The commentary text in this handbook is written to assist users in understanding and applying the provisions of NFPA 20. The commentary explains the reasoning behind the Standard's requirements and provides numerous examples, tables, photographs, and illustrations. The commentary text is printed in red type to distinguish it from the text of NFPA 20. Please note that the commentary is not part of NFPA 20 and therefore is not enforceable.

faster than normal, and since this is a relatively rare event, it is permitted for the discharge from the pressure relief valve to be piped back to the suction side of the pump.

4.18.1.1 Where a diesel engine fire pump is installed and where a total of 121 percent of the net rated shutoff (churn) pressure plus the maximum static suction pressure, adjusted for elevation, exceeds the pressure for which the system components are rated, a pressure relief valve shall be installed.

Prior to the 1996 edition, NFPA 20 required the installation of pressure relief valves for all diesel engine fire pumps. This requirement was based on the assumption that if engines ran too fast (a condition known as *overspeed*), the fire protection system would be exposed to pressures in excess of the pressure ratings of the system components. Because an overspeed shutdown device is required, the technical committee believes that a pressure relief valve is not needed on all diesel fire pump installations. Pumps that create pressures less than 175 psi (12.1 bar) at 120 percent of rated speed do not need a pressure relief valve. The following example illustrates the procedure used to determine if a pressure relief valve is needed.

EXAMPLE

A fire pump rated for 1500 gpm (5677 L/min) and 100 psi (6.9 bar) at 1750 rpm has a shutoff pressure of 120 psi (8.3 bar) to 140 psi (9.7 bar). The shutoff pressure produces 145 psi (10 bar) to 169 psi (11.7 bar) at 121 percent of rated pressure. If 45 psi (3.1 bar) static pressure is available from the city water supply, the total pressure when the pump is running at a churn pressure of 120 psi (8.3 bar) is 145 psi (10 bar).

$$120 \text{ psi } (8.3 \text{ bar}) \times 1.21 = 145 \text{ psi } (10 \text{ bar})$$

$$145 \text{ psi } (10 \text{ bar}) + 45 \text{ psi } (3.1 \text{ bar}) = 190 \text{ psi } (13.1 \text{ bar})$$

In this case, a pressure relief valve (see Exhibit II.4.10) is needed because the pressure is greater than 175 psi (12 bar).

Commentary art is set within red lines and labeled "Exhibit." The caption is printed in red ink. The commentary exhibits, including both drawings and photographs, provide detailed views of NFPA 20's concepts and are numbered sequentially throughout each chapter.

EXHIBIT II.8.1 Positive Displacement Pump with Electric Motor Driver. (Photograph courtesy of Pentair Water, North Aurora Operations)

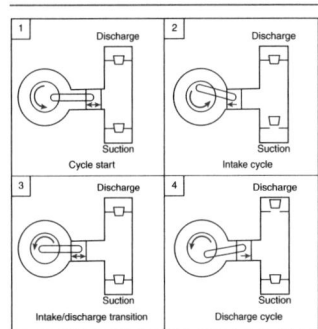

EXHIBIT II.8.2 Reciprocating Positive Displacement Pump—Pump Action.

the void volume is decreased on the discharge side of the pump, the liquid is "displaced" and exits the pump discharge port.

8.1.2* Suitability.

A.8.1.2 Special attention to the pump inlet piping size and length should be noted.

8.1.2.1 The positive displacement–type pump shall be listed for the intended application.

Stationary Fire Pumps Handbook

THIRD EDITION

Edited by

Jason R. Gamache
Fire Protection Engineer, NFPA

With the complete text of the 2010 edition of NFPA® 20, *Standard for the Installation of Stationary Pumps for Fire Protection*

National Fire Protection Association®
Quincy, Massachusetts

Product Manager: Debra Rose
Development Editor: Marla Marek
Project Editor: Irene Herlihy
Project Coordinator: Michael S. Barresi, Jr.
Permissions Editor: Josiane Domenici
Copy Editor: Janet Provost

Composition: Omegatype Typography, Inc.
Art Coordinator: Cheryl Langway
Cover Designer: Greenwood Associates
Manufacturing Manager: Ellen Glisker
Printer: R. R. Donnelley/Willard

Copyright © 2010
National Fire Protection Association®
One Batterymarch Park
Quincy, Massachusetts 02169–7471

All rights reserved.

Important Notices and Disclaimers: Publication of this handbook is for the purpose of circulating information and opinion among those concerned for fire and electrical safety and related subjects. While every effort has been made to achieve a work of high quality, neither the NFPA® nor the contributors to this handbook guarantee the accuracy or completeness of or assume any liability in connection with the information and opinions contained in this handbook. The NFPA and the contributors shall in no event be liable for any personal injury, property, or other damages of any nature whatsoever, whether special, indirect, consequential, or compensatory, directly or indirectly resulting from the publication, use of, or reliance upon this handbook.

This handbook is published with the understanding that the NFPA and the contributors to this handbook are supplying information and opinion but are not attempting to render engineering or other professional services. If such services are required, the assistance of an appropriate professional should be sought.

NFPA codes, standards, recommended practices, and guides ("NFPA Documents"), including the NFPA Documents that are the subject of this handbook, are made available for use subject to Important Notices and Legal Disclaimers, which appear at the end of this handbook and can also be viewed at *www.nfpa.org/disclaimers*.

Notice Concerning Code Interpretations: This third edition of *Stationary Fire Pumps Handbook* is based on the 2010 edition of NFPA® 20, *Installation of Stationary Pumps for Fire Protection*. All NFPA codes, standards, recommended practices, and guides ("NFPA Documents") are developed in accordance with the published procedures of the NFPA by technical committees comprised of volunteers drawn from a broad array of relevant interests. The handbook contains the complete text of NFPA 20, NFPA 24, and NFPA 291; relevant extracts from the text of NFPA 13, NFPA 14, NFPA 22, and NFPA 1; and any applicable Formal Interpretations issued by the NFPA. These NFPA Documents are accompanied by explanatory commentary and other supplementary materials.

The commentary and supplementary materials in this handbook are not a part of the NFPA Documents and do not constitute Formal Interpretations of the NFPA (which can be obtained only through requests processed by the responsible technical committees in accordance with the published procedures of the NFPA). The commentary and supplementary materials, therefore, solely reflect the personal opinions of the editor or other contributors and do not necessarily represent the official position of the NFPA or its technical committees.

The following are registered trademarks of the National Fire Protection Association®:

National Fire Protection Association®
NFPA®
Building Construction and Safety Code® and NFPA 5000®
NFPA 72®
Life Safety Code® and NFPA 101®
NEC®, National Electrical Code®, and NFPA 70®
Fire Protection Handbook®

NFPA No.: 20HB10
ISBN-10: 0-87765-854-4
ISBN-13: 978-0-87765-854-2
Library of Congress Card Control No.: 2009926677

Printed in the United States of America
09   10   11   12   13   5   4   3   2   1

Contents

Preface   vii
About the Contributors   ix
About the Editor   xi

PART I
An Overview of Water Supply Requirements, Suppression Systems, and Water Demand   1

Section 1   Standpipe Water Supply: Extracts from NFPA 14, 2007 Edition   5
Section 2   Sprinkler System Demand: Extracts from NFPA 13, 2010 Edition   21

PART II
NFPA 20, *Standard for the Installation of Stationary Pumps for Fire Protection*, 2010 Edition, with Commentary   29

1 Administration   31
   1.1 Scope   31
   1.2 Purpose   32
   1.3 Application   33
   1.4 Retroactivity   33
   1.5 Equivalency   34
   1.6 Units   34

2 Referenced Publications   37
   2.1 General   37
   2.2 NFPA Publications   37
   2.3 Other Publications   37
   2.4 References for Extracts in Mandatory Sections   38

3 Definitions   39
   3.1 General   39
   3.2 NFPA Official Definitions   39
   3.3 General Definitions   41

4 General Requirements   59
   4.1 Pumps   59
   4.2 Approval Required   59
   4.3 Pump Operation   60
   4.4 Fire Pump Unit Performance   62
   4.5 Certified Shop Test   63
   4.6 Liquid Supplies   63
   4.7 Pumps, Drivers, and Controllers   65
   4.8 Centrifugal Fire Pump Capacities   68
   4.9 Nameplate   68
   4.10 Pressure Gauges   69
   4.11 Circulation Relief Valve   70
   4.12 Equipment Protection   71
   4.13 Pipe and Fittings   73
   4.14 Suction Pipe and Fittings   76
   4.15 Discharge Pipe and Fittings   85
   4.16 Valve Supervision   88
   4.17 Protection of Piping Against Damage Due to Movement   89
   4.18 Relief Valves for Centrifugal Pumps   89
   4.19 Pumps Arranged in Series   94
   4.20 Water Flow Test Devices   95
   4.21 Steam Power Supply Dependability   100
   4.22 Shop Tests   101
   4.23 Pump Shaft Rotation   101
   4.24 Other Signals   102
   4.25 Pressure Maintenance (Jockey or Make-Up) Pumps   103
   4.26 Summary of Centrifugal Fire Pump Data   106
   4.27 Backflow Preventers and Check Valves   106
   4.28 Earthquake Protection   110
   4.29 Packaged Fire Pump Assemblies   110
   4.30 Pressure Actuated Controller Pressure Sensing Lines   112
   4.31 Break Tanks   114
   4.32 Field Acceptance Test of Pump Units   117

5 Fire Pumps for High-Rise Buildings   119
   5.1 General   120
   5.2 Types   120
   5.3 Equipment Access   120

Contents

- 5.4 Fire Pump Test Arrangement   120
- 5.5 Auxiliary Power   121
- 5.6 Fire Pump Backup   121
- 5.7 Water Supply Tanks   121

6 Centrifugal Pumps   125
- 6.1 General   125
- 6.2 Factory and Field Performance   131
- 6.3 Fittings   133
- 6.4 Foundation and Setting   137
- 6.5 Connection to Driver and Alignment   138

7 Vertical Shaft Turbine–Type Pumps   143
- 7.1 General   143
- 7.2 Water Supply   145
- 7.3 Pump   151
- 7.4 Installation   156
- 7.5 Driver   159
- 7.6 Operation and Maintenance   160

8 Positive Displacement Pumps   163
- 8.1 General   163
- 8.2 Foam Concentrate and Additive Pumps   167
- 8.3 Water Mist System Pumps   169
- 8.4 Fittings   169
- 8.5 Pump Drivers   172
- 8.6 Controllers   174
- 8.7 Foundation and Setting   174
- 8.8 Driver Connection and Alignment   174
- 8.9 Flow Test Devices   175

9 Electric Drive for Pumps   177
- 9.1 General   177
- 9.2 Normal Power   178
- 9.3 Alternate Power   181
- 9.4 Voltage Drop   183
- 9.5 Motors   184
- 9.6 On-Site Standby Generator Systems   187
- 9.7 Junction Boxes   189
- 9.8 Listed Electrical Circuit Protective System to Controller Wiring   190
- 9.9 Raceway Terminations   190

10 Electric-Drive Controllers and Accessories   193
- 10.1 General   193
- 10.2 Location   196
- 10.3 Construction   201
- 10.4 Components   205
- 10.5 Starting and Control   213
- 10.6 Controllers Rated in Excess of 600 V   220
- 10.7 Limited Service Controllers   223
- 10.8 Power Transfer for Alternate Power Supply   224
- 10.9 Controllers for Additive Pump Motors   231
- 10.10 Controllers with Variable Speed Pressure Limiting Control or Variable Speed Suction Limiting Control   231

11 Diesel Engine Drive   241
- 11.1 General   241
- 11.2 Engines   242
- 11.3 Pump Room   260
- 11.4 Fuel Supply and Arrangement   262
- 11.5 Engine Exhaust   266
- 11.6 Diesel Engine Driver System Operation   268

12 Engine Drive Controllers   271
- 12.1 Application   271
- 12.2 Location   273
- 12.3 Construction   273
- 12.4 Components   279
- 12.5 Battery Recharging   289
- 12.6 Battery Chargers   290
- 12.7 Starting and Control   293
- 12.8 Air-Starting Engine Controllers   303

13 Steam Turbine Drive   305
- 13.1 General   305
- 13.2 Turbine   308
- 13.3 Installation   310

14 Acceptance Testing, Performance, and Maintenance   313
- 14.1 Hydrostatic Tests and Flushing   314
- 14.2 Field Acceptance Tests   316
- 14.3 Manuals, Special Tools, and Spare Parts   338
- 14.4 Periodic Inspection, Testing, and Maintenance   339
- 14.5 Component Replacement   340

Annexes
- A Explanatory Material   345
- B Possible Causes of Pump Troubles   347
- C Informational References   353
- D Material Extracted by NFPA 70, Article 695   355

PART III
Private Water Supplies, Hydrants, Tanks, and Piping 357

Section 1 Extracts from NFPA 22, *Standard for Water Tanks for Private Fire Protection,* 2008 Edition 359
Section 2 Complete Text of NFPA 24, *Standard for the Installation of Private Fire Service Mains and Their Appurtenances,* 2010 Edition 389
Section 3 Section 18.4 and Annex I, NFPA 1, *Fire Code,* 2009 Edition 463
Section 4 Complete Text of NFPA 291, *Recommended Practice for Fire Flow Testing and Marking of Hydrants,* 2010 Edition 471

PART IV
Supplements 493

1 Fire Pump Installation from Design to Acceptance 495
2 Commissioning Forms for Fire Pumps 509
3 Article 695 from *NFPA 70®*, *National Electrical Code®*, 2008 edition 521
4 Technical/Substantive Changes from the 2007 Edition to the 2010 Edition of NFPA 20 527

NFPA 20 Index 545

Important Notices and Legal Disclaimers 555

Preface

Fire pumps have been used to supply water and pressure to fire protection systems for over 100 years. The first NFPA standard on automatic sprinkler systems was published in 1896 and included information on steam and rotary fire pumps that is still valid today. Among its requirements were 2½ in. outlets for testing purposes, equipment protection (the pump had to be located in a brick or stone enclosure and cut off from the main building by fire doors), and a weekly running test. The standard established the minimum size for fire pumps to be not less than 500 gpm rated capacity and required a 60 minute water supply. A spring-type pressure relief valve and pressure gauge were also required. These requirements amounted to less than one page of text for the installation of a fire pump.

These early pumps were not the primary water supply for sprinkler or standpipe systems and were started manually. Pumps were permitted to take suction by lift either from a connected water main or by means of connecting a primer pipe to a water tank of not less than 200 gallons capacity. The first pumps were usually powered by steam; gasoline engine–driven pumps were first mentioned in the standard in 1913. At first unreliable, these spark-ignited engines evolved into the reliable diesel engine–driven pumps of modern times.

Today, fire pumps are considered to be a primary component of the fire protection water supply and are started automatically. Modern fire pumps are connected to a reliable driver of either an electric motor or a diesel engine (some steam-driven units are still in service) and are designed to start and operate under the most demanding conditions. NFPA 20, *Standard for the Installation of Stationary Pumps for Fire Protection*, has undergone 30 revisions and has evolved into a comprehensive installation standard consisting of 14 chapters and 108 pages—far more comprehensive than the first standard on fire pumps.

In 1998 NFPA and the National Fire Sprinkler Association collaborated on the first edition of what was then called the *Fire Pump Handbook*. Kenneth E. Isman, P.E., of NFSA and Milosh Puchovsky, P.E., formerly of the NFPA staff, provided a depth of knowledge and expertise to create a handbook that became invaluable to those in the field. NFPA acknowledges the work of Mr. Isman and Mr. Puchovsky in their authorship of the first edition of the *Fire Pump Handbook,* portions of which have been used in the preparation of this book.

The purpose of this handbook, in addition to providing commentary on the requirements of NFPA 20, is to include in one document a complete handbook of all NFPA documents that establish water supply requirements for fixed suppression systems, regardless of the type of water supply. Part I discusses both automatic standpipe systems, which, due to their high flow and pressure demands, usually require the assistance of a fire pump, and sprinklers, which are the most common type of system installed and in many configurations create a pressure demand that necessitates a fire pump. Part II of this handbook contains the requirements for the installation of fire pumps from NFPA 20, with additional explanation in the form of commentary. Part III covers hydrant systems and how hydrant demand necessitates a fire pump installation and water tanks and private water supplies as they relate to the installation of fire pumps and suppression systems. Part III includes requirements and guidance from NFPA 22, *Standard for Water Tanks for Private Fire Protection*; NFPA 24, S*tandard for the Installation of Private Fire Service Mains and Their Appurtenances*; NFPA 1, *Fire Code*; and NFPA 291, *Recommended Practice for Fire Flow Testing and Marking of Hydrants*.

The handbook also includes four supplements: a sample project outline illustrating all of the considerations that must be taken into account during a fire pump installation; commissioning forms, which are intended to assist the user in managing a fire pump installation; extracted Article 695 from the *National Electrical Code®*; and a table of significant revisions from the 2007 to 2010 edition of NFPA 20.

I would like to thank all of the contributors to this project for their input and guidance on the preparation of this material.

Jason R. Gamache

About the Contributors

David B. Fuller (Chapter 5)

David B. Fuller is a Senior Engineering Technical Specialist for FM Global. Based in the company's Norwood, MA, offices, Fuller works in the Engineering Standards group and is responsible for providing technical property loss prevention guidance in areas of his expertise to FM Global's 1500 engineers. He is the company subject matter expert for fire protection system installation, maintenance, and testing; fire protection in cold storage facilities; fire pumps; and fire protection equipment corrosion. Fuller is a member of National Fire Protection Association and currently serves on the NFPA committees for sprinkler installation, fire pumps, inspection test and maintenance, and foam-water sprinklers. He received a bachelor's degree in Electrical Engineering from Northeastern University, Boston, MA.

David R. Hague (Chapter 4, Supplements 1–3)

David R. Hague, P.E., CFPS, CET, is Manager of the Engineering Technical Unit at Liberty Mutual Property (LMP) in Weston, MA, where he is responsible for the LMP Engineering Manual, training programs for internal and external customers, coordination of plan reviews and coordination of LMP's involvement in NFPA Technical Committees. He also manages the operation of the LMP Fire Lab in Wausau, WI. Hague is a member of the NFPA Technical Committees on Commissioning, Standpipe Systems, Fire Pumps, Combustible liquids and Protection of Records. Prior to joining LMP, Hague served as a principal engineer at the National Fire Protection Association, where he was responsible for NFPA standards on suppression systems and has written several books and developed several seminars on the subject of fire protection systems.

Bill M. Harvey (Chapters 11 and 13)

Bill M. Harvey, S.E.T., is CEO of Harvey & Associates, Inc., a fire sprinkler, fire alarm, and electrical contracting firm. Harvey is a 53-year veteran of the fire protection industry. Harvey serves on the NFPA 20 techinical committee, representing the American Fire Sprinkler Association as a principal member for 10 years. Harvey also serves on UL STP Committees, UL 448 and UL 1247 for Fire Pump and Diesel Engines Drivers. His work history includes 48 years in design, installation, and performance testing of fire pumps for fire protection systems.

John D. Jensen (Supplement 1)

John D. Jensen, P.E., is a licensed fire protection and mechanical engineer. He has more than 40 years' experience in the fire protection field. He was a member of the NFPA 20 Technical Committee on Fire Pumps from 1976 to 2006. Jensen has been chair of the committee for the last two editions of the standard. He has also served on the NFPA 22 technical committee from 1997 to 2006. Jensen instructed for the NFPA professional training seminars from 1987 until 2005, teaching courses in sprinkler systems, high-piled storage, fire alarm systems, fire pumps, and inspection, testing, and maintenance of water-based fire protection systems. He has been on the staff of the University of Idaho, teaching classes in Fire Protection and

Industrial Safety from 1980 to 2000. He has worked in the HPR Insurance business and as a consulting engineer. Jensen graduated from the University of Utah with a BSME degree.

John Kovacik (Chapters 9 and 10)

John Kovacik is Underwriters Laboratories' Primary Designated Engineer (PDE) for Circuit Breakers located at the Northbrook office of UL. He represents UL on a number of outside committees, including NFPA, NEMA, and IEC. Kovacik is UL's principal member on the National Electrical Code Technical Correlating Committee and is responsible for coordinating UL's participation in *NEC* activities. He is also a member of the NFPA 20 Technical Committee on Fire Pumps. Kovacik holds a BSEE from Bradley University.

James S. Nasby (Chapter 12)

James S. Nasby has been in responsible charge of fire pump controller design since 1972. He is a member of the NFPA 20 technical committee and former member of NFPA 110 and NFPA 70, the *NEC*®, committees for the past two cycles. He holds a BSEE from IIT and is a member of the NFPA, the SFPE, and the IEEE. He has contributed to the last two editions of the *Fire Protection Handbook* as well as the NFPA *Pumps for Fire Protection Systems* textbook. He also serves on UL Standards Technical Panels for Fire Pump Controllers, Fire Pump Motors, and Engine & Turbine Gen-Sets. His previous position was Director of Engineering for Master Control Systems, Inc., where he designed both diesel and electric fire pump controllers and battery chargers. He has given numerous seminars at NFPA and IEEE meetings on fire pumps and power supplies.

Gayle Pennel (Chapters 5–7, 14)

Gayle Pennel is a registered professional engineer and has worked at Schirmer Engineering Corporation since 1990. He specializes in water supplies and fire protection systems evaluation and design. Projects have ranged from multiple-zone high-rise buildings to exhibition centers, refineries, and failure analysis. He has over 40 years' fire protection experience and has served on the NFPA 20 Technical Committee on Fire Pumps for over ten years and also serves on the NFPA 25 technical committee.

Tom Reser (Chapter 8)

Tom Reser has been directly involved in both positive displacement and centrifugal pumps and systems since 1982. Starting in 1988, he has been applying positive displacement pumps in the fire protection industry. As a past owner of Edwards Mfg., a positive displacement pump manufacturer located in Oregon, he has extensive experience in special hazard foam and water mist systems that use positive displacement pumps. Reser has attended numerous rotating equipment technical equipment schools in the United States and holds a bachelor of science degree in marketing.

About the Editor

Jason R. Gamache is a fire protection engineer at the National Fire Protection Association, where he is responsible for NFPA standards on fire pumps, standpipes, dry and wet chemical extinguishing systems, explosion protection systems, venting systems for cooking appliances, laser fire protection, fixed guideway transit systems, cleanrooms, road tunnel and highway fire protection, electric generating plants, and recreational vehicles. He has written articles and presentations about these subjects and also provides informal interpretations for NFPA 13. He presently holds a bachelor's degree in mechanical engineering and a master's degree in fire protection engineering, both from Worcester Polytechnic Institute. While at WPI, Mr. Gamache worked on projects ranging from fire dynamic simulator modeling of train fires to the investigation of property protection in the building codes of Australia. He has also worked at Schirmer Engineering in New York as a code consultant, working on projects including code consultation, fire alarm testing, sprinkler design, egress analysis, risk assessment, site surveys, and due diligence reports.

PART I

An Overview of Water Supply Requirements, Suppression Systems, and Water Demand

Part I of this handbook outlines the hydraulic calculation requirements for the most common types of fire protection systems: standpipe systems and sprinkler systems. The water and pressure demands of standpipe and sprinkler systems drive the need for a fire pump. An important fact to remember is that a fire pump is a supplemental component of the water supply—a fire pump cannot create waterflow. However, a fire pump does transfer energy to the water, thus increasing the water pressure. This increase in pressure can then be used to move the water to the upper floors of a high-rise building, for example, or can be used to meet the high pressure demands of an early suppression fast response (ESFR) sprinkler system. Specific sections of NFPA 14, *Standard for the Installation of Standpipe and Hose Systems,* and NFPA 13, *Standard for the Installation of Sprinkler Systems,* are included to show how the water supply needs of sprinkler and standpipe systems are determined and to illustrate how the capacity of a fire pump is determined. Marginal icons indicate material of special interest with respect to design and calculations.

Once the need for a fire pump is established, the system designer or registered design professional (RDP) must then consider the requirements for the proper selection and installation of the fire pump and related equipment. Part II of this handbook includes a discussion of the requirements from NFPA 20, *Standard for the Installation for Stationary Pumps for Fire Protection.*

Background. Initially, when fire pumps were first used in 1907, they served only as a secondary water supply for sprinkler, standpipe, and hydrant systems and were started manually. Due to increased reliability of fire pumps, they now play a major role in fire protection. Most pumps provide the main supply of water and pressure and are started and controlled automatically.

Water supplies for fire protection systems generally consist of the following two types:

1. Public systems, which, in addition to supplying fire protection systems, serve domestic uses such as drinking and sanitation as well as commercial and industrial uses
2. Private systems, which serve only fire protection systems

In public systems, the system must be designed to handle the simultaneous demands of both the domestic and fire protection users. Where a public or private water supply system is deficient, a supplemental supply such as an elevated water tank and/or fire pump is needed. See Exhibit I.1 for an illustration of an elevated water tank.

EXHIBIT I.1 Elevated Water Tank.

EXHIBIT I.2 Suction Tank and Fire Pump.

A fire pump can be used in cases where flow, in addition to pressure, is lacking. For example, a ground-mounted suction tank can provide a sufficient amount of water, but without the fire pump to move that water to a fire protection system, the stored water is of little value. Therefore, in this case, the fire pump can be considered to be providing flow. See Exhibit I.2 for an illustration of a suction tank and fire pump.

In other situations, such as in a high-rise building, a city water supply usually lacks sufficient pressure to move water to the higher floors. In this case, the fire pump is used to supplement the available pressure. Many other examples of situations where a fire pump is needed are presented in Part I. The examples are not just cases where low pressure is experienced, as previously indicated, but also some design situations involving extra hazard occupancies or early suppression fast response (ESFR) sprinklers where high pressures and flows are required.

Standpipe Water Supply:

Extracts from NFPA 14, 2007 Edition

SECTION 1

Because their intended use is providing waterflow for manual fire fighting, standpipe systems present a substantial demand on water supplies. Historically, these systems have been designed to flow from 500 gpm (1893 L/min) at 50 psi (3.4 bar) beginning with the 1915 edition of NFPA 14, *Standard for the Installation of Standpipe and Hose Systems,* to as much as 1250 gpm (4731 L/min) at 100 psi (6.9 bar) in the 2007 edition. Several flow rates and pressures have been specified for standpipes between the first and current editions. The flow and particularly the pressure requirements in the current edition are based on the type of equipment used by the fire department in what is commonly referred to as a *standpipe pack.*

In preparation for fire operations in buildings equipped with standpipes, NFPA 13E, *Recommended Practice for Fire Department Operations in Properties Protected by Sprinkler and Standpipe Systems,* 2005 edition, states that fire departments should plan on providing "hose and nozzles of appropriate size and length along with proper accessory equipment for the anticipated fire conditions" (see NFPA 13E, 6.3.5). Following the tragic fire at One Meridian Plaza in Philadelphia, Pennsylvania, on February 23, 1991, discussions with several fire departments revealed that fire departments across the United States were using different tactics in fighting fires in buildings equipped with standpipe systems and were using a variety of equipment in their standpipe packs. As a result of this fire, NFPA surveyed approximately 200 fire departments across the country, including several fire departments in British Columbia, Canada. Those surveyed were asked what type of standpipe equipment was installed in their jurisdiction and what type of equipment was employed in their standpipe packs. (The results were compiled and are shown in Table A.7.8 from NFPA 14, 2007 edition, extracted on p. 12.)

The results of the survey indicated a wide variety of equipment configurations and, most importantly, indicated the use of combination-type nozzles that require a minimum of 100 psi (6.9 bar) to achieve an effective hose stream. Because of the pressure requirements of this very popular nozzle, the 1993 edition required 100 psi (6.9 bar) of pressure at the topmost outlet of the standpipe system. If additional pressure is needed, the fire department or the stationary fire pump can supplement the needed pressure beyond the minimum requirement.

The following section on design is extracted from Chapter 7 of NFPA 14, 2007 edition, and also provides commentary to explain the requirements.

DESIGN

7.1* General.

The design of the standpipe system is governed by building height, area per floor occupancy classification, egress system design, required flow rate and residual pressure, and the distance of the hose connection from the source(s) of the water supply.

A.7.1 The building height determines the number of vertical zones. The area of a floor or fire area and exit locations, as well as the occupancy classification, determines the number and locations of hose connections. Local building codes influence types of systems, classes of systems, and locations of hose connections. Pipe sizing is dependent on the number of hose connections flowing, the quantity of water flowed, the required residual pressure, and the vertical distance and horizontal distance of those hose connections from the water supplies.

For typical elevation drawings, see Figure A.7.1(a), Figure A.7.1(b), and Figure A.7.1(c). See Chapter 7 for general system requirements.

Because building height is not limited or restricted, a standpipe system may be required to provide the needed flow and pressure to floors in high-rise buildings that are beyond the reach of the local fire department. The definition of *high-rise building* from NFPA 14 deals with fire department vehicle access and is "a building where the floor of an occupiable story is greater than 75 ft (23 m) above the lowest level of fire department vehicle access."

A design scenario involving building heights beyond the reach of fire department access presents several problems that a designer must address. The first is how to move water to such heights, and the second is how to prevent dangerous and damaging pressures from reaching lower floors when high pressure is needed to reach upper floors. Another consideration is the reliability of critical systems, such as fuel and power, since loss of these systems on upper floors of a high-rise building leaves few fire-fighting options (see Part II of this handbook for more information on this topic).

A fire pump in this case will solve one pressure problem but will present another problem involving too much pressure at the lower points in the system. In lower levels of high-rise buildings, excessively high pressures are detrimental to the system components and also pres-

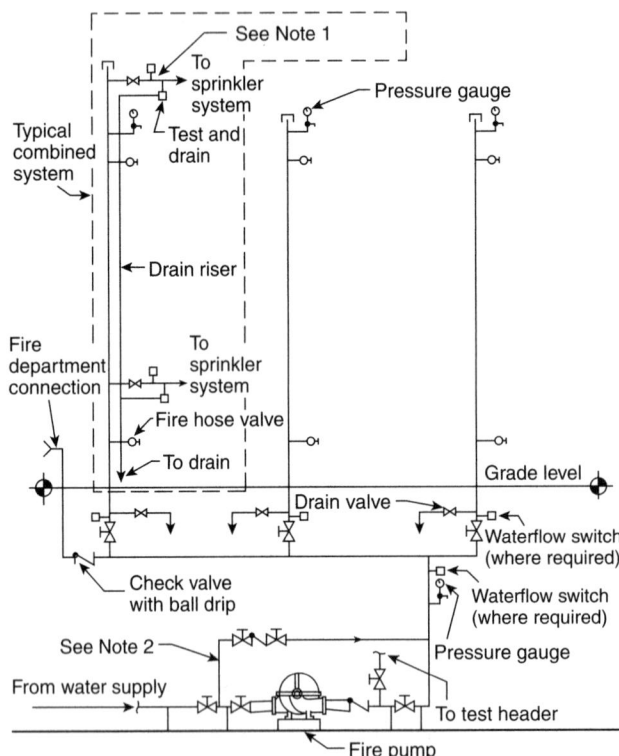

Notes:
1. Sprinkler floor assembly in accordance with NFPA 13, *Standard for the Installation of Sprinkler Systems.*
2. Bypass in accordance with NFPA 20, *Standard for the Installation of Stationary Pumps for Fire Protection.*

FIGURE A.7.1(a) *Typical Single-Zone System.*

Part I • Extracts from NFPA 14, 2007   7

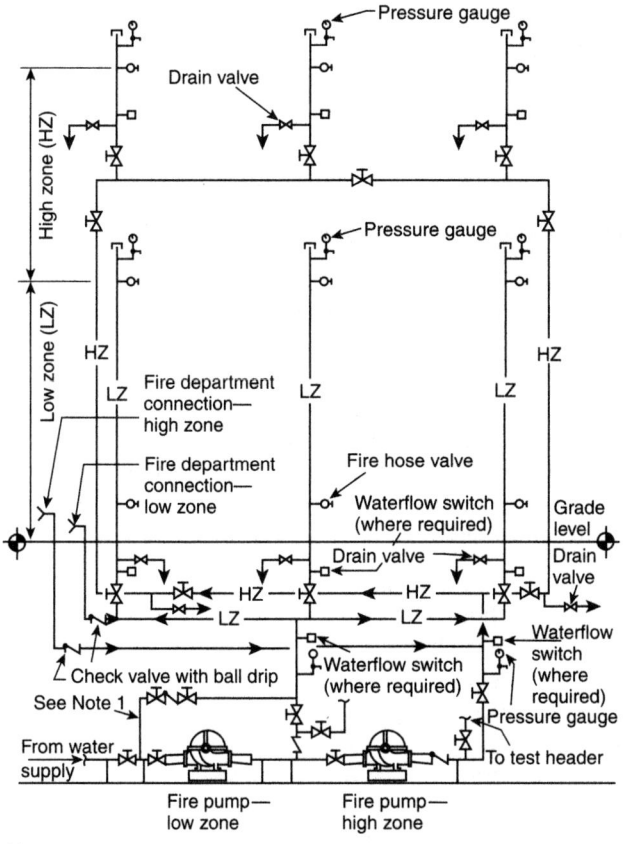

FIGURE A.7.1(b) Typical Two-Zone System.

FIGURE A.7.1(c) Typical Multizone System.

ent a safety issue to the end user—the fire fighter. In this instance, pressure-regulating devices must be employed.

7.1.1* When pressure-regulating devices are used, they shall be approved for installation within the maximum and minimum anticipated flow conditions.

A.7.1.1 It is important to determine the exact operating range to ensure that pressure-regulating devices function in accordance with the manufacturer's instructions for both maximum and minimum anticipated flow rates. Minimum flow can be from a single sprinkler for combined systems or flow from a 1½ in. (40 mm) hose connection on standpipe systems that do not supply sprinklers. This could require the use of two devices installed in parallel.

When selecting the appropriate pressure-regulating device, it is important to check the manufacturer's data to verify that the correct size of the device is selected so as to ensure that the flow and pressure settings are appropriate for the design condition. In cases where the pressure-regulating device serves a combined sprinkler and standpipe system, both the upper and lower values of flow and pressure must be evaluated. The lower value can be as little as 15 gpm (56.7 L/min), which is the flow from a ½ in. orifice (K-5.6) sprinkler at 7 psi (48.3 kPa). The higher value can be as much as 1750 gpm (5678 L/min), which is the maximum anticipated flow from a combined system with partial sprinkler protection [1250 gpm (3785 L/min)

Stationary Fire Pumps Handbook 2010

is the maximum anticipated flow from a standpipe system plus 500 gpm (1893 L/min), which is the required flow from an ordinary hazard sprinkler system; see Section 7.10 of NFPA 14.]. The manufacturer's literature should provide the limits of flow and pressure for each device.

7.2* Pressure Limitation.

The maximum pressure at any point in the system at any time shall not exceed 350 psi (24 bar).

A.7.2 The system pressure limits have been implemented to replace the former height units. Because the issue addressed by the height limits has always been maximum pressure, pressure limitations are a more direct method of regulation and allow flexibility in height units where pumps are used, because a pump curve with less excess pressure at churn yields lower maximum system pressures while achieving the required system demand.

The maximum system pressure normally is at pump churn. The measurement should include both the pump boost and city static pressures. The 350 psi (24 bar) limit was selected because it is the maximum pressure at which most system components are available, and it recognizes the need for a reasonable pressure unit.

Design pressures for standpipe systems were initially addressed by limiting the height of a standpipe system zone. The previous height limitation was 275 ft (84 m) with an exception limiting height to a maximum of 400 ft (122 m). With this limitation and a pressure requirement of 100 psi (6.9 bar) at the topmost outlet, pressures at the bottom of the riser could be expected to approach approximately 220 psi (15.2 bar), requiring pressure regulation of the hose valves on lower floors and the use of extra heavy valves and fittings to cope with pressures in excess of the 175 psi (12.1 bar) limitation on standard weight valves and fittings.

The 1993 edition of NFPA 14 removed the height limitation in favor of a pressure limitation of 350 psi (24 bar) based on the pressure ratings of available devices at the time and presented a logical pressure limitation. For the 2007 edition of NFPA 14, the Technical Committee on Standpipes opted to resort to a performance-based approach in referring to the pressure ratings of devices rather than a prescriptive pressure limitation, thus permitting pressures in excess of 350 psi (24 bar). However, this change was overturned at NFPA's Annual Meeting, and the 350 psi (24 bar) limitation was reinstated. Regardless of the exact pressure requirement in the standard, most system components are listed based on 300 psi (21 bar).

7.2.1 Maximum Pressure for Hose Connections.

7.2.1.1 Where the residual pressure at a 1½ in. (40 mm) outlet on a hose connection exceeds 100 psi (6.9 bar), an approved pressure-regulating device shall be provided to limit the residual pressure at the flow required by Section 7.10 to 100 psi (6.9 bar).

Paragraph 7.2.1.1 is intended to pertain to Class II standpipe systems for use by trained personnel. Pressures for a Class II standpipe system are generally limited to lower values due to the size of the hose, the nozzle type, and lower flows needed for these systems.

7.2.1.2 Where the static pressure at a hose connection exceeds 175 psi (12.1 bar), an approved pressure-regulating device shall be provided to limit static and residual pressures at the outlet of the hose connection to 100 psi (6.9 bar) for 1½ in. (40 mm) hose connections and 175 psi (12.1 bar) for other hose connections. The pressure on the inlet side of the pressure-regulating device shall not exceed the device's rated working pressure.

This paragraph limits pressures on the outlet side of 2½ in. (65 mm) hose valves to 175 psi (12.1 bar) for Class I standpipe systems and 100 psi (6.9 bar) on the 1½ in. (40 mm) hose connection installed on Class III standpipe systems. This pressure limitation is based upon the type of equipment used in each standpipe system type and the training of the intended user.

Although the pressure limitation is relatively high at 350 psi (24 bar), the designer or registered design professional (RDP) should be aware that pressures in the piping should never exceed that for which the pipe, fittings, or devices are rated.

Many fire departments lay a hose line from the pumper into the building and connect to an accessible valve outlet using a double female swivel where the building fire department connections are inaccessible or inoperable. To pressurize the standpipe, the hose valve is opened and the engine pumps into the system.

If the standpipe is equipped with pressure-reducing hose valves, the valve acts as a check valve, prohibiting pumping into the system when the valve is open.

A supplementary single-inlet fire department connection or hose valve with female threads at an accessible location on the standpipe allows pumping into that system.

7.2.2* When system pressure-regulating devices are used in lieu of providing separate pumps, multiple zones shall be permitted to be supplied by a single pump and pressure-regulating device(s) under the following conditions:

(1) Pressure-regulating device(s) shall be permitted to control pressure in the lower zone(s).
(2) A method to isolate the pressure-regulating device(s) shall be provided for maintenance and repair.
(3) Regulating devices shall be arranged so that the failure of any single device does not allow pressure in excess of 175 psi (12.1 bar) to more than two hose connections.
(4) An equally sized bypass around the pressure-regulating device(s), with a normally closed control valve, shall be installed.
(5) Pressure-regulating device(s) shall be installed not more than 7 ft 6 in. (2.31 m) above the floor.
(6) The pressure-regulating device shall be provided with inlet and outlet pressure gauges.
(7) The fire department connection(s) shall be connected to the system side of the outlet isolation valve.

It is the intent to always have the fire department connection installed downstream of pressure-regulating devices installed in accordance with 7.2.2.

(8) The pressure-regulating device shall be provided with a pressure relief valve in accordance with the manufacturers recommendations.
(9) Remote monitoring and supervision for detecting high pressure failure of the pressure-regulating device shall be provided in accordance with *NFPA 72, National Fire Alarm Code.*

A.7.2.2 A small diameter pressure-reducing device can be required to accommodate low flow conditions such as those created by the flow of a single sprinkler.

See Figure A.7.2.2(a) and Figure A.7.2.2(b) for methods to comply with this section.

The requirement to install a separate fire pump for each zone of a standpipe system was added in the 1973 edition of NFPA 14 and remains as a design option today. The 2007 edition permits the use of a pressure-reducing valve for the lower zone of a standpipe system, allowing the installation of a single fire pump in many instances.

NFPA 14 defines three types of pressure-regulating devices and permits the installation of any of these devices, provided that the device selected is listed in Section 7.2. Previously, the standard provided detailed instructions for the installation of pressure-regulating hose valves but did not provide any instructions for the installation of any other type of pressure-regulating device.

The design requirements for standpipe systems are complicated when considering zoning of standpipe systems in high-rise buildings. Unlike NFPA 13, *Standard for the Installation of Sprinkler Systems,* NFPA 14 permits system pressures up to the pressure rating of the system components while limiting hose valve outlet pressures to a range from a minimum

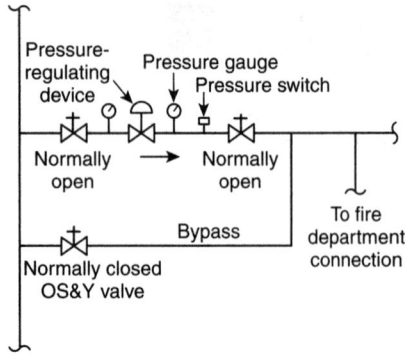

FIGURE A.7.2.2(a) *Pressure-Regulating Device Arrangement.*

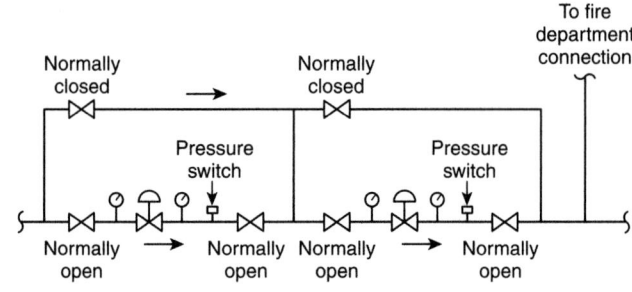

FIGURE A.7.2.2(b) *Dual Pressure-Regulating Device Arrangement.*

of 100 psi (6.9 bar) to a maximum of 175 psi (12.1 bar). In many cases this design results in a variety of pressure-regulating device settings from floor to floor, which must be carefully monitored during construction to ensure that each device is set properly for the elevation in which it is installed in the system. In some situations, the use of a master pressure-reducing valve to control pressures to the lower zone of a standpipe system may be convenient. Prior to the 2007 edition, the standard required a separate pump for each zone. In fact, a single pump could be used with a "master" pressure-regulating device protecting the lower zone, but the standard was silent on this issue. The current edition of NFPA 14 now permits this alternate arrangement, which is a common practice in the field. Such an application is consistent with the listing of the pressure-reducing valve.

 Further compounding this application is the flow range needed for combined sprinkler and standpipe systems. Many pressure-regulating devices on the market cannot provide the minimum and maximum flows needed in a combined system configuration. The estimated worst-case flow through a pressure-regulating device in a combined sprinkler and standpipe system with only partial sprinkler protection can range from a minimum of 15 gpm (57 L/min) (for a single sprinkler) to a maximum of 1750 gpm (6623 L/min) [full standpipe system flow of 1250 gpm (4731 L/min) plus the sprinkler discharge of 500 gpm (1893 L/min) as required by NFPA 14, 7.10.1.3.2]. This wide range results in the need to install two devices in parallel: one for low-flow and one for high-flow conditions (see Exhibit I.3).

EXHIBIT I.3 Parallel Systems

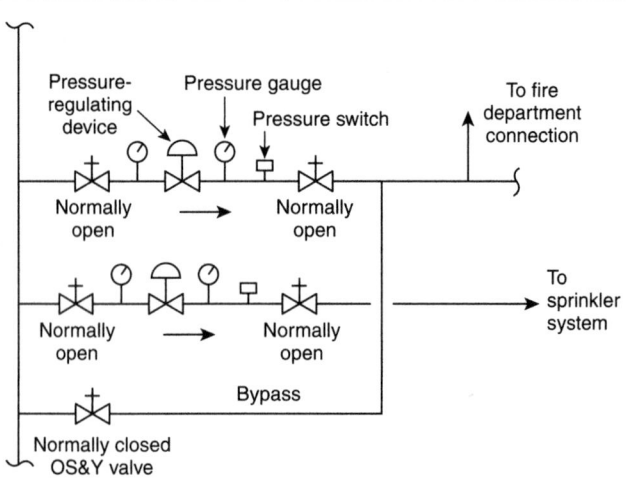

Previously, the NFPA 14 technical committee rejected a proposal to permit a "master" pressure-reducing valve as an option to the single pump per zone requirement. The Committee Statement in the Report on Comments, published by the NFPA, indicated a concern on the part of the committee about device reliability and the potential of leaving "entire segments of the standpipe system not being provided with proper water if the valve fails." The committee further stated that valves failing in an open position could present high pressures and a substantial risk to fire fighters. The installation requirements in this section are intended to address both situations. In NFPA 14, pressure gauges and a pressure switch are required for constant monitoring of the valve as shown in Figure A.7.2.2(a). A normally closed bypass valve is installed to permit bypassing the pressure-reducing valve in event of failure in the closed position and for maintenance and repair.

Additionally, to address concern of failure in the open position, Figure A.7.2.2(b) provides an example of redundancy where two valves can be installed in series. In this figure, since pressure-reducing valves generally require a 10 psi (0.7 bar) pressure differential, the primary valve should be set for an outlet pressure of 165 psi (11.4 bar) [10 psi (0.7 bar) below the standard pressure rating for most system components]. The second or redundant valve can be set for an outlet pressure of 155 psi (10.7 bar), thereby preventing exposure of the end user to dangerous pressures at any point in the lower standpipe system zone.

NFPA 25, *Standard for the Inspection, Testing, and Maintenance of Water-Based Fire Protection Systems,* includes weekly inspection requirements with quarterly and annual flow test requirements to provide additional insurance of the operation of these devices.

7.8* Minimum and Maximum Pressure Limits.

A.7.8 Where determining the pressure at the outlet of the remote hose connection, the pressure loss in the hose valve should be considered.

It is very important that fire departments choose an appropriate nozzle type for their standpipe fire-fighting operations. Constant pressure- (automatic-) type spray nozzles *(see NFPA 1964, Standard for Spray Nozzles)* should not be used for standpipe operations because many of these types require a minimum of 100 psi (6.9 bar) of pressure at the nozzle inlet to produce a reasonably effective fire stream. In standpipe operations, hose friction loss could prevent the delivery of 100 psi (6.9 bar) to the nozzle.

In high-rise standpipe systems with pressure-reducing hose valves, the fire department has little or no control over hose valve outlet pressure.

Many fire departments use combination (fog and straight stream) nozzles requiring 100 psi (6.9 bar) residual pressure at the nozzle inlet with 1½ in., 1¾ in., or 2 in. (40 mm, 44 mm, or 50 mm) hose in lengths of up to 150 ft (45.7 m). Some use 2½ in. (65 mm) hose with a smooth bore nozzle or a combination nozzle.

Some departments use 50 ft (15.2 m) of 2½ in. (65 mm) hose to a gated wye, supplying two 100 ft (30.5 m) lengths of 1½–2 in. (40–50 mm) hose with combination nozzles, requiring 120–149 psi (8.3–0.3 bar) at the valve outlet. *(See Table A.7.8.)*

Also see NFPA 1901, *Standard for Automotive Fire Apparatus.*

Given the various nozzle and hose configurations available, the fire department's selection of equipment that is compatible with the flows and pressures specified in NFPA 14 is imperative. The 100 psi (6.9 bar) pressure requirement was selected from a wide variety of proposed pressures ranging from 65 psi (4.5 bar) to 150 psi (10.3 bar) to meet most of the nozzle and hose configurations listed in the NFPA Standpipe Survey. Note that friction loss through the hose valve should be included in the calculations since the equivalent pipe length for a typical 2½ in. (65 mm) hose valve is 35 ft (10.7 m). When calculating 250 gpm (946 L/min) through 2½ in. (65 mm) pipe at 35 ft (10.7 m) long, the pressure loss is as follows:

Friction loss per foot of 2½ in. pipe while flowing at 250 gpm = x psi/ft × 35 ft = y psi

TABLE A.7.8 Hose Stream Friction Losses Summary

Calculation No.	Nozzle/Hose	Valve Outlet gpm	L/min	Flow psi	bar
1	2½ in. (65 mm) combination nozzle, with 150 ft (45.7 m) of 2½ in. (65 mm) hose	250	946	123	8.5
2	Two 1½ in. (40 mm) combination nozzles with 100 ft (30.5 m) of 1½ in. (40 mm) hose per nozzle, 2½ in. (65 mm) gated wye, and 50 ft (15.2 m) of 2½ in. (65 mm) hose	250	946	149	10.3
3	Same as calculation no. 2 with two 100 ft (30.5 m) lengths of 1½ in. (40 mm) hose	250	946	139	9.6
4	Same as calculation no. 3 with two 100 ft (30.5 m) lengths of 2 in. (50 mm) hose	250	946	120	8.3
5	1½ in. (40 mm) combination nozzle with 150 ft (45.7 m) of 2 in. (50 mm) hose	200	757	136	9.4
6	Same as calculation no. 5 with 1½ in. (40 mm) hose	200	757	168	11.6

Note: For a discussion of use by the fire department of fire department connections, see NFPA 13E, *Recommended Practice for Fire Department Operations in Properties Protected by Sprinkler and Standpipe Systems*.

7.8.1 Minimum Design Pressure for Hydraulically Designed Systems. Hydraulically designed standpipe systems shall be designed to provide the waterflow rate required by Section 7.10 at a minimum residual pressure of 100 (6.9 bar) at the outlet of the hydraulically most remote 2½ in. (65 mm) hose connection and 65 psi (4.5 bar) at the outlet of the hydraulically most remote 1½ in. (40 mm) hose station.

It is the intent that automatic and manual standpipe systems be hydraulically calculated.

FAQ ▶
Why is 100 psi (6.9 bar) required at the hydraulically most remote 2½ in. (65 mm) hose connection?

The 100 psi (6.9 bar) pressure requirement is based upon an in-depth study by the Technical Committee on Standpipes and selected based on the type of equipment employed by most fire departments today. Each fire department must carefully select the equipment used in the standpipe pack to ensure that it is compatible with the standpipe system. The 65 psi (4.5 bar) pressure requirement is intended for use in Class II standpipe systems, where the system is intended for use by industrial fire brigades only. The 65 psi (4.5 bar) pressure requirement was never an issue in the fires studied by the technical committee and, therefore, no changes to this requirement are necessary.

7.9 Standpipe System Zones.

Given the pressure limitation of 350 psi (24 bar) for systems with extra-heavy pattern fittings or 175 psi (12.1 bar) for systems utilizing standard weight fittings, the establishment of system zones becomes necessary to prevent the system design pressure from exceeding the aforementioned limitations. By sectionalizing systems into pressure zones and relay pumping, excessive pressures can be avoided. For example, assuming 12 ft (3.7 m) between floors in a high-rise building, the static head loss from floor to floor would be 5.196 psi (0.4 bar).

$$12 \text{ ft} \times 0.433 \text{ psi/ft} = 5.196 \text{ psi}$$

or in metrics,

$$3.7 \text{ m} \times 0.108 \text{ bar/m} = 0.4 \text{ bar}$$

Given a design limitation of 175 psi (12.1 bar) and a starting pressure of 100 psi (6.9 bar) at the topmost outlet, only 75 psi (5.2 bar) remain for friction loss and static head loss:

$$175 \text{ psi} - 100 \text{ psi} = 75 \text{ psi}$$

$$12.1 \text{ bar} - 6.9 \text{ bar} = 5.2 \text{ bar}$$

Given a head loss of 5.196 psi (0.358 bar) per floor, 14 floors can be supplied by a single zone:

$$\frac{75}{5.196} \text{ psi} = 14 \text{ floors}$$

$$\frac{5.2}{0.358} \text{ bar} = 14 \text{ floors}$$

Accounting for friction loss, approximately 12 floors can be supplied by a single zone. Using the maximum pressure permitted by the standard,

$$350 \text{ psi} - 100 \text{ psi} = 250 \text{ psi}$$

$$24.1 \text{ bar} - 6.9 \text{ bar} = 17.2 \text{ bar}$$

Applying the head loss of 5.196 psi per floor,

$$\frac{250}{5.196} \text{ psi} = 48 \text{ floors}$$

$$\frac{17.2}{0.358} \text{ bar} = 48 \text{ floors}$$

Based on this information, each zone of a standpipe system would be 12 or 48 floors in height based on the design pressure limitations used. The zones exposed to high pressure, of course, would require the use of pressure-regulating devices to deliver pressure at safe levels for the end user.

7.9.1 Each standpipe system zone created due to system component pressure limitations, requiring pumps, shall be provided with a separate pump.

The intent of 7.9.1 is to avoid the use of high-pressure pumps that may not be listed for fire protection use. Each standpipe system zone is required to have its own fire pump except as permitted by 7.2.2.

7.9.1.1 The requirement in 7.9.1 shall not preclude the use of pumps arranged in series.

Pumps arranged in series are needed to achieve the pressures necessary to deliver water to the upper floors of a high-rise building.

7.9.2 Where pumps supplying two or more zones are located at the same level, each zone shall have separate and direct supply piping of a size not smaller than the standpipe that it serves.

Subsection 7.9.2 prohibits using the lower zone standpipe system as a feed to the upper zone.

7.9.2.1 Zones with two or more standpipes shall have at least two direct supply pipes of a size not smaller than the largest standpipe that they serve.

The intent of 7.9.2.1 is to require redundancy in the supply pipe to the upper zone standpipe.

7.9.3 Where the supply for each zone is pumped from the next lower zone, and the standpipe or standpipes in the lower zone are used to supply the higher zone, such standpipes shall comply with the provisions for supply lines in 7.9.2.

7.9.3.1 At least two lines shall be provided between zones.

The intent of 7.9.3.1 is to require each supply pipe to be capable of supplying the upper zone fire pump.

7.9.3.2 One of the lines specified in 7.9.3.1 shall be arranged so that the supply can be automatically delivered from the lower to the higher zone.

7.9.4 For systems with two or more zones in which portions of the second and higher zones cannot be supplied using the residual pressure required by Section 7.8 by means of fire department pumpers through a fire department connection, an auxiliary means of supply shall be provided.

The intent of 7.9.4 is to require an auxiliary water supply when the ability of the fire apparatus to supply the system has been exceeded.

7.9.4.1 The auxiliary means shall be in the form of high-level water storage with additional pumping equipment or other means acceptable to the authority having jurisdiction.

Figure A.7.1(c) illustrates the method used to meet this requirement.

7.10 Flow Rates.

7.10.1 Class I and Class III Systems.

7.10.1.1* Flow Rate.

A.7.10.1.1 If a water supply system supplies more than one building or more than one fire area, the total supply can be calculated based on the single building or fire area requiring the greatest number of standpipes.

For a discussion of use by the fire department of fire department connections, see NFPA 13E, *Recommended Practice for Fire Department Operations in Properties Protected by Sprinkler and Standpipe Systems.*

Where buildings are attached, complete fire separation must be provided if the systems are intended to be calculated separately. Otherwise, the two buildings must be considered to be one and the standpipe systems calculated accordingly.

7.10.1.1.1 For Class I and Class III systems, the minimum flow rate for the hydraulically most remote standpipe shall be 500 gpm (1893 L/min), and the calculation procedure shall be in accordance with 7.10.1.2.

The hydraulically most demanding standpipe must be sized to handle this flow. The calculation is based on flowing the two topmost outlets simultaneously.

7.10.1.1.2* Where a horizontal standpipe on a Class I and Class III system supplies three or more hose connections on any floor, the minimum flow rate for the hydraulically most demanding horizontal standpipe shall be 750 gpm (2840 L/min), and the calculation procedure shall be in accordance with 7.10.1.2.

In the case of horizontal standpipes, any pipe supplying more than three hose connections should be considered to be a standpipe by definition. Therefore, each of these pipes in a standpipe system must be sized to flow a total of 750 gpm (2840 L/min).

A.7.10.1.1.2 The intent of this section is to provide a different flow requirement for large area low-rise buildings and other structures protected by horizontal standpipes.

Due to the problems associated with horizontal reach of hose lines in large area low-rise buildings, a higher initial flow is required.

7.10.1.1.3 The minimum flow rate for additional standpipes shall be 250 gpm (946 L/min) per standpipe, with the total not to exceed 1250 gpm (4731 L/min) or 1000 gpm (3785 L/min) for buildings sprinklered throughout.

Each standpipe must be sized to accommodate a flow of 500 gpm (1893 L/min); however, only the hydraulically most demanding standpipe is required to produce that flow in the calculations. Additional standpipes must produce at least 250 gpm (946 L/min) as a supplemental measure. Each standpipe in the system is not anticipated to flow the 500 gpm (1893 L/min).

Since a standpipe system is intended to provide a backup service to the sprinkler system, a lower flow rate is permitted when the building is provided with a sprinkler system that covers all floor areas. The higher flow is needed when the building is provided with a partial sprinkler system.

7.10.1.1.4 Flow rates for combined systems shall be in accordance with 7.10.1.3.

7.10.1.1.4.1 When the floor area exceeds 80,000 ft^2 (7432 m^2), the second most remote standpipe shall be designed to accommodate 500 gpm (1893 L/min).

The intention of 7.10.1.1.4.1 is to require a higher flow for large area buildings, because a larger quantity of water must be available to handle the larger fire area that is anticipated.

It is intended that buildings that exceed 80,000 ft^2 (7432 m^2) per floor be provided with minimum flow rates for additional standpipes of 500 gpm (1893 L/min) for the second standpipe and 250 gpm (946 L/min) for the third standpipe if the additional flow is required for an unsprinklered building.

7.10.1.2* Hydraulic Calculation Requirements.

The characteristics of the attached water supply must be determined prior to calculating any system. For municipal or private water distribution systems, a hydrant flow test in accordance with NFPA 291, *Recommended Practice for Fire Flow Testing and Marking of Hydrants*, must be completed not more than 1 year prior to the commencement of design of the standpipe system. In any case, waterflow test data should never be more than 5 years old. Where fire pumps are installed, see Part II of this handbook for determining fire pump output.

Since manual standpipe systems are intended to be supplied by a fire department pumper through the fire department connection (FDC), the output from the fire department pumper should be ascertained. Lacking this information, the design specifications required by NFPA 1901, *Standard for Automotive Fire Apparatus*, can be used. According to NFPA 1901, a fire department pumper must meet the following three performance points:

1. 100 percent rated capacity at 150 psi (10.3 bar)
2. 70 percent rated capacity at 200 psi (13.8 bar)
3. 50 percent rated capacity at 250 psi (17.2 bar)

For example, a 1500 gpm (5677 L/min) pumper must meet the following specifications:

1. 1500 gpm (5677 L/min) at 150 psi (10.3 bar)
2. 1050 gpm (3974 L/min) at 200 psi (13.8 bar)
3. 750 gpm (2840 L/min) at 250 psi (17.2 bar)

The calculation procedures set forth in Chapter 22, Plans and Calculations, of NFPA 13 should be used to calculate the required flow and pressure for a standpipe system.

The calculation must begin at the hydraulically most demanding point in the system. This point is usually furthest from the supply. NFPA 14 mandates the starting pressure and flow as 250 gpm (946 L/min) at 100 psi (6.9 bar). The accumulated system demand, accounting for friction loss through piping, fittings, and other devices, and pressure adjustments due to elevation changes, must be less than the available water supply. If not, the calculation must be modified, usually by increasing the diameter of some piping.

A.7.10.1.2 See Section 22.4 of NFPA 13, *Standard for the Installation of Sprinkler Systems.*

When performing a hydraulic design, the hydraulic characteristics of each water supply need to be known. The procedure for determining the hydraulic characteristics of permanent water supplies, such as pumps, is fairly straightforward and is described in NFPA 20, *Standard for the Installation of Stationary Pumps for Fire Protection*. The procedure for determining the hydraulic characteristics of fire apparatus supplying a standpipe system are similar. Lacking better information about local fire apparatus, a conservative design would accommodate a 1000 gpm (3785 L/min) fire department pumper performing at the level of design specifications set forth in NFPA 1901, *Standard for Automotive Fire Apparatus* (hereinafter referred to as NFPA 1901). NFPA 1901 specifies that fire department pumpers must be able to achieve three pressure/flow combinations. These are 100 percent of rated capacity at 150 psi (1034 kPa) net pump pressure, 70 percent of rated capacity at 200 psi (1379 kPa) net pump pressure, and 50 percent of rated capacity at 250 psi (1724 kPa) net pump pressure. Therefore, a 1000 gpm (3785 L/min) pumper can be expected to deliver no less than 1000 gpm (3785 L/min) at 150 psi (1034 kPa), 700 gpm (2650 L/min) at 200 psi (1379 kPa), and 500 gpm (1893 L/min) at 250 psi (1724 kPa). Residual supply pressure on the suction side of a pump from a municipal or other pressurized water supply can also be added.

To perform a hydraulic design, one should determine the minimum required pressure and flow at the hydraulically most remote hose connection and calculate this demand back through system piping to each water supply, accumulating losses for friction and elevation changes and adding flows for additional standpipes and sprinklers at each point where such standpipes or sprinklers connect to the hydraulic design path. When considering fire apparatus as a water supply, flows are calculated from system piping through the fire department connection and back through connecting hoses to the pump. If the pressure available at each supply source exceeds a standpipe system's pressure demand at the designated flow, the design is acceptable. Otherwise, the piping design or the water supply need to be adjusted.

The intent of the standard is to require that each vertical standpipe serving two or more hose connections be capable of individually flowing 500 gpm (1893 L/min) and 250 gpm (946 L/min) at each of the two hydraulically most demanding connections at the required residual pressure. Given the requirement in 7.10.1.1.4.1 for the hydraulically most remote standpipe to supply this pressure and flow rate and given the minimum standpipe sizes in Section 7.6, the ability of standpipes that are not hydraulically most remote to satisfy this requirement is implicit and should not require additional hydraulic calculations.

Note that a calculation of a manual standpipe system can be made based on the fire department pumper. See Section 22.4 of NFPA 13 for Hydraulic Calculation Procedures found in NFPA 13. Calculating the system demand back to the FDC is allowed for buildings where the building height is not beyond the pumping capacity of the fire department. In such cases, the standpipe system is calculated by providing 250 gpm (946 L/min) at each of the two topmost outlets on the hydraulically most demanding standpipe at 100 psi (6.9 bar) and adding 250 gpm (946 L/min) for each additional standpipe, up to the maximum flow required.

This procedure is basically the same as that required for automatic standpipe systems, but in this case the calculations terminate at the FDC. Manual standpipe systems, therefore, do not require a fire pump.

In cases where the sprinkler system portion of a combined sprinkler/standpipe system may dictate the need for a stationary fire pump, the stationary fire pump does not need to be sized to handle the standpipe system's demand. It cannot be overemphasized that this calculation method only applies to manual standpipe systems where the building height is not beyond the pumping capacity of the fire department.

7.10.1.2.1 Hydraulic calculations and pipe sizes for each standpipe shall be based on providing 250 gpm (946 L/min) at the two hydraulically most remote hose connections on the standpipe and at the topmost outlet of each of the other standpipes at the minimum residual pressure required by Section 7.8.

This requirement was introduced into the standard for the 1993 edition of NFPA 14. At the time, the technical committee elected to retain the 500 gpm (1893 L/min) requirement for the hydraulically most demanding riser plus an additional 250 gpm (946 L/min) for other risers unless the floor area dictated a higher flow rate. Most fire departments use a gated wye connection to supply the 250 gpm (946 L/min) flow with two 1¾ in. (45 mm) connections each flowing 125 gpm (474 L/min). In order to flow the required 500 gpm (1893 L/min), four hose lines attached to two connections each on different floors of the building would be necessary.

Where a standpipe system has risers that terminate at different floor levels, it is the intent of NFPA 14 to allow separate hydraulic calculations for the standpipes that exist on each level. In each case, flow only needs to be added for standpipes that exist on the floor level of the calculations.

7.10.1.2.2 Where a horizontal standpipe on a Class I and Class III system supplies three or more hose connections on any floor, hydraulic calculations and pipe sizes for each standpipe shall be based on providing 250 gpm (946 L/min) at the three hydraulically most remote hose connections on the standpipe and at the topmost outlet of each of the other standpipes at the minimum residual pressure required by Section 7.8.

7.10.1.2.3 Common supply piping shall be calculated and sized to provide the required flow rate for all standpipes connected to such supply piping, with the total not to exceed 1250 gpm (4731 L/min).

7.10.1.3 Combined Systems.

7.10.1.3.1 For a building protected throughout by an approved automatic sprinkler system, the system demand established by Section 7.7 and 7.10.1 also shall be permitted to serve the sprinkler system.

Figures A.6.3.5(a) and A.6.3.5(b) from NFPA 14 illustrate the proper method for connecting a sprinkler system to a standpipe. The combination sprinkler/standpipe riser must be sized to accommodate the flow and pressure from either system. In cases where the building is not completely sprinklered, the combination sprinkler/standpipe riser must be sized to accommodate both system demands simultaneously.

7.10.1.3.1.1 Where the sprinkler system water supply requirement, including the hose stream allowance as determined in accordance with NFPA 13, *Standard for the Installation of Sprinkler Systems,* exceeds the system demand established by Section 7.7 and 7.10.1, the larger of the two values shall be provided.

In most cases the standpipe system demand will be much greater than the sprinkler demand. Therefore, calculate the standpipe system first to establish the appropriate size of feedmain piping and stationary fire pump.

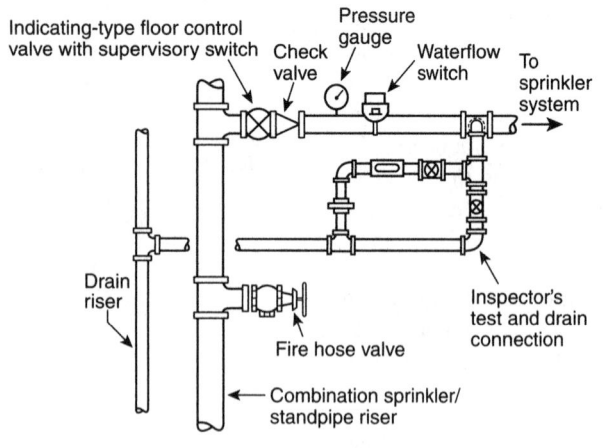

FIGURE A.6.3.5(a) *Acceptable Piping Arrangement for Combined Sprinkler/Standpipe System.*

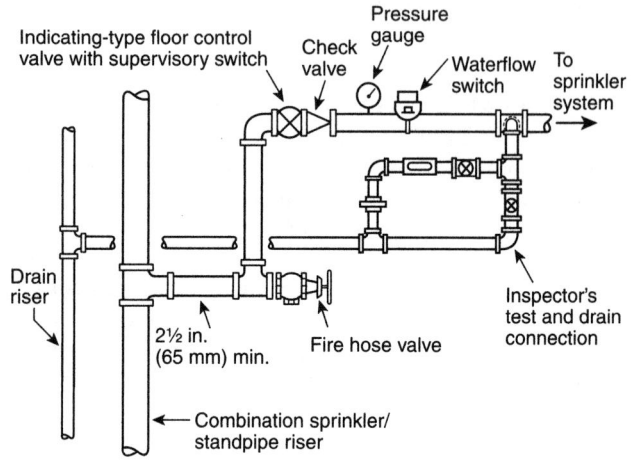

FIGURE A.6.3.5(b) *Combined Sprinkler/Standpipe System.*

7.10.1.3.1.2 A separate sprinkler demand shall not be required.

7.10.1.3.2 For a combined system in a building equipped with partial automatic sprinkler protection, the flow rate required by 7.10.1 shall be increased by an amount equal to the hydraulically calculated sprinkler demand or 150 gpm (568 L/min) for light hazard occupancies, or by 500 gpm (1893 L/min) for ordinary hazard occupancies, whichever is less.

 For situations where only partial sprinkler protection is provided, the standpipe system must be calculated as discharging simultaneously with the sprinkler system. Where complete sprinkler protection is provided, each system (sprinkler and standpipe) can be calculated as discharging separately.

WATER SUPPLY

9.1.5 Water supplies from the following sources shall be permitted:

(1) A public waterworks system where pressure and flow rate are adequate
(2) Automatic fire pumps connected to an approved water source in accordance with NFPA 20, *Standard for the Installation of Stationary Pumps for Fire Protection*
(3) Manually controlled fire pumps in combination with pressure tanks
(4) Pressure tanks installed in accordance with NFPA 22, *Standard for Water Tanks for Private Fire Protection*
(5) Manually controlled fire pumps operated by remote control devices at each hose station, supervised in accordance with *NFPA 72, National Fire Alarm Code,* at each hose station
(6) Gravity tanks installed in accordance with NFPA 22, *Standard for Water Tanks for Private Fire Protection*

SUMMARY

Part I, Section 1 of this handbook covers water supply requirements for standpipe systems and outlines the considerable flow and pressure requirements needed by standpipes in order to provide fire fighters with a sufficient tool to fight interior structure fires. The pressure and flow demands of standpipe systems often create the need for fire pumps, more so than most other fire protection systems. On most projects, the standpipe system is calculated first to establish the size of feedmain and supplymain piping, and the standpipe calculations are frequently used to determine the size and capacity of the fire pump.

REFERENCES CITED IN COMMENTARY

National Fire Protection Association, 1 Batterymarch Park, Quincy, MA 02169-7471.

NFPA 13, *Standard for the Installation of Sprinkler Systems,* 2010 edition.
NFPA 13E, *Recommended Practice for Fire Department Operations in Properties Protected by Sprinkler and Standpipe Systems,* 2005 edition.
NFPA 14, *Standard for the Installation of Standpipe and Hose Systems,* 2007 edition.
NFPA 25, *Standard for the Inspection, Testing, and Maintenance of Water-Based Fire Protection Systems,* 2008 edition.
NFPA 291, *Recommended Practice for Fire Flow Testing and Marking of Hydrants,* 2010 edition.
NFPA 1901, *Standard for Automotive Fire Apparatus,* 2009 edition.

Sprinkler System Demand:

Extracts from NFPA 13, 2010 Edition

SECTION 2

This section outlines the hydraulic requirements for sprinkler systems that establish the need for a fire pump. Sprinkler systems that are not combined with standpipe systems can create a significant water and pressure demand based on the nature of the protected hazard.

Section 11.1 from NFPA 13, *Standard for the Installation of Sprinkler Systems,* 2010 edition, establishes the basic hydraulic criteria that apply to most sprinkler systems. Storage occupancies are one of the few exceptions, which are covered in Chapters 12 through 20 of NFPA 13. This section of the handbook focuses on extra hazard hydraulic criteria, because the high flows and pressures needed for these hazards typically require a fire pump. The room design method and light and ordinary hazard design approaches are not covered here, as these design approaches do not require a high flow or pressure demand and usually do not drive the need for a fire pump. Examples of hydraulic criteria that drive the need for a fire pump are selected from Chapter 11, because this chapter is the most frequently used for sprinkler systems, and from Section 14.4, since this section describes the application of early suppression fast response (ESFR) sprinklers producing the highest system demand for storage applications (thus the need for a fire pump).

DESIGN APPROACHES

11.1.1 A building or portion thereof shall be permitted to be protected in accordance with any applicable design basis at the discretion of the designer.

The designer in most cases selects the criteria that are the least demanding in an effort to produce a cost-effective design. Some occupancies such as extra hazard require higher flow and pressure, and a cost effective design is difficult to accomplish. In any case, the designer has the responsibility to select the criteria that match the hazard. The authority having jurisdiction will only confirm that the design criteria is appropriate.

11.1.2* Adjacent Hazards or Design Methods. For buildings with two or more adjacent hazards or design methods, the following shall apply:

(1) Where areas are not physically separated by a barrier or partition capable of delaying heat from a fire in one area from fusing sprinklers in the adjacent area, the required sprinkler protection for the more demanding design basis shall extend 15 ft (4.6 m) beyond its perimeter.
(2) The requirements of 11.1.2(1) shall not apply where the areas are separated by a barrier partition that is capable of preventing heat from a fire in one area from fusing sprinklers in the adjacent area.

Varying hazards and the resultant design areas necessitate multiple submissions of hydraulic calculations to prove that the most demanding area was selected. The supplymains and feedmains should be sized based on the most demanding area chosen.

A.11.1.2 The situation frequently arises where a small area of a higher hazard is surrounded by a lesser hazard. For example, consider a 600 ft² (55.7 m²) area consisting of 10 ft (3.05 m) high on-floor storage of cartoned unexpanded plastic commodities surrounded by a plastic extruding operation in a 15 ft (4.57 m) high building. In accordance with Chapter 12, the density required for the plastic storage must meet the requirements for extra hazard (Group 1) occupancies. The plastic extruding operation should be considered an ordinary hazard (Group 2) occupancy. In accordance with Chapter 11, the corresponding discharge densities should be 0.3 gpm/ft^2 (12.2 mm/min) over 2500 ft^2 (232 m^2) for the storage and 0.2 gpm/ft^2 (8.1 mm/min) over 1500 ft^2 (139 m^2) for the remainder of the area. (*Also see Chapter 11 for the required minimum areas of operation.*)

If the storage area is not separated from the surrounding area by a wall or partition (*see 11.1.2*), the size of the operating area is determined by the higher hazard storage.

For example, the operating area is 2500 ft^2 (232 m^2). The system must be able to provide the 0.3 gpm/ft^2 (12.2 mm/min) density over the storage area and 15 ft (4.57 m) beyond. If part of the remote area is outside the 600 ft^2 (55.7 m^2) plus the 15 ft (4.57 m) overlap, only 0.2 gpm/ft^2 (8.1 mm/min) is needed for that portion.

If the storage is separated from the surrounding area by a floor-to-ceiling/roof partition that is capable of preventing heat from a fire on one side from fusing sprinklers on the other side, the size of the operating area is determined by the occupancy of the surrounding area. In this example, the design area is 1500 ft^2 (139 m^2). A 0.3 gpm/ft^2 (12.2 mm/min) density is needed within the separated area with 0.2 gpm/ft^2 (8.1 mm/min) in the remainder of the remote area.

11.1.3 For hydraulically calculated systems, the total system water supply requirements for each design basis shall be determined in accordance with the procedures of Section 22.4 unless modified by a section of Chapter 11 or Chapter 12.

11.1.4 Water Demand.

11.1.4.1* The water demand requirements shall be determined from the following:

(1) Occupancy hazard fire control approach and special design approaches of Chapter 11
(2) Storage design approaches of Chapter 12 through Chapter 20
(3) Special occupancy approaches of Chapter 21

A.11.1.4.1 See Section A.4.3.

11.1.4.2* The minimum water demand requirements for a sprinkler system shall be determined by adding the hose stream allowance to the water demand for sprinklers.

A.11.1.4.2 Appropriate area–density, other design criteria, and water supply requirements should be based on scientifically based engineering analyses that can include submitted fire testing, calculations, or results from appropriate computational models.

Recommended water supplies anticipate successful sprinkler operation. Because of the small but still significant number of uncontrolled fires in sprinklered properties, which have various causes, there should be an adequate water supply available for fire department use.

The hose stream demand required by this standard is intended to provide the fire department with the extra flow they need to conduct mop-up operations and final extinguishment of a fire at a sprinklered property. This is not the fire department manual fire flow, which is determined by other codes or standards. However, it is not the intent of this standard to require that the sprinkler demand be added to the fire flow demand. The sprinkler demand should be permitted to be considered in the fire flow required by other codes and standards.

11.1.5 Water Supplies.

11.1.5.1 The minimum water supply shall be available for the minimum duration specified in Chapter 11.

In addition to the sprinkler demand, hose demand must be included in the hydraulic calculations because flow from either building occupant hose or fire department hose may be employed simultaneously. This demand should be included in the calculations at the point of connection.

11.1.5.2* Tanks shall be sized to supply the equipment that they serve.

A.11.1.5.2 Where tanks serve sprinklers only, they can be sized to provide the duration required for the sprinkler system, ignoring any hose stream demands. Where tanks serve some combination of sprinklers, inside hose stations, outside hose stations or domestic/process use, the tank needs to be capable of providing the duration for the equipment that is fed from the tank, but the demands of equipment not connected to the tank can be ignored. Where a tank is used for both domestic/process water and fire protection, the entire duration demand of the domestic/process water does not need to be included in the tank if provisions are made to segregate the tank so that adequate fire protection water is always present or if provisions are made to automatically cut off the simultaneous use in the event of fire.

11.1.5.3* Pumps shall be sized to supply the equipment that they serve.

A.11.1.5.3 Where pumps serve sprinklers only, they can be sized to provide the flow required for the sprinkler system, ignoring any hose stream demands. Where pumps serve some combination of sprinklers, inside hose stations, or outside hose stations, the pump needs to be capable of providing the flow for the equipment that is fed from the pump, but the demands of equipment not connected to the pump can be ignored.

11.1.6 Hose Allowance.

11.1.6.1 An allowance for inside and outside hose shall not be required where tanks supply sprinklers only.

When sprinklers are supplied by water storage tanks, normally hose connections are not provided. Hose demand in these cases should not be included in the calculations to avoid unnecessarily oversizing the tank.

11.1.6.2 Systems with Multiple Hazard Classifications. For systems with multiple hazard classifications, the hose stream allowance and water supply duration shall be in accordance with one of the following:

(1) The water supply requirements for the highest hazard classification within the system
(2) The water supply requirements for each individual hazard classification shall be used in the calculations for the design area for that hazard.
(3)* For systems with multiple hazard classifications where the higher classification only lies within single rooms less than or equal to 400 ft^2 (37.2 m^2) in area with no such rooms adjacent, the water supply requirements for the principal occupancy shall be used for the remainder of the system.

A.11.1.6.2(3) When a light hazard occupancy, such as a school, contains separate ordinary hazard rooms no more than 400 ft^2 (37.2 m^2), the hose stream allowance and water supply duration would be that required for a light hazard occupancy.

11.1.6.3 Where pumps taking suction from a private fire service main supply sprinklers only, the pump need not be sized to accommodate inside and outside hose. Such hose allowance shall be considered in evaluating the available water supplies.

Where pumps serve sprinklers only, they can be sized to provide the flow required for the sprinkler system, ignoring any hose stream demands. Where pumps serve some combination of sprinklers, inside hose stations, or outside hose stations, the intent of 11.1.6.3 is to require the pump to be capable of providing the flow for the equipment that is fed from the pump, but the demands of equipment not connected to the pump can be ignored.

11.1.6.4* Water allowance for outside hose shall be added to the sprinkler requirement at the connection to the city main or a private fire hydrant, whichever is closer to the system riser.

A.11.1.6.4 When the hose demand is provided by a separate water supply, the sprinkler calculation does not include the outside hose demand.

11.1.6.5 Where inside hose connections are planned or are required, the following shall apply:

(1) A total water allowance of 50 gpm (189 L/min) for a single hose connection installation shall be added to the sprinkler requirements.
(2) A total water allowance of 100 gpm (379 L/min) for a multiple hose connection installation shall be added to the sprinkler requirements.
(3) The water allowance shall be added in 50 gpm (189 L/min) increments beginning at the most remote hose connection, with each increment added at the pressure required by the sprinkler system design at that point.

Paragraph 11.1.6.5 simply requires that the flow for a hose station be added to the calculations without consideration of pressure. This added demand creates additional friction loss in the system piping and, of course, produces a higher flow demand at the water supply.

11.1.6.6* When hose valves for fire department use are attached to wet pipe sprinkler system risers in accordance with 8.17.5.2, the following shall apply:

(1) The sprinkler system demand shall not be required to be added to standpipe demand as determined from NFPA 14, *Standard for the Installation of Standpipe and Hose Systems.*
(2) Where the combined sprinkler system demand and hose stream allowance of Table 11.2.3.1.2 exceeds the requirements of NFPA 14, *Standard for the Installation of Standpipe and Hose Systems,* this higher demand shall be used.
(3) For partially sprinklered buildings, the sprinkler demand, not including hose stream allowance, as indicated in Figure 11.2.3.1.1 shall be added to the requirements given in NFPA 14, *Standard for the Installation of Standpipe and Hose Systems.*

A.11.1.6.6 For fully sprinklered buildings, if hose valves or stations are provided on a combination sprinkler riser and standpipe for fire department use in accordance with NFPA 14, *Standard for the Installation of Standpipe and Hose Systems,* the hydraulic calculation for the sprinkler system is not required to include the standpipe allowance.

Only on rare occasions does the sprinkler demand need to be added to the standpipe system demand because most buildings are provided with a complete sprinkler system throughout. As a result, calculating a sprinkler system and a standpipe system discharging simultaneously is unusual. Sprinkler system demand approaches that of a standpipe system, beginning with Extra Hazard Group I criteria or higher. The following calculation is based on Table 11.2.3.1.2 from NFPA 13:

$$0.3 \text{ gpm} \times 2500 \text{ ft}^2 = 750 \text{ gpm} + 250 \text{ gpm (hose)} = 1000 \text{ gpm}$$

or in metrics,

$$12.2 \text{ L/min} \times 232 \text{ m}^2 = 2839 \text{ L/min} + 946 \text{ L/min (hose)} = 3785 \text{ L/min}$$

As stated previously in the standpipe section, standpipe system demand is typically 1000 gpm (3785 L/min) for buildings that are protected throughout by an automatic sprinkler system and 1250 gpm (4731 L/min) for buildings with only partial protection. However, consideration of the pressure requirement for standpipe systems is included in this calculation.

Even with the highest densities and spacing of sprinklers, the starting pressure for a sprinkler system when using the density/area curves is considerably below that of a standpipe system. Consider a sprinkler spacing of 130 ft² (12.1 m²) as permitted by Table 8.6.2.2.1(c) and a discharge density of 0.4 gpm/ft² (16.3 mm/min) from Table 11.2.3.1.2 from NFPA 13:

$$130 \text{ ft}^2 \times 0.4 \text{ gpm/ft}^2 = 52 \text{ gpm}$$

$$12.1 \text{ m}^2 \times 16.3 \text{ L/min/m}^2 = 197 \text{ L/min}$$

The most demanding sprinkler in this system must discharge 52 gpm (197 L/min) to meet this criterion. In order to achieve this discharge with a ½ in. orifice (K-5.6) sprinkler, a starting pressure of 86 psi (5.9 bar) would be required:

$$p = (Q/k)^2$$

$$p = (52 \text{ gpm}/5.6)^2$$

$$p = 86 \text{ psi (5.9 bar)}$$

where:

p = pressure (psi)
Q = flow (gpm)
k = K-factor of sprinkler

Depending on the amount of friction loss in the system piping, a standpipe system in most cases still presents the higher system demand.

ESFR SPRINKLERS

Extracts from Section 14.4 of NFPA 13 were selected as an example of sprinkler systems that may require fire pumps because of the high flow rates and pressures needed for the proper operation of ESFR sprinklers. A wide variety of criteria for storage applications exist that include standard spray sprinklers, large drop sprinklers, and rack sprinklers. None of these criteria, however, approach those of the ESFR sprinkler in terms of flow and pressure demands. The following is an example of the hydraulic demand for K-25.2 ESFR sprinklers, although other K factor configurations from Table 14.4.1 can produce flow and pressure demands that drive the need for fire pumps.

14.4 Early Suppression Fast-Response (ESFR) Sprinklers for Palletized or Solid Piled Storage of Class I Through Class IV Commodities.

14.4.1 Protection of palletized and solid-piled storage of Class I through Class IV commodities shall be in accordance with Table 14.4.1.

14.4.2 ESFR sprinkler systems shall be designed such that the minimum operating pressure is not less than that indicated in Table 14.4.1 for commodity, storage height, and building height involved.

Assuming a maximum ceiling/roof height of 45 ft (13.7 m), a maximum storage height of 40 ft (12.2 m), and the use of a K-25.2 ESFR sprinkler, a minimum operating pressure of 40 psi (2.8 bar) is required. Therefore,

$$Q = k \times \sqrt{p}$$

$$Q = 25.2 \times \sqrt{40}$$

$$Q = 159.37 \text{ gpm (603.2 L/min)}$$

14.4.3 The design area shall consist of the most hydraulically demanding area of 12 sprinklers, consisting of four sprinklers on each of three branch lines.

TABLE 14.4.1 ESFR Protection of Palletized and Solid-Piled Storage of Class I Through Class IV Commodities

Commodity	Maximum Storage Height ft	Maximum Storage Height m	Maximum Ceiling/Roof Height ft	Maximum Ceiling/Roof Height m	Nominal K-Factor	Orientation	Minimum Operating Pressure psi	Minimum Operating Pressure bar	Hose Stream Allowance gpm	Hose Stream Allowance L/min	Water Supply Duration (hours)
Class I, II, III, or IV, encapsulated (no open-top containers or solid shelves)	20	6.1	25	7.6	14.0 (201)	Upright or pendent	50	3.4	250	946	1
					16.8 (242)	Upright/Pendent	35	2.4			
					22.4 (322)	Pendent	25	1.7			
					25.2 (363)	Pendent	15	1.0			
	25	7.6	30	9.1	14.0 (201)	Upright or pendent	50	3.4			
					16.8 (242)	Upright/Pendent	35	2.4			
					22.4 (322)	Pendent	25	1.7			
					25.2 (363)	Pendent	15	1.0			
			32	9.8	14.0 (201)	Upright or pendent	60	4.1			
					16.8 (242)	Pendent	42	2.9			
	30	9.1	35	10.7	14.0 (201)	Upright or pendent	75	5.2			
					16.8 (242)	Upright/Pendent	52	3.6			
					22.4 (322)	Pendent	35	2.4			
					25.2 (363)	Pendent	20	1.4			
	35	10.7	40	12.2	14.0 (201)	Pendent	75	5.2			
					16.8 (242)	Upright/Pendent	52	3.6			
					22.4 (322)	Pendent	40	2.8			
					25.2 (363)	Pendent	25	1.7			
	35	10.7	45	13.7	22.4 (322)	Pendent	40	2.8			
					25.2 (363)	Pendent	40	2.8			
	40	12.2	45	13.7	22.4 (322)	Pendent	40	2.8			
					25.2 (363)	Pendent	40	2.8			

This subsection specifies the exact number of sprinklers that must be considered to be discharging in the calculation. Therefore,

$$159.2 \text{ gpm} \times 12 = 1910 \text{ gpm} (7230.8 \text{ L/min}) + 250 \text{ gpm for hose}$$
$$= 2160 \text{ gpm} (8175 \text{ L/min})$$

This flow is a minimum rate as it does not include over-discharge due to friction loss in the branch lines or between branch lines. Usually a sprinkler system discharges between 110 to 120 percent of the minimum specified discharge due to friction and balancing. The calculated flow would be needed at a pressure in excess of 59 psi (4.1 bar) as follows:

$$44 \text{ ft} (13.4 \text{ m}) \times 0.433 \text{ psi/ft} (0.097 \text{ bar/m})$$
$$= 19 \text{ psi} (1.3 \text{ bar}) + 40 \text{ psi} (2.8 \text{ bar}) \text{ (starting pressure)}$$
$$= 59 \text{ psi} (4.1 \text{ bar})$$

The 44 ft (13.4 m) elevation is estimated to be the elevation of the sprinkler piping [approximately 1 ft (0.3 m) below the roof height] and the 40 psi (2.8 bar) pressure is the minimum operating pressure for the K-25.2 sprinkler as referenced in Table 14.4.1. This estimated calculation does not include friction loss through the piping system, which can be substantial based on the overall size of the system. Therefore, initially for this system, a total system demand of 2160 gpm (8175 L/min) at a minimum pressure of 59 psi (4.1 bar) is needed. Most municipal water supply systems and private water supply systems cannot meet this demand without the assistance of a fire pump.

SUMMARY

Part I, Section 2, of this handbook covers extra hazard hydraulic criteria water supply requirements for sprinkler systems. In systems protecting extra hazard occupancies or for ESFR systems protecting storage of Class I through Class IV commodities, both high flows and sometimes high pressures are needed. In most cases a fire pump is necessary to meet the hydraulic demands of these systems.

REFERENCE CITED IN COMMENTARY

National Fire Protection Association, 1 Batterymarch Park, Quincy, MA 02169-7471.

NFPA 13, *Standard for the Installation of Sprinkler Systems,* 2010 edition.

PART II

NFPA® 20, *Standard for the Installation of Stationary Pumps for Fire Protection,* with Commentary

Part II of this handbook includes the complete text of the 2010 edition of NFPA 20, *Standard for the Installation of Stationary Pumps for Fire Protection.* The standard consists of mandatory core chapters (1 through 14) and nonmandatory annex material. The mandatory provisions found in Chapters 1 through 14 were prepared by the Technical Committee on Fire Pumps within the framework of NFPA's consensus codes and standards development process. Because these provisions are designed to be suitable for adoption into law, or for reference by other codes and standards, the text is concise, without extended explanation.

The material found in Annex A of the standard was also developed by the fire pump committee within NFPA's codes and standards process. The annex material is designed to assist users in interpreting the mandatory provisions. This text is not considered to be part of the requirements of the standard; it is advisory or informational. An asterisk (*) following a chapter paragraph number indicates that nonmandatory material pertaining to that paragraph appears in Annex A. For readers' convenience in this handbook, Annex A material has been repositioned to appear immediately following its base paragraph in the body of the standard.

The explanatory commentary accompanying the standard was prepared by the handbook editors. The commentary immediately follows the part of NFPA 20 it discusses and is easily identified by its brown type. Designed to help users understand and apply provisions of the standard, the commentary gives detailed explanations of the reasoning behind requirements, examples of calculations, applications of requirements, and illustrations of key fire pump components.

This edition of the handbook includes a frequently asked question (FAQ) feature. The marginal FAQs are based on the questions most commonly asked of the NFPA 20 staff. In addition, "design alert" and "calculation" icons indicate material of special interest.

Administration

CHAPTER 1

Chapter 1 of NFPA 20, *Standard for the Installation of Stationary Pumps for Fire Protection,* covers the administrative requirements for the selection and installation of fire pumps. In addition to the scope and purpose of the standard, Chapter 1 provides guidance on the ongoing use of existing fire pumps and the use of new equipment or technologies that may not be referenced in the standard. Chapter 1 also provides guidance on the appropriate units and trade sizes used in the standard.

1.1* Scope.

NFPA 20 is an installation standard for stationary fire pumps. As an installation standard, NFPA 20 does not specify when or if fire pumps must be installed to supplement an existing water supply. The purpose of the standard is to specify how to install a fire pump properly when one is needed and which components, equipment, and power supplies are acceptable for use in a fire pump installation. In other words, NFPA 20 indicates *how* to properly arrange and install a fire pump and its supporting equipment. The standard does not identify *when* a fire pump is required or needed.

◀ FAQ
Does NFPA 20 contain requirements mandating when a fire pump must be installed?

Stationary fire pumps, such as the centrifugal or positive displacement types, are pumps that are permanently installed in a building. The term *stationary* is used to distinguish between the types of pumps described in NFPA 20 and the motorized types used by public and private fire departments.

The need for a fire pump is usually determined through an analysis of the fire protection system being considered, the water supply required for the system, and the water supply available, as indicated in Part I of this handbook. Fire pumps are principally used where pressure from the attached water supply is insufficient. Where capacity of the water supply is also in question, means in addition to a fire pump, such as water tanks, also need to be considered.

◀ FAQ
How is the need for a fire pump determined?

When considering whether a fire pump is necessary, the reader should refer to the other National Fire Protection Association (NFPA) standards that address various types of water-based fire protection systems. These standards include the following documents:

NFPA 11, *Standard for Low-, Medium-, and High-Expansion Foam*

NFPA 13, *Standard for the Installation of Sprinkler Systems*

NFPA 13D, *Standard for the Installation of Sprinkler Systems in One- and Two-Family Dwellings and Manufactured Homes*

NFPA 13R, *Standard for the Installation of Sprinkler Systems in Residential Occupancies up to and Including Four Stories in Height*

NFPA 14, *Standard for the Installation of Standpipe and Hose Systems*

NFPA 15, *Standard for Water Spray Fixed Systems for Fire Protection*

NFPA 16, *Standard for the Installation of Foam-Water Sprinkler and Foam-Water Spray Systems*

NFPA 750, *Standard on Water Mist Fire Protection Systems*

Other NFPA documents address water supplies and their components and piping arrangements. These documents include the following:

NFPA 22, *Standard for Water Tanks for Private Fire Protection*
NFPA 24, *Standard for the Installation of Private Fire Service Mains and Their Appurtenances*

A.1.1 For more information, see NFPA 25, *Standard for the Inspection, Testing, and Maintenance of Water-Based Fire Protection Systems,* and *NFPA 70, National Electrical Code,* Article 695.

1.1.1 This standard deals with the selection and installation of pumps supplying liquid for private fire protection.

1.1.2 The scope of this document shall include liquid supplies; suction, discharge, and auxiliary equipment; power supplies, including power supply arrangements; electric drive and control; diesel engine drive and control; steam turbine drive and control; and acceptance tests and operation.

In addition to the selection of the fire pump, NFPA 20 includes minimum requirements for the attached water supply, power supply arrangement, type of driver, and acceptance test procedures.

1.1.3 This standard does not cover system liquid supply capacity and pressure requirements, nor does it cover requirements for periodic inspection, testing, and maintenance of fire pump systems.

For liquid supply capacity and pressure requirements, see the appropriate system installation standard such as NFPA 13. For inspection, testing, and maintenance requirements for existing fire pump installations, see NFPA 25, *Standard for the Inspection, Testing, and Maintenance of Water-Based Fire Protection Systems.*

1.1.4 This standard does not cover the requirements for installation wiring of fire pump units.

See *NFPA 70®, National Electrical Code®,* for requirements related to the installation of wiring for fire pumps.

1.2 Purpose.

The purpose of this standard is to provide a reasonable degree of protection for life and property from fire through installation requirements for stationary pumps for fire protection based upon sound engineering principles, test data, and field experience.

FAQ ▶
Is the purpose of a fire pump to create water or can a fire pump only boost pressure?

Fire pump installations are a critical and essential component of the water supply for a fire protection system, as they provide the necessary system flow and pressure. The performance of any water-based fire protection system is dependent on the availability, adequacy, and reliability of the water supply to which it is connected. Fire pumps installed in accordance with NFPA 20 and inspected, tested, and maintained in accordance with NFPA 25 ensure that the available water supply will have the necessary operating pressure and flow when needed during an emergency. An important point to recognize, however, is that fire pumps cannot create water; they can only boost or supply the pressure and flow in an available water source. Fire pumps do not increase the capacity of a water supply.

NFPA 20 provides the minimum requirements needed for the satisfactory operation of all types of listed fire pumps. Even though the overall level of performance intended might be enhanced by exceeding the minimum requirements stated, the provisions of NFPA 20 allow for proper and effective fire pump performance.

1.3 Application.

1.3.1 This standard shall apply to centrifugal single-stage and multistage pumps of the horizontal or vertical shaft design and positive displacement pumps of the horizontal or vertical shaft design.

NFPA 20 covers every type of listed fire pump. The most commonly used type in fire protection is centrifugal, which consists of horizontal split case, end suction, and in-line. Other types of fire pumps include piston plunger, positive displacement, rotary lobe, rotary vane, and vertical lineshaft turbine.

The title of the 1999 edition of the standard was revised to indicate that the standard dealt with all stationary pumps and not just centrifugal fire pumps. At that time, the standard began to include positive displacement pumps that pump water additives such as foam concentrate. Section 1.3 was added to indicate the intended application of both centrifugal and positive displacement fire pumps.

◀ FAQ
Which types of fire pumps does NFPA 20 cover?

1.3.2 Requirements are established for the design and installation of single-stage and multistage pumps, pump drivers, and associated equipment.

1.4 Retroactivity.

The provisions of this standard reflect a consensus of what is necessary to provide an acceptable degree of protection from the hazards addressed in this standard at the time the standard was issued.

The retroactivity statement in Section 1.4 appears in most NFPA documents. The section's purpose is to reinforce the premise that any fire pump installed in accordance with the applicable edition of NFPA 20 is considered to be compliant with the standard for its lifetime as long as the associated hazards remain unchanged and the pump is properly inspected, tested, and maintained. Therefore, an existing installation is not required to be reviewed for compliance with every new edition of the standard.

◀ FAQ
Are fire pumps prescribed by older editions considered unsafe?

Omission of the retroactivity statement would require the never-ending task of updating and revising a fire pump installation every time a new edition of NFPA 20 is published. Although newer editions contain information that may provide a greater level of safety or a more effective means of accomplishing a certain objective than that prescribed by older editions, the provisions of older editions should not necessarily be interpreted as unsafe. In those instances where a severe deficiency is discovered, Section 1.4 allows latitude for authorities having jurisdiction to require an upgrade.

1.4.1 Unless otherwise specified, the provisions of this standard shall not apply to facilities, equipment, structures, or installations that existed or were approved for construction or installation prior to the effective date of the standard. Where specified, the provisions of this standard shall be retroactive.

1.4.2 In those cases where the authority having jurisdiction determines that the existing situation presents an unacceptable degree of risk, the authority having jurisdiction shall be permitted to apply retroactively any portions of this standard deemed appropriate.

1.4.3 The retroactive requirements of this standard shall be permitted to be modified if their application clearly would be impractical in the judgment of the authority having jurisdiction, and only where it is clearly evident that a reasonable degree of safety is provided.

1.5 Equivalency.

Nothing in this standard is intended to prevent the use of systems, methods, or devices of equivalent or superior quality, strength, fire resistance, effectiveness, durability, and safety over those prescribed by this standard.

The increasing availability of specially listed materials and products, along with the continuing development of new pump-related technologies, necessitates some of the language in Section 1.5. This section allows for products or arrangements not specifically addressed by the standard to be used, provided it can be demonstrated that the use of these products or arrangements does not lower the level of safety provided by the standard or alter the intent of the standard. This statement also alerts the user of NFPA 20 that specific requirements or limitations are often associated with specialized products and that these limitations are not addressed by the standard. Therefore, the listing information and the relevant manufacturers' literature will need to be reviewed.

1.5.1 Technical documentation shall be submitted to the authority having jurisdiction to demonstrate equivalency.

1.5.2 The system, method, or device shall be approved for the intended purpose by the authority having jurisdiction.

Subsection 1.5.2 clarifies that the authority having jurisdiction determines whether equivalency has been demonstrated based on technical documentation that has been submitted.

1.6 Units.

1.6.1 Metric units of measurement in this standard are in accordance with the modernized metric system known as the International System of Units (SI).

1.6.2 *Liter* and *bar* in this standard are outside of but recognized by SI.

1.6.3 Units are listed in Table 1.6.3 with conversion factors.

TABLE 1.6.3 System of Units

Name of Unit	Unit Abbreviation	Conversion Factor
Meter(s)	m	1 ft = 0.3048 m
Foot (feet)	ft	1 m = 3.281 ft
Millimeter(s)	mm	1 in. = 25.4 mm
Inch(es)	in.	1 mm = 0.03937 in.
Liter(s)	L	1 gal = 3.785 L
Gallon(s) (U.S.)	gal	1 L = 0.2642 gal
Cubic decimeter(s)	dm^3	1 gal = 3.785 dm^3
Cubic meter(s)	m^3	1 ft^3 = 0.0283 m^3
Cubic foot (feet)	ft^3	1 m^3 = 35.31 ft^3
Pascal(s)	Pa	1 psi = 6894.757 Pa; 1 bar = 10^5 Pa
Pound(s) per square inch	psi	1 Pa = 0.000145 psi; 1 bar = 14.5 psi
Bar	bar	1 Pa = 10^{-5} bar; 1 psi = 0.0689 bar

Note: For additional conversions and information, see IEEE/ASTM SI10, *Standard for Use of the International System of Units (SI): The Modern Metric System.*

1.6.4 Conversion. The conversion procedure is to multiply the quantity by the conversion factor and then round the result to an appropriate number of significant digits.

1.6.5 Trade Sizes. Where industry utilizes nominal dimensions to represent materials, products, or performance, direct conversions have not been utilized and appropriate trade sizes have been included.

REFERENCES CITED IN COMMENTARY

National Fire Protection Association, 1 Batterymarch Park, Quincy, MA 02169-7471.

NFPA 11, *Standard for Low-, Medium-, and High-Expansion Foam,* 2005 edition.
NFPA 13, *Standard for the Installation of Sprinkler Systems,* 2010 edition.
NFPA 13D, *Standard for the Installation of Sprinkler Systems in One- and Two-Family Dwellings and Manufactured Homes,* 2010 edition.
NFPA 13R, *Standard for the Installation of Sprinkler Systems in Residential Occupancies up to and Including Four Stories in Height,* 2010 edition.
NFPA 14, *Standard for the Installation of Standpipe and Hose Systems,* 2007 edition.
NFPA 15, *Standard for Water Spray Fixed Systems for Fire Protection,* 2007 edition.
NFPA 16, *Standard for the Installation of Foam-Water Sprinkler and Foam-Water Spray Systems,* 2007 edition.
NFPA 22, *Standard for Water Tanks for Private Fire Protection,* 2008 edition.
NFPA 24, *Standard for the Installation of Private Fire Service Mains and Their Appurtenances,* 2010 edition.
NFPA 25, *Standard for the Inspection, Testing, and Maintenance of Water-Based Fire Protection Systems,* 2008 edition.
NFPA 70®, National Electrical Code®, 2008 edition.
NFPA 750, *Standard on Water Mist Fire Protection Systems,* 2006 edition.

Referenced Publications

CHAPTER 2

This chapter lists the mandatory referenced publications. Annex C lists nonmandatory referenced publications. Because the information is located immediately after Chapter 1, Administration, the user is presented with the complete list of publications needed for effective use of the standard before reading the specific requirements. The provisions of the publications that are mandated by NFPA 20, *Standard for the Installation of Stationary Pumps for Fire Protection,* are also requirements. Regardless of whether an actual requirement resides within NFPA 20 or is mandatorily referenced and appears only in the referenced publication, the requirement must be met to achieve compliance with NFPA 20.

2.1 General.

The documents or portions thereof listed in this chapter are referenced within this standard and shall be considered part of the requirements of this document.

2.2 NFPA Publications.

National Fire Protection Association, 1 Batterymarch Park, Quincy, MA 02169-7471.

NFPA 13, *Standard for the Installation of Sprinkler Systems,* 2010 edition.
NFPA 22, *Standard for Water Tanks for Private Fire Protection,* 2008 edition.
NFPA 24, *Standard for the Installation of Private Fire Service Mains and Their Appurtenances,* 2010 edition.
NFPA 25, *Standard for the Inspection, Testing, and Maintenance of Water-Based Fire Protection Systems,* 2008 edition.
NFPA 37, *Standard for the Installation and Use of Stationary Combustion Engines and Gas Turbines,* 2006 edition.
NFPA 51B, *Standard for Fire Prevention During Welding, Cutting, and Other Hot Work,* 2009 edition.
NFPA 70®, *National Electrical Code*®, 2008 edition.
NFPA 101®, *Life Safety Code*®, 2009 edition.
NFPA 110, *Standard for Emergency and Standby Power Systems,* 2010 edition.
NFPA 1963, *Standard for Fire Hose Connections,* 2009 edition.

2.3 Other Publications.

2.3.1 AGMA Publications.

American Gear Manufacturers Association, 1500 King Street, Suite 201, Alexandria, VA 22314-2730.

AGMA 390.03, *Handbook for Helical and Master Gears,* 1995.

2.3.2 ANSI Publications.

American National Standards Institute, Inc., 25 West 43rd Street, 4th Floor, New York, NY 10036.

ANSI B15.1, *Safety Standard for Mechanical Power Transmission Apparatus,* 2000.
ANSI/IEEE C62.1, *IEEE Standard for Gapped Silicon-Carbide Surge Arresters for AC Power Circuits,* 1989.
ANSI/IEEE C62.11, *IEEE Standard for Metal-Oxide Surge Arresters for Alternating Current Power Circuits (>1 kV),* 2005.
ANSI/IEEE C62.41, *IEEE Recommended Practice for Surge Voltages in Low-Voltage AC Power Circuits,* 1991.

2.3.3 HI Publications.

Hydraulic Institute, 6 Campus Drive, First Floor North, Parsippany, NJ 07054-4406.

Hydraulic Institute Standards for Centrifugal, Rotary and Reciprocating Pumps, 14th edition, 1983.
HI 3.6, *Rotary Pump Tests,* 1994.

2.3.4 IEEE Publications.

Institute of Electrical and Electronics Engineers, Three Park Avenue, 17th Floor, New York, NY 10016-5997.

IEEE/ASTM SI10, *Standard for Use of the International System of Units (SI): The Modern Metric System,* 2003.

2.3.5 NEMA Publications.

National Electrical Manufacturers Association, 1300 North 17th Street, Suite 1847, Rosslyn, VA 22209.

NEMA MG-1, *Motors and Generators,* 1998.

2.3.6 UL Publications.

Underwriters Laboratories Inc., 333 Pfingsten Road, Northbrook, IL 60062-2096.

ANSI/UL 142, *Standard for Steel Aboveground Tanks for Flammable and Combustible Liquids,* 2006.
ANSI/UL 508, *Standard for Industrial Control Equipment,* 2005.

2.3.7 Other Publications.

Merriam-Webster's Collegiate Dictionary, 11th edition, Merriam-Webster, Inc., Springfield, MA, 2003.

2.4 References for Extracts in Mandatory Sections.

NFPA 14, *Standard for the Installation of Standpipe and Hose Systems,* 2007 edition.
NFPA 37, *Standard for the Installation and Use of Stationary Combustion Engines and Gas Turbines,* 2006 edition.
NFPA 70®, National Electrical Code®, 2008 edition.
NFPA 110, *Standard for Emergency and Standby Power Systems,* 2010 edition.
NFPA 1451, *Standard for a Fire Service Vehicle Operations Training Program,* 2007 edition.
NFPA 5000®, Building Construction and Safety Code®, 2009 edition.

Definitions

CHAPTER 3

All definitions that apply to subjects covered throughout the standard are located in Chapter 3. Otherwise, where terms are not defined, the ordinary meanings as defined in *Merriam-Webster's Collegiate Dictionary* apply. For definitions of other fire protection–related terms that do not appear in either NFPA 20, *Standard for the Installation of Stationary Pumps for Fire Protection,* or *Merriam-Webster's Collegiate Dictionary,* see the *NFPA Glossary of Fire Protection Terms.*

3.1 General.

The definitions contained in this chapter shall apply to the terms used in this standard. Where terms are not defined in this chapter or within another chapter, they shall be defined using their ordinarily accepted meanings within the context in which they are used. *Merriam-Webster's Collegiate Dictionary,* 11th edition, shall be the source for the ordinarily accepted meaning.

3.2 NFPA Official Definitions.

3.2.1* Approved. Acceptable to the authority having jurisdiction.

A.3.2.1 Approved. The National Fire Protection Association does not approve, inspect, or certify any installations, procedures, equipment, or materials; nor does it approve or evaluate testing laboratories. In determining the acceptability of installations, procedures, equipment, or materials, the authority having jurisdiction may base acceptance on compliance with NFPA or other appropriate standards. In the absence of such standards, said authority may require evidence of proper installation, procedure, or use. The authority having jurisdiction may also refer to the listings or labeling practices of an organization that is concerned with product evaluations and is thus in a position to determine compliance with appropriate standards for the current production of listed items.

The term *approved* is not the same as the term *listed,* which is defined in 3.2.3. A component or equipment that is required to be approved is not necessarily also required to be listed. The term *listed* indicates that a specific component or piece of equipment has been evaluated by an approved testing laboratory. The term *approved,* which is defined in 3.2.1, indicates acceptance by the authority having jurisdiction. In general, all components in a fire pump installation are required by NFPA 20 to be approved, but only specific components, such as the fire pump itself, are required to be listed.

◀ **FAQ**
What is the difference between the terms *approved* and *listed*?

3.2.2* Authority Having Jurisdiction (AHJ). An organization, office, or individual responsible for enforcing the requirements of a code or standard, or for approving equipment, materials, an installation, or a procedure.

A.3.2.2 Authority Having Jurisdiction (AHJ). The phrase "authority having jurisdiction," or its acronym AHJ, is used in NFPA documents in a broad manner, since jurisdictions and approval agencies vary, as do their responsibilities. Where public safety is primary, the authority having jurisdiction may be a federal, state, local, or other regional department or individual such as a fire chief; fire marshal; chief of a fire prevention bureau, labor department, or health department; building official; electrical inspector; or others having statutory authority. For insurance purposes, an insurance inspection department, rating bureau, or other insurance company representative may be the authority having jurisdiction. In many circumstances, the property owner or his or her designated agent assumes the role of the authority having jurisdiction; at government installations, the commanding officer or departmental official may be the authority having jurisdiction.

The phrase *authority having jurisdiction (AHJ)* is intended to include anyone responsible for enforcing the requirements of NFPA 20. On many construction projects, multiple AHJs, such as the building owner's representative, a code official (e.g., a fire marshal), and an insurance representative, are present. All of the aforementioned AHJs are involved in the plan review and acceptance process and issue an approval or comment on plan reviews and acceptance tests.

3.2.3* Listed. Equipment, materials, or services included in a list published by an organization that is acceptable to the authority having jurisdiction and concerned with evaluation of products or services, that maintains periodic inspection of production of listed equipment or materials or periodic evaluation of services, and whose listing states that either the equipment, material, or service meets appropriate designated standards or has been tested and found suitable for a specified purpose.

Components critical to proper system operation are required to be listed. One testing laboratory uses the designation *classified* as an indication that a specific product meets its evaluation criteria. Materials with this designation meet the intent of *listed*.

A.3.2.3 Listed. The means for identifying listed equipment may vary for each organization concerned with product evaluation; some organizations do not recognize equipment as listed unless it is also labeled. The authority having jurisdiction should utilize the system employed by the listing organization to identify a listed product.

3.2.4 Shall. Indicates a mandatory requirement.

The term *shall* indicates a requirement of this standard and mandates that a specific provision of NFPA 20 be followed. When the term *shall* is used with a specific provision of the standard, compliance with that provision is mandatory. Any exceptions to a requirement of this standard are specifically stated. The use of the term *shall* in the exception indicates that it is only under the conditions specified that the exception is applicable. Mandatory requirements of NFPA 20 are found in the main body of the standard (see Chapters 1 through 14).

3.2.5 Should. Indicates a recommendation or that which is advised but not required.

The term *should* indicates a recommendation of the standard. When the term *should* is used with a provision of the standard, the provision is not meant to be a mandatory requirement. Recommended provisions are limited to Annexes A and B of NFPA 20. The term *should* identifies a good idea or a better practice. If the recommendation is not followed, the fire pump is still expected to perform satisfactorily. Any section number preceded by a letter (e.g., A.1.1) is an annex item. Terms such as *should* and *recommend* are prevalent in annex paragraphs.

3.2.6 Standard. A document, the main text of which contains only mandatory provisions using the word "shall" to indicate requirements and which is in a form generally suitable for mandatory reference by another standard or code or for adoption into law. Nonmandatory provisions shall be located in an appendix or annex, footnote, or fine-print note and are not to be considered a part of the requirements of a standard.

3.3 General Definitions.

3.3.1 Additive. A liquid such as foam concentrates, emulsifiers, and hazardous vapor suppression liquids and foaming agents intended to be injected into the water stream at or above the water pressure.

The term *additive* was added to the standard in the 1999 edition because the document scope was expanded to include positive displacement pumps, which are used for pumping foam concentrate and emulsifiers.

3.3.2 Aquifer. An underground formation that contains sufficient saturated permeable material to yield significant quantities of water.

3.3.3 Aquifer Performance Analysis. A test designed to determine the amount of underground water available in a given field and proper well spacing to avoid interference in that field. Basically, test results provide information concerning transmissibility and storage coefficient (available volume of water) of the aquifer.

A thorough analysis of the water supply for a fire protection system is critical to determine the supply's adequacy and reliability. See Part III of this handbook for water supply testing requirements.

3.3.4 Automatic Transfer Switch. See 3.3.48.2.1.

3.3.5 Branch Circuit. See 3.3.7.1.

3.3.6 Break Tank. A tank providing suction to a fire pump whose capacity is less than the fire protection demand (flow rate times flow duration).

A break tank, as its name suggests, provides a break or separation between a fire pump and a city water supply. This separation is needed for two purposes: (1) to offer backflow protection to avoid contamination of the city water supply and (2) to avoid seasonal fluctuations in pressure, thus preventing overpressurization of the fire protection system.

◀ FAQ
What is the purpose of a break tank?

The definition *break tank* was added in the 2007 edition to describe the component that is covered in Section 4.31. Since some municipalities do not permit the direct connection of a fire pump to a water main due to the potential for backflow and overpressurization, a break tank can be used to create this gap. Break tanks are also used to eliminate pressure fluctuations in the attached water supply or as backflow prevention devices.

3.3.7 Circuit.

3.3.7.1 Branch Circuit. The circuit conductors between the final overcurrent device protecting the circuit and the outlet(s). [70: Art. 100]

3.3.7.2 Fault Tolerant External Control Circuit. Those control circuits either entering or leaving the fire pump controller enclosure, which if broken, disconnected, or shorted will not prevent the controller from starting the fire pump from all other internal or external means and can cause the controller to start the pump under these conditions.

3.3.8 Circulation Relief Valve. See 3.3.55.5.1.

3.3.9 Corrosion-Resistant Material. Materials such as brass, copper, Monel®, stainless steel, or other equivalent corrosion-resistant materials.

The list of corrosion-resistant materials in this definition is not intended to be comprehensive. The causes of corrosion vary and the examples given in this definition are effective against certain types of water conditions that can cause corrosion. However, in the context of this standard, a corrosion-resistant material must resist the type of corrosion anticipated. (Monel® is a type of nickel alloy.)

3.3.10 Diesel Engine. See 3.3.13.1.

3.3.11 Disconnecting Means. A device, or group of devices, or other means by which the conductors of a circuit can be disconnected from their source of supply. [**70:** Art. 100]

Disconnecting means are usually devices used within fire pump controllers for electrically driven fire pumps. Means of disconnecting are usually accomplished through a circuit breaker or a locked-rotor overcurrent protection device. See 10.4.3, 10.6.3, 10.6.8, and 10.7.2.1.

3.3.12 Drawdown. The vertical difference between the pumping water level and the static water level.

The term *drawdown* is usually associated with vertical shaft turbine–type pumps that take water from a supply below the discharge flange of the pump, such as a well, river, or pond. When these pumps operate at peak capacity, the water level of the supply will usually drop a certain distance. The level of the water before the fire pump starts is referred to as the *static liquid level*. The level to which the water drops when the pump is operating at 150 percent of its rated capacity is referred to as the *pumping liquid level*. The difference between these two levels is referred to as the *drawdown*. See Figure A.7.2.2.1.

3.3.13 Engine.

3.3.13.1 Diesel Engine. An internal combustion engine in which the fuel is ignited entirely by the heat resulting from the compression of the air supplied for combustion. The oil-diesel engine, which operates on fuel oil injected after compression is practically completed, is the type usually used as a fire pump driver.

FAQ ▶
Are diesel engines the only type of internal combustion engine permitted by NFPA 20?

Diesel engines have proven to be very reliable and effective and are currently the only type of internal combustion engine permitted by NFPA 20 for driving fire pumps. Diesel engines used to power fire pumps must be listed for such use and, therefore, only specific engines are permitted. (See Chapter 9.) Spark-ignited internal combustion engines have not been permitted as a means of driving fire pumps since 1974. Exhibit II.3.1 illustrates a diesel engine–driven fire pump.

3.3.13.2 Internal Combustion Engine. Any engine in which the working medium consists of the products of combustion of the air and fuel supplied. This combustion usually is effected within the working cylinder but can take place in an external chamber.

EXHIBIT II.3.1 Diesel Engine–Driven Pump. (Photograph courtesy of Liberty Mutual Property)

The diesel engine is the only type of internal combustion engine permitted by NFPA 20 to drive fire pumps. See Chapter 11.

3.3.14 Fault Tolerant External Control Circuit. See 3.3.7.2.

3.3.15 Feeder. All circuit conductors between the service equipment, the source of a separately derived system, or other power supply source and the final branch-circuit overcurrent device. [**70:** Art. 100]

3.3.16 Fire Pump Alarm. A supervisory signal indicating an abnormal condition requiring immediate attention.

3.3.17 Fire Pump Controller. A group of devices that serve to govern, in some predetermined manner, the starting and stopping of the fire pump driver and to monitor and signal the status and condition of the fire pump unit.

Different types of controllers are available for electric motor– and diesel engine–driven fire pumps. Controllers must be listed for fire pump service and incorporate both automatic and manual features for pump operation. Controllers for electrically driven fire pumps are available for full voltage and reduced voltage starting. Limited-service controllers for electrically driven fire pumps are also qualified for certain applications but are not required by NFPA 20 to meet the same requirements as full-service controllers. Chapter 10 addresses controllers for electric motor–driven pumps, and Chapter 12 addresses controllers for diesel engine–driven pumps. Exhibit II.3.2 illustrates the internal components of an electric fire pump controller, and Exhibit II.3.3 illustrates the internal components of a diesel engine fire pump controller.

EXHIBIT II.3.2 Electric Fire Pump Controller. (Photograph courtesy of Liberty Mutual Property)

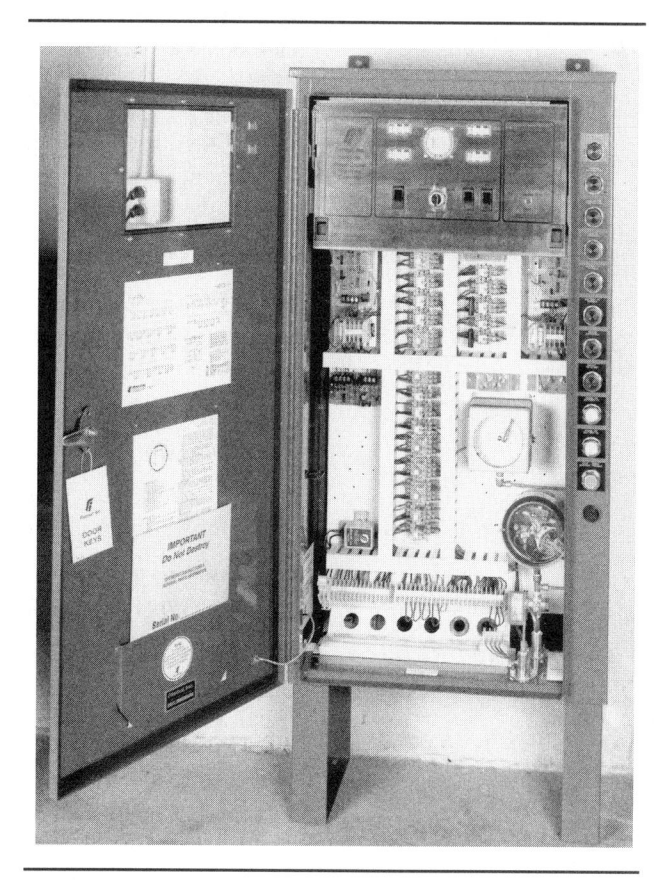

EXHIBIT II.3.3 Diesel Engine Fire Pump Controller. (Photograph courtesy of Firetrol Products, ASCO Power Technologies)

EXHIBIT II.3.4 *Fire Pump Unit. (Photograph courtesy of A-C Fire Pump Systems)*

3.3.18 Fire Pump Unit. An assembled unit consisting of a fire pump, driver, controller, and accessories.

Exhibit II.3.4 illustrates an example of a fire pump unit.

3.3.19 Flexible Connecting Shaft. A device that incorporates two flexible joints and a telescoping element.

Subsection 11.2.3 indicates where a flexible connecting shaft is required.

3.3.20 Flexible Coupling. A device used to connect the shafts or other torque-transmitting components from a driver to the pump, and that permits minor angular and parallel misalignment as restricted by both the pump and coupling manufacturers.

Subsections 6.5.1 and 11.2.3 indicate where a flexible coupling is needed.

3.3.21 Flooded Suction. The condition where water flows from an atmospheric vented source to the pump without the average pressure at the pump inlet flange dropping below atmospheric pressure with the pump operating at 150 percent of its rated capacity.

3.3.22 Groundwater. That water that is available from a well, driven into water-bearing subsurface strata (aquifer).

3.3.23* Head. A quantity used to express a form (or combination of forms) of the energy content of water per unit weight of the water referred to any arbitrary datum.

In the context of fire pumps, the pressure component is usually expressed in terms of feet (ft) or meters (m) of head rather than in pounds per square inch (psi) or bar.

A.3.3.23 Head. The unit for measuring head is the foot (meter). The relation between pressure expressed in pounds per square inch (bar) and pressure expressed in feet (meters) of head is expressed by the following formulas:

$$\text{Head in feet} = \frac{\text{Pressure in psi}}{0.433 \text{ specific gravity}}$$

$$\text{Head in feet} = \frac{\text{Pressure in bar}}{0.098 \text{ specific gravity}}$$

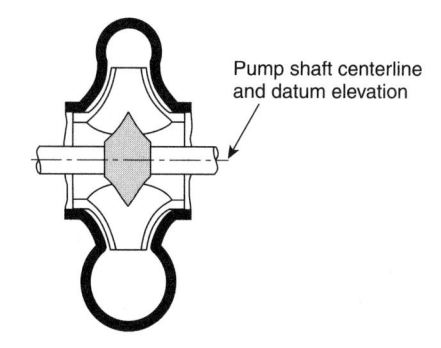

Horizontal double-suction pump

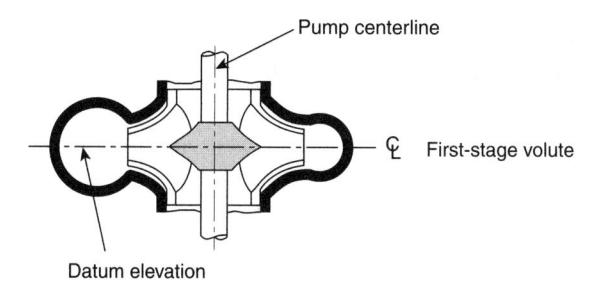

Vertical double-suction pump

Notes:
(1) For all types of horizontal shaft pumps (single-stage double-suction pump shown). Datum is same for multistage, single- (end) suction ANSI-type or any pump with a horizontal shaft.
(2) For all types of vertical shaft pumps (single-stage vertical double-suction pump shown). Datum is same for single- (end) suction, in-line, or any pump with a vertical shaft.

FIGURE A.3.3.23 *Datum Elevation of Two Stationary Pump Designs.*

In terms of foot-pounds (meter-kilograms) of energy per pound (kilogram) of water, all head quantities have the dimensions of feet (meters) of water. All pressure readings are converted into feet (meters) of the water being pumped. *(See Figure A.3.3.23.)*

3.3.23.1 *Net Positive Suction Head (NPSH)* (h_{sw}). The total suction head in feet (meters) of liquid absolute, determined at the suction nozzle, and referred to datum, less the vapor pressure of the liquid in feet (meters) absolute.

In NFPA 20, the two types of net positive suction head (NPSH) that need to be considered are the *NPSH required* and the *NPSH supplied*. The NPSH required by the pump is determined by the pump manufacturer and is a function of both the speed and capacity of the pump. Curves indicating the NPSH versus flow can be obtained from the manufacturer. The NPSH supplied is the pressure head at the pump inlet that causes water to flow through the pump into the eye of the impeller. The NPSH supplied is a function of the water supply.

For any pump installation, the NPSH supplied must be at least equal to the NPSH required for the operating conditions specified. If the NPSH supplied is less than the NPSH required, some of the incoming water vaporizes and forms bubbles. These bubbles can collapse within the pump, producing noise and vibration and causing significant damage to the pump. This phenomenon is referred to as *cavitation*.

3.3.23.2 Total Discharge Head (h_d). The reading of a pressure gauge at the discharge of the pump, converted to feet (meters) of liquid, and referred to datum, plus the velocity head at the point of gauge attachment.

Figures A.3.3.23.3.1 and A.3.3.23.3.2 illustrate the components of total discharge head.

3.3.23.3 Total Head.

3.3.23.3.1* Total Head (H), Horizontal Pumps. The measure of the work increase, per pound (kilogram) of liquid, imparted to the liquid by the pump, and therefore the algebraic difference between the total discharge head and the total suction head. Total head, as determined on test where suction lift exists, is the sum of the total discharge head and total suction lift. Where positive suction head exists, the total head is the total discharge head minus the total suction head.

A.3.3.23.3.1 Total Head (H), Horizontal Pumps. See Figure A.3.3.23.3.1. (Figure A.3.3.23.3.1 does not show the various types of pumps applicable.)

Total head is the energy (pressure) added to the water supply by the pump. It is expressed in feet (ft) or meters (m). For horizontal pumps, where a positive suction pressure exists, the total head is usually calculated by subtracting the pressure at the suction gauge from the pressure at the discharge gauge. Where the suction and discharge flanges are not of the same diameter, the velocity head for the volume of water passing through the flanges must be calculated using the formula identified in the definition of velocity head (see 3.3.23.6). Where the suction and discharge flanges have the same diameter, the incoming and outgoing velocities are not different.

3.3.23.3.2* Total Head (H), Vertical Turbine Pumps. The distance from the pumping liquid level to the center of the discharge gauge plus the total discharge head.

A.3.3.23.3.2 Total Head (H), Vertical Turbine Pumps. See Figure A.3.3.23.3.2.

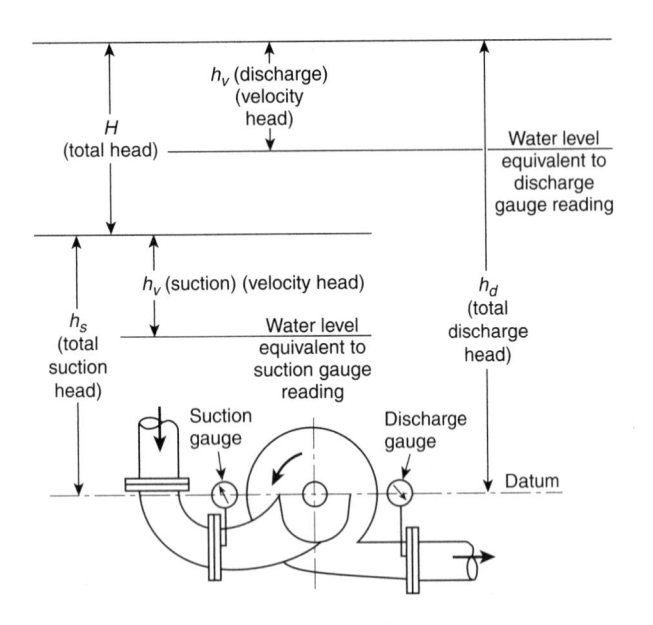

FIGURE A.3.3.23.3.1 *Total Head of All Types of Stationary (Not Vertical Turbine–Type) Fire Pumps.*

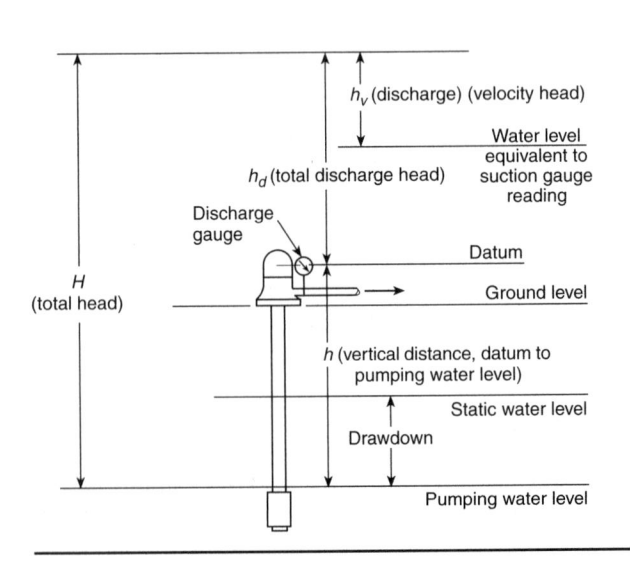

FIGURE A.3.3.23.3.2 *Total Head of Vertical Turbine–Type Fire Pumps.*

As noted earlier, total head is the energy (pressure) added to the water supply by the pump and is expressed in feet (ft) or meters (m). For vertical turbine pumps, a discharge gauge is located at the top of the pump column pipe. The reading of this gauge, therefore, indicates the pressure provided by the pump minus the pressure caused by moving the water from the pumping water level up to the discharge flange as well as the effects of friction loss though the pipe column.

3.3.23.4 Total Rated Head. The total head developed at rated capacity and rated speed for either a horizontal split-case or a vertical shaft turbine–type pump.

The total rated head is the amount of energy given to water as it passes through a specific pump when that pump is operating at its rated speed and at its rated flow.

3.3.23.5 Total Suction Head. Suction head exists where the total suction head is above atmospheric pressure. Total suction head, as determined on test, is the reading of a gauge at the suction of the pump, converted to feet (meters) of liquid, and referred to datum, plus the velocity head at the point of gauge attachment.

Figure A.3.3.23.3.1 provides a graphical description of total suction head.

3.3.23.6* Velocity Head (hv). Figured from the average velocity (v) obtained by dividing the flow in cubic feet per second (cubic meters per second) by the actual area of pipe cross section in square feet (square meters) and determined at the point of the gauge connection.

A.3.3.23.6 Velocity Head (h_v). Velocity head (hv) is expressed by the following formula:

$$h_v = \frac{v^2}{2g}$$

where:

v = velocity in the pipe [ft/sec (m/sec)]

g = acceleration due to gravity: 32.17 ft/sec^2 (9.807 m/sec^2) at sea level and 45 degrees latitude

3.3.24 High-Rise Building. A building where the floor of an occupiable story is greater than 75 ft (23 m) above the lowest level of fire department vehicle access. [*5000,* 2009]

The definition of high-rise building was added to the 2007 edition to assist the user in determining the appropriate protection criteria for fire pump rooms in high-rise buildings. Fire pump rooms must be protected by 2-hour fire-rated enclosures in high-rise buildings.

3.3.25 Internal Combustion Engine. See 3.3.13.2.

3.3.26 Isolating Switch. See 3.3.48.1.

3.3.27 Liquid. For the purposes of this standard, liquid refers to water, foam-water solution, foam concentrates, water additives, or other liquids for fire protection purposes.

3.3.28 Liquid Level.

3.3.28.1 Pumping Liquid Level. The level, with respect to the pump, of the body of liquid from which it takes suction when the pump is in operation. Measurements are made the same as with the static liquid level.

The term *pumping liquid level* is usually associated with vertical turbine–type pumps that take water or additives from an unpressurized source below the discharge flange of the pump, such as a well, river, or pond. See the commentary following the definition of static liquid level (see 3.3.28.2) for more information.

3.3.28.2 Static Liquid Level. The level, with respect to the pump, of the body of liquid from which it takes suction when the pump is not in operation. For vertical shaft turbine–type pumps, the distance to the liquid level is measured vertically from the horizontal centerline of the discharge head or tee.

The term *static liquid level* is usually associated with vertical turbine–type pumps that take water or additives from an unpressurized source below the discharge flange of the pump, such as a well, river, or pond. When these pumps operate at peak capacity, the liquid level of the supply usually drops a certain distance. The level of the liquid before the fire pump starts is referred to as the *static liquid level.* The level to which the liquid drops when the pump is operating at 150 percent of its rated capacity is referred to as the *pumping liquid level* (see 3.3.28.1). The difference between these two levels is referred to as the *drawdown* (see 3.3.12). See Figure A.7.2.2.1.

3.3.29 Loss of Phase. The loss of one or more, but not all, phases of the polyphase power source.

3.3.30 Manual Transfer Switch. See 3.3.48.2.2.

3.3.31 Maximum Pump Brake Horsepower. The maximum brake horsepower required to drive the pump at rated speed. The pump manufacturer determines this by shop test under expected suction and discharge conditions. Actual field conditions can vary from shop conditions.

When selecting a driver for a pump, the maximum brake horsepower demand of the pump when operating at its rated speed must be determined. The pump driver must be capable of supplying this horsepower. A pump's maximum brake horsepower can be obtained from the horsepower curve supplied by the pump manufacturer. The maximum demand of most pumps usually occurs between 140 and 170 percent of the pump's rated capacity.

Exhibit II.3.5 illustrates the range in brake horsepower required to operate a hypothetical 750 gpm (2839 L/min) rated horizontal split-case fire pump. Notice that as the pump produces larger flows, the brake horsepower required increases and the pressure produced by the pump decreases. Pump manufacturers provide similar curves for each of their pumps.

3.3.32 Motor.

3.3.32.1 Dripproof Guarded Motor. A dripproof machine whose ventilating openings are guarded in accordance with the definition for dripproof motor.

3.3.32.2 Dripproof Motor. An open motor in which the ventilating openings are so constructed that successful operation is not interfered with when drops of liquid or solid particles strike or enter the enclosure at any angle from 0 to 15 degrees downward from the vertical.

3.3.32.3 Dust-Ignition-Proof Motor. A totally enclosed motor whose enclosure is designed and constructed in a manner that will exclude ignitible amounts of dust or amounts that might affect performance or rating and that will not permit arcs, sparks, or heat otherwise generated or liberated inside of the enclosure to cause ignition of exterior accumulations or atmospheric suspensions of a specific dust on or in the vicinity of the enclosure.

3.3.32.4 Electric Motor. A motor that is classified according to mechanical protection and methods of cooling.

Electric motors are a reliable, effective means of supplying power to fire pumps; however, at the time of publication of this handbook, only listed motors were permitted to be used. Electric motors must be selected so that they are of sufficient size to turn the pump at its rated speed under the required range of operating conditions. When an electric motor is considered,

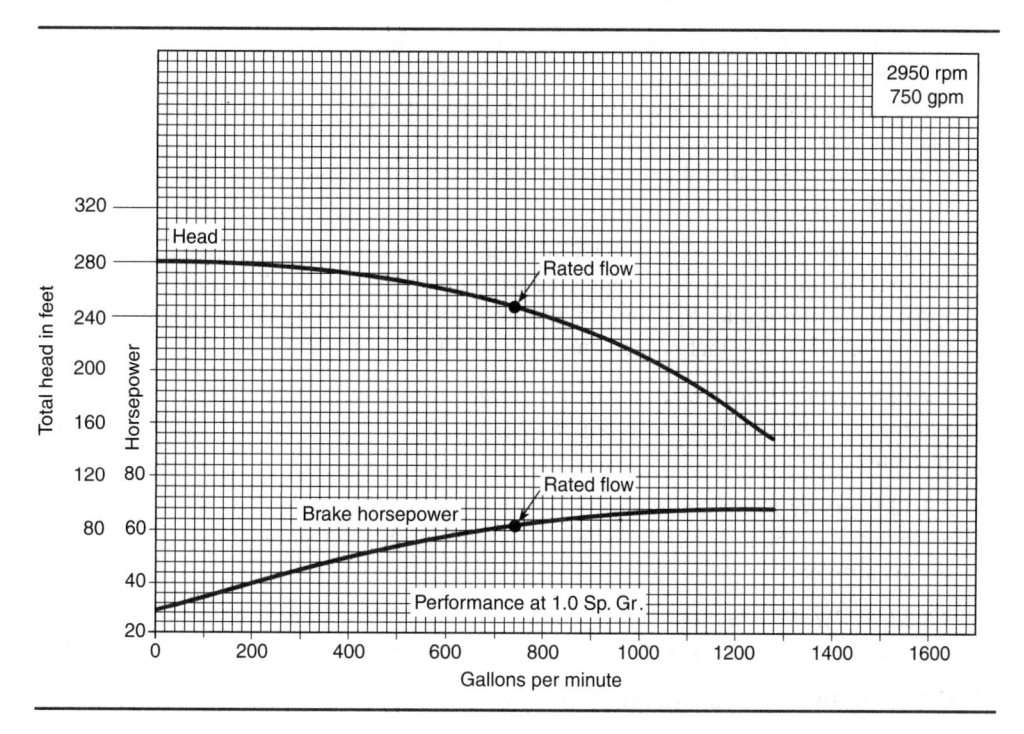

EXHIBIT II.3.5 Brake Horsepower Curve.

the power supply to the motor must be from a reliable source. Power supplies that can be compromised due to storms, fires, or other accidents can leave electric motors without power and render the fire pump useless. Where reliable sources cannot be secured, back-up supplies such as an emergency generator or diesel engine are necessary. See Chapter 9 for more information.

Electric motors used to drive fire pumps are required to withstand severe changes in operating conditions and should be arranged to accommodate large electric loads that might cause the motor to be destroyed. This strategy is contrary to the design and arrangement of most other electric motor applications. However, the reliability of the fire pump is the primary concern. The fire pump must continue to operate during a fire emergency for as long as is necessary. Exhibit II.3.6 shows a horizontal split-case fire pump driven by an electric motor.

3.3.32.5 Explosionproof Motor. A totally enclosed motor whose enclosure is designed and constructed to withstand an explosion of a specified gas or vapor that could occur within it and to prevent the ignition of the specified gas or vapor surrounding the motor by sparks, flashes, or explosions of the specified gas or vapor that could occur within the motor casing.

EXHIBIT II.3.6 Electric Motor–Driven Fire Pump. (Photograph courtesy of Peerless Pump Company)

3.3.32.6 Guarded Motor. An open motor in which all openings giving direct access to live metal or rotating parts (except smooth rotating surfaces) are limited in size by the structural parts or by screens, baffles, grilles, expanded metal, or other means to prevent accidental contact with hazardous parts. Openings giving direct access to such live or rotating parts shall not permit the passage of a cylindrical rod 0.75 in. (19 mm) in diameter.

3.3.32.7 Open Motor. A motor having ventilating openings that permit passage of external cooling air over and around the windings of the motor. Where applied to large apparatus without qualification, the term designates a motor having no restriction to ventilation other than that necessitated by mechanical construction.

3.3.32.8 Totally Enclosed Fan-Cooled Motor. A totally enclosed motor equipped for exterior cooling by means of a fan or fans integral with the motor but external to the enclosing parts.

3.3.32.9 Totally Enclosed Motor. A motor enclosed so as to prevent the free exchange of air between the inside and the outside of the case but not sufficiently enclosed to be termed airtight.

3.3.32.10 Totally Enclosed Nonventilated Motor. A totally enclosed motor that is not equipped for cooling by means external to the enclosing parts.

3.3.33 Net Positive Suction Head (NPSH) (*hsv*). See 3.3.23.1.

3.3.34 On-Site Power Production Facility. The normal supply of electric power for the site that is expected to be constantly producing power.

3.3.35 On-Site Standby Generator. A facility producing electric power on site as the alternate supply of electrical power. It differs from an on-site power production facility in that it is not constantly producing power.

3.3.36 Pressure-Regulating Device. A device designed for the purpose of reducing, regulating, controlling, or restricting water pressure. [14, 2007]

3.3.37 Pump.

3.3.37.1 Additive Pump. A pump that is used to inject additives into the water stream.

3.3.37.2 Can Pump. A vertical shaft turbine–type pump in a can (suction vessel) for installation in a pipeline to raise water pressure.

Vertical shaft turbine–type pumps are used primarily to pump water from nonpressurized water sources such as wells, ponds, and rivers. Can pumps are an application of vertical turbine–type pumps that boost pressure from a pressurized water source such as a public water main. A cylinder or canister is constructed into which the pressurized water flows. The can pump, when placed into this cylinder, boosts the pressure of the incoming water supply. A can pump is essentially a vertical turbine–type pump that is used as a booster pump. Exhibit II.3.7 illustrates a typical vertical can pump installation.

3.3.37.3 Centrifugal Pump. A pump in which the pressure is developed principally by the action of centrifugal force.

Water in centrifugal pumps enters the suction inlet and passes to the center of the impeller. Rotation of the impeller drives the water by centrifugal force to the rim, where it discharges.

3.3.37.4 End Suction Pump. A single suction pump having its suction nozzle on the opposite side of the casing from the stuffing box and having the face of the suction nozzle perpendicular to the longitudinal axis of the shaft.

Exhibit II.3.8 illustrates an end suction pump.

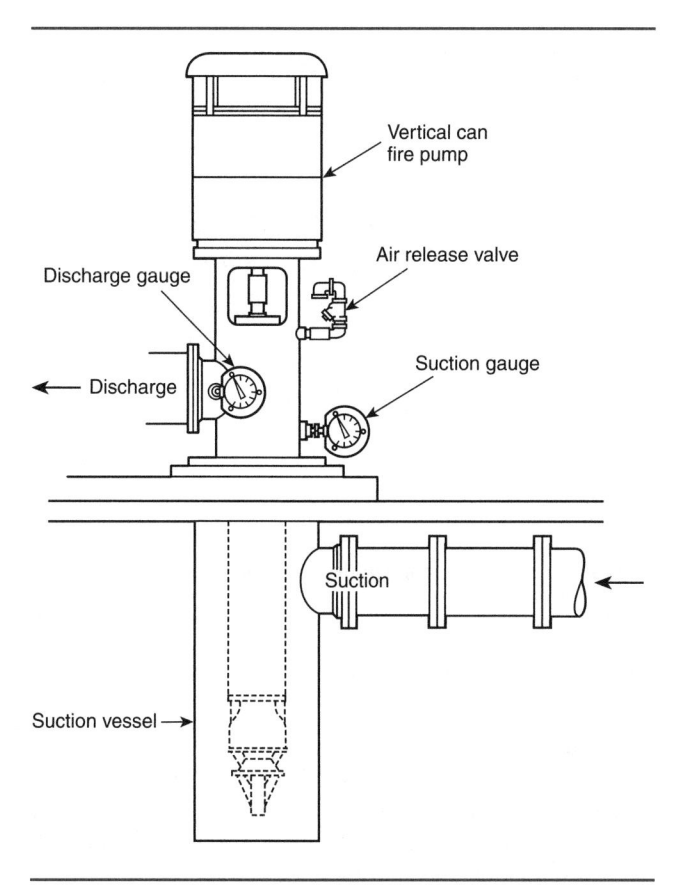

EXHIBIT II.3.7 Vertical Can Pump.

3.3.37.5 Fire Pump. A pump that is a provider of liquid flow and pressure dedicated to fire protection.

3.3.37.6 Foam Concentrate Pump. See 3.3.37.1, Additive Pump.

3.3.37.7 Gear Pump. A positive displacement pump characterized by the use of gear teeth and casing to displace liquid.

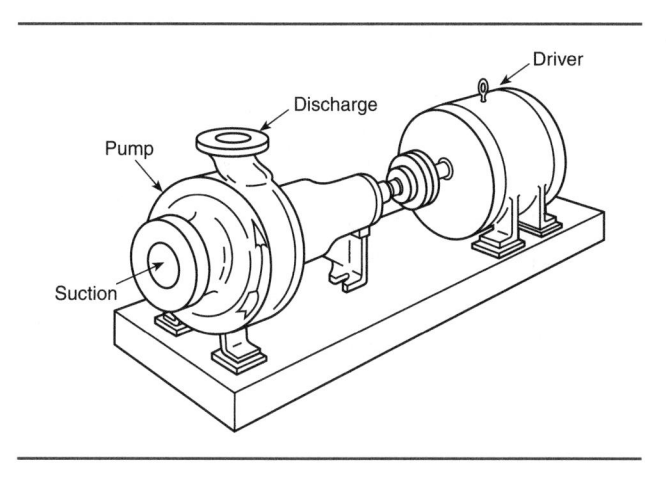

EXHIBIT II.3.8 End Suction Pump Driven by an Electric Motor.

EXHIBIT II.3.9 *Horizontal Split-Case Pump. (Courtesy of Hydraulics Institute)*

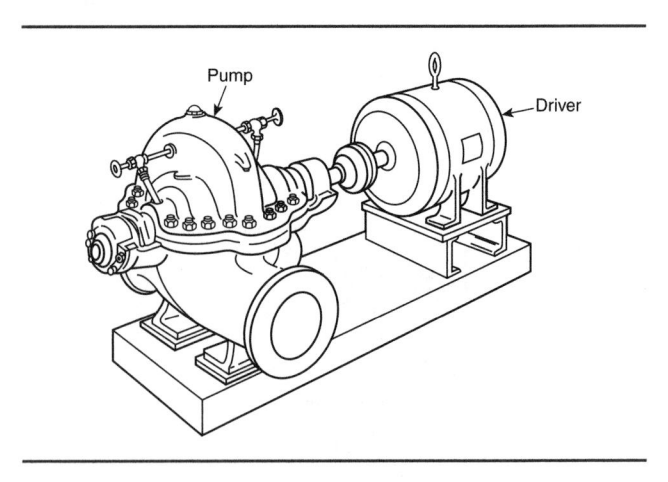

3.3.37.8 Horizontal Pump. A pump with the shaft normally in a horizontal position.

Chapter 6 addresses the installation of horizontal pumps.

3.3.37.9 Horizontal Split-Case Pump. A centrifugal pump characterized by a housing that is split parallel to the shaft.

With this type of pump, the case in which the shaft and impeller are housed is split in the middle and can be separated. This type of housing allows for easy access for repair or replacement of internal components. These pumps are not self-priming and must be supplied with water at the suction flange at some positive pressure. Exhibit II.3.9 illustrates a horizontal split-case pump. (See Chapter 6 for information on the installation of this type of pump.)

3.3.37.10 In-Line Pump. A centrifugal pump whose drive unit is supported by the pump having its suction and discharge flanges on approximately the same centerline.

Exhibit II.3.10 shows a diagram of an in-line pump. See Chapter 6 for more information.

EXHIBIT II.3.10 *In-Line Pump.*

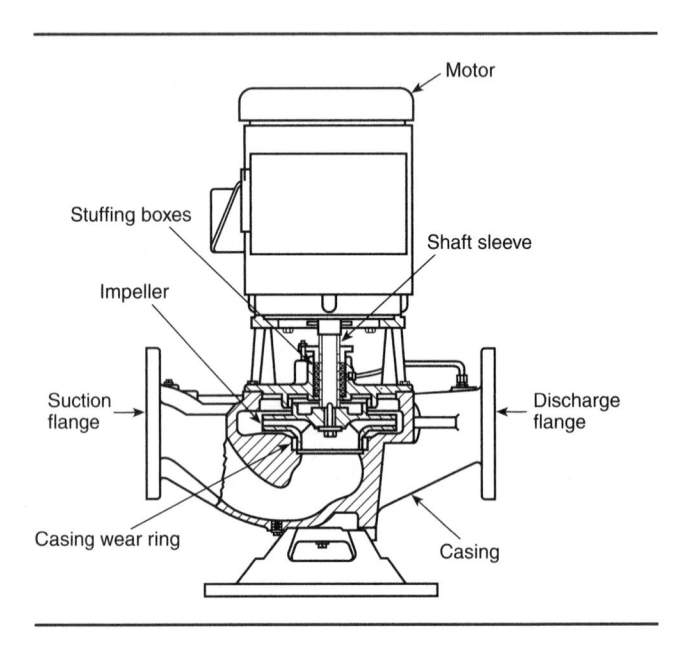

3.3.37.11 Packaged Fire Pump Assembly. Fire pump unit components assembled at a packaging facility and shipped as a unit to the installation site. The scope of listed components (where required to be listed by this standard) in a pre-assembled package includes the pump, driver, controller, and other accessories identified by the packager assembled onto a base with or without an enclosure.

See Section 4.29 for packaged fire pump assembly requirements and associated commentary.

3.3.37.12 Piston Plunger Pump. A positive displacement pump characterized by the use of a piston or plunger and a cylinder to displace liquid.

3.3.37.13 Positive Displacement Pump. A pump that is characterized by a method of producing flow by capturing a specific volume of fluid per pump revolution and reducing the fluid void by a mechanical means to displace the pumping fluid.

3.3.37.14 Pressure Maintenance (Jockey or Make-Up) Pump. A pump designed to maintain the pressure on the fire protection system(s) between preset limits when the system is not flowing water.

See Section 4.25 for pressure maintenance (jockey or make-up) pump requirements and associated commentary.

3.3.37.15 Rotary Lobe Pump. A positive displacement pump characterized by the use of a rotor lobe to carry fluid between the lobe void and the pump casing from the inlet to the outlet.

3.3.37.16 Rotary Vane Pump. A positive displacement pump characterized by the use of a single rotor with vanes that move with pump rotation to create a void and displace liquid.

3.3.37.17 Vertical Lineshaft Turbine Pump. A vertical shaft centrifugal pump with rotating impeller or impellers and with discharge from the pumping element coaxial with the shaft. The pumping element is suspended by the conductor system, which encloses a system of vertical shafting used to transmit power to the impellers, the prime mover being external to the flow stream.

Vertical lineshaft turbine pumps were originally designed to pump water from wells; however, the vertical lineshaft turbine pumps can be used wherever a nonpressurized water source serves as the water supply. These pumps are most often used where horizontal fire pumps are prohibited from use because horizontal pumps would operate under a suction lift condition.

Although vertical turbine pumps move water up to the discharge flange of the pump, they do not operate under a suction lift condition. The impellers for these pumps are located within the water supply and force water up to the discharge flange. Because the impellers sit in the water source, these pumps do not require priming (see Chapter 7). Exhibit II.3.11 shows a typical vertical lineshaft turbine pump.

3.3.38 Pumping Liquid Level. See 3.3.28.1.

3.3.39 Qualified Person. A person who, by possession of a recognized degree, certificate, professional standing, or skill, and who, by knowledge, training, and experience, has demonstrated the ability to deal with problems related to the subject matter, the work, or the project. [**1451,** 2007]

3.3.40 Series Fire Pump Unit. All fire pump units that operate in a series arrangement where the first fire pump takes suction directly from a water supply and each sequential pump takes suction from the preceding pump; pumps taking suction from tanks or break tanks are not considered series fire pump units even if fire pumps at lower elevations are used to refill the tanks or break tanks.

EXHIBIT II.3.11 Vertical Lineshaft Turbine Pump. (Photograph courtesy of Patterson Pump)

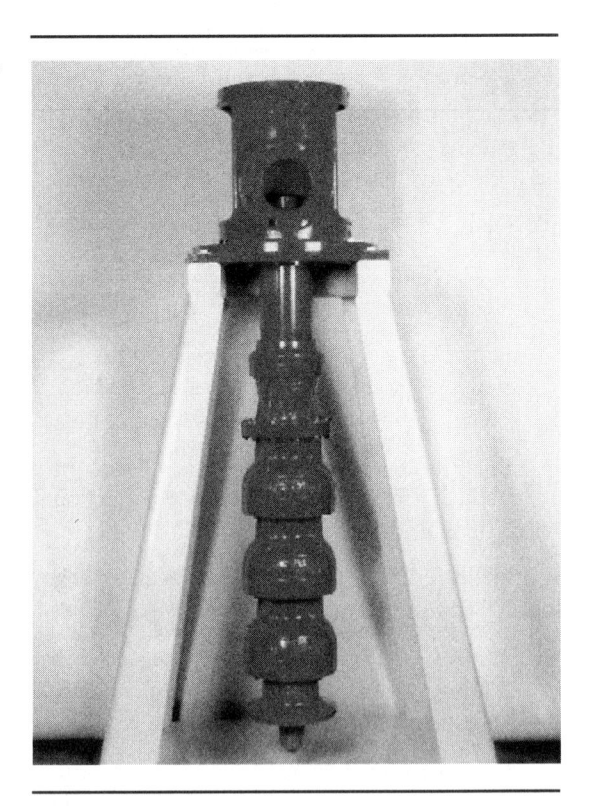

This definition does not include pumps that may operate in series but are separated by a distribution system that supplies other uses between the pumps. An example of a series pumping arrangement that is not included within this definition is a fire pump that takes suction from a municipal supply that is supplied by a pump. See also the commentary following 9.2.2.

3.3.41* Service. The conductors and equipment for delivering electric energy from the serving utility to the wiring system of the premises served. [**70:** Art. 100]

A.3.3.41 Service. For more information, see *NFPA 70, National Electrical Code,* Article 100.

3.3.42* Service Equipment. The necessary equipment, usually consisting of a circuit breaker(s) or switch(es) and fuse(s) and their accessories, connected to the load end of service conductors to a building or other structure, or an otherwise designated area, and intended to constitute the main control and cutoff of the supply. [**70:** Art. 100]

A.3.3.42 Service Equipment. For more information, see *NFPA 70, National Electrical Code,* Article 100.

See also Chapters 9 and 10.

3.3.43 Service Factor. A multiplier of an ac motor that, when applied to the rated horsepower, indicates a permissible horsepower loading that can be carried at the rated voltage, frequency, and temperature. For example, the multiplier 1.15 indicates that the motor is permitted to be overloaded to 1.15 times the rated horsepower.

The motor for a fire pump installation must be of sufficient capacity so that it is not overloaded beyond the limit of the service factor at the pump's maximum brake horsepower when operating at its rated speed. For example, a 75 hp (56 kW) motor with a service factor of 1.12

could safely meet the demand of 84 hp (63 kW) [75 hp (56 kW) × 1.12]. The service factor is a function of the type of motor and is determined by the manufacturer (see 9.5.2.2).

3.3.44 Set Pressure. As applied to variable speed pressure limiting control systems, the pressure that the variable speed pressure limiting control system is set to maintain.

3.3.45* Signal. An indicator of status.

3.3.46 Speed.

3.3.46.1 Engine Speed. The speed indicated on the engine nameplate.

3.3.46.2 Motor Speed. The speed indicated on the motor nameplate.

3.3.46.3 Rated Speed. The speed for which the fire pump is listed and that appears on the fire pump nameplate.

3.3.47 Static Liquid Level. See 3.3.28.2.

3.3.48 Switch.

3.3.48.1 Isolating Switch. A switch intended for isolating an electric circuit from its source of power. It has no interrupting rating, and it is intended to be operated only after the circuit has been opened by some other means.

Isolating switches typically are used within controllers for electric motor–driven pumps. They are used to isolate the power source from the controller and must be operable from the outside of the controller (see 10.4.2).

3.3.48.2 Transfer Switch.

3.3.48.2.1 Automatic Transfer Switch (ATS). Self-acting equipment for transferring the connected load from one power source to another power source. [**110,** 2010]

Where an emergency or backup power source is supplied for electric motors driving fire pumps, an automatic transfer switch is used to transfer power spontaneously from the primary supply to the backup supply when the primary power supply fails. The transfer switch may be provided as an integral part of the fire pump controller, or it may be installed as a separate piece of equipment. Where the transfer switch is used independently from the controller, the transfer switch is required to be listed. Where the transfer switch is integrated into the controller, the entire assembly is required to be listed (see Section 10.8).

3.3.48.2.2 Manual Transfer Switch. A switch operated by direct manpower for transferring one or more load conductor connections from one power source to another.

Where an emergency or backup power source is supplied for electric motors driving fire pumps, a manual power transfer switch allows for the manual (nonautomatic) transfer of power from the normal power supply to the secondary supply. However, NFPA 20 prohibits the use of manual transfer switches (see 10.8.2). In some facilities, a common practice is to have multiple sources of power to improve reliability. As a result, the intention of the Technical Committee on Fire Pumps is to prohibit the use of manual transfer switches.

◀ **FAQ**
Does NFPA 20 allow the use of manual transfer switches?

3.3.49 Total Discharge Head (*hd*). See 3.3.23.2.

3.3.50 Total Head (*H*), Horizontal Pumps. See 3.3.23.3.1.

3.3.51 Total Head (*H*), Vertical Turbine Pumps. See 3.3.23.3.2.

3.3.52 Total Rated Head. See 3.3.23.4.

3.3.53 Total Suction Head (*h$_s$*). See 3.3.23.5.

3.3.54 Total Suction Lift (h_l). Suction lift that exists where the total suction head is below atmospheric pressure. Total suction lift, as determined on test, is the reading of a liquid manometer at the suction nozzle of the pump, converted to feet (meters) of liquid, and referred to datum, minus the velocity head at the point of gauge attachment.

NFPA 20 no longer permits fire pumps to operate under a suction lift condition—that is, a negative pressure is not permitted at the suction flange. Prior to the 1974 edition of the standard, however, horizontal shaft pumps were permitted to take suction lift. The difficulty in taking a suction lift with a horizontal shaft fire pump is that the pump must remain primed with water at all times. The priming mechanism was required to have sufficient capacity to remove air from the pump and suction pipe within 3 minutes. At one point, NFPA 20 contained seven methods for priming pumps and suction pipes. This provision was removed from the standard beginning with the 1974 edition because failure of the priming mechanism results in failure of the pump to move water. Exhibit II.3.12 illustrates one priming method that was permitted in previous editions of the standard. This exhibit is included here for information only as some of these installations may still be in service.

3.3.55 Valve.

3.3.55.1 Dump Valve. An automatic valve installed on the discharge side of a positive displacement pump to relieve pressure prior to the pump driver reaching operating speed.

EXHIBIT II.3.12 Centrifugal Fire Pump Installation Where Pump Has Suction Lift. (Source: NFPA 20, 1970, Figure 100b)

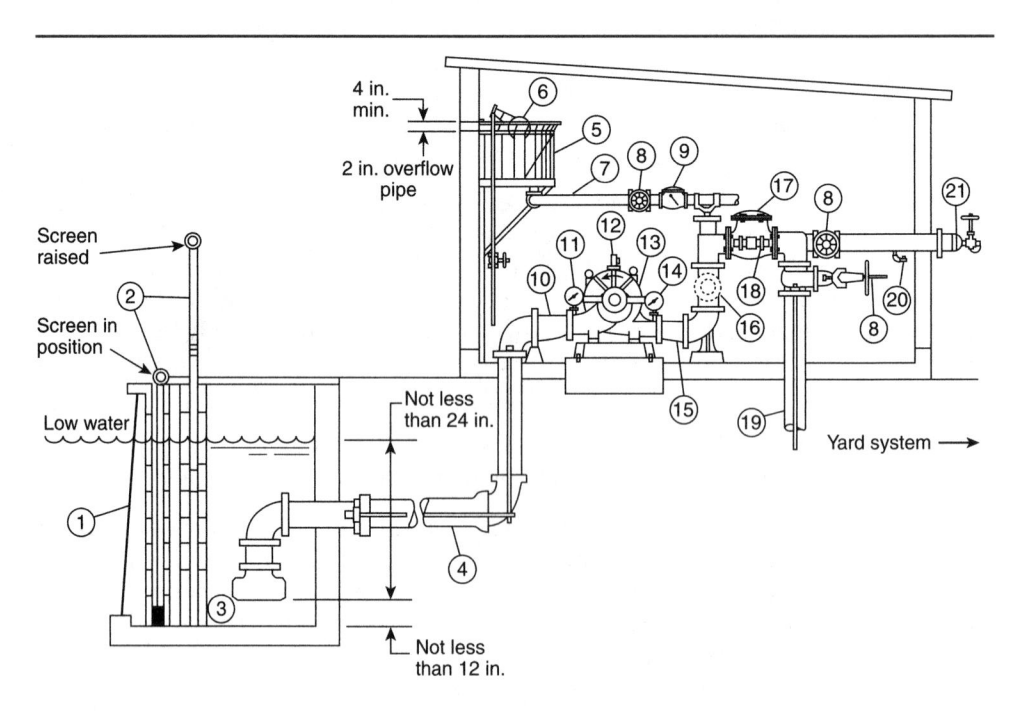

1. Trash rack (½ in. flat or ¾ in. round steel bars spaced 2 to 3 in. apart)
2. Double screens
3. Approved foot valve
4. Suction pipe from water supply to pump
5. Priming tank
6. Automatic float valve
7. Priming connection
8. OS&Y gate valves
9. Priming check valve
10. Eccentric reducer
11. Suction gauge
12. Umbrella cock
13. Horizontal fire pump
14. Discharge gauge
15. Concentric reducer
16. Relief valve (if required)
17. Discharge check valve
18. Priming bypass
19. Discharge pipe
20. Drain valve or ball drip
21. Hose valve manifold with hose valves

3.3.55.2 Low Suction Throttling Valve. A pilot-operated valve installed in discharge piping that maintains positive pressure in the suction piping, while monitoring pressure in the suction piping through a sensing line.

3.3.55.3 Pressure Control Valve. A pilot-operated pressure-reducing valve designed for the purpose of reducing the downstream water pressure to a specific value under both flowing (residual) and nonflowing (static) conditions. [**14,** 2007]

3.3.55.4 Pressure-Reducing Valve. A valve designed for the purpose of reducing the downstream water pressure under both flowing (residual) and nonflowing (static) conditions. [**14,** 2007]

3.3.55.5 Relief Valve. A device that allows the diversion of liquid to limit excess pressure in a system.

3.3.55.5.1 Circulation Relief Valve. A valve used to cool a pump by discharging a small quantity of water, this valve is separate from and independent of the main relief valve.

3.3.55.6 Unloader Valve. A valve that is designed to relieve excess flow below pump capacity at set pump pressure.

3.3.56 Variable Speed Pressure Limiting Control. A speed control system used to limit the total discharge pressure by reducing the pump driver speed from rated speed.

Variable speed pressure limiting control (VSPLC) systems are similar to conventional fire pump systems but with the addition of a pressure-sensing line that prevents the pump from overpressurizing the fire protection system. Prevention of overpressurization is accomplished by altering the speed of the driver. VSPLC can be used in place of a break tank or a pressure-reducing or pressure relief valve when the pressure varies in the attached water supply. The VSPLC can be driven by either an electric or a diesel engine.

3.3.57 Variable Speed Suction Limiting Control. A speed control system used to maintain a minimum positive suction pressure at the pump inlet by reducing the pump driver speed while monitoring pressure in the suction piping through a sensing line.

3.3.58 Velocity Head (h_v). See 3.3.23.6.

3.3.59 Wet Pit. A timber, concrete, or masonry enclosure having a screened inlet kept partially filled with water by an open body of water such as a pond, lake, or stream.

REFERENCES CITED IN COMMENTARY

National Fire Protection Association, 1 Batterymarch Park, Quincy, MA 02169-7471.

NFPA Glossary of Fire Protection Terms (available online at *www.nfpa.org*).

Merriam-Webster, Inc., 7 Federal Street, P.O. Box 281, Springfield, MA 01102.

Merriam-Webster's Collegiate Dictionary, 11th edition, 2003

General Requirements

CHAPTER 4

Chapter 4 contains general requirements for fire pumps. It is important to note that this standard does not establish the requirement to install a fire pump. Such requirements are based on the hydraulic demand of suppression systems. In some cases, where supplemental water supplies are needed, other types of water supplies can be used, such as elevated water tanks or pressure tanks. Where a fire pump is needed or desired, however, the design, installation, and acceptance testing requirements should be in accordance with this standard.

The chapter begins with design and operational requirements for fire pumps, including information related to required approvals, pump operation and performance, and water supplies. Limitations on fire pump rated capacities are also established. The chapter concludes with specifications and installation requirements for all pump components and equipment, except for pump controllers and drivers.

4.1 Pumps.

4.1.1 This standard shall apply to centrifugal single-stage and multistage pumps of the horizontal or vertical shaft design and positive displacement pumps of the horizontal or vertical shaft design.

Positive displacement pumps were added to the 1999 edition to provide guidance on the design and installation of pumps that are typically used for foam systems. Previously, no guidance for the design and installation of positive displacement pumps existed.

4.1.2 Other Pumps.

4.1.2.1 Pumps other than those specified in this standard and having different design features shall be permitted to be installed where such pumps are listed by a testing laboratory.

4.1.2.2 These pumps shall be limited to capacities of less than 500 gpm (1892 L/min).

4.2* Approval Required.

4.2.1 Stationary pumps shall be selected based on the conditions under which they are to be installed and used.

4.2.2 The pump manufacturer or its authorized representative shall be given complete information concerning the liquid and power supply characteristics.

Pump capacities are based on the calculated system demand. Pressure boost or output of the pump is determined by the pressures available from the attached water supply. The characteristics of that water supply—whether positive pressure or static—must be determined in order to select the correct type of pump and its performance characteristics.

The available power supply for electric pumps must be suitable for the fire pump controller. The power supply must be analyzed for reliability, capacity, and suitability. This information must be made available to the pump manufacturer or manufacturer's representative for analysis.

4.2.3 A complete plan and detailed data describing pump, driver, controller, power supply, fittings, suction and discharge connections, and liquid supply conditions shall be prepared for approval.

Ordinarily, upon purchase of a fire pump, the manufacturer or supplier provides product data for the pump, driver, controller, and all accessories to the buyer. This data package should include the pump's rated capacity and speed in addition to other performance characteristics. The data package should be submitted to the registered design professional (RDP) and authority having jurisdiction (AHJ) for approval before shipment of the pump and accessories. The reviewing authority should be contacted prior to submittal to determine the number of plans, calculations, or product data needed for review. The submitter should include an additional number of copies of the package to be returned documenting the approval or comments from the AHJ.

In addition to this data package, a complete plan of the proposed installation, drawn to scale, should be submitted, indicating as a minimum the information outlined in Exhibit II.4.1.

4.2.4 Each pump, driver, controlling equipment, power supply and arrangement, and liquid supply shall be approved by the authority having jurisdiction for the specific field conditions encountered.

4.3 Pump Operation.

4.3.1 In the event of fire pump operation, qualified personnel shall respond to the fire pump location to determine that the fire pump is operating in a satisfactory manner.

A qualified person is one who is familiar with the purpose and function of the fire pump and related equipment. This person should be trained in the operation of the pump. The weekly operating test is an excellent training opportunity to demonstrate the proper operation of pumping equipment. One of the reasons for placing a fire pump in a protected enclosure, as required in Section 4.12, is to provide protection for both the pump operator and the pumping equipment. Consideration should be given to locating the fire pump room at grade level with direct access to the exterior of the building to afford quick and easy egress for the pump operator and access to the pumping equipment for fire department personnel.

4.3.2 System Designer. The system designer shall be identified on the system design documents. Acceptable minimum evidence of qualifications or certification shall be provided when requested by the authority having jurisdiction. Qualified personnel shall include, but not be limited to, one or more of the following:

(1) Personnel who are factory trained and certified for fire pump system design of the specific type and brand of system being designed
(2)* Personnel who are certified by a nationally recognized fire protection certification organization acceptable to the authority having jurisdiction
(3) Personnel who are registered, licensed, or certified by a state or local authority

A.4.3.2(2) Nationally recognized fire protection certification programs include, but are not limited to, those programs offered by the International Municipal Signal Association (IMSA) and the National Institute for Certification in Engineering Technologies (NICET). Note: These organizations and the products or services offered by them have not been independently verified by the NFPA, nor have the products or services been endorsed or certified by the NFPA or any of its technical committees.

EXHIBIT II.4.1 *Checklist for Plan of Proposed Installation.*

PUMP INSTALLATION PLAN CHECKLIST

General
- Name of owner or occupant
- Location including street address
- Point of compass
- Name and address of designer and installing contractor
- Listed pump, make, model number, driver type, and rated capacity
- Type of system supplied by pump
- Design standard used including edition

Water Supply Characteristics
- Flow test data not more than 5 years old
- Underground main of adequate size
- Water storage tank of adequate capacity with automatic refill connection

Suction Piping
- Proper size
- Galvanized or painted on the inside for corrosion protection
- Isolation valve (OS&Y) in the proper location
- Backflow prevention or other device in proper location
- Elbows in the proper orientation or more than 10 pipe diameters away from suction flange of pump
- Eccentric reducer (if needed) installed correctly
- Pump bypass
- Suction and discharge pressure gauges
- Circulation relief valve

Discharge Piping
- Proper size
- Check valve
- Discharge isolation valve

Fire Pump Controller
- Listed for type of pump served
- If electric, type and arrangement of power supply

Water Flow Test Devices
- Test header or flowmeter
- Proper number of 2½ in. hose valves on test header
- Test header piping of proper size
- Proper size of flowmeter (if provided)

Jockey Pump
- Jockey pump bypasses fire pump
- Separate and dedicated sensing line for jockey pump
- Sensing lines ½ in. diameter and of brass, copper, or stainless steel piping
- No shutoff valves in sensing lines

Isolation Valves
- All isolation valves supervised in the open position
- Test header and flowmeter valves supervised in the closed position

Diesel Fire Pump
- Relief valve (if provided) with no isolation valves
- Two storage batteries provided with charger
- Cooling system from heat exchanger or cooling water supply from pump discharge
- Diesel tank located above ground
- Diesel tank of sufficient capacity 1 gallon per horsepower (gal/hp) plus 10 percent

Notes:
1. For all types of pumping equipment, a complete bill of material should be provided. This list should include the make, model, and part numbers of all components.
2. For all electrically operated pumps, provide an electrical schematic drawing that depicts which components are proposed for the power supply from the utility connection to the pump motor controller.
3. Ratings of all equipment and components and settings of breakers, fuses, switches, and transformers should be indicated. Size and length of all circuit conductors should also be noted.

4.3.2.1 Additional evidence of qualification or certification shall be permitted to be required by the AHJ.

4.3.3 System Installer. Installation personnel shall be qualified or shall be supervised by persons who are qualified in the installation, inspection, and testing of fire protection systems. Minimum evidence of qualifications or certification shall be provided when requested by the authority having jurisdiction. Qualified personnel shall include, but not be limited to, one or more of the following:

(1) Personnel who are factory trained and certified for fire pump system designed of the specific type and brand of system being designed

(2)* Personnel who are certified by a nationally recognized fire protection certification organization acceptable to the authority having jurisdiction
(3) Personnel who are registered, licensed, or certified by a state or local authority

A.4.3.3(2) See A.4.3.2(2).

4.3.3.1 Additional evidence of qualification or certification shall be permitted to be required by the AHJ.

4.3.4* Service Personnel Qualifications and Experience.

A.4.3.4 Service personnel should be able to do the following:

(1) Understand the requirements contained in this standard and in NFPA 25, *Standard for the Inspection, Testing, and Maintenance of Water-Based Fire Protection Systems,* and the fire pump requirements contained in *NFPA 70, National Electrical Code*
(2) Understand basic job site safety laws and requirements
(3) Apply troubleshooting techniques and determine the cause of fire protection system trouble conditions
(4) Understand equipment-specific requirements, such as programming, application, and compatibility
(5) Read and interpret fire protection system design documentation and manufacturers' inspection, testing, and maintenance guidelines
(6) Properly use tools and equipment required for testing and maintenance of fire protection systems and their components
(7) Properly apply the test methods required by this standard and NFPA 25, *Standard for the Inspection, Testing, and Maintenance of Water-Based Fire Protection Systems.*

4.3.4.1 Service personnel shall be qualified and experienced in the inspection, testing, and maintenance of fire protection systems. Qualified personnel shall include, but not be limited to, one or more of the following:

(1) Personnel who are factory trained and certified for fire pump system design of the specific type and brand of system being designed
(2)* Personnel who are certified by a nationally recognized fire protection certification organization acceptable to the authority having jurisdiction
(3) Personnel who are registered, licensed, or certified by a state or local authority
(4) Personnel who are employed and qualified by an organization listed by a nationally recognized testing laboratory for the servicing of fire protection systems

A.4.3.4.1(2) See A.4.3.2(2).

4.3.4.2 Additional evidence of qualification or certification shall be permitted to be required by the AHJ.

4.4 Fire Pump Unit Performance.

4.4.1* The fire pump unit, consisting of a pump, driver, and controller, shall perform in compliance with this standard as an entire unit when installed or when components have been replaced.

A.4.4.1 A single entity should be designated as having unit responsibility for the pump, driver, controller, transfer switch equipment, and accessories. *Unit responsibility* means the accountability to answer and resolve any and all problems regarding the proper installation, compatibility, performance, and acceptance of the equipment. Unit responsibility should not be construed to mean purchase of all components from a single supplier.

Unit responsibility should be the responsibility of the installer until the equipment is accepted and officially turned over to the building owner. A representative from the installing contractor (usually a designer, project manager, or engineer) is normally involved with the fire pump installation from purchase of the equipment to acceptance testing. On projects where a formal commissioning program is used, the responsibility for the fire pump may be given to a commissioning agent.

When individual components have been replaced, the contractor installing the replacement components must verify that the entire unit functions as intended—see NFPA 25, *Standard for the Inspection, Testing, and Maintenance of Water-Based Fire Protection Systems*, for component replacement testing requirements.

The standard does not require the purchase of all components of a fire pump from a single source. However, consideration should be given to the listing and compatibility of all system components.

4.4.2 The complete fire pump unit shall be field acceptance tested for proper performance in accordance with the provisions of this standard. *(See Section 14.2.)*

◀ FAQ
Must a fire pump, driver, and all related accessories/components be purchased from a single source?

Upon completion of the installation and prior to final acceptance, the installation contractor should coordinate the field acceptance test, which includes participation of all interested parties such as the pump manufacturer's representative and the fire pump controller manufacturer's representative, the building owner or his or her representative, the RDP, and the AHJ. The acceptance test should demonstrate to the building owner's representative and the AHJ that the pump performs as intended and complies with the requirements of this standard and any project specification. Ordinarily, the services of both the pump manufacturer and controller manufacturer are part of the fire pump purchase order and include the necessary labor hours to complete the acceptance test.

4.5 Certified Shop Test.

4.5.1 Certified shop test curves showing head capacity and brake horsepower of the pump shall be furnished by the manufacturer to the purchaser.

The certified shop test provides verification that the fire pump was tested by the manufacturer prior to shipment to the job site and performs properly under ideal conditions. This test report should be submitted to the RDP and AHJ for review and approval. The data collected during the acceptance test are compared to the certified shop test report to reveal differences in pump performance from the shop test to the acceptance test. Minor differences between these test results may be corrected by applying affinity laws as referenced in Chapter 14. See Chapter 14 for more information.

4.5.2 The purchaser shall furnish the data required in 4.5.1 to the authority having jurisdiction.

The certified shop test is performed when the fire pump and driver are assembled at the manufacturing facility. In some cases, the commissioning agent or project engineer may wish to witness the shop test. In such cases, the shop test should be coordinated with the manufacturer, purchaser, and commissioning agent. The certified shop test results should be submitted for review and approval as part of the commissioning process.

4.6 Liquid Supplies.

4.6.1* **Reliability.** The adequacy and dependability of the water source are of primary importance and shall be fully determined, with due allowance for its reliability in the future.

For a discussion regarding water supply adequacy and reliability, see Part III of this handbook.

A.4.6.1 For water supply capacity and pressure requirements, see the following documents:

(1) NFPA 13, *Standard for the Installation of Sprinkler Systems*
(2) NFPA 14, *Standard for the Installation of Standpipe and Hose Systems*
(3) NFPA 15, *Standard for Water Spray Fixed Systems for Fire Protection*
(4) NFPA 16, *Standard for the Installation of Foam-Water Sprinkler and Foam-Water Spray Systems*
(5) NFPA 24, *Standard for the Installation of Private Fire Service Mains and Their Appurtenances*

4.6.2* Sources.

A.4.6.2 Where the suction supply is from a factory-use water system, pump operation at 150 percent of rated capacity should not create hazardous process upsets due to low water pressure.

4.6.2.1 Any source of water that is adequate in quality, quantity, and pressure shall be permitted to provide the supply for a fire pump.

4.6.2.2 Where the water supply from a public service main is not adequate in quality, quantity, or pressure, an alternative water source shall be provided.

4.6.2.3 The adequacy of the water supply shall be determined and evaluated prior to the specification and installation of the fire pump.

4.6.2.3.1 Where the maximum flow available from the water supply cannot provide a flow of 150 percent of the rated flow of the pump, but the water supply can provide the greater of 100 percent of rated flow or the maximum flow demand of the fire protection system(s), the water supply shall be deemed to be adequate. In this case, the maximum flow shall be considered the highest flow that the water supply can achieve.

It is the intent of this standard to size fire pumps based on the flow requirements of the attached fire protection system. If the attached water supply cannot provide sufficient flow to meet 150 percent of the rated flow of the pump, such a situation should not be considered to be inadequate. This subparagraph is consistent with the requirements of 14.2.5.2.4.

4.6.2.3.2 Where the water supply cannot provide 150 percent of the rated flow of the pump, a placard shall be placed in the pump room indicating the minimum suction pressure that the fire pump is allowed to be tested at and also indicating the required flow rate.

4.6.2.4 For liquids other than water, the liquid source for the pump shall be adequate to supply the maximum required flow rate for any simultaneous demands for the required duration and the required number of discharges.

4.6.3 Level. The minimum water level of a well or wet pit shall be determined by pumping at not less than 150 percent of the fire pump rated capacity.

4.6.4* Stored Supply.

A.4.6.4 Water sources containing salt or other materials deleterious to the fire protection systems should be avoided.

Where the authority having jurisdiction approves the start of an engine-driven fire pump on loss of ac power supply, the liquid supply should be sufficient to meet the additional cooling water demand.

4.6.4.1 A stored supply plus reliable automatic refill shall be sufficient to meet the demand placed upon it for the design duration.

A stored supply of water for any fire protection system should be designed and installed in accordance with NFPA 22, *Standard for Water Tanks for Private Fire Protection*. The quantity of water is determined by multiplying the system demand in gallons per minute (liters per minute) [gpm (L/min)] by the discharge duration specified by the appropriate system installation standard. The size of the stored supply is then selected based on the capacity of the stored supply between the discharge and the overflow connections on the storage tank. An automatic refill connection is preferable and should be designed and installed with sufficient flow capacity to refill the stored supply within 8 hours in accordance with NFPA 22.

4.6.4.2 A reliable method of replenishing the supply shall be provided.

4.6.5 Head.

4.6.5.1 Except as provided in 4.6.5.2, the head available from a water supply shall be figured on the basis of a flow of 150 percent of rated capacity of the fire pump.

The capacity of a fire pump should be selected based upon the anticipated system demand (see Section 4.8). However, consideration should be given to performance of the pump when operating at overload (150 percent of rated capacity). Oversizing a fire pump can place an undue demand on the attached water supply, while undersizing the fire pump can cause difficulty in meeting system demand. Further, some water authorities prohibit flowing a quantity of water that causes residual pressures to drop below 20 psi (1.4 bar). Note that 4.14.3.1 limits suction pressure to not less than 0 psi (0 bar). Because of the aforementioned limitations, careful consideration of pump sizing must be exercised. Also important to note is that failure to operate a pump at 150 percent of rated capacity during the acceptance test does not constitute an unacceptable installation, provided that the pump meets the system demand (see 14.2.5.2.4).

4.6.5.2 Where the water supply cannot provide a flow of 150 percent of the rated flow of the pump, but the water supply can provide the flow demand of the fire protection system, the head available from the water supply shall be permitted to be calculated on the basis of the maximum flow available as allowed by 4.6.2.3.1.

This paragraph was revised to be consistent with 4.6.2.3.1 and reaffirms that the attached water supply need not be capable of flowing 150 percent of the fire pump rated capacity. The attached water supply must, however, be capable of providing sufficient flow to exceed the fire protection system demand.

Flow test data should never be more than 5 years old since many factors affecting the flow test results may change during that time. See Part III of this handbook for information regarding fire flow testing.

4.6.5.3 This head shall be as indicated by a flow test.

4.7 Pumps, Drivers, and Controllers.

4.7.1* Fire pumps shall be dedicated to and listed for fire protection service.

A.4.7.1 This subsection does not preclude the use of pumps in public and private water supplies that provide water for domestic, process, and fire protection purposes. Such pumps are not fire pumps and are not expected to meet all the requirements of this standard. Such pumps are permitted for fire protection if they are considered reliable by the analysis mandated in Section 4.6. Evaluating the reliability should include at least the levels of supervision and rapid response to problems as are typical in municipal water systems.

If a private development (campus) needs a fire protection pump, this is typically accomplished by installing a dedicated fire pump (in accordance with this standard) in parallel with a domestic pump or as part of a dedicated fire branch/loop off a water supply.

Subsection 4.7.1 is intended to prohibit the use of fire pumps for other purposes, as the annex material points out. However, other pumps such as those used for domestic or industrial purposes can be used to supply fire protection water, provided that they are evaluated for reliability and are properly supervised.

4.7.2 Acceptable drivers for pumps at a single installation shall be electric motors, diesel engines, steam turbines, or a combination thereof.

4.7.3* A pump shall not be equipped with more than one driver.

A.4.7.3 It is not the intent of this subsection to require replacement of dual driver installations made prior to the adoption of the 1974 edition of this standard.

4.7.4 Each fire pump shall have its own dedicated driver unless otherwise permitted in 8.5.3.1.

4.7.5 Each driver shall have its own dedicated controller.

4.7.6* The driver shall be selected in accordance with 9.5.2 (electric motors), 11.2.2 (diesel engines), or 13.1.2 (steam turbines) to provide the required power to operate the pump at rated speed and maximum pump load under any flow condition.

A.4.7.6 For centrifugal and turbine pumps, the maximum brake horsepower required to drive the pump typically occurs at a flow beyond 150 percent of the rated capacity. For positive displacement pumps, the maximum brake horsepower required to drive the pump typically occurs when the relief valve is flowing 100 percent of the rated pump capacity. Pumps connected to variable speed drivers may operate at lower speeds, but the driver needs to be selected based upon the power required to drive the pump at rated speed and maximum pump load under any flow condition.

4.7.7* Maximum Pressure for Centrifugal Pumps.

A.4.7.7 It is poor design practice to overdesign the fire pump and driver and then count on the pressure relief valve to open and relieve the excess pressure. A pressure relief valve is not an acceptable method of reducing system pressure under normal operating conditions.

4.7.7.1 The net pump shutoff (churn) pressure plus the maximum static suction pressure, adjusted for elevation, shall not exceed the pressure for which the system components are rated.

When selecting the fire pump output, pressure input from the attached water supply must be added to the pressure output of the pump in the tabulation. Pressure limitations for standard rated components in a fire protection system are 175 psi (12.1 bar) and up to 300 psi (24.1 bar) for extra heavy fittings. Pressures in excess of this limitation can damage system components. The maximum pressure permitted in a standpipe system per NFPA 14, *Standard for the Installation of Standpipe and Hose Systems,* currently is 350 psi (24.1 bar), which requires the use of extra-heavy fittings and components. A fire pump should be sized properly to provide sufficient pressure to meet the system demand without subjecting the system component(s) to pressures in excess of their rating.

4.7.7.2* Pressure relief valves and pressure regulating devices in the fire pump installation shall not be used as a means to meet the requirements of 4.7.7.1.

A pump should never be deliberately oversized so that a pressure-reducing valve or pressure relief valve is needed unless absolutely necessary. Pressure-reducing valves and pressure relief valves are maintenance intensive and should only be incorporated into the design of a fire pump system when necessary. These devices are not permitted to be installed in the fire pump system piping. Where high pressures are part of the system design, the pipe and fittings between the fire pump discharge flange and the discharge isolation valve must be rated for the pressures designed (i.e., extra heavy pattern). It is not the intent of this section to limit the use of pressure-reducing or pressure-regulating valves downstream of the discharge isolation valve. In many cases, high-pressure output is needed to meet the design requirements of systems installed in high-rise buildings and in standpipe systems in particular.

A.4.7.7.2 It is not the intent of this subsection to restrict the use of pressure reducing valves downstream of the discharge isolation valve for the purpose of meeting the requirements of 4.7.7.

4.7.7.3 Variable Speed Pressure Limiting Control.

4.7.7.3.1 Variable speed pressure limiting control drivers, as defined in this standard, shall be acceptable to limit system pressure.

Variable speed pressure limiting control or variable speed drive (VSD) was first introduced into NFPA 20 in the 2003 edition. As its name suggests, the pump speed is changed to prevent overpressurization of the fire protection system. A VSD fire pump and controller can be used when high flow and high pressure are needed but, due to the nature of fire pump performance, pressures at lower flows may be at levels in excess of the rating of system components. Examples of such systems are early suppression fast response (ESFR) sprinkler systems or standpipe systems. For example, a standpipe system with a system demand of 1000 gpm (3785 L/min) at 150 psi (10.3 bar) would require a fire pump rated for 1000 gpm (3785 L/min) at 160 psi (11.0 bar) [assuming 0 psi (0 bar) on the suction flange of the fire pump]. A fire pump of this rated capacity can produce as much as 224 psi (15.5 bar)—well in excess of the 175 psi (12.1 bar) rating of the system component. In this case, a VSD with a set pressure of 165 psi (11.4 bar) would prevent overpressurization of the system while meeting the fire protection system demand. Exhibit II.4.2 illustrates a typical VSD.

EXHIBIT II.4.2 Variable Speed Drive.

4.7.7.3.2* The set pressure plus the maximum pressure variance of the variable speed pressure limiting controlled systems during variable speed operation and adjusted for elevation shall not exceed the pressure rating of any system component.

Subparagraph 4.7.7.3.2 is intended to be a performance requirement to address the pressure tolerance rather than establishing a prescriptive requirement for a specific tolerance. This subparagraph and A.4.7.7.3.2 also provide guidance for the fire protection system designer on how to deal with the effects of the variable speed pressure tolerance.

A.4.7.7.3.2 This requirement is intended to take into consideration the set pressure tolerance performance of the variable speed pressure limiting control as stated by the manufacturer.

4.8* Centrifugal Fire Pump Capacities.

A.4.8 The performance of the pump when applied at capacities over 140 percent of rated capacity can be adversely affected by the suction conditions. Application of the pump at capacities less than 90 percent of the rated capacity is not recommended.

The selection and application of the fire pump should not be confused with pump operating conditions. With proper suction conditions, the pump can operate at any point on its characteristic curve from shutoff to 150 percent of its rated capacity.

Selecting a pump so that the system demand falls between 90 and 140 percent of rated capacity ensures that the pump is not oversized and conversely is not operating at its maximum output. While NFPA 20 recommends selecting a pump of a size such that the system demand falls between 90 and 140 percent, some insurance companies limit pump sizing to 120 percent of rated capacity. The intent of this limitation is to avoid operating the pump at or near the overload point, which over time may compromise pump performance. After years of service, a fire pump may begin to show signs of wear and may not be capable of reaching 150 percent of rated capacity. If such a pump were selected based on performance at 150 percent capacity, eventually, a deficiency would result. Selecting a fire pump of such a size that the system demand falls below 90 percent results in a fire pump that is too large for the system. The results of such a design include a potential overpressurization of the system and excessive cost.

4.8.1 A centrifugal fire pump for fire protection shall be selected so that the greatest single demand for any fire protection system connected to the pump is less than or equal to 150 percent of the rated capacity (flow) of the pump.

4.8.2* Centrifugal fire pumps shall have one of the rated capacities in gpm (L/min) identified in Table 4.8.2 and shall be rated at net pressures of 40 psi (2.7 bar) or more.

A.4.8.2 In countries that use the metric system, there do not appear to be standardized flow ratings for pump capacities; therefore, the metric conversions listed in Table 4.8.2 are soft conversions.

4.8.3 Centrifugal fire pumps with ratings over 5000 gpm (18,925 L/min) shall be subject to individual review by either the authority having jurisdiction or a listing laboratory.

4.9 Nameplate.

Pumps shall be provided with a nameplate.

The nameplate provides specific performance data on the fire pump. This information must be maintained for the life of the system. Without the nameplate, it may be difficult to deter-

TABLE 4.8.2 Centrifugal Fire Pump Capacities

gpm	L/min	gpm	L/min
25	95	1,000	3,785
50	189	1,250	4,731
100	379	1,500	5,677
150	568	2,000	7,570
200	757	2,500	9,462
250	946	3,000	11,355
300	1,136	3,500	13,247
400	1,514	4,000	15,140
450	1,703	4,500	17,032
500	1,892	5,000	18,925
750	2,839		

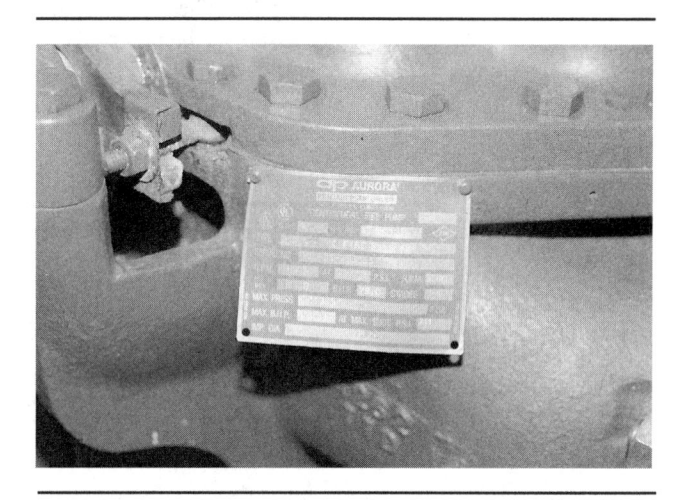

EXHIBIT II.4.3 Fire Pump Nameplate.

mine the performance characteristics of the pump. See Exhibit II.4.3 for an illustration of a typical fire pump nameplate.

4.10 Pressure Gauges.

Suction and discharge gauges are used to measure pressure on both sides of the fire pump when in operation and during the weekly running test. The gauges should be of sufficient quality to provide a reasonably accurate pressure reading since they are used for testing purposes on a weekly and annual basis. These gauges should be calibrated each year to maintain accuracy.

4.10.1 Discharge.

4.10.1.1 A pressure gauge having a dial not less than 3.5 in. (89 mm) in diameter shall be connected near the discharge casting with a nominal 0.25 in. (6 mm) gauge valve.

4.10.1.2 The dial shall indicate pressure to at least twice the rated working pressure of the pump but not less than 200 psi (13.8 bar).

4.10.1.3 The face of the dial shall read in bar, pounds per square inch, or both with the manufacturer's standard graduations.

4.10.2* Suction.

A.4.10.2 For protection against damage from overpressure, where desired, a gauge protector should be installed.

4.10.2.1 Unless the requirements of 4.10.2.4 are met, a gauge having a dial not less than 3.5 in. (89 mm) in diameter shall be connected to the suction pipe near the pump with a nominal 0.25 in. (6 mm) gauge valve.

4.10.2.1.1 Where the minimum pump suction pressure is below 20 psi (1.3 bar) under any flow condition, the suction gauge shall be a compound pressure and vacuum gauge.

4.10.2.2 The face of the dial shall read in inches of mercury (millimeters of mercury) or psi (bar) for the suction range.

4.10.2.3 The gauge shall have a pressure range two times the rated maximum suction pressure of the pump.

4.10.2.4 The requirements of 4.10.2 shall not apply to vertical shaft turbine–type pumps taking suction from a well or open wet pit.

4.11 Circulation Relief Valve.

For electrical drive fire pumps, a listed circulation relief valve is needed to provide cooling water when the pump is operating at churn. The pipe connection for this valve must be located on the discharge side of the pump to cause flow through the pump casing and should discharge outdoors or to a floor drain where discharge can be observed by the pump operator. This valve should be installed in the vertical position because installation in the horizontal position may cause the valve to fail at an accelerated rate due to obstructing material collecting in the valve seat. Failure or lack of this valve can result in overheating and subsequent damage to the fire pump. Exhibit II.4.4 illustrates a ¾ in. (19 mm) casing relief valve.

EXHIBIT II.4.4 Casing Relief Valve. (Photograph courtesy of Liberty Mutual Property)

4.11.1 Automatic Relief Valve.

4.11.1.1 Unless the requirements of 4.11.1.7 are met, each pump(s) shall have an automatic relief valve listed for the fire pump service installed and set below the shutoff pressure at minimum expected suction pressure.

4.11.1.2 The valve shall be installed on the discharge side of the pump before the discharge check valve.

4.11.1.3 The valve shall provide flow of sufficient water to prevent the pump from overheating when operating with no discharge.

4.11.1.4 Provisions shall be made for discharge to a drain.

4.11.1.5 Circulation relief valves shall not be tied in with the packing box or drip rim drains.

4.11.1.6 The automatic relief valve shall have a nominal size of 0.75 in. (19 mm) for pumps with a rated capacity not exceeding 2500 gpm (9462 L/min) and have a nominal size of 1 in. (25 mm) for pumps with a rated capacity of 3000 gpm to 5000 gpm (11,355 L/min to 18,925 L/min).

4.11.1.7 The requirements of 4.11.1 shall not apply to engine-driven pumps for which engine cooling water is taken from the pump discharge.

Diesel fire pumps that use water from the discharge side of a fire pump for engine cooling perform the same function as a circulation relief valve. See Chapter 11 for more details.

4.12* Equipment Protection.

Section 4.12 provides requirements and guidance on the proper placement of a fire pump and related equipment. In this case, equipment protection is also intended to provide a relatively safe environment for the equipment operator. In accordance with Section 4.3, a qualified fire pump operator will report to the fire pump room to confirm that the equipment is operating properly in the event of operation of the plant. By locating the fire pump room near an exterior wall and enclosing the equipment in a fire-rated room, the operator is provided with a means of egress and both the operator and the equipment are protected from fire exposure.

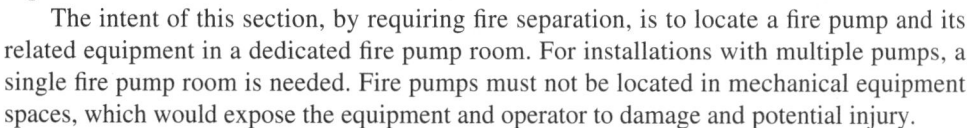

The intent of this section, by requiring fire separation, is to locate a fire pump and its related equipment in a dedicated fire pump room. For installations with multiple pumps, a single fire pump room is needed. Fire pumps must not be located in mechanical equipment spaces, which would expose the equipment and operator to damage and potential injury.

◀ FAQ
Does the standard require a separate fire pump room for each fire pump installation?

A.4.12 Special consideration needs to be given to fire pump installations installed belowgrade. Light, heat, drainage, and ventilation are several of the variables that need to be addressed. Some locations or installations might not require a pump house. Where a pump room or pump house is required, it should be of ample size and located to permit short and properly arranged piping. The suction piping should receive first consideration. The pump house should preferably be a detached building of noncombustible construction. A one-story pump room with a combustible roof, either detached or well cut off from an adjoining one-story building, is acceptable if sprinklered. Where a detached building is not feasible, the pump room should be located and constructed so as to protect the pump unit and controls from falling floors or machinery and from fire that could drive away the pump operator or damage the pump unit or controls. Access to the pump room should be provided from outside the building. Where the use of brick or reinforced concrete is not feasible, metal lath and plaster is recommended for the construction of the pump room. The pump room or pump house should not be used for storage purposes. Vertical shaft turbine–type pumps might necessitate a removable panel in the pump house roof to permit the pump to be removed for inspection or repair. Proper clearances to equipment should be provided as recommended by the manufacturer's drawings.

4.12.1* General Requirements. The fire pump, driver, controller, water supply, and power supply shall be protected against possible interruption of service through damage caused by explosion, fire, flood, earthquake, rodents, insects, windstorm, freezing, vandalism, and other adverse conditions.

A.4.12.1 A fire pump that is inoperative for any reason at any time constitutes an impairment to the fire protection system. It should be returned to service without delay.

Rain and intense heat from the sun are adverse conditions to equipment not installed in a completely protective enclosure. At a minimum, equipment installed outdoors should be shielded by a roof or deck.

4.12.1.1* Indoor Fire Pump Units.

A.4.12.1.1 Most fire departments have procedures requiring operation of a fire pump unit during an incident. Building designers should locate the fire pump room to be easily accessible during an incident.

4.12.1.1.1 Fire pump units serving high-rise buildings shall be protected from surrounding occupancies by a minimum of 2-hour fire-rated construction or physically separated from the protected building by a minimum of 50 ft (15.3 m).

A minimum of 50 ft (15.3 m) of separation or a 2-hour fire-rated enclosure will protect the building's fire pump and personnel from an exposure fire.

4.12.1.1.2 Indoor fire pump rooms in non-high-rise buildings or in separate fire pump buildings shall be physically separated or protected by fire-rated construction in accordance with Table 4.12.1.1.2.

TABLE 4.12.1.1.2 Equipment Protection

Pump Room/House	Building(s) Exposing Pump Room/House	Required Separation
Not sprinklered	Not sprinklered	2 hour fire-rated
Not sprinklered	Fully sprinklered	or
Fully sprinklered	Not sprinklered	50 ft (15.3 m)
Fully sprinklered	Fully sprinklered	1 hour fire-rated or 50 ft (15.3 m)

4.12.1.1.3 The location of and access to the fire pump room shall be preplanned with the fire department.

4.12.1.1.4* Except as permitted in 4.12.1.1.5, rooms containing fire pumps shall be free from storage, equipment, and penetrations not essential to the operation of the pump and related components.

It is the intent of the standard to prohibit the use of a fire pump room for the purposes of storage. It is not the intent of the standard to prohibit the installation of other types of equipment such as domestic water distribution equipment as indicated in 4.12.1.1.5. No materials or equipment that add to the combustibility of the space should be placed in the fire pump room. This provision is intended to provide protection to equipment as well as operating personnel.

A.4.12.1.1.4 Equipment that increases the fire hazard (such as boilers) and is not related to fire protection systems should not be in a fire pump room.

4.12.1.1.5 Equipment related to domestic water distribution shall be permitted to be located within the same room as the fire pump equipment.

4.12.1.2 Outdoor Fire Pump Units.

4.12.1.2.1 Fire pump units that are outdoors shall be located at least 50 ft (15.3 m) away from any buildings and other fire exposures exposing the building.

4.12.1.2.2 Outdoor installations shall be required to be provided with protection against possible interruption, in accordance with 4.12.1.

4.12.1.3 Fire Pump Buildings or Rooms with Diesel Engines. Fire pump buildings or rooms enclosing diesel engine pump drivers and day tanks shall be protected with an automatic sprinkler system installed in accordance with NFPA 13, *Standard for the Installation of Sprinkler Systems.*

4.12.2 Equipment Access.

Fire pump rooms should be accessible under fire conditions. Where a fire pump room cannot be located on grade with direct access from outside of the building, a protected space with a 2-hour fire rating must be provided.

4.12.2.1 Access to the fire pump room shall be pre-planned with the fire department.

4.12.2.1.1 Fire pump rooms not directly accessible from the outside shall be accessible through an enclosed passageway from an enclosed stairway or exterior exit. The enclosed passageway shall have a minimum 2-hour fire-resistance rating.

4.12.3 Heat.

4.12.3.1 An approved or listed source of heat shall be provided for maintaining the temperature of a pump room or pump house, where required, above 40°F (5°C).

4.12.3.2 The requirements of 11.6.5 shall be followed for higher temperature requirements for internal combustion engines.

4.12.4 Normal Lighting.
Artificial light shall be provided in a pump room or pump house.

4.12.5 Emergency Lighting.

4.12.5.1 Emergency lighting shall be provided in accordance with NFPA *101, Life Safety Code.*

4.12.5.2 Emergency lights shall not be connected to an engine starting battery.

4.12.6 Ventilation.
Provision shall be made for ventilation of a pump room or pump house.

4.12.7* Drainage.

A.4.12.7 Pump rooms and pump houses should be dry and free of condensate. To accomplish a dry environment, heat might be necessary.

4.12.7.1 Floors shall be pitched for adequate drainage of escaping water away from critical equipment such as the pump, driver, controller, and so forth.

4.12.7.2 The pump room or pump house shall be provided with a floor drain that will discharge to a frost-free location.

4.12.8 Guards.
Couplings and flexible connecting shafts shall be installed with a coupling guard in accordance with Section 8 of ANSI B15.1, *Safety Standard for Mechanical Power Transmission Apparatus.*

4.13 Pipe and Fittings.

4.13.1* Steel Pipe.

The type of steel piping used for fire pump installation should be identical to that specified for the suppression system. See Commentary Table II.4.1 for acceptable pipe types.

COMMENTARY TABLE II.4.1 *Pipe or Tube Materials and Dimensions*

Materials and Dimensions	Standard
Ferrous Piping (Welded and Seamless)	
Specification for black and hot-dipped zinc-coated (galvanized) welded and seamless steel pipe for fire protection use	ASTM A 795
Specification for welded and seamless steel pipe	ANSI/ASTM A 53
Wrought steel pipe	ANSI/ASME B36.10M
Specification for electric-resistance-welded steel pipe	ASTM A 135
Copper Tube (Drawn, Seamless)	
Specification for seamless copper tube	ASTM B 75
Specification for seamless copper water tube	ASTM B 88
Specification for general requirements for wrought seamless copper and copper-alloy tube	ASTM B 251
Fluxes for soldering applications of copper and copper-alloy tube	ASTM B 813
Brazing filler metal (classification BCuP-3 or BCuP-4)	AWS A5.8
Solder metal, Section 1: Solder alloys containing less than 0.2% lead and having solidus temperatures greater than 400°F	ASTM B 32
Alloy materials	ASTM B 446

Source: NFPA 13, 2010, Table 6.3.1.1.

A.4.13.1 The exterior of aboveground steel piping should be kept painted.

4.13.1.1 Steel pipe shall be used aboveground except for connection to underground suction and underground discharge piping.

4.13.1.2 Where corrosive water conditions exist, steel suction pipe shall be galvanized or painted on the inside prior to installation with a paint recommended for submerged surfaces.

In most cases, black steel, when used with any water, eventually corrodes to some extent. All suction pipes are recommended to be protected from corrosion by galvanizing or painting to protect the pump. Should the suction pipe corrode, rust or scale could dislodge from the inside wall of the pipe and enter the pump, potentially damaging the pump impeller.

4.13.1.3 Thick bituminous linings shall not be used.

4.13.2* Joining Method.

The type of fittings used for the fire pump installation should be identical to those specified for the suppression system. The fittings used in sprinkler systems have been evaluated for that use and have been found to be acceptable. Using the same type of fittings for a system's fire pump installation eliminates performance and compatibility issues. See Commentary Table II.4.2 for acceptable fitting types.

Chlorinated polyvinyl chloride (CPVC) and copper pipe and fittings should not be used on the suction side of a fire pump due to potential settlement of either the foundation on which the fire pump rests or the water supply pipe.

A.4.13.2 Flanges welded to pipe are preferred.

4.13.2.1 Sections of steel piping shall be joined by means of screwed, flanged mechanical grooved joints or other approved fittings.

4.13.2.2 Slip-type fittings shall be permitted to be used where installed as required by 4.14.6 and where the piping is mechanically secured to prevent slippage.

COMMENTARY TABLE II.4.2 *Fittings Materials and Dimensions*

Materials and Dimensions	Standard
Cast Iron	
Cast iron threaded fittings, Class 125 and 250	ASME B16.4
Cast iron pipe flanges and flanged fittings	ASME B16.1
Malleable Iron	
Malleable iron threaded fittings, Class 150 and 300 steel	ASME B16.3
Factory-made wrought steel buttweld fittings	ASME B16.9
Buttwelding ends for pipe, valves, flanges, and fittings	ASME B16.25
Specification for piping fittings of wrought carbon steel and alloy steel for moderate and elevated temperatures	ASTM A 234
Steel pipe flanges and flanged fittings	ASME B16.5
Forged steel fittings, socket welded and threaded copper	ASME B16.11
Wrought copper and copper alloy solder joint pressure fittings	ASME B16.22
Cast copper alloy solder joint pressure fittings	ASME B16.18

Source: NFPA 13, 2010, Table 6.4.1.

4.13.3 Concentrate and Additive Piping.

4.13.3.1 Foam concentrate or additive piping shall be a material that will not corrode in this service.

Pure foam concentrate is frequently corrosive; therefore, foam concentrate piping is normally installed with either stainless steel or brass pipe. Galvanized pipe should never be used for foam concentrate piping because the concentrate may cause the galvanizing to "flake" and detach from the pipe wall.

4.13.3.2 Galvanized pipe shall not be used for foam concentrate service.

4.13.4 Drain Piping. Drain pipe and its fittings that discharge to atmosphere shall be permitted to be constructed of metallic or polymeric materials.

This new subsection clarifies that drain piping can be constructed of metallic or plastic materials.

4.13.5 Piping, Hangers, and Seismic Bracing. Pipe, fittings, hangers, and seismic bracing for the fire pump unit, including the suction and discharge piping, shall comply with the applicable requirements of NFPA 13, *Standard for the Installation of Sprinkler Systems*.

This new subsection clarifies that NFPA 13, *Standard for the Installation of Sprinkler Systems*, should be referenced for seismic bracing.

4.13.6* Cutting and Welding. Torch cutting or welding in the pump house shall be permitted as a means of modifying or repairing pump house piping when it is performed in accordance with NFPA 51B, *Standard for Fire Prevention During Welding, Cutting, and Other Hot Work*.

A.4.13.6 When welding is performed on the pump suction or discharge piping with the pump in place, the welding ground should be on the same side of the pump as the welding.

4.14 Suction Pipe and Fittings.

4.14.1* Components.

A.4.14.1 The exterior of steel suction piping should be kept painted.

Buried iron or steel pipe should be lined and coated or protected against corrosion in conformance with AWWA C104, *Cement-Mortar Lining for Cast-Iron and Ductile-Iron Pipe and Fittings for Water,* or equivalent standards.

Underground pipe frequently is coated with a material that prevents the formation of debris or tuberculation. Steel pipe should never be used for underground piping, particularly on the suction side of the pump.

4.14.1.1 The suction components shall consist of all pipe, valves, and fittings from the pump suction flange to the connection to the public or private water service main, storage tank, or reservoir, and so forth, that feeds water to the pump.

4.14.1.2 Where pumps are installed in series, the suction pipe for the subsequent pump(s) shall begin at the system side of the discharge valve of the previous pump.

4.14.2 Installation. Suction pipe shall be installed and tested in accordance with NFPA 24, *Standard for the Installation of Private Fire Service Mains and Their Appurtenances.*

4.14.3 Suction Size.

4.14.3.1* Unless the requirements of 4.14.3.2 are met, the size of the suction pipe for a single pump or of the suction header pipe for multiple pumps (operating together) shall be such that, with all pumps operating at maximum flow (150 percent of rated capacity or the maximum flow available from the water supply as discussed in 4.6.2.3.1), the gauge pressure at the pump suction flanges shall be 0 psi (0 bar) or higher.

It is important to note that horizontal shaft fire pumps are not permitted to draft water. The suction pipe must also be large enough so that a substantial pressure loss does not occur when the pump is supplied by a water storage tank. For such applications, the suction pressure is permitted to drop to not less than –3 psi (–0.2 bar). This negative pressure is only permitted when the base of the tank is even with or higher than the pump and is intended to compensate for the friction loss through the suction pipe, fittings, and gate valve.

A.4.14.3.1 It is permitted that the suction pressure drop to –3 psi for a horizontal pump that is taking suction from a grade level storage tank where the pump room elevation and bottom of the water storage tank are at the same elevation. This negative suction pressure is to allow for the friction loss in the suction piping when the pump is operating at 150 percent capacity.

4.14.3.2 The requirements of 4.14.3.1 shall not apply where the supply is a suction tank with its base at or above the same elevation as the pump, where the gauge pressure at the pump suction flange shall be permitted to drop to –3 psi (–0.2 bar) with the lowest water level after the maximum system demand and duration have been supplied.

4.14.3.3 The suction pipe shall be sized such that, with the pump(s) operating at 150 percent of rated capacity, the velocity in that portion of the suction pipe located within 10 pipe diameters upstream of the pump suction flange does not exceed 15 ft/sec (4.57 m/sec).

The size of the suction pipe is based on limiting water velocity to not more than 15 ft/sec (4.57 m/sec) to limit turbulent flow in the pipe. Turbulent flow generates air bubbles in the water, which adversely affects pump efficiency. As the pump impeller is imparting energy to the water (which is relatively incompressible), the introduction of air bubbles (which are easily compressible) can create small regions of high pressure due to collapsed air bubbles and causes inefficient pump operation.

4.14.3.4 The size of that portion of the suction pipe located within 10 pipe diameters upstream of the pump suction flange shall be not less than that specified in Section 4.26.

4.14.4* Pumps with Bypass.

A.4.14.4 The following notes apply to Figure A.4.14.4:

(1) A jockey pump is usually required with automatically controlled pumps.
(2) If testing facilities are to be provided, also see Figure A.4.20.1.2(a) and Figure A.4.20.1.2(b).
(3) Pressure-sensing lines also need to be installed in accordance with 10.5.2.1 or 12.7.2.1. See Figure A.4.30(a) and Figure A.4.30(b).

4.14.4.1 Where the suction supply is of sufficient pressure to be of material value without the pump, the pump shall be installed with a bypass. *(See Figure A.4.14.4.)*

In most cases, a pump that is connected to a public or private water supply should include a bypass. Only in rare cases, where the pressure available is so low that the water supply is of no value without the pump, should a fire pump be installed without a bypass. When a pump is supplied by a suction tank, a bypass is not needed (see Figure A.6.3.1(b) because a suction

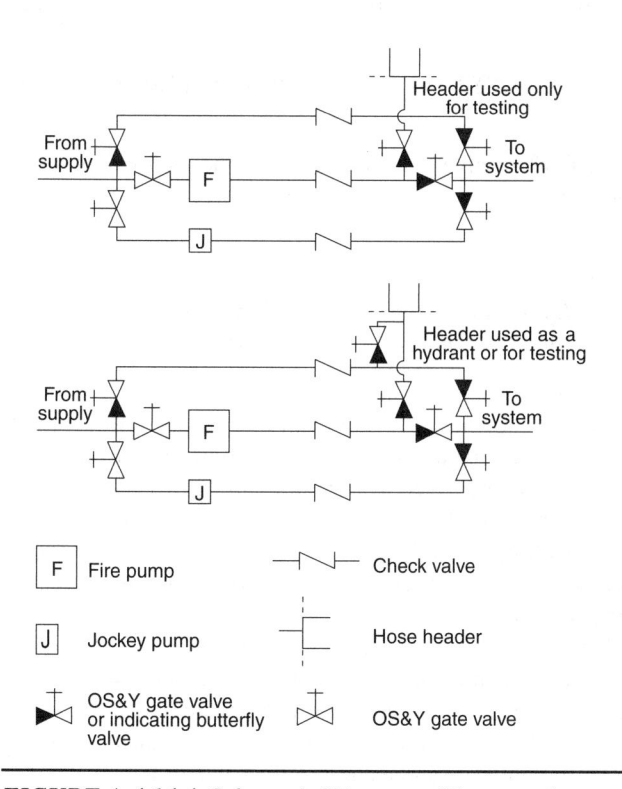

FIGURE A.4.14.4 *Schematic Diagram of Suggested Arrangements for a Fire Pump with a Bypass, Taking Suction from Public Mains.*

tank will not provide sufficient pressure to be of value without the fire pump operating. The valves on the bypass are required to be normally open so that the attached water supply is available automatically. In this case, "normally open" refers to the valve being in the open position at all times. The bypass valves should be closed only for system maintenance. In some cases, the function of the pump bypass is mistakenly thought to be used only when the fire pump is out of service. This function is not the intent of the standard, because the water from the bypass must be available automatically if the pump fails to start.

4.14.4.2 The size of the bypass shall be at least as large as the pipe size required for discharge pipe as specified in Section 4.26.

4.14.5* Valves.

A.4.14.5 Where the suction supply is from public water mains, the gate valve should be located as far as is practical from the suction flange on the pump. Where it comes from a stored water container, the gate valve should be located at the outlet of the container. A butterfly valve on the suction side of the pump can create turbulence that adversely affects the pump performance and can increase the possibility of blockage of the pipe.

The required placement of the valve in the suction line is intended to limit turbulence. While an outside screw and yoke (OS&Y) valve does not interfere with flow when it is fully open, the butterfly in a butterfly valve is still suspended in the pipe and causes turbulent flow. While previous editions of NFPA 20 prohibited the use of butterfly valves in the suction pipe, the standard now permits the use of this type of valve if the valve is located at a sufficient distance from the suction flange of the pump. By locating the butterfly valve 50 ft (15.3 m) or more away from the suction flange of the pump, the turbulence can be eliminated before the water reaches the suction flange.

4.14.5.1 A listed outside screw and yoke (OS&Y) gate valve shall be installed in the suction pipe.

4.14.5.2 No valve other than a listed OS&Y valve shall be installed in the suction pipe within 50 ft (15.3 m) of the pump suction flange.

4.14.6* Installation.

A.4.14.6 See Figure A.4.14.6. *(See Hydraulic Institute Standards for Centrifugal, Rotary and Reciprocating Pumps for additional information.)*

The layout and design of the suction pipe must be carefully considered in order to avoid the generation of air bubbles in the suction pipe and to avoid unbalanced flow into the pump suction flange. Where a reduction in pipe diameter is necessary, an eccentric reducer must be used to prevent the development of an air pocket, as illustrated in Figure A.4.14.6. The flat portion of the reducer must be on top with the actual reduction in pipe diameter taking place on the bottom of the fitting. Installing the reducer in this position will prevent an air pocket from forming. Placing an elbow with the centerline in the horizontal plane within 10 pipe diameters of the suction flange creates an unbalanced flow of water into the impeller. An unbalanced flow causes an axial load to be placed on the pump shaft and bearings, causing excessive wear to the bearings and subsequent damage to the pump over time and/or severe cavitation if not corrected. The elbow shown in Exhibit II.4.5 will eventually cause damage to the pump and may result in pump shaft-bearing failure. A butterfly valve installed within 10 pipe diameters will also cause turbulent flow and cavitation in the pump and should be avoided.

In the lower portion of Figure A.4.14.6, the correct method for designing a turn in the suction piping is illustrated. The suction elbow should always be of the long turn radius type.

FIGURE A.4.14.6 Right and Wrong Pump Suctions.

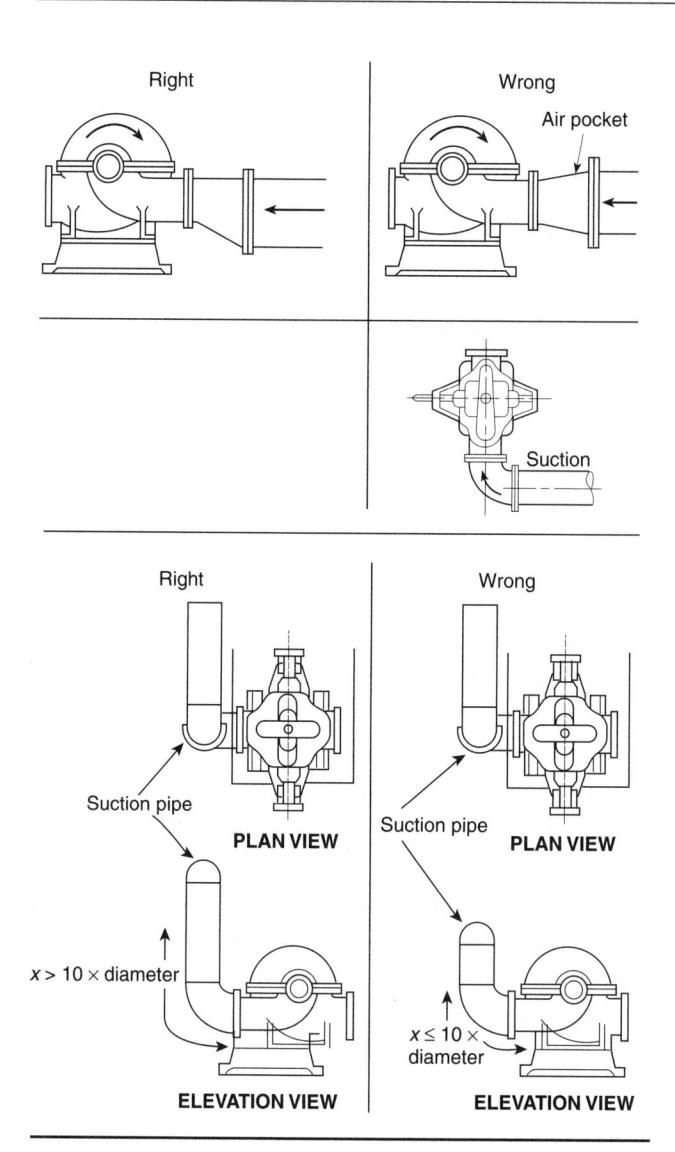

EXHIBIT II.4.5 Elbow in the Horizontal Position in a Fire Pump Suction Line.

An elbow in the vertical position (either directed upward or downward) should be used, and the elbow causing a turn in the pipe parallel with the pump shaft should only be installed when located greater than 10 pipe diameters away from the suction flange. For example, an elbow installed in the horizontal plane in a suction pipe with a diameter of 8 in. (200 mm) should be located 80 in. (2000 mm) away from the suction flange of the pump.

4.14.6.1 General. Suction pipe shall be laid carefully to avoid air leaks and air pockets, either of which can seriously affect the operation of the pump.

4.14.6.2 Freeze Protection.

4.14.6.2.1 Suction pipe shall be installed below the frost line or in frostproof casings.

4.14.6.2.2 Where pipe enters streams, ponds, or reservoirs, special attention shall be given to prevent freezing either underground or underwater.

4.14.6.3 Elbows and Tees.

4.14.6.3.1 Unless the requirements of 4.14.6.3.2 are met, elbows and tees with a centerline plane parallel to a horizontal split-case pump shaft shall not be permitted. *(See Figure A.4.14.6.)*

Any fitting that causes a change in the direction of flow, including elbows, tees, and crosses when installed in the horizontal plane, should be located a minimum of 10 pipe diameters away from the suction flange of the pump. While fittings in the vertical position are not considered to be a problem, such an installation should be avoided if possible to prevent turbulence.

4.14.6.3.2 The requirements of 4.14.6.3.1 shall not apply to elbows and tees with a centerline plane parallel to a horizontal split-case pump shaft where the distance between the flanges of the pump suction intake and the elbow and tee is greater than 10 times the suction pipe diameter.

4.14.6.3.3 Elbows with a centerline plane perpendicular to the horizontal split-case pump shaft shall be permitted at any location in the pump suction intake.

An elbow perpendicular to the pump shaft will not produce an unbalanced flow condition. Regardless of the position of the elbow in a fire pump suction pipe, when designing a change in direction in the suction pipe of a pump, the elbows used should be of the long radius type to permit as smooth a transition in the change in direction as possible. A long radius elbow provides a smoother turn in direction, further limiting turbulence.

FAQ ▶
Does NFPA 20 require the installation of a long radius elbow in the suction pipe?

Although a long radius elbow is *not* required by NFPA 20, the use of such a fitting reduces the likelihood of turbulence in the suction pipe. In all cases, even a long radius elbow should never be installed in the parallel plane with the pump shaft.

4.14.6.4 Eccentric Tapered Reducer or Increaser. Where the suction pipe and pump suction flange are not of the same size, they shall be connected with an eccentric tapered reducer or increaser installed in such a way as to avoid air pockets.

4.14.6.5 Strain Relief. Where the pump and its suction supply are on separate foundations with rigid interconnecting pipe, the pipe shall be provided with strain relief. *(See Figure A.6.3.1(a).)*

Strain relief can be accomplished by means of an expansion joint as depicted in Exhibit II.4.6 or a series of grooved couplings as shown in Exhibit II.4.7. When using grooved couplings

EXHIBIT II.4.6 Expansion Joint.

EXHIBIT II.4.7 Grooved Couplings

2010 Stationary Fire Pumps Handbook

for strain relief, consideration should be given to installing the pipe with the proper pipe end separation as recommended by the manufacturer. Deflection for most grooved couplings is on the order of 3½ in. to 4½ in. (89 mm to 114 mm) per 20 ft (6 m) length of pipe in sizes 6 in. to 8 in. (150 mm to 200 mm). A series of these grooved couplings in the suction pipe provides flexibility if the pump or tank foundations begin to settle and should also provide for limited movement due to expansion and contraction. The purpose of providing such strain relief is to prevent damage to the pump suction flange, since misalignment or stress placed on the pump suction flange can cause the flange to crack. The fire pump design should avoid the use of separate foundations for the pump and the driver, particularly in earthquake-prone areas.

Seismic design should follow the design requirements found in ASCE 7, *Minimum Design Loads for Buildings and Other Structures,* and NFPA 13. Prior to the design of the installation, the locally adopted building code must be checked for the seismic requirements for the fire pump piping installation.

4.14.7 Multiple Pumps. Where a single suction pipe supplies more than one pump, the suction pipe layout at the pumps shall be arranged so that each pump will receive its proportional supply.

Multiple pump installations are used to meet one of the design scenarios as follows:

1. Two or more pumps are installed in parallel to meet the minimum water demand of the fire protection system.
2. Two or more pumps are installed in series to meet the pressure demand of the fire protection system.
3. Two or more pumps are installed, and each of the pumps is individually required to meet the minimum water supply and pressure demand required by the fire protection system. One pump will serve as the primary source, the other(s) will serve as a backup or reserve supply.

In the first situation, the suction piping is required to be sized assuming that all of the installed pumps are operating simultaneously. In the second situation, the piping is sized for the flow of both pumps operating but the flow will equal that of a single pump. The third arrangement will have piping sized for either (but not both or multiple) of the pumps operating.

4.14.8* Suction Screening.

Suction screens are located in a suction pit at the pump end of a channel, reservoir, pond, or lake that allows water to flow from the source to the pump, as indicated in Figure A.7.2.2.2. The figure shows a vertical turbine–type pump. A horizontal pump may also be used, if the suction pressure is positive. Any debris entering the channel can accumulate in the area in front of the suction crib. If a fire occurs, it is possible that fire pump operation will draw some of this settled material onto the first screen, reducing the flow area and restricting flow to the fire protection system. The standard requires two screens to be installed to allow removal of the first screen if it becomes obstructed. The opening size in the screens is intended to be small to provide protection to the distribution portion of the fire protection system. For example, the openings would be restricted to less than ½ in. (15 mm) for a sprinkler having an orifice of ½ in. (15 mm). Caution should be exercised when removing a debris-loaded screen because the removal may cause the obstructing material to fall off and thereby obstruct the second screen.

A.4.14.8 In the selection of screen material, consideration should be given to prevention of fouling from aquatic growth. Antifouling is best accomplished with brass or copper wire.

4.14.8.1 Where the water supply is obtained from an open source such as a pond or wet pit, the passage of materials that might clog the pump shall be obstructed.

4.14.8.2 Double intake screens shall be provided at the suction intake.

4.14.8.3 Screens shall be removable, or an in situ cleaning shall be provided.

4.14.8.4 Below minimum water level, these screens shall have an effective net area of opening of 1 in.^{2} for each 1 gpm (170 mm^2 for each 1 L/min) at 150 percent of rated pump capacity.

The following example illustrates how to correctly size a suction screen for a fire pump.

EXAMPLE

A 750 gpm (2839 L/min) vertical shaft turbine pump requires a screen to protect the wet pit that takes water from an open lake or river. What size is needed for the screen?

Solution: Since the pump is rated for 750 gpm (2839 L/min), the pump delivers 1125 gpm (4258 L/min) at overload (150 percent of 750). Therefore, the screen must have an area of 1125 in.^{2} (0.72 m^2). Since 4.14.8.6 requires a 0.50 in. (12.7 mm) mesh and a No. 10 B&S gauge wire, the total screen area must be 1800 in.^{2} (1.1 m^2) (1.6 × 1125 in.^{2}). This area is equivalent to a screen having an area of 12.5 ft^2 (1.16 m^2).

FAQ ▶
Does NFPA 20 prohibit the installation of any device in the suction pipe of a fire pump?

The installation of devices in the suction pipe of a fire pump is restricted to the devices listed in 4.14.9.2. The installation of any device or fitting should be made at least 10 pipe diameters upstream of the pump suction flange to avoid unnecessary turbulence.

4.14.8.5 Screens shall be so arranged that they can be cleaned or repaired without disturbing the suction pipe.

4.14.8.6 Mesh screens shall be brass, copper, Monel, stainless steel, or other equivalent corrosion-resistant metallic material wire screen of 0.50 in. (12.7 mm) maximum mesh and No. 10 B&S gauge.

4.14.8.7 Where flat panel mesh screens are used, the wire shall be secured to a metal frame sliding vertically at the entrance to the intake.

4.14.8.8 Where the screens are located in a sump or depression, they shall be equipped with a debris-lifting rake.

4.14.8.9 Periodically, the system shall be test pumped, the screens shall be removed for inspection, and accumulated debris shall be removed.

4.14.8.10 Continuous slot screens shall be brass, copper, Monel, stainless steel, or other equivalent corrosion-resistant metallic material of 0.125 in. (3.2 mm) maximum slot and profile wire construction.

4.14.8.11 Screens shall have at least 62.5 percent open area.

4.14.8.12 Where zebra mussel infestation is present or reasonably anticipated at the site, the screens shall be constructed of a material with demonstrated resistance to zebra mussel attachment or coated with a material with demonstrated resistance to zebra mussel attachment at low velocities.

With the introduction of zebra mussels into many United States waterways, extra care should be exercised with the use of suction screens. While the introduction of zebra mussels was first observed in the western end of Lake Erie, the organisms have since spread to all of the Great Lakes and the Mississippi, Ohio, and Illinois Rivers, including approximately 22 major lakes and tributaries. The mussels can spread rapidly once established in a water source, with each female producing approximately 300,000 to 1,000,000 larvae annually. The larvae, in turn, can grow from the size of fine sand to an adult size of one-half foot in length in 2 to 3 years. They attach to many materials as well as to each other, forming obstructions to piping and discharge device orifices.

NFPA 20 requires screens to be constructed of brass, copper, Monel, stainless steel, or other corrosion-resistant material. Copper-based materials have been tested and proven resistant to the attachment of zebra mussels. Pure copper is resistant to attachments until corrosion products form on its surface. Some copper alloys, Monel, and stainless steel have been tested and found to offer no resistance to the attachment of zebra mussels. In areas where zebra mussel infestation is present, screens should be constructed with materials or coated with a corrosion-resistant material that has demonstrated long-term resistance to their attachment in low-velocity conditions. Materials that require regular maintenance to remove zebra mussels should not be used.

4.14.8.13 The overall area of the screen shall be 1.6 times the net screen opening area. *(See screen details in Figure A.7.2.2.2.)*

4.14.9* Devices in Suction Piping.

A.4.14.9 The term *device* as used in this subsection is intended to include, but not be limited to, devices that sense suction pressure and then restrict or stop the fire pump discharge. Due to the pressure losses and the potential for interruption of the flow to the fire protection systems, the use of backflow prevention devices is discouraged in fire pump piping. Where required, however, the placement of such a device on the discharge side of the pump is to ensure acceptable flow characteristics to the pump suction. It is more efficient to lose the pressure after the pump has boosted it, rather than before the pump boosts it. Where the backflow preventer is on the discharge side of the pump and a jockey pump is installed, the jockey pump discharge and sensing lines need to be located so that a cross-connection is not created through the jockey pump.

4.14.9.1 No device or assembly, unless identified in 4.14.9.2, that will stop, restrict the starting of, or restrict the discharge of a fire pump or pump driver shall be installed in the suction piping.

The purpose of the requirement in 4.14.9.1 is to prohibit the use of certain devices in the suction pipe of a fire pump, that would cause turbulence or excess friction loss, or cut off water flow to the pump when flowing at 150 percent of rated capacity. If the water supply to a fire pump is shut off while the pump is running, a catastrophic failure will occur. Only an OS&Y control valve is permitted to be installed in the suction pipe of a fire pump to allow isolation of the fire pump for maintenance and repair. Additional valves are permitted only where a piece of equipment such as a backflow prevention device is installed at least 10 pipe diameters upstream of the pump suction flange. All control valves, which should be of the approved indicating type, are required to be supervised, preferably by a constantly attended monitoring station. The problem is that they could be shut off before or during operation of the fire pump.

It is important to note that valves that have been shut off continue to be the leading cause of fire protection system failure. A vigorous inspection program and constant electronic surveillance of all water supply control valves are critical to the continued operation of any fire protection system, particularly water supplies such as fire pumps.

4.14.9.2 The following devices shall be permitted in the suction piping where the following requirements are met:

(1) Check valves and backflow prevention devices and assemblies shall be permitted where required by other NFPA standards or the authority having jurisdiction.

Check valves and backflow prevention devices are generally prohibited in the suction pipe due to concerns over turbulence and friction loss; however, some jurisdictions require the installation of one of these devices in the suction pipe when backflow contamination is a possibility.

Backflow preventers and their isolation valves contribute to turbulence and friction loss, usually more so than check valves. It is common for reduced pressure backflow preventers to cause more than 10 psi (0.7 bar) of friction loss at the overload point (150 percent of pump rated capacity).

For example, consider a fire protection system connected to a public main where the pump is at an elevation 20 ft (6.1 m) higher than the street main and at maximum pump flow the residual pressure in the public main is 20 psi (138 kPa). Taking into account a 9 psi (62 kPa) loss in pressure due to elevation (0.433 psi × 20 ft) (0.098 bar x 6.1 m = 0.6 bar = 60 kPA) and 5 psi (39.5 kPa) friction loss in the suction pipe, the pressure at the suction flange will be 6 psi (41.4 kPa) (20 psi – 9 psi – 5 psi = 6 psi). This situation is acceptable in accordance with 4.14.3.1. However, if a backflow preventer were installed in this suction pipe with a friction loss of 10 psi (69 kPa), the system would become inoperative (20 – 9 – 5 – 10 = –4 psi). The pump would not be able to obtain water at a positive pressure and would be severely damaged if operated at maximum flow.

For these reasons, the Technical Committee on Fire Pumps prefers that backflow preventers be installed on the discharge side of the pump. In the preceding example, the pump would receive water with adequate pressure at 6 psi (41.4 kPa), boost the pressure, and force it through the backflow preventer, where it would then lose 10 psi (69 kPa). Although this arrangement solves one concern, installation of backflow prevention on the discharge side of pumps can also cause other concerns. Three of the more common concerns are as follows:

1. *Maximum pressure.* Backflow preventers are listed for maximum pressures of 175 psi (1207 kPa). The pressure on the discharge side of the pump frequently exceeds this value.

2. *Water authorities.* Water authorities sometimes assume that pumps are a possible cause of contamination of the public water supply, and, therefore, they require the backflow preventer on the suction side. However, the water authorities are incorrect in this assumption. As indicated by the American Water Works Association, fire pumps do not cause contamination problems for potable water supplies. Their manual AWWA M-14, *Recommended Practice for Backflow Prevention and Cross-Connection Control,* states that "booster pumps do not affect the potability of the system." (See the definition of *Class 2 Fire Protection Systems* in AWWA M-14, 4th edition.)

3. *Jurisdiction.* In some communities, the backflow preventer is the traditional point where the jurisdiction of one labor trade group ends and another begins. Typically, work from the water supply to the backflow preventer is done by mechanical contractors, and work from the backflow preventer to the fire protection devices is done by fire protection contractors. If the backflow preventer is installed on the discharge side of the fire pump, the fire protection contractor in many jurisdictions is prohibited from working on the fire pump because of its location. This situation can result in the installation of the fire pump by inadequately trained personnel inexperienced with fire pump functions and operations. Installation of fire pumps by unqualified personnel should be a cause of concern.

NFPA 20 does not mandate the location of backflow preventers. Instead, a technical preference for the installation of the backflow preventer on the discharge side of the pump is indicated. However, this arrangement is not always possible, so Section 4.27 is included, which contains all of the regulations to follow if the backflow preventer is to be installed on the suction side.

(2) Where the authority having jurisdiction requires positive pressure to be maintained on the suction piping, a pressure sensing line for a low suction pressure control, specifically listed for fire pump service, shall be permitted to be connected to the suction piping.

Previously, NFPA 20 appeared to permit the installation of low suction throttling valves to be installed in the suction pipe of a fire pump. The 2003 edition of the standard clarified that the industry term for a flow control valve is actually a *low suction throttling valve* and that such a device is required to be installed on the discharge side of a fire pump with the sensing line connected to the suction piping. Now, only a pressure-sensing line serving a low suction throttling valve is referenced.

(3) Devices shall be permitted to be installed in the suction supply piping or stored water supply and arranged to activate a signal if the pump suction pressure or water level falls below a predetermined minimum.

This list item permits the installation of a device such as a pressure switch to monitor the pressure in the suction pipe (where the suction pipe is connected to municipal water supply). The intent of this requirement is to permit monitoring of the suction pressure to activate a signal indicating that the pressure has dropped below a predetermined level. This list item does not permit shutting off the pump due to low pressure. In the 2007 edition, the technical committee added a definition in Chapter 3 for the term *signal,* which reads "an indicator of status." The purpose of this new definition is to provide an indication that the suction pressure is low, but does not allow the pump to be shut off due to low suction pressure conditions.

(4) Suction strainers shall be permitted to be installed in the suction piping where required by other sections of this standard.
(5) Other devices specifically permitted or required by this standard shall be permitted.

4.14.10* Vortex Plate.
For pump(s) taking suction from a stored water supply, a vortex plate shall be installed at the entrance to the suction pipe. *[See Figure A.6.3.1(a).]*

When water discharges at or near the bottom of an abovegrade storage tank or holding basin, a vortex forms as the water level lowers, and without an anti-vortex plate, air will be introduced into the suction pipe. A similar phenomenon occurs when a sink full of water empties through its drain. If this vortex were allowed to develop in a suction tank, it would result in turbulence (air bubbles) in the suction pipe and cause damage to the pump. To prevent this from happening, vortex plates (called *anti-vortex plates* in NFPA 22) are placed in the bottom of the tank and connected to the discharge flange of the tank.

A vortex plate is a flat, square or round sheet of steel. The length of the sides or the diameter of the plate must be at least twice the diameter of the suction pipe, although it is recommended that the plate be at least 48 in. (1219 mm) along each side. A hole, the same size as the suction pipe, is placed in the center of the plate.

The plate is mounted parallel to the floor of the tank [the plate is located at least half the diameter of the suction pipe or at least 6 in. (152 mm) above the floor] and connected to a long turn elbow that connects to the discharge flange installed in the wall of the tank. Water that flows out of the tank drops under the plate, up through the hole, up into the elbow, and out the discharge flange into the suction pipe. By forcing the water to follow this path, the creation of a vortex is prevented, which is why the NFPA 22 technical committee changed the name of this device to anti-vortex plate. See Exhibit II.4.8 for vortex plate details.

A.4.14.10 For more information, see the *Hydraulic Institute Standards for Centrifugal, Rotary and Reciprocating Pumps.*

4.15 Discharge Pipe and Fittings.

4.15.1 The discharge components shall consist of pipe, valves, and fittings extending from the pump discharge flange to the system side of the discharge valve.

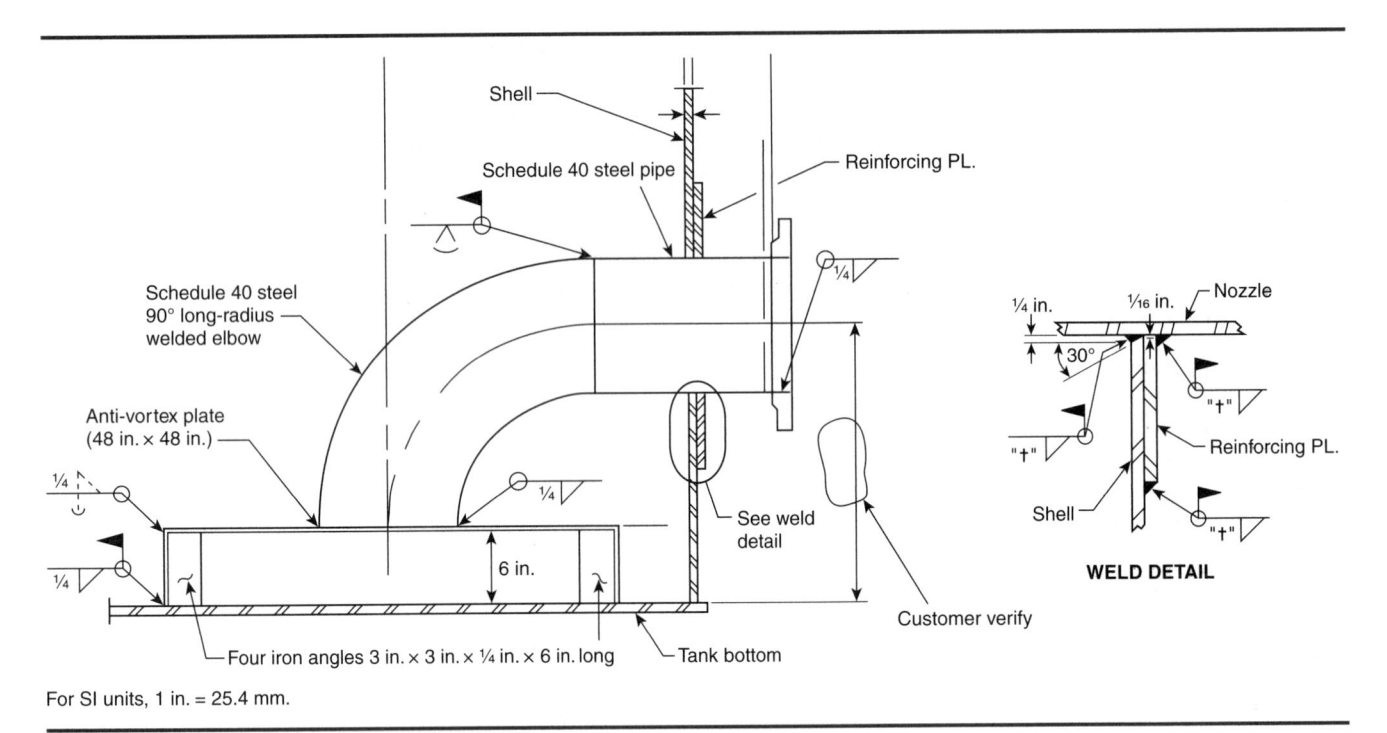

EXHIBIT II.4.8 Suction Nozzle with Anti-Vortex Plate for Welded Suction Tanks. [Source: NFPA 22, 2008, Figure B.1(o)]

4.15.2 The pressure rating of the discharge components shall be adequate for the maximum total discharge head with the pump operating at shutoff and rated speed but shall not be less than the rating of the fire protection system.

The standard makes reference to "total discharge head" rather than working pressure to require that discharge components be rated for the actual pressure as measured at the discharge gauge of the fire pump. The intent of this subsection is to require the installation of discharge components that are rated for the maximum pressure that occurs with a variable speed pressure-limiting control turned off and the pump running at rated speed. The maximum pressure is normally the shutoff or churn pressure developed by the pump, plus any suction head at the inlet to the pump.

4.15.3* Steel pipe with flanges, screwed joints, or mechanical grooved joints shall be used above ground.

A.4.15.3 Flanges welded to the pipe are preferred.

4.15.4 All pump discharge pipe shall be hydrostatically tested in accordance with NFPA 13, *Standard for the Installation of Sprinkler Systems*, and NFPA 24, *Standard for the Installation of Private Fire Service Mains and Their Appurtenances.*

4.15.5* The size of pump discharge pipe and fittings shall not be less than that given in Section 4.26.

The discharge pipe in this case is considered to be the pipe between the fire pump discharge flange and the discharge isolation valve specified in 4.15.7. Piping downstream of the discharge isolation valve is beyond the scope of NFPA 20 and should be designed and installed in accordance with the standard governing the system it supplies, such as NFPA 13, *Standard for the Installation of Sprinkler Systems*, or NFPA 14, *Standard for the Installation of Standpipe and Hose Systems.*

A.4.15.5 The discharge pipe size should be such that, with the pump(s) operating at 150 percent of rated capacity, the velocity in the discharge pipe does not exceed 20 ft/sec (6.1 m/sec).

4.15.6* A listed check valve or backflow preventer shall be installed in the pump discharge assembly.

The check valve or backflow preventer is necessary to prevent water from flowing back into the fire pump, thus allowing the fire pump to be removed and serviced without draining the entire fire protection system that it serves.

A.4.15.6 Large fire protection systems sometimes experience severe water hammer caused by backflow when the automatic control shuts down the fire pump. Where conditions can be expected to cause objectionable water hammer, a listed anti-water-hammer check valve should be installed in the discharge line of the fire pump. Automatically controlled pumps in tall buildings could give trouble from water hammer as the pump is shutting down.

Where a backflow preventer is substituted for the discharge check valve, an additional backflow preventer might be necessary in the bypass piping to prevent backflow through the bypass.

Where a backflow preventer is substituted for the discharge check valve, the connection for the sensing line is permitted to be between the last check valve and the last control valve if the pressure-sensing line connection can be made without altering the backflow valve or violating its listing. This method can sometimes be done by adding a connection through the test port on the backflow valve. In this situation, the discharge control valve is not necessary, because the last control valve on the backflow preventer serves this function.

Where a backflow preventer is substituted for the discharge check valve and the connection of the sensing line cannot be made within the backflow preventer, the sensing line should be connected between the backflow preventer and the pump's discharge control valve. In this situation, the backflow preventer cannot substitute for the discharge control valve because the sensing line must be able to be isolated.

4.15.7 A listed indicating gate or butterfly valve shall be installed on the fire protection system side of the pump discharge check valve.

Isolation valves on the discharge side of the fire pump do not affect the operation of the pump, so the kind of approved indicating valve that is installed does not matter. The exception to this statement is when two or more pumps are installed in series (see Exhibit II.12.4). In this design, the discharge from one pump leads to the suction of the next. Therefore, OS&Y gate valves must be used as both the suction valve and the discharge valve for pumps installed in series. See Exhibit II.4.9 for an example of an OS&Y gate valve.

4.15.8 Where pumps are installed in series, a butterfly valve shall not be installed between pumps.

4.15.9 Low Suction Pressure Controls.

4.15.9.1 Low suction throttling valves or variable speed suction limiting controls for pump drivers that are listed for fire pump service and that are suction pressure sensitive shall be permitted where the authority having jurisdiction requires positive pressure to be maintained on the suction piping.

Paragraph 4.15.9.1 was added to the 2003 edition to clarify that the low suction control valve is to be installed in the discharge piping and the sensing line for this device is connected to the suction piping.

4.15.9.2 When a low suction throttling valve is used, it shall be installed according to manufacturers' recommendations in the piping between the pump and the discharge check valve.

EXHIBIT II.4.9 OS&Y Gate Valve. (Photograph courtesy of Liberty Mutual Property)

4.15.9.3* The size of the low suction throttling valve shall not be less than that given for discharge piping in Section 4.26.

The intent of this paragraph is to require that the size of the low suction throttling valve be not less than that specified in Table 4.26 for discharge piping. This valve must be sized properly so that the system design is not compromised.

A.4.15.9.3 The friction loss through a low suction throttling valve must be taken into account in the design of the fire protection system.

4.15.10* Pressure Regulating Devices. No pressure regulating devices shall be installed in the discharge pipe except as permitted in this standard.

Subsection 4.15.10 was added to the 2003 edition to prohibit the use of pressure-regulating valves for the purpose of preventing water hammer during pump startup. The intent of the technical committee is to prevent the installation of any device in the discharge pipe unless permitted by NFPA 20. An important note is that the committee does not intend to prohibit the use of pressure-regulating devices downstream of the discharge isolation valve. Any pipe, fitting, valve, or other device downstream of the discharge isolation valve is beyond the scope of NFPA 20, and such pipe, fittings, valves, or other devices should be installed in accordance with the appropriate listing and installation standard such as NFPA 13 or NFPA 14.

A.4.15.10 See 4.7.7.2.

4.16* Valve Supervision.

A.4.16 Isolation valves and control valves are considered to be identical when used in conjunction with a backflow prevention assembly.

4.16.1 Supervised Open. Where provided, the suction valve, discharge valve, bypass valves, and isolation valves on the backflow prevention device or assembly shall be supervised open by one of the following methods:

(1) Central station, proprietary, or remote station signaling service
(2) Local signaling service that will cause the sounding of an audible signal at a constantly attended point

(3) Locking valves open
(4) Sealing of valves and approved weekly recorded inspection where valves are located within fenced enclosures under the control of the owner

The four methods of supervision represented in 4.16.1 help to ensure that a common mode of sprinkler system failure is reduced to a minimum. Approximately one-third of system failures are attributable to closed or partially closed sprinkler system control valves. The methods listed are in descending order of preference.

Central station, proprietary, or remote station supervision methods are preferred over the other methods listed. Supervisory signals received at such alarm facilities are usually forwarded to maintenance staff who are responsible for the protected property. This method also allows for much more efficient notification of the fire department that will respond (the signal is not considered a fire alarm signal and, therefore, does not require an immediate response). Early notification that a control valve has been closed for any reason may alter the fire department response and fire-fighting tactics. However, central station, proprietary, or remote station supervision methods have not been mandated by NFPA 20.

All bypass valves are required to be supervised open. Where a bypass is installed, water is expected to flow through the bypass into the fire protection system automatically, that is, without human intervention, in the event that the fire pump does not start. When the control valves are open, water can flow through the bypass automatically. A check valve, therefore, needs to be installed in the bypass line so that, when the pump is running, water does not circulate through the bypass back to the pump.

4.16.2 Supervised Closed. The test outlet control valves shall be supervised closed.

For test headers that are installed outside or where freezing is a danger, an isolation valve as required by 4.20.3.3.1 must be installed. This valve is normally closed and is only opened for testing purposes. The valve is required to be supervised in the closed position to prevent accidental freezing of the test header. The hose header is required to have an automatic drain valve where it is subject to freezing conditions.

4.17* Protection of Piping Against Damage Due to Movement.

A clearance of not less than 1 in. (25 mm) shall be provided around pipes that pass through walls or floors.

A.4.17 Pipe breakage caused by movement can be greatly lessened and, in many cases, prevented by increasing flexibility between major parts of the piping. One part of the piping should never be held rigidly and another free to move without provisions for relieving the strain. Flexibility can be provided by the use of flexible couplings at critical points and by allowing clearances at walls and floors. Fire pump suction and discharge pipes should be treated the same as sprinkler risers for whatever portion is within a building. *(See NFPA 13, Standard for the Installation of Sprinkler Systems.)*

Holes through pump room fire walls should be packed with mineral wool or other suitable material held in place by pipe collars on each side of the wall. Pipes passing through foundation walls or pit walls into ground should have clearance from these walls, but holes should be watertight. Space around pipes passing through pump room walls or pump house floors can be filled with asphalt mastic.

4.18 Relief Valves for Centrifugal Pumps.

4.18.1* General.

A.4.18.1 The pressure is required to be evaluated at 121 percent of the net rated shutoff pressure because the pressure is proportional to the square of the speed that the pump is turned. A diesel engine governor is required to be capable of limiting the maximum engine speed to 110 percent, creating a pressure of 121 percent. Since the only time that a pressure relief valve is required by the standard to be installed is when the diesel engine is turning faster than normal, and since this is a relatively rare event, it is permitted for the discharge from the pressure relief valve to be piped back to the suction side of the pump.

4.18.1.1 Where a diesel engine fire pump is installed and where a total of 121 percent of the net rated shutoff (churn) pressure plus the maximum static suction pressure, adjusted for elevation, exceeds the pressure for which the system components are rated, a pressure relief valve shall be installed.

Prior to the 1996 edition, NFPA 20 required the installation of pressure relief valves for all diesel engine fire pumps. This requirement was based on the assumption that if engines ran too fast (a condition known as *overspeed*), the fire protection system would be exposed to pressures in excess of the pressure ratings of the system components. Because an overspeed shutdown device is required, the technical committee believes that a pressure relief valve is not needed on all diesel fire pump installations. Pumps that create pressures less than 175 psi (12.1 bar) at 120 percent of rated speed do not need a pressure relief valve. The following example illustrates the procedure used to determine if a pressure relief valve is needed.

EXAMPLE

A fire pump rated for 1500 gpm (5677 L/min) and 100 psi (6.9 bar) at 1750 rpm has a shutoff pressure of 120 psi (8.3 bar) to 140 psi (9.7 bar). The shutoff pressure produces 145 psi (10 bar) to 169 psi (11.7 bar) at 121 percent of rated pressure. If 45 psi (3.1 bar) static pressure is available from the city water supply, the total pressure when the pump is running at a churn pressure of 120 psi (8.3 bar) is 145 psi (10 bar).

$$120 \text{ psi } (8.3 \text{ bar}) \times 1.21 = 145 \text{ psi } (10 \text{ bar})$$

$$145 \text{ psi } (10 \text{ bar}) + 45 \text{ psi } (3.1 \text{ bar}) = 190 \text{ psi } (13.1 \text{ bar})$$

In this case, a pressure relief valve (see Exhibit II.4.10) is needed because the pressure is greater than 175 psi (12 bar).

EXHIBIT II.4.10 *Pressure Relief Valve.*

4.18.1.2* Pressure relief valves shall be used only where specifically permitted by this standard.

Paragraph 4.18.1.2 states that main relief valves should only be used where specifically permitted by this standard. No section in this standard specifically permits the use of a main pressure relief valve. The use of a main pressure relief valve on an electrical fire pump is considered to be poor design and should be avoided. Several methods are available to cope with excessive pressures such as the following:

1. A break tank (see Section 4.31)
2. A variable speed pressure-limiting control device (see 11.2.4.3)
3. Other pressure-regulating devices downstream of the fire pump

◀ **FAQ**
Does NFPA 20 permit the installation of main pressure relief valves in electric fire pump systems?

A.4.18.1.2 In situations where the required system pressure is close to the pressure rating of the system components and the water supply pressure varies significantly over time, to eliminate system overpressurization, it might be necessary to use one of the following:

(1) A tank between the water supply and the pump suction, in lieu of directly connecting to the water supply piping
(2) A variable speed pressure limiting control device

4.18.1.3 Where an electric variable speed pressure limiting control driver is installed, and the maximum total discharge head adjusted for elevation with the pump operating at shutoff and rated speed exceeds the pressure rating of the system components, a pressure relief valve shall be installed.

4.18.2 Size. The relief valve size shall be determined by one of the methods specified in 4.18.2.1 or 4.18.2.2.

Sizing the relief valve piping hydraulically can minimize the loss of water to the fire protection system if the pressure relief valve fails to open and conserves water during the weekly operating test when the pump is running at churn. The intent of the technical committee is to allow a performance-based option for the sizing of the relief valve piping for this reason. The purpose of a performance-based design option is to establish a set of stated goals. In this case, the goal is conservation of water discharge from the pressure relief valve. Performance-based design options are used in place of prescriptive design options that in the case of pressure relief valves may not consider water conservation or short runs of piping where smaller pipe diameters can be capitalized through hydraulic calculations.

4.18.2.1* The relief valve shall be permitted to be sized hydraulically to discharge sufficient water to prevent the pump discharge pressure, adjusted for elevation, from exceeding the pressure rating of the system components.

A.4.18.2.1 See Figure A.4.18.2.1.

4.18.2.2 If the relief valve is not sized hydraulically, the relief valve size shall not be less than that given in Section 4.26. *(See also 4.18.7 and A.4.18.7 for conditions that affect size.)*

4.18.3 Location. The relief valve shall be located between the pump and the pump discharge check valve and shall be so attached that it can be readily removed for repairs without disturbing the piping.

4.18.4 Type.

4.18.4.1 Pressure relief valves shall be either a listed spring-loaded or a pilot-operated diaphragm type.

SAMPLE PRESSURE RELIEF VALVE CALCULATION
DISCHARGE TO ATMOSPHERE

Pressure rating of the system components				175
Maximum pump overspeed				105%
Pump size				1500
Rated pump pressure				100
			Maximum Pressure Static or Pump Overspeed	Normal Static or Rated Speed
Pump net pressure			112.5	102
Pump net churn pressure			132.3	120
Pump net pressure @ 150% of rated flow			71.7	65
Static pressure at pump suction			60	57
Available flow at pump suction			1320	1320
Residual pressure at pump suction			50	47
Maximum pump discharge pressure at churn			192.3	177.0
Pump flow rate at which the maximum discharge pressure does not exceed the pressure rating of the system components			1068.0	340.0
Estimated flow rate through the pressure relief valve			1795.5	1725.3
Pump discharge pressure at estimated flow			114.2	105.7
Pressure relief valve size				4
Pressure relief valve pipe size				4.026
Nozzle (pipe) discharge coefficient				0.9
C factor				120
Pressure relief valve Cv				240
Pressure Relief Valve Fittings	Type Fitting	Number	Equiv. Length	Total Equiv. Length
	45°	1	4	4
	ELLs	2	10	20
	LRE	0	6	0
Pressure relief valve pipe length				30
Total equivalent length				54
Calculated Results			Maximum	Normal
Friction loss per foot in pipe at estimated flow			0.7641	0.7098
Total loss in pressure relief valve piping			41.3	38.3
Friction loss in pressure relief valve at estimated flow			56.0	51.7
Pressure at pressure relief valve discharge			17.0	15.7
Elevation difference			0	0
Calculated discharge flow out of wide open pressure relief valve			1795	1726

FIGURE A.4.18.2.1 Sample Pressure Relief Valve Calculation.

4.18.4.2 Pilot-operated pressure relief valves, where attached to vertical shaft turbine pumps, shall be arranged to prevent relieving of water at water pressures less than the pressure relief setting of the valve.

4.18.5* Discharge.

A.4.18.5 The relief valve cone should be piped to a point where water can be freely discharged, preferably outside the building. If the relief valve discharge pipe is connected to an underground drain, care should be taken that no steam drains enter near enough to work back through the cone and into the pump room.

The discharge of the relief valve should preferably be piped to a water storage tank for safe discharge or to the outside to discharge at a safe point. If discharge to a tank or a safe point outside is not possible, then discharge to a drain of adequate size and capacity to accept the maximum flow from any discharge should be arranged.

4.18.5.1 The relief valve shall discharge into an open pipe or into a cone or funnel secured to the outlet of the valve.

4.18.5.2 Water discharge from the relief valve shall be readily visible or easily detectable by the pump operator.

4.18.5.3 Splashing of water into the pump room shall be avoided.

4.18.5.4 If a closed-type cone is used, it shall be provided with means for detecting motion of water through the cone.

Relief valve discharge is permitted to be piped back to the suction side of a fire pump (closed loop). Where this piping arrangement is used, a method to observe flow must be provided. This method is usually a closed waste cone (see Exhibit II.4.11). When this method is used, a circulation relief valve must be installed for cooling purposes. For relief valve arrangements (open and closed discharge), see Exhibits II.4.12 and II.4.13.

4.18.5.5 If the relief valve is provided with means for detecting motion (flow) of water through the valve, then cones or funnels at its outlet shall not be required.

4.18.6 Discharge Piping.

4.18.6.1 Except as permitted in 4.18.6.2, the relief valve discharge pipe shall be of a size not less than that given in Section 4.26.

4.18.6.2 The discharge pipe shall be permitted to be sized hydraulically to discharge sufficient water to prevent the pump discharge pressure, adjusted for elevation, from exceeding the pressure rating of the system components.

EXHIBIT II.4.11 Waste Cone Used for Observation of Water Flow. (Photograph courtesy of Liberty Mutual Property)

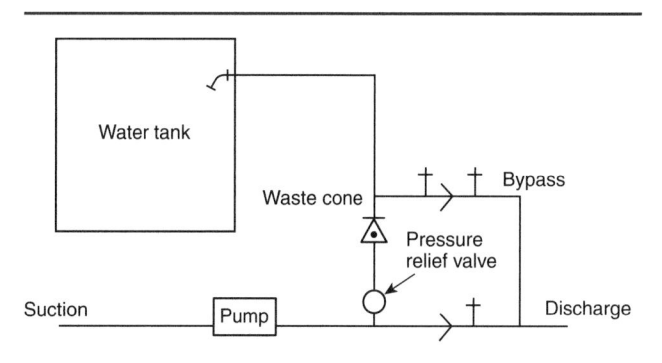

EXHIBIT II.4.12 Pressure Relief Valve Piping Arrangement to Tank (Open).

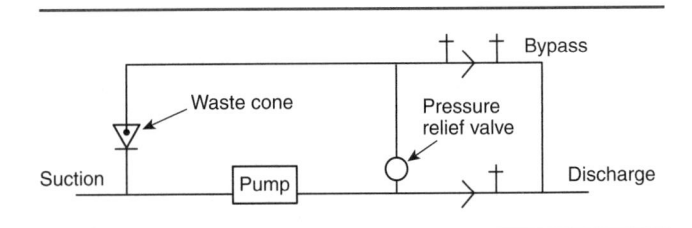

EXHIBIT II.4.13 *Pressure Relief Valve Piping Arrangement to Pump Suction (Closed).*

4.18.6.2.1 If the pipe employs more than one elbow, the next larger pipe size shall be used.

4.18.6.3 Relief valve discharge piping returning water back to the supply source, such as an aboveground storage tank, shall be run independently and not be combined with the discharge from other relief valves.

4.18.7* Discharge to Source of Supply. Where the relief valve is piped back to the source of supply, the relief valve and piping shall have sufficient capacity to prevent pressure from exceeding that for which system components are rated.

A.4.18.7 Where the relief valve discharges back to the source of supply, the back pressure capabilities and limitations of the valve to be used should be determined. It might be necessary to increase the size of the relief valve and piping above the minimum to obtain adequate relief capacity due to back pressure restriction.

4.18.7.1 Where a pressure relief valve has been piped back to suction, a circulation relief valve sized in accordance with 4.11.1.6 and discharged to atmosphere shall be provided downstream of the pressure relief valve.

4.18.8* Discharge to Suction Reservoir. Where the supply of water to the pump is taken from a suction reservoir of limited capacity, the drain pipe shall discharge into the reservoir at a point as far from the pump suction as is necessary to prevent the pump from drafting air introduced by the drain pipe discharge.

A.4.18.8 When discharge enters the reservoir below minimum water level, there is not likely to be an air problem. If it enters over the top of the reservoir, the air problem is reduced by extending the discharge to below the normal water level.

4.18.9 Shutoff Valve. A shutoff valve shall not be installed in the relief valve supply or discharge piping.

4.19 Pumps Arranged in Series.

4.19.1 Series Fire Pump Unit Performance.

4.19.1.1 A series fire pump unit (pumps, drivers, controllers, and accessories) shall perform in compliance with this standard as an entire unit.

4.19.1.2 Within 20 seconds after a demand to start, pumps in series shall supply and maintain a stable discharge pressure (±10 percent) throughout the entire range of operation.

4.19.1.2.1 The discharge pressure shall be permitted to be adjusted and restabilized whenever the flow condition changes.

4.19.1.3 The complete series fire pump unit shall be field acceptance tested for proper performance in accordance with the provisions of this standard. *(See Section 14.2.)*

4.19.2 Fire Pump Arrangement.

4.19.2.1 No more than three pumps shall be allowed to operate in series.

4.19.2.2 No pump in a series pump unit shall be shut down automatically for any condition of suction pressure.

4.19.2.3 No pressure reducing or pressure regulating valves shall be installed between fire pumps arranged in series.

4.19.2.4 The pressure at any point in any pump in a series fire pump unit, with all pumps running at shutoff and rated speed at the maximum static suction supply, shall not exceed any pump suction, discharge, or case working pressure rating.

4.20 Water Flow Test Devices.

See the commentary in Chapter 9 for a discussion of power supply reliability and the need for auxiliary power supplies. See also the commentary in 4.2.3 for the details needed for a proper plan review of a proposed fire pump power supply.

4.20.1 General.

Every fire pump installation needs a method for performing the acceptance test required in Chapter 14. NFPA 25 permits the following three methods for testing fire pumps:

   1. *Test header.* This device is connected to the discharge side of the pump and has a number of hose outlets as illustrated in Exhibit II.4.14. When testing the pump, the hose is connected to the outlets with water discharged in a safe location as illustrated in Exhibit II.4.15. Flow readings are usually taken from the end of the hose with a Pitot tube or other flow measuring device.

   2. *Flowmeter.* A special pipe is run from the discharge side of the pump back to the water supply (or to some other acceptable discharge point) with a flowmeter, control valve, and check valve in the line. When testing the pump, the control valve is opened partially (with the pump already running) to achieve the 100 percent flow condition. The valve is then opened more to achieve the 150 percent flow condition.

EXHIBIT II.4.14 Test Header. (Photograph courtesy of Liberty Mutual Property)

EXHIBIT II.4.15 Test Header with Hose Attached.

3. *Closed loop metering.* This method consists of a bypass line with a flowmeter, control valves, and a check valve, which goes directly from the pump discharge to the pump suction. This test is run the same way as method 2, but the water does not come from the supply. The water is recirculated in a small loop through the pump, as illustrated in Exhibit II.4.16. It should be noted that A.4.20.2.1.1 recommends that metering devices discharge to drain, to the water supply tank, or to the outside.

EXHIBIT II.4.16 Closed Loop Metering.

Because closed loop metering (method 3) does not test the water supply's ability to get water to the pump, NFPA 25 limits its application. Once every 3 years, testing must be done in accordance with method 1 or 2. This requirement means that the designer must make sure that equipment to perform the annual flow test in accordance with method 1 or 2 is included in the original design or, 3 years after installation, the building owner will not be able to comply with NFPA 25.

FAQ ▶
If the fire pump has no means to discharge water, must a test header or flowmeter be provided for a test that cannot be completed?

Under no circumstances can a fire pump be placed into service without an acceptance test. NFPA 25 requires an annual test of the fire pump by means of observing discharge of water at least once every 3 years. A flowmeter in a closed loop arrangement permits testing the pump without discharging water; however, water discharge is required at least once every 3 years.

4.20.1.1* A fire pump installation shall be arranged to allow the test of the pump at its rated conditions as well as the suction supply at the maximum flow available from the fire pump.

A.4.20.1.1 The two objectives of running a pump test are to make sure that the pump itself is still functioning properly and to make sure that the water supply can still deliver the correct amount of water to the pump at the correct pressure. Some arrangements of test equipment do not permit the water supply to be tested. Every fire pump installation needs to have at least one arrangement of test equipment where the water supply can be tested. Inspection, testing, and maintenance standards (NFPA 25, *Standard for the Inspection, Testing, and Maintenance of Water-Based Fire Protection Systems*) require the pump test to be run at least once every 3 years using a method that tests the water supply's ability to provide water to the pump.

4.20.1.2* Where water usage or discharge is not permitted for the duration of the test specified in Chapter 14, the outlet shall be used to test the pump and suction supply and determine that the system is operating in accordance with the design.

A.4.20.1.2 Outlets can be provided through the use of standard test headers, yard hydrants, wall hydrants, or standpipe hose valves.

The following notes apply to Figure A.4.20.1.2(a) and Figure A.4.20.1.2(b):

(1) The distance from the flowmeter to either isolation valve should be as recommended by the meter manufacturer.
(2) There should be a distance of not less than 5 diameters of suction pipe for top or bottom suction connection to the fire pump suction flange. There should be a distance of not less than 10 diameters of suction pipe for side connection (not recommended) to the fire pump suction flange.
(3) Automatic air release should be provided if piping forms an inverted "U," trapping air.
(4) The fire protection system should have outlets available to test the fire pump and suction supply piping. *(See A.4.20.3.1.)*
(5) The closed loop meter arrangement will test only net pump performance. It does not test the condition of the suction supply, valves, piping, and so forth.
(6) Return piping should be arranged so that no air can be trapped that would eventually end up in the eye of the pump impeller.
(7) Turbulence in the water entering the pump should be avoided to eliminate cavitation, which would reduce pump discharge and damage the pump impeller. For this reason, side connection is not recommended.
(8) Prolonged recirculation can cause damaging heat buildup, unless some water is wasted.
(9) The flowmeter should be installed according to manufacturer's instructions.
(10) Pressure sensing lines also need to be installed in accordance with 10.5.2.1. *[See Figure A.4.30(a) and Figure A.4.30(b).]*

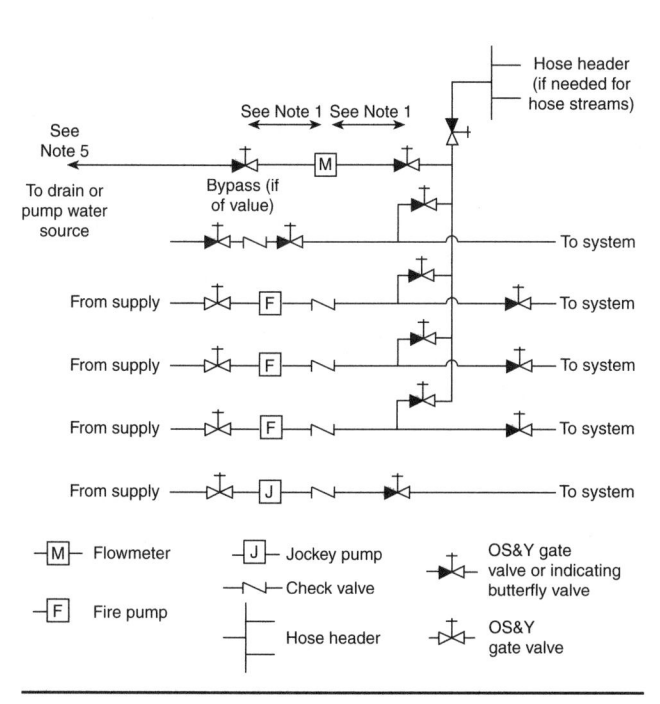

FIGURE A.4.20.1.2(a) Preferred Arrangement for Measuring Fire Pump Water Flow with Meter for Multiple Pumps and Water Supplies. Water is permitted to discharge to a drain or to the fire pump water source. (See the text for information on the notes.)

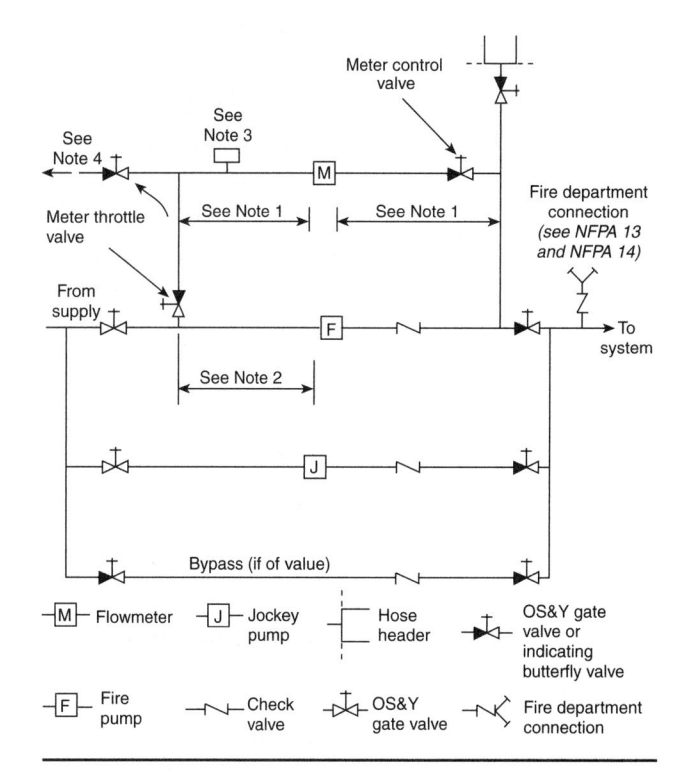

FIGURE A.4.20.1.2(b) Typical Arrangement for Measuring Fire Pump Water Flow with Meter. Discharge from the flowmeter is recirculated to the fire pump suction line. (See the text for information on the notes.)

4.20.1.3 The flow shall continue until the flow has stabilized. *(See 14.2.5.4.)*

4.20.1.4 Where a test header is installed, it shall be installed on an exterior wall or in another location outside the pump room that allows for water discharge during testing in accordance with 14.2.7.2.

4.20.2 Meters and Testing Devices.

4.20.2.1 Testing Devices.

4.20.2.1.1* Metering devices or fixed nozzles for pump testing shall be listed.

A.4.20.2.1.1 Metering devices should discharge to drain.

In the case of a limited water supply, the discharge should be back to the water source (e.g., suction tank, small pond). If this discharge enters the source below minimum water level, it is not likely to create an air problem for the pump suction. If it enters over the top of the source, the air problem is reduced by extending the discharge to below the normal water level.

4.20.2.2 Metering devices or fixed nozzles shall be capable of water flow of not less than 175 percent of rated pump capacity.

4.20.2.3 All of the meter system piping shall be permitted to be sized hydraulically but shall not be smaller than as specified by the meter manufacturer.

Sizing the meter piping hydraulically can reduce the cost of installation without sacrificing system performance. The intent of the technical committee is to allow a performance-based option for sizing the flowmeter piping.

4.20.2.4 If the meter system piping is not sized hydraulically, then all of the meter system piping shall be sized as specified by the meter manufacturer but not less than the meter device sizes shown in Section 4.26.

4.20.2.5 For nonhydraulically sized piping, the minimum size meter for a given pump capacity shall be permitted to be used where the meter system piping does not exceed 100 ft (30.5 m) equivalent length.

4.20.2.6 For nonhydraulically sized piping, where meter system piping exceeds 100 ft (30.5 m), including length of straight pipe plus equivalent length in fittings, elevation, and loss through meter, the next larger size of piping shall be used to minimize friction loss.

4.20.2.7 The primary element shall be suitable for that pipe size and pump rating.

4.20.2.8 The readout instrument shall be sized for the pump-rated capacity. *(See Section 4.26.)*

4.20.2.9 When discharging back into a tank, the discharge nozzle(s) or pipe shall be located at a point as far from the pump suction as is necessary to prevent the pump from drafting air introduced by the discharge of test water into the tank.

4.20.3 Hose Valves.

4.20.3.1* General.

A.4.20.3.1 The hose valves should be attached to a header or manifold and connected by suitable piping to the pump discharge piping. The connection point should be between the discharge check valve and the discharge gate valve. Hose valves should be located to avoid any possible water damage to the pump driver or controller, and they should be outside the pump room or pump house. If there are other adequate pump testing facilities, the hose valve header can be omitted when its main function is to provide a method of pump and suction supply testing. Where the hose header also serves as the equivalent of a yard hydrant, this omission should not reduce the number of hose valves to less than two.

4.20.3.1.1 Hose valves shall be listed.

4.20.3.1.2 The number and size of hose valves used for pump testing shall be as specified in Section 4.26.

4.20.3.1.3 Where outlets are being utilized as a means to test the fire pump in accordance with 4.20.1.1, one of the following methods shall be used:

(1)* Hose valves mounted on a hose valve header with supply pipe sized in accordance with 4.20.3.4 and Section 4.26
(2) Wall hydrants, yard hydrants, or standpipe outlets of sufficient number and size to allow testing of the pump

A.4.20.3.1.3(1) Outlets are typically provided through a standard test header. The test header is usually connected to the pump system between the discharge check valve and the discharge control valve for the pump so that the fire protection system can be isolated from the pump during testing if desired. However, the objective of testing the pump can be achieved with other arrangements as well.

4.20.3.2 Thread Type. Thread types shall be in compliance with one of the following:

(1) Hose valve(s) shall have the NH standard external thread for the valve size specified, as stipulated in NFPA 1963, *Standard for Fire Hose Connections.*
(2) Where local fire department connections do not conform to NFPA 1963, the authority having jurisdiction shall designate the threads to be used.

4.20.3.3 Location.

4.20.3.3.1 Where the hose valve header is located outside or at a distance from the pump and there is danger of freezing, a listed indicating butterfly or gate valve and drain valve or ball drip shall be located in the pipeline to the hose valve header.

4.20.3.3.2 The valve required in 4.20.3.3.1 shall be at a point in the line close to the pump. *(See Figure A.6.3.1(a).)*

4.20.3.4 Pipe Size. The pipe size shall be in accordance with one of the following two methods:

(1) Where the pipe between the hose valve header and the connection to the pump discharge pipe is over 15 ft (4.5 m) in length, the next larger pipe size than that required by 4.20.3.1.3 shall be used.
(2)* This pipe is permitted to be sized by hydraulic calculations based on a total flow of 150 percent of rated pump capacity, including the following:
  (a) This calculation shall include friction loss for the total length of pipe plus equivalent lengths of fittings, control valve, and hose valves, plus elevation loss, from the pump discharge flange to the hose valve outlets.
  (b) The installation shall be proven by a test flowing the maximum water available.

When using the second method for testing a pump that produces pressures in excess of the ratings of standard weight fittings, the designer's use of extra heavy pattern fittings in the test header piping is imperative. The reason for the use of extra heavy fittings is because pressure-regulating devices must be connected downstream of the test header connection. Therefore, pressure-regulating devices may offer no protection for test header pipe, fittings, and hose valves. This result would not be the case for pressure relief valves.

SAMPLE PUMP TEST HEADER SIZE CALCULATION

Pump size		1500		
Number of test hose streams		6		
Size of hose		2½		
Feet of hose per test hose		50		
Nozzle size		1.75		
Nozzle coefficient		0.97		
Pump test header pipe size		8.071		
C factor		120		

	Type Fitting	Number	Equiv. Length	Total Equiv. Length
Pump Test Header Pipe Fittings	45°	1	9	9
	E	1	18	18
	LRE	0	13	0
	T	1	35	35
	BV	0	12	0
	GV	1	4	4
	SW	1	45	45

Pump test header pipe length	30	
Total equivalent length	141	
Maximum test flow	2250	
Friction loss per ft in pipe	0.0392	
Total loss in pump test header pipe		5.5
Flow in each hose	375	
Friction loss in 100 ft of hose	28.125	
Total friction loss in hose		14.1
Equivalent pipe length 2½ in. valve	7	
Friction loss in 2½ in. pipe	0.4561	
Friction loss through 2½ in. valve		3.2
Required pitot pressure		18
Elevation difference		0
Required pump discharge		40.8

FIGURE A.4.20.3.4(2) Sample Pump Test Header Calculation.

A.4.20.3.4(2) See Figure A.4.20.3.4(2).

4.21 Steam Power Supply Dependability.

4.21.1 Steam Supply.

4.21.1.1 Careful consideration shall be given in each case to the dependability of the steam supply and the steam supply system.

4.21.1.2 Consideration shall include the possible effect of interruption of transmission piping either on the property or in adjoining buildings that could threaten the property.

4.22 Shop Tests.

Shop test results are used for comparison to acceptance testing results during the field acceptance test (for annual testing requirements, see NFPA 25). See Chapter 14 for more information regarding shop and acceptance tests.

4.22.1 General. Each individual pump shall be tested at the factory to provide detailed performance data and to demonstrate its compliance with specifications.

4.22.2 Preshipment Tests.

4.22.2.1 Before shipment from the factory, each pump shall be hydrostatically tested by the manufacturer for a period of not less than 5 minutes.

4.22.2.2 The test pressure shall not be less than one and one-half times the sum of the pump's shutoff head plus its maximum allowable suction head, but in no case shall it be less than 250 psi (17.24 bar).

4.22.2.3 Pump casings shall be essentially tight at the test pressure.

4.22.2.4 During the test, no objectionable leakage shall occur at any joint.

4.22.2.5 In the case of vertical turbine–type pumps, both the discharge casting and pump bowl assembly shall be tested.

4.23* Pump Shaft Rotation.

Pump shaft rotation shall be determined and correctly specified when fire pumps and equipment involving that rotation are ordered.

The rotation of a fire pump dictates the entire layout of piping and equipment in the fire pump room. The correct rotation must be determined before attempting a piping layout and design, and the rotation should be specified on the fire pump purchase order. It is important to note that not all fire pump manufacturers view rotation from the same perspective. Therefore, the manufacturer's catalog should be consulted for the proper procedure for determining pump rotation. Some diesel engine drives are only available using a single rotation direction.

A.4.23 Pumps are designated as having right-hand, or clockwise (CW), rotation or left-hand, or counterclockwise (CCW), rotation. Diesel engines are commonly stocked and supplied with clockwise rotation.
  Pump shaft rotation can be determined as follows:

(1) *Horizontal Pump Shaft Rotation.* The rotation of a horizontal pump can be determined by standing at the driver end and facing the pump. *[See Figure A.4.23(a).]* If the top of the shaft revolves from the left to the right, the rotation is right-handed, or clockwise (CW). If the top of the shaft revolves from right to left, the rotation is left-handed, or counterclockwise (CCW).
(2) *Vertical Pump Shaft Rotation.* The rotation of a vertical pump can be determined by looking down at the top of the pump. If the point of the shaft directly opposite revolves from left to right, the rotation is right-handed, or clockwise (CW). *[See Figure A.4.23(b).]* If the point of the shaft directly opposite revolves from right to left, the rotation is left-handed, or counterclockwise (CCW).

See Exhibit II.4.17 for an example of pump shaft rotation.

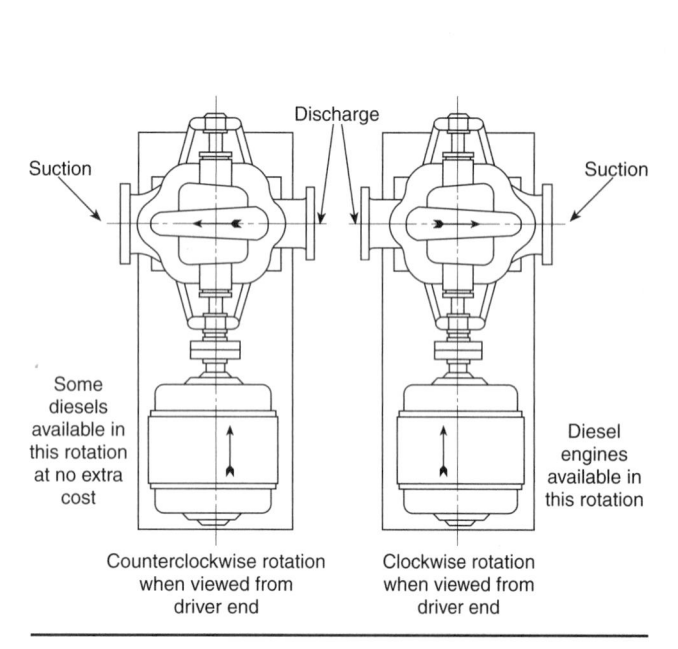

FIGURE A.4.23(a) Horizontal Pump Shaft Rotation.

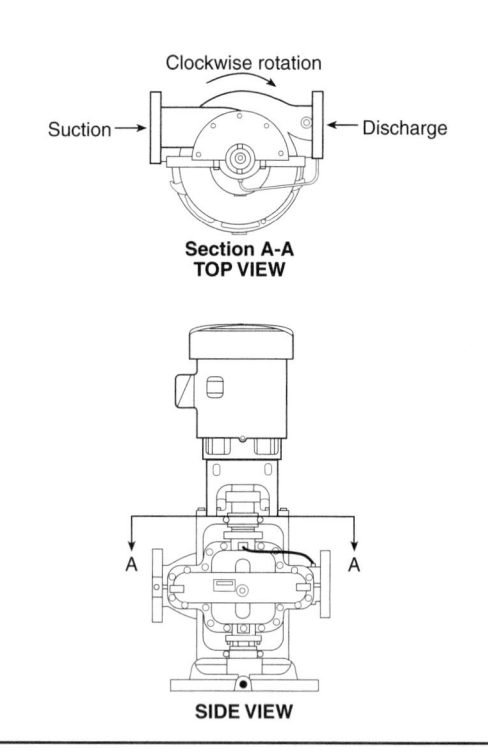

FIGURE A.4.23(b) Vertical Pump Shaft Rotation.

EXHIBIT II.4.17 Pump Shaft Rotation. (Photograph courtesy of Liberty Mutual Property)

4.24* Other Signals.

Where required by other sections of this standard, signals shall call attention to improper conditions in the fire pump equipment.

A.4.24 In addition to those conditions that require signals for pump controllers and engines, there are other conditions for which such signals might be recommended, depending upon local conditions. Some of these conditions are as follows:

(1) Low pump room temperature
(2) Relief valve discharge
(3) Flowmeter left on, bypassing the pump
(4) Water level in suction supply below normal
(5) Water level in suction supply near depletion
(6) Steam pressure below normal

Such additional signals can be incorporated into the trouble signals already provided on the controller, or they can be independent.

4.25* Pressure Maintenance (Jockey or Make-Up) Pumps.

Depending on factors such as the nature of the water supply or leakage through the system piping, pressure drops or fluctuations can occur within the fire protection system. These pressure fluctuations can cause the fire pump to operate even though the fire protection system has not been activated—that is, a sprinkler did not actuate or a standpipe valve was not opened. It is not the intent to operate the fire pump frequently for short durations to maintain system pressure. Fire pumps are not designed nor are they intended for this purpose. Subsection 4.25.6 prohibits the use of fire pumps as pressure maintenance devices.

The benefits of a pressure maintenance pump go beyond maintaining system pressure. A pressure maintenance pump will provide a higher pressure for the first sprinklers that actuate and may improve their effectiveness, thus limiting the total number of sprinklers that actuate. The pump will also reduce the likelihood of a pressure surge and thereby improve the performance of waterflow alarms by avoiding false alarms created by pressure surges.

It is important to note that pressure maintenance pumps (also called make-up pumps or jockey pumps) are not required by NFPA 20. NFPA 20 provides only the requirements for jockey pumps when they are installed; however, some method of maintaining system pressure, other than through the fire pump(s), should be provided. The pump, as illustrated in Exhibit II.4.18, is the most frequently used method to maintain system pressure. For systems that use the underground piping for both domestic and fire protection purposes, the operating pressure must be coordinated and the domestic pumps can operate as pressure maintenance pumps.

Since pressure maintenance pumps are not required, this section has been revised for the 2010 edition to indicate that valves and other components are not required to be listed. The suction isolation valve no longer needs to be an OS&Y and isolation valves are not required to be supervised. These changes were made because failure of the pressure maintenance pump will not compromise the fire protection features of the fire pump system. Therefore, the level of protection required of a fire pump is not necessary for a pressure maintenance pump.

A.4.25 Pressure maintenance (jockey or make-up) pumps should be used where it is desirable to maintain a uniform or relatively high pressure on the fire protection system.

A domestic water pump in a dual-purpose water supply system can function as a means of maintaining pressure.

4.25.1 Pressure maintenance pumps shall not be required to be listed. Pressure maintenance pumps shall be approved.

4.25.1.1* The pressure maintenance pump shall be sized to replenish the fire protection system pressure due to allowable leakage and normal drops in pressure.

This paragraph was revised for the 2010 edition to include the use of pressure maintenance pumps for stabilizing the air pressure in fire protection systems for situations such as thermal changes and expansion of liquid to fill voids left by escaping pressurized air.

EXHIBIT II.4.18 Pressure Maintenance Pump.

A.4.25.1.1 The sizing of the pressure maintenance pump requires a thorough analysis of the type and size of system the pressure maintenance pump will serve. Pressure maintenance pumps on fire protection systems that serve large underground mains need to be larger than pressure maintenance pumps that serve small aboveground fire protection systems. Underground mains are permitted by NFPA 24, *Standard for the Installation of Private Fire Service Mains and Their Appurtenances,* to have some leakage *(see 10.10.2.2.4 of NFPA 24 for allowable leakage rates)* while aboveground piping systems are required to be tight when new and should not have significant leakage.

For situations where the pressure maintenance pump serves only aboveground piping for fire sprinkler and standpipe systems, the pressure maintenance pump should be sized to provide a flow less than a single fire sprinkler. The main fire pump should start and run (providing a pump running signal) for any waterflow situation where a sprinkler has opened, which will not happen if the pressure maintenance pump is too large.

One guideline that has been successfully used to size pressure maintenance pumps is to select a pump that will make up the allowable leakage rate in 10 minutes or 1 gpm (3.8 L/min), whichever is larger.

4.25.2 Pressure maintenance pumps shall have rated capacities not less than any normal leakage rate.

Pressure maintenance pumps, also called jockey pumps, are generally low-flow, high-pressure pumps. For sprinkler systems, jockey pumps are usually sized to flow an amount of water less than or equal to that required by a single sprinkler. This way, if a sprinkler opens on the system, the jockey pump will not be able to keep up with system demand, the pressure will continue to fall, and the fire pump will start usually after an additional 5 psi (0.3 bar) has been lost to friction following the start of the jockey pump. See Chapter 14 Annex A material for recommendations on starting pressures for jockey pumps and fire pumps. The pressure differential between the pressure maintenance pump and the fire pump should be a minimum of 10 psi (0.7 bar) to avoid false starting of the fire pump due to pressure fluctuations in the system piping.

A general rule of thumb for sizing jockey pumps has been to take 1 percent of the fire pump rated capacity and 10 psi (0.7 bar) greater than the pressure rating of the fire pump. For example, a fire pump with a rated capacity of 1000 gpm (3785 L/min) at 100 psi (6.9 bar) should be provided with a jockey pump of 10 gpm (37.8 L/min) at 110 psi (7.6 bar) rated capacity. The only exception to this general rule is when older underground systems leak excessively. In that case, the jockey pump capacity should be increased further based on the leakage rate of the underground system.

4.25.3 Pressure maintenance pumps shall have discharge pressure sufficient to maintain the desired fire protection system pressure.

4.25.4* Excess Pressure.

A.4.25.4 A centrifugal-type pressure maintenance pump is preferable.
The following notes apply to a centrifugal-type pressure maintenance pump:

(1) A jockey pump is usually required with automatically controlled pumps.
(2) Jockey pump suction can come from the tank filling supply line. This situation would allow high pressure to be maintained on the fire protection system even when the supply tank is empty for repairs.
(3) Pressure sensing lines also need to be installed in accordance with 10.5.2.1. *[See Figure A.4.30(a) and Figure A.4.30(b).]*

4.25.4.1 Where a centrifugal-type pressure maintenance pump has a total discharge pressure with the pump operating at shutoff exceeding the working pressure rating of the fire protection equipment, or where a turbine vane (peripheral) type of pump is used, a relief valve sized to prevent overpressuring of the system shall be installed on the pump discharge to prevent damage to the fire protection system.

4.25.4.2 Running period timers shall not be used where jockey pumps are utilized that have the capability of exceeding the working pressure of the fire protection systems.

4.25.5 Piping and Components for Pressure Maintenance Pumps.

4.25.5.1 Steel pipe shall be used for suction and discharge piping on pressure maintenance pumps, which includes packaged prefabricated systems.

4.25.5.2 Valves and components for the pressure maintenance pump shall not be required to be listed.

4.25.5.3 An isolation valve shall be installed on the suction side of the pressure maintenance pump to isolate the pump for repair.

4.25.5.4 A check valve and isolation valve shall be installed in the discharge pipe.

4.25.5.5* Indicating valves shall be installed in such places as needed to make the pump, check valve, and miscellaneous fittings accessible for repair.

A.4.25.5.5 See Figure A.4.25.5.5.

4.25.5.6 The pressure sensing line for the pressure maintenance pump shall be in accordance with Section 4.30.

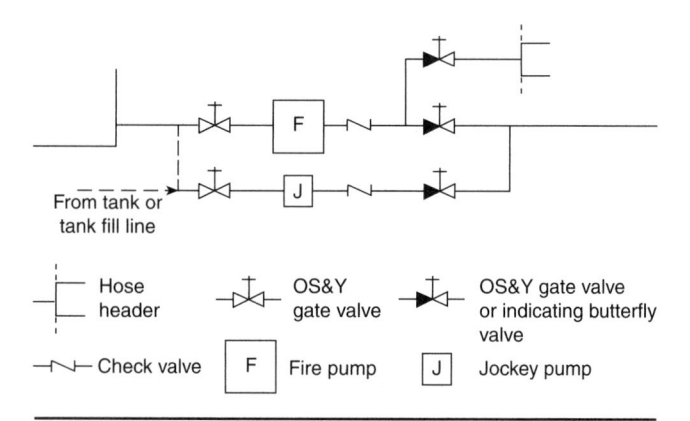

FIGURE A.4.25.5.5 *Jockey Pump Installation with Fire Pump.*

4.25.5.7 The isolation valves serving the pressure maintenance pump shall not be required to be supervised.

4.25.6 The primary or standby fire pump shall not be used as a pressure maintenance pump.

4.25.7 The controller for a pressure maintenance pump shall be listed but shall not be required to be listed for fire pump service.

4.25.8 The pressure maintenance pump is not required to have secondary or standby power.

4.26 Summary of Centrifugal Fire Pump Data.

The sizes indicated in Table 4.26(a) and Table 4.26(b) shall be used as a minimum.

4.27 Backflow Preventers and Check Valves.

4.27.1 Check valves and backflow prevention devices and assemblies shall be listed for fire protection service.

4.27.2 Relief Valve Drainage.

4.27.2.1 Where the backflow prevention device or assembly incorporates a relief valve, the relief valve shall discharge to a drain appropriately sized for the maximum anticipated flow from the relief valve.

Relief valves on reduced pressure backflow preventers (shown in Exhibit II.4.19) can discharge a great deal of water. The drain for these devices needs to be adequately sized to handle this flow. When reduced pressure backflow preventers are used, the pressure maintenance pump sensing line connection should be made on the supply side of the reduced pressure backflow preventers. If the pressure maintenance pump is permitted to pressurize the system side of a reduced pressure backflow preventer, the pressure relief valve remains open, continually discharging water into the drain.

The air gap on a reduced pressure backflow preventer is necessary for two reasons. First, without an air gap, the backflow preventer becomes a potential cross connection that could

TABLE 4.26(a) *Summary of Centrifugal Fire Pump Data (U.S. Customary)*

Pump Rating (gpm)	Suction*† (in.)	Discharge* (in.)	Relief Valve (in.)	Relief Valve Discharge (in.)	Meter Device (in.)	Number and Size of Hose Valves (in.)	Hose Header Supply (in.)
25	1	1	¾	1	1¼	1—1½	1
50	1½	1¼	1¼	1½	2	1—1½	1½
100	2	2	1½	2	2½	1—2½	2½
150	2½	2½	2	2½	3	1—2½	2½
200	3	3	2	2½	3	1—2½	2½
250	3½	3	2	2½	3½	1—2½	3
300	4	4	2½	3½	3½	1—2½	3
400	4	4	3	5	4	2—2½	4
450	5	5	3	5	4	2—2½	4
500	5	5	3	5	5	2—2½	4
750	6	6	4	6	5	3—2½	6
1000	8	6	4	8	6	4—2½	6
1250	8	8	6	8	6	6—2½	8
1500	8	8	6	8	8	6—2½	8
2000	10	10	6	10	8	6—2½	8
2500	10	10	6	10	8	8—2½	10
3000	12	12	8	12	8	12—2½	10
3500	12	12	8	12	10	12—2½	12
4000	14	12	8	14	10	16—2½	12
4500	16	14	8	14	10	16—2½	12
5000	16	14	8	14	10	20—2½	12

Notes:
(1) The pressure relief valve shall be permitted to be sized in accordance with 4.18.2.1.
(2) The pressure relief valve discharge shall be permitted to be sized in accordance with 4.18.6.2.
(3) The flowmeter device shall be permitted to be sized in accordance with 4.19.2.2.
(4) The hose header supply shall be permitted to be sized in accordance with 4.19.3.4.
*Actual diameter of pump flange is permitted to be different from pipe diameter.
†Applies only to that portion of suction pipe specified in 4.14.3.4.

contaminate the water supply from the drain. Second, the drain sometimes leads to an unheated area, in which case the cold temperatures could freeze the inside of the backflow preventer if it is not separated by an air gap.

4.27.2.2 An air gap shall be provided in accordance with the manufacturer's recommendations.

4.27.2.3 Water discharge from the relief valve shall be readily visible or easily detectable.

4.27.2.4 Performance of the requirements in 4.27.2.1 through 4.27.2.3 shall be documented by engineering calculations and tests.

4.27.3 Devices in Suction Piping. Where located in the suction pipe of the pump, check valves and backflow prevention devices or assemblies shall be located a minimum of 10 pipe diameters from the pump suction flange.

TABLE 4.26(b) *Summary of Centrifugal Fire Pump Data (Metric)*

Pump Rating (L/min)	Suction*† (mm)	Discharge* (mm)	Relief Valve (mm)	Relief Valve Discharge (mm)	Meter Device (mm)	Number and Size of Hose Valves (mm)	Hose Header Supply (mm)
95	25	25	19	25	32	1—38	25
189	38	32	32	38	50	1—38	38
379	50	50	38	50	65	1—65	65
568	65	65	50	65	75	1—65	65
757	75	75	50	65	75	1—65	65
946	85	75	50	65	85	1—65	75
1,136	100	100	65	85	85	1—65	75
1,514	100	100	75	125	100	2—65	100
1,703	125	125	75	125	100	2—65	100
1,892	125	125	100	125	125	2—65	100
2,839	150	150	100	150	125	3—65	150
3,785	200	150	150	200	150	4—65	150
4,731	200	200	150	200	150	6—65	200
5,677	200	200	150	200	200	6—65	200
7,570	250	250	150	250	200	6—65	200
9,462	250	250	200	250	200	8—65	250
11,355	300	300	200	300	200	12—65	250
13,247	300	300	200	300	250	12—65	300
15,140	350	300	200	350	250	16—65	300
17,032	400	350	200	350	250	16—65	300
18,925	400	350	200	350	250	20—65	300

Notes:
(1) The pressure relief valve shall be permitted to be sized in accordance with 4.18.2.1.
(2) The pressure relief valve discharge shall be permitted to be sized in accordance with 4.18.6.2.
(3) The flow meter device shall be permitted to be sized in accordance with 4.19.2.2.
(4) The hose header supply shall be permitted to be sized in accordance with 4.19.3.4.
*Actual diameter of pump flange is permitted to be different from pipe diameter.
†Applies only to that portion of suction pipe specified in 4.14.3.4.

Installation of a backflow preventer on the suction side of a fire pump should be avoided if possible because the water discharging from a backflow prevention device is extremely turbulent. Before this water enters the fire pump, turbulent flow must be minimized. The distance of 10 pipe diameters is considered acceptable in the water control industries for performing this task. Exhibit II.4.20 illustrates a double check backflow preventer assembly that should have been located 10 pipe diameters from the fire pump suction flange.

4.27.3.1 Where a backflow preventer with butterfly control valves is installed in the suction pipe, the backflow preventer is required to be at least 50 ft (15.2 m) from the pump suction flange (as measured along the route of pipe) in accordance with 4.14.5.2.

4.27.4 Evaluation.

4.27.4.1 Where the authority having jurisdiction requires the installation of a backflow prevention device or assembly in connection with the pump, special consideration shall be given to the increased pressure loss resulting from the installation.

EXHIBIT II.4.20 Double Check Valve Assembly Installed on Suction Side of Fire Pump.

EXHIBIT II.4.19 Reduced Pressure Backflow Preventer.

Water must enter the suction flange of the pump at a positive pressure. See the commentary following 4.14.3 for more information on this subject. The friction loss from the backflow prevention device and two accompanying isolation OS&Y gate valves is considerable and must be included in the system calculations.

4.27.4.2 Where a backflow prevention device is installed, the final arrangement shall provide effective pump performance with a minimum suction pressure of 0 psi (0 bar) at the gauge at 150 percent of rated capacity.

4.27.4.3 If available suction supplies do not permit the flowing of 150 percent of rated pump capacity, the final arrangement of the backflow prevention device shall provide effective pump performance with a minimum suction pressure of 0 psi (0 bar) at the gauge at the maximum allowable discharge.

4.27.4.4 The discharge shall exceed the fire protection system design flow.

4.27.4.5 Determination of effective pump performance shall be documented by engineering calculations and tests.

4.28 Earthquake Protection.

The potential for seismic activity is higher than previously thought for most areas of the United States and for a significant portion of the world. Fire protection systems, including fire pump installations, need to be designed so that they can function during and following a seismic event, because post-earthquake fires are relatively common.

In general, fire protection systems are designed to handle horizontal forces equal to one-half of their weight (0.5 g, where g is force of gravity). This design handles most of the areas where earthquakes are an infrequent, but still possible, occurrence. In some areas, the building code allows systems to be designed to withstand horizontal forces less than one-half of the weight of the component. In these situations, the lower value specified by the building code is acceptable. For more active and intense earthquake areas, fire protection systems need to be designed to withstand larger horizontal and vertical forces. In this case, the design factors will be provided in the building code. NFPA 13 should be used for guidance related to seismic support. The building codes specify very strict earthquake design parameters of vital buildings such as hospitals, fire stations, schools, and other similar occupancies.

4.28.1* Unless the requirements of 4.28.2 are met and where local codes require seismic design, the fire pump, driver, diesel fuel tank (where installed), and fire pump controller shall be attached to their foundations with materials capable of resisting lateral movement of horizontal forces equal to one-half the weight of the equipment.

A.4.28.1 NFPA 13, *Standard for the Installation of Sprinkler Systems,* contains specific requirements for seismic design of fire protection systems. Tables are available to determine the relative strength of many common bracing materials and fasteners.

4.28.2 The requirements of 4.28.1 shall not apply where the authority having jurisdiction requires horizontal force factors other than 0.5; in such cases, NFPA 13, *Standard for the Installation of Sprinkler Systems,* shall apply for seismic design.

4.28.3 Pumps with high centers of gravity, such as vertical in-line pumps, shall be mounted at their base and braced above their center of gravity in accordance with the requirements of 4.28.1 or 4.28.2, whichever is applicable.

4.28.4 Where the system riser is also a part of the fire pump discharge piping, a flexible pipe coupling shall be installed at the base of the system riser.

4.29 Packaged Fire Pump Assemblies.

See Exhibit II.4.21 for an illustration of a packaged fire pump system.

4.29.1 A packaged pump house and/or skid unit[s] shall include detailed design information acceptable to the authority having jurisdiction.

4.29.2 All electrical components, clearances, and wiring shall meet the minimum requirements of the applicable *NFPA 70, National Electrical Code,* articles.

4.29.3 Packaged and prefabricated skid unit(s) shall meet all the requirements in this standard, including those described in Sections 4.12 through 4.17.

EXHIBIT II.4.21 Packaged Fire Pump System.

4.29.4 Careful consideration shall be given to the possible effects of system component damage during shipment to the project site.

4.29.4.1 The structural integrity shall be maintained with minimal flexing and movement.

4.29.4.2 The necessary supports and restraints shall be installed to prevent damage and breakage during transit.

4.29.5 The packaged fire pump shall have the correct lifting points marked to ensure safe rigging of the unit.

4.29.6 All packaged pump house and/or pump skids shall meet the requirements of Section 4.28 through 4.28.4.

4.29.7 Suction and discharge piping shall be thoroughly inspected, including checking all flanged and mechanical connections per manufacturers' recommendations, after the pump house or skid unit is set in place on the permanent foundation.

4.29.8 The units shall be properly anchored and grouted in accordance with Section 6.4.

4.29.9* The interior floor shall be solid with grading to provide for proper drainage for the fire pump components. The structural frame for a packaged pump shall be mounted to a properly engineered footing designed to withstand the live loads of the packaged unit and the applicable wind loading requirements. The foundation footings shall include the necessary anchor points required to secure the package to the foundation.

A.4.29.9 Figure A.4.29.9 illustrates a typical foundation detail for a packaged fire pump assembly.

4.29.10 A high skid-resistant, solid structural plate floor with grout holes shall be permitted to be used where protected from corrosion and drainage is provided for all incidental pump room spillage or leakage.

Stationary Fire Pumps Handbook 2010

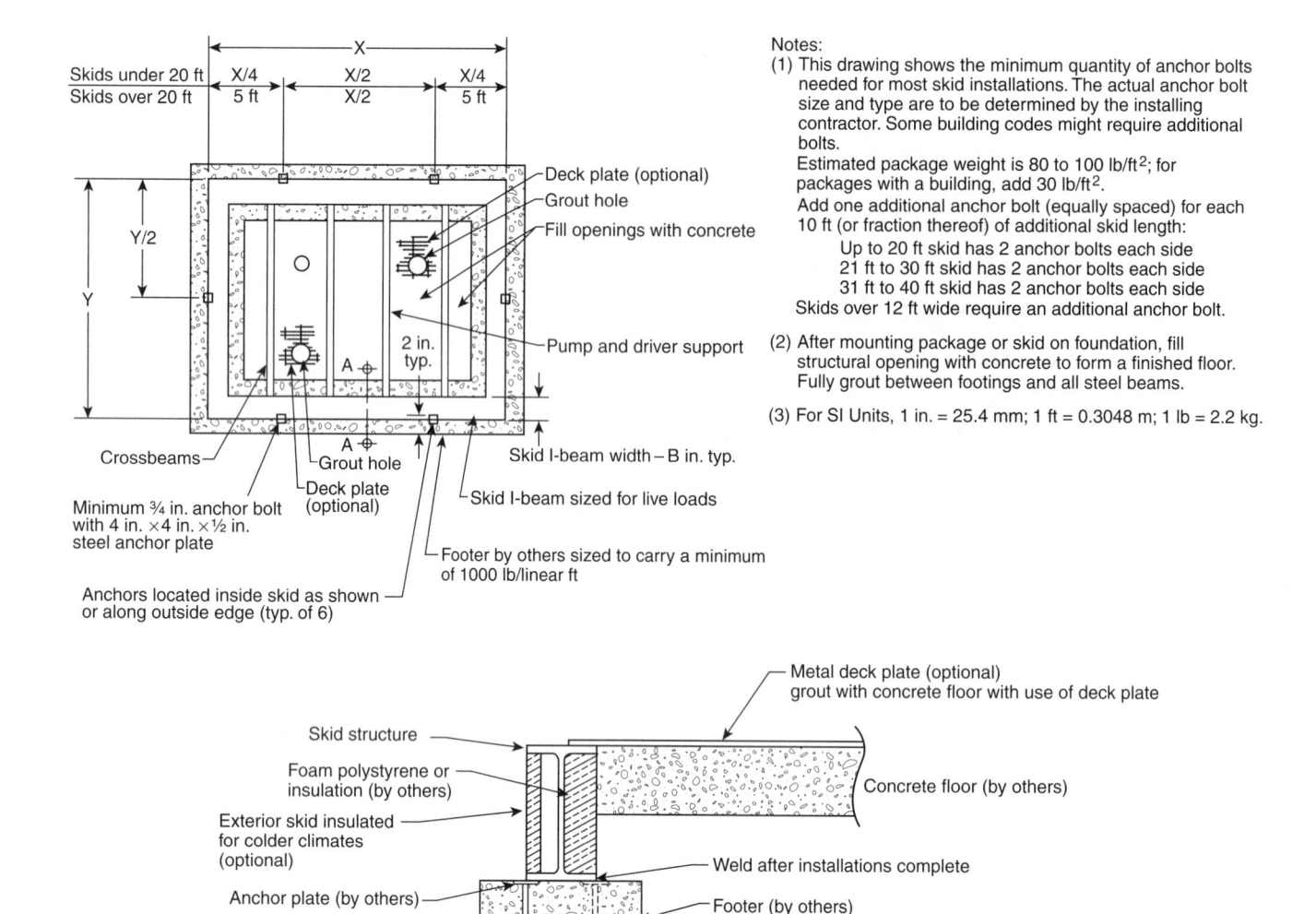

FIGURE A.4.29.9 Typical Foundation Detail for Packaged Fire Pump Assembly.

4.30* Pressure Actuated Controller Pressure Sensing Lines.

The pressure sensing line specified in Section 4.30 is intended to control starting of the pump motor and the running cycle of the fire pump. The sensing line specified in Section 4.30 is in addition to the pressure sensing lines required by 10.5.2.1.7.5 and 11.2.4.3.4. The pressure sensing lines required by 10.5.2.1.7.5 and 11.2.4.3.4 are intended to control the variable speed pressure-limiting control function of the fire pump controller.

A.4.30 See Figure A.4.30(a) and Figure A.4.30(b).

4.30.1 For all pump installations, including jockey pumps, each controller shall have its own individual pressure sensing line.

Each pump, including the pressure maintenance pump, must have its own dedicated sensing line. Each pump, including the pressure maintenance pump in multiple pump installations,

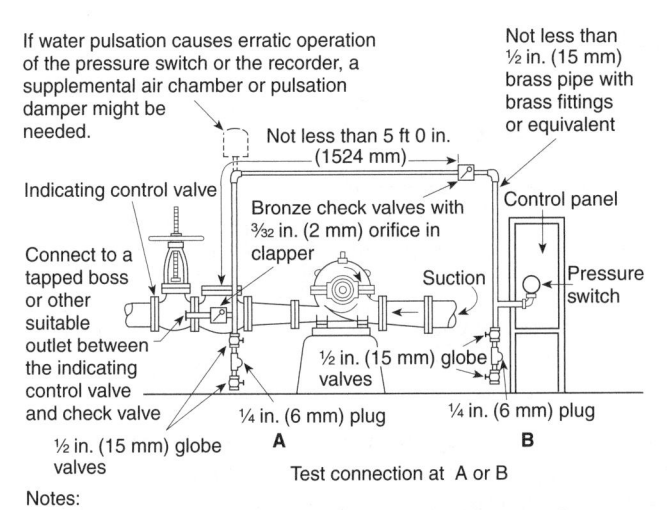

FIGURE A.4.30(a) *Piping Connection for Each Automatic Pressure Switch (for Diesel Fire Pump and Jockey Pumps).*

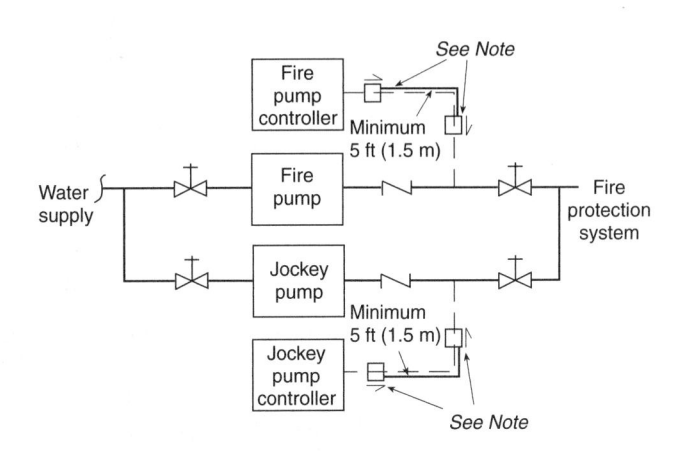

FIGURE A.4.30(b) *Piping Connection for Pressure-Sensing Line (Diesel Fire Pump).*

must also be provided with a separate dedicated sensing line. Providing separate dedicated pressure sensing lines adds reliability to the starting and control of the pumps. For example, if one line becomes obstructed, the other line is not affected. See Exhibit II.4.22 for an illustration of combined sensing lines.

◀ **FAQ**
Can the sensing line for a fire pump and pressure maintenance pump be combined and installed as a single pipe?

4.30.2 The pressure sensing line connection for each pump, including jockey pumps, shall be made between that pump's discharge check valve and discharge control valve.

4.30.3* The pressure sensing line shall be brass, rigid copper pipe Types K, L, or M, or Series 300 stainless steel pipe or tube, and the fittings shall be of ½ in. (15 mm) nominal size.

EXHIBIT II.4.22 *Combined Sensing Lines.*

A.4.30.3 The use of soft copper tubing is not permitted for a pressure sensing line because it is easily damaged.

4.30.4 Check Valves or Ground-Face Unions.

4.30.4.1 Where the requirements of 4.30.4.2 are not met, there shall be two check valves installed in the pressure sensing line at least 5 ft (1.52 m) apart with a nominal 0.09375 in. (2.4 mm) hole drilled in the clapper to serve as dampening. *[See Figure A.4.30(a) and Figure A.4.30(b).]*

The check valves referred to in 4.30.4.1 are intended to absorb the pressure surge created when the fire pump starts. Without this feature, the pressure switch inside the fire pump controller could be damaged.

4.30.4.2 Where the water is clean, ground-face unions with noncorrosive diaphragms drilled with a nominal 0.09375 in. (2.4 mm) orifice shall be permitted in place of the check valves.

Paragraph 4.30.4.2 is intended to refer to water that is free from obstructing material. In most cases, potable water is considered to be acceptable for this application.

4.30.5 Shutoff Valve. There shall be no shutoff valve in the pressure sensing line.

Because shutoff valves are not permitted, an important verification is that the pressure sensing line is connected to the pump side of the discharge control valve, thus permitting the sensing line to be serviced without draining the entire system.

4.30.6 Pressure Switch Actuation. Pressure switch actuation at the low adjustment setting shall initiate the pump starting sequence (if the pump is not already in operation).

4.31 Break Tanks.

Where a break tank is used to provide the pump suction water supply, the installation shall comply with this section.

As defined in 3.3.6, a break tank is not intended to provide a storage capacity equal to the entire system demand as specified by a fire protection system installation standard such as NFPA 13 or NFPA 14. The break tank is also not intended to be a pressurized source except for the pressure that is developed by the depth of water in the tank. The tank is normally equipped with an automatic fill connection to a source of water that is considered reliable in normal circumstances. For example, in an area subject to earthquakes, a city supply would not be considered reliable and the tank would be sized for the required total facility water supply.

The fire pump technical committee added information on the use of break tanks for the 2007 edition of NFPA 20 because guidance was needed on the design and testing of this type of water tank, which may not have the capacity to meet the full fire protection system demand. This application should not be substituted for the requirements of NFPA 22, which mandate that tanks be sized for full system demand. The primary purpose of a break tank should be to address situations where the municipality or water purveyor prohibits the direct connection of a fire pump to the water main because of pressure drawdown or in cases where the pressure fluctuation causes overpressurization of the fire protection system. See Exhibit II.4.23 for a typical break tank installation.

4.31.1 Application. Break tanks shall be used for one or more of the following reasons:

(1) As a backflow prevention device between the water supply and the fire pump suction pipe

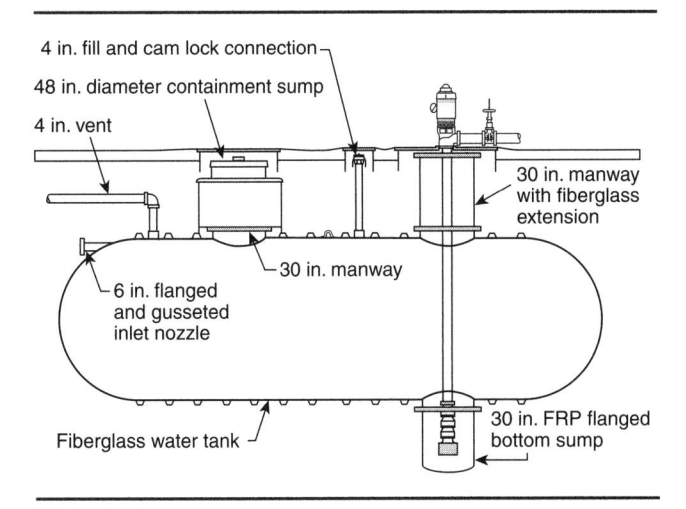

EXHIBIT II.4.23 Break Tank Installation.

(2) To eliminate fluctuations in the water supply pressure and provide a steady suction pressure to the fire pump
(3) To provide a quantity of stored water on site where the normal water supply will not provide the required quantity of water required by the fire protection system

In 4.31.1, the fire protection system demand comes into play and the automatic fill line is sized according to the system demand, *not* the capacity of the fire pump at overload (150 percent of pump rated capacity).

Refer to the two examples in the commentary following 4.30.2. In the first example, the fire protection system total water demand is greater than the capacity of the break tank. Therefore, the tank requires two automatic fill lines, each having the flow capacity of 1500 gpm (5677 L/min) [suggested size of 6 in. (150 mm) diameter pipe].

In the second example, the storage capacity of the break tank is greater than the 30-minute minimum fire protection water supply and only one automatic fill line is required having a capacity of 1100 gpm (4163 L/min) or 110 percent of the fire protection system demand.

4.31.2 Break Tank Size. The tank shall be sized for a minimum duration of 15 minutes with the fire pump operating at 150 percent of rated capacity.

Subsection 4.31.2 indicates that a break tank is not intended to meet the minimum system flow duration of 30 minutes as established by NFPA 13 and NFPA 14. Instead, a break tank must be sized for 15 minutes of discharge of the fire pump at 150 percent of fire pump flow (the overload point). This quantity of water exceeds the required fire protection system demand in gallons per minute. An important consideration is that the break tank is sized based upon the size of the fire pump, not the fire protection system demand, and the two different quantities should not be confused.

For example, if the fire protection demand is 1000 gpm (3785 L/min) and the fire pump is rated at 1000 gpm (3785 L/min) and at 1500 gpm (5677 L/min) at overload, then the quantity of water available for 15 minutes would be 22,500 gal (85,163 L) or 22.5 minutes at 1000 gpm (3785 L/min).

As a second example, if the demand is 1000 gpm (3785 L/min) and the fire pump is rated at 1500 gpm (5677 L/min) [2250 gpm (8208 L/min) at overload], then the quantity of water available would be 33,750 gal (127,744 L) or 33.75 minutes at 1000 gpm (3785 L/min).

FAQ▶
Does NFPA 20 specify the flow rate of an automatic fill connection for all water storage tanks?

Note that the intent of NFPA 20 is to establish the requirements for the application, sizing, and refill of break tanks only. For capacity and filling requirements for water storage tanks for private fire protection systems, see NFPA 22.

4.31.3 Refill Mechanism. The refill mechanism shall be listed and arranged for automatic operation.

4.31.3.1 If the break tank capacity is less than the maximum system demand for 30 minutes, the refill mechanism shall meet the requirements in 4.31.3.1.1 through 4.31.3.1.5.

4.31.3.1.1 Dual automatic refill lines, each capable of refilling the tank at a minimum rate of 150 percent of the fire pump(s) capacity, shall be installed.

4.31.3.1.2 If available supplies do not permit refilling the tank at a minimum rate of 150 percent of the rated pump capacity, each refill line shall be capable of refilling the tank at a rate that meets or exceeds 110 percent of the maximum fire protection system design flow.

4.31.3.1.3 A manual tank fill bypass designed for and capable of refilling the tank at a minimum rate of 150 percent of the fire pump(s) capacity shall be provided.

4.31.3.1.4 If available supplies do not permit refilling the tank at a minimum rate of 150 percent of the rated pump capacity, the manual fill bypass shall be capable of refilling the tank at a rate that meets or exceeds 110 percent of the maximum fire protection system design flow.

4.31.3.1.5 A local visible and audible low liquid level signal shall be provided in the vicinity of the tank fill mechanism.

In addition to the local audible and visual signal, a supervisory signal is suggested to be transmitted to the constantly attended location.

4.31.3.2 If the break tank is sized to provide a minimum duration of 30 minutes of the maximum system demand, the refill mechanism shall meet the requirements in 4.31.3.2.1 through 4.31.3.2.5.

4.31.3.2.1 The refill mechanism shall be designed for and capable of refilling the tank at 110 percent of the rate required to provide the total fire protection system demand [110% × (Total Demand—Tank Capacity) / Duration].

The formula in 4.31.3.2.1 should be used to determine the required refill capacity. See the following sample calculation for a fire pump supplying a standpipe system.

EXAMPLE

Assuming a standpipe system demand of 750 gpm (2839 L/min) for a duration as specified in NFPA 14 of 30 minutes and a break tank of 1000 gal (3785 L) capacity,

$$\text{Total demand} = 750 \text{ gpm} \times 30 \text{ min} = 22{,}500 \text{ gal}$$

$$(2839 \text{ L/min} \times 30 \text{ min} = 85{,}163 \text{ L})$$

$$\frac{1.1(22{,}500 - 1000)}{30} = 788 \text{ gpm refill rate}$$

$$\frac{1.1(85{,}163 - 2839)}{30} = 3019 \text{ L/min refill rate}$$

4.31.3.2.2 A manual tank fill bypass shall be designed for and capable of refilling the tank at 110 percent of the rate required to provide the total fire protection system demand [110% × (Total Demand—Tank Capacity) / Duration].

4.31.3.2.3 The pipe between the municipal connection and the automatic fill valve shall be installed in accordance with NFPA 24, *Standard for the Installation of Private Fire Service Mains and Their Appurtenances.*

4.31.3.2.4 The automatic filling mechanism shall be maintained at a minimum temperature of 40°F (4.4°C).

4.31.3.2.5 The automatic filling mechanism shall activate a maximum of 6 in. (152 mm) below the overflow level.

4.31.4 Installation Standard. The break tank shall be installed in accordance with NFPA 22, *Standard for Water Tanks for Private Fire Protection.*

4.32 Field Acceptance Test of Pump Units.

Upon completion of the entire fire pump installation, an acceptance test shall be conducted in accordance with the provisions of this standard. *(See Chapter 14.)*

REFERENCES CITED IN COMMENTARY

National Fire Protection Association, 1 Batterymarch Park, Quincy, MA 02169-7471.

NFPA 13, *Standard for the Installation of Sprinkler Systems,* 2010 edition.
NFPA 14, *Standard for the Installation of Standpipe and Hose Systems,* 2007 edition.
NFPA 22, *Standard for Water Tanks for Private Fire Protection,* 2008 edition.
NFPA 25, *Standard for the Inspection, Testing, and Maintenance of Water-Based Fire Protection Systems,* 2008 edition.

American Society of Civil Engineers, 1801 Alexander Bell Drive, Reston, VA 20191-4400.

ASCE 7, *Minimum Design Loads for Buildings and Other Structures,* 2005 edition.

American Society of Mechanical Engineers, Three Park Avenue, New York, NY 10016-5990.

ASME B16.1, *Cast Iron Pipe Flanges and Flanged Fittings, Classes 25, 125, and 250,* 1998.
ASME B16.3, *Malleable Iron Threaded Fittings, Classes 150 and 300,* 1998.
ASME B16.4, *Cast Iron Threaded Fittings, Classes 125 and 250,* 1998.
ASME B16.5, *Pipe Flanges and Flanged Fittings,* 1996.
ASME B16.9, *Factory-Made Wrought Steel Buttwelding Fittings,* 2001.
ASME B16.11, *Forged Steel Fittings, Socket-Welding and Threaded,* 1996.
ASME B16.18, *Cast Copper Alloy Solder Joint Pressure Fittings,* 1994.
ASME B16.22, *Wrought Copper and Copper Alloy Solder Joint Pressure Fittings,* 1995.
ASME B16.25, *Buttwelding Ends,* 1997.
ANSI/ASME B36.10M, *Welded and Seamless Wrought Steel Pipe,* 2000.

American Society for Testing and Materials, 100 Barr Harbor Drive, P.O. Box C700, West Conshohocken, PA 19428-2959.

ANSI/ASTM A 53, *Standard Specification for Pipe, Steel, Black and Hot-Dipped, Zinc-Coated, Welded and Seamless,* 2001.
ASTM A 135, *Standard Specification for Electric-Resistance-Welded Steel Pipe,* 2001.
ASTM A 234, *Standard Specification for Piping Fittings of Wrought-Carbon Steel and Alloy Steel for Moderate and High Temperature Service,* 2001.

ASTM A 795, *Standard Specification for Black and Hot-Dipped Zinc-Coated (Galvanized) Welded and Seamless Steel Pipe for Fire Protection Use,* 2000.

ASTM B 32, *Standard Specification for Solder Metal,* 2000. ASTM B 75, *Standard Specification for Seamless Copper Tube,* 1999.

ASTM B 88, *Standard Specification for Seamless Copper Water Tube,* 1999.

ASTM B 251, *Standard Specification for General Requirements for Wrought Seamless Copper and Copper-Alloy Tube,* 1997.

ASTM B 446, *Standard Specification for Nickel-Chromium-Molybdenum-Columbium Alloy (UNSN 06625) and Nickel-Chromium-Molybdenum-Silicon Alloy (UNSN 06219) Rod and Bar,* 2000.

ASTM B 813, *Standard Specification for Liquid and Paste Fluxes for Soldering Applications of Copper and Copper-Alloy Tube,* 2000.

American Water Works Association, 6666 West Quincy Avenue, Denver, CO 80235.

AWWA M-14, *Recommended Practice for Backflow Prevention and Cross-Connection Control,* 4th edition.

American Welding Society, 550 N.W. LeJeune Road, Miami, FL 33126.

AWS A5.8, *Specification for Filler Metals for Brazing and Braze Welding,* 1992.

Fire Pumps for High-Rise Buildings

CHAPTER 5

Chapter 5 is a new chapter in the 2010 edition delineating special requirements for fire pumps in high-rise buildings. A special chapter on high-rise buildings was removed from NFPA 20 in 1996 after a review revealed that only one requirement in the chapter was unique to high-rise buildings—standby power was required whenever a fire pump served a portion of a building that was beyond the pumping capacity of the fire department. This requirement was moved to another chapter and the high-rise chapter was eliminated.

The chapter was reinstated in the 2010 edition because some of the most complex fire pump installations exist in high-rise buildings. This chapter does not, however, provide a complete guideline for fire protection design in high-rise buildings. Tanks, standpipes, and fire pumps are highly interrelated and must be coordinated and require the use of multiple standards.

The NFPA 20 technical committee understands that this new high-rise chapter has significant implications for high-rise fire protection and may also impact cost and some current design practices. The requirements were developed after reviewing current code requirements, current design practices, fire-fighting operations, maintenance implications, and overall reliability and risk exposure.

NFPA 20 is a minimum requirements standard with an attempted balance between installation cost and risk built into the standard's requirements. Typically high-value and high-risk facilities are built with higher performance requirements than required in NFPA 20. It is also appropriate within minimum requirements standards such as NFPA 20 to provide for higher performance requirements, especially whenever the known life safety risk is high.

The following characteristics of high-rise buildings are the basis of the requirements in this new chapter:

1. High-rise buildings require some of the most complex fire pump installations.
   a. Most series fire pump arrangements are found in high-rise buildings.
   b. High-rise buildings may contain multiple vertical zone standpipe systems, tanks, and automatic refill valves.
2. More than in any other occupancy, high-rise occupants are dependent on the building fire pump to function reliably during a fire. High-rise evacuation plans, which can incorporate "refuge floors," depend on the building's automatic and manual fire-fighting systems to be fully operational.
3. High-rise buildings are high-value buildings.
4. Direct access to the fire pump(s) is needed during fire-fighting operations.
5. The occupancy fire load can be significant in many high-rise buildings.
6. For buildings that exceed 300 ft (91 m) in height, fire-fighting and fire department operations are totally reliant on fire pumps within the building.
7. For buildings that exceed 500 ft (152.4 m) in height, it is impractical to evacuate the occupants, and occupants must be "protected in place."

The requirements in this chapter are intended to provide a highly reliable fire pump installation as part of the fire protection system(s). For high-rise buildings with floors above the

◀ FAQ
Why was a chapter on fire pumps in high-rise buildings, removed from earlier editions of NFPA 20, reinstated in the 2010 edition?

◀ FAQ
What is the intent of this chapter?

pumping capacity of the fire department, the fire pump and water supply arrangement requirements are intended to be such that even with any single piece of equipment impaired, the full fire protection demand can be met.

5.1 General.

5.1.1 Application.

5.1.1.1 This chapter applies to all fire pumps within a building wherever a building is defined as high-rise per 3.3.24.

5.1.1.2 The provisions of all other chapters of this standard shall apply unless specifically addressed by this chapter.

5.2 Types.

5.2.1 Fire pumps used for high-rise application shall be of a type addressed by Chapter 6 or 7 (centrifugal pumps) of this standard.

Fire pumps are not specifically listed for high-rise buildings. Fire pumps used in high-rise buildings are standard fire pumps, which typically have a high pressure rating.

5.2.2 Fire pumps complying with Chapter 8 (positive displacement pumps) shall be allowed for local applications.

This chapter is not intended to prevent the use of local application fire protection systems that may require positive displacement pumps. The use of positive displacement pumps supplying systems such as water mist for the primary protection of a building is not appropriate.

5.3 Equipment Access.

Location and access to the fire pump room shall be preplanned with the fire department.

In general, preplanning and coordination with the fire service are critical to successfully fighting a fire in a high-rise building. The fire service should have specific knowledge of the location, type, pressures, and operation of the building fire pumps.

5.4 Fire Pump Test Arrangement.

Where the water supply to a fire pump is a tank, a listed flowmeter or a test header discharging back into the tank with a calibrated nozzle(s) arranged for the attachment of a pressure gauge to determine pitot pressure shall be permitted.

Holding a pitot tube in a hose stream discharging into a water tank in tight spaces is difficult, can create safety issues, and, therefore, should be avoided. Permanently affixed flowmeters or calibrated nozzles with fixed attachment points avoid this problem.

High-rise buildings tend to be located in congested areas where testing may present a hazard to the public and test times may be restricted. Annual flow testing in accordance with NFPA 25, *Standard for the Inspection, Testing, and Maintenance of Water-Based Fire Protection Systems,* verifies that the pump system is fully functional. Omission of this testing can result in an unreliable pump. There are many high-rise buildings where pump flow testing is difficult or impractical. Locating the pump test header on an outside wall facilitates testing and allows quicker setup and takedown time while minimizing hose usage. Special consideration should be given to the pressure rating of the test header and the pressure developed by the pump, which in most cases will exceed 175 psi (12.1 bar).

5.5 Auxiliary Power.

Where electric motors are used and the height of the structure is beyond the pumping capability of the fire department apparatus, a reliable emergency source of power in accordance with Section 9.6 shall be provided for the fire pump installation.

While it is appropriate to provide standby power for all electric motor fire pumps serving a high-rise building, it is specifically required if the fire department cannot provide an alternative source of fire protection water.

5.6* Fire Pump Backup.

Fire pumps serving zones that are partially or wholly beyond the pumping capability of the fire department apparatus shall be provided with an auxiliary means that is capable of providing the full fire protection demand.

A.5.6 If backup fire pumps are required, they can be arranged to prevent both pumps from starting and running simultaneously and, if done, the following arrangement is suggested:

(1) Turning off or disconnecting power to the primary fire pump controller should not prevent the redundant fire pump from starting or running.
(2) Turning off or disconnecting power to the redundant fire pump controller should not prevent the primary fire pump from starting or running.
(3) Once the primary fire pump is locked out, it should remain locked out until manual reset.
(4) Once the redundant fire pump is running, it should remain running until manual reset.
(5) Either controller should always be capable of being operated by local manual starting regardless of any lockout that has occurred.
(6) A local visual alarm and remote contacts should be provided to indicate that the primary fire pump has been locked out.

A fully independent and automatic backup fire pump unit(s) arranged so that all zones can be maintained in full service with any one pump out of service should be considered one such means.

5.7 Water Supply Tanks.

5.7.1 Water tanks shall be installed in accordance with NFPA 22, *Standard for Water Tanks for Private Fire Protection*.

The auxiliary means of providing the full fire protection demand allows for service and maintenance without the loss of protection. Exhibit II.5.1 shows one possible means of meeting this requirement using a redundant fire pump fed from a water tank. Gravity feed from a higher level system might be another way to meet this requirement.

5.7.2 When a water tank serves domestic and fire protection systems, the domestic supply connection shall be connected above the level required for fire protection demand.

EXHIBIT II.5.1 Redundant Fire Pump Fed from a Water Tank. (Courtesy of Schirmer Engineering Corp.)

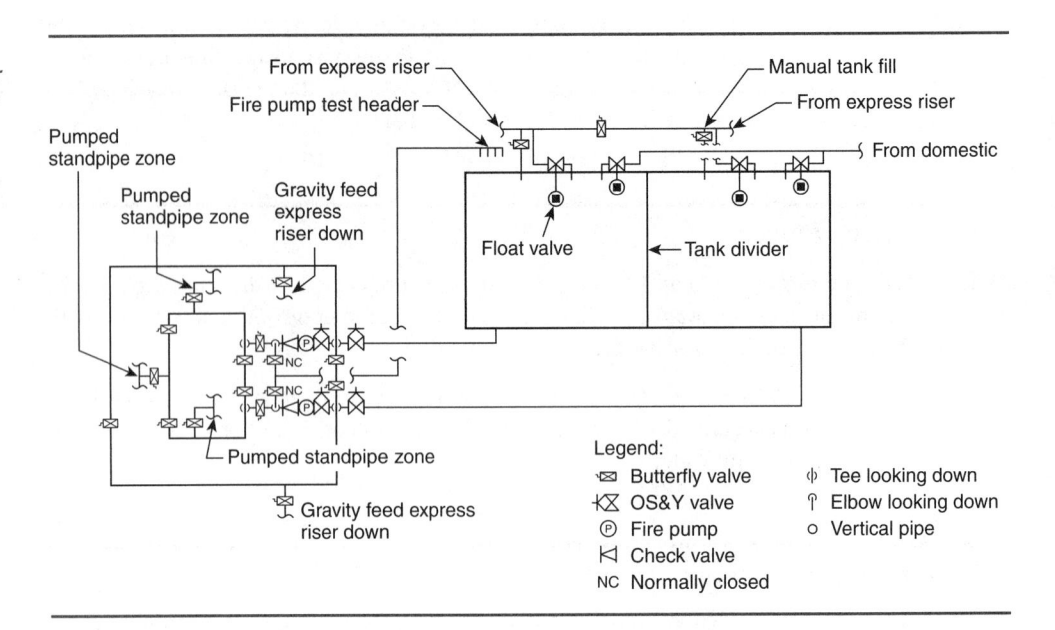

Connecting the domestic water above the level required for fire protection demand prevents domestic water usage from depleting the fire protection water reserve.

5.7.3 Water tanks supplying suction to fire pumps serving zones that are partially or wholly beyond the pumping capability of the fire department apparatus shall meet the requirements in 5.7.3.1 through 5.7.3.5 with a minimum of two automatic fill valves with separate piping connected to the zone below or primary water supply to the building. A manual fill valve shall also be provided.

This subsection ensures the following:

1. The stored water supply will meet the full fire protection water demand.
2. The tanks are automatically refilled at a minimum rate of the fire protection water demand.
3. With any single tank or tank compartment (and automatic refill valve for that tank) out of service, the full fire protection water demand can still be met through stored water and automatic tank refill.

5.7.3.1 Two or more water tanks shall be provided. Alternatively, a water tank shall be permitted to be divided into compartments such that the compartments function as individual tanks.

This paragraph allows maintenance and servicing of a tank without impairing the building fire protection.

5.7.3.2 Water tank(s) shall be sized for the full fire protection demand and arranged so that at least 50 percent of the fire protection demand is stored with any one compartment or tank out of service.

5.7.3.3 An automatic refill valve shall be provided for each tank or tank compartment.

5.7.3.4 A manual refill valve shall be provided.

5.7.3.5 Each refill valve shall be sized and arranged to independently supply the system fire protection demand.

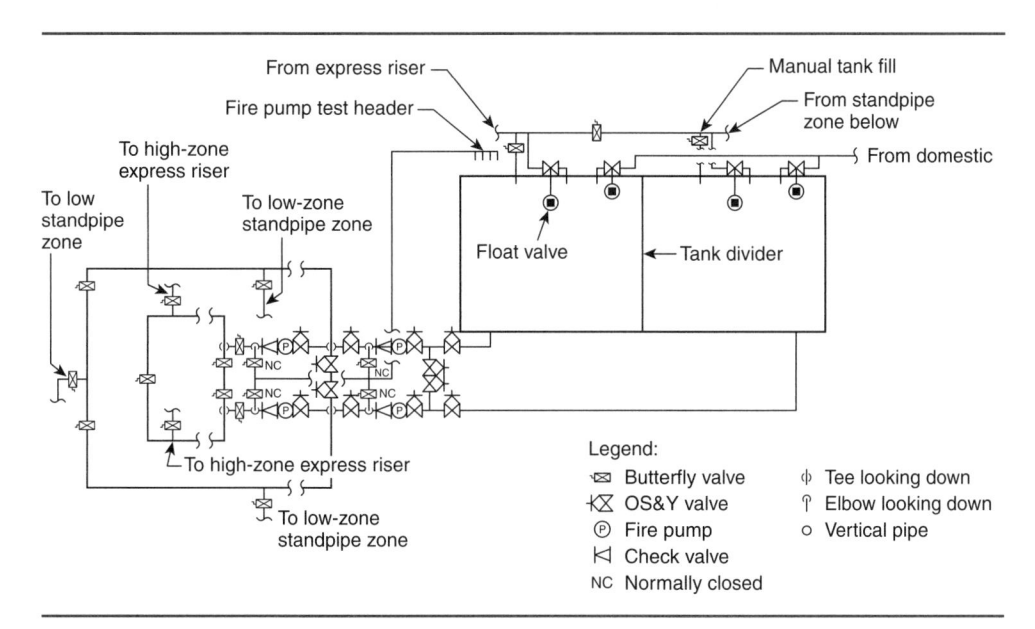

EXHIBIT II.5.2 Divided Water Tank. (Courtesy of Schirmer Engineering Corp.)

Exhibit II.5.1 shows one way of meeting the requirements of Section 5.7 through 5.7.3.5 with a divided water tank with refill valves supplying a two zone standpipe system with backup fire pumps. Exhibit II.5.2 shows a divided water tank with refill valves supplying a two-zone standpipe system, with the lower zone gravity fed and the upper zone supplied by a fire pump with a backup fire pump.

The refill valves do not have to be designed to operate simultaneously. When operating independently, the refill rate must be the minimum of the system's fire protection demand. Under conditions where the refill valves are operating simultaneously, the total refill must be the minimum of the system's fire protection demand.

Exhibit II.5.3 shows a tank with duplicate automatic refill.

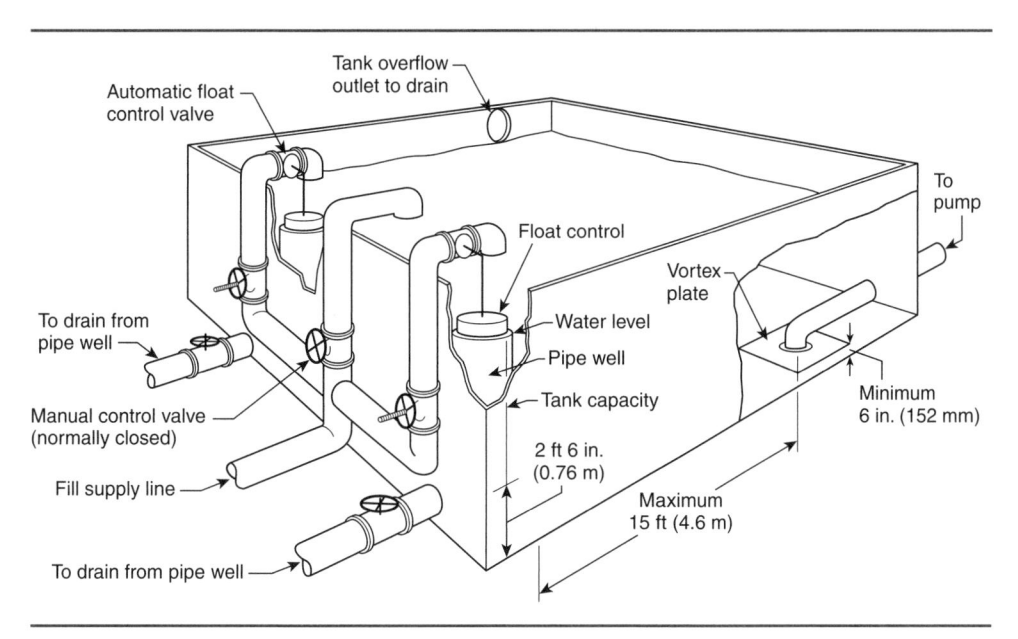

EXHIBIT II.5.3 Tank with Duplicate Automatic Refill. (Courtesy of FM Global)

REFERENCE CITED IN COMMENTARY

National Fire Protection Association, 1 Batterymarch Park, Quincy MA 02169-7471.

NFPA 25, *Standard for the Inspection, Testing, and Maintenance of Water-Based Fire Protection Systems,* 2008 edition.

Centrifugal Pumps

CHAPTER 6

Chapter 6 covers the basic physical and performance characteristics of centrifugal fire pumps for fire protection service. Centrifugal fire pumps are generally compact, reliable, and easy to maintain. The hydraulic characteristics of the pump and variety of drivers available have made the centrifugal pump the most common of all pumps for fire protection.

A desirable feature of the centrifugal pump is its relationship between flow and pressure at a constant speed—that is, when flow is decreased, pressure increases. A centrifugal pump functions by converting kinetic energy to pressure and velocity energy through centrifugal force. As water enters the eye of the impeller, centrifugal force from the rotation of the impeller forces water through the impeller to the discharge outlet of the pump.

Centrifugal fire pumps are not appropriate for suction lift applications and are limited to applications where the water source can provide a positive pressure to the suction side of the pump.

6.1 General.

6.1.1* Types.

An impeller is the rotating element of the pump that imparts energy in the form of increased pressure to the water passing through the pump. Exhibit II.6.1 illustrates a typical centrifugal pump impeller.

◀ **FAQ**
What is an impeller?

An overhung impeller is located at the end of the impeller shaft. The shaft is supported only on one side of the impeller.

◀ **FAQ**
What is overhung impeller design?

The shaft on an impeller between bearings design extends through the impeller and is supported on both sides of the impeller.

◀ **FAQ**
What is impeller between bearings design?

A.6.1.1 See Figure A.6.1.1(a) through Figure A.6.1.1(h).

EXHIBIT II.6.1 Centrifugal Pump Impeller.

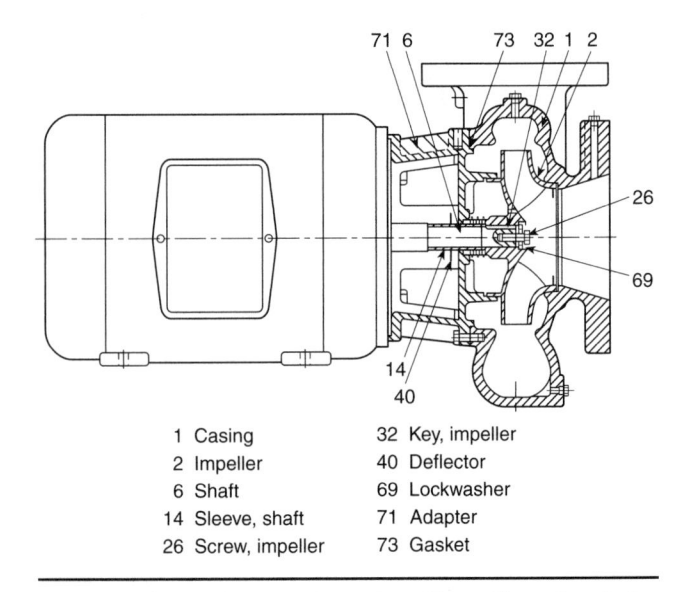

1 Casing	32 Key, impeller	
2 Impeller	40 Deflector	
6 Shaft	69 Lockwasher	
14 Sleeve, shaft	71 Adapter	
26 Screw, impeller	73 Gasket	

FIGURE A.6.1.1(a) *Overhung Impeller—Close Coupled Single Stage—End Suction.*

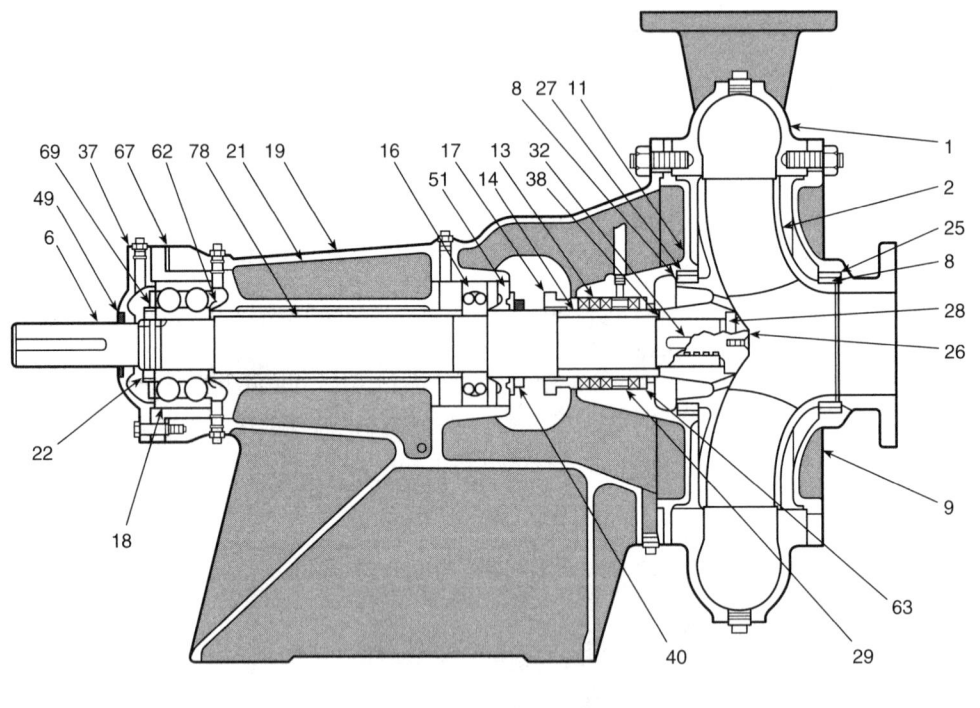

1 Casing	16 Bearing, inboard	27 Ring, stuffing-box cover	49 Seal, bearing cover, outboard
2 Impeller	17 Gland	28 Gasket	51 Retainer, grease
6 Shaft, pump	18 Bearing, outboard	29 Ring, lantern	62 Thrower (oil or grease)
8 Ring, impeller	19 Frame	32 Key, impeller	63 Busing, stuffing-box
9 Cover, suction	21 Liner, frame	37 Cover, bearing, outboard	67 Shim, frame liner
11 Cover, stuffing-box	22 Locknut, bearing	38 Gasket, shaft sleeve	69 Lockwasher
13 Packing	25 Ring, suction cover	40 Deflector	78 Spacer, bearing
14 Sleeve, shaft	26 Screw, impeller		

FIGURE A.6.1.1(b) *Overhung Impeller—Separately Coupled Single Stage—Frame Mounted.*

1 Casing	13 Packing	40 Deflector	
2 Impeller	14 Sleeve, shaft	71 Adapter	
11 Cover, seal chamber	17 Gland, packing	73 Gasket, casing	

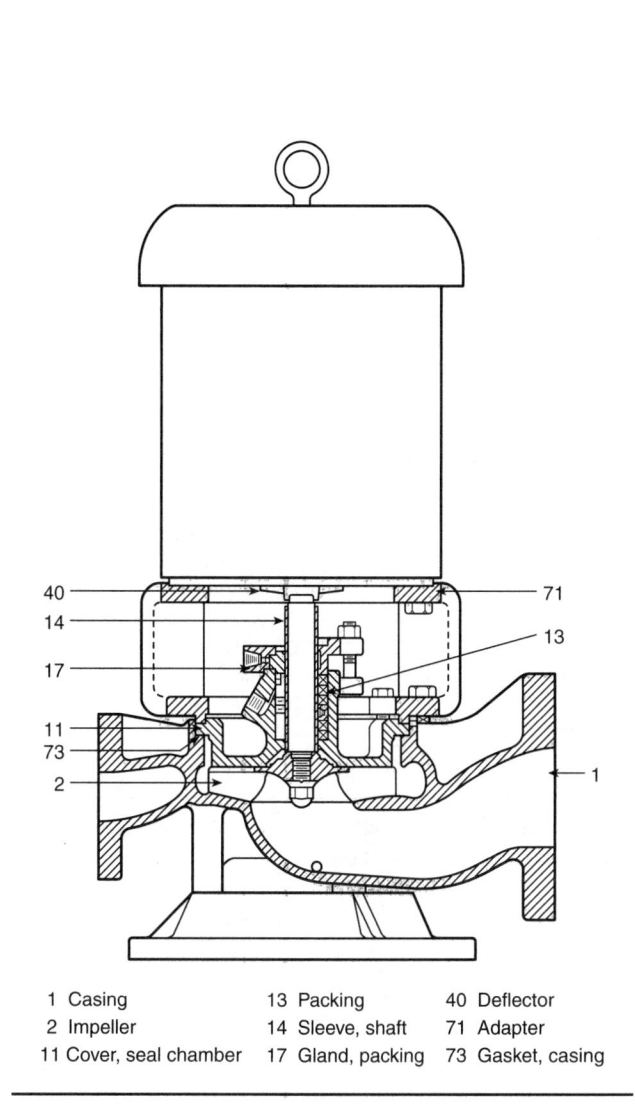

FIGURE A.6.1.1(c) *Overhung Impeller—Close Coupled Single Stage—In-Line (Showing Seal and Packaging).*

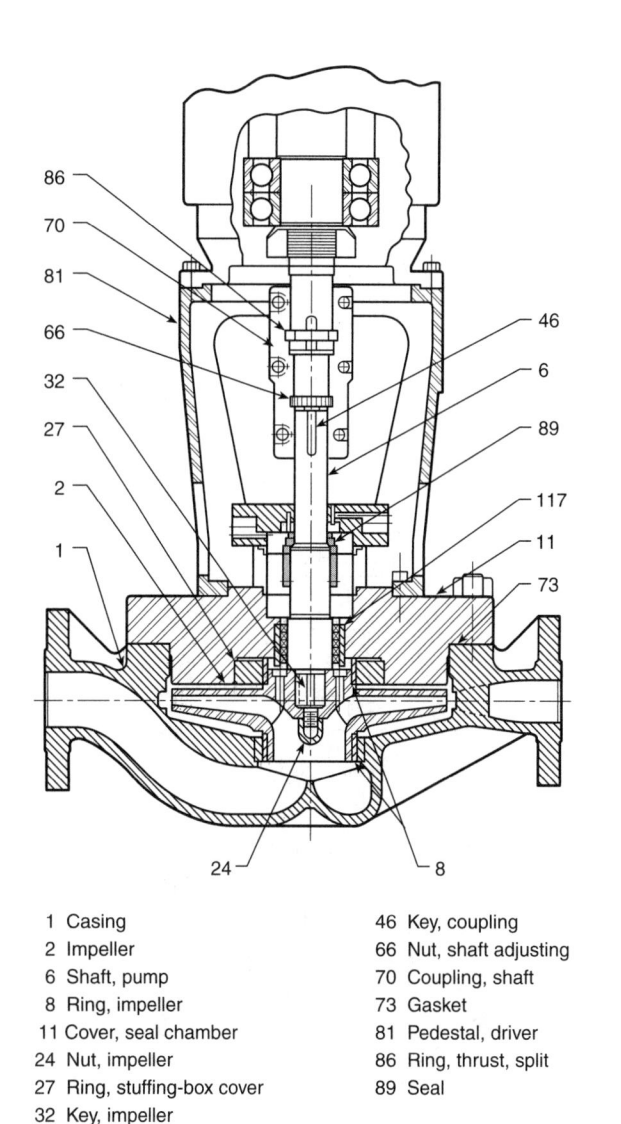

1 Casing	46 Key, coupling
2 Impeller	66 Nut, shaft adjusting
6 Shaft, pump	70 Coupling, shaft
8 Ring, impeller	73 Gasket
11 Cover, seal chamber	81 Pedestal, driver
24 Nut, impeller	86 Ring, thrust, split
27 Ring, stuffing-box cover	89 Seal
32 Key, impeller	

FIGURE A.6.1.1(d) *Overhung Impeller—Separately Coupled Single Stage—In-Line—Rigid Coupling.*

6.1.1.1 Centrifugal pumps shall be of the overhung impeller design and the impeller between bearings design.

6.1.1.2 The overhung impeller design shall be close coupled or separately coupled single- or two-stage end-suction-type *[see Figure A.6.1.1(a) and Figure A.6.1.1(b)]* or in-line-type *[see Figure A.6.1.1(c) through Figure A.6.1.1(e)]* pumps.

The driver on a *close coupled* pump is attached to the pump housing, thereby allowing a short shaft between the driver and the pump. On a *separately coupled* pump, the pump and the driver have separate drive shafts that must be connected. A *multistage pump* has multiple impellers, and as water passes from one impeller to the next, the pressure increases. Exhibit II.6.2 illustrates a typical two-stage centrifugal pump.

An end suction–type pump is a device in which water enters the end of the pump, perpendicular to the plane of the impeller.

◀ **FAQ**
What do the terms *close coupled, separately coupled,* and *multistage pump* mean?

◀ **FAQ**
What is an end suction–type pump?

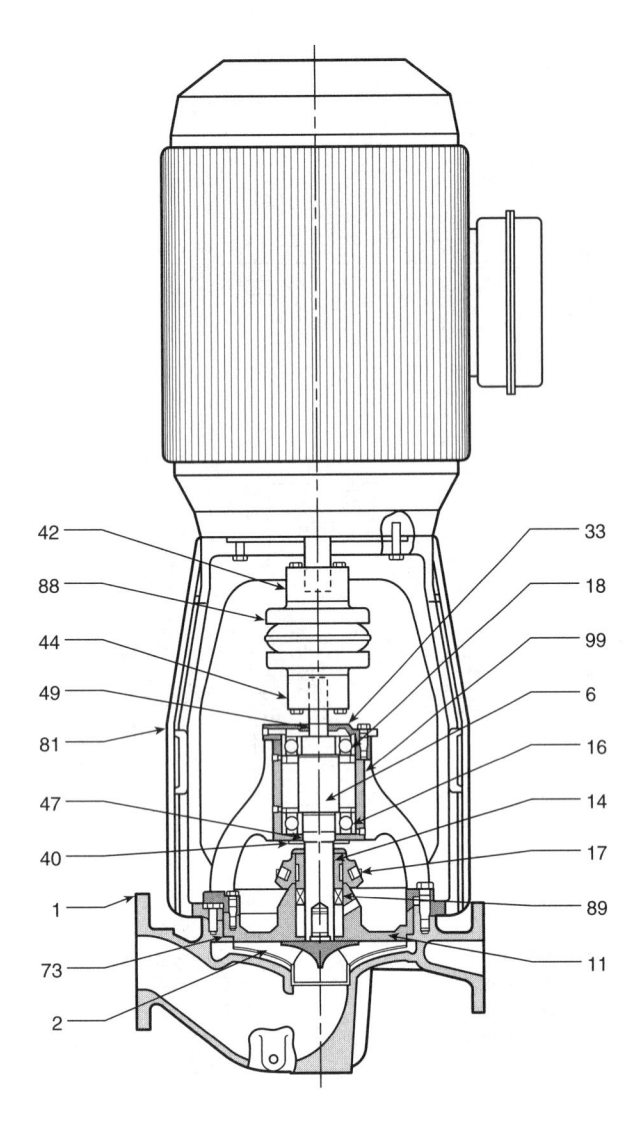

1A Casing, lower half	22 Locknut
1B Casing, upper half	31 Housing, bearing inboard
2 Impeller	32 Key, impeller
6 Shaft	33 Housing, bearing outboard
7 Ring, casing	35 Cover, bearing inboard
8 Ring, impeller	37 Cover, bearing outboard
14 Sleeve, shaft	40 Deflector
16 Bearing, inboard	65 Seal, mechanical stationary element
18 Bearing, outboard	
20 Nut, shaft sleeve	80 Seal, mechanical rotating element

FIGURE A.6.1.1(f) *Impeller Between Bearings—Separately Coupled—Single Stage—Axial (Horizontal) Split Case.*

1 Casing	42 Coupling half, driver
2 Impeller	44 Coupling half, pump
6 Shaft, pump	47 Seal, bearing cover, inboard
11 Cover, seal chamber	49 Seal, bearing cover, outboard
14 Sleeve, shaft	73 Gasket
16 Bearing, inboard	81 Pedestal, driver
17 Gland	88 Spacer, coupling
18 Bearing, outboard	89 Seal
33 Cap, bearing, outboard	99 Housing, bearing
40 Deflector	

FIGURE A.6.1.1(e) *Overhung Impeller—Separately Coupled Single Stage—In-Line—Flexible Coupling.*

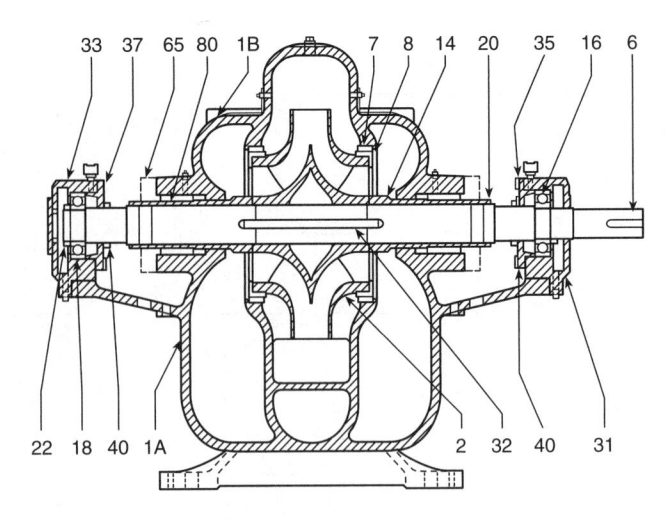

Part II • NFPA 20   129

#	Part	#	Part	#	Part
1	Casing	16	Bearing, inboard, sleeve	33	Housing, bearing, outboard
2	Impeller	18A	Bearing, outboard, sleeve	37	Cover, bearing, outboard
6	Shaft	18B	Bearing, outboard, ball	40	Deflector
7	Ring, casing	20	Nut, shaft sleeve	50	Locknut, coupling
8	Ring, impeller	22	Locknut, bearing	60	Ring, oil
11	Cover, stuffing-box	31	Housing, bearing, inboard		
14	Sleeve, shaft	32	Key, impeller		

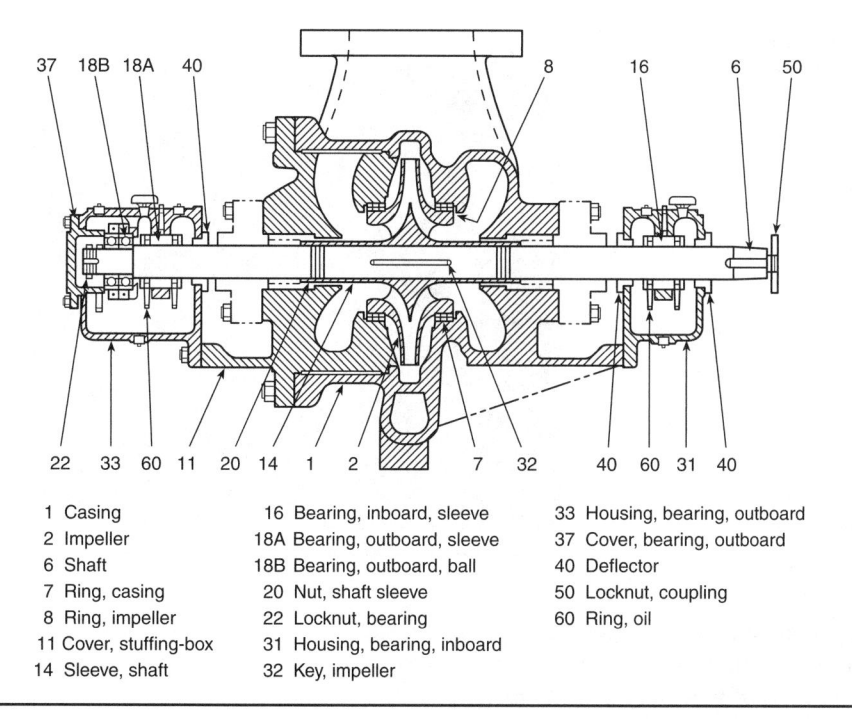

FIGURE A.6.1.1(g) *Impeller Between Bearings—Separately Coupled—Single Stage—Radial (Vertical) Split Case.*

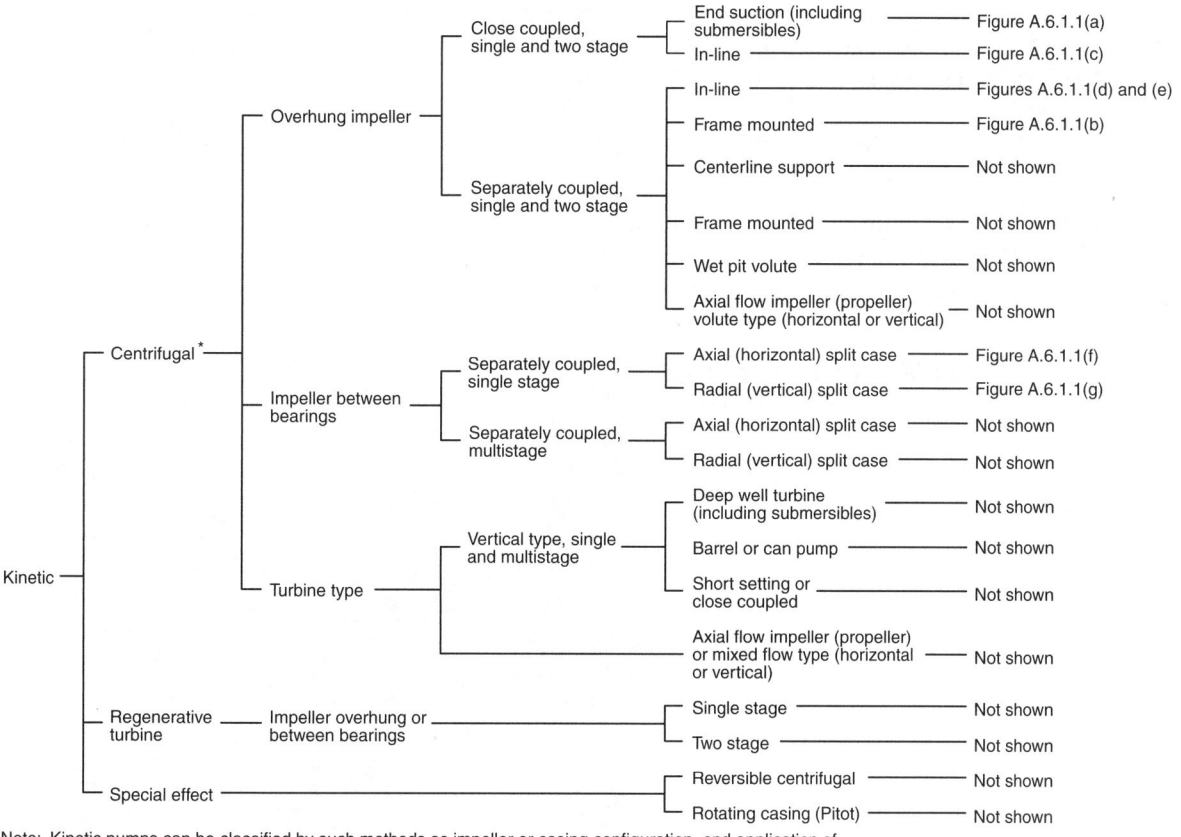

Kinetic
- Centrifugal*
  - Overhung impeller
    - Close coupled, single and two stage
      - End suction (including submersibles) — Figure A.6.1.1(a)
      - In-line — Figure A.6.1.1(c)
    - Separately coupled, single and two stage
      - In-line — Figures A.6.1.1(d) and (e)
      - Frame mounted — Figure A.6.1.1(b)
      - Centerline support — Not shown
      - Frame mounted — Not shown
      - Wet pit volute — Not shown
      - Axial flow impeller (propeller) volute type (horizontal or vertical) — Not shown
  - Impeller between bearings
    - Separately coupled, single stage
      - Axial (horizontal) split case — Figure A.6.1.1(f)
      - Radial (vertical) split case — Figure A.6.1.1(g)
    - Separately coupled, multistage
      - Axial (horizontal) split case — Not shown
      - Radial (vertical) split case — Not shown
  - Turbine type
    - Vertical type, single and multistage
      - Deep well turbine (including submersibles) — Not shown
      - Barrel or can pump — Not shown
      - Short setting or close coupled — Not shown
    - Axial flow impeller (propeller) or mixed flow type (horizontal or vertical) — Not shown
- Regenerative turbine
  - Impeller overhung or between bearings
    - Single stage — Not shown
    - Two stage — Not shown
- Special effect
  - Reversible centrifugal — Not shown
  - Rotating casing (Pitot) — Not shown

Note: Kinetic pumps can be classified by such methods as impeller or casing configuration, end application of the pump, specific speed, or mechanical configuration. The method used in this chart is based primarily on mechanical configuration.

*Includes radial, mixed flow, and axial flow designs.

FIGURE A.6.1.1(h) *Types of Stationary Pumps.*

EXHIBIT II.6.2 Two-Stage Centrifugal Pump.

6.1.1.3 The impeller between bearings design shall be separately coupled single-stage or multistage axial (horizontal) split-case-type *[see Figure A.6.1.1(f)]* or radial (vertical) split-case-type *[see Figure A.6.1.1(g)]* pumps.

FAQ ▶
What is an axial (horizontal) split-case-type pump?

An axial (horizontal) split-case-type pump is a device in which the impeller shaft is in a horizontal plane and the impeller rotates in a vertical plane. Exhibit II.6.3 illustrates a typical single-stage axial split-case-type centrifugal pump.

FAQ ▶
What is a radial (vertical) split-case-type of pump?

A radial (vertical) split-case-type pump is a device in which the impeller shaft is in a vertical plane and the impeller rotates in a horizontal plane. Exhibit II.6.4 illustrates a typical radial split-case-type centrifugal pump.

6.1.2* Application. Centrifugal pumps shall not be used where a static suction lift is required.

FAQ ▶
Why are centrifugal pumps not allowed where a suction lift is required?

A centrifugal pump is not self-priming—that is, it will not pump unless water covers the impeller. Once primed, a centrifugal pump can theoretically operate with a suction lift up to atmospheric pressure. Earlier editions of NFPA 20 allowed centrifugal pumps to operate under suction

EXHIBIT II.6.3 Single-Stage Axial (Horizontal) Split-Case-Type Centrifugal Pump. (Photograph courtesy of Schirmer Engineering Corp.)

EXHIBIT II.6.4 Radial (Vertical) Split-Case-Type Centrifugal Pump. (Photograph courtesy of Schirmer Engineering Corp.)

lift when they were provided with a tank of priming water. This arrangement proved problematic. It did not prove to be sufficiently reliable and was removed from the standard beginning with the 1974 edition. (Problems included leaky foot valves that caused the loss of priming water.)

Unlike positive displacement pumps, which continue to build pressure when operating against a closed system (until something breaks or the power source can no longer drive the pump), a centrifugal pump reaches a maximum pressure and can operate against a closed system indefinitely, provided adequate water is discharged to cool the pump. When operating against a closed system, the energy the pump is imparting to the water is converted to heat, so water must be discharged to keep the pump from overheating.

A.6.1.2 The centrifugal pump is particularly suited to boost the pressure from a public or private supply or to pump from a storage tank where there is a positive static head.

6.2* Factory and Field Performance.

A.6.2 Listed pumps can have different head capacity curve shapes for a given rating. Figure A.6.2 illustrates the extremes of probable curve shapes. Shutoff head will range from a minimum of 101 percent to a maximum of 140 percent of rated head. At 150 percent of rated capacity, head will range from a minimum of 65 percent to a maximum of just below rated head. Pump manufacturers can supply expected curves for their listed pumps.

The suction pressure from the water supply must be added to the fire pump characteristic curve to determine the fire pump discharge curve. See Exhibit II.6.5.

Fire protection systems can be designed to operate at any point along the pump discharge curve; however, all system components must be rated for the maximum pressure developed by the fire pump.

◀ **FAQ**
How does a fire pump characteristic curve impact fire protection system design?

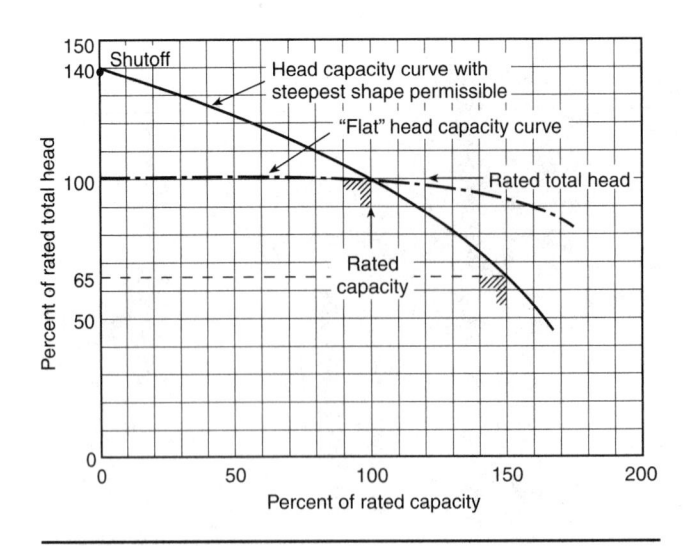

FIGURE A.6.2 *Pump Characteristics Curves.*

6.2.1 Pumps shall furnish not less than 150 percent of rated capacity at not less than 65 percent of total rated head.

NFPA 20 defines the following three operating points that a fire pump must meet:

1. The shutoff head is limited to 140 percent of the rated head.
2. The rated head at the rated flow is noted on the pump nameplate and is commonly referred to as the rated capacity.

EXHIBIT II.6.5 *Sample Fire Pump Discharge Curve.*

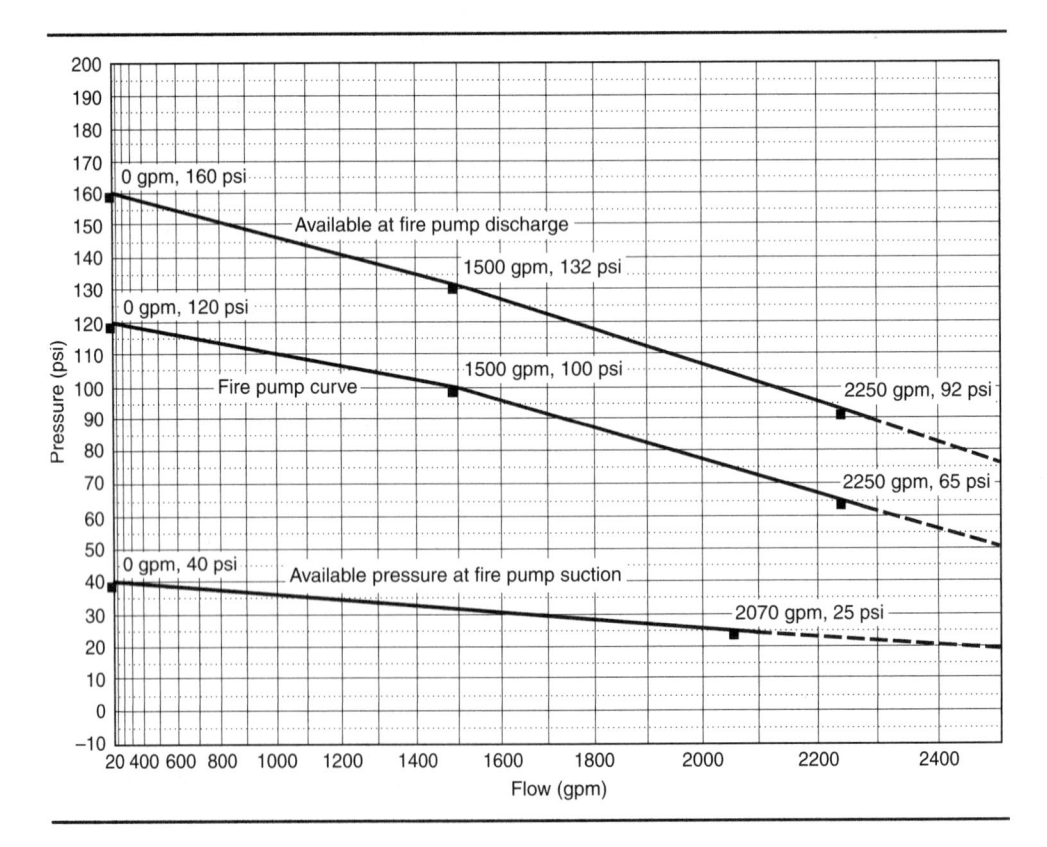

3. The pump must provide a minimum of 65 percent of the rated head at 150 percent of the rated capacity.

Design is permitted along any point of a fire pump curve between shutoff and 150 percent of rated flow. NFPA 25, *Standard for the Inspection, Testing, and Maintenance of Water-Based Fire Protection Systems,* allows 5 percent degradation in performance before requiring an investigation.

The actual fire pump curve is typically below the maximum limits established by NFPA 20 (140 percent of rated pressure at churn) when flowing less than 100 percent of rated flow, and above the minimum limits established by NFPA 20 (65 percent of rated pressure at 150 percent of rated flow) when flowing above the rated flow. Although NPFA 20 requires the pump performance to be measured against the original pump curve when evaluating pump performance, the test results are sometimes compared against the NFPA 20 limiting values, especially when the original pump curve is not readily available. When designing to flow rates that exceed 100 percent of the rated flow, designers should design to the NFPA 20 limits (65 percent of rated pressure at 150 percent of rated flow) unless they are sure that the actual pump curve will be used to evaluate the pump performance whenever the fire pump is tested.

6.2.2 The shutoff head shall not exceed 140 percent of rated head for any type pump. *(See Figure A.6.2.)*

Historically, vertical turbine pumps have a higher churn pressure than horizontal centrifugal pumps. Earlier editions of NFPA 20 limited the horizontal centrifugal pump's shutoff head to 120 percent of the rated head. Although the distinction was removed from NFPA 20 in the 1990 edition, most centrifugal fire pumps still churn at less than 120 percent of the rated pressure.

The fire pump characteristic curve must meet the minimum and maximum flow and pressure requirements of this section, as indicated in Figure A.6.2. The fire pump output can vary significantly. For an example of fire pump outputs, see Commentary Table II.6.1.

The pump capacity should be carefully selected to meet the fire protection system demand with a reasonable cushion of approximately 10 psi (0.6 bar) or more but should not be oversized to risk the overpressurization of the system. To plot a fire pump curve using the sample fire pump discharge curve in Figure A.6.2, first plot the rated capacity of the pump, which should be a known value. In this case, the value is 1500 gpm (5677 L/min) at 100 psi (6.8 bar). The maximum pump output, commonly referred to as *overload,* can then be plotted as 2250 gpm (8516 L/min at 65 psi 4.5 bar) (150 percent of rated flow at 65 percent of rated pressure). The shutoff or churn point can be estimated at midrange or 120 percent of rated pressure at 0 gpm (0 bar) flow or may be obtained from the manufacturer's shop test curve if available. In this case, 120 psi (8.3 bar) was used.

For pumps attached to a city water supply, the city water flow test data must be adjusted to the fire pump curve (elevation difference and friction loss). These data should be plotted next and are shown on the sample fire pump discharge curve in Exhibit II.6.5 as 40 psi (2.8 bar) static pressure and 2070 gpm (7835 L/min) at 25 psi (1.7 bar) residual pressure. To determine the combined fire pump/city water supply curve, the pressures (churn, rated capacity, and overload) from the city water supply must be added to each point on the fire pump curve. Doing this calculation on the sample curve results in the following data: 160 psi (11 bar) at 0 flow (churn), 132 psi (9.1 bar) at 1500 gpm (5677 L/min) (rated capacity), and 92 psi (6.3 bar) at 2250 gpm (8516 L/min) (overload). This procedure can be used for any size fire pump that is connected to a city water supply.

6.3 Fittings.

6.3.1* Where necessary, the following fittings for the pump shall be provided by the pump manufacturer or an authorized representative:

COMMENTARY TABLE II.6.1 *Fire Pump Selection Table*

Item	Value	Comments
1. Maximum system demand flow (gpm)	1700	Determined by hydraulic calculations based on design criteria.
2. Maximum system demand pressure (psi)	107	Determined by hydraulic calculations based on design criteria.
3. Pump suction pressure at zero (0) flow (psi)	40	Based on a flow test adjusted for the friction loss to the fire pump suction. The friction loss from the flow test to the pump suction was assumed to be 0 psi for this example.
4. Pump suction pressure at maximum system demand flow (psi)	30	Based on a flow test curve at a flow of 1700 gpm. The friction loss from the flow test to the pump suction was assumed to be 0 psi for this example.
5. Pump flow rating (gpm)	1500	Maximum system demand must be between 0% and 150% of pump's rated capacity.
6. Net pump pressure rating (psi)	100	Size so the discharge pressure (net pump pressure plus suction pressure) provides adequate pressure for the fire protection system(s).
7. Net pump churn pressure (psi)	120	Review pump curves from manufacturer to approximate churn pressure (120% of rated pressure was used for this example).
8. Net pump pressure at 150% of rated flow (psi)	65	Must be a minimum of 65% of rated pressure. Actual pump curve may be higher.
9. Net pump pressure at maximum system demand flow (psi)	92	From pump curve at maximum system demand flow (1700 gpm for this example).
10. Pump discharge pressure at 0 flow (psi)	160	Items 3 + 7: Pump suction pressure plus net pump pressure at 0 flow. The pump discharge pressure after adjusting for elevation cannot exceed the rating of the system components.
11. Pump discharge pressure at maximum system demand flow (psi)	122	Items 4 + 9: Must be higher than the maximum system demand plus any required safety factors.

(1) Automatic air release valve

See 6.3.3 for more information on automatic air release.

(2) Circulation relief valve

A circulation relief valve (see Exhibit II.6.6) allows a small amount of water to discharge from the pump to keep the fire pump from overheating when operating at no flow or very low flows.

(3) Pressure gauges

The pressure gauge on the suction side of the fire pump should be a compound gauge to register negative pressures whenever the water supply is a ground level storage tank or has a low residual pressure. Liquid-filled gauges dampen pressure fluctuations, making the gauges easier to read during pump testing.

A.6.3.1 See Figure A.6.3.1(a) and Figure A.6.3.1(b).

6.3.2 Where necessary, the following fittings shall be provided:

(1) Eccentric tapered reducer at suction inlet

When the pump suction pipe is larger than the pump suction flange, an eccentric tapered reducer (see Exhibit II.6.7) is required to minimize the possibility of air pockets forming at the pump suction.

EXHIBIT II.6.6 *Circulation Relief Valve.*

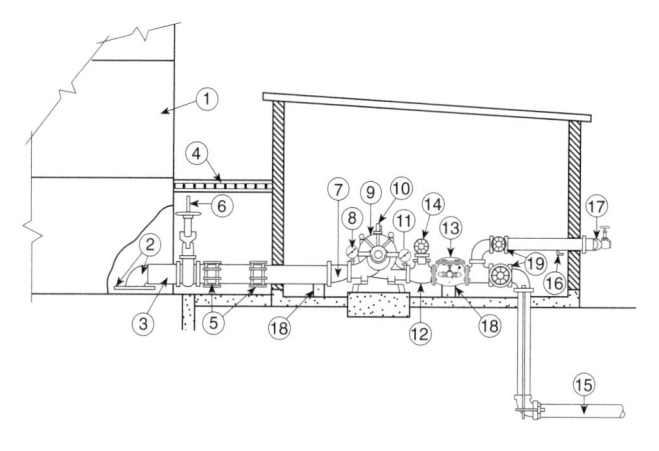

1. Aboveground suction tank
2. Entrance elbow and square steel vortex plate with dimensions at least twice the diameter of the suction pipe. Distance above the bottom of tank is one-half the diameter of the suction pipe with minimum of 6 in. (152 mm).
3. Suction pipe
4. Frostproof casing
5. Flexible couplings for strain relief
6. OS&Y gate valve *(see 4.14.5 and A.4.14.5)*
7. Eccentric reducer
8. Suction gauge
9. Horizontal split-case fire pump
10. Automatic air release
11. Discharge gauge
12. Reducing discharge tee
13. Discharge check valve
14. Relief valve (if required)
15. Supply pipe for fire protection system
16. Drain valve or ball drip
17. Hose valve manifold with hose valves
18. Pipe supports
19. Indicating gate or indicating butterfly valve

FIGURE A.6.3.1(a) *Horizontal Split-Case Fire Pump Installation with Water Supply Under a Positive Head.*

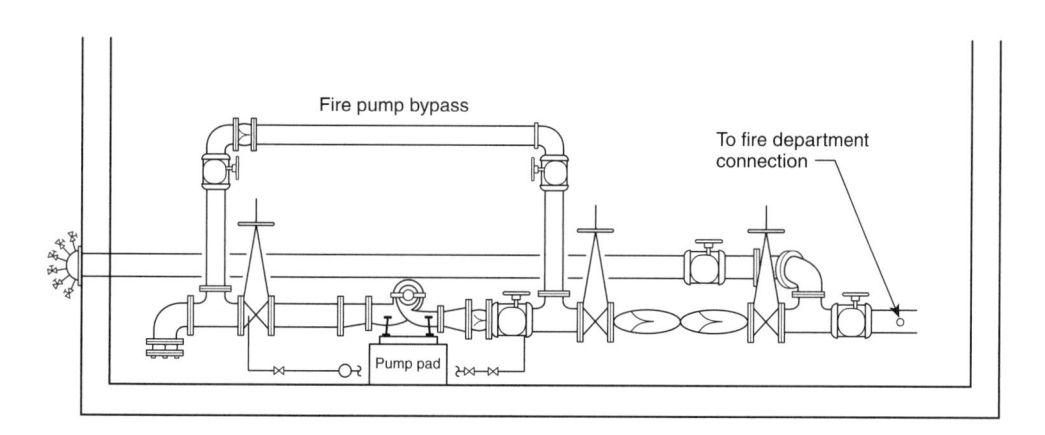

FIGURE A.6.3.1(b) *Backflow Preventer Installation.*

(2) Hose valve manifold with hose valves

See 4.20.3 for more information on hose valves. Exhibit II.6.8 illustrates a hose valve manifold.

(3) Flow measuring device

See 4.20.2 for more information on meters. A water flowmeter (see Exhibit II.6.9) must be listed and rated for a minimum of 175 percent of the rated pump capacity. The provision of

EXHIBIT II.6.7 Eccentric Reducer Installation.

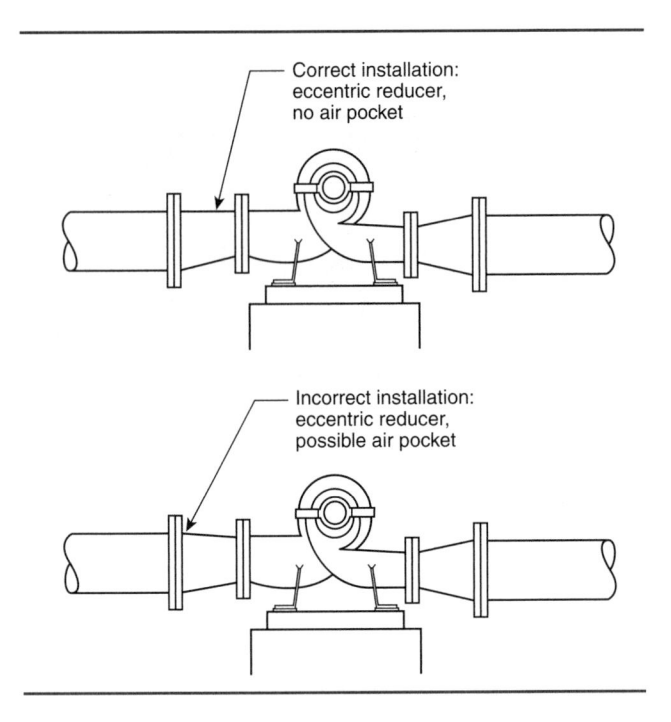

EXHIBIT II.6.8 Hose Valve Manifold.

EXHIBIT II.6.9 Flowmeter.

a hose valve manifold downstream of a flowmeter is preferable so that the flowmeter can be calibrated and the accuracy of the flowmeter verified.

(4) Relief valve and discharge cone

See Section 4.18 for the allowable use of pressure relief valves. Paragraph 4.18.5.5 requires a means of detecting flow and a circulation relief valve (see Exhibit II.6.10).

(5) Pipeline strainer

Earlier editions of NFPA 20 required a strainer in the suction lines of pumps that required removal of the driver to remove rocks or debris from the pump impeller. This requirement

EXHIBIT II.6.10 Pressure Relief Valve Piped Back to Suction.

was removed from the 2007 edition. See 7.2.2.2.4, 7.3.4, 8.4.5, 11.2.8.5.3.2, and 11.2.8.6.2 for current strainer requirements.

6.3.3 Automatic Air Release.

6.3.3.1 Unless the requirements of 6.3.3.2 are met, pumps that are automatically controlled shall be provided with a listed float-operated air release valve having a nominal 0.50 in. (12.7 mm) minimum diameter discharged to atmosphere.

Air in the impeller can cause cavitation, damage the impeller, and negatively impact the pump performance. Exhibit II.6.11 illustrates an automatic air release valve.

6.3.3.2 The requirements of 6.3.3.1 shall not apply to overhung impeller–type pumps with top centerline discharge or that are vertically mounted to naturally vent the air.

6.4 Foundation and Setting.

6.4.1* Overhung impeller and impeller between bearings design pumps and driver shall be mounted on a common grouted base plate.

A.6.4.1 Flexible couplings are used to compensate for temperature changes and to permit end movement of the connected shafts without interfering with each other.

6.4.2 Pumps of the overhung impeller close coupled in-line type *[see Figure A.6.1.1(c)]* shall be permitted to be mounted on a base attached to the pump mounting base plate.

6.4.3 The base plate shall be securely attached to a solid foundation in such a way that proper pump and driver shaft alignment is ensured.

6.4.4* The foundation shall be sufficiently substantial to form a permanent and rigid support for the base plate.

A.6.4.4 A substantial foundation is important in maintaining alignment. The foundation preferably should be made of reinforced concrete.

6.4.5 The base plate, with pump and driver mounted on it, shall be set level on the foundation.

6.5* Connection to Driver and Alignment.

A.6.5 A pump and driver shipped from the factory with both machines mounted on a common base plate are accurately aligned before shipment. All base plates are flexible to some extent and, therefore, should not be relied upon to maintain the factory alignment. Realignment is necessary after the complete unit has been leveled on the foundation and again after the grout has set and foundation bolts have been tightened. The alignment should be checked after the unit is piped and rechecked periodically. To facilitate accurate field alignment, most manufacturers either do not dowel the pumps or drivers on the base plates before shipment or, at most, dowel the pump only.

After the pump and driver unit has been placed on the foundation, the coupling halves should be disconnected. The coupling should not be reconnected until the alignment operations have been completed.

The purpose of the flexible coupling is to compensate for temperature changes and to permit end movement of the shafts without interference with each other while transmitting power from the driver to the pump.

The two forms of misalignment between the pump shaft and the driver shaft are as follows:

(1) *Angular misalignment*—shafts with axes concentric but not parallel
(2) *Parallel misalignment*—shafts with axes parallel but not concentric

The faces of the coupling halves should be spaced within the manufacturer's recommendations and far enough apart so that they cannot strike each other when the driver rotor is moved hard over toward the pump. Due allowance should be made for wear of the thrust bearings. The necessary tools for an approximate check of the alignment of a flexible coupling are a straight edge and a taper gauge or a set of feeler gauges.

A check for angular alignment is made by inserting the taper gauge or feelers at four points between the coupling faces and comparing the distance between the faces at four points spaced at 90-degree intervals around the coupling. *[See Figure A.6.5(a).]* The unit will be in angular alignment when the measurements show that the coupling faces are the same distance apart at all points.

A check for parallel alignment is made by placing a straight edge across both coupling rims at the top, bottom, and both sides. *[See Figure A.6.5(b).]* The unit will be in parallel alignment when the straight edge rests evenly on the coupling rim at all positions.

Allowance might be necessary for temperature changes and for coupling halves that are not of the same outside diameter. Care should be taken to have the straight edge parallel to the axes of the shafts.

Angular and parallel misalignment are corrected by means of shims under the motor mounting feet. After each change, it is necessary to recheck the alignment of the coupling halves. Adjustment in one direction can disturb adjustments already made in another direction. It should not be necessary to adjust the shims under the pump.

The permissible amount of misalignment will vary with the type of pump, driver, and coupling manufacturer, model, and size.

The best method for putting the coupling halves in final accurate alignment is by the use of a dial indicator.

When the alignment is correct, the foundation bolts should be tightened evenly but not too firmly. The unit can then be grouted to the foundation. The base plate should be completely filled with grout, and it is desirable to grout the leveling pieces, shims, or wedges in place. Foundation bolts should not be fully tightened until the grout has hardened, usually about 48 hours after pouring.

FIGURE A.6.5(a) Checking Angular Alignment. (Courtesy of Hydraulic Institute, www.pumps.org.)

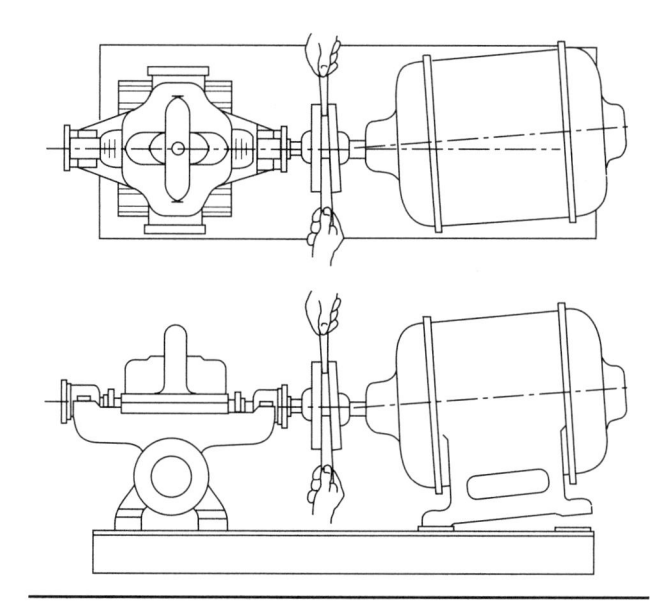

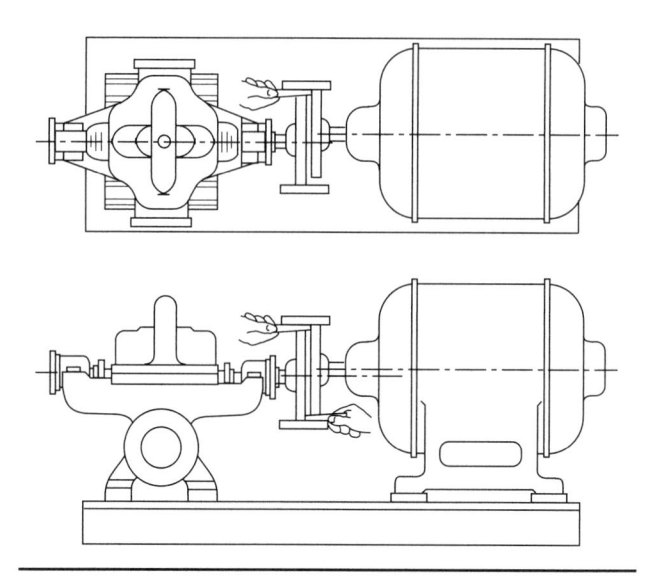

FIGURE A.6.5(b) Checking Parallel Alignment. (Courtesy of Hydraulic Institute, www.pumps.org.)

After the grout has set and the foundation bolts have been properly tightened, the unit should be checked for parallel and angular alignment, and, if necessary, corrective measures taken. After the piping of the unit has been connected, the alignment should be checked again.

The direction of driver rotation should be checked to make certain that it matches that of the pump. The corresponding direction of rotation of the pump is indicated by a direction arrow on the pump casing.

The coupling halves can then be reconnected. With the pump properly primed, the unit should be operated under normal operating conditions until temperatures have stabilized. It then should be shut down and immediately checked again for alignment of the coupling. All alignment checks should be made with the coupling halves disconnected and again after they are reconnected.

After the unit has been in operation for about 10 hours or 3 months, the coupling halves should be given a final check for misalignment caused by pipe or temperature strains. If the alignment is correct, both pump and driver should be dowelled to the base plate. Dowel location is very important, and the manufacturer's instructions should be followed, especially if the unit is subject to temperature changes.

The unit should be checked periodically for alignment. If the unit does not stay in line after being properly installed, the following are possible causes:

(1) Settling, seasoning, or springing of the foundation and pipe strains distorting or shifting the machine
(2) Wearing of the bearings
(3) Springing of the base plate by heat from an adjacent steam pipe or from a steam turbine
(4) Shifting of the building structure due to variable loading or other causes
(5) If the unit and foundation are new, need for the alignment to be slightly readjusted from time to time

6.5.1 Coupling Type.

6.5.1.1 Separately coupled–type pumps with electric motor drivers shall be connected by a flexible coupling or flexible connecting shaft.

The couplings referred to in 6.5.1.1 connect the pump to the driver—either an electric motor or a diesel engine. The two types of couplings used are a flexible coupling, as illustrated in Exhibit II.6.12, or a flexible shaft, as illustrated in Exhibit II.6.13.

6.5.1.2 All coupling types shall be listed for the service referenced in 6.5.1.1.

6.5.2 Pumps and drivers on separately coupled–type pumps shall be aligned in accordance with the coupling and pump manufacturers' specifications and the *Hydraulic Institute Standards for Centrifugal, Rotary and Reciprocating Pumps. (See A.6.5.)*

EXHIBIT II.6.12 *Flexible Coupling.*

EXHIBIT II.6.13 *Flexible Connecting Shaft.*

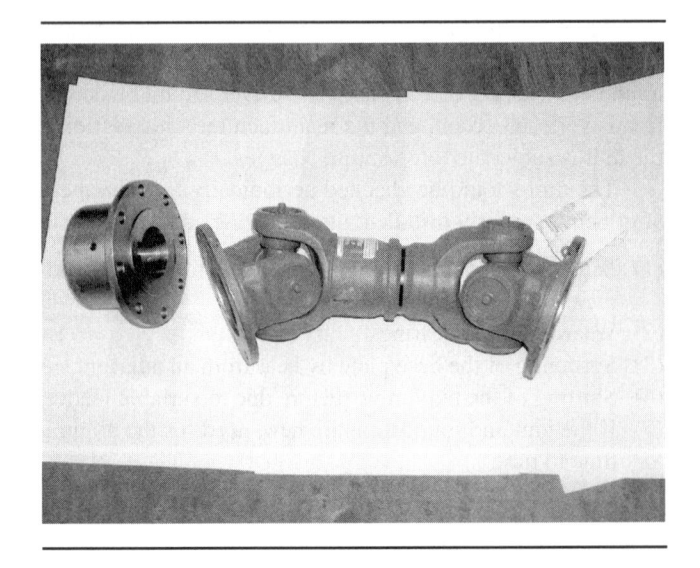

REFERENCES CITED IN COMMENTARY

National Fire Protection Association, 1 Batterymarch Park, Quincy, MA 02169-7471.

NFPA 25, *Standard for the Inspection, Testing, and Maintenance of Water-Based Fire Protection Systems,* 2008 edition.

Vertical Shaft Turbine–Type Pumps

CHAPTER 7

The vertical turbine pump is designed to provide suction lift from a static source of water. Therefore, this pump differs in its design and installation requirements from the other pumps covered in this handbook thus far. The design and installation of a vertical turbine pump depend on the construction of either a well or a wet pit. In either case, this chapter provides specific requirements for the design of the well or the wet pit as well as for the necessary accessories for the installation and proper operation of the pump.

7.1* General.

A.7.1 Satisfactory operation of vertical turbine–type pumps is dependent to a large extent upon careful and correct installation of the unit; therefore, it is recommended that this work be done under the direction of a representative of the pump manufacturer.

The designer of the pump installation should work with the AHJ and the designer of the fire protection system to determine the pressure and volume of water required by the system to control any anticipated fire. The designer of the pump installation should obtain the assistance of the pump manufacturer's representative and the water source (well, tank, water main, or reservoir) supplier. This collaboration ensures that the pump installation meets all of the requirements of the fire protection system without overpressurizing the system piping, while at the same time providing the required pressure and volume of water required by the system installation standard. The pump installation designer should perform the function of a system integrator, preparing technical specifications for the selection of properly sized and coordinated equipment, such as discharge pressure, driveshaft, diesel engine or electric motor, pump cooling and/or driver cooling system, relief valves, test equipment, and so forth.

7.1.1* Suitability. Where the water supply is located below the discharge flange centerline and the water supply pressure is insufficient for getting the water to the fire pump, a vertical shaft turbine–type pump shall be used.

A.7.1.1 The vertical shaft turbine–-type pump is particularly suitable for fire pump service where the water source is located below ground and where it would be difficult to install any other type of pump below the minimum water level. It was originally designed for installation in drilled wells but is permitted to be used to lift water from lakes, streams, open swamps, and other subsurface sources. Both oil-lubricated enclosed-line-shaft and water-lubricated open-line-shaft pumps are used. *(See Figure A.7.1.1.)* Some health departments object to the use of oil-lubricated pumps; such authorities should be consulted before proceeding with oil-lubricated design.

Horizontal pumps are limited by 4.14.3.2 to operate with a suction lift of −3 psi (−0.2 bar) at the lowest water level and at 150 percent of the rated flow. Therefore, the vertical shaft turbine–type pump is the best choice for locations that cannot provide positive pressure to the suction side of the fire pump. As shown in Figures A.7.2.2.1 and A.7.2.2.2, vertical shaft

Water-lubricated, open lineshaft pump, surface discharge, threaded column and bowls

Oil-lubricated, enclosed lineshaft pump, underground discharge, flanged column and bowls

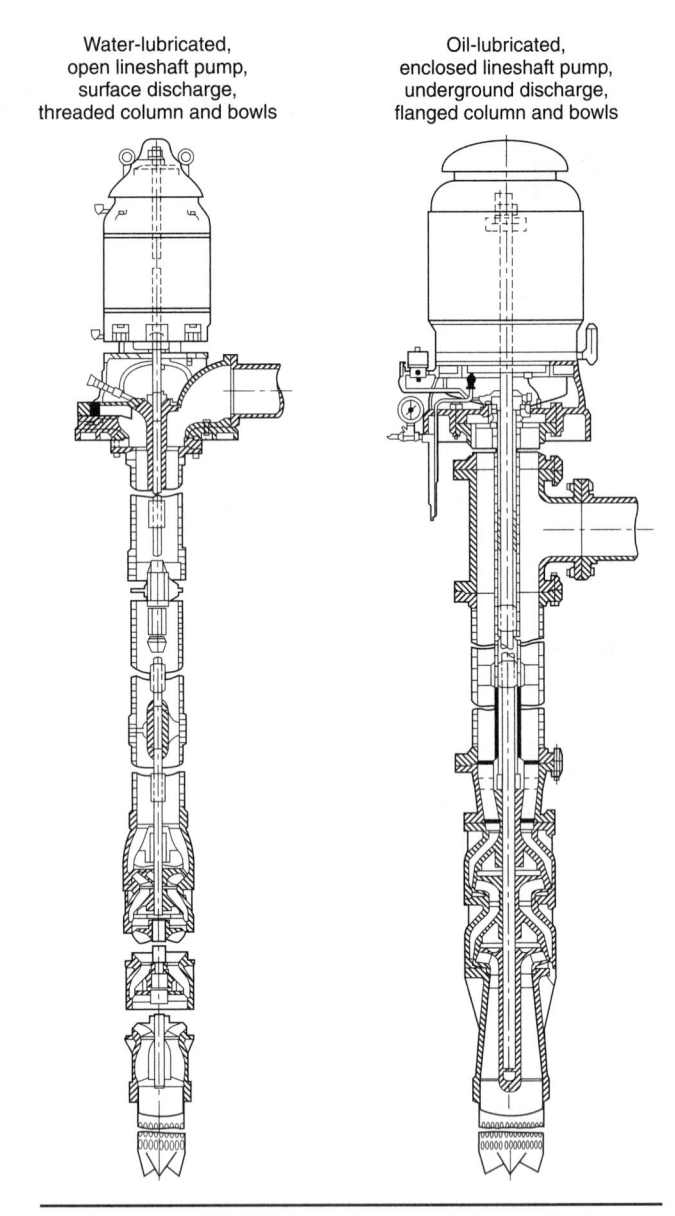

FIGURE A.7.1.1 *Water-Lubricated and Oil-Lubricated Shaft Pumps.*

turbine–type pumps have their impellers submerged in the water supply. The operating impellers force the water up the pipe column to the discharge flange and into the system piping. (See Exhibit II.3.11 for an example of a vertical lineshaft turbine pump.)

7.1.2 Characteristics.

Fire pumps are designed and built so that they can operate over a range of pressure and flow conditions in relation to their rating. Subsection 7.1.2 identifies the endpoints of this range, and all listed pumps must have characteristic curves somewhere within this range. The shutoff head corresponds to the maximum pressure the pump can produce when the pump is operating in a churn condition—that is, when no water is flowing. When the pump operates at churn pressure, the shutoff head or pressure cannot exceed 140 percent of the pump's pressure rating. For example, a pump with a rating of 100 psi (6.7 bar) at 1000 gpm (3785 L/min) cannot produce a shutoff head in excess of 140 psi (9.7 bar). A pump can be designed and manufactured so that the shutoff head falls below this value, such as at 100 psi (6.7 bar).

The other endpoint identifies the minimum pressure the pump can produce. This minimum pressure occurs when the pump is operating at 150 percent of its capacity and does not fall below 65 percent of the pump's rated pressure. Consider a pump with a rating of 125 psi at 1000 gpm (8.6 bar) (3785 L/min). When the flow reaches 150 percent of the pump's capacity or 1500 gpm (5677 L/min), the minimum pressure cannot fall below 81.25 psi (5.6 bar). Any pressure between 81.25 psi (5.6 bar) and 125 psi (8.6 bar) would be acceptable. Each manufacturer's pump can be different, but all must meet the minimum criteria.

7.1.2.1 Pumps shall furnish not less than 150 percent of rated capacity at a total head of not less than 65 percent of the total rated head.

7.1.2.2 The total shutoff head shall not exceed 140 percent of the total rated head on vertical turbine pumps. *(See Figure A.6.2.)*

7.2 Water Supply.

7.2.1 Source.

7.2.1.1* The water supply shall be adequate, dependable, and acceptable to the authority having jurisdiction.

Water supplies need to be thoroughly evaluated. Without a sufficient and reliable water supply, the entire fire protection system is in jeopardy. A fire pump augments an existing water supply by increasing the pressure. The fire pump cannot create water; the pump only increases pressure. The amount of water necessary for a fire protection system in terms of flow, pressure, and capacity is obtained through other NFPA standards, such as NFPA 13, *Standard for the Installation of Sprinkler Systems,* or NFPA 14, *Standard for the Installation of Standpipe and Hose Systems;* authorities having jurisdiction (e.g., insurance companies); and building codes. The amount of water available from a supply can be evaluated using NFPA 291, *Recommended Practice for Fire Flow Testing and Marking of Hydrants.* When the municipal water supply is inadequate, the supply must be augmented using tanks, reservoirs, lakes, and so forth.

◀ **FAQ**
Does installation of a fire pump increase the amount of water in the water supply?

The components of a fire protection system are limited by their pressure rating. The use of the correct components in the design of a fire pump system is important to eliminate overpressure problems. The pump churn or shutoff pressure is often a restricting factor that is missed in the design process and then has to be compensated for by adding pressure relief or control devices that have less reliability than the fire pump or fire protection system components. Where vertical shaft turbine–type fire pumps draw water from naturally occurring water sources such as wells, lakes, rivers, or ponds, consideration should be given to factors such as seasonal fluctuation of water level, the presence of snow and ice, the accumulation of trash, and chemical and/or organic factors that may cause corrosion and/or tuberculation.

The minimum water level required for vertical turbine pumps to draw suction must be considered for aboveground and belowground tanks. Generally, a suction pit needs to be constructed below the tank, so that the total capacity of the tank can be used by the pump. The depth of the suction pit depends on the length of the pump, which is determined by the required pressure and pumping volume of the assembly.

◀ **FAQ**
How is the depth of a suction pit determined?

A.7.2.1.1 Stored water supplies from reservoirs or tanks supplying wet pits are preferable. Lakes, streams, and groundwater supplies are acceptable where investigation shows that they can be expected to provide a suitable and reliable supply.

Lakes, rivers, and open top reservoirs are subject to particulate matter and trash that may clog suction screens. They are also subject to chemical and/or organic factors that may cause corrosion and/or tuberculation.

7.2.1.2* The acceptance of a well as a water supply source shall be dependent upon satisfactory development of the well and establishment of satisfactory aquifer characteristics.

A.7.2.1.2 The authority having jurisdiction can require an aquifer performance analysis. The history of the water table should be carefully investigated. The number of wells already in use in the area and the probable number that can be in use should be considered in relation to the total amount of water available for fire protection purposes.

Wells, lakes, streams, tanks, and any other water supply source must be carefully analyzed for suitability and serviceability. Data no older than 10 years should be used to determine the minimum conditions regarding the water level expected. Some of the factors to be considered are the water level in the source and how it changes from year to year and during the year, seasonal changes in water elevation, ground freezing conditions in cold climates, trash and debris in the water source, and how the pump suction is designed to use all of the water in a storage tank or reservoir.

7.2.2 Pump Submergence.

7.2.2.1* Well Installations.

A.7.2.2.1 See Figure A.7.2.2.1.

7.2.2.1.1 Proper submergence of the pump bowls shall be provided for reliable operation of the fire pump unit. Submergence of the second impeller from the bottom of the pump bowl assembly shall be not less than 10 ft (3.2 m) below the pumping water level at 150 percent of rated capacity. (*See Figure A.7.2.2.1.*)

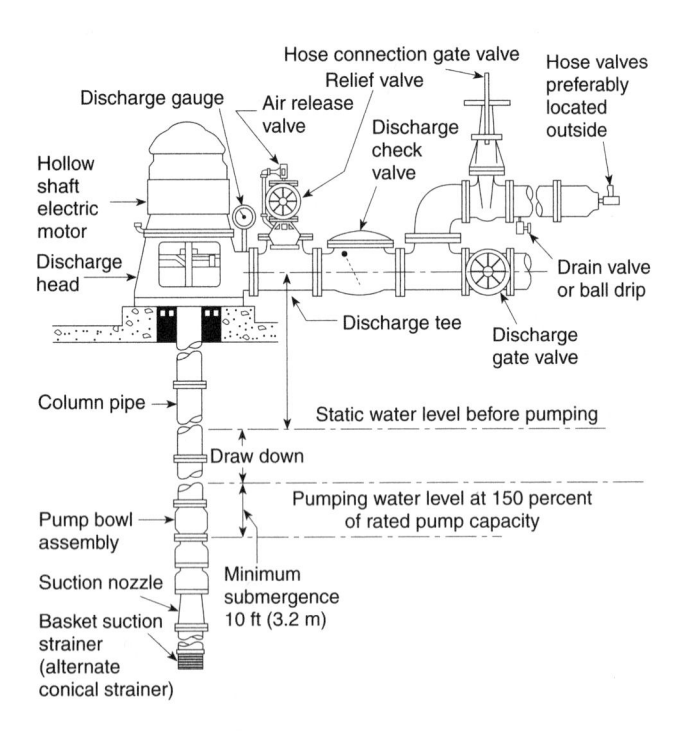

Note: The distance between the bottom of the strainer and the bottom of the wet pit should be one-half of the pump bowl diameter but not less than 12 in. (305 mm).

FIGURE A.7.2.2.1 Vertical Shaft Turbine–Type Pump Installation in a Well.

7.2.2.1.2* The submergence shall be increased by 1 ft (0.3 m) for each 1000 ft (305 m) of elevation above sea level.

The increase specified in 7.2.2.1.2 is in addition to any other increases that are required by the standard. The effect of not including this increase may cause cavitation in the pump when operating at higher pump flows.

A.7.2.2.1.2 The acceptability of a well is determined by a 24-hour test that flows the well at 150 percent of the pump flow rating. This test should be reviewed by qualified personnel (usually a well drilling contractor or a person having experience in hydrology and geology). The adequacy and reliability of the water supply are critical to the successful operation of the fire pump and fire protection system.

A 10 ft (3.05 m) submergence is considered the minimum acceptable level to provide proper pump operation in well applications. The increase of 1 ft (0.30 m) for each 1000 ft (305 m) increase in elevation is due to loss of atmospheric pressure that accompanies elevation. Therefore, the net positive suction head (NPSH) available must be considered in selection of the pump. For example, to obtain the equivalent of 10 ft (3.05 m) of NPSH available at an elevation of 1000 ft (305 m), approximately 11 ft (3.35 m) of water is required.

Several other design parameters need to be considered in the selection of a vertical turbine pump, including the following:

(1) *Lineshaft lubrication when the pump is installed in a well.* Bearings are required to have lubrication and are installed along the lineshaft to maintain alignment. Lubrication fluid is usually provided by a fluid reservoir located aboveground, and the fluid is supplied to each bearing by a copper tube or small pipe. This lubrication fluid should use a vegetable-based material that is approved by the federal Clean Water Act to minimize water contamination.

(2) *Determination of the water level in the well.* When a vertical turbine pump is tested, the water level in the well needs to be known so that the suction pressure can be determined. Often the air line for determining the depth is omitted, so testing of the pump for performance is not possible. The arrangement of this device is shown in Figure A.7.3.5.3, and its installation should be included in the system design.

7.2.2.2* Wet Pit Installations.

A.7.2.2.2 The velocities in the approach channel or intake pipe should not exceed approximately 2 ft/sec (0.7 m/sec), and the velocity in the wet pit should not exceed approximately 1 ft/sec (0.3 m/sec). *(See Figure A.7.2.2.2.)*

The ideal approach is a straight channel coming directly to the pump. Turns and obstructions are detrimental because they can cause eddy currents and tend to initiate deep-cored vortices. The amount of submergence for successful operation will depend greatly on the approaches of the intake and the size of the pump.

The *Hydraulic Institute Standards for Centrifugal, Rotary and Reciprocating Pumps* recommends sump dimensions for flows 3000 gpm (11,355 L/min) and larger. The design of sumps for pumps with discharge capacities less than 3000 gpm (11,355 L/min) should be guided by the same general principles shown in the *Hydraulic Institute Standards for Centrifugal, Rotary and Reciprocating Pumps*.

The quality of the water source is the most challenging part of the equation if a pond, lake, or other natural source is used as a water supply. Open ponds and lakes may contain fish, debris, leaves, or other floating materials that may cause problems with the pump's water supply.

The function of the intake structure, whether it is an open channel, a fully wetted tunnel, a sump, or a tank, is to supply an evenly distributed flow to the pump suction. An uneven distribution of flow, such as that caused by strong local currents, can result in the formation of surface or submerged vortices. With certain low values of submergence, an uneven dis-

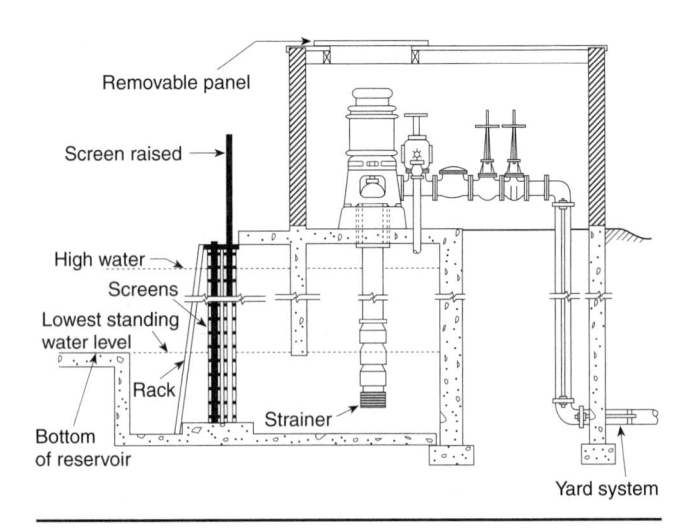

FIGURE A.7.2.2.2 *Vertical Shaft Turbine–Type Pump Installation in a Wet Pit.*

tribution of flow may introduce air into the pump, causing reduced capacity and increased vibration and additional noise. Uneven flow distribution can also increase or decrease the power consumption with a change in total developed head.

Calculation of low average velocity, as required by 7.2.2.2, is not the sole basis for judging the suitability of an intake structure. High velocities of isolated currents and swirls may be present; however, in general, overall very low average velocities are present.

Water should not flow past one pump suction bell, suction pipe, or other intake to reach the next intake. If the intakes must be placed in line with the flow, the construction of an open front cell around each pump intake or placement of turning vanes under the intake may be necessary to deflect the water upward. Streamlining should be used to reduce alternating vortices in the wake of an intake or other obstructions in the stream flow.

The amount of submergence available is only one factor affecting vortex-free operation. The pump can have adequate submergence and still have submerged vortices that may have an adverse effect on pump operation. Successful vortex-free operation depends greatly on the approach upstream of the sump. The design of any suction inlet structure should use the principles of the *Hydraulics Institute Standards for Centrifugal, Rotary and Reciprocating Pumps.*

7.2.2.2.1 To provide submergence for priming, the elevation of the second impeller from the bottom of the pump bowl assembly shall be such that it is below the lowest pumping water level in the open body of water supplying the pit.

The minimum required submergence is normally obtained from the pump manufacturer or its representative. Note that if the water depth is shallow, a larger suction bell can reduce the required submergence, as can a round plate attached to the suction bell. The intent of this requirement is to maintain the water velocities from 2 ft/sec (0.7 m/sec) to 4 ft/sec (1.2 m/sec).

7.2.2.2.2 For pumps with rated capacities of 2000 gpm (7570 L/min) or greater, additional submergence is required to prevent the formation of vortices and to provide required net positive suction head (NPSH) in order to prevent excessive cavitation.

7.2.2.2.3 The required submergence shall be obtained from the pump manufacturer.

7.2.2.2.4 The distance between the bottom of the strainer and the bottom of the wet pit shall be at least one-half of the pump bowl diameter but not less than 12 in. (305 mm).

7.2.3 Well Construction.

7.2.3.1 It shall be the responsibility of the groundwater supply contractor to perform the necessary groundwater investigation to establish the reliability of the supply, to develop a well to produce the required supply, and to perform all work and install all equipment in a thorough and workmanlike manner.

Reliability of the water supply is critical. The groundwater investigation required by 7.2.3.1 must be performed prior to well construction and purchase of the pump. A very important consideration in the well analysis is the well's water level with the pump operating at churn (zero flow), because the fire protection components may not be installed to withstand the developed pressures if a large drawdown of the water level occurs at zero flow and 150 percent flow. The well casing may have to be perforated to allow additional flow into the casing and reduce the drawdown. The concept of using wells is much more complex than using an underground tank as a water source.

7.2.3.2 The vertical turbine–type pump is designed to operate in a vertical position with all parts in correct alignment.

7.2.3.3 To support the requirements of 7.2.3.1, the well shall be of ample diameter and sufficiently plumb to receive the pump.

The vertical shaft turbine–type pump should be installed vertically, plumbed for straightness, and checked for clearance between the pump bowls and well casing before the pump is started. The longer the pump column, the more critical the pump's straightness becomes. The depth of the well is no longer limited when the well is used as a fire protection water supply. Therefore, as the depth increases, the requirements for vertical alignment become more critical.

The vertical shaft turbine–type pump should be mounted on a machined, level, grouted baseplate. The mating surfaces between the baseplate and discharge head need to be clean and free of dirt, rust, burrs, or scale.

7.2.4 Unconsolidated Formations (Sands and Gravels).

7.2.4.1 All casings shall be of steel of such diameter and installed to such depths as the formation could justify and as best meet the conditions.

7.2.4.2 Both inner and outer casings shall have a minimum wall thickness of 0.375 in. (9.5 mm).

7.2.4.3 Inner casing diameter shall be not less than 2 in. (51 mm) larger than the pump bowls.

7.2.4.4 The outer casing shall extend down to approximately the top of the water-bearing formation.

7.2.4.5 The inner casing of lesser diameter and the well screen shall extend as far into the formation as the water-bearing stratum could justify and as best meets the conditions.

The outer casing must protect the pump bowls and pump column from damage under all conditions of service. The purpose of the outer casing is to keep the well walls from collapsing onto the lineshaft and pump.

7.2.4.6 The well screen is a vital part of the construction, and careful attention shall be given to its selection.

7.2.4.7 The well screen shall be the same diameter as the inner casing and of the proper length and percent open area to provide an entrance velocity not exceeding 0.15 ft/sec (46 mm/sec).

7.2.4.8 The screen shall be made of a corrosion- and acid-resistant material, such as stainless steel or Monel.

7.2.4.9 Monel shall be used where it is anticipated that the chloride content of the well water will exceed 1000 parts per million.

7.2.4.10 The screen shall have adequate strength to resist the external forces that will be applied after it is installed and to minimize the likelihood of damage during the installation.

7.2.4.11 The bottom of the well screen shall be sealed properly with a plate of the same material as the screen.

If the screen fails or the screen size is incorrect, sand or larger rock particles could enter the well and cause damage to the pump impellers, bearings, and other pump components. Proper selection of the screen material and size is as critical to the successful operation of the pump as is low entrance velocity of the water.

7.2.4.12 The sides of the outer casing shall be sealed by the introduction of neat cement placed under pressure from the bottom to the top.

The well must be sealed by cement to prevent the ingress of sand or rock particles. This installation is usually a one-time event. Extreme care should be used in this operation and should be accomplished by experienced personnel.

7.2.4.13 Cement shall be allowed to set for a minimum of 48 hours before drilling operations are continued.

7.2.4.14 The immediate area surrounding the well screen not less than 6 in. (152 mm) shall be filled with clean and well-rounded gravel.

7.2.4.15 This gravel shall be of such size and quality as will create a gravel filter to ensure sand-free production and a low velocity of water leaving the formation and entering the well.

7.2.4.16 Tubular Wells.

7.2.4.16.1 Wells for fire pumps not exceeding 450 gpm (1703 L/min) developed in unconsolidated formations without an artificial gravel pack, such as tubular wells, shall be acceptable sources of water supply for fire pumps not exceeding 450 gpm (1703 L/min).

7.2.4.16.2 Tubular wells shall comply with all the requirements of 7.2.3 and 7.2.4, except compliance with 7.2.4.11 through 7.2.4.15 shall not be required.

7.2.5* Consolidated Formations. Where the drilling penetrates unconsolidated formations above the rock, surface casing shall be installed, seated in solid rock, and cemented in place.

The purpose of a surface casing is to prevent sand and rock particles from unconsolidated areas above rock formations from entering the well, in addition to preventing the well walls from collapsing.

A.7.2.5 Where wells take their supply from consolidated formations such as rock, the specifications for the well should be decided upon by the authority having jurisdiction after consultation with a recognized groundwater consultant in the area.

It is critical that the water supply obtained from a well be of sufficient capacity and that it be reliable. Consideration should also be given to seasonal and yearly fluctuations in the water supply.

When installing vertical shaft turbine–type pumps in wells, consideration must be given to the formation of the well before installation. Installing a unit in a crooked well may bind and distort the pump column or pump–motor assembly, leading to potential malfunction.

Well straightness should be within 1 in. (25.4 mm) per 100 ft (30.5 m) and without double bend. The well should be "gauged" prior to installation by lowering a dummy assembly on a cable into the well. This assembly should be slightly longer and larger in diameter than the actual pump or pump–motor assembly. Gauging is also important when a stepped well casing is used—that is, when the lower part of the well casing has a smaller inside diameter than the upper part.

Wells that have not been properly constructed or developed or that produce sand can be detrimental to the pump. If a well is suspected of producing an excessive amount of sand, a unit other than the fire pump must be used to clear the well. (See also 7.2.6.) If corrosives are present, pump material selection must be made to accommodate this condition. Special materials such as nickel-aluminum-bronze alloy, stainless steel, or other special alloys can be specified.

7.2.6 Developing a Well.

7.2.6.1 Developing a new well and cleaning it of sand or rock particles (not to exceed 5 ppm) shall be the responsibility of the groundwater supply contractor.

7.2.6.2 Such development shall be performed with a test pump and not a fire pump.

7.2.6.3 Freedom from sand shall be determined when the test pump is operated at 150 percent of rated capacity of the fire pump for which the well is being prepared.

7.2.7* Test and Inspection of Well.

A.7.2.7 Before the permanent pump is ordered, the water from the well should be analyzed for corrosiveness, including such items as pH, salts such as chlorides, and harmful gases such as carbon dioxide (CO_2) or hydrogen sulfide (H_2S). If the water is corrosive, the pumps should be constructed of a suitable corrosion-resistant material or covered with special protective coatings in accordance with the manufacturers' recommendations.

7.2.7.1 A test to determine the water production of the well shall be made.

7.2.7.2 An acceptable water measuring device such as an orifice, a venturi meter, or a calibrated pitot tube shall be used.

7.2.7.3 The test shall be witnessed by a representative of the customer, contractor, and authority having jurisdiction, as required.

7.2.7.4 The test shall be continuous for a period of at least 8 hours at 150 percent of the rated capacity of the fire pump with 15-minute-interval readings over the period of the test.

7.2.7.5 The test shall be evaluated with consideration given to the effect of other wells in the vicinity and any possible seasonal variation in the water table at the well site.

7.2.7.6 Test data shall describe the static water level and the pumping water level at 100 percent and 150 percent, respectively, of the rated capacity of the fire pump for which the well is being prepared.

7.2.7.7 All existing wells within a 1000 ft (305 m) radius of the fire well shall be monitored throughout the test period.

7.3 Pump.

7.3.1* Vertical Turbine Pump Head Component.

A.7.3.1 See Figure A.7.3.1.

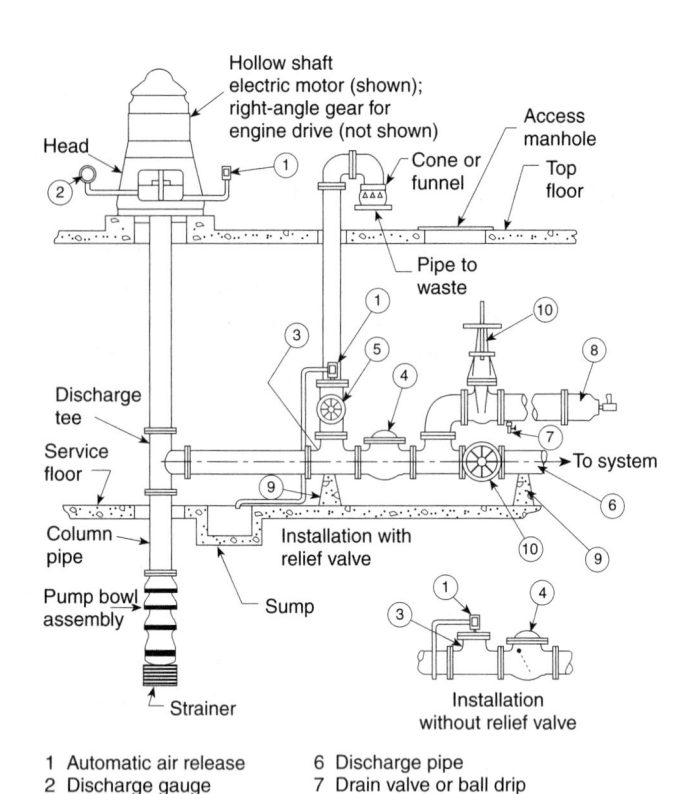

FIGURE A.7.3.1 Belowground Discharge Arrangement.

7.3.1.1 The pump head shall be either the aboveground or belowground discharge type.

7.3.1.2 The pump head shall be designed to support the driver, pump, column assembly, bowl assembly, maximum down thrust, and the oil tube tension nut or packing container.

7.3.2 Column.

7.3.2.1* The pump column shall be furnished in sections not exceeding a nominal length of 10 ft (3 m), shall be not less than the weight specified in Table 7.3.2.1(a) and Table 7.3.2.1(b), and shall be connected by threaded-sleeve couplings or flanges.

A.7.3.2.1 In countries that utilize the metric system, there do not appear to be standardized flow ratings for pump capacities; therefore, a soft metric conversion is utilized.

7.3.2.2 The ends of each section of threaded pipe shall be faced parallel and machined with threads to permit the ends to butt so as to form accurate alignment of the pump column.

7.3.2.3 All column flange faces shall be parallel and machined for rabbet fit to permit accurate alignment.

7.3.2.4 Where the static water level exceeds 50 ft (15.3 m) below ground, oil-lubricated-type pumps shall be used. *(See Figure A.7.1.1.)*

7.3.2.5 Where the pump is of the enclosed line shaft oil-lubricated type, the shaft-enclosing tube shall be furnished in interchangeable sections not over 10 ft (3 m) in length of extra-strong pipe.

TABLE 7.3.2.1(a) Pump Column Pipe Weights (U.S. Customary)

Nominal Size (in.)	Outside Diameter (O.D.) (in.)	Weight per Unit Length (Plain Ends) (lb/ft)
6	6.625	18.97
7	7.625	22.26
8	8.625	24.70
9	9.625	28.33
10	10.75	31.20
12	12.75	43.77
14	14.00	53.57

TABLE 7.3.2.1(b) Pump Column Pipe Weights (Metric)

Nominal Size (mm)	Outside Diameter (O.D.) (mm)	Weight per Unit Length (Plain Ends) (kg/m)
150	161	28.230
200	212	36.758
250	264	46.431
300	315	65.137
350	360	81.209

7.3.2.6 An automatic sight feed oiler shall be provided on a suitable mounting bracket with connection to the shaft tube for oil-lubricated pumps. *(See Figure A.7.1.1.)*

7.3.2.7 The pump line shafting shall be sized so critical speed shall be 25 percent above and below the operating speed of the pump.

When a vertical shaft turbine–type pump ramps up to speed, it generally passes through one or more critical speeds. Critical speeds correspond to the speeds of the rotating pump shaft. At critical speeds the pump shaft reaches its natural frequency of vibration and becomes unstable. A pump passing through a critical speed has no detrimental effect on the pump because it passes through this point in 1 second or less. However, running the pump at a critical speed is not recommended for any pump; most pumps operate between two critical speeds. Experience has shown that a 25 percent margin is a safe range for a pump to operate. If a higher percentage was used, the results would be the gross oversizing of the pump shaft and much closer bearing spacing than is required in present pump designs.

7.3.2.8 Operating speed shall include all speeds from shutoff to the 150 percent point of the pump, which vary on engine drives.

7.3.2.9 Operating speed for variable speed pressure limiting control drive systems shall include all speeds from rated to minimum operating speed.

7.3.3 Bowl Assembly.

7.3.3.1 The pump bowl shall be of close-grained cast iron, bronze, or other suitable material in accordance with the chemical analysis of the water and experience in the area.

The proper material must be specified when the pump is ordered. Local conditions determine what material is most suitable for the application.

7.3.3.2 Impellers shall be of the enclosed type and shall be of bronze or other suitable material in accordance with the chemical analysis of the water and experience in the area.

Enclosed impeller design is specified because, with the hydraulic design of vertical shaft turbine–type pumps, shaft stretch can impact the operating efficiency of an open impeller. The efficiency of open impeller designs decreases as the depth of the pumping water level increases. Impellers constructed of bronze are well suited for most water supplies.

7.3.4 Suction Strainer.

7.3.4.1 A cast or heavy fabricated, corrosion-resistant metal cone or basket-type strainer shall be attached to the suction manifold of the pump.

7.3.4.2 The suction strainer shall have a free area of at least four times the area of the suction connections, and the openings shall be sized to restrict the passage of a 0.5 in. (12.7 mm) sphere.

The openings in the strainer are sized to prevent clogging of the strainer itself and to allow for proper rotation of the pump impeller. The free area of the strainer serves to reduce the entrance water velocity. The 0.5 in. (12.7 mm) sphere size serves to minimize obstruction of a fire protection system's components such as sprinklers.

7.3.4.3 For installations in a wet pit, this suction strainer shall be required in addition to the intake screen. *(See Figure A.7.2.2.2.)*

The intake screen usually allows passage of objects with sphere sizes larger than 0.5 in. (12.7 mm). Screens serve to protect the pump from objects falling into the wet pit or biological growth that may occur in an open pond or stream. These objects may clog the impeller or suction strainer.

7.3.5 Fittings.

7.3.5.1 The following fittings shall be required for attachment to the pump:

(1) Automatic air release valve as specified in 7.3.5.2
(2) Water level detector as specified in 7.3.5.3
(3) Discharge pressure gauge as specified in 4.10.1
(4) Relief valve and discharge cone where required by 4.18.1
(5) Hose valve header and hose valves as specified in 4.20.3 or metering devices as specified in 4.20.2

The installation must also include piping for interconnection of the fittings per the design.

The installation of a flowmeter is preferred to measure the flow during the acceptance testing required by Chapter 14 and the periodic testing required by NFPA 25, *Standard for the Inspection, Testing, and Maintenance of Water-Based Fire Protection Systems*, although the installation of a test header can be beneficial for use as a fire hydrant.

7.3.5.2 Automatic Air Release.

NFPA 20 requires an air release valve because it is critical that air within the pump column and pump head be vented when the pump is started. The presence of trapped air can damage system components and impair operation of the fire protection system.

It is equally important that when the pump stops, a vacuum condition does not exist within the pump column. For this reason the air release valve must be a type that also allows air to enter the pump column when the pump stops. See Exhibit II.7.1 for an illustration of an air release valve.

7.3.5.2.1 A nominal 1.5 in. (38 mm) pipe size or larger automatic air release valve shall be provided to vent air from the column and the discharge head upon the starting of the pump.

7.3.5.2.2 This valve shall also admit air to the column to dissipate the vacuum upon stopping of the pump.

7.3.5.2.3 This valve shall be located at the highest point in the discharge line between the fire pump and the discharge check valve.

EXHIBIT II.7.1 Air Release Valve.

7.3.5.3* Water Level Detection. Water level detection shall be required for all vertical turbine pumps installed in wells to monitor the suction pressure available at the shutoff, 100 percent flow, and 150 percent flow points, to determine if the pump is operating within its design conditions.

NFPA 20 requires water level detectors because the water level must be determined in order to calculate the net pump discharge pressure. Water level detectors of other than the air line type can be used if they are approved for the intended use. The electronic type is preferred for pumps taking suction from belowgrade tanks and suction cribs.

A.7.3.5.3 Water level detection using the air line method is as follows:

(1) A satisfactory method of determining the water level involves the use of an air line of small pipe or tubing of known vertical length, a pressure or depth gauge, and an ordinary bicycle or automobile pump installed as shown in Figure A.7.3.5.3. The air line pipe should be of known length and extend beyond the lowest anticipated water level in the well, to ensure more reliable gauge readings, and should be properly installed. An air pressure gauge is used to indicate the pressure in the air line. *(See Figure A.7.3.5.3.)*

(2) The air line pipe is lowered into the well, a tee is placed in the line above the ground, and a pressure gauge is screwed into one connection. The other connection is fitted with an ordinary bicycle valve to which a bicycle pump is attached. All joints should be made carefully and should be airtight to obtain correct information. When air is forced into the line by means of the bicycle pump, the gauge pressure increases until all of the water has been expelled. When this point is reached, the gauge reading becomes constant. The maximum maintained air pressure recorded by the gauge is equivalent to that necessary to support a column of water of the same height as that forced out of the air line. The length of this water column is equal to the amount of air line submerged.

(3) Deducting this pressure converted to feet (meters) (pressure in psi × 2.31 = pressure in feet, and pressure in bar × 10.3 = pressure in meters) from the known length of the air line will give the amount of submergence.

Example: The following calculation will serve to clarify Figure A.7.3.5.3.

Assume a length (L) of 50 ft (15.2 m).

The pressure gauge reading before starting the fire pump (p_1) = 10 psi (0.68 bar). Then A = 10 × 2.31 = 23.1 ft (0.68 × 10.3 = 7.0 m). Therefore, the water level in the well before starting the pump would be $B = L - A$ = 50 ft − 23.1 ft = 26.9 ft ($B = L - A$ = 15.2 m − 7 m = 8.2 m).

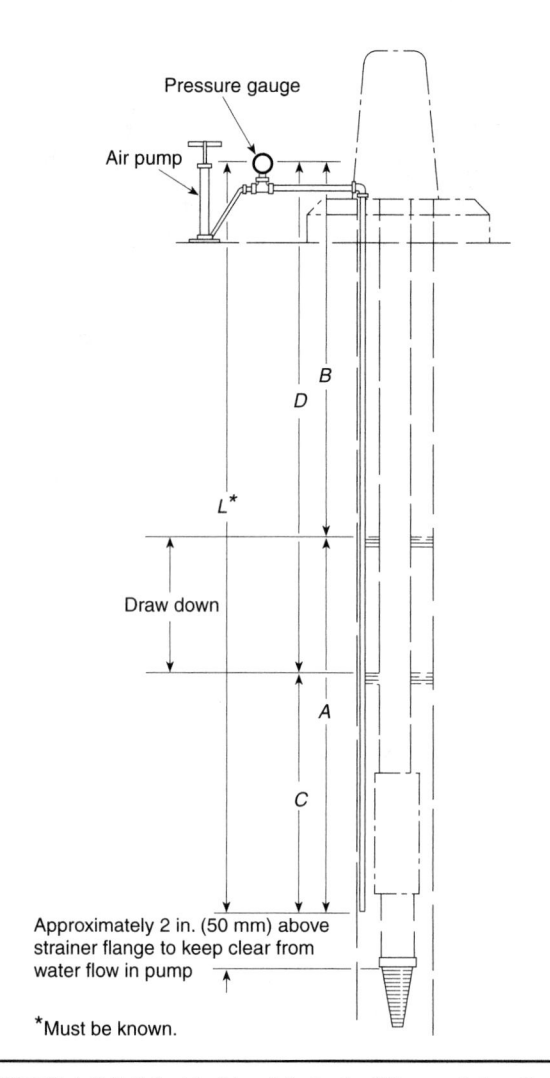

FIGURE A.7.3.5.3 *Air Line Method of Determining Depth of Water Level.*

The pressure gauge reading when pump is running (p_2) = 8 psi (0.55 bar). Then C = 8 × 2.31 = 18.5 ft (0.55 × 10.3 = 5.6 m). Therefore, the water level in the well when the pump is running would be $D = L - C$ = 50 ft − 18.5 ft = 31.5 ft ($D = L - C$ = 15.2 m − 5.6 m = 9.6 m).

The draw down can be determined by any of the following methods:

(1) $D - B$ = 31.5 ft − 26.9 ft = 4.6 ft (9.6 m − 8.2 m = 1.4 m)
(2) $A - C$ = 23.1 ft − 18.5 ft = 4.6 ft (7.0 m − 5.6 m = 1.4 m)
(3) $p_1 - p_2$ = 10 − 8 = 2 psi = 2 × 2.31 = 4.6 ft (0.68 − 0.55 = 0.13 bar = 0.13 × 10.3 = 1.4 m)

7.3.5.3.1 Each well installation shall be equipped with a suitable water level detector.

7.3.5.3.2 If an air line is used, it shall be brass, copper, or series 300 stainless steel.

7.3.5.3.3 Air lines shall be strapped to column pipe at 10 ft (3 m) intervals.

7.4* Installation.

Exhibit II.7.2 shows the process of installing a vertical turbine fire pump. Exhibit II.7.3 shows the vertical turbine fire pump in a concrete reservoir.

A.7.4 Several methods of installing a vertical pump can be followed, depending upon the location of the well and facilities available. Since most of the unit is underground, extreme care should be used in assembly and installation, thoroughly checking the work as it progresses. The following simple method is the most common:

(1) Construct a tripod or portable derrick and use two sets of installing clamps over the open well or pump house. After the derrick is in place, the alignment should be checked carefully with the well or wet pit to avoid any trouble when setting the pump.
(2) Attach the set of clamps to the suction pipe on which the strainer has already been placed and lower the pipe into the well until the clamps rest on a block beside the well casing or on the pump foundation.
(3) Attach the clamps to the pump stage assembly, bring the assembly over the well, and install pump stages to the suction pipe, until each piece has been installed in accordance with the manufacturer's instructions.

EXHIBIT II.7.2 Vertical Turbine Fire Pump Installaton. (Top) Unloading a vertical turbine fire pump. (Bottom) Lowering a vertical turbine fire pump through the fire pump house roof hatch. (Photographs courtesy of Schirmer Engineering Corp.)

EXHIBIT II.7.3 Vertical Turbine Fire Pump in Concrete Reservoir prior to Removing the Concrete Core. (Photograph courtesy of Schirmer Engineering Corp.)

7.4.1 Pump House.

7.4.1.1 The pump house shall be of such design as will offer the least obstruction to the convenient handling and hoisting of vertical pump parts.

The pump house normally includes a roof hatch at least 4 ft × 4 ft (1219 mm × 1219 mm) in size to provide access to the pump room so that the lineshaft can be removed for maintenance. When establishing clearances, *NFPA 70®*, *National Electrical Code®*, requirements as well as normal working distances need to be taken into consideration. External access needs to be large enough to remove a pump driver or other component. Heating, cooling, and combustion air units (for diesel engines) need to be included and sized for the location of the unit. Floor drainage and spill protection must also be included in any design to protect the environment.

7.4.1.2 The requirements of Sections 4.12 and 11.3 shall also apply.

7.4.2 Outdoor Setting.

7.4.2.1 If in special cases the authority having jurisdiction does not require a pump room and the unit is installed outdoors, the driver shall be screened or enclosed and adequately protected against tampering.

The design of the pump and related equipment must still comply with the requirements of 4.12.1.

7.4.2.2 The screen or enclosure required in 7.4.2.1 shall be easily removable and shall have provision for ample ventilation.

7.4.3 Foundation.

In situations where sound pressure levels are too high because of the occupancy, additional measures, such as using isolation pads and other means, must be utilized to separate the fire pump and driver from the building.

In earthquake-prone areas, the design of the foundation is required to be by a licensed professional engineer certified in design of earthquake-resistant structures. The provisions for earthquake design are found in the applicable building code and NFPA 13.

7.4.3.1 Certified dimension prints shall be obtained from the manufacturer.

7.4.3.2 The foundation for vertical pumps shall be substantially built to carry the entire weight of the pump and driver plus the weight of the water contained in it.

7.4.3.3 Foundation bolts shall be provided to firmly anchor the pump to the foundation.

7.4.3.4 The foundation shall be of sufficient area and strength that the load per square inch (square millimeter) on concrete does not exceed design standards.

7.4.3.5 The top of the foundation shall be carefully leveled to permit the pump to hang freely over a well pit on a short-coupled pump.

7.4.3.6 On a well pump, the pump head shall be positioned plumb over the well, which is not necessarily level.

7.4.3.7 Sump or Pit.

7.4.3.7.1 Where the pump is mounted over a sump or pit, I-beams shall be permitted to be used.

7.4.3.7.2 Where a right-angle gear is used, the driver shall be installed parallel to the beams.

7.5 Driver.

7.5.1 Method of Drive.

7.5.1.1 The driver provided shall be so constructed that the total thrust of the pump, which includes the weight of the shaft, impellers, and hydraulic thrust, can be carried on a thrust bearing of ample capacity so that it will have an average life rating of 5 years continuous operation.

7.5.1.2 All drivers shall be so constructed that axial adjustment of impellers can be made to permit proper installation and operation of the equipment.

7.5.1.3 Vertical shaft turbine pumps shall be driven by a vertical hollow shaft electric motor or vertical hollow shaft right-angle gear drive with diesel engine or steam turbine except as permitted in 7.5.1.4.

7.5.1.4 The requirements of 7.5.1.3 shall not apply to diesel engines and steam turbines designed and listed for vertical installation with vertical shaft turbine–type pumps, which shall be permitted to employ solid shafts and shall not require a right-angle gear drive but shall require a nonreverse ratchet.

7.5.1.5 Motors shall be of the vertical hollow-shaft type and comply with 9.5.1.9.

7.5.1.6 Mass Elastic System.

7.5.1.6.1 For drive systems that include a right angle gear drive, the pump manufacturer shall provide a complete mass elastic system torsional analysis to ensure there are no damaging stresses or critical speeds within 25 percent above and below the operating speed of the pump and drive.

7.5.1.6.2 The torsional analysis specified in 7.5.1.6.1 shall include the mass elastic characteristics for a wetted pump with the specific impeller trim, coupling, right-angle gear, flexible connecting shaft, and engine, plus the excitation characteristics of the engine.

7.5.1.6.3 For variable speed vertical hollow shaft electric motors, the pump manufacturer shall provide a complete mass elastic system torsional analysis to ensure there are no damaging stresses or critical speeds within 25 percent above and below the operating speed of the pump and drive.

7.5.1.7 Gear Drives.

7.5.1.7.1 Gear drives and flexible connecting shafts shall be acceptable to the authority having jurisdiction.

7.5.1.7.2 Gear drives shall be of the vertical hollow-shaft type, permitting adjustment of the impellers for proper installation and operation of the equipment.

7.5.1.7.3 The gear drive shall be equipped with a nonreverse ratchet.

7.5.1.7.4 All gear drives shall be listed and rated by the manufacturer at a load equal to the maximum horsepower and thrust of the pump for which the gear drive is intended.

7.5.1.7.5 Water-cooled gear drives shall be equipped with a visual means to determine whether water circulation is occurring.

7.5.1.8 Flexible Connecting Shafts.

7.5.1.8.1 The flexible connecting shaft shall be listed for this service.

7.5.1.8.2 The operating angle for the flexible connecting shaft shall not exceed the limits specified by the manufacturer for the speed and horsepower transmitted under any static or operating conditions.

The operating angle is usually set for a minimum offset angle of 1 degree up to a maximum of 3 degrees. This angle is to prevent failure of the universal joints during operation. In order to conserve space, the flexible connecting shaft should be as short as possible.

7.5.2 Controls. The controllers for the motor, diesel engine, or steam turbine shall comply with specifications for either electric-drive controllers in Chapter 10 or engine drive controllers in Chapter 12.

FAQ ▶
When should the use of a variable speed controller be considered?

Coordination of the variable speed controller manufacturer, engineers, electric power supplier, owner, installing contractor, and AHJ is essential in the proper selection of the controller and the required features. A full-service controller is preferred for electric motor drives. Use of a variable speed controller should be considered when high discharge pressures are required and electric costs are excessive.

7.5.3 Variable Speed Vertical Turbine Pumps.

7.5.3.1 The pump supplier shall inform the controller manufacturer of any and all critical resonant speeds within the operating speed range of the pump, which is from zero up to full speed.

7.5.3.2 When water-lubricated pumps with line shaft bearings are installed, the pump manufacturer shall inform the controller manufacturer of the maximum allowed time for water to reach the top bearing under the condition of the lowest anticipated water level of the well or reservoir.

7.6 Operation and Maintenance.

7.6.1 Operation.

7.6.1.1* Before the unit is started for the first time after installation, all field-installed electrical connections and discharge piping from the pump shall be checked.

The following method may be used to mount and align vertical hollow shaft drivers:

1. Remove the coupling from the top of the hollow shaft and mount the driver on top of the discharge head/driver stand.
2. For designs requiring the pump head to be installed prior to mounting the driver, lower the hollow shaft driver with care over the head shaft to be sure the latter is not damaged.
3. Check the driver for correct rotation, as given in the manufacturer's installation instructions.
4. Install the head shaft, if not already done, and check it for centering in the hollow shaft. If the head shaft is off-center, check for runout in head shaft misalignment from the discharge head to the driver or check to see whether the suspended pump is out of plumb. Shims can be placed under the discharge head to center the head shaft, but shims should not be placed between the motor and the discharge head.
5. Install the driver coupling and check the nonreverse ratchet for operability.
6. Install the coupling gib key and the adjusting nut, and raise the shaft assembly with the impeller(s) to the correct running position in accordance with the manufacturer's instructions.
7. Secure the adjusting nut to the clutch, and double-check the driver hold-down bolts for tightness.
8. Most hollow shaft drivers have register fits. Therefore, further recentering of these drivers normally is not required, nor are dowels recommended.
9. Always consult the manufacturer's manual for proper installation and operation.

A.7.6.1.1 The setting of the impellers should be undertaken only by a representative of the pump manufacturer. Improper setting will cause excessive friction loss due to the rubbing of impellers on pump seals, which results in an increase in power demand. If the impellers are adjusted too high, there will be a loss in capacity, and full capacity is vital for fire pump service. The top shaft nut should be locked or pinned after proper setting.

7.6.1.2 With the top drive coupling removed, the drive shaft shall be centered in the top drive coupling for proper alignment and the motor shall be operated momentarily to ensure that it rotates in the proper direction.

7.6.1.3 With the top drive coupling reinstalled, the impellers shall be set for proper clearance according to the manufacturer's instructions.

7.6.1.4* With the precautions of 7.6.1.1 through 7.6.1.3 taken, the pump shall be started and allowed to run.

A.7.6.1.4 Pumping units are checked at the factory for smoothness of performance and should operate satisfactorily on the job. If excessive vibration is present, the following conditions could be causing the trouble:

(1) Bent pump or column shaft
(2) Impellers not properly set within the pump bowls
(3) Pump not hanging freely in the well
(4) Strain transmitted through the discharge piping

Excessive motor temperature is generally caused either by a maintained low voltage of the electric service or by improper setting of impellers within the pump bowls.

7.6.1.5 The operation shall be observed for vibration while running, with vibration limits according to the *Hydraulic Institute Standards for Centrifugal, Rotary and Reciprocating Pumps.*

The responsible installing contractor should generate a written report of the testing and running for inclusion in the acceptance test report.

7.6.1.6 The driver shall be observed for proper operation.

7.6.2 Maintenance.

7.6.2.1 The manufacturer's instructions shall be carefully followed in making repairs and dismantling and reassembling pumps.

7.6.2.2 When spare or replacement parts are ordered, the pump serial number stamped on the nameplate fastened to the pump head shall be included in order to make sure the proper parts are provided.

The pump manufacturer must know the serial number of the pump in order to provide the proper replacement parts. Consult the operating manual for information regarding ordering parts.

7.6.2.3 Ample head room and access for removal of the pump shall be maintained.

Accessibility for pump repairs and maintenance should be considered when the pump house or pump room is designed. This information is available from all major pump manufacturers to assist the designer with proper design.

REFERENCES CITED IN COMMENTARY

National Fire Protection Association, 1 Batterymarch Park, Quincy, MA 02169-7471.

NFPA 13, *Standard for the Installation of Sprinkler Systems,* 2010 edition.
NFPA 14, *Standard for the Installation of Standpipe and Hose Systems,* 2007 edition.
NFPA 25, *Standard for the Inspection, Testing, and Maintenance of Water-Based Fire Protection Systems,* 2008 edition.
NFPA 70®, National Electrical Code®, 2008 edition.
NFPA 291, *Recommended Practice for Fire Flow Testing and Marking of Hydrants,* 2010 edition.

Hydraulic Institute, 6 Campus Drive, First Floor North, Parsippany, NJ 07054-4406.

Hydraulic Institute Standards for Centrifugal, Rotary and Reciprocating Pumps, 14th edition, 1983.

Positive Displacement Pumps

CHAPTER 8

The requirements in Chapter 8 for positive displacement fire pumps are relatively new for NFPA 20, first appearing in the 1999 edition. Positive displacement pumps, sometimes referred to as PD pumps, are more apt to be found as part of water mist and foam-based fire protection system installations. Note that, in most instances, NFPA 20 uses the term *liquid* rather than *water* due to the type of pumps addressed in Chapter 8. The following relevant NFPA installation standards reference NFPA 20 for the installation of positive displacement pumps:

   NFPA 11, *Standard for Low-, Medium-, and High-Expansion Foam*
   NFPA 16, *Standard for the Installation of Foam-Water Sprinkler and Foam-Water Spray Systems*
   NFPA 750, *Standard on Water Mist Fire Protection Systems*

8.1* General.

A.8.1 All the requirements in Chapter 4 might not apply to positive displacement pumps.

In cases where requirements in Chapter 8 conflict with requirements in Chapter 4, the requirements in Chapter 8 take precedence for positive displacement fire pumps.

8.1.1 Types. Positive displacement pumps shall be as defined in 3.3.37.13.

A positive displacement pump is defined in 3.3.37.13 as "a pump that is characterized by a method of producing flow by capturing a specific volume of fluid per pump revolution and reducing the fluid void by a mechanical means to displace the pumping fluid." Exhibit II.8.1 illustrates one style of positive displacement pump and is not meant to illustrate the only acceptable pump design or style.

All fire pumps create pressure by imparting energy into the fluid the pump is pumping. A positive displacement pump differs from (traditional) centrifugal pumps because a positive displacement pump imparts energy into the fluid by "pushing" the fluid, whereas a centrifugal pump imparts energy into the fluid through spinning action. In a common example, displacement forces keep a heavy boat from sinking through the pressure created by the displacement of water by the bottom of the boat.

Two common displacement techniques are used with which a positive displacement pump imparts energy into the liquid. The first technique is through a two-way piston stroke in a chamber (see Exhibit II.8.2), which clearly differs from the mechanism used in the common centrifugal pump.

The second technique is through a series of gears or vanes rotating in a chamber (see Exhibit II.8.3). As the gears or vanes rotate inside the pump casing, a void is created by the gears or vanes passing by the suction port. Liquid is captured by the rotating elements, and as

◀ **FAQ**
When the requirements in Chapter 8 conflict with those in Chapter 4, which chapter takes precedence?

◀ **FAQ**
How do positive displacement pumps differ from (traditional) centrifugal pumps?

◀ **FAQ**
Certain positive displacement pumps involve a spinning action to create pressure. Why isn't this type of pump addressed in Chapter 6 on centrifugal pumps?

EXHIBIT II.8.1 Positive Displacement Pump with Electric Motor Driver. (Photograph courtesy of Pentair Water, North Aurora Operations)

EXHIBIT II.8.2 Reciprocating Positive Displacement Pump—Pump Action.

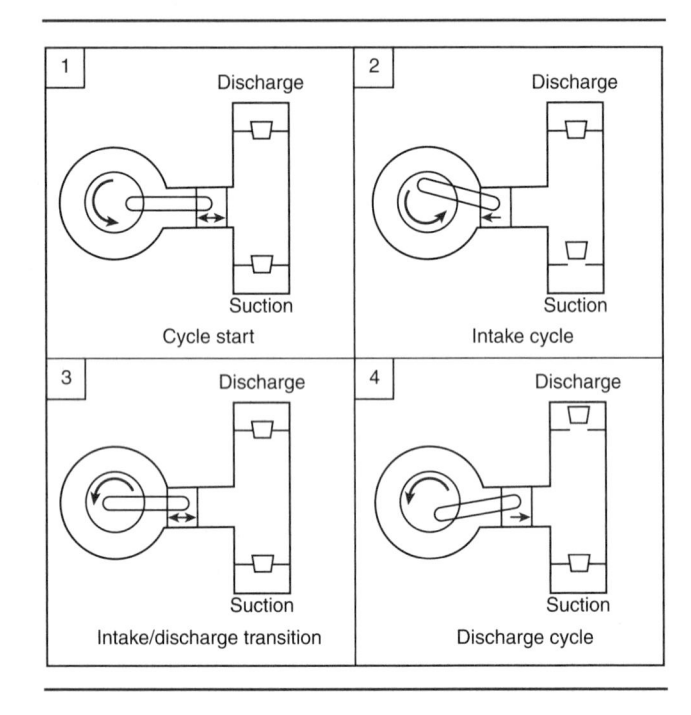

the void volume is decreased on the discharge side of the pump, the liquid is "displaced" and exits the pump discharge port.

8.1.2* Suitability.

A.8.1.2 Special attention to the pump inlet piping size and length should be noted.

8.1.2.1 The positive displacement–type pump shall be listed for the intended application.

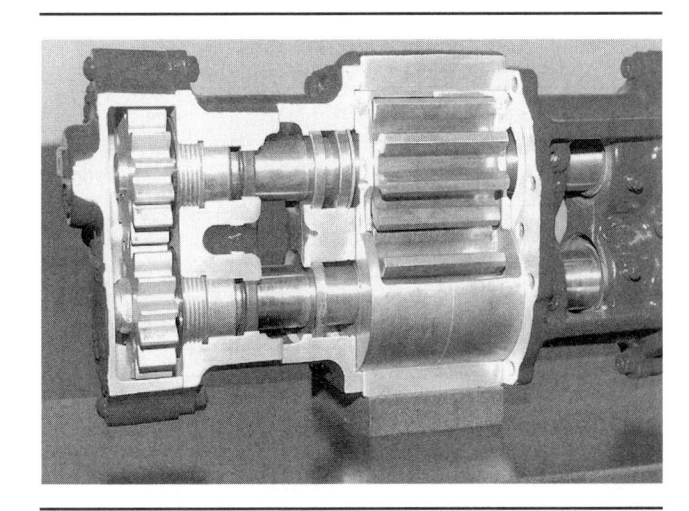

EXHIBIT II.8.3 *One Type of a Rotary Gear Positive Displacement Pump—Interior Cross Section.*

In order for the pump installation to comply with NFPA 20, the positive displacement pump and the system type must be listed for fire protection use as follows:

1. Water mist pumps:
   a. UL Listing Category—ZDPJ
   b. FM Approval Standard—5560
2. Foam concentrate pumps:
   a. UL Listing Category—GKWT
   b. FM Approval Standard—1313

8.1.2.2* The listing shall verify the characteristic performance curves for a given pump model.

A.8.1.2.2 This material describes a sample pump characteristic curve and gives an example of pump selection methods. Characteristic performance curves should be in accordance with HI 3.6, *Rotary Pump Tests*.

Example: An engineer is designing a foam-water fire protection system. It has been determined, after application of appropriate safety factors, that the system needs a foam concentrate pump capable of 45 gpm at the maximum system pressure of 230 psi. Using the performance curve *(see Figure A.8.1.2.2)* for pump model "XYZ-987," this pump is selected for the application. First, find 230 psi on the horizontal axis labeled "Differential pressure," then proceed vertically to the flow curve to 45 gpm. It is noted that this particular pump produces 46 gpm at a standard motor speed designated "rpm-2." This pump is an excellent fit for the application. Next, proceed to the power curve for the same speed of rpm-2 at 230 psi and find that it requires 13.1 hp to drive the pump. An electric motor will be used for this application, so a 15 hp motor at rpm-2 is the first available motor rating above this minimum requirement.

Positive displacement pump performance curves determined through the pump's listing and approval tests must be provided by the manufacturer so that the fire protection system designer can properly select the correct size positive displacement pump. See the example in A.8.1.2.2 for more details.

8.1.3 Application.

8.1.3.1 Positive displacement pumps shall be permitted to pump liquids for fire protection applications.

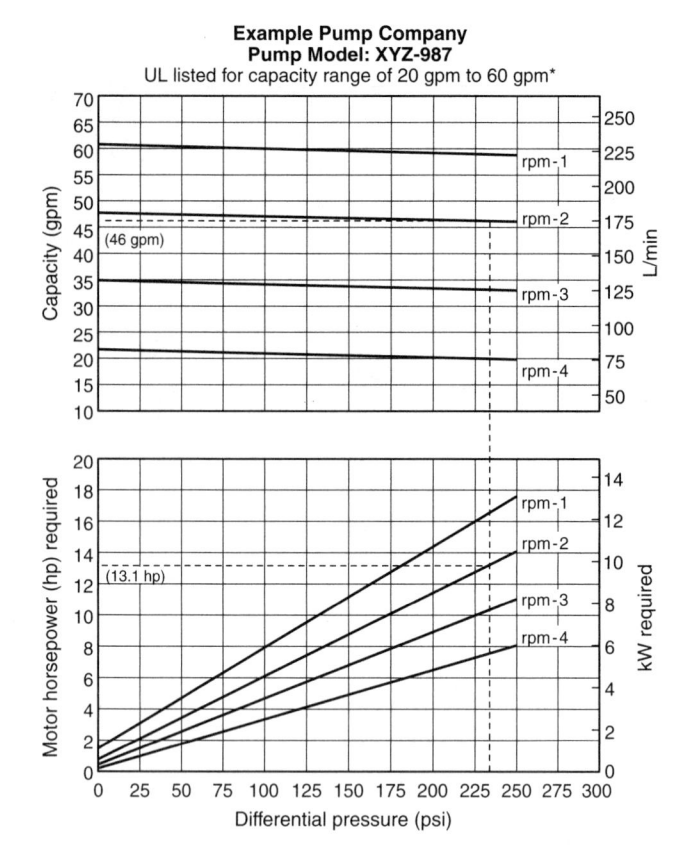

Example Pump Company
Pump Model: XYZ-987
UL listed for capacity range of 20 gpm to 60 gpm*

*Conforms to requirements of Chapter 8 on positive displacement foam concentrate and additive pumps.

FIGURE A.8.1.2.2 *Example of Positive Displacement Pump Selection.*

Paragraph 8.1.3.1 was changed for the 2007 edition to recognize that the requirements of Chapter 8 apply to any stationary positive displacement fire pump installation that must comply with NFPA 20, regardless of the liquid that the pump handles.

8.1.3.2 The selected pump shall be appropriate for the viscosity of the liquid.

The selected pump is required to be verified for the maximum viscosity of the liquid to be pumped. Liquid viscosity maximums are a part of the listings and approvals process and are posted in UL category GKWT and ZDPJ and FM Approvals.

All pumps have maximum viscosities listed as part of the listing process. Since foam concentrates and additives have widely ranging viscosities, it is important that users of the standard refer to the listing itself, which shows the maximum viscosity for a given pump.

8.1.4 Pump Seals.

8.1.4.1 The seal type acceptable for positive displacement pumps shall be either mechanical or lip seal.

8.1.4.2 Packing shall not be used.

Note the difference in packing requirements between the positive displacement pump and the common horizontal split-case centrifugal pump covered in Chapters 4 and 6. Packing seals, found on the common horizontal split-case centrifugal pump, are prohibited for positive displacement pumps.

8.1.5* Pump Materials. Materials used in pump construction shall be selected based on the corrosion potential of the environment, fluids used, and operational conditions. *(See 3.3.9 for corrosion-resistant materials.)*

Foam concentrates and solutions can be significantly more corrosive to common fire protection system materials than domestic water.

A.8.1.5 Positive displacement pumps are tolerance dependent. Corrosion can affect pump performance and function. *(See HI 3.5, Standard for Rotary Pumps for Nomenclature, Design, Application and Operation.)*

8.1.6 Dump Valve.

A dump valve is required to be provided for diesel-driven foam concentrate or additive pumps to allow the positive displacement pump to bleed off excess pressure during the start cycle. It is not necessary for a dump valve to be used for electric-driven foam concentrate or additive pumps.

Electric-driven foam pumps have never required a dump valve. The requirement originated because of the crank cycle of diesel engines trying to start the engine against a pressurized system. Dump valves are only needed for diesel-driven foam or additive pumps.

8.1.6.1 A dump valve shall be provided on all closed head systems to allow the positive displacement pump to bleed off excess pressure and achieve operating speed before subjecting the driver to full load.

The driver for a positive displacement pump is not generally sized to overcome the back pressure created if none of the fire protection system discharge devices have activated while starting. Back pressure can result in a phenomenon similar to the "banana in the tailpipe" prank, whereby a car engine is prevented from being started by blocking the exhaust pipe. The requirement in 8.1.6.1 is meant to address the same problem.

8.1.6.2 The dump valve shall operate only for the duration necessary for the positive displacement pump to achieve operating speed.

8.1.6.3 Dump Valve Control.

8.1.6.3.1 Automatic Operation. When an electrically operated dump valve is used, it shall be controlled by the positive displacement pump controller.

Requiring that the dump valve control be part of the fire pump controller ensures that the fire pump operator can monitor all fire pump functions in one location.

◀ **FAQ**
Why does NFPA 20 require the use of a dump valve control?

8.1.6.3.2 Manual Operation. Means shall be provided at the controller to ensure dump valve operation during manual start.

8.1.6.4 Dump valves shall be listed.

8.1.6.5 Dump valve discharge shall be permitted to be piped to the liquid supply tank, pump suction, drain, or liquid supply.

When the discharge is piped to the pump suction, the design must take into consideration that the maximum suction pressures are not exceeded.

8.2 Foam Concentrate and Additive Pumps.

The requirements in Section 8.2 are intended to apply to any positive displacement pump in a fire protection system that is pumping any liquid other than water-only solutions.

8.2.1 Additive Pumps. Additive pumps shall meet the requirements for foam concentrate pumps.

8.2.2* Net Positive Suction Head. Net positive suction head (NPSH) shall exceed the pump manufacturer's required NPSH plus 5 ft (1.52 m) of liquid.

The net positive suction head (NPSH), which is defined in 3.3.23.1, is the total liquid head-absolute (in feet) at the suction nozzle minus the head pressure of the liquid (in feet). To convert pressure in psi to head in feet, multiply the pressure in psi by 2.31 and divide by the specific gravity of the liquid.

A.8.2.2 Specific flow rates should be determined by the applicable NFPA standard. Viscose concentrates and additives have significant pipe friction loss from the supply tank to the pump suction.

8.2.3 Seal Materials. Seal materials shall be compatible with the foam concentrate or additive.

Incompatibility of the seal material with the foam or additive being pumped could lead to degradation of the seal and inadequate performance of the pump.

8.2.4* Dry Run. Foam concentrate pumps shall be capable of dry running for 10 minutes without damage.

The time of 10 minutes specified in 8.2.4 permits the fire pump operator time to shut down the positive displacement pump after the foam/additive tank has been depleted and before damage occurs to the pump. Positive displacement pumps typically rely on the fluid being pumped to provide pump lubrication.

A.8.2.4 This requirement does not apply to water mist pumps.

8.2.5* Minimum Flow Rates. Pumps shall have foam concentrate flow rates to meet the maximum foam flow demand for their intended service.

The required flow rate is determined through other NFPA fire protection system installation standards, such as the following:

NFPA 11, *Standard for Low-, Medium-, and High-Expansion Foam*
NFPA 13, *Standard for the Installation of Sprinkler Systems*
NFPA 14, *Standard for the Installation of Standpipe and Hose Systems*
NFPA 15, *Standard for Water Spray Fixed Systems for Fire Protection*
NFPA 16, *Standard for the Installation of Foam-Water Sprinkler and Foam-Water Spray Systems*

A.8.2.5 Generally, pump capacity is calculated by multiplying the maximum water flow by the percentage of concentration desired. To that product is added a 10 percent "over demand" to ensure that adequate pump capacity is available under all conditions.

The "over demand" is not fixed at 10 percent but is intended to be a guide.

8.2.6* Discharge Pressure. The discharge pressure of the pump shall exceed the maximum water pressure under any operating condition at the point of foam concentrate injection.

A.8.2.6 Generally, concentrate pump discharge pressure is required to be added to the maximum water pressure at the injection point plus 25 psi (2 bar).

The required positive displacement pump discharge pressure is based on the calculated discharge pressure of the foam system, which in turn is determined through other NFPA fire protection system installation standards such as NFPA 11 and NFPA 16. The required discharge pressure of the foam concentrate pump has an added safety factor to ensure that foam concentrate enters the system.

8.3 Water Mist System Pumps.

8.3.1* Positive displacement pumps for water shall have adequate capacities to meet the maximum system demand for their intended service.

A.8.3.1 It is not the intent of this standard to prohibit the use of stationary pumps for water mist systems.

A listed (traditional) centrifugal fire pump meeting the system demand is permitted to be used in a water mist fire protection system. It is more common to see a positive displacement pump(s) supplying a water mist fire protection system due to the relatively higher pressures that positive displacement pumps can produce.

◀ FAQ
Can a (traditional) centrifugal fire pump be used in a water mist system?

8.3.2 NPSH shall exceed the pump manufacturer's required NPSH plus 5 ft (1.52 m) of liquid.

The NPSH is the total liquid head-absolute (in feet) at the suction nozzle minus the head pressure of the liquid (in feet). To convert pressure in psi to head in feet, multiply the pressure in psi by 2.31 and divide by the specific gravity of the liquid.

8.3.3 The inlet pressure to the pump shall not exceed the pump manufacturer's recommended maximum inlet pressure.

8.3.4 When the pump output has the potential to exceed the system flow requirements, a means to relieve the excess flow such as an unloader valve or orifice shall be provided.

In order to meet the pressure requirements, a larger pump that produces excess flow might be needed. A common fire protection goal of reducing water discharge in the protected space often leads to the use of water mist fire protection systems.

8.3.5 Where the pump is equipped with an unloader valve, it shall be in addition to the safety relief valve as outlined in 8.4.2.

8.4 Fittings.

8.4.1 Gauges. A compound suction gauge and a discharge pressure gauge shall be furnished.

A compound suction gauge is a gauge that is capable of reading negative and positive pressures. The discharge pressure gauge is not required to be a compound gauge due to the assumption that the discharge pressures will never be negative. Both gauges are needed to indicate the pressure boost across the pump.

8.4.2* General Information for Relief Valves.

A.8.4.2 Positive displacement pumps are capable of quickly exceeding their maximum design discharge pressure if operated against a closed discharge system. Other forms of protective

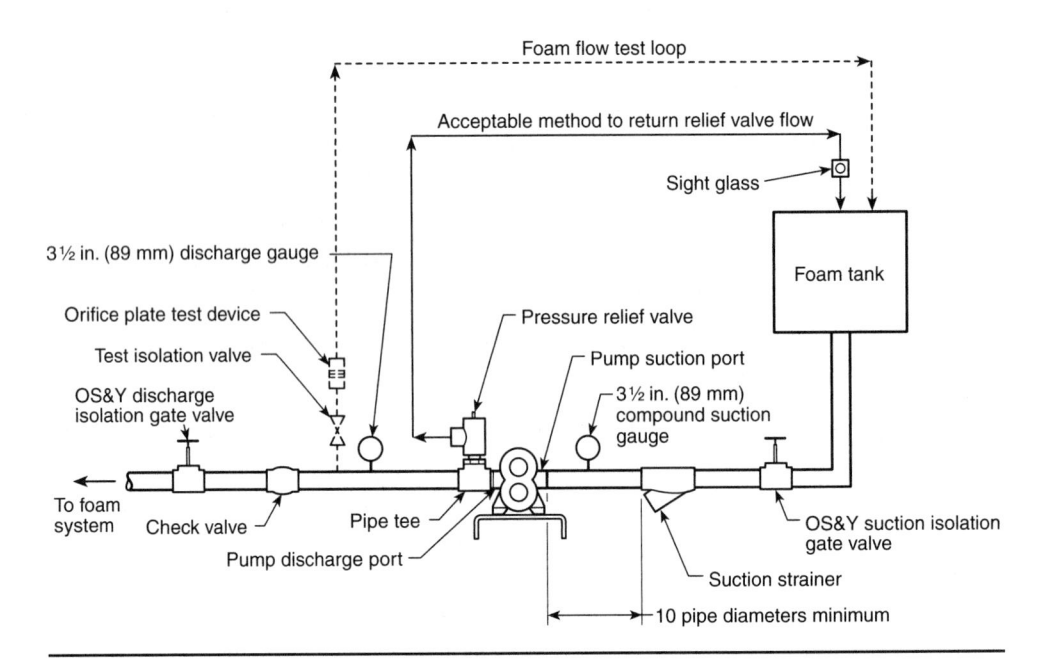

FIGURE A.8.4.2 *Typical Foam Pump Piping and Fittings.*

devices (e.g., automatic shutdowns, rupture discs) are considered a part of the pumping system and are generally beyond the scope of the pump manufacturer's supply. These components should be safely designed into and supplied by the system designer and/or by the user. *(See Figure A.8.4.2 for proposed schematic layout of pump requirements.)*

8.4.2.1 All pumps shall be equipped with a listed safety relief valve capable of relieving 100 percent of the rated pump capacity at a pressure not exceeding 125 percent of the relief valve set pressure.

Without a properly sized pressure relief valve, positive displacement pumps continue to increase pressure when pumping against a fire protection system that has few or no openings until the driver stalls or the fire protection system fails due to exceeding component failure pressures.

8.4.2.2 The pressure relief valve shall be set such that the pressure required to discharge the rated pump capacity is at or below the lowest rated pressure of any component.

8.4.2.3 The relief valve shall be installed on the pump discharge to prevent damage to the fire protection system.

8.4.3* Relief Valves for Foam Concentrate Pumps. For foam concentrate pumps, safety relief valves shall be piped to return the valve discharge to the concentrate supply tank.

The relief valve discharge needs to be piped back to the supply so that the foam/additive supply is not depleted. The quantity of foam concentrate is based on a specific discharge duration and should not be depleted by pressure relief valve discharge. The routine testing of the pump installation with no flow is frequently performed where the relief valve discharges.

A.8.4.3 Only the return to source and external styles should be used when the outlet line can be closed for more than a few minutes. Operation of a pump with an integral relief valve and a closed outlet line will cause overheating of the pump and a foamy discharge of fluid after the outlet line is reopened.

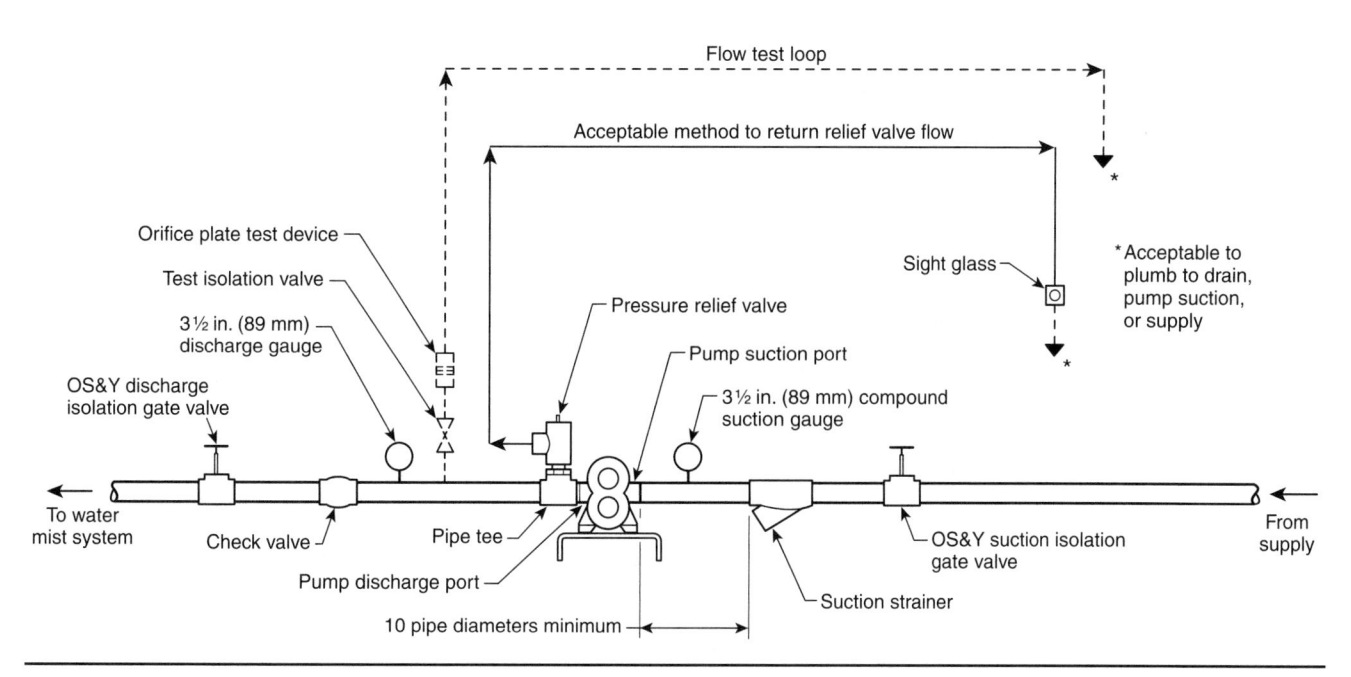

FIGURE A.8.4.4 *Typical Water Mist System Pump Piping and Fittings.*

8.4.4* Relief Valves for Water Mist Pumps.

A.8.4.4 Backpressure on the discharge side of the pressure relief valve should be considered. *(See Figure A.8.4.4 for proposed schematic layout of pump requirement.)*

8.4.4.1 For positive displacement water mist pumps, safety relief valves shall discharge to a drain or to a water supply at atmospheric pressure.

8.4.4.2 A means of preventing overheating shall be provided when the relief valve is plumbed to discharge to the pump suction.

A pressure relief valve installed as a circulation relief valve and discharging to a drain could be used to prevent overheating.

8.4.5* Suction Strainer.

Positive displacement pumps are more susceptible to damage from debris in the pumped fluid and, therefore, a strainer in the suction piping is required for all positive displacement pumps.

A.8.4.5 Strainer recommended mesh size is based on the internal pump tolerances. *(See Figure A.8.4.5 for standard mesh sizes.)*

Mesh	20	40	60	80	100
Opening (in.)	0.034	0.015	0.0092	0.007	0.0055
Opening (μ)	860	380	230	190	140

FIGURE A.8.4.5 *Standard Mesh Sizes.*

8.4.5.1 Pumps shall be equipped with a removable and cleanable suction strainer installed at least 10 pipe diameters from the pump suction inlet.

8.4.5.2 Suction strainer pressure drop shall be calculated to ensure that sufficient NPSH is available to the pump.

See 3.3.23.1 for the definition of NPSH and the commentary following 8.2.2 for more detailed information on the term.

8.4.5.3 The net open area of the strainer shall be at least four times the area of the suction piping.

8.4.5.4 Strainer mesh size shall be in accordance with the pump manufacturer's recommendation.

8.4.6 Water Supply Protection. Design of the system shall include protection of potable water supplies and prevention of cross connection or contamination.

The system designer must plan for backflow prevention devices in the design of the system. Many water purveyors require protection of water supplies when chemicals such as foam/additives are used in a fire protection system. NFPA 20 permits the use of a backflow prevention device or a break tank (see Chapter 4) when water supply protection is required by local authorities.

8.5 Pump Drivers.

8.5.1* The driver shall be sized for and have enough power to operate the pump and drive train at all design points.

A.8.5.1 Positive displacement pumps are typically driven by electric motors, internal combustion engines, or water motors.

FAQ ▶
Does the motor driving a positive displacement pump have to be installed in accordance with Chapter 9?

A positive displacement pump is considered a type of fire pump. Therefore, the intent is that all of the NFPA 20 requirements apply to a fire pump installation involving a positive displacement pump (see Exhibit II.8.4), unless specific NFPA 20 provisions dictate otherwise. The driver must comply with Chapter 9, Chapter 11, or Chapter 13.

8.5.2 Reduction Gears.

8.5.2.1 If a reduction gear is provided between the driver and the pump, it shall be listed for the intended use. Reduction gears shall meet the requirements of AGMA 390.03, *Handbook for Helical and Master Gears.*

If gears are designed improperly, or if gears are selected on speed change alone, the reduction gear can become a failure point, resulting in loss of the fire pump. The Technical Committee on Fire Pumps has received reports that gear failures have occurred in fire pump installations.

8.5.2.2 Gears shall be AGMA Class 7 or better, and pinions shall be AGMA Class 8 or better.

8.5.2.3 Bearings shall be in accordance with AGMA standards and applied for an L10 life of 15,000 hours.

8.5.2.4 For drive systems that include a gear case, the pump manufacturer shall provide a complete mass elastic system torsional analysis to ensure there are no damaging stresses or critical speeds within 25 percent above and below the operating speed of the pump(s) and driver.

EXHIBIT II.8.4 Pre-Manufactured Fire Pump Unit with Positive Displacement Pump and Diesel Engine Driver. (Photograph courtesy of Pentair Water, North Aurora Operations)

8.5.2.4.1 For variable speed drives, the analysis of 8.5.2.4 shall include all speeds down to 25 percent below the lowest operating speed obtainable with the variable speed drive.

8.5.3 Common Drivers.

8.5.3.1 A single driver shall be permitted to drive more than one positive displacement pump.

Subsection 8.5.3 overrides the requirements in 4.7.3 and permits one motor/engine to drive multiple pumps. The flow demands of a water mist fire protection system, combined with low operating flow range of a positive displacement pump, commonly require that more than one pump be driven off the same motor/engine driver. Manufacturers typically supply this arrangement in a pre-assembled unit.

8.5.3.2 Redundant pump systems shall not be permitted to share a common driver.

The intention of NFPA 20 is that the requirement in 8.5.3.2 does not conflict with the requirement in 8.5.3.1. NFPA 750 requires that a redundant pump installation be provided for a water mist fire protection system in case a failure occurs in the primary pump installation. For example, if two pumps are needed to meet the system demand, then four pumps are needed to comply with NFPA 750. If electric motors are installed as drivers, at least two motors are required—one motor for the primary set of pumps as allowed by 8.5.3.1 and another motor for the redundant (emergency) set of pumps as allowed by 8.5.3.1. The requirement in 8.5.3.2 prohibits a single motor to drive all four pumps. See Exhibit II.8.5 for an example of acceptable common driver arrangements compared to unacceptable arrangements.

◀ FAQ
Does the requirement specified in 8.5.3.2 conflict with the requirement in 8.5.3.1?

EXHIBIT II.8.5 Acceptable and Unacceptable Common Driver Arrangements.

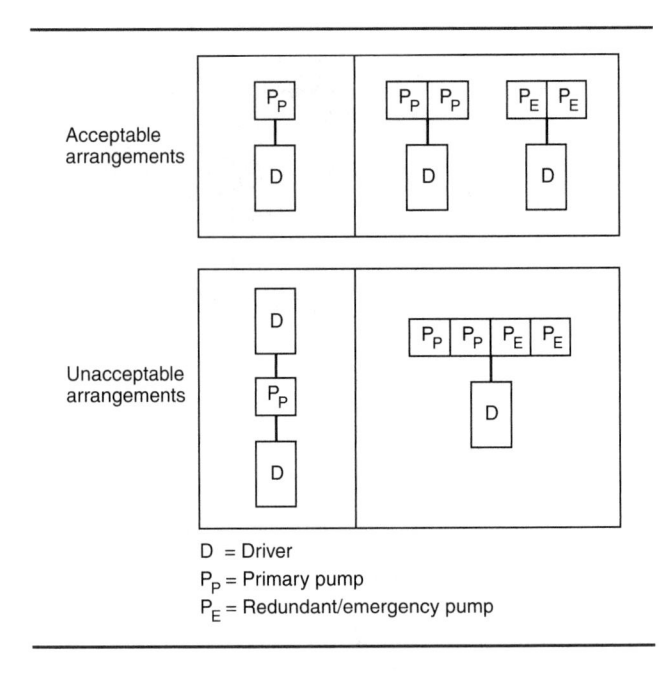

D = Driver
P_P = Primary pump
P_E = Redundant/emergency pump

8.6* Controllers.

See Chapters 10 and 12 for requirements for controllers.

The requirements in NFPA 20 for controllers of positive displacement pumps are the same as those for (traditional) centrifugal pumps.

A.8.6 These controllers can incorporate means to permit automatic unloading or pressure relief when starting the pump driver.

8.7 Foundation and Setting.

The requirements in NFPA 20 for foundations and setting of positive displacement pumps are similar to those for (traditional) centrifugal pumps.

8.7.1 The pump and driver shall be mounted on a common grouted base plate.

8.7.2 The base plate shall be securely attached to a solid foundation in such a way that proper pump and driver shaft alignment will be maintained.

8.7.3 The foundation shall provide a solid support for the base plate.

8.8 Driver Connection and Alignment.

8.8.1 The pump and driver shall be connected by a listed, closed coupled, flexible coupling or timing gear type of belt drive coupling.

FAQ ▶
Does NFPA 20 permit the use of a belt drive to connect a positive displacement pump to a driver?

A positive displacement pump is permitted to be connected to the driver through a belt drive, which is not permitted for any other type of pump complying with NFPA 20.

8.8.2 The coupling shall be selected to ensure that it is capable of transmitting the horsepower of the driver and does not exceed the manufacturer's maximum recommended horsepower and operating speed.

8.8.3 Pumps and drivers shall be aligned once final base plate placement is complete.

8.8.4 Alignment shall be in accordance with the coupling manufacturer's specifications.

8.8.5 The operating angle for the flexible coupling shall not exceed the recommended tolerances.

8.9 Flow Test Devices.

8.9.1 A positive displacement pump installation shall be arranged to allow the test of the pump at its rated conditions as well as the suction supply at the maximum flow available from the pump.

A means of testing the suction supply by actually flowing liquid through the pump from the supply, not just through a test loop, must be available. Due to the viscosity of foam/additive solutions, significant friction loss can occur in the suction supply piping.

8.9.2 Additive pumping systems shall be equipped with a flow meter or orifice plate installed in a test loop back to the additive supply tank.

A test loop is required to discharge back to the foam/additive tank so that the foam/additive tank is not depleted during routine testing of the foam/additive supply.

8.9.3 Water pumping systems shall be equipped with a flowmeter or orifice plate installed in a test loop back to the water supply, tank, inlet side of the water pump, or drain.

REFERENCES CITED IN COMMENTARY

National Fire Protection Association, 1 Batterymarch Park, Quincy, MA 02169-7471.

NFPA 11, *Standard for Low-, Medium-, and High-Expansion Foam,* 2005 edition.
NFPA 13, *Standard for the Installation of Sprinkler Systems,* 2010 edition.
NFPA 14, *Standard for the Installation of Standpipe and Hose Systems,* 2007 edition.
NFPA 15, *Standard for Water Spray Fixed Systems for Fire Protection,* 2007 edition.
NFPA 16, *Standard for the Installation of Foam-Water Sprinkler and Foam-Water Spray Systems,* 2007 edition.
NFPA 750, *Standard on Water Mist Fire Protection Systems,* 2006 edition.

Electric Drive for Pumps

CHAPTER 9

Chapter 9 describes the sources and transmission of electric power for motors driving stationary fire pumps. Also covered are the performance and installation criteria for motors that provide the prime mover for stationary fire pumps. The transmission of power includes all equipment between the source(s) and the fire pump motor, with the exception of the controllers, transfer switches, and accessories, which are covered in Chapter 10. The focus of this chapter is the proper installation of electrical equipment associated with a stationary fire pump.

Chapter 9 also provides information to aid in specifying, designing, and operating this equipment. Article 695 of *NFPA 70*®, *National Electrical Code*® (*NEC*®), also contains the installation requirements for fire pumps (see Supplement 3). Chapter 9 is the source for these requirements, which are then extracted into Article 695. However, it is important to note that not all of the installation requirements for electrical equipment associated with a stationary fire pumps are located in Chapter 9. Article 695 must be reviewed to determine all of the installation requirements applicable to a specific fire pump installation.

◀ FAQ
What is the relationship between Chapter 9 of NFPA 20 and Article 695 of the *NEC*?

9.1 General.

9.1.1 This chapter covers the minimum performance and testing requirements of the sources and transmission of electrical power to motors driving fire pumps.

Although performance and testing are critical, the emphasis of Chapter 9 is the proper installation of the sources and the equipment used to transmit power to the fire pump motor. The objective is continuity of power to drive the fire pump motor in the event of a fire.

9.1.2 This chapter also covers the minimum performance requirements of all intermediate equipment between the source(s) and the pump, including the motor(s) but excepting the electric fire pump controller, transfer switch, and accessories *(see Chapter 10)*.

All equipment necessary to ensure the minimum performance of sources of electric power to the fire pump motor is subject to the requirements of this chapter.

9.1.3 All electrical equipment and installation methods shall comply with *NFPA 70, National Electrical Code,* Article 695, and other applicable articles.

Some of the other applicable *NEC* articles that are important include Article 230, Services, and Article 700, Emergency Systems.

9.1.4* All power supplies shall be located and arranged to protect against damage by fire from within the premises and exposing hazards.

Damage to equipment involved in the transmission of power to the fire pump as a result of exposure to fire could render the fire pump inoperable.

◀ FAQ
What *NEC* articles other than 695 most affect the installation of fire pumps?

A.9.1.4 Where the power supply involves an on-site power production facility, the protection is required for the facility in addition to the wiring and equipment.

The requirement for protection in 9.1.4 also applies to the building housing the on-site power production facility.

9.1.5 All power supplies shall have the capacity to run the fire pump on a continuous basis.

> **FAQ ▶**
> As fire pumps operate infrequently, is any of the fire pump equipment rated for less than continuous duty?

Although fire pumps are not expected to run continuously, there is no reliable method for sizing the power supply based on duty cycle and ensuring adequate pumping capacity. Duty cycle in this context is the proportion of time during which the fire pump is operated. Duty cycle can be expressed as a ratio or as a percentage. The higher the duty cycle, the shorter the useful life, if all other things are equal. If a pump had a life expectancy of a certain number of hours based on a specific duty cycle, that same pump's life expectancy would be about half that number of hours if the duty cycle doubled. Because the number of hours a pump operates is application specific, sizing the power supply to run the fire pump on a continuous basis (100 percent duty cycle) ensures adequate pumping capacity. For example, operation of the fire pump where there is a signal for a specific fire occurrence may not be a one-time event. In some instances fire pumps may operate for days in order to prevent rekindling of a fire after the main extinguishing is completed.

9.1.6 All power supplies shall comply with the voltage drop requirements of Section 9.4.

In some cases, the capacity of a power supply may need to be increased in order to meet the voltage drop requirements of Section 9.4.

9.1.7* Phase converters shall not be used to supply power to a fire pump.

> **FAQ ▶**
> Are phase converters acceptable for use in a fire pump circuit?

Phase converters are normally used to provide three-phase power where none is available, such as in rural locations. However, a phase converter is not considered a reliable source of power for a fire pump motor because of the imbalance in the voltage between the phases when there is no load on the equipment. Additionally, a phase converter would need to be constantly energized; otherwise, the fire pump controller would not be in the standby condition. This is not always possible or practical. When three-phase power is not available, a single-phase fire pump controller and motor would need to be used. Single-phase fire pump controllers are available to 10 hp max at 240 volts.

A.9.1.7 Phase converters that take single-phase power and convert it to three-phase power for the use of fire pump motors are not recommended because of the imbalance in the voltage between the phases when there is no load on the equipment. If the power utility installs phase converters in its own power transmission lines, such phase converters are outside the scope of this standard and need to be evaluated by the authority having jurisdiction to determine the reliability of the electric supply.

9.2 Normal Power.

9.2.1 An electric motor–driven fire pump shall be provided with a normal source of power as a continually available source.

If a normal source of power is not continually available, an alternate source of power must be provided as described in Section 9.3.

9.2.2 The normal source of power required in 9.2.1 and its routing shall be arranged in accordance with one of the following:

(1) Service connection dedicated to the fire pump installation

The requirement in 9.2.2(1) means that the service must supply no loads other than those associated with the fire pump. This requirement is normally met by providing a separate service or a tap ahead of the main service for the building in which the conductors supplying power to the fire pump enter. Frequently, the fire department shuts down power to the building during a fire event to provide for fire-fighter safety. In such situations, power is still needed for the fire pump.

(2) On-site power production facility connection dedicated to the fire pump installation

An on-site standby or emergency generator does not satisfy the requirement in 9.2.2(2). See 3.3.34 and 3.3.35 for the definitions of on-site power production facility and on-site standby generator. An on-site standby generator differs from an on-site power production facility in that it is not constantly producing power and is therefore not considered a private power production facility. In many cases, on-site power production sources are electric generating stations dedicated to a particular facility or to a particular facility campus-style distribution system. Information on fire protection systems for on-site generating stations can be found in NFPA 850, *Recommended Practice for Fire Protection for Electric Generating Plants and High Voltage Direct Current Converter Stations.*

(3) Dedicated feeder connection derived directly from the dedicated service to the fire pump installation

The dedicated service and feeder connection referred to in 9.2.2(3) may not supply loads other than those associated with the fire pump.

◀ **FAQ**
Can a fire pump feeder circuit supply loads other than those associated with the fire pump?

(4) As a feeder connection where all of the following conditions are met:

The feeder connection referred to in 9.2.2(4) may supply loads other than those associated with the fire pump. This practice is acceptable and not considered to reduce the reliability of the power supplied to the fire pump because a backup source of power is required in 9.2.2(4)(b) and all feeder circuits supplying fire pump(s) must be selectively coordinated.

◀ **FAQ**
Can a fire pump feeder circuit supply loads other than those associated with the fire pump if an alternate source of power is provided for the fire pump motor?

(a) The protected facility is part of a multibuilding campus-style arrangement.
(b) A backup source of power is provided from a source independent of the normal source of power.

In order to satisfy the requirement in 9.2.2(4)(b), the independent source must be a separate service, a separate feeder, or an on-site generator. If a separate feeder is provided, it cannot derive its power from the service providing the feeder connection for the normal supply.

◀ **FAQ**
What is meant by the term *independent feeder*?

(c) It is impractical to supply the normal source of power through the arrangement in 9.2.2(1), 9.2.2(2), or 9.2.2(3).
(d) The arrangement is acceptable to the authority having jurisdiction.
(e) The overcurrent protection device(s) in each disconnecting means is selectively coordinated with any other supply side overcurrent protective device(s).

Selective coordination in the context of this chapter ensures that when an overcurrent condition occurs in a fire pump branch circuit, power is not interrupted to any loads served by upstream protective devices.

(5) Dedicated transformer connection directly from the service meeting the requirements of Article 695 of *NFPA 70, National Electrical Code*

A fire pump may be powered by a transformer provided that the transformer is dedicated to the fire pump and does not serve any other loads not associated with the fire pump. The exception to this is stated in *NEC* 695.5(C): "Where a feeder source is provided in accordance with 695.3(B)(2), transformers supplying the fire pump system shall be permitted to supply

◀ **FAQ**
What is meant by a "dedicated transformer connection"?

other loads." Section 695.3(B)(2) applies to multibuilding, campus-style complexes with fire pumps at one or more buildings. In this application, two or more feeder sources are permitted as one power source or as more than one power source where such feeders are connected to or derived from separate utility services. If any of these feeders are supplied by a transformer, that transformer may supply other loads.

9.2.3 For fire pump installations using the arrangement of 9.2.2(1), 9.2.2(2), 9.2.2(3), or 9.2.2(5) for the normal source of power, no more than one disconnecting means and associated overcurrent protection device shall be installed in the power supply to the fire pump controller.

FAQ ▶
Why does NFPA 20 place a limit on the number of disconnecting means allowed upstream of the fire pump controller?

The one upstream disconnect permitted in 9.2.3 affords disconnection of the fire pump feeder circuit for maintenance, for example, and is not intended to be operated under fire conditions. Additional disconnecting means would reduce the reliability of the power supplied to the fire pump, as these disconnects can be remote from the pump room and are not easily located in the event they must be operated. Supervising multiple disconnects can be difficult to manage, especially in a fire event.

9.2.3.1 Where the disconnecting means permitted by 9.2.3 is installed, the disconnecting means shall meet all of the following:

(1) They shall be identified as being suitable for use as service equipment.
(2) They shall be lockable in the closed position.

FAQ ▶
Why does NFPA 20 require the disconnect external to the fire pump controller to be capable of being locked in the closed position?

As generally provided by the manufacturer, most disconnects are suitable for locking in the "off" position, but not the "on" position. In order to provide the capability of locking in both positions, a locking accessory with such capability may need to be field installed on the disconnect device. Locking in the closed position prevents inadvertent disconnection of power to the fire pump motor by untrained and/or unauthorized personnel.

(3)* They shall be located remote from other building disconnecting means.

FAQ ▶
Why does NFPA 20 require the disconnects external to the fire pump controller to be located remote from other building disconnecting means?

The disconnects are required to be located remotely so that they are not confused with disconnects for building loads and inadvertently opened by the fire department staff in the event of a fire.

A.9.2.3.1(3) The disconnecting means should be located such that inadvertent simultaneous operation is not likely.

It is important to avoid opening of multiple disconnects at the same time, resulting in disabling fire protection in more than one location or building.

(4)* They shall be located remote from other fire pump source disconnecting means.

A.9.2.3.1(4) The disconnecting means should be located such that inadvertent simultaneous operation is not likely.

(5) They shall be marked "Fire Pump Disconnecting Means" in letters that are no less than 1 in. (25 mm) in height and that can be seen without opening enclosure doors or covers.

Disconnects need to be readily identifiable for maintenance and fire service personnel.

9.2.3.2 Where the disconnecting means permitted by 9.2.3 is installed, a placard shall be placed adjacent to the fire pump controller stating the location of this disconnecting means and the location of any key needed to unlock the disconnect.

FAQ ▶
Why does NFPA 20 require a placard adjacent to the fire pump controller?

A placard provides for easy identification of the one disconnect in the event the power provided to the controller needs to be disconnected.

9.2.3.3 Where the disconnecting means permitted by 9.2.3 is installed, the disconnect shall be supervised in the closed position by one of the following methods:

(1) Central station, proprietary, or remote station signal device
(2) Local signaling service that will cause the sounding of an audible signal at a constantly attended location
(3) Locking the disconnecting means in the closed position
(4) Sealing of disconnecting means and approved weekly recorded inspections where the disconnecting means are located within fenced enclosures or in buildings under the control of the owner

All of the methods listed in 9.2.3.3(1) through (4) reduce the likelihood that the one disconnecting means will be open when a fire event occurs and power needs to be supplied to the fire pump motor.

◀ FAQ
What is the purpose of the supervision methods for the disconnecting means external to the fire pump controller?

9.2.3.4 Where the overcurrent protection permitted by 9.2.3 is installed, the overcurrent protection device shall be rated to carry indefinitely the sum of the locked rotor current of the fire pump motor(s) and the pressure maintenance pump motor(s) and the full-load current of the associated fire pump accessory equipment.

The requirement in 9.2.3.4 ensures that in the event that an overload occurs in the fire pump branch circuit, the circuit breaker in the fire pump controller is the only protective device to disconnect power to the fire pump motor. This arrangement is necessary as the overload may only be temporary. In this case, the circuit breaker in the fire pump controller may be reset and the fire pump motor restarted if needed. If protective devices upstream of the fire pump controller open during an overload condition, resetting or replacing them in a reasonable time to place the fire pump back in operation to successfully extinguish the fire is virtually impossible.

This requirement in the 2007 edition of NFPA 20 stated that "the overcurrent protection device shall be selected or set to carry indefinitely . . . ." The words "or set" have been removed. The effect of this change is to require a circuit breaker continuous ampere rating to be sized at the "carry indefinitely current." Previously an adjustable trip circuit breaker could have its trip setting adjusted to the "carry indefinitely current." This change in sizing procedure will result in larger amp rated circuit breakers in many applications as a circuit breaker's trip settings are normally higher than its continuous ampere rating.

When sizing the overcurrent protection based on the requirements of 9.2.3.4, compliance with the selective coordination requirement in 9.2.2(4)(e) also needs to be met.

9.3 Alternate Power.

9.3.1 Except for an arrangement described in 9.3.3, at least one alternate source of power shall be provided where the height of the structure is beyond the pumping capacity of the fire department apparatus.

9.3.2* Other Sources. Except for an arrangement described in 9.3.3, at least one alternate source of power shall be provided where the normal source is not reliable.

A reliable power source is not defined by NFPA 20. Therefore, guidance is provided in A.9.3.2 to determine if the normal power source can be considered reliable.

◀ FAQ
Are any guidelines available for determining what constitutes a reliable power source?

A.9.3.2 A reliable power source possesses the following characteristics:

(1) The source power plant has not experienced any shutdowns longer than 4 continuous hours in the year prior to plan submittal. NFPA 25, *Standard for the Inspection, Testing, and Maintenance of Water-Based Fire Protection Systems,* requires special undertakings (i.e., fire watches) when a water-based fire protection system is taken out of service for longer than 4 hours. If the normal source power plant has been intentionally shut down for longer than 4 hours in the past, it is reasonable to require a backup source of power.

(2) No power outages have been experienced in the area of the protected facility caused by failures in the power grid that were not due to natural disasters or electric grid management failure. The standard does not require that the normal source of power is infallible. NFPA 20 does not intend to require a back-up source of power for every installation using an electric motor–driven fire pump. Should the normal source of power fail due to a natural disaster (hurricane) or due to a problem with electric grid management (regional blackout), the fire protection system could be supplied through the fire department connection. However, if the power grid is known to have had problems in the past (i.e., switch failures or animals shorting a substation), it is reasonable to require a backup source of power.

(3) The normal source of power is not supplied by overhead conductors outside the protected facility. Fire departments responding to an incident at the protected facility will not operate aerial apparatus near live overhead power lines, without exception. A backup source of power is required in case this scenario occurs and the normal source of power must be shut off. Additionally, many utility providers will remove power to the protected facility by physically cutting the overhead conductors. If the normal source of power is provided by overhead conductors, which will not be identified, the utility provider could mistakenly cut the overhead conductor supplying the fire pump.

(4) Only the disconnect switches and overcurrent protection devices permitted by 9.2.3 are installed in the normal source of power. Power disconnection and activated overcurrent protection should only occur in the fire pump controller. The provisions of 9.2.2 for the disconnect switch and overcurrent protection essentially require disconnection and overcurrent protection to occur in the fire pump controller. If unanticipated disconnect switches or overcurrent protection devices are installed in the normal source of power that do not meet the requirements of 9.2.2, the normal source of power must be considered not reliable and a back-up source of power is necessary.

Typical methods of routing power from the source to the motor are shown in Figure A.9.3.2. Other configurations are also acceptable. The determination of the reliability of a service is left up to the discretion of the authority having jurisdiction.

FAQ ▶
What is meant by the terms *separate service* and *tap ahead of the main service*?

As shown in Figure A.9.3.2, Arrangement A is generally referred to as a *separate service* and is completely separated from the building service and Arrangement B as a *tap ahead of the main service* for the premises being protected. Arrangement B depicts a common service point and a separate tap for the fire pump. The transformer in Arrangement B may or may not be provided. The transformer is only needed if the service voltage is higher than the utilization voltage.

9.3.3 An alternate source of power is not required where a backup engine-driven or backup steam turbine–driven fire pump is installed in accordance with this standard.

The use of a backup engine-driven or backup steam turbine–driven fire pump increases the reliability of the fire suppression system for the premises being served and, therefore, an alternate source of power is not required.

9.3.4 Where provided, the alternate source of power shall be supplied from one of the following sources:

(1) A generator installed in accordance with Section 9.6
(2) One of the sources identified in 9.2.2(1), 9.2.2(2), 9.2.2(3), or 9.2.2(5) where the power is provided independent of the normal source of power

FAQ ▶
What is meant by "power is provided independent of the normal source of power"?

The phrase "power is provided independent of the normal source of power" in 9.3.4(2) means the alternate sources of power are not derived from the service providing the normal source of power.

9.3.5 Where provided, the alternate supply shall be arranged so that the power to the fire pump is not disrupted when overhead lines are de-energized for fire department operations.

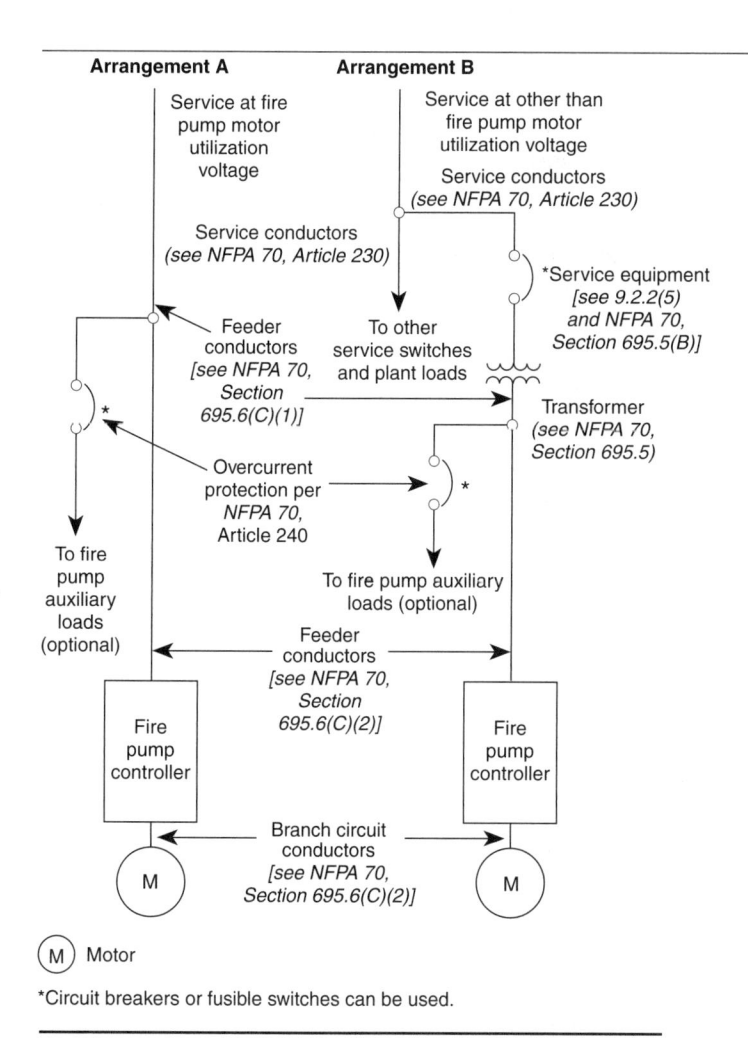

FIGURE A.9.3.2 *Typical Power Supply Arrangements from Source to Motor.*

The alternate power supply must not be transmitted through overhead power lines in close proximity to the building being supplied so that fire department operations do not result in de-energization of these lines.

9.3.6 Two or More Alternate Sources. Where the alternate source consists of two or more sources of power and one of the sources is a dedicated feeder derived from a utility service separate from that used by the normal source, the disconnecting means, overcurrent protective device, and conductors shall not be required to meet the requirements of Section 9.2 and shall be permitted to be installed in accordance with *NFPA 70, National Electrical Code.*

The alternate source may consist of two or more sources of power. When one of those sources is a dedicated feeder derived from a utility service separate from that used by the normal source, the reliability of the power available to the fire pump controller is increased significantly and the special requirements of Section 9.2 are considered unnecessary.

9.4* Voltage Drop.

A.9.4 Normally, conductor sizing is based on appropriate sections of *NFPA 70, National Electrical Code,* Article 430, except larger sizes could be required to meet the requirements of *NFPA 70*, Section 695.7 (NFPA 20, Section 9.4). Transformer sizing is to be in accordance with *NFPA 70*, Section 695.5(a), except larger minimum sizes could be required to meet the requirements of *NFPA 70*, Section 695.7.

9.4.1 Unless the requirements of 9.4.2 or 9.4.3 are met, the voltage at the controller line terminals shall not drop more than 15 percent below normal (controller-rated voltage) under motor-starting conditions.

FAQ ▶ Why is voltage drop critical in the operation of a fire pump?

Voltage drop is critical during motor starting and running conditions, as insufficient voltage could prevent the motor from permeating excessive loading due to obstruction(s) in the pump and achieving full pump speed. Excessive voltage drop could also result in the contacts in the magnetic contactor chattering, since the holding coils of magnetically operated devices are only required to hold the contacts closed at a minimum of 85 percent of rated voltage. Voltage drop upon starting is measured on the controller line terminals and as an option during running at the motor terminals.

9.4.2 The requirements of 9.4.1 shall not apply to emergency-run mechanical starting. *(See 10.5.3.2.)*

FAQ ▶ Why do the voltage drop requirements of Section 9.4 not apply to the emergency-run mechanical starting operation of the fire pump controller?

Many controllers are of the reduced voltage–type start construction. These controllers are often used when the capacity and voltage regulation of the source is not sufficient to start the motor under full voltage conditions. The emergency-run mechanical start operation of these controllers is accomplished by starting the motor at full line voltage (bypassing the reduced voltage start operation of the controller). However, the capacity of the source for a reduced voltage controller is not sufficient to meet the voltage drop requirements of Section 9.4. As such, the voltage drop requirements cannot be imposed on emergency-run mechanical starting operation of the controller.

9.4.3 The requirements of 9.4.1 shall not apply to the bypass mode of a variable speed pressure limiting control *(see 10.10.3),* provided a successful start can be demonstrated on the standby gen-set.

The voltage drop requirements of Section 9.4 do not apply to the bypass mode of a variable speed pressure limiting control for the same reasons they do not apply to the emergency-run mechanical starting operation of the fire pump controller. The requirement that a successful start can be demonstrated on the standby gen-set, if provided, is necessary because the capacity of the standby generator may not be sufficient to allow for starting of the fire pump motor. If this cannot be demonstrated, the bypass mode is of no use in the event of failure of the normal power source.

9.4.4 The voltage at the motor terminals shall not drop more than 5 percent below the voltage rating of the motor when the motor is operating at 115 percent of the full-load current rating of the motor.

FAQ ▶ Is voltage drop critical at any time other than motor starting?

Voltage drop is critical during motor operation at the maximum allowable service factor of 1.15 because insufficient voltage could prevent the motor from driving the pump to deliver maximum waterflow and achieving full pump speed.

9.5 Motors.

9.5.1 General.

9.5.1.1 All motors shall comply with NEMA MG-1, *Motors and Generators,* shall be marked as complying with NEMA Design B standards, and shall be specifically listed for fire pump service. *(See Table 9.5.1.1.)*

TABLE 9.5.1.1 Horsepower and Locked Rotor Current Motor Designation for NEMA Design B Motors

Rated Horsepower	Locked Rotor Current Three-Phase 460 V (A)	Motor Designation (NFPA 70, Locked Rotor Indicating Code Letter) "F" to and Including
5	46	J
7½	64	H
10	81	H
15	116	G
20	145	G
25	183	G
30	217	G
40	290	G
50	362	G
60	435	G
75	543	G
100	725	G
125	908	G
150	1085	G
200	1450	G
250	1825	G
300	2200	G
350	2550	G
400	2900	G
450	3250	G
500	3625	G

See Exhibit II.9.1 for a typical fire pump motor nameplate.

9.5.1.2 The requirements of 9.5.1.1 shall not apply to direct-current, high-voltage (over 600 V), large-horsepower [over 500 hp (373 kW)], single-phase, universal-type, or wound-rotor motors, which shall be permitted to be used where approved.

Paragraph 9.5.1.1 does not apply to the motors specified in 9.5.1.2. Currently no published standards cover these motors for fire pump applications.

◀ **FAQ**
Are any types of fire pump motors not required to be listed?

9.5.1.3 Part-winding motors shall have a 50–50 winding ratio in order to have equal currents in both windings while running at nominal speed.

The requirement in 9.5.1.3 specifies the criteria needed for proper sizing of contactors in a part-winding fire pump controller.

9.5.1.4 Motors used with variable speed controllers shall additionally meet the applicable requirements of NEMA MG-1, *Motors and Generators*, Part 31 and shall be marked for inverter duty.

Standard NEMA Design B motors may not be suitable for use with variable speed controllers, as their performance may be adversely affected by the output voltage waveforms produced by these controllers. Only those motors marked for inverter duty are suitable for use with variable speed controllers.

◀ **FAQ**
Why aren't all fire pump motors suitable for use with variable speed fire pump controllers?

EXHIBIT II.9.1 Fire Pump Motor Nameplate. (Courtesy of Marathon Electric)

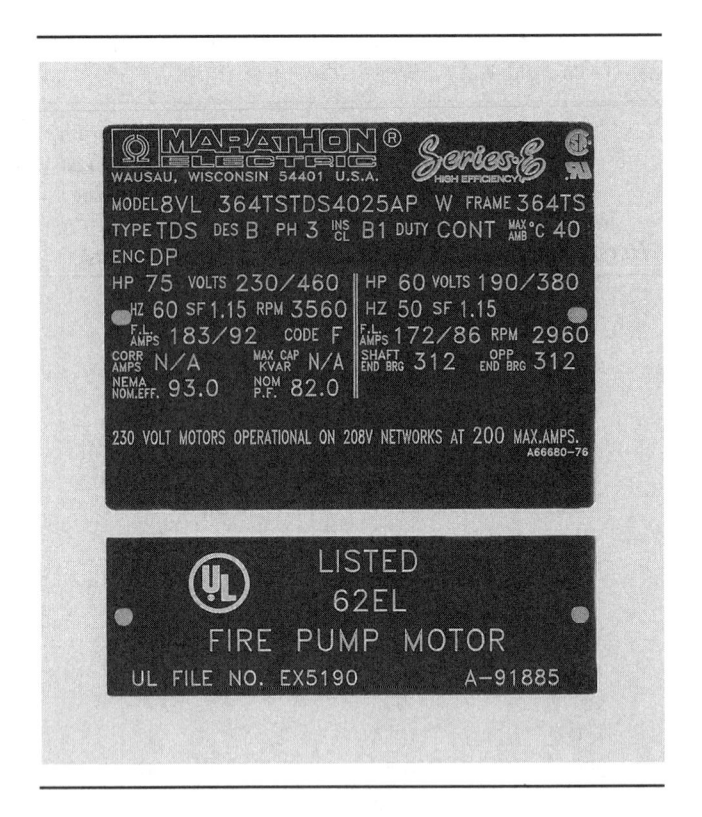

9.5.1.5* The corresponding values of locked rotor current for motors rated at other voltages shall be determined by multiplying the values shown by the ratio of 460 V to the rated voltage in Table 9.5.1.1.

A.9.5.1.5 The locked rotor currents for 460 V motors are approximately six times the full-load current.

9.5.1.6 Code letters of motors for all other voltages shall conform with those shown for 460 V in Table 9.5.1.1.

9.5.1.7 All motors shall be rated for continuous duty.

9.5.1.8 Electric motor–induced transients shall be coordinated with the provisions of 10.4.3.3 to prevent nuisance tripping of motor controller protective devices.

FAQ ▶
Why are electric motor–induced transient currents a concern in the operation of a fire pump motor?

Electric motor–induced transient currents are normal during motor starting. Some of these currents may contain enough energy to cause the fire pump circuit breaker specified in 10.4.3 to trip. The instantaneous setting of this circuit breaker should be high enough to prevent the electric motor–induced transient currents from tripping the circuit breaker. The instantaneous setting of the circuit breaker is allowed to be as high as 20 times the full-load current rating of the motor.

9.5.1.9 Motors for Vertical Shaft Turbine–Type Pumps.

9.5.1.9.1 Motors for vertical shaft turbine–type pumps shall be dripproof, squirrel-cage induction type.

9.5.1.9.2 The motor shall be equipped with a nonreverse ratchet.

Pumps must turn in only one direction and the same rotation for proper operation.

9.5.2 Current Limits.

9.5.2.1 The motor capacity in horsepower shall be such that the maximum motor current in any phase under any condition of pump load and voltage unbalance shall not exceed the motor-rated full-load current multiplied by the service factor.

9.5.2.2 The following shall apply to the service factor:

(1) The maximum service factor at which a motor shall be used is 1.15.

The requirement in 9.5.2.2(1) only applies to motors with a service factor of 1.15. Fire pump motors are allowed but not required to have a service factor of 1.15.

◀ FAQ
What is the required service factor for fire pump motors?

(2) Where the motor is used with a variable speed pressure limiting controller, the service factor shall not be used.

In order to operate an inverter duty motor in excess of its full-load current rating, such a motor must be operated in excess of its rated frequency. In fire pump applications, operating a motor above its rated frequency is not desirable. Therefore, the service factor of an inverter duty rated motor does not need to be used in a variable speed pressure limiting application.

◀ FAQ
Why is the service factor not utilized for motors used with a variable speed pressure limiting controller?

9.5.2.3 These service factors shall be in accordance with NEMA MG-1, *Motors and Generators*.

9.5.2.4 General-purpose (open and dripproof) motors, totally enclosed fan-cooled (TEFC) motors, and totally enclosed nonventilated (TENV) motors shall not have a service factor larger than 1.15.

9.5.2.5 Motors used at altitudes above 3300 ft (1000 m) shall be operated or derated according to NEMA MG-1, *Motors and Generators*, Part 14.

9.5.3 Marking.

9.5.3.1 Marking of motor terminals shall be in accordance with NEMA MG-1, *Motors and Generators*, Part 2.

9.5.3.2 A motor terminal connecting diagram for multiple lead motors shall be furnished by the motor manufacturer.

Achieving proper wiring of multiple lead motors (more than three conductors) is difficult without a wiring diagram that identifies the proper connections. This issue is generally confined to part-winding and wye-delta motors.

9.6 On-Site Standby Generator Systems.

9.6.1 Capacity.

9.6.1.1 Where on-site generator systems are used to supply power to fire pump motors to meet the requirements of 9.3.2, they shall be of sufficient capacity to allow normal starting and running of the motor(s) driving the fire pump(s) while supplying all other simultaneously operated load(s) while meeting the requirements of Section 9.4.

The requirement in 9.6.1.1 prevents undersizing of the generator under all conditions of loading.

9.6.1.2 A tap ahead of the on-site generator disconnecting means shall not be required.

An on-site generator may supply loads other than those associated with the fire pump. Although the normal source power supply is required to be a service connection dedicated to the

◀ FAQ
Why is a tap ahead of the on-site generator disconnecting means not required?

fire pump installation when the source supplies loads other than those associated with the fire pump, this requirement does not apply to the alternate source power supply. As such, the fire pump load may be connected on the load side of the generator disconnecting means.

9.6.2* Power Sources.

A.9.6.2 Where a generator is installed to supply power to loads in addition to one or more fire pump drivers, the fuel supply should be sized to provide adequate fuel for all connected loads for the desired duration. The connected loads can include such loads as emergency lighting, exit signage, and elevators.

9.6.2.1 On-site standby generator systems shall comply with Section 9.4 and shall meet the requirements of Level 1, Type 10, Class X systems of NFPA 110, *Standard for Emergency and Standby Power Systems.*

> **FAQ ▶** Why were the requirements of Level 1, Type 10, Class X systems of NFPA 110 chosen to define the performance of on-site generators?

Adherence to the requirements of NFPA 110, *Standard for Emergency and Standby Power Systems,* for Level 1, Type 10, Class X systems ensures that the alternate power source will be available within 10 seconds. This time period is considered necessary to provide adequate water supply in the event of a fire.

9.6.2.2 The engine shall run and continue to produce rated nameplate power without shutdown or de-rate for alarms and warnings, or failed engine sensors, except for overspeed shutdown.

Level 1 emergency power supply systems (EPSS) as defined by NFPA 110 and Articles 700, 701, and 702 of the *NEC* require the emergency backup unit (generator, or other) to operate with limited safety or warning shutdowns. This requirement is applicable to the standby generator used to provide an alternative power source. Certain municipalities have required adherence to a level greater than the Level 1 EPSS, not allowing the engine to shut down or derate power in the event of a failed sensor on ECM engines, high water temperature, or low oil pressure. This requirement ensures that engines used as the electrical backup to electric drivers have a level of performance similar to what is required in Chapter 11, which includes a requirement that is more stringent than what is required for a Level 1 EPSS.

9.6.2.3 The fuel supply capacity shall be sufficient to provide 8 hours of fire pump operation at 100 percent of the rated pump capacity in addition to the supply required for other demands.

9.6.3 Sequencing.
Automatic sequencing of the fire pumps shall be permitted in accordance with 10.5.2.5.

9.6.4 Transfer of Power.
Transfer of power to the fire pump controller between the normal supply and one alternate supply shall take place within the pump room.

> **FAQ ▶** What is meant by "transfer of power . . . shall take place within the pump room"?

The requirement in 9.6.4 specifies the location for the transfer of power to the fire pump motor. The transfer switch that has the fire pump controller connected to its load terminals must be located in the pump room. As multiple transfer switches may be available for one fire pump motor, the requirement only applies to the transfer switch immediately upstream of the fire pump motor. This limitation is specified by the reference in the requirement to "one alternate supply."

9.6.5* Protective Devices.
Protective devices installed in the on-site power source circuits at the generator shall allow instantaneous pickup of the full pump room load and shall comply with *NFPA 70, National Electrical Code,* Section 700.27.

The wording of 9.6.5 was revised to clarify that these protective devices are not optional. Article 445 of the *NEC* requires that the load side of the generator be supplied with overcurrent protection. The reference to Section 700.27 of the *NEC* was added to ensure that these protective devices are selectively coordinated with all overcurrent devices in the alternate power circuit.

A.9.6.5 Generator protective devices are to be sized to permit the generator to allow instantaneous pickup of the full pump room load. This includes starting any and all connected fire pumps in the across-the-line (direct on line) full voltage starting mode. This is always the case when the fire pump(s) is started by use of the emergency-run mechanical control in 10.5.3.2.

The requirement in 9.6.5 ensures that no nuisance tripping of the generator protective device(s) occurs when the full pump room load is connected to the generator by the transfer switch. The circuit breaker in the fire pump controller should be the only protective device to disconnect power to the fire pump motor under any conditions of loading other than short circuit. If protective devices upstream of the fire pump controller open when the full pump room load is transferred to the generator or during an overload condition, resetting or replacing them in a reasonable period of time to place the fire pump back in operation to successfully extinguish the fire is virtually impossible.

9.7 Junction Boxes.

Where fire pump wiring to or from a fire pump controller is routed through a junction box, the following requirements shall be met:

Junction boxes are commonly mounted on fire pump controllers for the purposes of terminating cables. In some cases, these junction boxes are used to provide compliance with Section 9.8.

(1) The junction box shall be securely mounted.
(2)* Mounting and installation of a junction box shall not violate the enclosure type rating of the fire pump controller(s).

If the junction box is mounted to the fire pump controller enclosure, care needs to be taken to ensure that the mounting process does not affect the environmental rating of the fire pump controller enclosure. Modification of the controller enclosure may be necessary in order to mount the junction box. For example, cutting or drilling an opening in the controller enclosure that would allow the entrance of water or expose live electrical parts could pose a shock hazard.

◀ **FAQ**
If mounting the junction box requires modification of the fire pump controller enclosure, what concerns need to be addressed?

A.9.7(2) See also 10.3.3.

(3)* Mounting and installation of a junction box shall not violate the integrity of the fire pump controller(s) and shall not affect the short circuit rating of the controller(s).

If the junction box is mounted to the fire pump controller enclosure, care needs to be taken to ensure that the mounting process does not affect the strength and rigidity of the enclosure and its ability to contain an internal disturbance due to a short circuit within or downstream of the controller. Modification of the enclosure may be necessary in order to mount the junction box. An example would be cutting or drilling an opening in the controller enclosure that would allow flammable material to escape during a short circuit occurrence.

A.9.7(3) See 10.1.2.1, controller short circuit (withstand) rating.

(4) As a minimum, a Type 2, dripproof enclosure (junction box) shall be used. The enclosure shall be listed to match the fire pump controller enclosure type rating.

A Type 2 enclosure provides a degree of protection for the enclosed equipment against incidental contact with falling dirt, falling liquids, and light splashing. As the junction box and

◀ **FAQ**
Does a junction box need a specific environmental rating?

controller enclosure are in the same environment, the junction box needs to have an environmental rating at least equivalent to that of the controller enclosure. A junction box with an environmental rating more stringent than that of the controller enclosure is acceptable. (Refer to Commentary Table II.10.1 following 10.3.3 to determine the degree of protection against specific environmental conditions provided by an enclosure based on its type rating.)

(5) Terminals, junction blocks, and splices, where used, shall be listed.

9.8 Listed Electrical Circuit Protective System to Controller Wiring.

9.8.1* Where single conductors (individual conductors) are used, they shall be terminated in a separate junction box. Single conductors (individual conductors) shall not enter the fire pump enclosure separately.

FAQ ▶
What special requirements apply when single conductors (individual conductors) are used to provide power to the fire pump controller?

NFPA 20 does not permit the individual conductors of an electrical circuit protective system to terminate in the fire pump controller enclosure. The possibility exists that flammable gases may enter the controller enclosure through the individual conductors.

A.9.8.1 Cutting slots or rectangular cutouts in a fire pump controller will violate the enclosure type rating and the controller's short circuit (withstand) rating and will void the manufacturer's warranty. See also *NFPA 70, National Electrical Code,* Articles 300.20 and 322, for example, for further information.

FAQ ▶
Is the modification of a fire pump controller enclosure permitted in order to comply with 9.8.1?

If cutting slots or cutouts in the controller enclosure is absolutely necessary, care needs to be taken to ensure that these slots or cutouts do not affect the environmental rating and the strength and rigidity of the fire pump controller enclosure. The manufacturer of the controller should be consulted for guidance on this issue before any modifications are made to the controller enclosure.

9.8.2* The raceway between a junction box and the fire pump controller shall be sealed at the junction box end with an identified compound and in accordance with the instructions of the electrical circuit protective systems if provided.

A.9.8.2 When so required, this seal is to prevent flammable gases from entering the fire pump controller.

The provision in 9.8.2 was revised to require sealing of the raceway between a junction box and the fire pump controller in all installations where a junction box is provided. Previously, sealing was only necessary where required by the manufacturer of a listed electrical circuit protective system or by the *NEC*, or by the listing.

FAQ ▶
What type of sealing means is acceptable in order to comply with 9.8.2?

Types of acceptable sealing compounds are specified by the manufacturer of the listed electrical circuit protective system. Normally, acceptable sealing compounds are those that do not suffer degradation from exposure to gases and do not harden and/or become brittle with age.

9.8.3 Standard wiring between the junction box and the controller is acceptable.

The provision in 9.8.3 is made on the basis that the requirement of 9.8.2 is met.

9.9 Raceway Terminations.

9.9.1 Listed conduit hubs shall be used to terminate raceway (conduit) to the fire pump controller.

Conduit hubs may be marked with a type rating, as in the case of the controller enclosure. However, most hubs are marked with designations such as "raintight" and "watertight." In order to determine an equivalent type rating for these hubs, refer to Commentary Table II.10.1 to determine the degree of protection against specific environmental conditions provided by an enclosure based on its type rating. Some product standards for enclosures allow an enclosure rated Type 3, 3S, 3SX, 3X, 4, and 4X to be marked "raintight" and an enclosure rated Type 4 or 4X to be marked "watertight."

◀ FAQ
What types of conduit hubs are acceptable in order to comply with 9.9.1?

9.9.2 The type rating of the conduit hub(s) shall be at least equal to that of the fire pump controller.

9.9.3 The installation instructions of the manufacturer of the fire pump controller shall be followed.

9.9.4 Alterations to the fire pump controller, other than conduit entry as allowed by *NFPA 70, National Electrical Code,* shall be approved by the authority having jurisdiction.

REFERENCES CITED IN COMMENTARY

National Fire Protection Association, 1 Batterymarch Park, Quincy, MA 02169-7471.

NFPA 70®, National Electrical Code®, 2008 edition.
NFPA 110, *Standard for Emergency and Standby Power Systems,* 2010 edition.
NFPA 850, *Recommended Practice for Fire Protection for Electric Generating Plants and High Voltage Direct Current Converter Stations,* 2005 edition.

Electric-Drive Controllers and Accessories

CHAPTER 10

Chapter 10 describes the attributes of electric motor–driven fire pump controllers, including controllers with power transfer switches and separate power transfer switches, all for use in electric fire pump motor circuits. Also covered in this chapter are accessories used with the controllers to ensure the minimum performance required by NFPA 20. The focus of Chapter 10 is the design, construction, and performance of the controllers and their accessories. This chapter also provides information to aid in specifying, installing, and operating this equipment.

10.1 General.

10.1.1 Application.

10.1.1.1 This chapter covers the minimum performance and testing requirements for controllers and transfer switches for electric motors driving fire pumps.

10.1.1.2 Accessory devices, including fire pump alarm and signaling means, are included where necessary to ensure the minimum performance of the equipment mentioned in 10.1.1.1.

All equipment necessary to ensure the minimum performance of the fire pump controller and/or transfer switch, whether installed as part of such equipment or separately, is subject to the requirements of this chapter.

10.1.2 Performance and Testing.

10.1.2.1 Listing. All controllers and transfer switches shall be specifically listed for electric motor–driven fire pump service.

Equipment that is listed meets the criteria specified in the definition for this term in 3.2.3. The listing identification for this equipment must include a reference to "Fire Pump(s)" or "Fire Pump Service." Typical listing agencies for fire pump equipment include Underwriters Laboratories Inc. and FM Global.

10.1.2.2* Marking.

See Exhibit II.10.1 for a typical fire pump controller nameplate.

A.10.1.2.2 The phrase *suitable for use* means that the controller and the transfer switch have been prototype tested and have demonstrated by those tests their short-circuit withstandability and interrupting capacity at the stated magnitude of short-circuit current and voltage available at their line terminals. *(See ANSI/UL 508, Standard for Safety Industrial Control Equipment, and ANSI/UL 1008, Standard for Safety Automatic Transfer Switches.)*

A short-circuit study should be made to establish the available short-circuit current at the controller in accordance with IEEE 141, *Electric Power Distribution for Industrial Plants,* IEEE 241, *Electric Systems for Commercial Buildings,* or other acceptable methods.

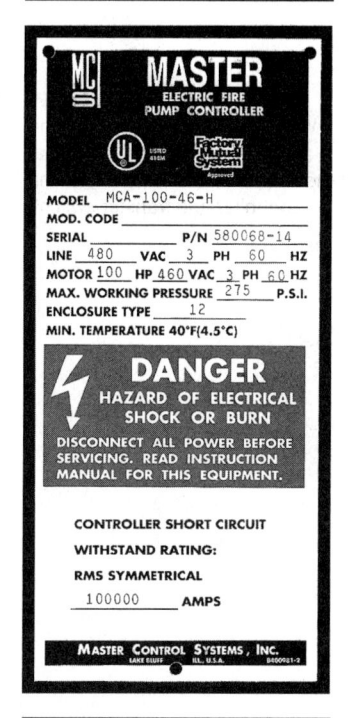

EXHIBIT II.10.1 Fire Pump Controller Nameplate. (Courtesy of Master Control Systems, Inc.)

◀ **FAQ**
What is meant by the term *listed* and who are some of the typical listing agencies?

A short-circuit study is necessary to ensure that equipment will not present a risk of fire or shock hazard if a short circuit occurs in the fire pump motor circuit.

After the controller and transfer switch have been subjected to a high fault current, they might not be suitable for further use without inspection or repair. *(See NEMA ICS 2.2, Maintenance of Motor Controllers After a Fault Condition.)*

FAQ ▶
What if no evidence exists that a fire pump controller and transfer switch subjected to a high fault current present a risk of fire or shock hazard?

Equipment must be thoroughly inspected, and repaired if necessary, after a short circuit has occurred. Unlike overload currents, a short circuit can cause damage to equipment that would render it inoperable or unsafe for further use. While the short-circuit event might not present a risk of fire or shock hazard, the equipment may sustain damage that would require repair before further use.

10.1.2.2.1 The controller and transfer switch shall be suitable for the available short-circuit current at the line terminals of the controller and transfer switch.

FAQ ▶
Why is it necessary to know the available short-circuit current at the line terminals of the fire pump controller and transfer switch?

The compatibility of the short-circuit current rating marked on the equipment (required in 10.1.2.2.2) with the available short-circuit current at the line terminals of the equipment is covered in 10.1.2.2.1. The available short-circuit current determined by a short-circuit study, as described in 10.1.2.2, must not exceed the short-circuit current rating marked on the equipment.

10.1.2.2.2 The controller and transfer switch shall be marked "Suitable for use on a circuit capable of delivering not more than ____ amperes RMS symmetrical at ____ volts ac," or "____ amperes RMS symmetrical at ____ volts ac short-circuit current rating," or equivalent, where the blank spaces shown shall have appropriate values filled in for each installation.

The values to be inserted in the blank spaces in 10.1.2.2.2 are determined through prototype testing, which is described in A.10.1.2.2. Other marking that conveys the same information may be used. See Exhibit II.10.2 for typical combination fire pump controller/transfer switch nameplates. Exhibit II.10.2 (left) illustrates a marking where the combination fire pump

EXHIBIT II.10.2 Combination Fire Pump Controller/Transfer Switch Nameplates. (Courtesy of Master Control Systems, Inc.)

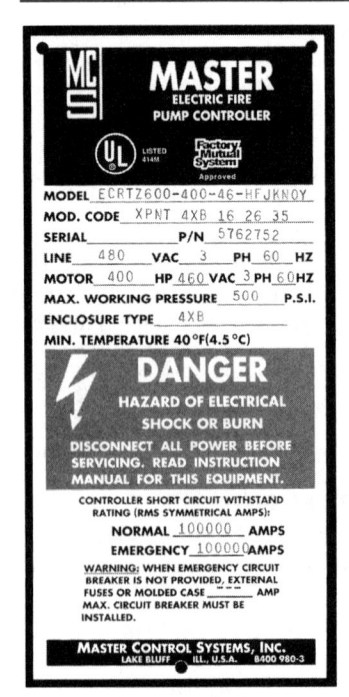

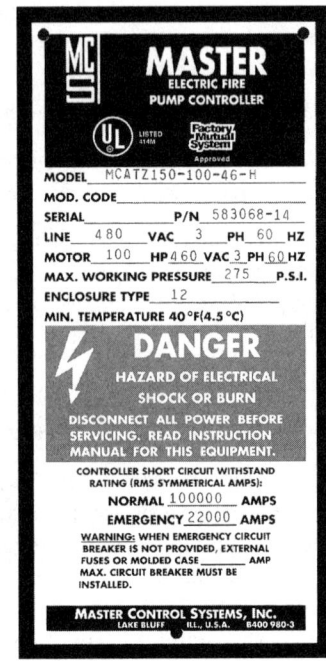

controller/transfer switch has the same short-circuit rating for both the normal and alternate power circuits. Exhibit II.10.2 (right) illustrates a marking where the combination fire pump controller/transfer switch has different short-circuit ratings for the normal and alternate power circuits.

10.1.2.3 Preshipment. All controllers shall be completely assembled, wired, and tested by the manufacturer before shipment from the factory. Controllers shipped in sections shall be completely assembled, wired, and tested by the manufacturer before shipment from the factory. Such controllers shall be reassembled in the field, and the proper assembly shall be verified by the manufacturer or designated representative.

This paragraph was revised to recognize that some fire pump controllers due to size and construction cannot be shipped as a complete assembly and need to be shipped in sections. Proper reassembly at the installation site is ensured by requiring the reassembly to be verified by the manufacturer or designated representative.

Fire pump controllers are generally not assembled in the field from individual parts or subassemblies. Complete assembly prior to shipment ensures proper coordination and collective performance of the components used in the manufacture of the controller. Large equipment, such as controllers with high horsepower ratings and/or those with transfer switches, resistor banks, and variable speed pressure limiting control, is impractical to ship as a complete assembly. Such equipment is assembled, wired, and tested by the manufacturer before shipment from the factory. However, the equipment is subsequently disassembled, shipped in pieces, reassembled at the installation site, and tested again to ensure proper coordination and collective performance of the components. The tests at the installation site are not necessarily all of those conducted at the factory before shipment. Generally, some functional tests and a dielectric strength test are conducted at the installation site.

◀ FAQ
Is a fire pump controller allowed to be assembled in the field from individual parts or subassemblies?

10.1.2.4 Service Equipment Listing. All controllers and transfer switches shall be listed as "suitable for use as service equipment" where so used.

Controllers and transfer switches that provide power to the fire pump motor by connection to a separate service or tap ahead of a service disconnect for the premises must be listed as "suitable for use as service equipment." Previous editions of NFPA 20 permitted only these two types of connection schemes and, as such, all controllers and transfer switches were required to be listed as "suitable for use as service equipment." The standard was subsequently revised for the 2007 edition to permit controllers and transfer switches to be supplied from feeder circuits under certain conditions. Controllers and transfer switches used in these applications need not be suitable for use as service equipment. Therefore, the phrase *where so used* was added to require the listing as "suitable for use as service equipment" only when controllers and transfer switches are connected to a separate service or tap ahead of a service disconnect.

◀ FAQ
When is a fire pump controller or transfer switch required to be listed as "suitable for use as service equipment"?

10.1.2.5 Additional Marking.

10.1.2.5.1 All controllers shall be marked "Electric Fire Pump Controller" and shall show plainly the name of the manufacturer, identifying designation, maximum operating pressure, enclosure type designation, and complete electrical rating.

Most controllers are marked with the minimum operating temperature of their environment. Controllers that are pressure actuated are connected to pipes containing water. Consequently, the controller must not be located in an area where temperatures at or below freezing may occur.

◀ FAQ
What is the significance of marking the minimum operating temperature on a fire pump controller?

10.1.2.5.2 Where multiple pumps serve different areas or portions of the facility, an appropriate sign shall be conspicuously attached to each controller indicating the area, zone, or portion of the system served by that pump or pump controller.

10.1.2.6 Service Arrangements. It shall be the responsibility of the pump manufacturer or its designated representative to make necessary arrangements for the services of a manufacturer's representative when needed for service and adjustment of the equipment during the installation, testing, and warranty periods.

Although service and adjustment of the equipment is normally performed by the manufacturer of the controller and transfer switch, ultimately the pump manufacturer is responsible for ensuring that this work is performed.

10.1.2.7 State of Readiness. The controller shall be in a fully functional state within 10 seconds upon application of ac power.

The requirement for state of readiness relates to controllers that are computer controlled. For such equipment, boot-up time may be needed before the controller is fully functional, and any delay in being fully functional must be minimized in order to ensure that the performance of the fire pump is not diminished.

10.1.3* Design. All electrical control equipment design shall comply with *NFPA 70, National Electrical Code,* Article 695, and other applicable documents.

The importance of Article 695 of *NFPA 70®, National Electrical Code® (NEC®)*, is emphasized in 10.1.3 (see also Supplement 3). NFPA 20 covers the minimum performance and testing requirements for controllers and transfer switches, while Article 695 covers the installation requirements of such equipment. Some of the requirements of Article 695 are extracted from NFPA 20 because those requirements are under the purview of the NFPA 20 Technical Committee on Fire Pumps. However, all of the installation requirements for controllers, transfer switches, and fire pump accessory equipment are located in Article 695.

A.10.1.3 All electrical control equipment design should also follow the guidelines within NEMA ICS 14, *Application Guide for Electric Pump Controllers.*

NEMA ICS 14, *Application Guide for Electric Pump Controllers,* provides practical information concerning the general technical considerations in the installation of electric fire pump controllers. The guide is intended to be used by specifiers, purchasers, installers, and owners of fire pump controllers.

10.2 Location.

10.2.1* Controllers shall be located as close as is practical to the motors they control and shall be within sight of the motors.

A.10.2.1 If the controller must be located outside the pump room, a glazed opening should be provided in the pump room wall for observation of the motor and pump during starting. The pressure control pipe line should be protected against freezing and mechanical injury.

10.2.2 Controllers shall be located or protected so that they will not be damaged by water escaping from pumps or pump connections.

FAQ ▶
Are any restrictions or guidance given on where a fire pump controller can be installed?

The location of the controller is critical with regard to exposure to water. Although the controller enclosure is rated Type 2, the enclosure protects equipment only from contact with water dripping from a source located above the controller. The spray pattern of valves should be determined, as well as potential leak points of pump connections.

10.2.3 Current-carrying parts of controllers shall be not less than 12 in. (305 mm) above the floor level.

10.2.4 Working clearances around controllers shall comply with *NFPA 70, National Electrical Code,* Article 110.

Providing adequate working clearances is especially important in those pump rooms where space is limited. Subsection 10.2.4 references Article 110 of the *NEC*, 2008 edition. The relevant paragraphs from Section 110.26 of the *NEC*, which contain the working clearance requirements, are extracted following this paragraph. The accompanying commentary is modified from the *National Electrical Code® Handbook*, 2008 edition.

> **110.26 Spaces About Electrical Equipment**
>
> Sufficient access and working space shall be provided and maintained about all electric equipment to permit ready and safe operation and maintenance of such equipment.
>
> **(A) Working Space** Working space for equipment operating at 600 volts, nominal, or less to ground and likely to require examination, adjustment, servicing, or maintenance while energized shall comply with the dimensions of 110.26(A)(1), (A)(2), and (A)(3) or as required or permitted elsewhere in this *Code*.
>
> **(1) Depth of Working Space.** The depth of the working space in the direction of live parts shall not be less than that specified in Table 110.26(A)(1) unless the requirements of 110.26(A)(1)(a), (A)(1)(b), or (A)(1)(c) are met. Distances shall be measured from the exposed live parts or from the enclosure or opening if the live parts are enclosed.

TABLE 110.26(A)(1) Working Spaces

Nominal Voltage to Ground	Minimum Clear Distance		
	Condition 1	Condition 2	Condition 3
0–150	914 mm (3 ft)	914 mm (3 ft)	914 mm (3 ft)
151–600	914 mm (3 ft)	1.07 m (3 ft 6 in.)	1.22 m (4 ft)

Note: Where the conditions are as follows:

Condition 1—Exposed live parts on one side of the working space and no live or grounded parts on the other side of the working space, or exposed live parts on both sides of the working space that are effectively guarded by insulating materials.

Condition 2—Exposed live parts on one side of the working space and grounded parts on the other side of the working space. Concrete, brick, or tile walls shall be considered as grounded.

Condition 3—Exposed live parts on both sides of the working space.

> (a) *Dead-Front Assemblies.* Working space shall not be required in the back or sides of assemblies, such as dead-front switchboards or motor control centers, where all connections and all renewable or adjustable parts, such as fuses or switches, are accessible from locations other than the back or sides. Where rear access is required to work on nonelectrical parts on the back of enclosed equipment, a minimum horizontal working space of 762 mm (30 in.) shall be provided.
>
> (b) *Low Voltage.* By special permission, smaller working spaces shall be permitted where all exposed live parts operate at not greater than 30 volts rms, 42 volts peak, or 60 volts dc.
>
> (c) *Existing Buildings.* In existing buildings where electrical equipment is being replaced, Condition 2 working clearance shall be permitted between dead-front switchboards, panelboards, or motor control centers located across the aisle from each other where conditions of maintenance and supervision ensure that written procedures have been adopted to prohibit equipment on both sides of the aisle from being open at the same time and qualified persons who are authorized will service the installation.

(2) Width of Working Space. The width of the working space in front of the electrical equipment shall be the width of the equipment or 762 mm (30 in.), whichever is greater. In all cases, the work space shall permit at least a 90 degree opening of equipment doors or hinged panels.

(3) Height of Working Space. The work space shall be clear and extend from the grade, floor, or platform to the height required by 110.26(E). Within the height requirements of this section, other equipment that is associated with the electrical installation and is located above or below the electrical equipment shall be permitted to extend not more than 150 mm (6 in.) beyond the front of the electrical equipment.

(B) Clear Spaces. Working space required by this section shall not be used for storage. When normally enclosed live parts are exposed for inspection or servicing, the working space, if in a passageway or general open space, shall be suitably guarded.

The key to understanding Section 110.26 is the division of requirements for spaces about electrical equipment into two separate and distinct categories: working space and dedicated equipment space. The term *working space* generally applies to the protection of the worker, and *dedicated equipment space* applies to the space reserved for future access to electrical equipment and to protection of the equipment from intrusion by nonelectrical equipment. The performance requirements for all spaces about electrical equipment are set forth in the first sentence. Storage of materials that blocks access or prevents safe work practices must be avoided at all times.

The intent of 110.26(A) is to provide enough space for personnel to perform any of the operations listed without jeopardizing worker safety. These operations include examination, adjustment, servicing, and maintenance of equipment. Examples of such equipment include panelboards, switches, circuit breakers, controllers, and controls on heating and air-conditioning equipment. It is important to understand that the word *examination,* as used in 110.26(A), includes such tasks as checking for the presence of voltage using a portable voltmeter.

Minimum working clearances are not required if the equipment is such that it is not likely to require examination, adjustment, servicing, or maintenance while energized. However, "sufficient" access and working space are still required by the opening paragraph of Section 110.26.

Included in the clearance requirements in 110.26(A) is the step-back distance from the face of the equipment. Table 110.26(A)(1) provides requirements for clearances away from the equipment, based on the circuit voltage to ground and whether grounded or ungrounded objects are in the step-back space or exposed live parts across from each other. The voltages to ground consist of two groups: 0 to 150, inclusive, and 151 to 600, inclusive. Examples of common electrical supply systems covered in the 0 to 150 volts to ground group include 120/240-volt, single-phase, 3-wire; and 208Y/120-volt, 3-phase, 4-wire. Examples of common electrical supply systems covered in the 151 to 600 volts to ground group include 240-volt, 3-phase, 3-wire; 480Y/277-volt, 3-phase, 4-wire; and 480-volt, 3-phase, 3-wire (ungrounded and corner grounded). Remember, where an ungrounded system is utilized, the voltage to ground (by definition) is the greatest voltage between the given conductor and any other conductor of the circuit. For example, the voltage to ground for a 480-volt ungrounded delta system is 480 volts. See Exhibit II.10.3 for the general working clearance requirements for each of the three conditions listed in Table 110.26(A)(1). Distances are measured from the live parts if the live parts are exposed, or from the enclosure front if the live parts are enclosed. If any assemblies, such as switchboards or motor-control centers, are accessible from the back and exposed live parts, the working clearance dimensions would be required at the rear of the equipment, as illustrated. Note that for Condition 3, where an enclosure is on opposite sides of the working space, the clearance for only one working space is required.

The intent of 110.26(A)(1)(a) is to point out that work space is required only from the side(s) of the enclosure that requires access. The general rule still applies: Equipment that re-

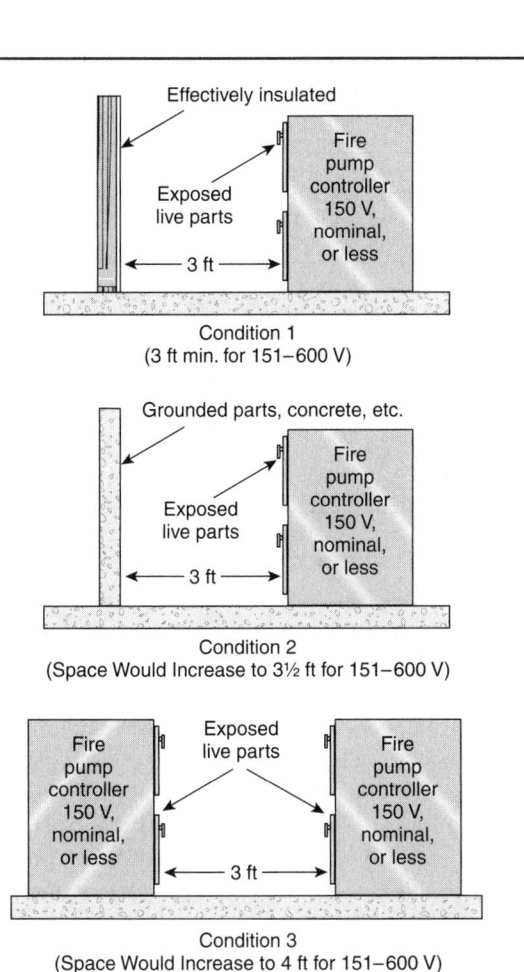

EXHIBIT II.10.3 General Working Clearance Requirements from NEC Table 110.26(A)(1). (Source: Adapted from the National Electrical Code® Handbook, NFPA, Quincy, MA, 2008, Exhibit 110.9)

Condition 1
(3 ft min. for 151–600 V)

Condition 2
(Space Would Increase to 3½ ft for 151–600 V)

Condition 3
(Space Would Increase to 4 ft for 151–600 V)

quires front, rear, or side access for electrical activities described in 110.26(A) must meet the requirements of Table 110.26(A)(1). In many cases, equipment of "dead-front" assemblies requires only front access. For equipment that requires rear access for nonelectrical activity, however, a reduced working space of at least 30 in. (762 mm) must be provided.

Section 110.26(A)(1)(c) permits some relief for installations that are being upgraded. Where assemblies such as dead-front switchboards, panelboards, or motor-control centers are replaced in an existing building, the working clearance allowed is that required by Table 110.26(A)(1), Condition 2. The reduction from a Condition 3 to a Condition 2 clearance is allowed only where a written procedure prohibits facing doors of equipment from being open at the same time and where only authorized and qualified persons service the installation. Exhibit II.10.4 illustrates this relief for existing buildings.

Regardless of the width of the electrical equipment, the working space cannot be less than 30 in. (762 mm) wide in accordance with 110.26(A)(2). This space allows an individual to have at least shoulder-width space in front of the equipment. The 30 in. (762 mm) measurement can be made from either the left or the right edge of the equipment and can overlap other electrical equipment, provided the other equipment does not extend beyond the clearance required by Table 110.26(A)(1). If the equipment is wider than 30 in. (762 mm), the left-to-right space must be equal to the width of the equipment. Exhibit II.10.5 illustrates the 30 in. (762 mm) wide front working space, which is not required to be directly centered on the electrical equipment if space is sufficient for safe operation and maintenance of such equipment.

EXHIBIT II.10.4 Permitted Reduction from a Condition 3 to a Condition 2 Clearance According to NEC 110.26(A)(1)(c). (Source: Adapted from the National Electrical Code® Handbook, NFPA, Quincy, MA, 2008, Exhibit 110.11)

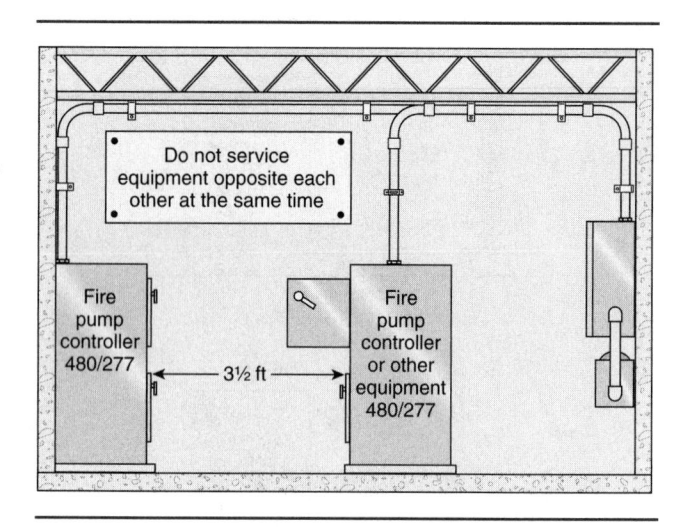

Sufficient depth in the working space also must be provided to allow a panel or a door to open at least 90 degrees. If doors or hinged panels are wider than 3 ft (914 mm), more than a 3 ft (914 mm) deep working space must be provided to allow a full 90 degree opening.

In addition to requiring a working space to be clear from the floor to a height of 6½ ft (1981 mm) or to the height of the equipment, whichever is greater, 110.26(A)(3) permits electrical equipment located above or below other electrical equipment to extend into the working space not more than 6 in. (152 mm). This requirement allows the placement of a 12 in. (305 mm) × 12 in. (305 mm) wireway on the wall directly above or below a 6 in. (152 mm) deep panelboard without impinging on the working space or compromising practical working clearances. This requirement continues to prohibit large differences in depth of equipment below or above other equipment that specifically requires working space. In order to minimize the amount of space required for electrical equipment, it was not uncommon to find installations of large freestanding, dry-type transformers within the required work space for a wall-mounted panelboard. Clear access to the panelboard is compromised by the location of the transformer with its grounded enclosure and this type of installation and is clearly not permitted by 110.26(A)(3). Electrical equipment that produces heat or that otherwise requires ventilation also must comply with 110.3(B) and Section 110.13.

EXHIBIT II.10.5 The 30 in. (762 mm) Wide Front Working Space. (Source: Adapted from the National Electrical Code® Handbook, NFPA, Quincy, MA, 2008, Exhibit 110.12)

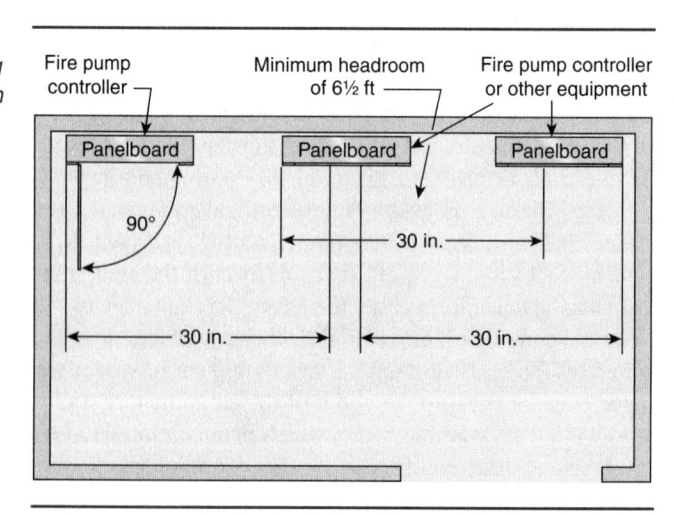

Section 110.26(B), as well as the rest of Section 110.26, does not prohibit the placement of panelboards in corridors or passageways. For that reason, when the covers of corridor-mounted panelboards are removed for servicing or other work, access to the area around the panelboard should be guarded or limited to protect unqualified persons using the corridor.

10.3 Construction.

10.3.1 Equipment. All equipment shall be suitable for use in locations subject to a moderate degree of moisture, such as a damp basement.

Controllers listed for fire pump service fulfill the requirement in 10.3.1 and are suitable for use in environments that may be damp.

◀ FAQ
What ensures that the enclosure for fire pump equipment provides the minimum amount of protection required for the enclosed equipment?

10.3.2 Mounting. All equipment shall be mounted in a substantial manner on a single noncombustible supporting structure.

Some controllers generate a considerable amount of heat when operating—in some cases at a level to ignite combustible materials. Some controllers have open bottoms. The use of a noncombustible supporting structure to close the bottom maintains enclosure integrity.

10.3.3 Enclosures.

See Commentary Table II.10.1 to determine the degree of protection against specific environmental conditions provided by an enclosure based on its type rating.

10.3.3.1* The structure or panel shall be securely mounted in, as a minimum, a National Electrical Manufacturers Association (NEMA) Type 2, dripproof enclosure(s) or an enclosure(s) with an ingress protection (IP) rating of IP31.

A Type 2 enclosure provides a degree of protection against incidental contact with the enclosed equipment from falling dirt, falling liquids, and light splashing. This paragraph was revised to recognize the ingress protection (IP) rating system for enclosures. An enclosure with an IP rating of IP31 is considered to provide protection equivalent to the protection provided by a Type 2 enclosure.

A.10.3.3.1 For more information, see NEMA 250, *Enclosures for Electrical Equipment.*

Controllers listed for fire pump service fulfill the requirement for NEMA Type 2. Some controllers may be marked with ratings other than Type 1. These ratings provide protection for the equipment within a level at least equivalent to the protection provided by a Type 2 enclosure. These enclosure types and the protection they provide are defined in NEMA 250, *Enclosures for Electrical Equipment.* Type 1 enclosures do not provide adequate protection for controllers used in fire pump applications because they are located in areas where they are exposed to dripping or splashing water.

10.3.3.2 Where the equipment is located outside, or where special environments exist, suitably rated enclosures shall be used.

Refer to NEMA 250 to determine the correct enclosure-type rating required. Each enclosure type is associated with specific degrees of protection provided. The NEMA standard describes the protection provided by an enclosure according to type number. All controllers listed for fire pump service are marked with one or more type ratings.

◀ FAQ
How should equipment used in a special environment be protected?

10.3.3.3 The enclosure(s) shall be grounded in accordance with *NFPA 70, National Electrical Code,* Article 250.

COMMENTARY TABLE II.10.1 Enclosure Selection

For Outdoor Use

Provides a Degree of Protection Against the Following Environmental Conditions	\multicolumn{10}{c}{Enclosure-Type Number}
	3	3R	3S	3X	3RX	3SX	4	4X	6	6P
Incidental contact with the enclosed equipment	X	X	X	X	X	X	X	X	X	X
Rain, snow, and sleet	X	X	X	X	X	X	X	X	X	X
Sleet*	—	—	X	—	—	X	—	—	—	—
Windblown dust	X	—	X	X	—	X	X	X	X	X
Hosedown	—	—	—	—	—	—	X	X	X	X
Corrosive agents	—	—	—	X	X	X	—	X	—	X
Temporary submersion	—	—	—	—	—	—	—	—	X	X
Prolonged submersion	—	—	—	—	—	—	—	—	—	X

For Indoor Use

Provides a Degree of Protection Against the Following Environmental Conditions	\multicolumn{10}{c}{Enclosure-Type Number}
	1	2	4	4X	5	6	6P	12	12K	13
Incidental contact with the enclosed equipment	X	X	X	X	X	X	X	X	X	X
Falling dirt	X	X	X	X	X	X	X	X	X	X
Falling liquids and light splashing	—	X	X	X	X	X	X	X	X	X
Circulating dust, lint, fibers, and flyings	—	—	X	X	—	X	X	X	X	X
Settling airborne dust, lint, fibers, and flyings	—	—	X	X	X	X	X	X	X	X
Hosedown and splashing water	—	—	X	X	—	X	X	—	—	—
Oil and coolant seepage	—	—	—	—	—	—	—	X	X	X
Oil or coolant spraying and splashing	—	—	—	—	—	—	—	—	—	X
Corrosive agents	—	—	—	X	—	—	X	—	—	—
Temporary submersion	—	—	—	—	—	X	X	—	—	—
Prolonged submersion	—	—	—	—	—	—	X	—	—	—

*Mechanism shall be operable when ice covered.

Note: The term *raintight* is typically used in conjunction with Enclosure Types 3, 3S, 3SX, 3X, 4, 4X, 6, and 6P. The term *rainproof* is typically used in conjunction with Enclosure Types 3R and 3RX. The term *watertight* is typically used in conjunction with Enclosure Types 4, 4X, 6, 6P. The term *driptight* is typically used in conjunction with Enclosure Types 2, 5, 12, 12K, and 13. The term *dusttight* is typically used in conjunction with Enclosure Types 3, 3S, 3SX, 3X, 5, 12, 12K, and 13.

Source: *NFPA 70®*, *National Electrical Code®*, 2008, Table 110.20.

10.3.4 Connections and Wiring.

10.3.4.1 All busbars and connections shall be readily accessible for maintenance work after installation of the controller.

Enclosures for equipment must contain adequate space to allow field-wiring connections to be made without interfering with any of the components within the enclosure. When the connections are made, no obstructions are to be placed in close proximity that prevent the connections from being readily accessible once the equipment is placed into service.

10.3.4.2 All busbars and connections shall be arranged so that disconnection of the external circuit conductors will not be required.

10.3.4.3 Means shall be provided on the exterior of the controller to read all line currents and all line voltages with an accuracy within ±5 percent of motor nameplate voltage and current.

The requirement in 10.3.4.3 allows for reading all line currents and all line voltages with the equipment energized and the enclosure door(s) closed. Some controllers may have meters mounted on the outside of the controller, while others may have external connection points for portable meters.

◀ FAQ
How can line voltages and currents be measured without the risk of electric shock?

The requirement in 10.3.4.3 was new to NFPA 20 in the 2007 edition. Previous editions of NFPA 20 required that provisions be made within the controller to permit the use of test instruments for measuring all line voltages and currents without disconnecting any conductors within the controller. Using this provision, measurements of both current and voltage were performed inside the enclosure with the fire pump controller energized. This was essentially "hot" work as the person making the measurements was exposed to energized live parts. The new requirement in 10.3.4.3 for a means on the exterior of the controller to read all line currents and all line voltages mitigated this hazard and provided a safer environment for measuring all line currents and all line voltages. Considering all of the heightened concern regarding the hazards of arc flash, it made no sense to allow personnel to be exposed to shock and arc flash hazards when NFPA 20 mandated a safer method that mitigates these hazards. The requirement that provisions be made within the controller to permit the use of test instruments for measuring all line voltages and currents without disconnecting any conductors within the controller was deleted.

10.3.4.4 Continuous-Duty Basis.

10.3.4.4.1 Unless the requirements of 10.3.4.4.2 are met, busbars and other wiring elements of the controller shall be designed on a continuous-duty basis.

10.3.4.4.2 The requirements of 10.3.4.4.1 shall not apply to conductors that are in a circuit only during the motor starting period, which shall be permitted to be designed accordingly.

Some factory-installed conductors, such as those connecting starting resistors, reactors, or an autotransformer, are generally smaller in size than the conductors carrying the line currents of the motor. This difference in size is permissible as these conductors only carry current during the motor starting period.

10.3.4.5 Field Connections.

10.3.4.5.1 A fire pump controller shall not be used as a junction box to supply other equipment.

Allowing the fire pump controller to be used as a junction box to supply other equipment could result in the fire pump controller being rendered inoperable if a short circuit occurred in the equipment being supplied by the fire pump controller.

10.3.4.5.2 No undervoltage, phase loss, frequency sensitive, or other device(s) shall be field installed that automatically or manually prohibits electrical actuation of the motor contactor.

The protection of equipment from undervoltage, phase loss, or other abnormal conditions is common in industrial control systems. Fire pump controllers must allow the fire pump motor to start even if such conditions exist. Damage to equipment from such conditions is tolerated in the event a fire occurs. Energizing the motor contactor coil is necessary to initiate motor starting. The presence of equipment in the coil circuit sensing abnormal conditions reduces the reliability of the motor contactor. The text of 10.3.4.5.2 was clarified in the 2010 edition so that the requirement applies to any field-installed device that automatically or manually prohibits electrical actuation of the motor contactor.

10.3.4.6 Electrical supply conductors for pressure maintenance (jockzey or make-up) pump(s) shall not be connected to the fire pump controller.

Although jockey or make-up pump(s) are allowed to be supplied by the same source supplying the fire pump motor, a means for connection upstream of the fire pump controller must be provided. This arrangement allows for servicing the jockey or make-up pump(s) without disturbing the wiring connections in the fire pump controller.

10.3.5 Protection of Control Circuits.

10.3.5.1 Circuits that are necessary for proper operation of the controller shall not have overcurrent protective devices connected in them.

FAQ ▶
Why are overcurrent protective devices (fuses or circuit breakers) not found in some of the control circuits of the fire pump controller?

The requirement in 10.3.5.1 generally applies to the coil circuit of the motor contactor, but may include other circuits where failures would prevent the controller from starting. Fire pump controllers must allow the fire pump motor to start even if conditions exist in these circuits that would normally open a fuse or trip a circuit breaker. The presence of a protective device in these circuits reduces the reliability of the fire pump controller.

10.3.5.2 The secondary of the transformer and control circuitry shall be permitted to be ungrounded except as required in 10.6.5.4.

10.3.6* External Operation. All switching equipment for manual use in connecting or disconnecting or starting or stopping the motor shall be externally operable.

A.10.3.6 For more information, see *NFPA 70, National Electrical Code.*

All manual switches in the fire pump motor circuit must be readily accessible to allow for prompt energizing of the fire pump motor circuit.

10.3.7 Electrical Diagrams and Instructions.

10.3.7.1 An electrical schematic diagram shall be provided and permanently attached to the inside of the controller enclosure.

Proper maintenance of the controller requires knowledge of the electrical equipment and circuitry within the fire pump controller.

10.3.7.2 All the field wiring terminals shall be plainly marked to correspond with the field connection diagram furnished.

FAQ ▶
What ensures that the electrical connections of field conductors are being terminated properly?

How to properly connect a controller in the field is not always obvious. The combination of properly identified field-wiring terminals and a field connection diagram ensures correct electrical connections.

10.3.7.3* Complete instructions covering the operation of the controller shall be provided and conspicuously mounted on the controller.

A.10.3.7.3 Pump operators should be familiar with instructions provided for controllers and should observe in detail all their recommendations.

The identification of the steps needed for proper operation of the controller is important. These steps must be easily understood by the operator, whether the controller is being operated in a fire condition or for startup/periodic testing.

10.3.8 Marking.

10.3.8.1 Each motor control device and each switch and circuit breaker shall be marked to plainly indicate the name of the manufacturer, the designated identifying number, and the electrical rating in volts, horsepower, amperes, frequency, phases, and so forth, as appropriate.

10.3.8.2 The markings shall be so located as to be visible after installation.

In the event that replacement of critical components is necessary, the clear identification of what is being replaced is important so that the same or equivalent component is installed.

◀ FAQ
Why is the marking of control devices, switches, and circuit breakers with so much identification information important?

10.4 Components.

10.4.1* Voltage Surge Arrester.

A.10.4.1 Operation of the surge arrester should not cause either the isolating switch or the circuit breaker to open. Arresters in ANSI/IEEE C62.11, *IEEE Standard for Metal-Oxide Surge Arresters for AC Power Circuits*, are normally zinc-oxide without gaps.

10.4.1.1 Unless the requirements of 10.4.1.3 or 10.4.1.4 are met, a voltage surge arrester complying with ANSI/IEEE C62.1, *IEEE Standard for Gapped Silicon-Carbide Surge Arresters for AC Power Circuits,* or ANSI/IEEE C62.11, *IEEE Standard for Metal-Oxide Surge Arresters for Alternating Current Power Circuits (>1 kV),* shall be installed from each phase to ground. *(See 10.3.3.3.)*

Surge arresters are provided in controllers to prevent power line surges from damaging components in the controller and/or rendering them inoperable. Typical failures due to a power line surge are burnout of indicating lamps and dielectric breakdown of the magnetic contactor holding coil.

10.4.1.2 The surge arrester shall be rated to suppress voltage surges above line voltage.

10.4.1.3 The requirements of 10.4.1.1 and 10.4.1.2 shall not apply to controllers rated in excess of 600 V. *(See Section 10.6.)*

10.4.1.4 The requirements of 10.4.1.1 and 10.4.1.2 shall not apply where the controller can withstand without damage a 10 kV impulse in accordance with ANSI/IEEE C62.41, *IEEE Recommended Practice for Surge Voltages in Low-Voltage AC Power Circuits.*

Either suitable surge arresters must be provided or the controller must be tested to determine that its construction is able to provide protection equivalent to that of a surge arrester.

◀ FAQ
Are surge arresters required in all fire pump controllers?

10.4.2 Isolating Switch.

10.4.2.1 General.

10.4.2.1.1 The isolating switch shall be a manually operable motor circuit switch or a molded case switch having a horsepower rating equal to or greater than the motor horsepower.

The requirement in 10.4.2.1.1 ensures that the isolating switch is capable of starting and stopping the fire pump motor, unless the switch is interlocked with the circuit breaker as permitted in 10.4.2.4.2.1.

◀ FAQ
Is the isolating switch required to be interlocked with the fire pump circuit breaker?

10.4.2.1.2* A molded case switch having an ampere rating not less than 115 percent of the motor rated full-load current and also suitable for interrupting the motor locked rotor current shall be permitted.

A.10.4.2.1.2 For more information, see *NFPA 70, National Electrical Code.*

A molded case switch used as an isolating switch need not be rated in horsepower if it has a suitable ampere rating and has been evaluated for switching the locked rotor current of the fire pump motor. The source of the 115 percent requirement is Section 430.110 of the *NEC*. This sizing rule accounts for motors running continuously in excess of their rated current.

10.4.2.1.3 A molded case isolating switch shall be permitted to have self-protecting instantaneous short-circuit overcurrent protection, provided that this switch does not trip unless the circuit breaker in the same controller trips.

FAQ ▶
Is the isolating switch permitted to have any integral instantaneous short-circuit overcurrent protection?

Molded case switches are generally the same construction as circuit breakers except the thermal (inverse-time) elements are removed. The instantaneous (magnetic) elements that remain are intended to provide only short-circuit protection. However, their trip current setting must be greater than the instantaneous trip current setting of the circuit breaker in the fire pump controller, which ensures that the fire pump circuit breaker is the primary overcurrent protective device for the fire pump motor.

These instantaneous elements also allow the fire pump controller to achieve a higher short-circuit rating than the interrupting capability of the circuit breaker in the fire pump controller. This higher short-circuit rating of the circuit breaker is accomplished by testing the combination of molded case switch and circuit breaker in a fire pump controller to demonstrate the controller can interrupt currents in excess of the interrupting capability of the circuit breaker.

10.4.2.2 Externally Operable. The isolating switch shall be externally operable.

Isolating switches must be readily accessible to allow for prompt energizing of the fire pump motor circuit. External operation is necessary to permit nonelectrically qualified personnel to safely energize and de-energize the fire pump motor circuit.

10.4.2.3* Ampere Rating. The ampere rating of the isolating switch shall be at least 115 percent of the full-load current rating of the motor.

A.10.4.2.3 For more information, see *NFPA 70, National Electrical Code.*

FAQ ▶
How is the ampere rating of the isolating switch determined?

The source of the 115 percent requirement is Section 430.110 of the *NEC*. This sizing rule accounts for motors running continuously in excess of their rated current.

10.4.2.4 Warning.

10.4.2.4.1 Unless the requirements of 10.4.2.4.2 are met, the following warning shall appear on or immediately adjacent to the isolating switch:

WARNING
DO NOT OPEN OR CLOSE THIS SWITCH WHILE
THE CIRCUIT BREAKER (DISCONNECTING MEANS)
IS IN CLOSED POSITION.

The isolating switch is not required to have a short-circuit interrupting rating. Only the fire pump controller circuit breaker is capable of manually closing on and interrupting a short-circuit current. Attempting to do so with an isolating switch could result in a fire and/or shock hazard.

10.4.2.4.2 Instruction Label. The requirements of 10.4.2.4.1 shall not apply where the requirements of 10.4.2.4.2.1 and 10.4.2.4.2.2 are met.

10.4.2.4.2.1 Where the isolating switch and the circuit breaker are so interlocked that the isolating switch can be neither opened nor closed while the circuit breaker is closed, the warning label shall be permitted to be replaced with an instruction label that directs the order of operation.

FAQ ▶
Is the interlocking mechanism for the isolating switch and circuit breaker interlocked with the controller enclosure door and, if so, is it provided with a means to circumvent it to allow for testing?

The interlock referenced in 10.4.2.4.2.1 ensures that the circuit breaker, and not the isolating switch, "makes" and "breaks" the current flowing in the fire pump power circuit. Most controllers are provided with interlocking mechanisms having either one or two operating handles. In most cases, these mechanisms are interlocked with the controller enclosure door and provided with a means to circumvent the interlock to allow for testing and maintenance with the controller energized.

10.4.2.4.2.2 This label shall be permitted to be part of the label required by 10.3.7.3.

10.4.2.5 Operating Handle.

10.4.2.5.1 Unless the requirements of 10.4.2.5.2 are met, the isolating switch operating handle shall be provided with a spring latch that shall be so arranged that it requires the use of the other hand to hold the latch released in order to permit opening or closing of the switch.

The use of a spring latch requires a two-hand operation to close or open the isolating switch. This safety feature minimizes inadvertent opening or closing of the switch. Instructions for operating the latch are normally included with the warning marking required in 10.4.2.4.1.

10.4.2.5.2 The requirements of 10.4.2.5.1 shall not apply where the isolating switch and the circuit breaker are so interlocked that the isolating switch can be neither opened nor closed while the circuit breaker is closed.

10.4.3 Circuit Breaker (Disconnecting Means).

10.4.3.1* General. The motor branch circuit shall be protected by a circuit breaker that shall be connected directly to the load side of the isolating switch and shall have one pole for each ungrounded circuit conductor.

An acceptable fire pump circuit breaker is either a circuit breaker listed for fire pump use or an instantaneous trip circuit breaker provided with a shunt trip device actuated by a separate device that provides the motor locked rotor protection as described in 10.4.4.

◀ **FAQ**
What types of circuit breakers are acceptable in a fire pump controller?

A.10.4.3.1 For more information, see *NFPA 70, National Electrical Code,* Article 100.

10.4.3.2 Mechanical Characteristics. The circuit breaker shall have the following mechanical characteristics:

(1) It shall be externally operable. *(See 10.3.6.)*
(2) It shall trip free of the handle.
(3) A nameplate with the legend "Circuit breaker—disconnecting means" in letters not less than ⅜ in. (10 mm) high shall be located on the outside of the controller enclosure adjacent to the means for operating the circuit breaker.

10.4.3.3* Electrical Characteristics.

A.10.4.3.3 Attention should be given to the type of service grounding to establish circuit breaker interrupting rating based on grounding type employed.

10.4.3.3.1 The circuit breaker shall have the following electrical characteristics:

(1) A continuous current rating not less than 115 percent of the rated full-load current of the motor

The source of the 115 percent requirement is Section 430.110 of the *NEC*. This sizing rule accounts for motors running continuously in excess of their rated current. This requirement is also applied to the isolating switch in 10.4.2.3.

◀ **FAQ**
How is the ampere rating of the circuit breaker determined?

(2) Overcurrent-sensing elements of the nonthermal type

The overcurrent protective function of the circuit breaker must be capable of being reset immediately without delay. Overcurrent-sensing elements of the thermal type would require cooling before the protective mechanism could be reset. If the circuit breaker trips while the pump motor is running under fire conditions, resetting the circuit breaker to attempt to place the motor back into service should be accomplished without interruption.

(3) Instantaneous short-circuit overcurrent protection

FAQ ▶
Why is a thermal-magnetic circuit breaker not allowed to provide overcurrent protection for the fire pump motor?

The short-circuit protective function of the circuit breaker must be capable of operating immediately when current through the circuit breaker exceeds the value of the instantaneous trip setting specified in 10.4.3.3.1(6).

(4)*An adequate interrupting rating to provide the suitability rating 10.1.2.2 of the controller

FAQ ▶
How can the interrupting rating of the circuit breaker be less than the short-circuit rating of the fire pump controller?

Fire pump controllers and transfer switches are type tested to determine their suitability or short-circuit rating. The short-circuit rating marked on such equipment may be the same or higher than the interrupting rating of the circuit breaker. The ultimate short-circuit rating of the equipment is dependent upon the ability of the circuit breaker to interrupt current without introducing a risk of electrical shock or fire during short-circuit conditions. The type tests are performed on combinations of circuit breakers and other devices within the controller.

A.10.4.3.3.1(4) The interrupting rating can be less than the suitability rating where other devices within the controller assist in the current-interrupting process.

(5) Capability of allowing normal and emergency starting and running of the motor without tripping *(see 10.5.3.2)*

Fire pump motors exhibit high inrush currents under starting conditions. These currents are maximized (as high as locked rotor) and sustained when the pump encounters debris in the water supply line. It is important that these inrush currents do not cause tripping of the circuit breaker.

(6) An instantaneous trip setting of not more than 20 times the full-load current

FAQ ▶
Why is the instantaneous trip setting of the fire pump breaker allowed to be a maximum of 20 times the motor full-load current?

The limit of 20 times for the instantaneous trip setting of the fire pump breaker was increased several code cycles ago from the value allowed in *NEC* Article 430 for ordinary motor branch circuits. This increase minimized the possibility of the circuit breaker tripping under motor starting conditions. The potential for motor damage at this higher trip setting is minimal, and any damage would be tolerable as starting and running the motor is most critical.

10.4.3.3.2* Current limiters, where integral parts of the circuit breaker, shall be permitted to be used to obtain the required interrupting rating, provided all the following requirements are met:

(1) The breaker shall accept current limiters of only one rating.
(2) The current limiters shall hold 300 percent of full-load motor current for a minimum of 30 minutes.
(3) The current limiters, where installed in the breaker, shall not open at locked rotor current.
(4) A spare set of current limiters of correct rating shall be kept readily available in a compartment or rack within the controller enclosure.

A.10.4.3.3.2 Current limiters are melting link-type devices that, where used as an integral part of a circuit breaker, limit the current during a short circuit to within the interrupting capacity of the circuit breaker.

FAQ ▶
Why are current limiters allowed when fuses may not be used to provide overcurrent protection in the fire pump controller?

Current limiters were very common before circuit breakers with high current-interrupting ratings were available. Current limiters are intended to provide short-circuit protection only, and they must not open under any conditions of motor overload.

10.4.4 Locked Rotor Overcurrent Protection. The only other overcurrent protective device that shall be required and permitted between the isolating switch and the fire pump motor shall be located within the fire pump controller and shall possess the following characteristics:

(1) For a squirrel-cage or wound-rotor induction motor, the device shall be as follows:

(a) Of the time-delay type having a tripping time between 8 seconds and 20 seconds at locked rotor current
(b) Calibrated and set at a minimum of 300 percent of motor full-load current
(2) For a direct-current motor, the device shall be as follows:
(a) Of the instantaneous type
(b) Calibrated and set at a minimum of 400 percent of motor full-load current
(3)* There shall be visual means or markings clearly indicated on the device that proper settings have been made.

A.10.4.4(3) It is recommended that the locked rotor overcurrent device not be reset more than two consecutive times if tripped due to a locked rotor condition without the motor first being inspected for excessive heating and to alleviate or eliminate the cause preventing the motor from attaining proper speed.

Motors that are subjected to more than two consecutive starts under locked rotor conditions may incur damage that would affect the motor's performance or render it inoperable.
See Formal Interpretation 83–1.

(4) It shall be possible to reset the device for operation immediately after tripping, with the tripping characteristics thereafter remaining unchanged.
(5) Tripping shall be accomplished by opening the circuit breaker, which shall be of the external manual reset type.

Formal Interpretation

NFPA 20
Stationary Pumps for Fire Protection
2010 Edition

Reference: 10.4.3, 10.4.4
F.I. 83–1

Question 1: Is it the intent to allow continuous 300 percent of full load current electrical overloading of the fire pump feeder circuits, including transformers, disconnects or other devices on this circuit?

Answer:
a) Relative to protective devices in the fire pump feeder circuit, such devices shall not open under locked rotor currents (see 9.3.2.2).

b) Relative to the isolating means and the circuit breaker of the fire pump controller, it is the intent of 10.4.3 to permit 300 percent of full load motor current to flow continuously through these devices until an electrical failure occurs. [This statement also applies to the motor starter of the fire pump controller, but this device is not in the feeder (see Section 3.3).]

c) Relative to all devices other than those cited above, refer to NFPA 70 for sizing.

Question 2: If the answer to Question 1 is no, what is meant by "setting the circuit breaker at 300 percent of full load current"?

Answer: The phrase "setting the circuit breaker at 300 percent of full load current" means that the circuit breaker will not open (as a normal operation) at 300 percent of full load current. It does not mean that the circuit breaker can pass 300 percent of full load current without ultimately failing from overheating.

Question 3: What is meant by "calibrated up to and set at 300 percent" of motor full load current?

Answer: Question 2 answers the "set at 300 percent" of motor full load current. "Calibrated up to 300 percent" of motor full load current means that calibration at approximately 300 percent is provided by the manufacturer of the circuit breaker.

Issue Edition: 1983
Reference: 6–3.5, 7–4.3
Date: January 1983

Copyright © 2009 All Rights Reserved
NATIONAL FIRE PROTECTION ASSOCIATION

FAQ ▶
Why are the requirements for locked rotor overcurrent protection specified independently of the requirements for circuit breakers?

Locked rotor overcurrent protection is normally provided as part of a circuit breaker listed for fire pump service, but may be provided as a separate device used in conjunction with an instantaneous trip circuit breaker. Calibrated and set at a minimum of 300 percent (or 400 percent for a dc motor) of motor full-load current means the locked rotor overcurrent protection must allow the motor to run continuously at a minimum of 300 percent of motor full-load current. The locked rotor overcurrent protection must be capable of being reset immediately. Overcurrent-sensing elements of the thermal type require cooling before the protective mechanism can be reset. If the locked rotor overcurrent protective device trips while the pump motor is running under fire conditions, resetting it to attempt to place the motor back into service should be accomplished without delay.

The 300 percent trip setting reflects a design point from which a trip curve could be drawn up to the locked rotor current of the motor (600 percent) for the locked rotor overcurrent protective device. Drawing a steady-state 300 percent of nameplate full-load current for a fire pump motor is virtually impossible. As the motor experiences an overload condition, its current draw increases steadily until it reaches a point near the locked rotor value. At that point tripping of the overcurrent protective device occurs within 8 to 20 seconds as defined in 10.4.4.(1)(a). The 300 percent trip setting is a design point only used by the manufacturers and the testing laboratories in the listing/approval process. Field verification is not recommended.

10.4.5 Motor Starting Circuitry.

10.4.5.1 Motor Contactor. The motor contactor shall be horsepower rated and shall be of the magnetic type with a contact in each ungrounded conductor.

10.4.5.1.1 Running contactors shall be sized for both the locked rotor currents and the continuous running currents encountered.

10.4.5.1.2 Starting contactors shall be sized for both the locked rotor current and the acceleration (starting) encountered.

The sizing criteria for the selection of running and starting contactors in a fire pump controller are provided in 10.4.5.1.1 and 10.4.5.1.2. Across-the-line starting fire pump controllers generally employ one contactor that needs to meet the sizing criteria specified for a running contactor. Reduced-voltage starting fire pump controllers employ multiple contactors and the sizing criteria that an individual contactor needs to meet is based on its function in the controller.

10.4.5.2 Timed Acceleration.

10.4.5.2.1 For electrical operation of reduced-voltage controllers, timed automatic acceleration of the motor shall be provided.

10.4.5.2.2 The period of motor acceleration shall not exceed 10 seconds.

The requirement in 10.4.5.2.2 ensures that the pump is running at full speed within 10 seconds of the signal to start.

10.4.5.3 Starting Resistors. Starting resistors shall be designed to permit one 5-second starting operation every 80 seconds for a period of not less than 1 hour.

Starting resistors that are not sized properly based on duty cycle are subject to overheating and burnout. Duty cycle in this context is the proportion of time during which the motor-starting resistor is dissipating its maximum rated power. Duty cycle can be expressed as a ratio or as a percentage. The higher the duty cycle, the shorter the useful life is, if all other things are equal.

10.4.5.4 Starting Reactors and Autotransformers.

10.4.5.4.1 Starting reactors and autotransformers shall comply with the requirements of ANSI/UL 508, *Standard for Industrial Control Equipment,* Table 92.1.

10.4.5.4.2 Starting reactors and autotransformers over 200 hp shall be permitted to be designed to Part 3 of ANSI/UL 508, *Standard for Industrial Control Equipment,* Table 92.1, in lieu of Part 4.

Starting reactors and autotransformers over 200 hp are not required to meet the more stringent performance criteria of ANSI/UL 508, *Standard for Industrial Control Equipment,* Table 92.1, Part 4, because of the requirement in 10.4.5.2.2 limiting motor acceleration time. The criteria in Part 4 of ANSI/UL 508 include a higher duty cycle.

10.4.5.5 Soft Start Units.

10.4.5.5.1 Soft start units shall be horsepower rated or specifically designed for the service.

Soft start units that are rated in amperes must have a rating at least equivalent to the motor full-load current.

◀ FAQ
Do soft start units for fire pump service need to be rated in horsepower?

10.4.5.5.2 The bypass contactor shall comply with 10.4.5.1.

The bypass contactor must have the same characteristics as the main contactor because it may be required to start and/or stop the motor if the soft start unit fails or requires maintenance.

10.4.5.5.3 Soft start units shall comply with the duty cycle requirements in accordance with 10.4.5.4.1 and 10.4.5.4.2.

Soft start units are categorized as reduced voltage starting controllers, similar to resistor, reactor, and autotransformer controllers and, therefore, must meet the same performance criteria as these controllers.

◀ FAQ
Are soft start units considered to be reduced voltage starting controllers?

10.4.5.6 Operating Coils. For controllers of 600 V or less, the operating coil(s) for any motor contactor(s) and any bypass contactor(s), if provided, shall be supplied directly from the main power voltage and not through a transformer.

Supplying the operating coil directly from the main power voltage maximizes the reliability of contactor operation.

◀ FAQ
Why is the operating coil for the motor contactor not allowed to be supplied through a transformer?

10.4.5.7* Single-Phase Sensors in Controller.

A.10.4.5.7 The signal should incorporate local visible indication and contacts for remote indication. The signal can be incorporated as part of the power available indication and loss of phase signal. *(See 10.4.6.1 and 10.4.7.2.2.)*

10.4.5.7.1 Sensors shall be permitted to prevent a three-phase motor from starting under single-phase condition.

10.4.5.7.2 Such sensors shall not cause disconnection of the motor if it is running at the time of single-phase occurrence.

10.4.5.7.3 Such sensors shall be monitored to provide a local visible signal in the event of malfunction of the sensors.

Attempting to start a three-phase motor at rest under a single-phase condition could result in permanent damage to the motor. If the motor is at rest, starting the motor is not possible. However, if the motor is turning at the time of the single-phase occurrence, the motor may continue to run while supplied from single-phase power. In this case attempting to keep the motor running (and the pump delivering water to the system) at the risk of damaging the motor is preferred. Because the sensors can prevent the motor from running, it is necessary for their operation to be monitored in an effort to avoid inadvertent disconnection of the motor.

10.4.6* Signal Devices on Controller.

A.10.4.6 The pilot lamp for signal service should have operating voltage less than the rated voltage of the lamp to ensure long operating life. When necessary, a suitable resistor or potential transformer should be used to reduce the voltage for operating the lamp.

10.4.6.1 Power Available Visible Indicator.

10.4.6.1.1 A visible indicator shall monitor the availability of power in all phases at the line terminals of the motor contactor or of the bypass contactor, if provided.

A digital display or message board would satisfy the requirement in 10.4.6.1.1.

10.4.6.1.2 If the visible indicator is a pilot lamp, it shall be accessible for replacement.

10.4.6.1.3 When power is supplied from multiple power sources, monitoring of each power source for phase loss shall be permitted at any point electrically upstream of the line terminals of the contactor, provided all sources are monitored.

> **FAQ ▶**
> How is power monitored when the fire pump motor is supplied from multiple sources?

Each power source available to the motor, not just the connected power source, is required to be monitored for phase loss at all times. The line terminals of the contactor are specified so that phase loss can be monitored without the motor running.

10.4.6.2 Phase Reversal.

10.4.6.2.1 Phase reversal of the power source to which the line terminals of the motor contactor are connected shall be indicated by a visible indicator.

A digital display or message board would satisfy the requirement in 10.4.6.2.1.

10.4.6.2.2 When power is supplied from multiple power sources, monitoring of each power source for phase reversal shall be permitted at any point electrically upstream of the line terminals of the contactor, provided all sources are monitored.

> **FAQ ▶**
> How is phase reversal monitored when the fire pump motor is supplied from multiple sources?

Each power source available to the motor, not just the connected power source, is required to be monitored for phase reversal at all times. The line terminals of the contactor are specified so that phase reversal can be monitored without the motor running.

10.4.7* Fire Pump Alarm and Signal Devices Remote from Controller.

A.10.4.7 Where unusual conditions exist whereby pump operation is not certain, a "failed-to-operate" fire pump alarm is recommended. In order to supervise the power source for the fire pump alarm circuit, the controller can be arranged to start upon failure of the supervised alarm circuit power.

10.4.7.1 Where the pump room is not constantly attended, audible or visible signals powered by a source not exceeding 125 V shall be provided at a point of constant attendance.

10.4.7.2 These fire pump alarms and signals shall indicate the information in 10.4.7.2.1 through 10.4.7.2.4.

10.4.7.2.1 Pump or Motor Running. The signal shall actuate whenever the controller has operated into a motor-running condition. This signal circuit shall be energized by a separate reliable supervised power source or from the pump motor power, reduced to not more than 125 V.

Energizing the signal from the pump motor power is preferable since the determination of what constitutes a reliable power source is sometimes difficult.

10.4.7.2.2 Loss of Phase.

10.4.7.2.2.1 The fire pump alarm shall actuate whenever any phase at the line terminals of the motor contactor is lost.

10.4.7.2.2.2 All phases shall be monitored. Such monitoring shall detect loss of phase whether the motor is running or at rest.

Some devices do not detect phase loss when the motor is running. The motor may incur damage if allowed to run for long periods under single-phase power. If no fire condition exists, identifying the existence of a single-phase condition is important so that the motor may be shut down.

10.4.7.2.2.3 When power is supplied from multiple power sources, monitoring of each power source for phase loss shall be permitted at any point electrically upstream of the line terminals of the contactor, provided all sources are monitored.

Each power source available to the motor, not just the connected power source, is required to be monitored for phase loss at all times. The line terminals of the contactor are specified so that phase loss can be monitored without the motor running.

◀ FAQ
How is loss of phase monitored when the fire pump motor is supplied from multiple sources?

10.4.7.2.3 Phase Reversal. *(See 10.4.6.2.)* This fire pump alarm circuit shall be energized by a separate reliable supervised power source or from the pump motor power, reduced to not more than 125 V. The fire pump alarm shall actuate whenever the three-phase power at the line terminals of the motor contactor is reversed.

Energizing the signal from the pump motor power is preferable, since determining what constitutes a reliable power source is sometimes difficult.

10.4.7.2.4 Controller Connected to Alternate Source. Where two sources of power are supplied to meet the requirements of 9.3.2, this signal shall indicate whenever the alternate source is the source supplying power to the controller. This signal circuit shall be energized by a separate, reliable, supervised power source, reduced to not more than 125 V.

Whether the motor is being supplied by the primary or alternate source of power is important to know from a remote location. Energizing the signal from the pump motor power is preferable, since determining what constitutes a reliable power source is sometimes difficult.

10.4.8 Controller Contacts for Remote Indication. Controllers shall be equipped with contacts (open or closed) to operate circuits for the conditions in 10.4.7.2.1 through 10.4.7.2.3 and when a controller is equipped with a transfer switch in accordance with 10.4.7.2.4.

The requirement for contacts to operate circuits for the condition specified in 10.4.7.2.4 is covered in 10.8.3.14. Paragraph 10.8.3.14 requires auxiliary open or closed contacts mechanically operated by the fire pump transfer switch mechanism to be provided for remote indication.

10.5 Starting and Control.

10.5.1* Automatic and Nonautomatic.

A.10.5.1 The following definitions are derived from *NFPA 70, National Electrical Code:*

(1) *Automatic.* Self-acting, operating by its own mechanism when actuated by some impersonal influence, as, for example, a change in current strength, pressure, temperature, or mechanical configuration.
(2) *Nonautomatic.* Action requiring intervention for its control. As applied to an electric controller, nonautomatic control does not necessarily imply a manual controller, but only that personal intervention is necessary.

10.5.1.1 An automatic controller shall be self-acting to start, run, and protect a motor.

10.5.1.2 An automatic controller shall be arranged to start the driver upon actuation of a pressure switch or nonpressure switch actuated in accordance with 10.5.2.1 or 10.5.2.2.

The standard was revised in 2003 to allow an automatic controller to be actuated by means other than a pressure switch. An example of a nonpressure switch–actuated controller would be one actuated by a flow switch, smoke detector, or heat sensor.

◀ FAQ
What is a nonpressure switch–actuated controller?

10.5.1.3 An automatic controller shall be operable also as a nonautomatic controller.

10.5.1.4 A nonautomatic controller shall be actuated by manually initiated electrical means and by manually initiated mechanical means.

An example of a manually initiated electrical means would be a start push button energizing the coil of a magnetic contactor.

10.5.2 Automatic Controller.

10.5.2.1* Water Pressure Control.

Water pressure control is the most common method for starting and stopping a fire pump. The pressure switch in both the pressure maintenance pump and fire pump controllers is set to start the pressure maintenance pump first to handle minor variations in pressure due to small leaks in the piping or due to temperature changes without starting the fire pump. In cases where a sprinkler activates, the pressure drop due to waterflow cannot be compensated for by the pressure maintenance pump and the fire pump then starts. The pressure maintenance pump stops when the normal system pressure is restored. The fire pump may be stopped automatically by the pressure switch if the fire pump is not the sole source of water supply to the fire protection system. If the fire pump is the sole source of water supply to the fire protection system, it cannot be stopped automatically. See Exhibit II.10.6 for an example of a controller with pressure switch.

A.10.5.2.1 Installation of the pressure-sensing line between the discharge check valve and the control valve is necessary to facilitate isolation of the jockey pump controller (and sensing line) for maintenance without having to drain the entire system. *[See Figure A.4.30(a) and Figure A.4.30(b).]*

10.5.2.1.1 Pressure-Actuated Switches.

10.5.2.1.1.1 There shall be provided a pressure-actuated switch or electronic pressure sensor having adjustable high- and low-calibrated set-points as part of the controller.

In the 2003 edition of NFPA 20 this requirement read ". . . having independent high- and low-calibrated adjustments . . ." The language was revised for the 2007 edition to clarify that the requirement is for the pressure-actuated switch to have individual, adjustable high- and low-pressure set points. In some pressure switches the set points are not independently adjustable of one another. In the 2010 edition of NFPA 20 this requirement was revised to reflect current technology and practice in permitting the use of electronic pressure transducers.

EXHIBIT II.10.6 *Controller with Pressure Switch.*

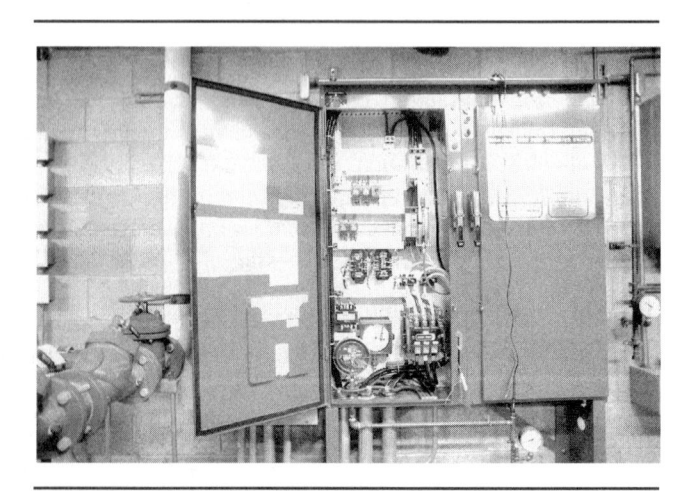

10.5.2.1.1.2 The requirements of 10.5.2.1.1.1 shall not apply in a nonpressure-actuated controller, where the pressure-actuated switch shall not be required.

10.5.2.1.2 There shall be no pressure snubber or restrictive orifice employed within the pressure switch or pressure responsive means.

10.5.2.1.3 There shall be no valve or other restrictions within the controller ahead of the pressure switch or pressure responsive means.

Restrictions of any type are not permitted within the controller ahead of the pressure switch or pressure transducer as they are an unnecessary item in the critical starting path of fire pump controllers. Their presence would reduce the reliability of the fire pump controller.

10.5.2.1.4 This switch shall be responsive to water pressure in the fire protection system.

10.5.2.1.5 The pressure-sensing element of the switch shall be capable of withstanding a momentary surge pressure of 400 psi (27.6 bar) or 133 percent of fire pump controller rated operating pressure, whichever is higher, without losing its accuracy.

Tests are necessary to verify compliance with the requirement in 10.5.2.1.5.

10.5.2.1.6 Suitable provision shall be made for relieving pressure to the pressure-actuated switch to allow testing of the operation of the controller and the pumping unit. *[See Figure A.4.30(a) and Figure A.4.30(b).]*

Testing the controller for operation through the pressure-actuated switch is normally accomplished by opening a solenoid-operated or manual drain valve in the water supply line to the controller pressure switch. This opening drops the pressure in the system below the lower set point of the pressure switch, which initiates the start-and-run sequence for the fire pump motor.

◀ **FAQ**
How is a controller tested for responsiveness to a pressure drop in a nonfire situation?

10.5.2.1.7 Water pressure control shall be in accordance with 10.5.2.1.7.1 through 10.5.2.1.7.5.

10.5.2.1.7.1 Pressure switch actuation at the low adjustment setting shall initiate pump starting sequence (if pump is not already in operation).

10.5.2.1.7.2* A listed pressure recording device shall be installed to sense and record the pressure in each fire pump controller pressure-sensing line at the input to the controller.

A.10.5.2.1.7.2 The pressure recorder should be able to record a pressure at least 150 percent of the pump discharge pressure under no-flow conditions. In a high-rise building, this requirement can exceed 400 psi (27.6 bar). This pressure recorder should be readable without opening the fire pump controller enclosure. This requirement does not mandate a separate recording device for each controller. A single multichannel recording device can serve multiple sensors.

The pressure recorder (see Exhibit II.10.7, top) is not required to be supplied as part of the controller. However, in most cases the pressure recorder is an integral part of the controller. The pressure recorder provides a permanent record of the performance of system pressure before, during, and after a fire event. This information is valuable to those conducting a fire investigation.

◀ **FAQ**
Why is a pressure recorder required?

Many controllers are now available with USB flashdrive log recorders (see Exhibit II.10.7, bottom). These controllers are supplied with a host USB port and can be set to log all events to a USB flashdrive.

10.5.2.1.7.3 The recorder shall be capable of operating for at least 7 days without being reset or rewound.

The recorder must be able to record or store at least 7 days' worth of data. This clause is often misinterpreted to mean the recorder must be able to operate off a backup source of power for at least 7 days, which is not the case.

◀ **FAQ**
What is meant by the phrase *capable of operating for at least 7 days*?

EXHIBIT II.10.7 Pressure Recorder (top) and USB Flashdrive Log Recorder (bottom).

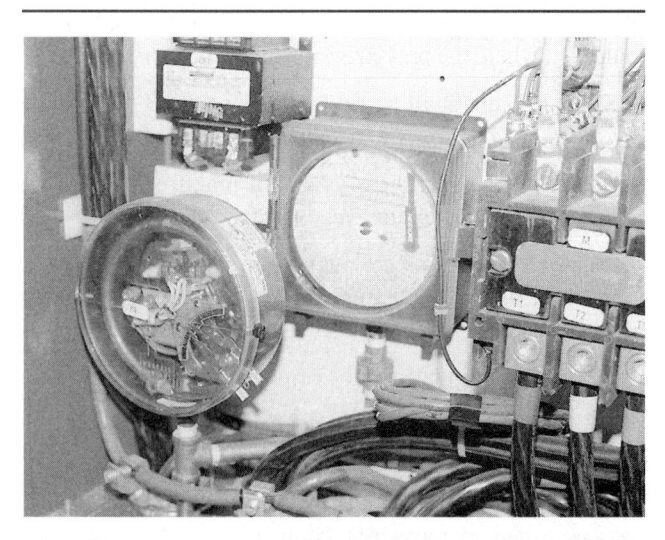

10.5.2.1.7.4 The pressure-sensing element of the recorder shall be capable of withstanding a momentary surge pressure of at least 400 psi (27.6 bar) or 133 percent of fire pump controller rated operating pressure, whichever is greater, without losing its accuracy.

Tests are necessary to verify compliance with the requirement in 10.5.2.1.7.4.

10.5.2.1.7.5 For variable speed pressure limiting control, a ½ in. (15 mm) nominal size inside diameter pressure line shall be connected to the discharge piping at a point recommended by the variable speed control manufacturer. The connection shall be between the discharge check valve and the discharge control valve.

FAQ ▶
Is the location of the pressure-sensing element of the recorder in the water supply piping different for variable speed pressure limiting fire pump controllers?

The pressure-sensing line specified in 10.5.2.1.7.5 is in addition to the pressure-sensing line required by Section 4.30. The pressure-sensing line required by 10.5.2.1.7.5 is needed for control of the variable speed pressure limiting control function of the fire pump controller.

The requirement in 10.5.2.1.7.5 was added to the 2007 edition of NFPA 20 to supplement new requirements for variable speed pressure limiting electric fire pump controllers. Reading the pressure on the system side of the check valve allows such a controller to determine when to shut down on the minimum run timer.

In the 2010 edition of NFPA 20 the location of the connection of the pressure-sensing line was clarified for variable speed pressure limiting fire pump controllers.

10.5.2.2 Nonpressure Switch–Actuated Automatic Controller.

10.5.2.2.1 Nonpressure switch–actuated automatic fire pump controllers shall commence the controller's starting sequence by the automatic opening of a remote contact(s).

10.5.2.2.2 The pressure switch shall not be required.

10.5.2.2.3 There shall be no means capable of stopping the fire pump motor except those on the fire pump controller.

The standard was revised in the 2003 edition to allow an automatic controller to be actuated by means other than a pressure switch. An example of a nonpressure switch–actuated controller would be one actuated by a flow switch, smoke detector, or heat sensor. Initiating the controller's starting sequence by the automatic opening of a remote contact(s) allows the fire pump to start in the event that breakage or disconnection occurs in the circuit connecting the remote contacts.

10.5.2.3 Fire Protection Equipment Control.

10.5.2.3.1 Where the pump supplies special water control equipment (deluge valves, dry pipe valves, etc.), it shall be permitted to start the motor before the pressure-actuated switch(es) would do so.

10.5.2.3.2 Under such conditions the controller shall be equipped to start the motor upon operation of the fire protection equipment.

10.5.2.3.3 Starting of the motor shall be initiated by the opening of the control circuit loop containing this fire protection equipment.

The control circuit of this special fire protection equipment is allowed to bypass the pressure-actuated switch of the controller. As in the case of other remote control stations, initiating the controller's starting sequence by opening the control circuit containing this special fire protection equipment allows the fire pump to start in the event that breakage or disconnection of conductors occurs in this control circuit.

10.5.2.4 Manual Electric Control at Remote Station. Where additional control stations for causing nonautomatic continuous operation of the pumping unit, independent of the pressure-actuated switch, are provided at locations remote from the controller, such stations shall not be operable to stop the motor.

When additional control stations are used to initiate the controller's starting sequence, the controller circuitry must be arranged to lock out these stations so that they are prevented from causing the controller to shut down the fire pump motor. Stopping the motor from a remote location is undesirable because it is difficult to determine if all fires have been extinguished.

10.5.2.5 Sequence Starting of Pumps.

See Exhibit II.10.8 for a diagram of fire pumps in series.

10.5.2.5.1 The controller for each unit of multiple pump units shall incorporate a sequential timing device to prevent any one driver from starting simultaneously with any other driver.

10.5.2.5.2 Each pump supplying suction pressure to another pump shall be arranged to start within 10 seconds before the pump it supplies.

In 10.5.2.5.2, the 2010 edition of NFPA 20 has clarified the time interval for starting pumps in series in order to ensure adequate water flow—especially in high-rise buildings, where the ability of one pump to feed another is critical.

10.5.2.5.3 If water requirements call for more than one pumping unit to operate, the units shall start at intervals of 5 to 10 seconds.

EXHIBIT II.10.8 *Fire Pumps in Series. [Source: NFPA 14, 2007, Figure A.7.1(b)]*

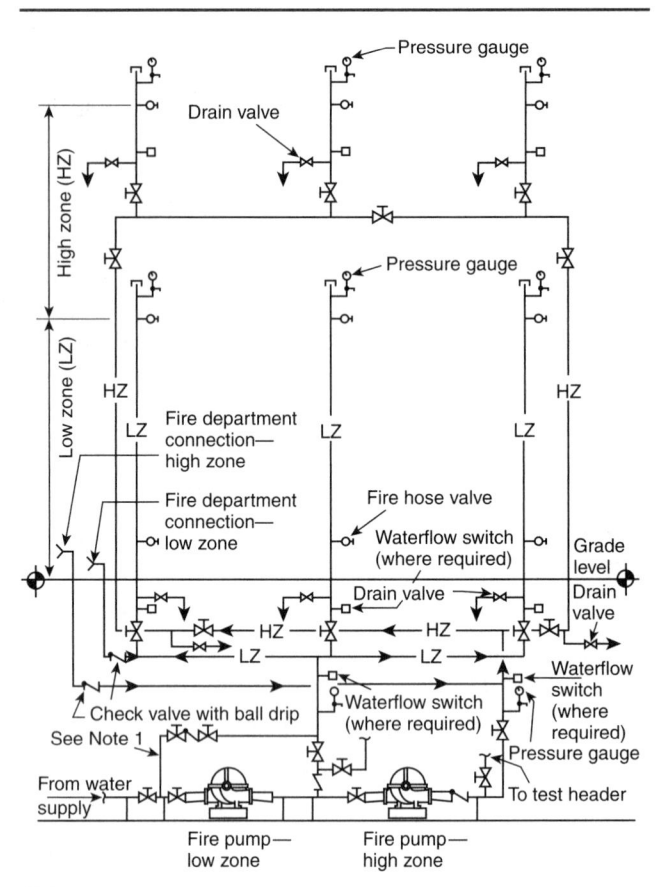

Notes:
1. Bypass in accordance with NFPA 20, *Standard for the Installation of Stationary Pumps for Fire Protection.*
2. High zone pump can be arranged to take suction directly from source of supply.

10.5.2.5.4 Failure of a leading driver to start shall not prevent subsequent pumping units from starting.

FAQ ▶
What is the starting sequence for a system of multiple pumps?

Some installations require multiple pumps, such as in high-rise buildings, in order to deliver adequate water pressure to the sprinklers. Starting such pumps simultaneously is undesirable. There is no reason to start a pump unless adequate water supply is available on its suction side. The 5- to 10-second delay between pumping unit starts allows each pump supplying suction pressure to another pump to operate near or at full speed to provide an adequate water supply. If a leading pump driver fails to start, the operation of subsequent pumping units is imperative to deliver water to the fire-fighting system.

10.5.2.6 External Circuits Connected to Controllers.

10.5.2.6.1 External control circuits that extend outside the fire pump room shall be arranged so that failure of any external circuit (open, ground-fault, or short circuit) shall not prevent operation of pump(s) from all other internal or external means.

10.5.2.6.2 Breakage, disconnecting, shorting of the wires, ground fault, or loss of power to these circuits shall be permitted to cause continuous running of the fire pump but shall not prevent the controller(s) from starting the fire pump(s) due to causes other than these external circuits.

External control circuits that extend outside the fire pump room are vulnerable to damage. It is critical that any such damage not prevent the controller from starting the fire pump when the pump start sequence is initiated by other means, such as the pressure-actuated switch. Any failure of these external control circuits must be allowed to start and run the fire pump motor continuously.

10.5.2.6.3 All control conductors within the fire pump room that are not fault tolerant as described in 10.5.2.6.1 and 10.5.2.6.2 shall be protected against mechanical injury.

Control conductors within the fire pump room are susceptible to damage in the same manner as control conductors outside the fire pump room. *Fault tolerant external control circuit,* as defined in 3.3.7.2, means protected against ground-fault or short circuit. Installation of these conductors in one of the wiring methods permitted by 695.6(E) of the 2008 edition of the *NEC* is one means of providing protection against mechanical injury.

◀ FAQ
What is the meaning of the term *fault tolerant*?

10.5.3 Nonautomatic Controller.

10.5.3.1 Manual Electric Control at Controller.

10.5.3.1.1 There shall be a manually operated switch on the control panel so arranged that, when the motor is started manually, its operation cannot be affected by the pressure-actuated switch.

10.5.3.1.2 The arrangement shall also provide that the unit will remain in operation until manually shut down.

The manually operated switch is one method of testing the controller for proper operation. The switch can also be used to start the fire pump motor in the event that a failure of the automatic starting circuitry occurs.

10.5.3.2* Emergency-Run Mechanical Control at Controller.

A.10.5.3.2 The emergency-run mechanical control provides means for externally and manually closing the motor contactor across-the-line to start and run the fire pump motor. It is intended for emergency use when normal electric/magnetic operation of the contactor is not possible.

When so used on controllers designed for reduced-voltage starting, the 15 percent voltage drop limitation in Section 9.4 is not applicable.

The 15 percent voltage drop limitation under motor-starting conditions is waived for reduced-voltage starting controllers because they are normally used in applications where the power supply cannot deliver locked rotor current when the motor is started at full line voltage. The emergency-run mechanical control bypasses the reduced-voltage starting components of the controller to apply full line voltage to the motor. Although the 15 percent voltage drop limitation is waived, reduced-voltage starting controllers are required to start and run the motor through the use of the emergency-run mechanical control.

◀ FAQ
Is compliance with the 15 percent voltage drop limitation in Section 9.4 required when the fire pump motor is started using the emergency-run mechanical control?

10.5.3.2.1 The controller shall be equipped with an emergency-run handle or lever that operates to mechanically close the motor-circuit switching mechanism.

10.5.3.2.1.1 This handle or lever shall provide for nonautomatic continuous running operation of the motor(s), independent of any electric control circuits, magnets, or equivalent devices and independent of the pressure-activated control switch.

10.5.3.2.1.2 Means shall be incorporated for mechanically latching or holding the handle or lever for manual operation in the actuated position.

10.5.3.2.1.3 The mechanical latching shall not be automatic, but at the option of the operator.

10.5.3.2.2 The handle or lever shall be arranged to move in one direction only from the off position to the final position.

10.5.3.2.3 The motor starter shall return automatically to the off position in case the operator releases the starter handle or lever in any position but the full running position.

The emergency-run mechanical control closes the contacts of the magnetic contactor in the event that control power is lost. The control is normally designed as an over-center mechanism or with a mechanical or magnetic assist to prevent "teasing" of the contactor contacts by the operator. Teasing is an undesirable effect and can result in damage to the contactor.

10.5.4 Methods of Stopping. Shutdown shall be accomplished by the methods in 10.5.4.1 and 10.5.4.2.

10.5.4.1 Manual. Manual shutdown shall be accomplished by operation of a pushbutton on the outside of the controller enclosure that, in the case of automatic controllers, shall return the controller to the full automatic position.

10.5.4.2 Automatic Shutdown After Automatic Start. Where provided, automatic shutdown after automatic start shall comply with the following:

(1) Unless the requirements of 10.5.4.2(3) are met, automatic shutdown shall be permitted only where the controller is arranged for automatic shutdown after all starting and running causes have returned to normal.
(2) A running period timer set for at least 10 minutes running time shall be permitted to commence at initial operation.
(3) The requirements of 10.5.4.2(1) shall not apply and automatic shutdown shall not be permitted where the pump constitutes the sole supply of a fire sprinkler or standpipe system or where the authority having jurisdiction has required manual shutdown.

FAQ ▶
When the fire pump motor is started automatically and all starting and running causes have returned to normal, does the pump have a minimum run time before automatic shutdown occurs?

The controller is not allowed to shut down automatically after an automatic start unless a need exists to run the pump during a fire event. The running period timer setting runs the pump for at least 10 minutes after an automatic start. The running period prevents unnecessary stops and starts of the fire pump motor. In cases where the pump is the only supply of water, stopping the motor automatically is not permitted.

10.6 Controllers Rated in Excess of 600 V.

10.6.1 Control Equipment. Controllers rated in excess of 600 V shall comply with the requirements of Chapter 10, except as provided in 10.6.2 through 10.6.8.

10.6.2 Provisions for Testing.

10.6.2.1 The provisions of 10.3.4.3 shall not apply.

10.6.2.2 An ammeter(s) shall be provided on the controller with a suitable means for reading the current in each phase.

10.6.2.3 An indicating voltmeter(s), deriving power of not more than 125 V from a transformer(s) connected to the high-voltage supply, shall also be provided with a suitable means for reading each phase voltage.

FAQ ▶
How is voltage and current measured in a controller operating in excess of 600 V without introducing a risk of electric shock?

Measuring voltage and current within equipment operating in excess of 600 V poses a high risk of electric shock. Meters on the outside of the controller enclosure enable measuring voltage and current without such risk. The voltage at the voltmeter terminals must be provided through a stepdown transformer to isolate voltages in excess of 600 V from the controller enclosure and operator.

10.6.3 Disconnecting Under Load.

10.6.3.1 Provisions shall be made to prevent the isolating switch from being opened under load.

Controllers in excess of 600 V are generally provided with an isolating switch that is not rated to interrupt current. In such cases the isolating switch must be prevented from being opened or closed when the contactor is closed, by interlocking the contactor with the isolating switch. Although not rated to interrupt current, the isolating switch is generally tested to determine whether it is capable of making and breaking the current of the primary winding of the control power transformer. See 10.6.5.2.

10.6.3.2 A load-break disconnecting means shall be permitted to be used in lieu of the isolating switch if the fault closing and interrupting ratings equal or exceed the requirements of the installation.

If a switch is used that can safely close on the available fault current and interrupt the locked rotor current of the motor, no interlocking is needed.

10.6.4 Pressure-Actuated Switch Location. Special precautions shall be taken in locating the pressure-actuated switch called for in 10.5.2.1 to prevent any water leakage from coming in contact with high-voltage components.

Water leakage on high-voltage components in the controller can cause a risk of electric shock for someone coming in contact with the controller enclosure. The pressure-actuated switch (see Exhibit II.10.9) is normally located at the bottom of the enclosure and in a barriered compartment for controllers rated in excess of 600 V.

10.6.5 Low-Voltage Control Circuit.

10.6.5.1 The low-voltage control circuit shall be supplied from the high-voltage source through a stepdown transformer(s) protected by high-voltage fuses in each primary line.

10.6.5.2 The transformer power supply shall be interrupted when the isolating switch is in the open position.

De-energizing the control circuit power prevents the contactor contacts from being closed by the operating coil of the contactor with the isolating switch open. This requirement

◀ **FAQ**
When is interlocking between the isolating switch and contactor in a controller rated over 600 V not required?

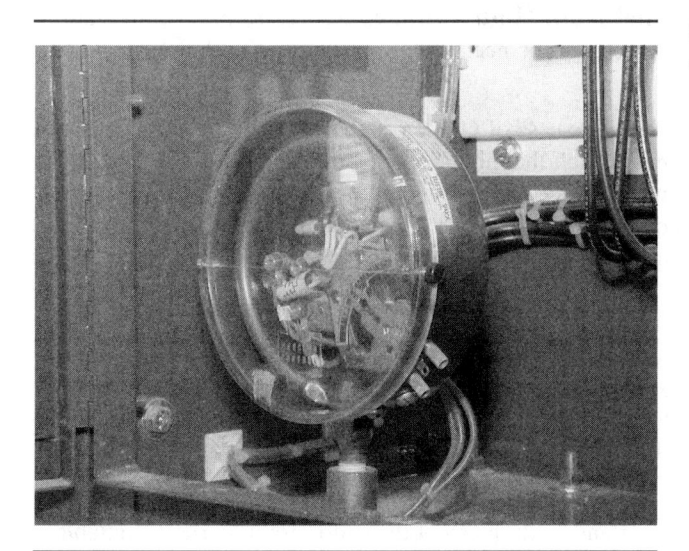

EXHIBIT II.10.9 Pressure-Actuated Switch.

supplements the interlocking mentioned in the commentary to 10.6.3.1 to prevent the isolating switch from operating under load. The contacts of the isolating switch are generally used to make and break the current of the primary winding of the control power transformer.

10.6.5.3 The secondary of the transformer and control circuitry shall otherwise comply with 10.3.5.

10.6.5.4 One secondary line of the high voltage transformer or transformers shall be grounded unless all control and operator devices are rated for use at the high (primary) voltage.

Low voltage in this case means the control circuit operates at 600 V or less. Operating the control circuit at 600 V or less reduces the risk of electric shock for operators of equipment and maintenance personnel. One secondary line of the transformer is normally grounded as control and operator devices are seldom rated for use at the high (primary) voltage. In the 2010 edition of NFPA 20, 10.6.5.4 was changed to clarify that the requirement applies to high (medium) voltage transformers.

10.6.5.5 Current Transformers. Unless rated at the incoming line voltage, the secondaries of all current transformers used in the high voltage path shall be grounded.

Unless the current transformer is rated for use at the incoming line (medium) voltage, the secondaries of all current transformers should be grounded to reduce the risk of electric shock to anyone who may come in contact with the fire pump controller.

10.6.6 Indicators on Controller.

10.6.6.1 Specifications for controllers rated in excess of 600 V differ from those in 10.4.6.

10.6.6.2 A visible indicator shall be provided to indicate that power is available.

10.6.6.3 The current supply for the visible indicator shall come from the secondary of the control circuit transformer through resistors, if found necessary, or from a small-capacity stepdown transformer, which shall reduce the control transformer secondary voltage to that required for the visible indicator.

The voltage at the visible indicator must be provided through a stepdown transformer or similar means to isolate voltages in excess of 600 V from the controller enclosure and operator.

10.6.6.4 If the visible indicator is a pilot lamp, it shall be accessible for replacement.

10.6.7 Protection of Personnel from High Voltage. Necessary provisions shall be made, including such interlocks as might be needed, to protect personnel from accidental contact with high voltage.

Contact with equipment operating in excess of 600 V poses a high risk of electric shock. The equipment must be constructed such that high-voltage circuits are de-energized when the equipment enclosure is opened. Intentional grounding of the power circuits may be necessary to provide isolation from the high-voltage source of supply. Protection from contact with high-voltage circuits during operation of the equipment is also critical. Such protection includes the necessary means to isolate voltages in excess of 600 V from the controller enclosure and operator.

10.6.8 Disconnecting Means. A contactor in combination with current-limiting motor circuit fuses shall be permitted to be used in lieu of the circuit breaker (disconnecting means) required in 10.4.3.1 if all of the following requirements are met:

(1) Current-limiting motor circuit fuses shall be mounted in the enclosure between the isolating switch and the contactor and shall interrupt the short-circuit current available at the controller input terminals.

(2) These fuses shall have an adequate interrupting rating to provide the suitability rating *(see 10.1.2.2)* of the controller.
(3) The current-limiting fuses shall be sized to hold 600 percent of the full-load current rating of the motor for at least 100 seconds.
(4) A spare set of fuses of the correct rating shall be kept readily available in a compartment or rack within the controller enclosure.

Controllers in excess of 600 V, unlike those rated 600 V or less, may be supplied with fuses that provide short-circuit protection and minimal locked rotor current protection. For this construction, closing on a short circuit will always be performed by the high-voltage contactor that is rated for such duty. The location of the fuses allows them to be replaced with de-energized fuseholders. The sizing of the fuses to hold 600 percent of the full-load current rating of the motor for at least 100 seconds is necessary to prevent opening of the fuses due to motor overload. The spare set of fuses is necessary in the event of the opening of one or more fuses and minimizes the amount of time the fire pump is out of service.

◀ **FAQ**
Why is a contactor permitted to be used in lieu of a circuit breaker in a controller operating in excess of 600 V?

10.6.9 Locked Rotor Overcurrent Protection.

10.6.9.1 Tripping of the locked rotor overcurrent device required by 10.4.4 shall be permitted to be accomplished by opening the motor contactor coil circuit(s) to drop out the contactor.

This paragraph is a companion requirement to 10.6.8. If a controller is provided with fuses in lieu of a circuit breaker, the only means to de-energize the motor under overload conditions is through the opening of the contactor that is rated for such duty.

◀ **FAQ**
If high-voltage fuses are provided in lieu of a circuit breaker in a controller operating in excess of 600 V, how are motor overload currents interrupted?

10.6.9.2 Means shall be provided to restore the controller to normal operation by an external manually reset device.

This paragraph is also a companion requirement to 10.6.8. Resetting a tripped circuit breaker is normally accomplished through the use of the circuit breaker operating handle mechanism that is accessible from outside the controller enclosure. If a controller is provided with fuses in lieu of a circuit breaker, a separate device to reset the locked rotor overcurrent protection must be provided external to the controller enclosure.

◀ **FAQ**
If high-voltage fuses are provided in lieu of a circuit breaker in a controller operating in excess of 600 V, how is the locked rotor overcurrent protection reset?

10.6.10 Emergency-Run Mechanical Control at Controller.

10.6.10.1 The controller shall comply with 10.5.3.2.1 and 10.5.3.2.2 except that the mechanical latching can be automatic.

Automatic mechanical latching is preferred in order to ensure a positive closing of the high-voltage contactor. As mentioned in the commentary to 10.5.3.2.3, teasing of the contacts is an undesirable effect. In controllers rated in excess of 600 V, there is zero tolerance for this action.

10.6.10.2 Where the contactor is latched in, the locked rotor overcurrent protection of 10.4.4 shall not be required.

This paragraph is a companion requirement to 10.6.9.1. If the high-voltage contactor is the only means to de-energize the motor under overload conditions, when this contactor is latched in, the overload protection is rendered inoperative. If a motor started by the emergency-run mechanical control assumes an overload condition, the motor may be disconnected by the opening of the fuses, which is allowed after 100 seconds at 600 percent of the full-load current rating of the motor. See 10.6.8(3).

◀ **FAQ**
When the motor is started by the emergency-run mechanical control, the motor overload protection is rendered inoperative. In this situation what protects the motor from overcurrent?

10.7* Limited Service Controllers

A.10.7 The use of limited service controllers for fire service is permitted for special situations where acceptable to the authority having jurisdiction. NFPA 20 permits limited service controllers

(LSCs) to have no isolating switch and use thermally responsive overcurrent protective devices that limit their application. These compromises have introduced the following:

(1) The controller circuit breaker can trip if the fire is near (cannot be reset because the breaker is hot, even when the manual/emergency handle is used).
(2) The reset time is substantially longer if the breaker trips due to distressed pump, and so forth. Tripping consistency and reset times are compromised on "hot-starts."
(3) The down time is substantially longer (no fire protection) if the breaker needs service/replacement due to no isolating switch. Most LSCs are suitable for use as service equipment (SUSE rated) and are so used.
(4) Sizing of breakers is different and can significantly exceed the 8 seconds to 20 seconds locked rotor current (LRC) trip time of a full-service fire pump controller depending on hot or cold starts.

10.7.1 Limitations. Limited service controllers consisting of automatic controllers for across-the-line starting of squirrel-cage motors of 30 hp or less, 600 V or less, shall be permitted to be installed where such use is acceptable to the authority having jurisdiction.

FAQ ▶
How is a limited service controller different from a full service fire pump controller?

The use of limited service controllers was restricted by their maximum allowable horsepower rating (30 hp). Limited service controllers were introduced many years ago as an economical alternative for premises that would otherwise have no fire suppression equipment. Although these controllers have numerous limitations compared to fire pump controllers, they did provide a degree of protection against the spread of fire when applied properly. They should not be confused with pressure maintenance (jockey or make-up) pump controllers. Pressure maintenance pump controllers are used with pumps to maintain system pressure in the water supply in the event of fluctuations in the system pressure.

10.7.2 Requirements. The provisions of Sections 10.1 through 10.5 shall apply, unless specifically addressed in 10.7.2.1 through 10.7.2.4.

10.7.2.1 In lieu of 10.4.3.3.1(2) and 10.4.4, the locked rotor overcurrent protection shall be permitted to be achieved by using an inverse time nonadjustable circuit breaker having a standard rating between 150 percent and 250 percent of the motor full-load current.

10.7.2.2 In lieu of 10.1.2.5.1, each controller shall be marked "Limited Service Controller" and shall show plainly the name of the manufacturer, the identifying designation, and the complete electrical rating.

Limited service controllers are not identified as fire pump controllers, although they perform the same function as full service fire pump controllers. This marking provides a distinction between the two types of controllers.

10.7.2.3 The controller shall have a short-circuit current rating not less than 10,000 A.

Tests are necessary to verify compliance with the requirement in 10.7.2.3.

10.7.2.4 The manually operated isolating switch specified in 10.4.2 shall not be required.

10.8* Power Transfer for Alternate Power Supply.

A.10.8 Typical fire pump controller and transfer switch arrangements are shown in Figure A.10.8. Other configurations can also be acceptable.

10.8.1 General.

10.8.1.1 Where required by the authority having jurisdiction or to meet the requirements of 9.3.2 where an on-site electrical power transfer device is used for power source selection, such switch shall comply with the provisions of Section 10.8 as well as Sections 10.1, 10.2, and 10.3 and 10.4.1.

FIGURE A.10.8 Typical Fire Pump Controller and Transfer Switch Arrangements.

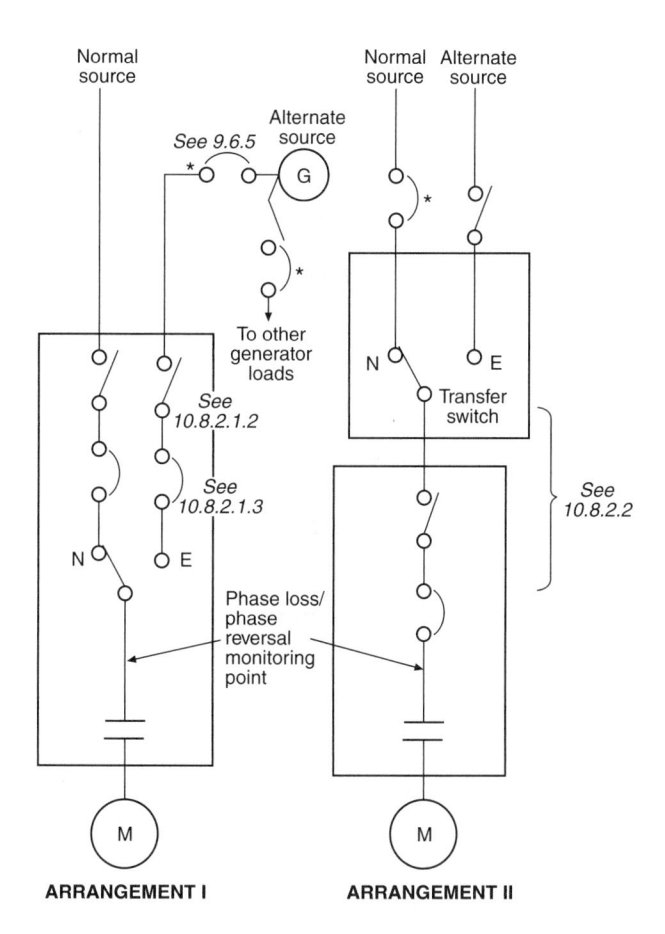

*Circuit breakers or fusible switches can be used.

- (M) Motor
- (G) Generator
- E Emergency
- N Normal

ARRANGEMENT I / **ARRANGEMENT II**

EXHIBIT II.10.10 Transfer Switch.

A transfer switch is used when the normal power source is not reliable (see Exhibit II.10.10). The function of the transfer switch is to transfer the fire pump motor to an alternate source of power when the power from the normal source is interrupted.

10.8.1.2 Manual transfer switches shall not be used to transfer power between the normal supply and the alternate supply to the fire pump controller.

FAQ ▶
What type of transfer switch is allowed in fire pump applications?

All transfer switches used in fire pump applications must be of the automatic type.

10.8.1.3 No remote device(s) shall be installed that will prevent automatic operation of the transfer switch.

The installation of any device (remote sensing or other) that inhibits the operation of the transfer switch reduces the reliability of the transfer operation from the normal to the alternate source of power.

10.8.2* Fire Pump Controller and Transfer Switch Arrangements.

A.10.8.2 The compartmentalization or separation is to prevent propagation of a fault in one compartment to the source in the other compartment.

10.8.2.1 Arrangement I (Listed Combination Fire Pump Controller and Power Transfer Switch).

10.8.2.1.1 Self-Contained Power Switching Assembly. Where the power transfer switch consists of a self-contained power switching assembly, such assembly shall be housed in a barriered compartment of the fire pump controller or in a separate enclosure attached to the controller and marked "Fire Pump Power Transfer Switch."

This type of arrangement is generally listed as one assembly, a fire pump controller, although the enclosure contains both a fire pump controller and a transfer switch. The compartment or enclosure housing the fire pump controller is marked "Fire Pump Controller," and the portion containing the transfer switch is marked accordingly.

10.8.2.1.2 Isolating Switch.

10.8.2.1.2.1 An isolating switch, complying with 10.4.2, located within the power transfer switch enclosure or compartment shall be provided ahead of the alternate input terminals of the transfer switch.

The isolating switch connected to the alternate power source terminals of the transfer switch provides for disconnection of the transfer switch and components in the fire pump motor branch circuit for maintenance and repair.

10.8.2.1.2.2 The isolating switch shall be suitable for the available short circuit of the alternate source.

Transfer switches may have different short-circuit ratings on the normal and alternate sources. This situation is typical when the transfer switch alternate source is a premises-owned generator. The normal source is generally a utility, and the available short-circuit current at the normal source terminals of the transfer switch would be somewhat or significantly higher than the available short-circuit current at the alternate source terminals.

In the 2010 edition of NFPA 20, the requirements for equipment in the alternate source side of the transfer switch when supplied by a second utility were revised such that the requirements are now identical to the those for equipment in the normal source side of the transfer switch. The essence of the change is that the transfer switch emergency side is now required to be provided with both an isolation switch and a circuit breaker. Previous editions of NFPA 20 included a requirement that if the alternate source was supplied by a generator with a capacity that exceeded 225 percent of the fire pump motor's rated full-load current, the combination fire pump controller/transfer switch had to be equipped with a circuit breaker and isolating switch on the alternate side input. This requirement was difficult, if not impossible, to enforce. Most fire pump controllers are shipped for installation at a site where the generator capacity is not known or is incorrectly stated. Determining if a generator's capacity exceeds 225 percent of the fire pump motor's rated full-load current requires several pieces of information that are not always available when the controller is assembled.

The revision also eliminated the need to determine whether the alternate source is supplied by one or more upstream transfer switches that can singly or in combination feed utility or on-site generated power to the fire pump controller. That determination is difficult to make, as it requires knowledge prior to installation of the types of power sources capable of supplying the alternate source side of the combination fire pump controller/transfer switch. There was also a void in the previous editions of the standard in that combination fire pump controller/transfer switches that were not supplied with a circuit breaker on the transfer switch emergency side were not required to be marked or identified that they are restricted in their use. Finally, generator capacity of 225 percent of the fire pump motor's rated full-load current was an arbitrarily chosen value. There is no proof that motor overload currents up to 225 percent of motor full-load current are free from producing hazards such as overheating of conductors.

10.8.2.1.3 Circuit Breaker. The transfer switch emergency side shall be provided with a circuit breaker complying with 10.4.3 and 10.4.4.

10.8.2.1.4 Cautionary Marking. The fire pump controller and transfer switch *(see 10.8.2.1)* shall each have a cautionary marking to indicate that the isolation switch for both the controller and the transfer switch is opened before servicing the controller, transfer switch, or motor.

When a transfer switch is used, both the normal and alternate source must be disconnected for servicing the controller, transfer switch, or motor. The disconnection is accomplished through the isolating switches required for both normal and alternate sources.

◀ **FAQ**
Why is an isolating switch required for both the normal and alternate input terminals of the transfer switch?

10.8.2.2 Arrangement II (Individually Listed Fire Pump Controller and Power Transfer Switch). The following shall be provided:

(1) A fire pump controller power transfer switch complying with Sections 9.6 and 10.8 and a fire pump controller shall be provided.

The "power transfer switch" described in this arrangement is commonly referred to as a *separately mounted fire pump transfer switch* and is not part of the fire pump controller.

◀ **FAQ**
What is meant by the term *separately mounted fire pump transfer switch*?

(2) An isolating switch, or service disconnect where required, ahead of the normal input terminals of the transfer switch shall be provided.

The isolating switch specified in 10.8.2.2(2) is the same isolating switch required in 10.8.2.1.2 for a listed combination fire pump controller and power transfer switch.

(3) The transfer switch overcurrent protection shall be selected or set to indefinitely carry the locked rotor current of the fire pump motor where the alternate source is supplied by a second utility.

The overcurrent protection specified in 10.8.2.2(3) is the same as that required for a transfer switch required by the *NEC*. Because it is not the overcurrent protection provided by the circuit breaker in the fire pump controller, its selection is critical in that it must not trip unless the circuit breaker in the fire pump controller trips.

◀ **FAQ**
How is overcurrent protection provided for a separately mounted fire pump transfer switch?

(4) An isolating switch ahead of the alternate source input terminals of the transfer switch shall meet the following requirements:
    (a) The isolating switch shall be lockable in the on position.

Stationary Fire Pumps Handbook 2010

(b) A placard shall be externally installed on the isolating switch stating "Fire Pump Isolating Switch," with letters at least 1 in. (25 mm) in height.
(c) A placard shall be placed adjacent to the fire pump controller stating the location of the isolating switch and the location of the key (if the isolating switch is locked).
(d) The isolating switch shall be supervised to indicate when it is not closed, by one of the following methods:
  i. Central station, proprietary, or remote station signal service
  ii. Local signaling service that will cause the sounding of an audible signal at a constantly attended point
  iii. Locking the isolating switch closed
  iv. Sealing of isolating switches and approved weekly recorded inspections where isolating switches are located within fenced enclosures or in buildings under the control of the owner
(e) This supervision shall operate an audible and visible signal on the transfer switch and permit monitoring at a remote point where required.

FAQ ▶
Are special requirements given for the isolating switch ahead of the alternate source input terminals on a separately mounted fire pump transfer switch?

The requirements in 10.8.2.2(4) are similar to those applied to the remote disconnecting means permitted in 9.2.3. These additional requirements are necessary for the alternate source isolating switch provided for the separately mounted transfer switch, as this isolating switch may be remote from the fire pump controller and not readily accessible.

10.8.2.3 Transfer Switch. Each fire pump shall have its own dedicated transfer switch(es) where a transfer switch(es) is required.

FAQ ▶
What is meant by a "dedicated transfer switch" in a fire pump circuit?

A transfer switch is not allowed to supply more than one fire pump unless it is through one or more downstream transfer switches—that is, a fire pump circuit may have additional transfer switches upstream from the dedicated transfer switch, and those transfer switches may supply other fire pumps and building loads. Only the dedicated transfer switch is subject to the requirements of NFPA 20. All other transfer switches upstream of the dedicated transfer switch are outside the scope of NFPA 20.

10.8.3 Power Transfer Switch Requirements.

10.8.3.1 Listing. The power transfer switch shall be specifically listed for fire pump service.

FAQ ▶
What are the differences between a listed transfer switch and a transfer switch listed for fire pump service?

Not all listed transfer switches are suitable for use in a fire pump branch circuit. Only those that are listed and marked as being suitable for fire pump service are acceptable in this application. Additional requirements are applied to transfer switches that are intended to be used in a fire pump application. Those additional requirements are specified in 10.8.3.3 to 10.8.3.14.

10.8.3.2 Suitability. The power transfer switch shall be suitable for the available short-circuit currents at the transfer switch normal and alternate input terminals.

10.8.3.3 Electrically Operated and Mechanically Held. The power transfer switch shall be electrically operated and mechanically held.

10.8.3.4 Horsepower or Ampere Rating.

10.8.3.4.1 Where rated in horsepower, the power transfer switch shall have a horsepower rating at least equal to the motor horsepower.

10.8.3.4.2 Where rated in amperes, the power transfer switch shall have an ampere rating not less than 115 percent of the motor full-load current and also be suitable for switching the motor locked rotor current.

10.8.3.5 Manual Means of Operation.

10.8.3.5.1 A means for safe manual (nonelectrical) operation of the power transfer switch shall be provided.

10.8.3.5.2 This manual means shall not be required to be externally operable.

10.8.3.6 Undervoltage-Sensing Devices. Unless the requirements of 10.8.3.6.5 are met, the requirements of 10.8.3.6.1 through 10.8.3.6.4 shall apply. Turning off the normal source isolating switch or the normal source circuit breaker shall not inhibit the transfer switch from operating as required by 10.8.3.6.1 through 10.8.3.6.4.

10.8.3.6.1 The power transfer switch shall be provided with undervoltage-sensing devices to monitor all ungrounded lines of the normal power source.

10.8.3.6.2 Where the voltage on any phase at the load terminals of the circuit breaker within the fire pump controller falls below 85 percent of motor-rated voltage, the power transfer switch shall automatically initiate starting of the standby generator, if provided and not running, and initiate transfer to the alternate source.

10.8.3.6.3 Where the voltage on all phases of the normal source returns to within acceptable limits, the fire pump controller shall be permitted to be retransferred to the normal source.

10.8.3.6.4 Phase reversal of the normal source power *(see 10.4.6.2)* shall cause a simulated normal source power failure upon sensing phase reversal.

10.8.3.6.5 The requirements of 10.8.3.6.1 through 10.8.3.6.4 shall not apply where the power transfer switch is electrically upstream of the fire pump controller circuit breaker, and voltage shall be permitted to be sensed at the input to the power transfer switch in lieu of at the load terminals of the fire pump controller circuit breaker.

10.8.3.7 Voltage- and Frequency-Sensing Devices. Unless the requirements of 10.8.3.7.3 are met, the requirements of 10.8.3.7.1 and 10.8.3.7.2 shall apply.

10.8.3.7.1 Voltage- and frequency-sensing devices shall be provided to monitor at least one ungrounded conductor of the alternate power source.

10.8.3.7.2 Transfer to the alternate source shall be inhibited until there is adequate voltage and frequency to serve the fire pump load.

10.8.3.7.3 Where the alternate source is provided by a second utility power source, the requirements of 10.8.3.7.1 and 10.8.3.7.2 shall not apply, and undervoltage-sensing devices shall monitor all ungrounded conductors in lieu of a frequency-sensing device.

When the alternate source is provided by a second utility, the concern of having adequate voltage and frequency to serve the fire pump load does not exist. The alternate source is always energized and available to drive the fire pump motor.

10.8.3.8 Visible Indicators. Two visible indicators shall be provided to externally indicate the power source to which the fire pump controller is connected.

10.8.3.9 Retransfer.

10.8.3.9.1 Means shall be provided to delay retransfer from the alternate power source to the normal source until the normal source is stabilized.

Damage could occur to the fire pump motor if the voltage and frequency to serve the fire pump load is not adequate. The delay in retransfer allows the normal source to achieve rated and consistent voltage and frequency.

◀ FAQ
Why is the delay of retransfer from the alternate power source to the normal source necessary until the normal source is stabilized?

10.8.3.9.2 This time delay shall be automatically bypassed if the alternate source fails.

If the alternate source fails during the period in which the transfer switch is attempting to return to the normal power source, it is important to continue driving the fire pump even if the normal power source does not have adequate voltage and frequency to serve the fire pump load.

10.8.3.10 In-Rush Currents. Means shall be provided to prevent higher than normal in-rush currents when transferring the fire pump motor from one source to the other.

10.8.3.11 Overcurrent Protection. The power transfer switch shall not have integral short circuit or overcurrent protection.

FAQ ▶
Why are fire pump transfer switches not allowed to have integral overcurrent protection?

If an overcurrent condition exists in the fire pump branch circuit, only the circuit breaker in the fire pump controller is permitted to operate to interrupt the overcurrent. Because of the special characteristics of this circuit breaker per 10.4.3.3 and associated locked rotor overcurrent protection per 10.4.4, any short-circuit or overcurrent protection integral to the transfer switch could render the fire pump inoperable in conditions where its operation is critical and it otherwise should be allowed to operate.

10.8.3.12 Additional Requirements. The following shall be provided:

(1) A device to delay starting of the alternate source generator to prevent nuisance starting in the event of momentary dips and interruptions of the normal source
(2) A circuit loop to the alternate source generator whereby either the opening or closing of the circuit will start the alternate source generator (when commanded by the power transfer switch) *(see 10.8.3.6)*
(3) A means to prevent sending of the signal for starting of the alternate source generator when commanded by the power transfer switch, if the alternate isolating switch or the alternate circuit breaker is in the open or tripped position

10.8.3.12.1 The alternate isolating switch and the alternate circuit breaker shall be monitored to indicate when one of them is in the open or tripped position, as specified in 10.8.3.12(3).

The monitoring required in 10.8.3.12.1 should eliminate the concern of starting the alternate source generator when it is unable to supply power to the fire pump motor because the alternate isolating switch or the alternate circuit breaker is in the open or tripped position. However, if the alternate isolating switch or circuit breaker is not closed before the need to signal the generator to start exists, there is no reason to send the signal to start.

Previous editions of NFPA 20 included an exception to the requirement in 10.8.3.12.1 that monitoring of both the alternate isolating switch and alternate circuit breaker was not required when they were interlocked. In the 2010 edition of NFPA 20 this requirement was deleted. It is necessary to monitor the position of the alternate isolating switch as it is the component that may defeat starting of the generator. It is also necessary to monitor the position of the alternate circuit breaker as the power available/power failure contacts will normally be supervising only the normal source side circuit breaker. There would be no warning of either a tripped or a manually opened breaker on the alternate side without this monitoring.

10.8.3.12.2 Supervision shall operate an audible and visible signal on the fire pump controller/automatic transfer switch combination and permit monitoring at a remote point where required.

10.8.3.13 Momentary Test Switch. A momentary test switch, externally operable, shall be provided on the enclosure that will simulate a normal power source failure.

10.8.3.14 Remote Indication. Auxiliary open or closed contacts mechanically operated by the fire pump power transfer switch mechanism shall be provided for remote indication in accordance with 10.4.8.

The remote indication requirement in 10.4.8 states that where two sources of power are supplied to meet the requirements of 9.3.2, controllers with transfer switches or separately mounted power transfer switches must be equipped with contacts (open or closed) to operate circuits for the condition where the alternate source is the source supplying power to the controller.

10.9 Controllers for Additive Pump Motors.

In previous editions of NFPA 20, this section was titled "Controllers for Foam Concentrate Pump Motors." The title was revised to clarify that the requirements apply to all controllers used in systems where additives are provided to aid in fire suppression.

10.9.1 Control Equipment.

Controllers for additive pump motors shall comply with the requirements of Sections 10.1 through 10.5 (and Section 10.8, where required) unless specifically addressed in 10.9.2 through 10.9.5.

10.9.2 Automatic Starting.

In lieu of the pressure-actuated switch described in 10.5.2.1, automatic starting shall be capable of being accomplished by the automatic opening of a closed circuit loop containing this fire protection equipment.

Controllers for additive pump motors are generally started by remote means and conditions other than a drop in water pressure due to a sprinkler system activation. The closed circuit loop ensures that the pump will start even if a failure of the starting circuit and/or starting circuit components exists.

10.9.3 Methods of Stopping.

10.9.3.1 Manual shutdown shall be provided.

10.9.3.2 Automatic shutdown shall not be permitted.

In fire suppression systems where an additive is employed, manual shutdown is the required method of stopping the pump. No automatic detection and signaling means are available to determine whether a need for this type of fire suppression system to operate still exists. Automatic shutdown of fire suppression systems is not permitted because malfunction of an automatic shutdown function can occur.

10.9.4 Lockout.

10.9.4.1 Where required, the controller shall contain a lockout feature where used in a duty-standby application.

10.9.4.2 Where supplied, this lockout shall be indicated by a visible indicator and provisions for annunciating the condition at a remote location.

The lockout feature allows for stopping the pump in the event the source of additive has been depleted. Generally, the operation of the fire pump in the event that the additive is no longer available is undesirable. For example, attempting to suppress a chemical-based fire with water only could result in spreading or increasing the size of the fire.

10.9.5 Marking.

The controller shall be marked "Additive Pump Controller."

10.10* Controllers with Variable Speed Pressure Limiting Control or Variable Speed Suction Limiting Control.

Section 10.10 was added to NFPA 20 in the 2007 edition. These added requirements reflect the latest technology in fire suppression systems. In the 2010 edition of NFPA 20, this section

◀ FAQ
Do the requirements of Section 10.9 apply to controllers for foam concentrate pumps?

◀ FAQ
Is there a means available to stop the pump in controllers for additive pump motors should the additive tank run dry?

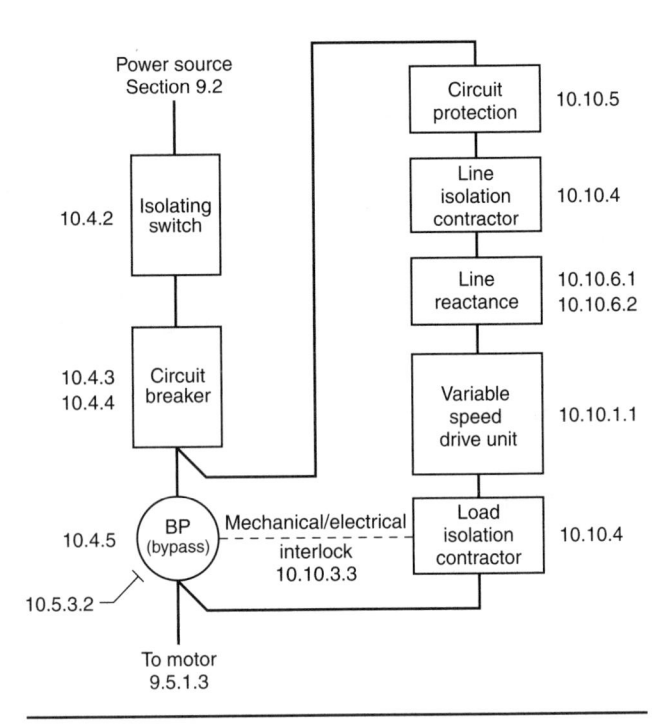

FIGURE A.10.10 *Variable Speed Pressure Limiting Control.*

was expanded to include variable speed suction-limiting control systems. These speed control systems are used to maintain a minimum positive suction pressure at the pump inlet by reducing the pump driver speed from rated speed while monitoring pressure in the suction piping through a sensing line.

A.10.10 See Figure A.10.10.

10.10.1 Control Equipment.

See Exhibit II.10.11 for a typical variable speed pressure limiting fire pump controller with key components identified.

10.10.1.1 Controllers equipped with variable speed pressure limiting control or variable speed suction limiting control shall comply with the requirements of Chapter 10, except as provided in 10.10.1 through 10.10.11.

FAQ ▶
When are controllers equipped with variable speed pressure limiting control used?

Controllers equipped with variable speed pressure limiting control are used in systems where overpressurizing the piping, fittings, connection points, and/or sprinklers could result in damage to one or more of these components. See Exhibit II.10.12 for a controller equipped with variable speed pressure limiting control.

10.10.1.2 Controllers with variable speed pressure limiting control or variable speed suction limiting control shall be listed for fire service.

10.10.1.3 The variable speed pressure limiting control or variable speed suction limiting control shall have a horsepower rating at least equal to the motor horsepower or, where rated in amperes, shall have an ampere rating not less than the motor full-load current.

The variable speed pressure limiting control replaces the motor contactor specified in 10.4.5.1. As such, its rating requirements are similar to those for the magnetic contactor. The variable speed pressure limiting control is generally a variable speed drive control, and many of these

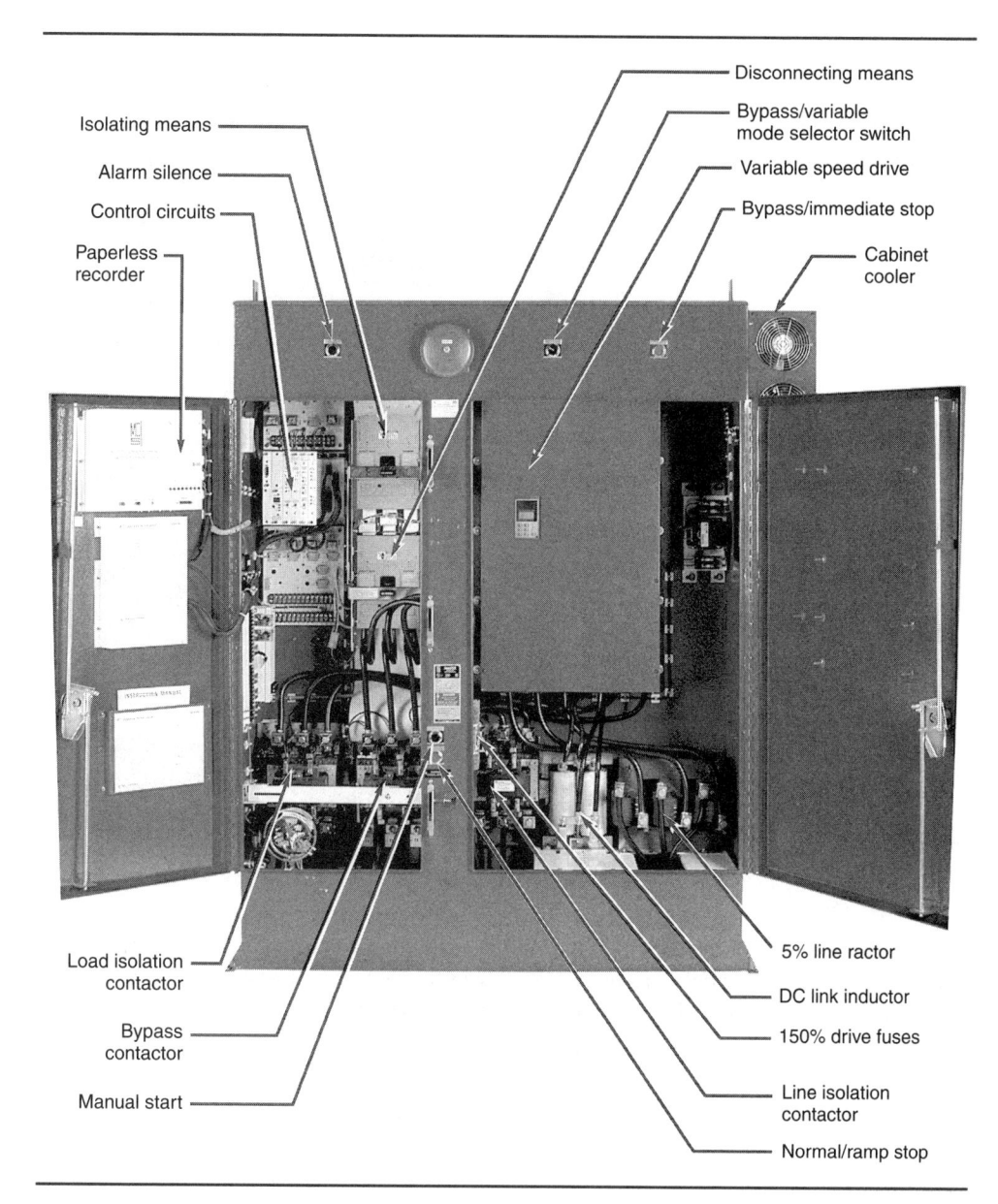

EXHIBIT II.10.11 Variable Speed Pressure Limiting Fire Pump Controller. (Photograph courtesy of Master Control Systems, Inc.)

devices specify their motor ratings in amperes. Therefore, an allowance is made for those devices rated only in amperes.

10.10.2 Additional Marking.

In addition to the markings required in 10.1.2.5.1, the controller shall be marked with the maximum ambient temperature rating.

Controllers equipped with variable speed pressure limiting control are generally provided with solid state components that are temperature sensitive to heat. A critical practice is that the controllers are employed within their maximum operating temperature.

10.10.3* Bypass Operation.

A.10.10.3 The bypass path constitutes all of the characteristics of a non-variable speed fire pump controller.

EXHIBIT II.10.12 Controller Equipped with Variable Speed Pressure Limiting Control. (Photograph courtesy of Master Control Systems, Inc.)

FAQ ▶
Why are controllers equipped with variable speed pressure limiting control provided with a bypass path?

In the event of failure of the variable speed pressure limiting control, a means to energize and operate the pump motor needs to be provided, and the bypass path serves that purpose. As such, all components in the bypass path must comply with all of the starting, stopping, and control requirements of a standard fire pump controller.

10.10.3.1* Upon failure of the variable speed pressure limiting control to keep the system pressure at or above the set pressure of the variable speed pressure limiting control system, the controller shall bypass and isolate the variable speed pressure limiting control system and operate the pump at rated speed.

In the event of failure of the variable speed pressure limiting control, the controller is expected to operate as a standard across-the-line fire pump controller. In this mode the possibility that the system could be overpressurized exists. However, the continued operation of the fire pump to suppress the fire is critical even if damage to the system is a risk.

A.10.10.3.1 The bypass contactor should be energized only when there is a pump demand to run and the variable speed pressure limiting control or variable speed suction limiting control is in the fault condition.

The bypass contactor should not be used to test the system or for any other reason where no fire condition exists.

10.10.3.1.1 Low Pressure. If system pressure remains below the set pressure for more than 15 seconds, the bypass operation shall occur.

The time of 15 seconds was chosen as a reasonable amount of time to allow the variable speed pressure limiting control to pressurize the system to the set pressure.

10.10.3.1.2* Drive Not Operational. If the variable speed drive indicates that it is not operational within 5 seconds, the bypass operation shall occur.

A.10.10.3.1.2 Variable speed drive units (VSDs) should have a positive means of indicating that the drive is operational within a few seconds after power application. If the VSD fails, there is no need to wait for the low pressure bypass time of 10.10.3.1.1.

The time of 5 seconds was chosen as a reasonable amount of time to allow the variable speed pressure limiting control to operate. If the variable speed pressure limiting control is not operational within 5 seconds, pressurizing the system to the set pressure in the next 10 seconds is unlikely.

10.10.3.1.3* Means shall be provided to prevent higher than normal in-rush currents when transferring the fire pump motor from the variable speed mode to the bypass mode.

A.10.10.3.1.3 A motor running at a reduced frequency cannot be connected immediately to a source at line frequency without creating high transient currents that can cause tripping of the fire pump circuit breaker. It is also important to take extra care not to connect (back feed) line frequency power to the VSD since this will damage the VSD, and, more importantly, can cause the fire pump circuit breaker to trip, which takes the pump out of service.

Reducing the speed of the fire pump motor or stopping it altogether before transferring the fire pump motor from the variable speed mode to the bypass mode is desirable. The maximum speed of transfer without tripping the fire pump breaker can be difficult to determine. A time delay is normally incorporated to allow the motor to slow down. The adjustment of this time delay may need to be done in the field to verify that tripping of the fire pump breaker does not occur.

10.10.3.2 When the variable speed pressure limiting control is bypassed, the unit shall remain bypassed until manually restored.

If the variable speed pressure limiting control is in a fault condition, there is no reason to attempt to restore it to service as the chances of doing so are unlikely. In this case the unit is required to remain bypassed until manually restored. If a variable speed fire pump controller is running in the bypass mode, it is doing so either by way of manual intervention (mode switch) or by way of automatic switch to bypass upon failure of the variable frequency drive to maintain the required pressure. Allowing automatic stopping in either case can and will leave a property or premise unprotected.

10.10.3.3 The bypass contactors shall be operable using the emergency-run handle or lever defined in 10.5.3.2.

The bypass contactors perform a function similar to that provided by the emergency-run mechanical control. As such, the bypass contactors need to be operated in a manner similar to that provided for the emergency-run mechanical control. This operation is accomplished through the emergency-run handle or lever specified in 10.5.3.2.

◀ FAQ
What performs the function of the emergency-run mechanical control in a variable speed pressure limiting fire pump controller?

10.10.4 Isolation.

10.10.4.1 The variable speed drive shall be line and load isolated when not in operation.

The variable speed pressure limiting control is generally a solid-state device that is sensitive to electrical impulses and voltage surges from the connected source(s). Therefore, isolating them from the connected source when not in use is important. This feature extends the life of the components, minimizes their failure, and increases their reliability.

10.10.4.2 The variable speed drive load isolation contactor and the bypass contactor shall be mechanically and electrically interlocked to prevent simultaneous closure.

When the bypass path is used, it is preferable to prevent back electromotive force (EMF) from the motor being fed into the variable speed pressure limiting control and to isolate the variable speed path from the utility power source.

10.10.5* Circuit Protection.

A.10.10.5 The intent is to prevent tripping of the fire pump controller circuit-breaker due to a variable speed drive failure and thus maintain the integrity of the bypass circuit.

10.10.5.1 Separate variable speed drive circuit protection shall be provided between the line side of the variable speed drive and the load side of the circuit breaker required in 10.4.3.

The specified location of the protective device for the variable speed pressure limiting control provides for protection of sensitive control equipment while allowing the controller to be fully functional in the event that the fire pump motor needs to be operated. Refer to Figure A.10.10 for the location of this protective device, referred to as *circuit protection.*

10.10.5.2 The circuit protection required in 10.10.5.1 shall be coordinated such that the circuit breaker in 10.4.3 does not trip due to a fault condition in the variable speed circuitry.

FAQ ▶
Are there special requirements for the overcurrent protection of the variable speed pressure limiting control?

The selection of the rating of the protective device for the variable speed pressure limiting control is critical in that the device should be expected to operate to protect the sensitive control equipment but not allow sufficient energy that will result in tripping of the fire pump circuit breaker. Generally, a fast-acting semiconductor fuse satisfies this condition.

10.10.6 Power Quality.

10.10.6.1 Power quality correction equipment shall be located in the variable speed circuit. As a minimum, 5 percent line reactance shall be provided.

FAQ ▶
Is power quality correction needed when a variable speed pressure limiting control is used?

The variable speed drive used for the pressure limiting control may produce harmonics, voltage transients, and other non-sinusoidal waveforms. These conditions affect the quality of the power supplied to the fire pump motor and result in motor inefficiency and/or possible degradation of the motor insulation system. In these cases, power quality correction equipment is needed. Line reactors are the most common devices for counteracting the waveform distortion caused by the variable speed drive. Power quality correction equipment also prevents harmonics and waveform distortions from affecting other equipment not necessarily associated with the fire pump.

10.10.6.2 Coordination shall not be required where the system voltage does not exceed 480 V and cable lengths between the motor and controller do not exceed 100 ft (30.5 m) *(see 10.10.6.3).*

FAQ ▶
Is cable length an issue when a variable speed pressure limiting control is used?

Voltage transients are generally not a problem when the supply voltage does not exceed 480 V and cable lengths between the motor and controller do not exceed 100 ft (30.5 m). In these cases, power quality correction equipment is normally not needed.

10.10.6.3* Where higher system voltages or longer cable lengths exist, the cable length and motor requirements shall be coordinated.

A.10.10.6.3 As the motor cable length between the controller and motor increases, the VSD high frequency switching voltage transients at the motor will increase. To prevent the transients from exceeding the motor insulation ratings, the motor manufacturer's recommended cable lengths must be followed.

10.10.7 Local Control.

10.10.7.1 All control devices required to keep the controller in automatic operation shall be within lockable enclosures.

Restricting access to setting or adjustment of control devices required to keep the controller in automatic operation is necessary, because in the event of a fire no automatic means to start and run the fire pump are available.

In the 2010 edition of NFPA 20, the term *cabinet* was changed to *enclosure.* The term *cabinet,* as it applies to fire pump controllers and their associated accessories, was used incorrectly in the context of the numerous places it appeared in NFPA 20. *Cabinet* is a specific type of enclosure defined in *NFPA 70.* Fire pump controllers and associated accessories are provided in various types of enclosure constructions, some of which may not meet the literal definition of *cabinet.* Enclosures that are free-standing or provided with a lift-off cover are examples of cabinets. Although the term *cabinet* appeared in several editions of NFPA 20, it did not accurately define the enclosure constructions currently provided in today's marketplace. The correct term is *enclosure,* which is also defined in *NFPA 70.*

10.10.7.2 The variable speed pressure sensing element connected in accordance with 10.5.2.1.7.5 shall only be used to control the variable speed drive.

10.10.7.3 Means shall be provided to manually select between variable speed and bypass mode.

Manual selection between variable speed and bypass mode is necessary in the event that a failure of the circuitry to automatically close the bypass contactor occurs.

10.10.7.4 Common pressure control shall not be used for multiple pump installations. Each controller pressure sensing control circuit shall operate independently.

System pressure can fluctuate considerably from one point in a multiple pump system to another. Allowing each controller pressure-sensing control circuit to operate independently results in a more uniform system pressure and prevents excessive pressure due to one or more pumps operating at a higher speed than necessary.

◀ **FAQ**
In multiple pump installations employing variable speed pressure limiting control, how is system pressure monitored?

10.10.8 Indicating Devices on Controller.

10.10.8.1 Drive Failure. A visible indicator shall be provided to indicate when the variable speed drive has failed.

10.10.8.2 Bypass Mode. A visible indicator shall be provided to indicate when the controller is in bypass mode.

10.10.8.3 Variable Speed Pressure Limiting Control Overpressure. Visible indication shall be provided on all controllers equipped with variable speed pressure limiting control to actuate at 115 percent of set pressure.

The visible indicator alerts personnel that the system is experiencing an overpressure condition and some precautions may need to be taken if this condition persists. Under such circumstances shutting down the fire pump is not desirable. Preventing the spread of fire is more critical than preventing the system from being overpressurized.

10.10.9 Controller Contacts for Remote Indication. Controllers shall be equipped with contacts (open or closed) to operate circuits for the conditions in 10.10.8.

Subsection 10.10.9 requires auxiliary open or closed contacts operated by the variable speed pressure limiting control to be provided for remote indication. The function of these contacts is similar to those required in 10.4.8.

10.10.10 System Performance.

10.10.10.1* The controller shall be provided with suitable adjusting means to account for various field conditions.

FAQ ▶
What types of field adjustments are available for variable speed pressure limiting controllers?

The variable speed drive used for the pressure limiting control generally has many adjustments that affect its performance. These adjustments allow the controller to be coordinated with the motor to provide maximum motor efficiency. These adjustments cover a myriad of performance factors such as voltage ramping for acceleration and deceleration, frequency control, and voltage sag.

A.10.10.10.1 This allows for field adjustments to reduce hunting, excessive overshooting, or oscillating.

10.10.10.2 Operation at reduced speed shall not result in motor overheating.

FAQ ▶
Some motors overheat when operated at less than their rated frequency. How is this avoided when using variable speed pressure limiting controllers?

Some motors tend to operate at higher-than-normal temperatures when the frequency of the applied voltage is less than the rated frequency of the motor. These higher-than-normal temperatures could result in degradation of the motor insulation system and reduced efficiency of operation. These conditions are generally eliminated by using motors marked for inverter duty. This problem was first recognized in the 2007 edition of NFPA 20 in 9.5.1.4, which requires motors used with variable speed controllers to be suitable for inverter duty.

10.10.10.3 The maximum operating frequency shall not exceed line frequency.

A properly designed system allows for the pump to deliver maximum flow at rated speed. As such, operating the fire pump motor in excess of its rated or line frequency is not necessary. Doing so may cause overheating and damage to the motor.

10.10.11 Critical Settings. Means shall be provided and permanently attached to the inside of the controller enclosure to record the following settings:

(1) Variable speed pressure limiting set point setting
(2) Pump start pressure
(3) Pump stop pressure

FAQ ▶
What are the critical pressure settings that may be field adjustable in variable speed pressure limiting controllers?

Recording the settings specified in 10.10.11 is critical as the need for future adjustment of these settings is common for controllers with variable speed pressure limiting control. The maintenance of these settings for proper operation of the system is important.

10.10.12 Variable Speed Drives for Vertical Pumps.

In the 2010 edition of NFPA 20, this new subsection was added to recognize the application of variable speed drives to vertical turbine pumps. This application is primarily for well pumps.

10.10.12.1 The pump supplier shall inform the controller manufacturer of any and all critical resonant speeds within the operating speed range of the pump, which is from zero speed up to full speed. The controller shall avoid operating at or ramping through these speeds. The controller shall make use of skip frequencies with sufficient bandwidth to avoid exciting the pump into resonance.

10.10.12.2 When water-lubricated pumps with line shaft bearings are installed, the pump manufacturer shall inform the controller manufacturer of the maximum allowed time for water to reach the top bearing under the condition of the lowest anticipated water level of the well or reservoir. The controller shall provide a ramp up speed within this time period.

10.10.12.3 The ramp down time shall be approved or agreed to by the pump manufacturer.

10.10.12.4 Any skip frequencies employed and their bandwidth shall be included along with the information required in 10.10.11.

10.10.12.5 Ramp up and ramp down times for water-lubricated pumps shall be included along with the information required in 10.10.11.

REFERENCES CITED IN COMMENTARY

National Fire Protection Association, 1 Batterymarch Park, Quincy, MA 02169-7471.

Earley, M. W., et al., eds., *National Electrical Code® Handbook,* 2008 edition.
NFPA 14, *Standard for the Installation of Standpipe and Hose Systems,* 2007 edition.
NFPA 70®, National Electrical Code®, 2008 edition.

National Electrical Manufacturers Association, 1300 North 17th Street, Suite 1752, Rosslyn, VA 22209.

NEMA ICS 14, *Application Guide for Electric Pump Controllers,* 2001 edition.
NEMA 250, *Enclosures for Electrical Equipment,* 1991 edition.

Underwriters Laboratories Inc., 333 Pfingsten Road, Northbrook, IL 60062-2096.

ANSI/UL 508, *Standard for Industrial Control Equipment,* 2006 edition.

Diesel Engine Drive CHAPTER 11

Chapter 11 describes the attributes for the requirements of diesel engine drives. Included in this chapter are requirements for diesel engines, fire pump and engine protection, diesel engine fuel supply and arrangement, engine exhaust, and driver system operation.

Diesel engine–driven fire pumps are often used where there is insufficient or unreliable electrical power for an electric driven fire pump. They are used in conjunction with, or in addition to, motor driven fire pumps as the best combination for reliable pumping systems. When combined with on-site water storage—such as a ground storage tank, underground reservoir, tower, pond, or well—a diesel-driven fire pump is completely self-contained. Since diesel engine–driven fire pumps are completely independent of external utilities, diesel drive units are well suited for remote locations or areas without a reliable power grid. Properly designed and installed units can survive extended periods in the absence of ac power. Some installations can remain in service indefinitely when the controller, engine, and other battery loads are low enough for the weekly run function to replenish the batteries. As with all fire protection equipment, the functionality of the fire pump system is only as good as the performance of the periodic maintenance and inspection programs. Refer to Chapter 8 of NFPA 25, *Standard for the Inspection, Testing, and Maintenance of Water-Based Fire Protection Systems,* for the total requirements for periodic care and maintenance schedules.

11.1 General.

Diesel engines are required to drive fire pumps at the pumps' rated speed without any delay. Safety or warning devices (such as low water, high temperature, or low oil pressure) associated with the engine are not allowed to stop the operation of the fire pump driver. Engine overspeed is the only safety feature that is required to stop the engine.

11.1.1 This chapter provides requirements for minimum performance of diesel engine drivers.

11.1.2 Accessory devices, such as monitoring and signaling means, are included where necessary to ensure minimum performance of the aforementioned equipment.

11.1.3* Engine Type.

A.11.1.3 The compression ignition diesel engine has proved to be the most dependable of the internal combustion engines for driving fire pumps.

11.1.3.1 Diesel engines for fire pump drive shall be of the compression ignition type.

The compression ignition diesel engine accomplishes combustion by spraying atomized fuel into a cylinder of compressed, heated air. The fuel spray is provided by an injector(s) controlled by the electronic fuel management control (electronic control module) (ECM) system at a precise time and of a specific volume. The cylinder air is compressed by the upstroke of the piston, which provides high compression that heats the air 2°F (1.1°C) for each pound per square inch (psi) of compression. This very simple ignition process has proven to be extremely reliable.

Diesel engines are designed with the necessary strength to accommodate high compression ratios and to provide long life of the fire pump. A typical diesel engine drive is shown in Exhibit II.11.1.

11.1.3.2 Spark-ignited internal combustion engines shall not be used.

There are still spark ignition engines operating throughout the world. The spark engines for gasoline and natural gas were dropped from the 1974 edition of NFPA 20. There are no equipment listings for ignition spark internal combustion engines and controls. Owing to the safety and environmental concerns about gasoline engines, it is advisable to replace these units with up-to-date diesel-driven units and controllers.

11.2 Engines.

11.2.1 Listing. Engines shall be listed for fire pump service.

FAQ▶
What information is provided on the engine's nameplate?

Listed engines carry the performance ratings and the listing laboratory's mark. This information is displayed on a nameplate mounted on the engine to aid the owner and authority having jurisdiction when reviewing the system components.

11.2.2 Engine Ratings.

11.2.2.1 Engines shall have a nameplate indicating the listed horsepower rating available to drive the pump.

Exhibit II.11.2 shows an example of a listed engine nameplate, in which the engine is listed for a speed range with a lower limit (1760 rpm) and an upper limit (2100 rpm). This rpm range means that the engine may be operated at any rpm between the lower and upper limits. For example, if the engine were operated at 2000 rpm, the maximum corresponding horsepower generated would be 329.

EXHIBIT II.11.1 Diesel Engine Drive. (Photograph courtesy of Mechanical Designs Ltd.)

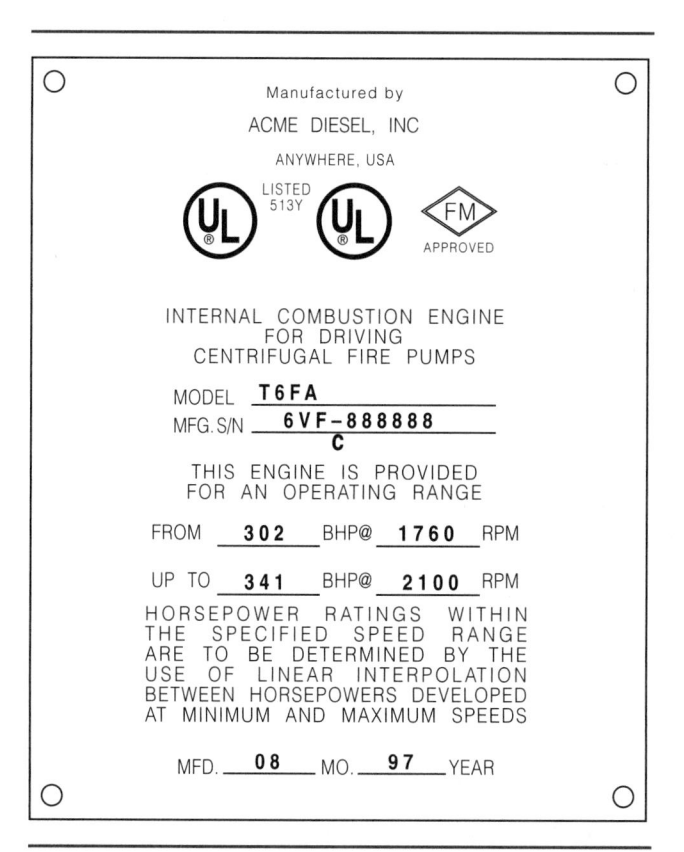

EXHIBIT II.11.2 Listed Engine Nameplate.

```
Manufactured by
ACME DIESEL, INC
ANYWHERE, USA

    UL LISTED 513Y    UL    FM APPROVED

INTERNAL COMBUSTION ENGINE
     FOR DRIVING
 CENTRIFUGAL FIRE PUMPS

MODEL    T6FA
MFG. S/N  6VF-888888
              C
THIS ENGINE IS PROVIDED
FOR AN OPERATING RANGE

FROM   302   BHP@  1760  RPM

UP TO  341   BHP@  2100  RPM
HORSEPOWER RATINGS WITHIN
THE SPECIFIED SPEED RANGE
ARE TO BE DETERMINED BY THE
USE OF LINEAR INTERPOLATION
BETWEEN HORSEPOWERS DEVELOPED
AT MINIMUM AND MAXIMUM SPEEDS

MFD.  08  MO.  97  YEAR
```

11.2.2.2* The horsepower capability of the engine, when equipped for fire pump service, shall have a 4-hour minimum horsepower rating not less than 10 percent greater than the listed horsepower on the engine nameplate.

A.11.2.2.2 For more information, see SAE J-1349, *Engine Power Test Code—Spark Ignition and Compression Engine*. The 4-hour minimum power requirement in NFPA 20 has been tested and witnessed during the engine listing process.

11.2.2.3 Engines shall be acceptable for horsepower ratings listed by the testing laboratory for standard SAE conditions.

11.2.2.4* A deduction of 3 percent from engine horsepower rating at standard SAE conditions shall be made for diesel engines for each 1000 ft (300 m) of altitude above 300 ft (91 m).

Each engine is rated for a certain number of horsepower at standard SAE conditions. The performance of any internal combustion engine is affected by the quantity of oxygen molecules in the combustion chamber. The oxygen is needed to support the burning of the fuel, which creates expanding gases that are then converted to mechanical energy. Higher altitude and/or ambient temperature causes the expansion of air, resulting in fewer molecules of oxygen per volume of air.

A.11.2.2.4 See Figure A.11.2.2.4.

11.2.2.5* A deduction of 1 percent from engine horsepower rating as corrected to standard SAE conditions shall be made for diesel engines for every 10°F (5.6°C) above 77°F (25°C) ambient temperature.

Because engine performance is affected by the combustion air temperature, the power rating of the engine must be corrected to the actual installation conditions, as in the following example.

◀ FAQ
How is the engine horsepower rating corrected to the installation conditions?

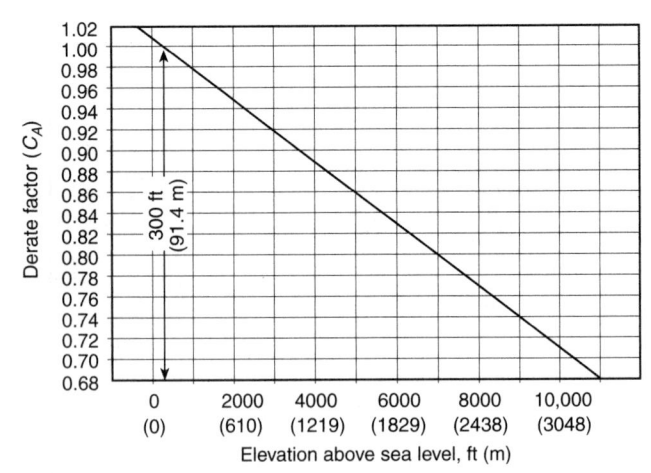

Note: The correction equation is as follows:

Corrected engine horsepower = $(C_A + C_T - 1) \times$ listed engine horsepower

where:

C_A = derate factor for elevation
C_T = derate factor for temperature

FIGURE A.11.2.2.4 *Elevation Derate Curve.*

EXAMPLE

An engine rated at 95 hp is being used to drive a fire pump in Yellowstone National Park, where the elevation is 7000 ft (2134 m) above sea level and the ambient temperature could be 105°F (40.5°C). What is the adjusted rating of this engine?

Solution: See Figures A.11.2.2.4 and A.11.2.2.5 and use the following formula:

Corrected engine horsepower $EH_{Corr} = [(C_A + C_T) - 1] \times$ listed engine horsepower

where:

C_A = derate factor for elevation
C_T = derate factor for temperature

From Figure A.11.2.2.4, the derate factor for altitude (C_A) is approximately 0.85 for an elevation of 7000 ft (2134 m). From Figure A.11.2.2.5, the derate factor for temperature (C_T) is approximately 0.973 for an ambient temperature of 105°F (40.5°C). Using the following correction equation, as indicated by the figures, the adjusted horsepower rating can be determined.

$$EH_{Corr} = [(C_A + C_T) - 1] \times \text{listed engine horsepower}$$
$$= [(0.85 + 0.973) - 1] \times 302$$
$$= 248.6$$

Therefore, an engine rated at 302 bhp at sea level is derated to 248 bhp at an elevation of 7000 ft (2134 m) and a temperature of 105°F (49.5°C).

A.11.2.2.5 Pump room temperature rise should be considered when determining the maximum ambient temperature specified. *(See Figure A.11.2.2.5.)*

11.2.2.6 Where right-angle gear drives *(see 11.2.3.2)* are used between the vertical turbine pump and its driver, the horsepower requirement of the pump shall be increased to allow for power loss in the gear drive.

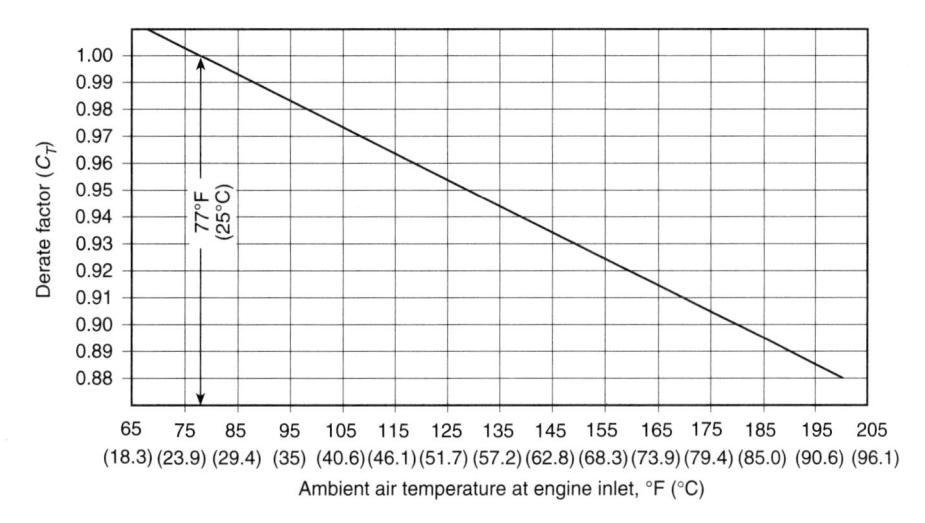

Note: The correction equation is as follows:

Corrected engine horsepower = $(C_A + C_T - 1) \times$ listed engine horsepower

where:

C_A = derate factor for elevation

C_T = derate factor for temperature

FIGURE A.11.2.2.5 *Temperature Derate Curve.*

To ensure that the engine has adequate horsepower, the power loss through the right-angle gear must be considered as additional load and added to the maximum pump load. The gear manufacturer provides the power loss of the right angle gear drive.

11.2.2.7 After the requirements of 11.2.2.1 through 11.2.2.6 have been complied with, engines shall have a 4-hour minimum horsepower rating equal to or greater than the brake horsepower required to drive the pump at its rated speed under any conditions listed for environmental conditions under pump load.

Typically, the shorter the performance duration, the higher the manufacturer rates the engine. Ratings of less than 4 hours continuous are not acceptable for fire pump drivers.

11.2.3 Engine Power Connection to Pump.

The fire pump and diesel engine are both rendered useless if the connecting drive fails. A flexible coupling or flexible-connecting shaft (universal driveshaft) that has been tested and listed is required by NFPA 20 for connecting a fire pump to a diesel engine.

The drive system between the pump and the engine should be as simple as possible with no means of disconnecting the pump and engine shafts without dismantling the system. This type of drive arrangement is considered to provide the greatest reliability. The manufacturer's requirements for the installation and alignment must be used to ensure proper operation of drive couplings and/or drive shafts.

11.2.3.1 Horizontal Shaft Pumps.

The installation and setup of the engine and gear drive must follow the instructions and requirements set forth in the installation data sheets and/or O&M manual provided by the manufacturers of the connecting shafts, the engine, and the gear drive. Special installation requirements must be followed to ensure a proper installation.

11.2.3.1.1 Engines shall be connected to horizontal shaft pumps by means of a flexible coupling or flexible connecting shaft listed for this service.

11.2.3.1.2 The flexible coupling shall be directly attached to the engine flywheel adapter or stub shaft. *(See Section 6.5.)*

11.2.3.2 Vertical Shaft Turbine–Type Pumps.

11.2.3.2.1 Unless the requirements of 11.2.3.2.2 are met, engines shall be connected to vertical shaft pumps by means of a right-angle gear drive with a listed flexible connecting shaft that will prevent undue strain on either the engine or gear drive. *(See Section 7.5.)*

11.2.3.2.2 The requirements of 11.2.3.2.1 shall not apply to diesel engines and steam turbines designed and listed for vertical installation with vertical shaft turbine–type pumps, which shall be permitted to employ solid shafts and shall not require a right-angle drive but shall require a nonreverse ratchet.

11.2.4 Engine Speed Controls.

11.2.4.1 Speed Control Governor.

11.2.4.1.1 Engines shall be provided with a governor capable of regulating engine speed within a range of 10 percent between shutoff and maximum load condition of the pump.

11.2.4.1.2 The governor shall be field adjustable and set and secured to maintain rated pump speed at maximum pump load.

The adjustment and setting of the speed control governor in the field should be performed by a qualified diesel engine technician or the manufacturer's representative. The engine manufacturer and/or the local authority having jurisdiction should approve the technician's qualifications prior to working on the engine speed controls.

FAQ ▶
What is meant by the term *droop*?

The older diesel engine governors used on fire pump drives allowed a speed variance as the load on the engine changed. As the load on the engine, or power demand, is increased, the engine slows down. This effect is called *droop*. Speed variation has a significant effect on the performance of the fire pump. Paragraph 11.2.4.1.1 advises that this droop be limited to a maximum of 10 percent for mechanical governors. Because of the droop effect, the engine speed should only be adjusted when the pump is operating at its maximum horsepower. The newest electronic governors do not droop nearly as much and, therefore, the engines do not slow down under an increase or decrease in load. All of the listed engine manufacturers have switched to the electronic governors for the new models.

11.2.4.2* Electronic Fuel Management Control.

The ECM-equipped diesel engines require periodic and annual testing, inspection, and service. Refer to 8.3.3.8 of NFPA 25 for test and inspection requirements.

A.11.2.4.2 Traditionally, engines have been built with mechanical devices to control the injection of fuel into the combustion chamber. To comply with requirements for reduced exhaust emissions, many engine manufacturers have incorporated electronics to control the fuel injection process, thus eliminating levers and linkages. Many of the mechanically controlled engines are no longer manufactured.

11.2.4.2.1 Alternate Electronic Control Module. Engines that incorporate an electronic control module (ECM) to accomplish and control the fuel injection process shall have an alternate ECM permanently mounted and wired so the engine can produce its full rated power output in the event of a failure of the primary ECM.

11.2.4.2.2 ECM Voltage Protection. ECMs shall be protected from transient voltage spikes and reverse dc current.

11.2.4.2.3 ECM Selector Switch.

11.2.4.2.3.1 Operation. The transition from the primary ECM to the alternate ECM shall be accomplished automatically upon failure of the primary ECM, and a hand/automatic switch without an off position shall be provided.

11.2.4.2.3.2 Supervision. A visual indicator shall be provided on the engine instrument panel and a supervisory signal shall be provided to the controller when the ECM selector switch is positioned to the alternate ECM.

11.2.4.2.3.3 Contacts.

(A) The contacts for each circuit shall be rated for both the minimum and maximum current and voltage.
(B) The total resistance of each ECM circuit through the selector switch shall be approved by the engine manufacturer.

11.2.4.2.3.4 Enclosure.

(A) The selector switch shall be enclosed in a NEMA Type 2 dripproof enclosure.
(B) Where special environments exist, suitably rated enclosures shall be used.

11.2.4.2.3.5 Mounting.

(A) The selector switch and enclosure shall be engine mounted.
(B) The selector switch enclosure and/or the selector switch inside shall be isolated from engine vibration to prevent any deterioration of contact operation.

11.2.4.2.4* Engine Power Output. The ECM (or its connected sensors) shall not, for any reason, intentionally cause a reduction in the engine's ability to produce rated power output.

A.11.2.4.2.4 ECMs can be designed by engine manufacturers to monitor various aspects of engine performance. A stressed engine condition (such as high cooling water temperature) is usually monitored by the ECM and is built into the ECM control logic to reduce the horsepower output of the engine, thus providing a safeguard for the engine. Such engine safeguards are not permitted for ECMs in fire pump engine applications.

11.2.4.2.5 ECM Sensors. Any sensor necessary for the function of the ECM that affects the engine's ability to produce its rated power output shall have a redundant sensor that shall operate automatically in the event of a failure of the primary sensor.

11.2.4.2.6 ECM Engine Supervision. A common supervisory signal shall be provided to the controller in the event of any of the following:

(1) Fuel injection failure
(2) Low fuel pressure
(3) Any primary sensor failure

11.2.4.2.7 ECM and Engine Power Supply.

11.2.4.2.7.1 In the standby mode, the engine standby battery shall be used to power the ECM.

11.2.4.2.7.2 These engines shall not require more than 0.5 ampere from the battery or battery charger while the engine is not running.

11.2.4.3 Variable Speed Pressure Limiting Control or Variable Speed Suction Limiting Control (Optional).

The addition of the variable speed pressure limiting control feature for a diesel engine has increased the tools available to the fire protection professional in the design of fire protection systems. The pressure limiting feature can be used to limit the available pressure to more workable levels when a variable pressure supply is used to provide suction to the fire pump, or when the fire protection system needs large quantities of water at high pressures because the discharge pressure curve from the fire pump drops off quickly as the flow increases. Similar results are available from electric motor–driven fire pumps. See Chapters 9 and 10 for details.

11.2.4.3.1 Variable speed pressure limiting control or variable speed suction limiting control systems used on diesel engines for fire pump drive shall be listed for fire pump service and be capable of limiting the pump output total rated head (pressure) or suction pressure by reducing pump speed.

11.2.4.3.2 Variable speed control systems shall not replace the engine governor as defined in 11.2.4.1.

11.2.4.3.3 In the event of a failure of the variable speeds control system, the engine shall be fully functional with the governor defined in 11.2.4.1.

11.2.4.3.4 A pressure sensing line shall be provided to the engine with a ½ in. (12.7 mm) nominal size inside diameter line, from a connection between the pump discharge flange and the discharge check valve.

The pressure sensing line required in 11.2.4.3.4 is needed in addition to the one specified in Section 4.30. The pressure sensing line required by Section 4.30 is intended to control starting and running time for the fire pump, while the pressure sensing line required in 11.2.4.3.4 is used to control the variable speed pressure limiting control function in the fire pump controller.

11.2.4.4 Engine Overspeed Shutdown Control.

11.2.4.4.1 Engines shall be provided with an overspeed shutdown device.

11.2.4.4.2 The overspeed device shall be arranged to shut down the engine in a speed range of 10 to 20 percent above rated engine speed and to be manually reset.

11.2.4.4.3 A means shall be provided to indicate an overspeed trouble signal to the automatic engine controller such that the controller cannot be reset until the overspeed shutdown device is manually reset to normal operating position.

11.2.4.4.4 Means shall be provided for verifying overspeed switch and circuitry shutdown function.

11.2.4.5 Engine Running and Crank Termination Control.

For automated operation of the engine, the engine needs to provide a reliable signal to the controller indicating that it has started and achieved sufficient revolutions per minute and that starter assistance is no longer required or useful. Generators or alternators do not provide the reliability required for this signal, which is necessary to ensure the starter is not damaged.

11.2.4.5.1 Engines shall be provided with a speed-sensitive switch to signal engine running and crank termination.

11.2.4.5.2 Power for this signal shall be taken from a source other than the engine generator or alternator.

11.2.5 Instrumentation.

11.2.5.1 Instrument Panel.

11.2.5.1.1 All engine instruments shall be placed on a panel secured to the engine or inside an engine base plate-mounted controller.

11.2.5.1.2 The engine instrument panel shall not be used as a junction box or conduit for any ac supply.

11.2.5.2 Tachometer.

11.2.5.2.1 A tachometer shall be provided to indicate revolutions per minute of the engine, including zero, at all times.

11.2.5.2.2 The tachometer shall be the totalizing type, or an hour meter shall be provided to record total time of engine operation.

11.2.5.2.3 Tachometers with digital display shall be permitted to be blank when the engine is not running.

11.2.5.3 Oil Pressure Gauge. Engines shall be provided with an oil pressure gauge to indicate lubricating oil pressure.

11.2.5.4 Temperature Gauge. Engines shall be provided with a temperature gauge to indicate engine coolant temperature at all times.

11.2.6 Wiring Elements.

11.2.6.1 Automatic Controller Wiring in Factory.

11.2.6.1.1* All connecting wires for automatic controllers shall be harnessed or flexibly enclosed, mounted on the engine, and connected in an engine junction box to terminals numbered to correspond with numbered terminals in the controller.

The interconnecting wiring between the engine control panel termination strip and the matching field connection strip in the controller should be wired as required by the equipment manufacturers' data sheets or the NFPA electrical codes and standards.

The power-carrying conductors should be sized to meet the requirements for the current and voltage between the engine and the automatic controller.

The minimum size wire is normally No. 10 AWG CHHN stranded wire for up to 25 ft (7.62 m) runs for the battery charging circuits. Undersized wiring components can cause failure of the circuit and possible failure of the starting system. Wiring schematics are included on the inside of the controller to assist in the installation of the wiring.

A.11.2.6.1.1 A harness on the enclosure will ensure ready wiring in the field between the two sets of terminals.

11.2.6.1.2 All wiring on the engine, including starting circuitry, shall be sized on a continuous-duty basis.

11.2.6.2* Automatic Control Wiring in the Field.

A.11.2.6.2 Terminations should be made using insulated ring-type compression connectors for post-type terminal blocks. Saddle-type terminal blocks should have the wire stripped with about $1/16$ in. (1.6 mm) of bare wire showing after insertion in the saddle, to ensure that no insulation is below the saddle. Wires should be tugged to ensure adequate tightness of the termination.

11.2.6.2.1 Interconnections between the automatic controller and the engine junction box shall be made using stranded wire sized on a continuous-duty basis.

11.2.6.2.2 The dc interconnections between the automatic controller and engine junction box and any ac power supply to the engine shall be routed in separate conduits.

11.2.6.3 Battery Cables.

11.2.6.3.1 Battery cables shall be sized in accordance with the engine manufacturer's recommendations considering the cable length required for the specific battery location.

11.2.7 Starting Methods.

11.2.7.1 Starting Devices. Engines shall be equipped with a reliable starting device and shall accelerate to rated output speed within 20 seconds.

The main battery contactors are normally open relays used for connecting and disconnecting one of the two battery units to one starter. To ensure that each battery is maintained at full charge, the two battery units must be isolated from each other. A main battery contactor is installed in the cable of each battery to the starter.

The controller provides a voltage signal to a selected main battery contactor, energizing its coil, which causes the relay to close and send battery potential to the starter, thus providing cranking power to start the engine. The main battery contactor is also equipped with a snap-action lever to allow the operator to manually crank the engine in the event of a circuit failure within the controller. Some engines are equipped with two starters, one for each battery unit. Therefore, main battery contactors are not required on the engines.

Main battery contactors or dual starter systems are intended to utilize only one battery unit at a time to crank an engine. However, in the event of an emergency, and where cranking the engine is difficult, both main battery contactors or both starters might need to be used simultaneously.

11.2.7.2 Electric Starting. Where electric starting is used, the electric starting device shall take current from a storage battery(ies).

11.2.7.2.1 Batteries.

11.2.7.2.1.1 Each engine shall be provided with two storage battery units.

11.2.7.2.1.2 Lead-acid batteries shall be furnished in a dry charge condition with electrolyte liquid in a separate container.

To ensure that the batteries are ready for in-service testing of the pump, battery electrolyte must be added prior to putting the pump in service so that the batteries have time to charge before testing begins.

11.2.7.2.1.3 Nickel-cadmium or other kinds of batteries shall be permitted to be installed in lieu of lead-acid batteries provided they meet the engine manufacturer's requirements.

11.2.7.2.1.4 At 40°F (4.5°C), each battery unit shall have twice the capacity sufficient to maintain the cranking speed recommended by the engine manufacturer through a 3-minute attempt-to-start cycle, which is six consecutive cycles of 15 seconds of cranking and 15 seconds of rest.

FAQ ▶
Why is the battery sized for twice the capacity?

Because battery output goes down with temperature, batteries must be sized to meet the engine requirements at 40°F (4.5°C) or lower for extreme conditions. Exhibit II.11.3 shows how a battery's performance is affected by temperature. By sizing the battery for twice the capacity, it has remaining capacity to start the engine even if it has gone through a full attempt-to-start cycle and shut down for overcrank. See 12.8.2 for requirements concerning starting equipment arrangement.

11.2.7.2.1.5* Batteries shall be sized on a calculated capacity of 72 hours of standby power followed by three 15–second attempt-to-start cycles per battery unit, without ac power being available for battery charging.

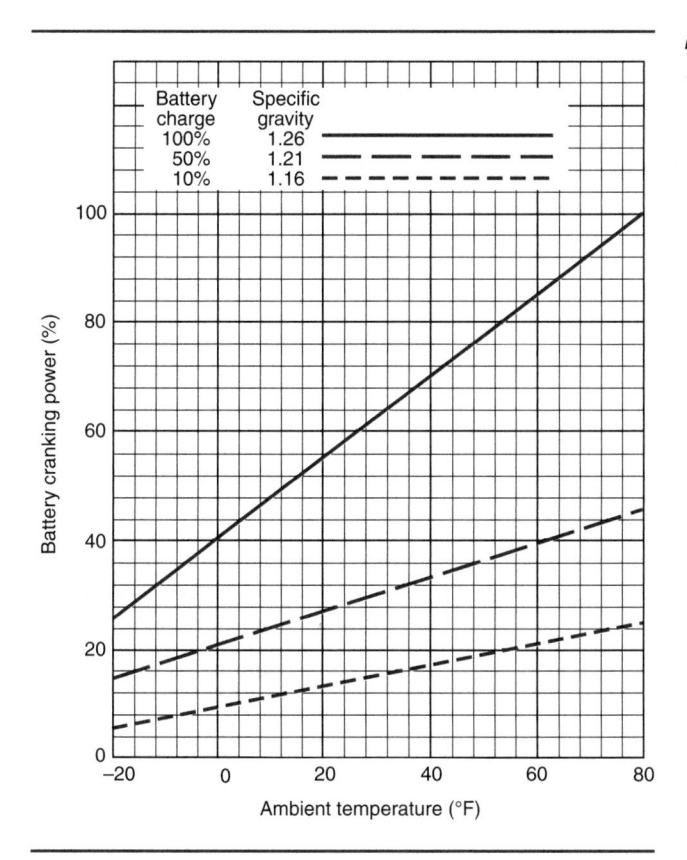

EXHIBIT II.11.3 Battery Performance Versus Temperature.

A.11.2.7.2.1.5 The 72-hour requirement is intended to apply when batteries are new. Some degradation is expected as batteries age.

11.2.7.2.2* Battery Isolation.

A.11.2.7.2.2 Manual mechanical operation of the main battery contactor will bypass all of the control circuit wiring within the controller.

In the event of a failure of control circuits within the controller or the interconnecting wires between the controller and engine, the main battery contactor provides a means of locally cranking the engine for emergency operation. Engines equipped with two starters, in lieu of main battery contactors, provide control switches on the engine instrument panel for local cranking during this type of emergency.

11.2.7.2.2.1 Engines with only one cranking motor shall include a main battery contactor installed between each battery and the cranking motor for battery isolation.

(A) Main battery contactors shall be listed for fire pump driver service.
(B) Main battery contactors shall be rated for the cranking motor current.
(C) Main battery contactors shall be capable of manual mechanical operation including positive methods such as spring-loaded, over-center operator to energize the starting motor in the event of controller circuit failure.

11.2.7.2.2.2 Engines with two cranking motors shall have one cranking motor dedicated to each battery.

(A) Each cranking motor shall meet the cranking requirements of a single cranking motor system.

(B) To activate cranking, each cranking motor shall have an integral solenoid relay to be operated by the pump set controller.

(C) Each cranking motor integral solenoid relay shall be capable of being energized from a manual operator listed and rated for the cranking motor solenoid relay and include spring-loaded, over-center operation to energize the starting motor in the event of controller circuit failure.

11.2.7.2.3 Battery Loads.

11.2.7.2.3.1 Nonessential loads shall not be powered from the engine starting batteries.

11.2.7.2.3.2 Essential loads, including the engine, controller, and all pump equipment combined, shall not exceed 0.5 ampere each for a total of 1.5 amperes, on a continuous basis.

11.2.7.2.4* Battery Location.

Exhibit II.11.4 shows two storage battery units for a diesel engine–driven pump. Note that the battery units are rack-supported and located adjacent to the diesel engine.

When calculating cable length for sizing purposes, all cable for the circuit must be included (positive cable, negative cable, battery jumper, cable from main battery contactor to starter, ground, and so forth).

A.11.2.7.2.4 Location at the side of and level with the engine is recommended to minimize lead length from battery to starter.

11.2.7.2.4.1 Storage batteries shall be rack supported above the floor, secured against displacement, and located where they will not be subject to excessive temperature, vibration, mechanical injury, or flooding with water.

11.2.7.2.4.2 Current-carrying parts shall not be less than 12 in. (305 mm) above the floor level.

11.2.7.2.4.3 Storage batteries shall be readily accessible for servicing.

11.2.7.2.4.4 Storage batteries shall not be located in front of the engine-mounted instruments and controls.

EXHIBIT II.11.4 Installed Battery Units for a Diesel Engine.

11.2.7.3 Hydraulic Starting.

11.2.7.3.1 Where hydraulic starting is used, the accumulators and other accessories shall be enclosed or so protected that they are not subject to mechanical injury.

11.2.7.3.2 The enclosure shall be installed as close to the engine as practical so as to prevent serious pressure drop between the engine and the enclosure.

11.2.7.3.3 The diesel engine as installed shall be without starting aid except that a thermostatically controlled electric water jacket heater shall be employed.

11.2.7.3.4 The diesel as installed shall be capable of carrying its full rated load within 20 seconds after cranking is initiated with the intake air, room ambient temperature, and all starting equipment at 32°F (0°C).

11.2.7.3.5 Hydraulic starting means shall comply with the following conditions:

(1) The hydraulic cranking device shall be a self-contained system that will provide the required cranking forces and engine starting revolutions per minute (rpm) as recommended by the engine manufacturer.
(2) Electrically operated means shall automatically provide and maintain the stored hydraulic pressure within the predetermined pressure limits.
(3) The means of automatically maintaining the hydraulic system within the predetermined pressure limits shall be energized from the main bus and the final emergency bus if one is provided.
(4) Means shall be provided to manually recharge the hydraulic system.
(5) The capacity of the hydraulic cranking system shall provide not fewer than six cranking cycles of not less than 15 seconds each.
(6) Each cranking cycle—the first three to be automatic from the signaling source—shall provide the necessary number of revolutions at the required rpm to permit the diesel engine to meet the requirements of carrying its full rated load within 20 seconds after cranking is initiated with intake air, room ambient temperature, and hydraulic cranking system at 32°F (0°C).
(7) The capacity of the hydraulic cranking system sufficient for three starts under conditions described in 11.2.7.3.5(5) shall be held in reserve and arranged so that the operation of a single control by one person will permit the reserve capacity to be employed.
(8) All controls for engine shutdown in the event of overspeed shall be 12 V dc or 24 V dc source to accommodate controls supplied on engine, and the following also shall apply:
   (a) In the event of such failure, the hydraulic cranking system shall provide an interlock to prevent the engine from recranking.
   (b) The interlock shall be manually reset for automatic starting when engine failure is corrected.

11.2.7.4 Air Starting.

11.2.7.4.1 In addition to the requirements of Section 11.1 through 11.2.7, 11.2.8.1, 11.2.8 through 11.6.2, 11.6.4, and 11.6.6, the requirements of 11.2.7.4 shall apply.

11.2.7.4.2 Automatic Controller Connections in Factory.

11.2.7.4.2.1 All conductors for automatic controllers shall be harnessed or flexibly enclosed, mounted on the engine, and connected in an engine junction box to terminals numbered to correspond with numbered terminals in the controller.

11.2.7.4.2.2 These requirements shall ensure ready connection in the field between the two sets of terminals.

11.2.7.4.3 Signal for Engine Running and Crank Termination.

11.2.7.4.3.1 Engines shall be provided with a speed-sensitive switch to signal running and crank termination.

11.2.7.4.3.2 Power for this signal shall be taken from a source other than the engine compressor.

11.2.7.4.4* Air Starting Supply.

A.11.2.7.4.4 Automatic maintenance of air pressure is preferable.

11.2.7.4.4.1 The air supply container shall be sized for 180 seconds of continuous cranking without recharging.

11.2.7.4.4.2 There shall be a separate, suitably powered automatic air compressor or means of obtaining air from some other system, independent of the compressor driven by the fire pump engine.

11.2.7.4.4.3 Suitable supervisory service shall be maintained to indicate high and low air pressure conditions.

11.2.7.4.4.4 A bypass conductor with a manual valve or switch shall be installed for direct application of air from the air container to the engine starter in the event of control circuit failure.

11.2.8 Engine Cooling System.

The cooling system for modern engines is very complex. The system not only must remove waste heat from the internal parts of the engine but must also accomplish the following:

1. Remove entrapped air from the coolant
2. Allow for coolant expansion
3. Allow for a specific volume of coolant loss
4. Maintain minimum flow under varying conditions

Therefore, the cooling system is installed and provided as part of the listed engine.

11.2.8.1 The engine cooling system shall be included as part of the engine assembly and shall be one of the following closed-circuit types:

(1) A heat exchanger type that includes a circulating pump driven by the engine, a heat exchanger, and an engine jacket temperature regulating device

A heat exchanger is the type of cooling system traditionally used for fire pump drivers. It takes water from the discharge piping of the fire pump and uses it to cool the engine. The water from the pump is passed through the engine heat exchanger, where it absorbs heat from the engine coolant. The water is then piped to a drain. In installations where the water is potable, the cooling water line is required to be equipped with listed backflow devices before being discharged to a drain or the building exterior. This is required because of the inhibitors that are part of the antifreeze solution.

(2) A radiator type that includes a circulating pump driven by the engine, a radiator, an engine jacket temperature regulating device, and an engine-driven fan for providing positive movement of air through the radiator

Engine-mounted radiator cooling does not require water from the fire pump. It does, however, require a considerable volume of air to pass through the pump room. Provision for sufficient airflow for radiator cooling through the pump room requires special attention. See 11.3.2 for more information on pump room ventilation.

11.2.8.2 A means shall be provided to maintain 120°F (49°C) at the combustion chamber.

In a diesel engine, combustion of the fuel is caused by the rapid compression of air in the combustion chamber or cylinder. When an engine starts, the temperature in the cylinder is greatly affected by the temperature of the block and cylinder head. Therefore, maintaining the engine at 120°F (49°C), required in 11.2.8.2, greatly improves the ability of the engine to start quickly.

Most modern engines require the coolant system to operate under pressure. This pressure is typically achieved by using a preset pressure relief cap on the coolant and fill openings. (See 11.2.8.3.)

The coolant provides a way to remove heat from the engine and to lubricate seals and prevent corrosion. Therefore, the coolant must be compatible with engine seals and gaskets. Be sure to follow the engine manufacturer's recommendations for antifreeze and additives.

When air bubbles form in the engine coolant, the coolant's ability to remove heat is reduced. The pressure cap and antifreeze raise the boiling point of the coolant, provide the heat transfer necessary, and increase the engine's reliability. Exhibit II.11.5 shows how antifreeze not only lowers the freezing point but also raises the boiling point of water.

◀ **FAQ**
Why must the temperature be maintained at 120°F (49°C)?

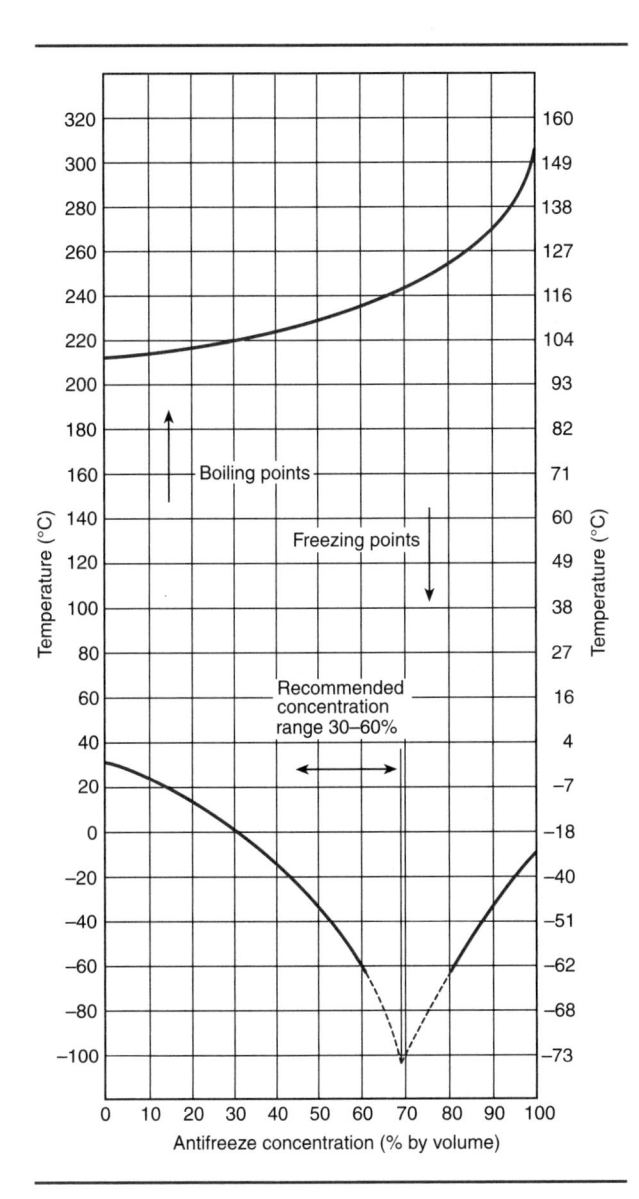

EXHIBIT II.11.5 Coolant Freezing and Boiling Temperature Versus Antifreeze Concentration (Sea Level).

The installer or mechanic should always mix the coolant and the water prior to filling the engine water jacket. The power switch to the electric jacket heater or block heaters must be in the off or open position until the coolant mix is completely installed. Keeping the power off the block heater will ensure the heater is not destroyed. Energizing the heater without completely filling the cooling system with the coolant and water mix will cause immediate failure of the electric heater.

11.2.8.3 An opening shall be provided in the circuit for filling the system, checking coolant level, and adding make-up coolant when required.

11.2.8.4 The coolant shall comply with the recommendation of the engine manufacturer.

11.2.8.5* Heat Exchanger Water Supply.

A.11.2.8.5 See Figure A.11.2.8.5. Water supplied for cooling the heat exchanger is sometimes circulated directly through water-jacketed exhaust manifolds and/or engine aftercoolers in addition to the heat exchangers.

11.2.8.5.1 The cooling water supply for a heat exchanger–type system shall be from the discharge of the pump taken off prior to the pump discharge check valve.

11.2.8.5.2 The cooling water flow required shall be set based on the maximum ambient cooling water.

The water supply for heat exchanger–cooled engines must provide an adequate volume to sufficiently cool the engine. The efficiency of the heat exchanger depends on the temper-

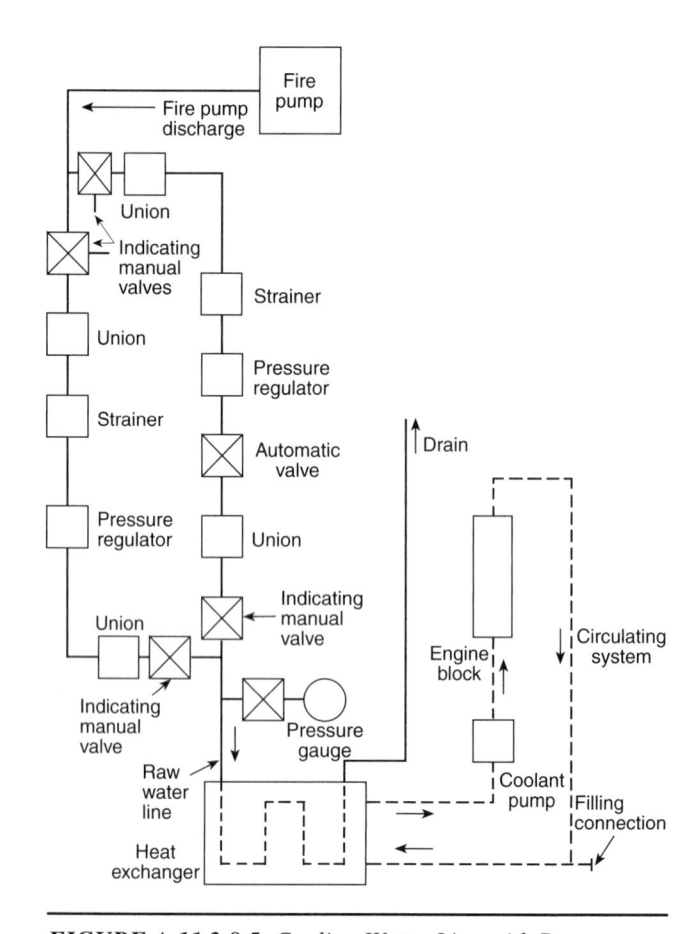

FIGURE A.11.2.8.5 Cooling Water Line with Bypass.

ature of the cooling water entering it. Refer to the engine manufacturer's instructions for the recommended water flow through the heat exchanger, based on the cooling water inlet temperature.

11.2.8.5.3 Heat Exchanger Water Supply Components.

Engine heat exchangers have low-pressure limits 15 psi (1 bar) to 60 psi (5 bars). The pressure regulator in the water supply and the bypass piping is provided to reduce the pump outlet pressure to a level below the heat exchanger limit.

11.2.8.5.3.1 Threaded rigid piping shall be used for this connection.

11.2.8.5.3.2 The pipe connection in the direction of flow shall include an indicating manual shutoff valve, an approved flushing-type strainer in addition to the one that can be a part of the pressure regulator, a pressure regulator, an automatic valve, and a second indicating manual shutoff valve or a spring-loaded check valve.

11.2.8.5.3.3 The indicating manual shutoff valves shall have permanent labeling with minimum ½ in. (12.7 mm) text that indicates the following: For the valve in the heat exchanger water supply, "Normal/Open" for the normal open position when the controller is in the automatic position and "Caution: Nonautomatic/Closed" for the emergency or manual position.

11.2.8.5.3.4 The pressure regulator shall be of such size and type that it is capable of and adjusted for passing approximately 120 percent of the cooling water required when the engine is operating at maximum brake horsepower and when the regulator is supplied with water at the pressure of the pump when it is pumping at 150 percent of its rated capacity.

11.2.8.5.3.5 Automatic Valve.

(A) An automatic valve listed for fire protection service shall permit flow of cooling water to the engine when it is running.
(B) Energy to operate the automatic valve shall come from the diesel driver or its batteries and shall not come from the building.
(C) The automatic valve shall be normally closed.
(D) The automatic valve shall not be required on a vertical shaft turbine–type pump or any other pump when there is no pressure in the discharge when the pump is idle.

11.2.8.5.3.6 A pressure gauge shall be installed in the cooling water supply system on the engine side of the last valve in the heat exchanger water supply and bypass supply.

11.2.8.5.3.7 Potable Water Separation (Optional.) Where two levels of separation for possible contaminants to the ground or potable water source are required by the authority having jurisdiction, dual spring-loaded check valves or backflow preventers shall be installed.

(A)* The spring-loaded check valve(s) shall replace the second indicating manual shutoff valve(s) in the cooling loop assembly as stated in 11.2.8.5.3.3.
(B)* If backflow preventers are used, the devices shall be listed for fire protection service and installed in parallel in the water supply and water supply bypass assembly.
(C) Where the authority having jurisdiction requires the installation of backflow prevention devices in connection with the engine, special consideration shall be given to the increased pressure loss, which will require that the cooling loop pipe size be evaluated and documented by engineering calculations to demonstrate compliance with the engine manufacturer's recommendation.

A.11.2.8.5.3.7(A) See Figure A.11.2.8.5.3.7(A).

A.11.2.8.5.3.7(B) See Figure A.11.2.8.5.3.7(B).

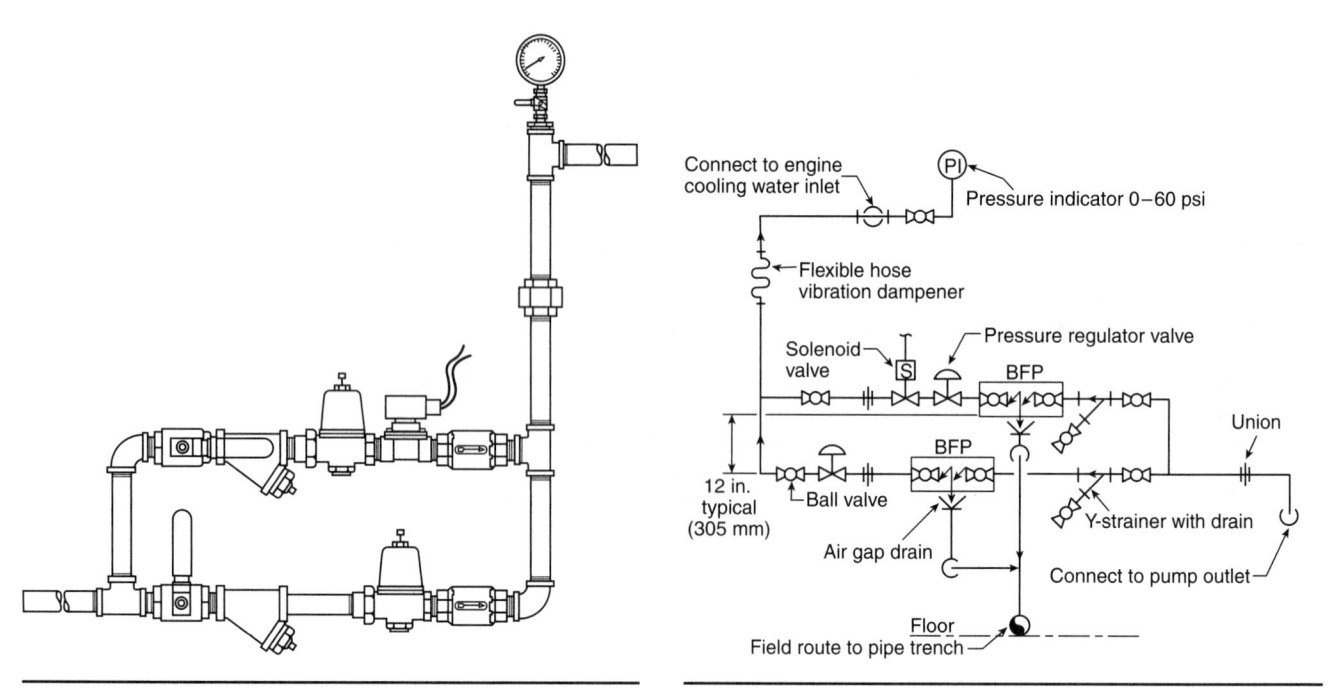

FIGURE A.11.2.8.5.3.7(A) *Spring-Loaded Check Valve Arrangement.*

FIGURE A.11.2 8.5.3.7(B) *Backflow Preventer Arrangement.*

11.2.8.6* Heat Exchanger Water Supply Bypass.

A.11.2.8.6 Where the water supply can be expected to contain foreign materials, such as wood chips, leaves, lint, and so forth, the strainers required in 11.2.8.5 should be of the duplex filter type. Each filter (clean) element should be of sufficient filtering capacity to permit full water flow for a 3-hour period. In addition, a duplex filter of the same size should be installed in the bypass line. *(See Figure A.11.2.8.5.)*

11.2.8.6.1 A threaded rigid pipe bypass line shall be installed around the heat exchanger water supply.

11.2.8.6.2 The pipe connection in the direction of flow shall include an indicating manual shutoff valve, an approved flushing-type strainer in addition to the one that can be a part of the pressure regulator, a pressure regulator, and an indicating manual shutoff valve or a spring-loaded check valve.

11.2.8.6.3 The indicating manual shutoff valves shall have permanent labeling with minimum ½ in. (12.7 mm) text that indicates the following: For the valve in the heat exchanger water supply bypass, "Normal/Closed" for the normal closed position when the controller is in the automatic position and "Emergency/Open" for manual operation or when the engine is overheating.

The water supply bypass is required to ensure the engine always has an adequate flow of cooling water even during maintenance of the cooling water line. If the flow through the water supply begins to deteriorate, the reading on the pressure gauge goes down—indicating a potential overheating condition.

11.2.8.7 Heat Exchanger Waste Outlet.

11.2.8.7.1 An outlet shall be provided for the wastewater line from the heat exchanger, and the discharge line shall not be less than one size larger than the inlet line.

The wastewater line piping must be sized to handle the flow of cooling water leaving the discharge outlet. The size of the diesel engine drive will determine the size of the piping and drain. Small diesel engines may only require 12 gpm (45 L/min) to 15 gpm (57 L/min) to keep the engine temperature at the correct level. However a large-tier three-engine arrangement may require as much as 30 gpm (114 L/min) to control the engine temperature at an expectable level. Contact the engine manufacturer's data sheet for the required flow needed to cool the engine. Local environmental rules should also be followed for the disposal of the wastewater.

11.2.8.7.2 The outlet line shall be as short as practical, shall provide discharge into a visible open waste cone, and shall have no valves in it.

11.2.8.7.3 The outlet shall be permitted to discharge to a suction reservoir provided a visual flow indicator and temperature indicator are installed.

11.2.8.7.4 When the waste outlet piping is longer than 15 ft (4.6 m) and/or its outlet discharges are more than 4 ft (1.2 m) higher than the heat exchanger, the pipe size shall be increased by at least one size.

11.2.8.8 Radiators.

11.2.8.8.1 The heat from the primary circuit of a radiator shall be dissipated by air movement through the radiator created by a fan included with, and driven by, the engine.

11.2.8.8.2 The radiator shall be designed to limit maximum engine operating temperature with an inlet air temperature of 120°F (49°C) at the combustion air cleaner inlet.

11.2.8.8.3 The radiator shall include the plumbing to the engine and a flange on the air discharge side for the connection of a flexible duct from the discharge side to the discharge air ventilator.

The use of a radiator-cooled engine requires the engineer to consider several other areas of concern related to the complete pump room layout. Due to the amount of air removed from the room during operation of the engine, the number of air changes must be considered during the cooler months of the year when the temperature outside drops below freezing, 32°F (0°C).

11.2.8.8.4 Fan.

11.2.8.8.4.1 The fan shall push the air through the radiator to be exhausted from the room via the air discharge ventilator.

11.2.8.8.4.2 To ensure adequate airflow through the room and the radiator, the fan shall be capable of a 0.5 in. water column (13 mm water column) restriction created by the combination of the air supply and the discharge ventilators in addition to the radiator, fan guard, and other engine component obstructions.

11.2.8.8.4.3 The fan shall be guarded for personnel protection.

11.2.9 Engine Lubrication.

This subsection does not cover the requirements for complete engine maintenance. Refer to NFPA 25, Section 8.5, Maintenance, and Table 8.5.3, for the requirements for all periodically required maintenance items.

11.2.9.1 The engine manufacturer's recommendations for oil heaters shall be followed.

Certain countries require the employment of an oil heater in the lube oil reservoir. The manufacturer's selection and installation instructions must be followed in the application of a lube oil heater. A thermo-well installation may be required to prevent the "cooking" of the lubrication oil. If the crankcase oil heater damages the oil, the engine may fail during emergency operation.

11.3* Pump Room.

A.11.3 The engine-driven pump can be located with an electric-driven fire pump(s) in a pump house or pump room that should be entirely cut off from the main structure by noncombustible construction. The fire pump house or pump room can contain facility pumps and/or equipment as determined by the authority having jurisdiction.

11.3.1 The floor or surface around the pump and engine shall be pitched for adequate drainage of escaping water away from critical equipment, such as pump, engine, controller, fuel tank, and so forth.

11.3.2* Ventilation.

A.11.3.2 For optimum room ventilation, the air supply ventilator and air discharge should be located on opposite walls.

When calculating the maximum temperature of the pump room, the radiated heat from the engine, the radiated heat from the exhaust piping, and all other heat-contributing sources should be considered.

If the pump room is to be ventilated by a power ventilator, reliability of the power source during a fire should be considered. If the power source is unreliable, the temperature rise calculation should assume the ventilator is not operable.

Air consumed by the engine for combustion should be considered as part of the air changes in the room.

Pump rooms with heat exchanger–cooled engines typically require more air changes than engine air consumption can provide. To control the temperature rise of the room, additional air flow through the room is normally required.*[See Figure A.11.3.2(a).]*

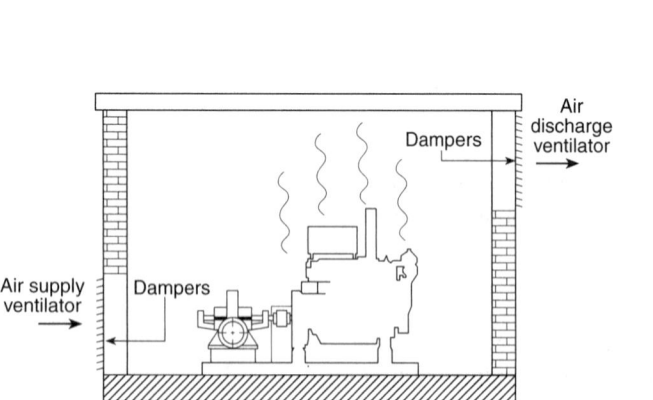

FIGURE A.11.3.2(a) Typical Ventilation System for a Heat Exchanger–Cooled Diesel-Driven Pump.

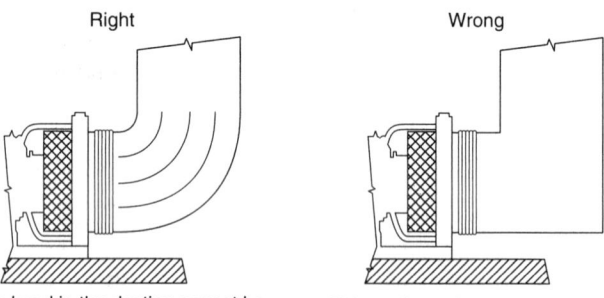

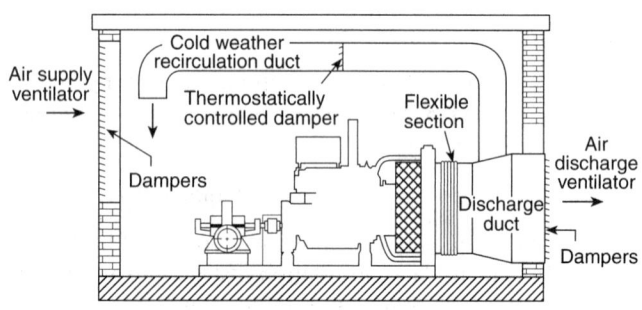

FIGURE A.11.3.2(b) Typical Ventilation System for a Radiator-Cooled Diesel-Driven Pump.

Pump rooms with radiator-cooled engines could have sufficient air changes due to the radiator discharge and engine consumption. *[See Figure A.11.3.2(b).]*

11.3.2.1 Ventilation shall be provided for the following functions:

(1) To control the maximum temperature to 120°F (49°C) at the combustion air cleaner inlet with engine running at rated load
(2) To supply air for engine combustion
(3) To remove any hazardous vapors
(4) To supply and exhaust air as necessary for radiator cooling of the engine when required

11.3.2.2 The ventilation system components shall be coordinated with the engine operation.

11.3.2.3* Air Supply Ventilator.

A.11.3.2.3 When motor-operated dampers are used in the air supply path, they should be spring operated to the open position and motored closed. Motor-operated dampers should be signaled to open when or before the engine begins cranking to start.

It is necessary that the maximum air flow restriction limit for the air supply ventilator be compatible with listed engines to ensure adequate air flow for cooling and combustion. This restriction typically includes louvers, bird screens, dampers, ducts, and anything else in the air supply path between the pump room and the outdoors.

Motor-operated dampers are recommended for the heat exchanger–cooled engines to enhance convection circulation.

Gravity-operated dampers are recommended for use with radiator-cooled engines to simplify their coordination with the air flow of the fan.

Another method of designing the air supply ventilator in lieu of dampers is to use a vent duct (with rain cap), the top of which extends through the roof or outside wall of a pump house and the bottom of which is approximately 6 in. (152.4 mm) off the floor of the pump house. This passive method reduces heat loss in the winter. Sizing of this duct must meet the requirements of 11.3.2.1.

11.3.2.3.1 The air supply ventilator shall be considered to include anything in the air supply path to the room.

11.3.2.3.2 The total air supply path to the pump room shall not restrict the flow of the air more than 0.2 in. water column (5.1 mm water column).

11.3.2.4* Air Discharge Ventilator.

A.11.3.2.4 When motor-operated dampers are used in the air discharge path, they should be spring operated to the open position, motored closed, and signaled to open when or before the engine begins cranking to start.

Prevailing winds can work against the air discharge ventilator. Therefore, the winds should be considered in determining the location for the air discharge ventilator. *(See Figure A.11.3.2.4 for the recommended wind wall design.)*

For heat exchanger–cooled engines, an air discharge ventilator with motor-driven dampers designed for convection circulation is preferred in lieu of a power ventilator. This arrangement requires the size of the ventilator to be larger, but it is not dependent on a power source that might not be available during the pump operation.

For radiator-cooled engines, gravity-operated dampers are recommended. Louvers and motor-operated dampers are not recommended due to the restriction to air flow they create and the air pressure against which they must operate.

The maximum air flow restriction limit for the air discharge ventilator is necessary to be compatible with listed engines to ensure adequate air flow cooling.

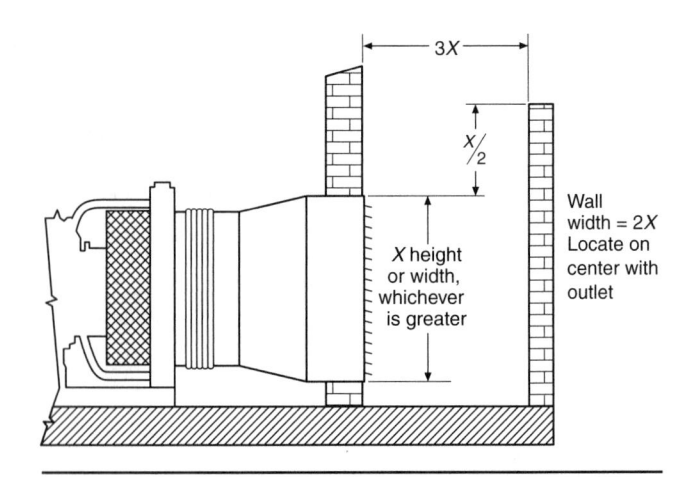

FIGURE A.11.3.2.4 *Typical Wind Wall.*

11.3.2.4.1 The air discharge ventilator shall be considered to include anything in the air discharge path from the engine to the outdoors.

11.3.2.4.2 The air discharge ventilator shall allow sufficient air to exit the pump room to satisfy 11.3.2.

11.3.2.4.3 Radiator-Cooled Engines.

11.3.2.4.3.1 For radiator-cooled engines, the radiator discharge shall be ducted outdoors in a manner that will prevent recirculation.

11.3.2.4.3.2 The duct shall be attached to the radiator via a flexible section.

11.3.2.4.3.3 The air discharge path for radiator-cooled engines shall not restrict the flow of air more than 0.3 in. water column (7.6 mm water column).

11.3.2.4.3.4 A recirculation duct shall be permitted for cold weather operation provided that the following requirements are met:

(1) The recirculation airflow shall be regulated by a thermostatically controlled damper.
(2) The control damper shall fully close in a failure mode.
(3) The recirculated air shall be ducted to prevent direct recirculation to the radiator.
(4) The recirculation duct shall not cause the temperature at the combustion air cleaner inlet to rise above 120°F (49°C).

11.3.3 The entire pump room shall be protected with fire sprinklers in accordance with NFPA 13, *Standard for the Installation of Sprinkler Systems,* as an Extra Hazard Group 2 space.

11.4 Fuel Supply and Arrangement.

11.4.1 General.

11.4.1.1 Plan Review. Before any fuel system is installed, plans shall be prepared and submitted to the authority having jurisdiction for agreement on suitability of the system for prevailing conditions.

11.4.1.2 Tank Construction.

11.4.1.2.1* Tanks shall be designed and constructed in accordance with recognized engineering standards such as ANSI/UL 142, *Standard for Steel Aboveground Tanks for Flammable and Combustible Liquids.*

A.11.4.1.2.1 Dikes are generally not necessary due to the requirement for double-wall tanks with monitoring.

11.4.1.2.2 Tanks shall be securely mounted on noncombustible supports.

11.4.1.2.3 Tanks used in accordance with the rules of this standard shall be limited in size to 1320 gal (4996 L). For situations where fuel tanks in excess of 1320 gal (4996 L) are being used, the rules of NFPA 37, *Standard for the Installation and Use of Stationary Combustion Engines and Gas Turbines,* shall apply.

11.4.1.2.4 Fuel tanks shall be enclosed with a wall, curb, or dike sufficient to hold the entire capacity of the tank.

When calculating the size of the containment wall or dike, the engineer must consider the overhead flow from the required sprinkler system. Sprinkler systems for pump rooms are calculated at 0.25 gpm/ft² (10.19 L/min · m²) over the entire area. The sprinkler flow duration for the hazard must be used when calculating the size of the containment. In most applications, a listed double-wall fuel tank requires less space within the pump room and is much safer. The minimum oversizing of the containment should be at least 110 percent of the total tank capacity.

11.4.1.2.5 Each tank shall have suitable fill, drain, and vent connections.

11.4.1.2.6 Fill pipes that enter the top of the tank shall terminate within 6 in. (152 mm) of the bottom of the tank and shall be installed or arranged so that vibration is minimized.

11.4.1.2.7 The fuel tank shall have one 2 in. (50.8 mm) NPT threaded port in the top, near the center, of the tank to accommodate the low fuel level switch.

11.4.1.2.8 Tank Venting.

11.4.1.2.8.1 Normal vents shall be sized in accordance with ANSI/UL 142, *Standard for Steel Aboveground Tanks for Flammable and Combustible Liquids,* or other approved standards. Alternatively, the normal vent shall be at least as large as the largest filling or withdrawal connection, but in no case shall it be less than 1¼ in. (32 mm) nominal inside diameter.

11.4.1.2.8.2 Vent piping shall be arranged so that the vapors are discharged upward or horizontally away from adjacent walls and so that vapors will not be trapped by eaves or other obstructions. Outlets shall terminate at least 5 ft (1.5 m) from building openings.

11.4.1.2.9 Engine Supply Connection.

11.4.1.2.9.1 The fuel supply pipe connection shall be located on a side of the tank.

11.4.1.2.9.2 The engine fuel supply (suction) pipe connection shall be located on the tank so that 5 percent of the tank volume provides a sump volume not usable by the engine.

11.4.2* Fuel Supply Tank and Capacity.

A.11.4.2 The quantity 1 gal per hp (5.07 L per kW) is equivalent to 1 pint per hp (0.634 L per kW) per hour for 8 hours. Where prompt replenishment of fuel supply is unlikely, a reserve supply should be provided along with facilities for transfer to the main tanks.

11.4.2.1* Fuel supply tank(s) shall have a capacity at least equal to 1 gal per hp (5.07 L per kW), plus 5 percent volume for expansion and 5 percent volume for sump.

The following is an example of how to size a fuel storage tank.

◀ FAQ
How is a fuel storage tank sized?

EXAMPLE

How large must the fuel tank be for a 120 hp engine driving a fire pump?

Solution: The initial calculations based on 11.4.2.1 are 1 gal per hp × 120 hp = 120 gal. Total capacity is determined by adding the allowances for fuel expansion and sump to the answer obtained in the preceding line:

$$120 + [120 \text{ gal} \times (5\% + 5\%)] = 120 + 12 = 132 \text{ gal}$$

Therefore, the minimum-sized fuel tank for a 120 hp engine driving a fire pump is 132 gal.

The 5 percent sump in the fuel tank is provided for any sediment or condensation that may be in the tank. By positioning the supply connection to the engine on the side of the tank at an elevation no lower than the engine fuel pump the plumbing to the engine is always under a small head pressure. In the event of a leak, fuel will escape and be observed. This method is preferred to air being drawn in, resulting in the possible loss of engine prime, if the supply line is under negative pressure.

A.11.4.2.1 Where the authority having jurisdiction approves the start of the fire pump on loss of ac power supply, provisions should be made to accommodate the additional fuel needed for this purpose.

11.4.2.2 Whether larger-capacity fuel supply tanks are required shall be determined by prevailing conditions, such as refill cycle and fuel heating due to recirculation, and shall be subject to special conditions in each case.

11.4.2.3 The fuel supply tank and fuel shall be reserved exclusively for the fire pump diesel engine.

11.4.2.4 A means shall be provided within the tank for low fuel level signal initiation.

11.4.2.5 There shall be a separate fuel line and separate fuel supply tank for each engine.

11.4.2.6 Tank Level Indication.

11.4.2.6.1 Means other than sight tubes for continuous indicating of the amount of fuel in each storage tank shall be provided.

11.4.2.6.2 A fuel level indicator shall be provided to activate at the two-thirds tank level.

11.4.3* Fuel Tank Supply Location.

A.11.4.3 Diesel fuel storage tanks preferably should be located inside the pump room or pump house, if permitted by local regulations. Fill and vent lines in such case should be extended to outdoors. The fill pipe can be used for a gauging well where practical.

Research has identified nothing in NFPA 30, *Flammable and Combustible Liquids Code,* or NFPA 37, *Standard for the Installation and Use of Stationary Combustion Engines and Gas Turbines,* that prohibits the outlet connection to the engine from the diesel tank from being in the location required by NFPA 20.

The applicable code is NFPA 37, not NFPA 30. The scope of NFPA 30 clearly states that if the installation meets the criteria in NFPA 37, then it satisfies the requirements of NFPA 30.

Therefore, NFPA 37 applies for the fuel tank for the fire pump, as it is considered to be part of the installation of the internal combustion engine. Subsection 6.3.2 of NFPA 37 deals with fuel tanks inside structures for fuels other than Class I liquids. Sections 6.6, 6.7, and 6.8 of NFPA 37 deal with filling, venting, and connections between the engine and the fuel tank, and these sections send the reader back to NFPA 30 for the requirements. A review of the tank chapter in NFPA 30 for fixed tanks with capacity of 119 gallons or more finds no requirement stating that the connection to the engine has to be from the top of the tank, if the tank is on the floor on legs, or otherwise above ground.

11.4.3.1 Diesel fuel supply tanks shall be located above ground in accordance with municipal or other ordinances and in accordance with requirements of the authority having jurisdiction and shall not be buried.

11.4.3.2 In zones where freezing temperatures [32°F (0°C)] are possible, the fuel supply tanks shall be located in the pump room.

11.4.3.3 The supply tank shall be located so the fuel supply pipe connection to the engine is no lower than the level of the engine fuel transfer pump.

11.4.3.4 The engine manufacturer's fuel pump static head pressure limits shall not be exceeded when the level of fuel in the tank is at a maximum.

11.4.3.5 If a double-wall tank is installed, the interstitial space between the shells of the diesel fuel storage tank shall be monitored for leakage and annunciated by the engine drive controller. The signal shall be of the supervisory type.

11.4.4* Fuel Piping.

A.11.4.4 NFPA 31, *Standard for the Installation of Oil-Burning Equipment,* can be used as a guide for diesel fuel piping. Figure A.11.4.4 shows a suggested diesel engine fuel system.

11.4.4.1 Flame-resistant reinforced flexible hose listed for this service with threaded connections shall be provided at the engine for connection to fuel system piping.

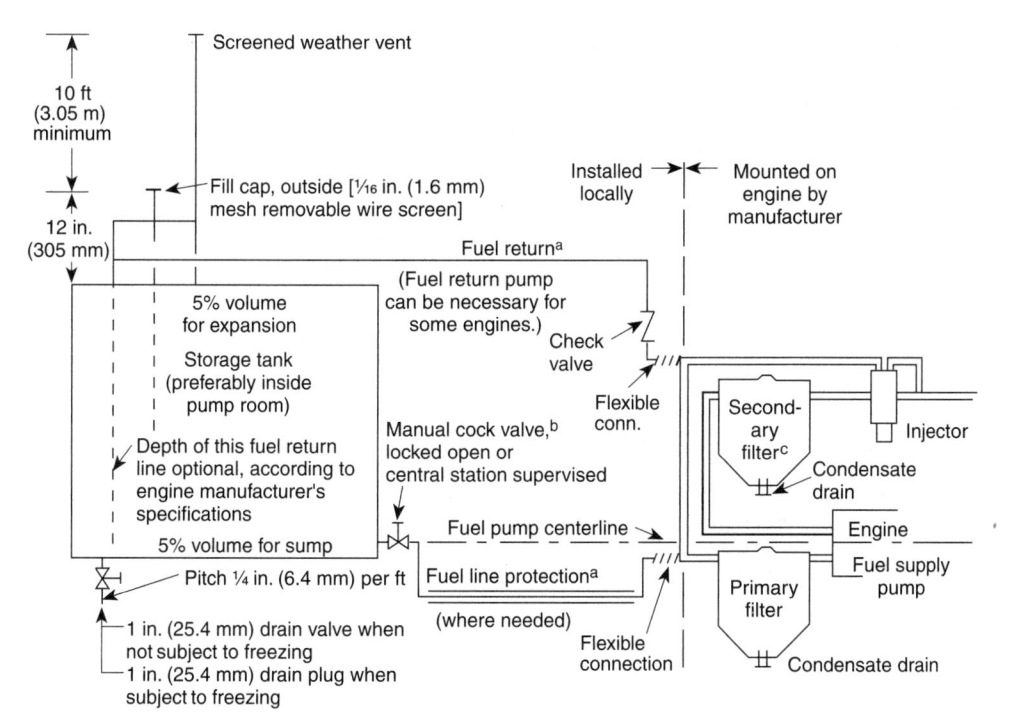

[a]Size fuel piping according to engine manufacturer's specifications.
[b]Excess fuel can be returned to fuel supply pump suction, if recommended by engine manufacturer.
[c]Secondary filter behind or before engine fuel pump, according to engine manufacturer's specifications.

FIGURE A.11.4.4 Fuel System for Diesel Engine–Driven Fire Pump.

11.4.4.2 Fuel piping shall not be galvanized steel or copper.

It is very important to follow the requirements set forth in two NFPA publications—NFPA 37, *Standard for the Installation and Use of Stationary Combustion Engines and Gas Turbines,* and NFPA 30, *Flammable and Combustible Liquids Code.*

All piping materials must be black steel or stainless steel pipe and fittings. The use of galvanized piping creates a problem with scaling of the coating and possibly clogging the fuel lines or injectors. Copper materials cause microbiologically influenced corrosion (MIC) growth in the fuel system and storage tank and can cause failure of the fuel system.

11.4.4.3 The fuel return line shall be installed according to the engine manufacturer's recommendation.

11.4.4.4 There shall be no shutoff valve in the fuel return line to the tank.

11.4.4.5* Fuel Line Protection. A guard, pipe protection, or approved double-walled pipe shall be provided for all exposed fuel lines.

A.11.4.4.5 A means, such as covered floor trough, angle, channel steel, or other adequate protection cover(s) (mechanical or nonmechanical), should be used on all fuel line piping "exposed to traffic," to prevent damage to the fuel supply and return lines between the fuel tank and diesel driver.

11.4.4.6 Fuel Solenoid Valve. Where an electric solenoid valve is used to control the engine fuel supply, it shall be capable of manual mechanical operation or of being manually bypassed in the event of a control circuit failure.

11.4.5* Fuel Type.

A.11.4.5 The pour point and cloud point should be at least 10°F (5.6°C) below the lowest expected fuel temperature. *(See 4.12.2 and 11.4.3.)*

11.4.5.1* The type and grade of diesel fuel shall be as specified by the engine manufacturer.

A.11.4.5.1 Biodiesel and other alternative fuels are not recommended for diesel engines used for fire protection because of the unknown storage life issues. It is recommended that these engines use only petroleum fuels.

11.4.5.2 In areas where local air quality management regulations allow only the use of DF #1 fuel, and no diesel fire pump driver is available listed for use with DF #1 fuel, an engine listed for use with DF #2 shall be permitted to be used but shall have the nameplate rated horsepower derated 10 percent, provided the engine manufacturer approves the use of DF #1 fuel.

11.4.5.3 The grade of fuel shall be indicated on the engine nameplate required in 11.2.2.1.

11.4.5.4 The grade of fuel oil shall be indicated on the fuel tank by letters that are a minimum of 6 in. (152 mm) in height and in contrasting color to the tank.

11.4.5.5 Residual fuels, domestic heating furnace oils, and drained lubrication oils shall not be used.

11.5 Engine Exhaust.

11.5.1 Exhaust Manifold. Exhaust manifolds and turbochargers shall incorporate provisions to avoid hazard to the operator or to flammable material adjacent to the engine.

11.5.2* Exhaust Piping.

A.11.5.2 A conservative guideline is that, if the exhaust system exceeds 15 ft (4.5 m) in length, the pipe size should be increased one pipe size larger than the engine exhaust outlet size for each 5 ft (1.5 m) in added length.

11.5.2.1 Each pump engine shall have an independent exhaust system.

11.5.2.2 A flexible connection with a section of stainless steel, seamless or welded corrugated (not interlocked), not less than 12 in. (305 mm) in length shall be made between the engine exhaust outlet and exhaust pipe.

11.5.2.3 The exhaust pipe shall not be any smaller in diameter than the engine exhaust outlet and shall be as short as possible.

11.5.2.4 The exhaust pipe shall be covered with high-temperature insulation or otherwise guarded to protect personnel from injury.

11.5.2.5 The exhaust pipe and muffler, if used, shall be suitable for the use intended, and the exhaust back pressure shall not exceed the engine manufacturer's recommendations.

11.5.2.6 Exhaust pipes shall be installed with clearances of at least 9 in. (229 mm) to combustible materials.

11.5.2.7 Exhaust pipes passing directly through combustible roofs shall be guarded at the point of passage by ventilated metal thimbles that extend not less than 9 in. (229 mm) above and 9 in. (229 mm) below roof construction and are at least 6 in. (152 mm) larger in diameter than the exhaust pipe.

11.5.2.8 Exhaust pipes passing directly through combustible walls or partitions shall be guarded at the point of passage by one of the following methods:

(1) Metal ventilated thimbles not less than 12 in. (305 mm) larger in diameter than the exhaust pipe
(2) Metal or burned clay thimbles built in brickwork or other approved materials providing not less than 8 in. (203 mm) of insulation between the thimble and construction material

11.5.2.9* Exhaust emission after treatment devices that have the potential to adversely impact the performance and reliability of the engine shall not be permitted.

A.11.5.2.9 Exhaust emission after treatment devices are typically dependent upon high exhaust temperature to burn away collected materials to prevent clogging. Due to the lower exhaust temperatures produced by the engine when operating at pump shutoff during weekly operation, there is a high possibility the after treatment device will accumulate collected material and will not be capable of flowing the volume of exhaust in the event the engine is required to produce full rated power for an emergency.

11.5.2.10 Where required by the authority having jurisdiction, the installation of an exhaust emission after treatment device shall be of the active regeneration type with a pressure limiting device that shall permit the engine exhaust to bypass the after treatment device when the engine manufacturer's maximum allowed exhaust backpressure is exceeded.

11.5.3 Exhaust Discharge Location.

11.5.3.1 Exhaust from the engine shall be piped to a safe point outside the pump room and arranged to exclude water.

11.5.3.2 Exhaust gases shall not be discharged where they will affect persons or endanger buildings.

11.5.3.3 Exhaust systems shall terminate outside the structure at a point where hot gases, sparks, or products of combustion will discharge to a safe location. [37:8.2.3.1]

11.5.3.4 Exhaust system terminations shall not be directed toward combustible material or structures, or into atmospheres containing flammable gases, flammable vapors, or combustible dusts. [37:8.2.3.2]

11.5.3.5 Exhaust systems equipped with spark-arresting mufflers shall be permitted to terminate in Division 2 locations as defined in Article 500 of *NFPA 70, National Electrical Code.* [37:8.2.3.3]

11.6* Diesel Engine Driver System Operation.

A.11.6 Internal combustion engines necessarily embody moving parts of such design and in such number that the engines cannot give reliable service unless given diligent care. The manufacturer's instruction book covering care and operation should be readily available, and pump operators should be familiar with its contents. All of its provisions should be observed in detail.

11.6.1 Weekly Run.

11.6.1.1 Engines shall be designed and installed so that they can be started no less than once a week and run for no less than 30 minutes to attain normal running temperature.

Running the engine for a minimum of 30 minutes provides sufficient time for the engine to reach its full operating temperature and for condensation to evaporate from the crankcase. Any irregularities in the engine operation should be investigated and repaired immediately. Running the engine for this length of time also gives the operator sufficient time to check out the pump operations, cooling loop operation, the battery chargers, and all other pump room components requiring an inspection. Section 8.5, Maintenance, and Table 8.5.3 of NFPA 25 describes the schedules and the specific requirements for the periodically required maintenance items. Additionally, Section 8.6, Component Replacement Testing Requirements in NFPA 25, provides the necessary testing required after repairs or upgrades are completed on any pump room component.

11.6.1.2 Engines shall run smoothly at rated speed, except for engines addressed in 11.6.1.3.

11.6.1.3 Engines equipped with variable speed pressure limiting control shall be permitted to run at reduced speeds provided factory-set pressure is maintained and they run smoothly.

11.6.2* Engine Maintenance. Engines shall be designed and installed so that they can be kept clean, dry, and well lubricated to ensure adequate performance.

A.11.6.2 See NFPA 25, *Standard for the Inspection, Testing, and Maintenance of Water-Based Fire Protection Systems,* for proper maintenance of engine(s), batteries, fuel supply, and environmental conditions.

11.6.3 Battery Maintenance.

11.6.3.1 Storage batteries shall be designed and installed so that they can be kept charged at all times.

11.6.3.2 Storage batteries shall be designed and installed so that they can be tested frequently to determine the condition of the battery cells and the amount of charge in the battery.

Lead-acid batteries last between 24 and 30 months of service. See NFPA 25 for the recommended replacement schedule.

11.6.3.3 Only distilled water shall be used in battery cells.

11.6.3.4 Battery plates shall be kept submerged at all times.

If the battery requires an excessive amount of replacement water, the battery should be checked to make sure it is not overcharging. Also, the battery plates need to be covered with water to help eliminate the possibility of battery explosions. Exposed plates may create an arc or spark that can ignite the hydrogen gasses that are present during the starting and charging cycles.

11.6.3.5 The automatic feature of a battery charger shall not be a substitute for proper maintenance of battery and charger.

11.6.3.6 The battery and charger shall be designed and installed so that periodic inspection of both battery and charger is physically possible.

11.6.3.6.1 This inspection shall determine that the charger is operating correctly, the water level in the battery is correct, and the battery is holding its proper charge.

11.6.4* Fuel Supply Maintenance.

A.11.6.4 Active systems that are permanently added to fuel tanks for removing water and particulates from the fuel can be acceptable, provided the following apply:

(1) All connections are made directly to the tank and are not interconnected with the engine or its fuel supply and return piping in any way.
(2) There are no valves or other devices added to the engine or its fuel supply and return piping in any way.

11.6.4.1 The fuel storage tanks shall be designed and installed so that they can be kept as full and maintained as practical at all times but never below 66 percent (two-thirds) of tank capacity.

11.6.4.2 The tanks shall be designed and installed so that they can always be filled by means that will ensure removal of all water and foreign material.

11.6.5* Temperature Maintenance.

A.11.6.5 Proper engine temperature, per 11.2.8.2 and 11.6.5.1, maintained through the use of a supplemental heater has many benefits, as follows:

(1) Quick starting (a fire pump engine might have to carry a full load as soon as it is started)
(2) Reduced engine wear
(3) Reduced drain on batteries
(4) Reduced oil dilution
(5) Reduced carbon deposits, so that the engine is far more likely to start every time

11.6.5.1 The temperature of the pump room, pump house, or area where engines are installed shall be designed so that the temperature is maintained at the minimum recommended by the engine manufacturer and is never less than the minimum recommended by the engine manufacturer.

Fuel tanks installed outdoors generate a greater volume of water from condensation than tanks installed inside the fire pump room. Tank maintenance should be increased for outdoor tanks to ensure water in the tank is kept to a minimum.

11.6.6 Emergency Starting and Stopping.

All personnel should be factory trained and experienced on starting and stopping the engine in all modes of operation. Follow the manufacturer's operating requirements for starting and stopping.

11.6.6.1 The sequence for emergency manual operation, arranged in a step-by-step manner, shall be posted on the fire pump engine.

11.6.6.2 It shall be the engine manufacturer's responsibility to list any specific instructions pertaining to the operation of this equipment during the emergency operation.

REFERENCES CITED IN COMMENTARY

National Fire Protection Association, 1 Batterymarch Park, Quincy, MA 02169-7471.

NFPA 25, *Standard for the Inspection, Testing, and Maintenance of Water-Based Fire Protection Systems,* 2008 edition.

NFPA 30, *Flammable and Combustible Liquids Code,* 2007 edition.

NFPA 37, *Standard for the Installation and Use of Stationary Combustion Engines and Gas Turbines,* 2006 edition.

Engine Drive Controllers

CHAPTER 12

Diesel engine–driven fire pumps are often used when independent self-contained fire protection is required. Diesel-driven fire pump controllers maximize this self-sufficiency in two ways: by having the engine starting powered by batteries, and by being independent of the ac power supply mains. Further, these controllers contain extensive monitoring, supervision, and annunciation provisions as well as contacts for remote alarm operation. Annunciation options might include pump room or pump house conditions such as temperature, reservoir level, or valve positions, among others. Substantial redundant circuitry is employed as well as means for manual intervention should it be required.

This chapter specifies the requirements for the application, location, construction, and internal components of engine drive controllers and also delineates their relevant operating requirements, including starting and stopping of the engine. New to Chapter 12 are the requirements for the static ac mains–powered battery charger. These requirements were formerly in Chapter 11.

The commentary contained in this chapter is intended to help explain the meaning, and often the reason, for the chapter's requirements. A more thorough discussion of the theory, parameters, and practical application consideration of fire pump controllers can be found in the National Fire Protection Association book on fire pumps, *Pumps for Fire Protection Systems*.

12.1 Application.

12.1.1 This chapter provides requirements for minimum performance of automatic/nonautomatic diesel engine controllers for diesel engine–driven fire pumps.

Although not available for many years, a nonautomatic (manual only) engine fire pump controller, which was a small, simple switch box that energized the fuel and cranking circuits, was once utilized. Most modern engines still incorporate this switch (in addition to the manual devices on the engine itself) and are called "combined" automatic/nonautomatic controllers. These controllers are the type commonly used for automatic fire protection and, while equipped to be used either in manual or in automatic (standby) mode, automatic is the method normally used. Exhibit II.12.1 shows a typical solid-state automatic/nonautomatic diesel engine–driven fire pump controller.

12.1.2 Accessory devices, such as fire pump alarm and signaling means, are included where necessary to ensure minimum performance of the equipment mentioned in 12.1.1.

This provision addresses items separate from the controller itself. One example is remote alarm equipment.

12.1.3 General.

12.1.3.1 All controllers shall be specifically listed for diesel engine–driven fire pump service.

EXHIBIT II.12.1 Diesel Drive Controller. (Photograph courtesy of Master Control Systems, Inc.)

See 3.2.3 and A.3.2.3 for the definition of *listed*. "Specifically listed" means that the equipment has been examined for compliance to these requirements. Note that most, but not all, fire pump controllers are both Underwriters Laboratories (UL) listed *and* Factory Mutual (FM) approved. Further, some options or modifications specified on a purchase order form may void one or another listing.

12.1.3.2 All controllers shall be completely assembled, wired, and tested by the manufacturer before shipment from the factory.

FAQ ▶
How can the buyer be sure the "listed" equipment has been adequately tested?

The amount of factory testing can vary substantially among the various listed manufacturers. Also, most, if not all, manufacturers can provide copies of test data sheets, and many can also accommodate "witness testing," although this is typically an extra cost item.

12.1.3.3 Markings.

12.1.3.3.1 All controllers shall be marked "Diesel Engine Fire Pump Controller" and shall show plainly the name of the manufacturer, the identifying designation, rated operating pressure, enclosure type designation, and complete electrical rating.

FAQ ▶
What is the best way to confirm that the controller received is appropriate?

It is very important to verify that the controller is a listed and approved controller in the proper category. The words "Diesel Engine Fire Pump Controller" should be on the same nameplate as the listing mark. There have been pump controllers sold and marked as "Fire Pump Controller" but have only general industrial or control panel listing. While this violates listing agency procedures or rules, it has to be reported before action can occur. See Exhibit II.12.2 for an example of a suitable fire pump controller nameplate.

12.1.3.3.2 Where multiple pumps serving different areas or portions of the facility are provided, an appropriate sign shall be conspicuously attached to each controller indicating the area, zone, or portion of the system served by that pump or pump controller.

Fire pumps are provided in different locations for a number of reasons, including the following:

1. Plant expansion
2. Fire water drawn from different sources or water mains

EXHIBIT II.12.2 Diesel Drive Controller Nameplate. (Photograph courtesy of Master Control Systems, Inc.)

3. Independent zones in a high-rise building, such as the upper zone as fed by a separate fire water reservoir (e.g., a swimming pool)
4. Some large processing plants
5. Some multi-building campuses, especially where a fire pump may not supply the campus fire water loop (main)

It is important that each controller have its own independent pressure-sensing line in accordance with Section 4.30. It is specifically required that jockey pumps (pressure maintenance pumps) also have their own independent pressure-sensing line. When pressure-sensing lines are combined, the system can and usually does malfunction until the required separate pressure sensing lines are installed. See Figure A.4.30(b) and 12.7.2.1.

12.1.4 It shall be the responsibility of the pump manufacturer or its designated representative to make necessary arrangements for the services of a controller manufacturer's representative, where needed, for services and adjustment of the equipment during the installation, testing, and warranty periods.

Typically, the installing fire pump contractor arranges for the services identified in 12.1.4.

12.2 Location.

12.2.1* Controllers shall be located as close as is practical to the engines they control and shall be within sight of the engines.

A.12.2.1 If the controller must be located outside the pump room, a glazed opening should be provided in the pump room wall for observation of the motor and pump during starting. The pressure control pipeline should be protected against freezing and mechanical injury.

It is important to see and hear whether an engine and pump are functioning as expected.

12.2.2 Controllers shall be so located or so protected that they will not be damaged by water escaping from pumps or pump connections.

Fire pump installations are sometimes located on lower floor levels or below grade. In the event of fire, water from fire-fighting operations can collect at these low points. By raising the current-carrying components above the floor, the likelihood of an electric hazard caused by water coming in contact with "hot" components is reduced. The electrical equipment associated with fire pump controllers is not waterproof or weatherproof, which is another reason that the equipment should be elevated above the floor.

◀ **FAQ**
How can controllers be protected from water damage?

12.2.3 Current carrying parts of controllers shall not be less than 12 in. (305 mm) above the floor level.

While the 12 in. (305 mm) requirement is adequate in most cases, it is recommended that a "housekeeping pad" be used for floor mounted controllers to keep them out of puddles or casual water in the pump room. Extra protection may be needed when controllers are mounted in pits or where sump pumps are needed to keep out water from the pump installation.

12.2.4 Working clearances around controllers shall comply with *NFPA 70, National Electrical Code*, Article 110.

12.3 Construction.

12.3.1 Equipment.

12.3.1.1* All equipment shall be suitable for use in locations subject to a moderate degree of moisture, such as a damp basement.

Where conditions warrant, a more robust and protective enclosure, such as Type 4, may be appropriate. This is also true for humidistatically controlled cabinet space heaters.

A.12.3.1.1 In areas affected by excessive moisture, heat can be useful in reducing the dampness.

This paragraph refers to cabinet (enclosure) space heaters. See the commentary following 12.3.3.1.2 for more information on cabinet heaters.

12.3.1.2 Reliability of operation shall not be adversely affected by normal dust accumulations.

Note that Type 12 or better enclosures provide dust protection.

12.3.2 Mounting. All equipment not mounted on the engine shall be mounted in a substantial manner on a single noncombustible supporting structure.

12.3.3 Enclosures.

12.3.3.1* Mounting.

A.12.3.3.1 For more information, see NEMA 250, *Enclosures for Electrical Equipment.*

Also refer to ANSI/UL 50, *Enclosures for Electrical Equipment.*

12.3.3.1.1 The structure or panel shall be securely mounted in, as a minimum, a NEMA Type 2 dripproof enclosure(s) or an enclosure(s) with an ingress protection (IP) rating of IP 31.

Commentary Table II.12.1 lists common enclosure types.

Note that the "NEMA" type designation is self-certified by the manufacturer and thus is not necessarily a reliable assurance as to compliance. However, when a controller is marked with a UL "Type" designation, such as "Type 4 Enclosure," then the construction of the enclosure is included in the UL investigation (conformity assessment) of the product and is also part of the periodic follow-up inspection. This designation does give reasonable assurance of

COMMENTARY TABLE II.12.1 *Fire Pump Controller Enclosure Types*

Enclosure Type*	Where	Protection Against:
Type 2	Indoor	Falling water and dirt
Type 12	Indoor	Dripping water and dust (was splash and oil resistant)
Type 13	Indoor	Spraying water (and oil) and dust
Type 3	Outdoor	Rain, dust, and ice
Type 3R	Outdoor	Falling rain (allows water, dust, and dirt to enter)
Type 4	Outdoor	Rain, dust, ice, splashing and hose-directed water (hose-tight)
Type 4XA	Outdoor	Same as NEMA 4 but corrosion resistant (special paint finish)
Type 4XB	Outdoor	Same as NEMA 4 but also corrosion resistant of Type 304 stainless steel
Type 4XC	Outdoor	Same as NEMA 4 but also corrosion resistant of Type 316 stainless steel
Type 4XCL	Outdoor	Same as NEMA 4XC but Type 316 low carbon stainless steel

Note: The NEMA 4X, A, B, and C designations are unofficial and are for reference only.

the compliance of the enclosure construction. For example, NEMA "X" is not the same as UL Type "X" regarding assurance or confidence in construction.

The "IP 31" rating is new to the 2010 edition of NFPA 20. It is an IEC rating that provides the same protection from dripping (vertical) water and slightly smaller openings than NEMA 2 enclosures.

12.3.3.1.2 Where the equipment is located outside or special environments exist, suitably rated enclosures shall be used.

For outdoor installations, the enclosure must be an outdoor rated type, such as Type 3, Type 4, or Type 4X. Moreover, a sunscreen or roof is highly advised to keep the controller internal temperature from exceeding the rating of the components, especially since the battery charger gives off heat when charging the batteries. For corrosive environments, including salt air, one of the Type 4X enclosures should be employed.

To protect outdoor equipment from cooler temperatures at night, an optional enclosure space heater should be employed, preferably one that is thermostatically controlled, or better yet, one that is controlled by a humidistat. This minimizes running the heater(s) unnecessarily. Better-built controllers run the heater at half voltage to lower the surface temperature of the heater. This is easily accomplished, for example, by using 240 Vac rated heaters at 120 Vac. The power is one-fourth of the rated power and, thus, the surface temperature is substantially lower.

◀ FAQ
What is the most efficient way to protect outdoor equipment where temperatures drop significantly at night?

12.3.3.2 Grounding. The enclosures shall be grounded in accordance with *NFPA 70, National Electrical Code,* Article 250.

Grounding is important to protect the equipment from lightning and/or other surges, but it is also an essential safety feature for personnel. Local codes must be followed. Most enclosures are grounded by way of conduits or raceways and the pressure-sensing line.

12.3.4 Locked Lockable Cabinet. All switches required to keep the controller in the automatic position shall be within locked enclosures having breakable glass panels.

In an emergency, the operator can break the glass window and switch the controller from automatic to manual mode and start the diesel engine manually. Likewise, an operator can break the glass window and switch the control switch to the "off" position to perform an emergency shutdown. Some modern controllers make use of glass rods or similar devices as an alternative to the usual break glass window. Note that the old phrase *locked cabinet* has been changed to the more correct *lockable enclosure*.

12.3.5 Connections and Wiring.

12.3.5.1 Field Wiring.

12.3.5.1.1 All wiring between the controller and the diesel engine shall be stranded and sized to carry the charging or control currents as required by the controller manufacturer.

Article 695 in *NFPA 70®, National Electrical Code®*, governs wiring methods for fire pump controllers. However, note that some of the material in Article 695 is extract text from NFPA 20. See Supplement 3 for the entire text of Article 695.

Note that 12.3.5.1.1 specifies stranded wire so as to better withstand engine vibration, and it allows for the use of ring lugs. Ring lugs are used because if a screw should come loose, as happens occasionally, the connection may still be good enough for emergency running of the pump.

◀ FAQ
Why does 12.3.5.1.1 specify stranded wire, when most wiring between the controller and the diesel engine is solid wire?

12.3.5.1.2 Such wiring shall be protected against mechanical injury.

Best wiring practice makes use of high-quality crimped ring lugs for connection to the engine junction terminal box terminal strip and the controller terminal strip(s).

Stationary Fire Pumps Handbook 2010

12.3.5.1.3 Controller manufacturer's specifications for distance and wire size shall be followed.

In some cases, the wire size recommended by the controller manufacturer will be larger than the minimum size specified in the *National Electrical Code*. Use of the larger wire size is intended to reduce the voltage drop in the wiring from the battery charger to the batteries.

12.3.5.2 Wiring Elements. Wiring elements of the controller shall be designed on a continuous-duty basis.

12.3.5.3 Field Connections.

12.3.5.3.1 A diesel engine fire pump controller shall not be used as a junction box to supply other equipment.

Only wiring directly involved with the fire pump engine and fire pump controller, and associated alarm and signals, is allowed in the fire pump controller—no other circuitry is allowed. This requirement is to both minimize needing access to the controller, which should be locked, and to minimize any hazard or threat posed to the controller by "foreign" signals, voltages, or circuitry.

12.3.5.3.2 No undervoltage, phase loss, frequency sensitive, or other device(s) shall be field installed that automatically or manually prohibits electrical actuation of the motor contactor.

In the 2010 edition of NFPA 20, the wording change from *sensor(s)* to *device(s)* makes the requirement more general. Adding "field" installed clarifies the intent of no field modifications to the controller design or construction. Modifying the controller in any way could void the listing(s) and likely void the manufacturer's warranty.

12.3.5.3.3 Electrical supply conductors for pressure maintenance (jockey or make-up) pump(s) shall not be connected to the diesel engine fire pump controller.

This paragraph reinforces the requirement of 12.3.5.3.1. Specifically, anything connected to the controller ac power (mains) circuit or circuitry is capable of tripping the upstream branch circuit breaker (or fuse), leaving the controller without ac power and the batteries without a charger.

12.3.5.3.4 Diesel engine fire pump controllers shall be permitted to supply essential and necessary ac and/or dc power to operate pump room dampers and engine oil heaters and other associated required engine equipment only when provided with factory-equipped dedicated field terminals and overcurrent protection.

FAQ ▶
Why is the diesel engine fire pump controller allowed to provide power to pump room dampers and engine oil heaters in this situation?

This allowance appears to be an exception to 12.3.5.3.1. However, the difference is a factory-equipped arrangement that has gone through the approval process with the listing agency. Done properly, the controller's overcurrent protection (fuses or circuit breakers) will be coordinated with upstream protection and prevent tripping that protection if a short-circuit or fault occurs in the auxiliary equipment or circuit.

12.3.6 Electrical Diagrams and Instructions.

12.3.6.1 A field connection diagram shall be provided and permanently attached to the inside of the enclosure.

The diagram required in 12.3.6.1 is used by the owner and contractor to identify internal circuits of the controller. See Exhibits II.12.3 and II.12.4 for typical engine-to-controller wiring diagrams.

12.3.6.2 The field connection terminals shall be plainly marked to correspond with the field connection diagram furnished.

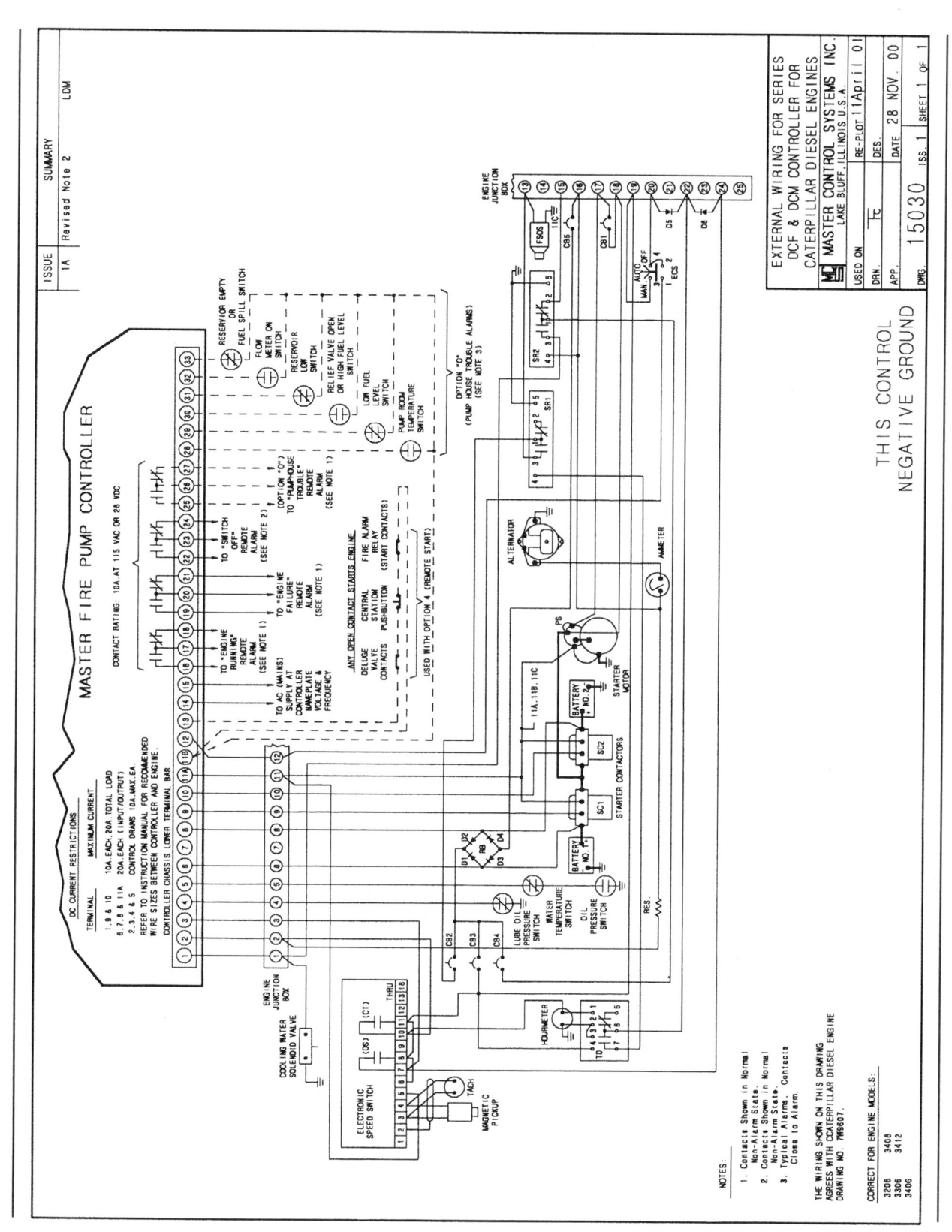

EXHIBIT II.12.3 External Wiring Diagram for Caterpillar Energize to Stop Diesel Engine. (Diagram courtesy of Master Control Systems, Inc.)

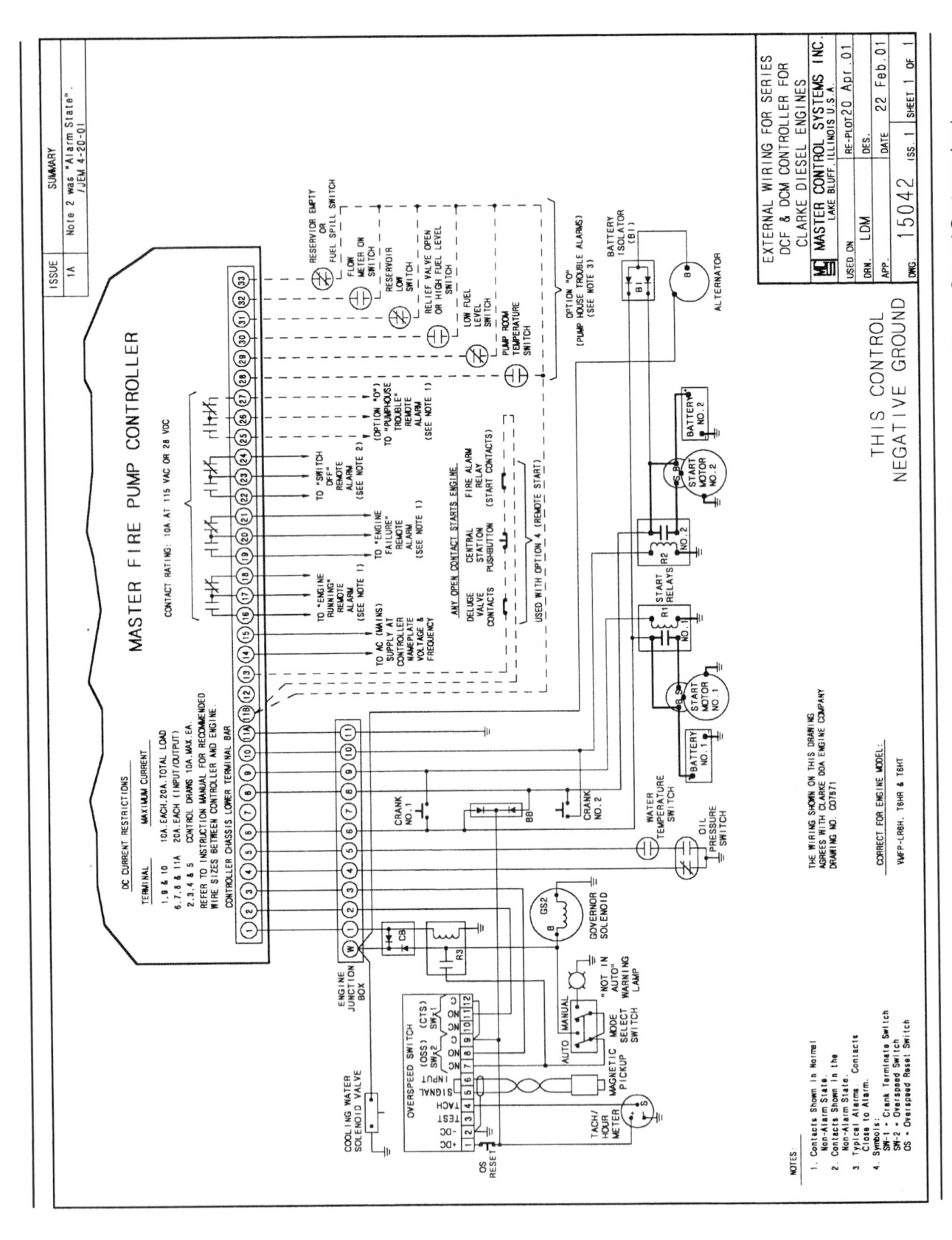

EXHIBIT II.12.4 External Wiring Diagram for Clarke Dual Starter Motor Diesel Engine. (Diagram courtesy of Master Control Systems, Inc.)

All required field connection terminals must be shown on the drawing(s) and permanently attached to the inside of the controller in addition to being shown in the instruction manual.

12.3.6.3 For external engine connections, the field connection terminals shall be commonly numbered between the controller and the engine terminals.

12.3.7 Marking.

12.3.7.1 Each operating component of the controller shall be plainly marked with the identification symbol that appears on the electrical schematic diagram.

This marking is to make field replacement of components and assemblies easier and reduce likelihood of mistakes.

12.3.7.2 The markings shall be located so as to be visible after installation.

Labels and designations must not be covered over by other components or assemblies.

12.3.8* Instructions. Complete instructions covering the operation of the controller shall be provided and conspicuously mounted on the controller.

The instruction plate should include clear and concise instructions for emergency starting and stopping and also how to set the controller's selector, control, and/or mode switch(es) for automatic (standby) fire protection service mode.

◀ **FAQ**
What steps should the instructions include to be considered "complete"?

A.12.3.8 Pump operators should be familiar with instructions provided for controllers and should observe in detail all their recommendations.

12.4 Components.

Exhibit II.12.5 shows the components of a relay-style diesel drive fire pump controller, while Exhibit II.12.17 on p. 302 shows components for a modern solid-state controller that employs hard-wired logic modules and relays for control, reporting, and communication.

12.4.1 Indicators on Controller.

12.4.1.1 All visible indicators shall be plainly visible.

12.4.1.2* Visible indication shall be provided to indicate that the controller is in the automatic position. If the visible indicator is a pilot lamp, it shall be accessible for replacement.

Visible indication is necessary whether the controller is in a dark room, a brightly lit pump house, or outdoors. This applies to alarm, trouble, and status indicators, as well as digital displays.

Legacy, or traditional, controllers commonly used incandescent lamp (pilot light) type indicators. Modern controllers use LED-type indicators and/or digital displays in one form or another. See Exhibits II.12.6 and II.12.17 for examples.

A.12.4.1.2 It is recommended that the pilot lamp for signal service have operating voltage less than the rated voltage of the lamp to ensure long operating life. When necessary, a suitable resistor should be used to reduce the voltage for operating the lamp.

This recommendation was important for legacy controllers, which made use of filament-type pilot lights. Better-designed controllers used 14 volt and 28 volt rated lamps for 12 Vdc and 24 Vdc rated systems, respectively.

The life of filament (incandescent) lamps is inversely proportional to the 12th power of the voltage. Hence, operating a 14 volt lamp at 13.2 Vdc (typical battery charger float

◀ **FAQ**
How can the pilot lamp be kept operable as long as possible?

280 Part II • NFPA 20

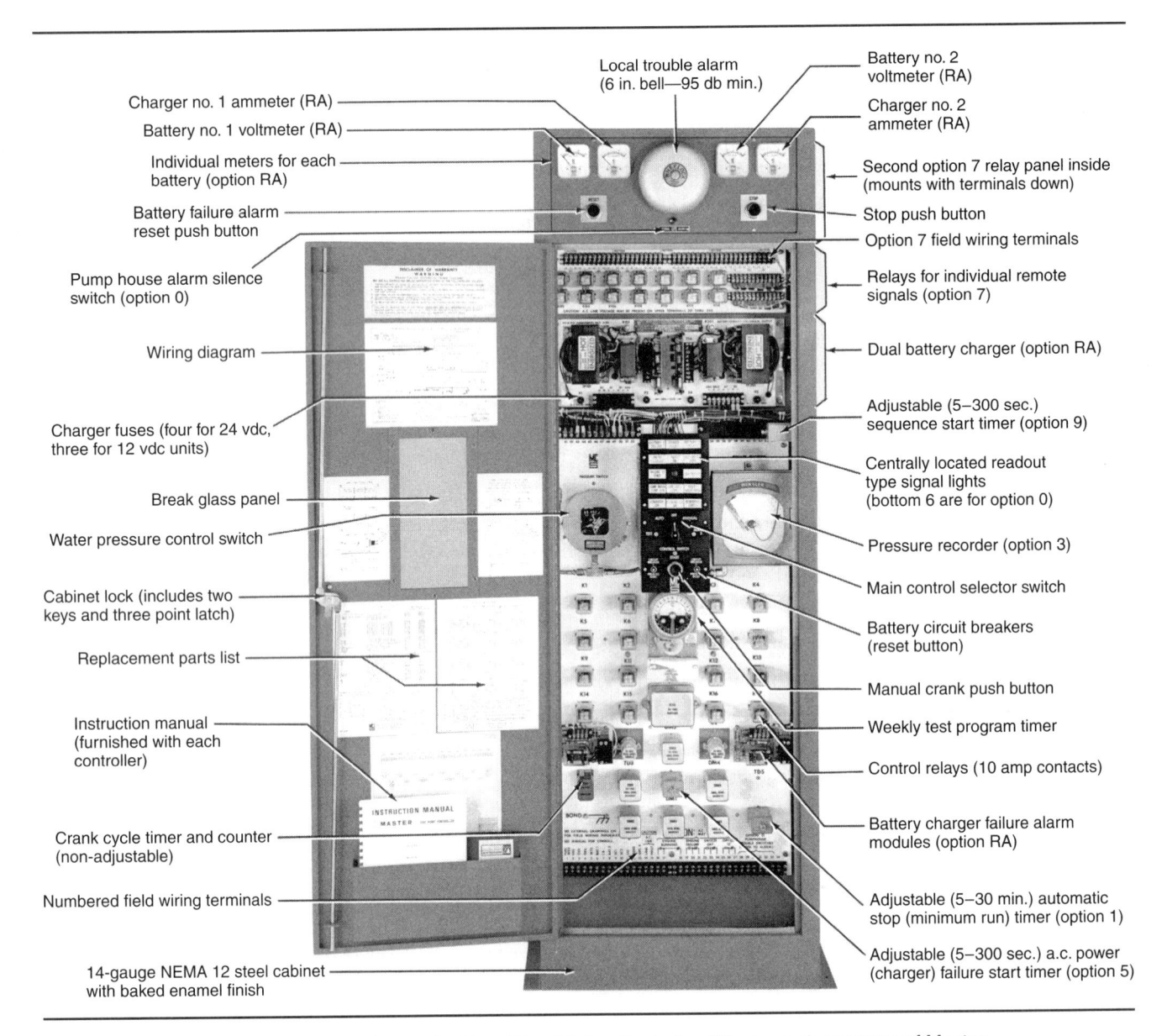

Labels around the controller image (clockwise from top):
- Local trouble alarm (6 in. bell—95 db min.)
- Battery no. 2 voltmeter (RA)
- Charger no. 2 ammeter (RA)
- Second option 7 relay panel inside (mounts with terminals down)
- Stop push button
- Option 7 field wiring terminals
- Relays for individual remote signals (option 7)
- Dual battery charger (option RA)
- Adjustable (5–300 sec.) sequence start timer (option 9)
- Centrally located readout type signal lights (bottom 6 are for option 0)
- Pressure recorder (option 3)
- Main control selector switch
- Battery circuit breakers (reset button)
- Manual crank push button
- Weekly test program timer
- Control relays (10 amp contacts)
- Battery charger failure alarm modules (option RA)
- Adjustable (5–30 min.) automatic stop (minimum run) timer (option 1)
- Adjustable (5–300 sec.) a.c. power (charger) failure start timer (option 5)
- 14-gauge NEMA 12 steel cabinet with baked enamel finish
- Numbered field wiring terminals
- Crank cycle timer and counter (non-adjustable)
- Instruction manual (furnished with each controller)
- Replacement parts list
- Cabinet lock (includes two keys and three point latch)
- Water pressure control switch
- Break glass panel
- Charger fuses (four for 24 vdc, three for 12 vdc units)
- Wiring diagram
- Pump house alarm silence switch (option 0)
- Battery failure alarm reset push button
- Individual meters for each battery (option RA)
- Battery no. 1 voltmeter (RA)
- Charger no. 1 ammeter (RA)

EXHIBIT II.12.5 Components of a Relay Logic–Type Diesel Drive Controller. (Photograph courtesy of Master Control Systems, Inc.)

voltage) is a 9.1 percent reduction in voltage, which results in approximately a 314 percent increase in lamp life.

12.4.1.3 Separate visible indicators and a common audible fire pump alarm capable of being heard while the engine is running and operable in all positions of the main switch except the off position shall be provided to immediately indicate the following conditions:

The audible device must be capable of being heard. Traditionally, controllers used alarm bells or horns, which could usually be heard over the noise of a running engine. The exception was 3,000 RPM engines, which are not as prevalent as they once were. It is important to note that solid-state devices such as piezoelectric beepers are often not audible over the sound of a running engine, and that the type rating of the enclosure is affected by which audible device is used. For example, Type 4 (NEMA) alarm devices are not as common as lower-rated types.

2010 Stationary Fire Pumps Handbook

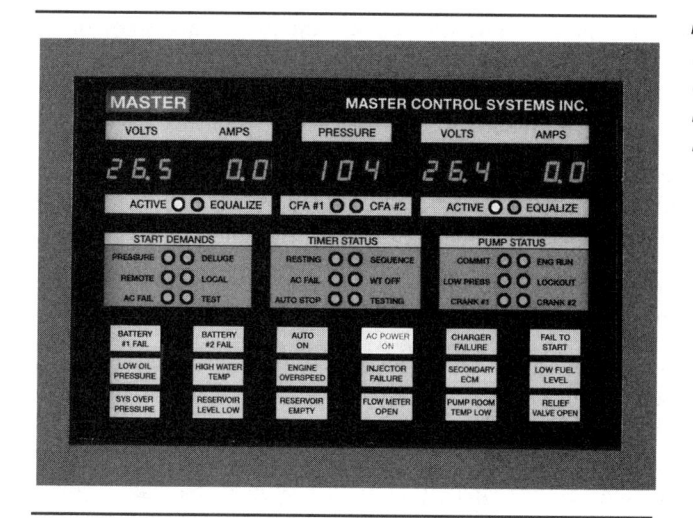

EXHIBIT II.12.6 Diesel Controller Door Display Panel. (Photograph courtesy of Master Control Systems, Inc.)

(1) Critically low oil pressure in the lubrication system. The controller shall provide means for testing the position of the pressure switch contacts without causing fire pump alarms.

The oil pressure referred to here is lube oil pressure and not fuel oil pressure. In most, if not all, cases the testing is accomplished by switching the controller to the manual mode or one of the two manual modes (MAN1 or MAN2) and not cranking the engine. The low oil pressure switch should be closed in manual mode, and the controller visible indicator should be lit. However, the audible device and remote failure contacts should remain in their "normal" (non-alarm) state. Typically, a delay time of a few seconds is used after the controller receives the cranking terminating (engine running) signal from the engine speed switch prior to actuating the audible device and alarm contacts. This delay is to prevent false alarms. See also 12.7.5.2(4).

(2) High engine jacket coolant temperature.

Similar to the low lube oil pressure alarm, the controller will not shut down the engine due to a high water alarm condition if it is running under a true demand, such as pressure loss, remote start, and so forth. The controller will shut down the engine if it is running under test. Note that some engines add an oil pressure switch in series with the high temperature switch to avoid false alarms after a high horsepower demand running. See also 12.7.5.2(4) for further requirements and information.

(3) Failure of engine to start automatically.

This condition occurs if the engine fails to start after six cranking attempts or if the engine (speed switch) fails to notify the controller that the engine is running. See 12.7.4 for more information. The controller treats this condition as a shutdown requirement, even though the engine may not be running. The reason for this shutdown is to kill the water and fuel solenoids to avoid further battery drain and also to avoid cooling water drain, which can be significant over an extended period, such as a three-day weekend.

(4) Shutdown from overspeed.

Typically the engine overspeed switch is set to trip at approximately 20 percent of the engine's rated running speed.

Although the general philosophy is that the fire pump and the engine are "sacrificial" if an overspeed condition is detected by the engine speed switch and transmitted to the controller, the controller will immediately shut down the engine. The reason is that a true engine overspeed condition is considered an "imminent destruction" condition. It is also a personnel safety consideration, since engines sometimes throw parts in the process of overspeed destruction. By shutting down the engine, there may also be a possibility of "manual intervention" to correct or temporarily remedy the situation.

12.4.1.4 Separate visible indicators and a common audible signal capable of being heard while the engine is running and operable in all positions of the main switch except the off position shall be provided to immediately indicate the following conditions:

(1) Battery failure or missing battery. Each controller shall be provided with a separate visible indicator for each battery.

The "battery failure" alarm occurs when the engine cranking voltage is too low to effectively crank (turn over) the engine. In addition to initiating the signal, the controller must also switch to the other battery set.

FAQ ▶
Why is the oldest battery in the engine controller suddenly stronger than the newer ones when cranking the engine?

As batteries age, their actual capacity (amp-hours) increases somewhat, then declines over the life of the battery. Once their actual capacity falls back to its initial capacity, the capacity curve will show that this capacity is declining rapidly and will soon be inadequate. Battery replacement is advised at this time.

(2) Battery charger failure. Each controller shall be provided with a separate visible indicator for battery charger failure and shall not require the audible signal for battery charger failure.

The controller is required to trip this signal if the battery charger either fails to maintain the charge in the battery, or fails to recharge the battery as required. See Sections 12.5 and 12.6 for more requirements.

(3) Low air or hydraulic pressure. Where air or hydraulic starting is provided *(see 11.2.7 and 11.2.7.4)*, each pressure tank shall provide to the controller separate visible indicators to indicate low pressure.

This condition is analogous to the battery failure alarm. Note that while not commonly used, a controller for either hydraulically or pneumatically started engines will usually make use of two batteries for controller power, although these batteries may be fairly small. In any event, the controller usually will still be provided with the battery failure and charger failure alarms, since both the battery and charger remain critical to the operation of the fire pump.

(4) System overpressure, for engines equipped with variable speed pressure limiting controls, to actuate at 115 percent of set pressure.

Pressure limiting diesel drivers use feedback from the pump discharge pressure to modulate the speed governor setting. This signal is used to indicate a problem or manual intervention resulting in pressure higher than the desired set point value. This signal also indicates that the main pressure relief valve may be open and flowing large quantities of water, or that it may not be working at all. See Section 4.18 for more information on relief valves.

(5) ECM selector switch in alternate ECM position (for engines with ECM control only).

Note that this position should also indicate when the alternate ECM is selected automatically by the engine.

(6) Fuel injection malfunction (for engines with ECM only).

This signal also occurs when the ECM detects an internal failure.

(7) Low fuel level. Signal at two-thirds tank capacity.

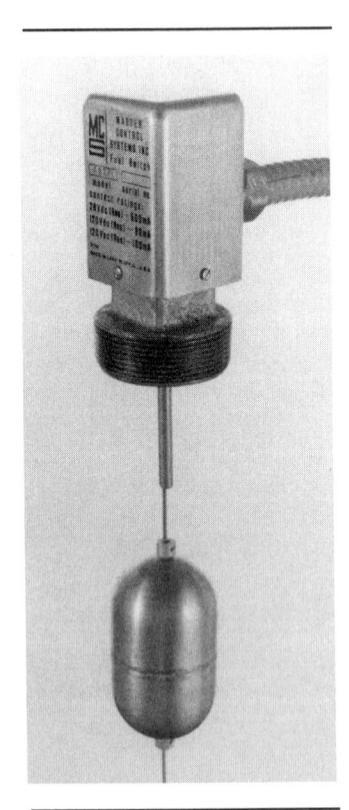

EXHIBIT II.12.7 Low Fuel Level Switch. (Photograph courtesy of Master Control Systems, Inc.)

This requirement was added in the 2003 edition. Prior to that edition, signaling at two-thirds tank capacity was a popular option but not mandatory. Exhibit II.12.7 shows a low fuel level switch.

2010 Stationary Fire Pumps Handbook

(8) Low air pressure (air-starting engine controllers only). The air supply container shall be provided with a separate visible indicator to indicate low air pressure.
(9) Low engine temperature.

Subparagraph 12.4.1.4(9) is new to the 2010 edition. The low engine temperature alarm picks up the low engine temperature sensor added to the engine as mentioned in Chapter 11. This signal indicates loss of ac (mains) power to the engine heater or a malfunction of the heater, that it became unplugged or disconnected, or that it has a kinked or plugged hose(s).

Failure to promptly investigate and correct any of the conditions indicated in 12.4.1.4 is likely to cause unsatisfactory pump performance or failure of the fire protection system. Although not specifically required by this chapter, some installations also include monitoring of the following conditions:

1. Low pump room temperature
2. Relief valve discharge
3. Low reservoir level
4. Low city water pressure
5. Control valve closure

12.4.1.5 No audible signal silencing switch or valve, other than the controller main switch, shall be permitted for the conditions reflected in 12.4.1.3 and 12.4.1.4.

The words "or valve" were added in the 2010 edition to address controllers that operate pneumatically or use fluidic logic rather than traditional electrical circuitry.

12.4.1.5.1 A separate signal silencing switch shall be used for the conditions of 12.4.1.4(5), 12.4.1.4(7), and 12.4.1.4(8).

These three alarms—ECM selector switch in alternate ECM position, low fuel level, and low air pressure—may exist for extended periods of time. Making these alarms silenceable is for attended pump rooms and also to avoid unnecessary battery drain during power outage periods and/or to avoid someone taking the controller out of service to quiet the alarm.

12.4.1.5.2* The controller shall automatically return to the nonsilenced state when the alarm(s) have cleared (returned to normal). This switch shall be clearly marked as to its function.

A.12.4.1.5.2 This automatic reset function can be accomplished by the use of a silence switch of the automatic reset type or of the self supervising type.

12.4.1.5.3 Where audible signals for the additional conditions listed in A.4.24 are incorporated with the engine fire pump alarms specified in 12.4.1.3, a silencing switch or valve for the additional A.4.24 audible signals shall be provided at the controller.

The conditions mentioned in A.4.24 are "pump house (pump room) trouble" conditions. They include the following:

1. Low pump room temperature (freeze alarm)
2. Relief valve discharge (relief valve open)
3. Flowmeter left on, bypassing the pump
4. Water level in suction supply below normal (reservoir low)
5. Water level in suction supply near depletion (reservoir empty)
6. Steam pressure below normal

These trouble condition alarms are all optional signals that the controller can monitor when so equipped. Other useful signals often used (monitored) are the following:

1. Low suction pressure
2. Fuel spill (primarily used in California)
3. High fuel level
4. Reservoir temperature low (freeze alarm)
5. Pump room intrusion (pump room door open)
6. Low discharge pressure

Exhibit II.12.8 shows pump room signal contacts for connection to the controller on the left, while contacts for remote annunciation are shown on the right.

12.4.1.5.4 The circuit shall be arranged so that the audible signal will be actuated if the silencing switch or valve is in the silent position when the supervised conditions are normal.

See 12.4.1.5 and A.12.4.1.5.2.

12.4.2 Signal Devices Remote from Controller.

12.4.2.1 Where the pump room is not constantly attended, audible or visible signals powered by a source other than the engine starting batteries and not exceeding 125 V shall be provided at a point of constant attendance.

Paragraph 12.4.2.1 requires the fire pump controller to be constantly monitored or supervised. This is not a requirement of the controller, although the controller is required to provide the contacts necessary to fulfill this requirement.

FAQ ▶
What is an example of a constantly attended location?

A boiler room where operating personnel are required 24 hours a day is one type of a constantly attended location. Because most new installations are not able to satisfy the provisions for a constantly attended location, a remote alarm panel is often installed at a place of constant attendance, such as a guard house, or a 24-hour phone switchboard. Although these units are termed *alarm sets* or *alarm panels*, they are usually not built nor sold as *NFPA 72*–style alarm equipment.

As an alternative, the authority having jurisdiction may require an off-site fire alarm monitoring station or alarm service. Note that a fire alarm monitoring station is often preferred. These signals are not considered "fire alarms" as defined in *NFPA 72®*, *National Fire Alarm and Signaling Code.* They would be considered "supervisory" or trouble signals instead. The monitoring station would normally notify the plant personnel, rather than the fire department, when these signals occur. A notable exception can be the "engine running" signal. This signal is often used to meet the required weekly line test function of *NFPA 72* since this signal will, or should, occur during the weekly half-hour engine test run. When an engine running signal occurs at any other than the expected weekly test run, the alarm company may signal the fire department of a potential fire.

The 125 V maximum requirement is to avoid imposing higher voltages on the controller remote alarm contacts of 12.4.3. This voltage is often the maximum voltage rating of these contacts.

12.4.2.2 Remote Indication. Controllers shall be equipped to operate circuits for remote indication of the conditions covered in 12.4.1.3, 12.4.1.4, and 12.4.2.3.

The controller is required to have at least three sets of contacts for engine running, switch-off, and engine (or controller) failure. Many controllers have more than these three as standard, and all controllers are available with additional or individual alarm contacts for the various conditions being supervised by the controller.

12.4.2.3 The remote panel shall indicate the following:

(1) The engine is running (separate signal).

It is good practice for the weekly test run to be attended to monitor starting and running conditions of the engine and controller. It is imperative that appropriate personnel be immediately

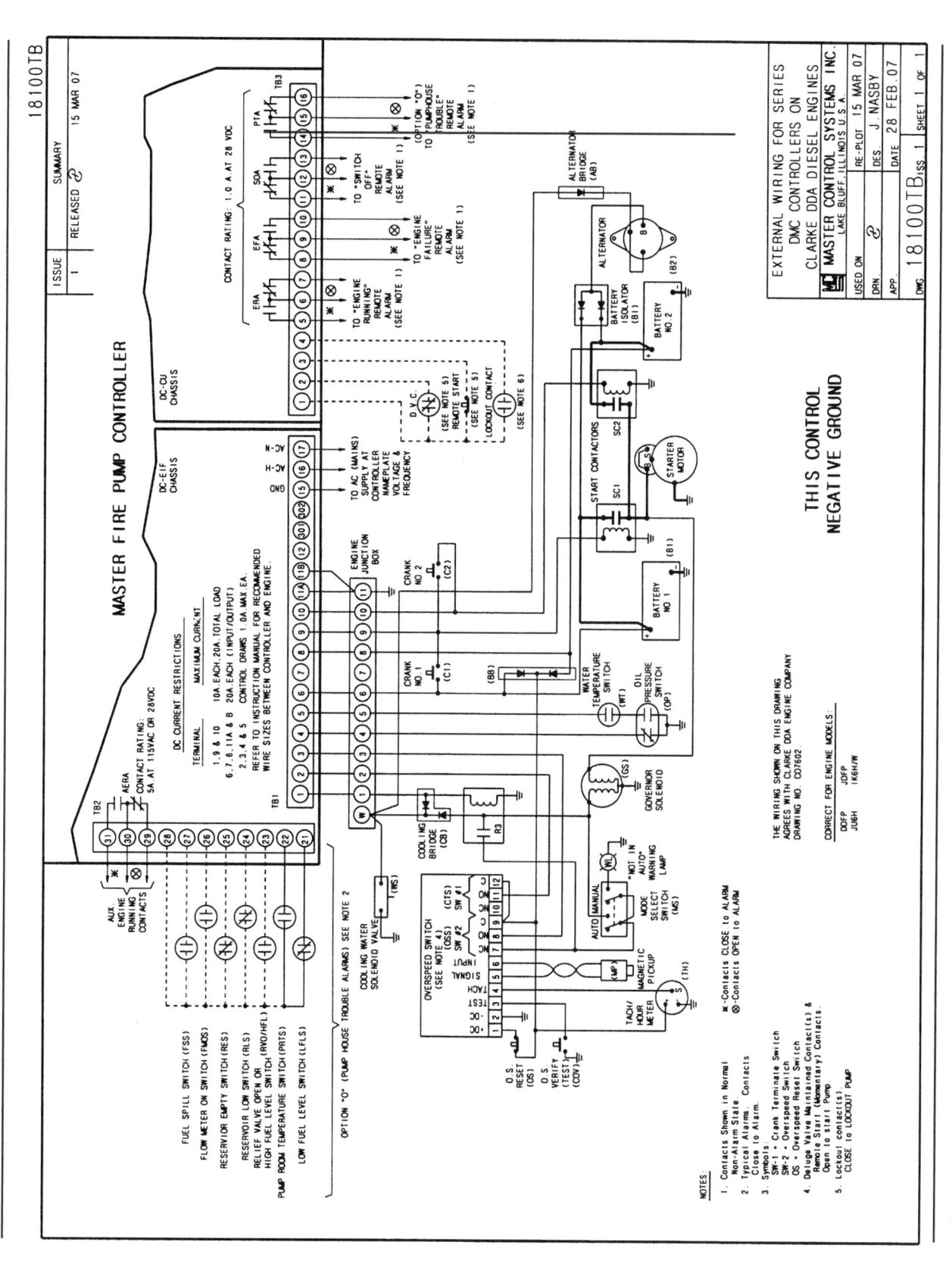

EXHIBIT II.12.8 Controller Wiring Diagram with Alarm Signals. (Diagram courtesy of Master Control Systems, Inc.)

dispatched to the pump room any other time the engine is running to determine the cause and to monitor the engine and controller operation since it is very likely that the engine and controller are supplying, or attempting to supply, fire water for a fire.

(2) The controller main switch has been turned to the off or manual position (separate signal).

If the controller control or selector switch is not in the "auto" automatic mode position, the controller and engine and, thus, the fire pump is "out of service." For sole source pump installations, which is often the situation, this means that the protected premises (building, plant, etc.) has no fire protection. This condition may require a fire watch or even a fire pump truck to be connected to the building until fire protection is restored.

(3)* There is trouble on the controller or engine (separate or common signals). *(See 12.4.1.4 and 12.4.1.5.)*

A.12.4.2.3(3) The following signals should be monitored remotely from the controller:

(1) A common signal can be used for the following trouble indications: the items in 12.4.1.4(1) through 12.4.1.4(7) and loss of output of battery charger on the load side of the dc overcurrent protective device.
(2) If there is no other way to supervise loss of power, the controller can be equipped with a power failure circuit, which should be time delayed to start the engine upon loss of current output of the battery charger.
(3) The arrangement specified in A.12.4.2.3(3)(3)(2) is only permitted where approved by the authority having jurisdiction in accordance with Section 1.5 and allows, upon loss of the ac power supply, the batteries to maintain their charge, activates ventilation in case conditions require cooling the engine, and/or maintains engine temperature in case conditions require heating the engine. *(See also A.4.6.4 and A.11.4.2.1.)*

It is suggested that when the engine is allowed by the authority having jurisdiction to be run in order to charge the batteries or keep the pump room temperature from freezing, it not be run any longer than necessary.

Some authorities having jurisdiction require ac power (mains) pump starting where hot processes are involved, such as with furnaces and salt baths. This is to bring the fire water supply on line since plantwide ac (mains) power loss will result in the loss of cooling fans, blowers, and/or pumps, which presents an increased likelihood of a fire.

12.4.3 Controller Contacts for Remote Indication. Controllers shall be equipped with open or closed contacts to operate circuits for the conditions covered in 12.4.2.

These contacts are part of the fire pump controller. Modern controllers employ contacts rated for 1.0 amp or less at 125 Vac or 28 Vdc. See also the commentary to 12.4.2.1 and 12.4.4.

12.4.4* Pressure Recorder.

The pressure recording device can be examined after operation of the fire pump. The recorder registers when the pump was in operation and what the output pressure was during operation. This recording can be used to help evaluate the performance of the pump.

FAQ ▶
Why is a pressure recorder needed?

A pressure recorder is of paramount importance for two main reasons—first, to record the weekly test run of the engine and pump, and second and most important, to record pressure events and excursions (abnormalities) during a fire for forensic analysis.

After a fire, forensic specialists can extract vital information about the operation or failure of equipment such as the pump and sprinklers, as well as the progress of the water supply during a fire. It also is important to provide definitive evidence that the engine, pump, and controller are being exercised (tested) weekly.

A.12.4.4 The pressure recorder should be able to record a pressure at least 150 percent of the pump discharge pressure under no-flow conditions. In a high-rise building, this requirement can exceed 400 psi (27.6 bar). This requirement does not mandate a separate recording device for each controller. A single multichannel recording device can serve multiple sensors.

12.4.4.1 A listed pressure recording device shall be installed to sense and record the pressure in each fire pump controller pressure-sensing line at the input to the controller.

Too often, a multiple fire pump installation has only a single pressure recorder. Fortunately, most modern controllers come with built-in pressure recorders as standard. It is paramount that each controller have a dedicated recorder to verify weekly tests and provide forensic data should a fire occur. Exhibit II.12.9 shows a traditional circular chart-type recorder, while Exhibit II.12.10 shows the control panel and 160-character display for a modern digital alarm, status, and pressure recorder.

12.4.4.2 The recorder shall be capable of operating for at least 7 days without being reset or rewound.

Older recorders were of the spring wound type. See Exhibit II.12.9, for example. The circular charts were good for 7 days. Most modern recorders are either battery operated or draw their power from the engine batteries. Exhibit II.12.11 shows a typical chart from the recorder shown in Exhibit II.12.9. The saw tooth pattern is the cycling of the jockey (pressure maintenance) pump.

12.4.4.3 The pressure-sensing element of the recorder shall be capable of withstanding a momentary surge pressure of at least 400 psi (27.6 bar) or 133 percent of fire pump controller rated operating pressure, whichever is higher, without losing its accuracy.

This requirement is important because there often are severe hydraulic transients and surges. These can occur on pump shutdown when the main check valve slams shut, or on start-up when the engine rapidly accelerates to full speed and the pump attempts to accelerate tons of water. Severe transients also occur on vertical pumps, especially deep well pumps, when the water column hits the air relief valve. See also A.12.4.4 and explanatory material.

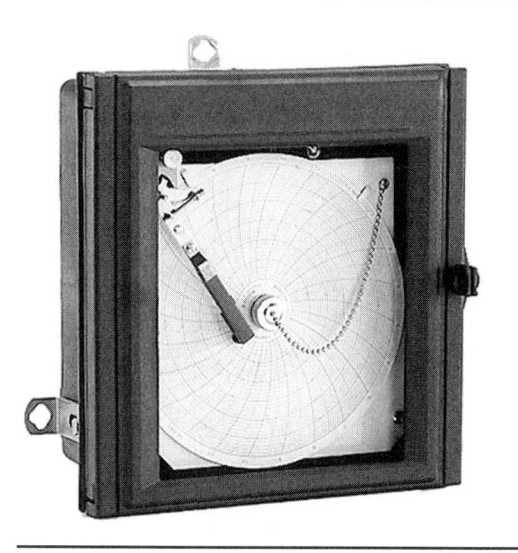

EXHIBIT II.12.9 Circular Chart Pressure Recorder.

EXHIBIT II.12.10 Digital Recorder Display and Control Panel. (Photograph courtesy of Master Control Systems, Inc.)

EXHIBIT II.12.11 Circular Pressure Chart.

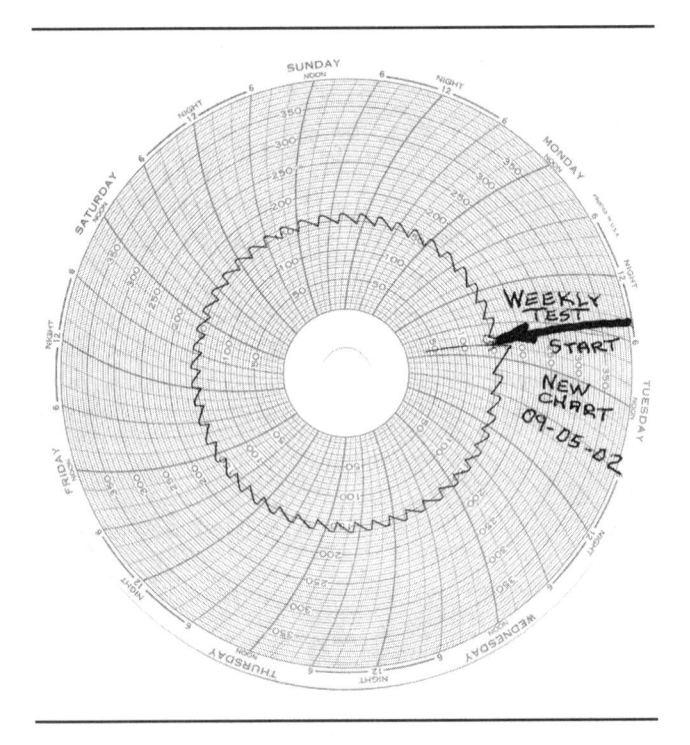

12.4.4.4 The pressure recording device shall be spring wound mechanically or driven by reliable electrical means.

After spring-wound recorders became obsolete, pressure recorders were often ac powered, and later powered by dry cell batteries.

12.4.4.5 The pressure recording device shall not be solely dependent upon alternating current (ac) electric power as its primary power source.

The ac powered chart recorders were equipped with a "spring backup" to keep the clock motor running when ac power was lost.

A diesel engine drive is used because it is independent of ac power. Therefore, the pressure recorder must also be designed to operate without ac power or without ac power for at least 24 hours.

12.4.4.6 Upon loss of ac electric power, the electric-driven recorder shall be capable of at least 24 hours of operation.

◀ **FAQ**
Why must a pressure recorder be designed to operate without ac power?

The spring had to be good for at least 1 full day of recording without ac (mains) power. Now almost all modern controllers make use of digital pressure recorders.

12.4.4.7 In a non-pressure-actuated controller, the pressure recorder shall not be required.

An example of a "non-pressure-actuated" controller is one used for such non-pressurized systems, such as deluge or monitor systems. These controllers have no pressure switch, no pressure-sensing line, and, thus, have no need for a pressure recorder. Most modern controllers, however, still have a digital recorder for the purpose of recording alarms and status change events.

12.4.5 Voltmeter. A voltmeter with an accuracy of ±5 percent shall be provided for each battery bank to indicate the voltage during cranking or to monitor the condition of batteries used with air-starting engine controllers.

12.5* Battery Recharging.

A.12.5 A single charger that automatically alternates from one battery to another can be used on two battery installations.

The type of charger referenced in A.12.5 was used until the 1970s, before controllers with built-in chargers were commonly available. These chargers had a relay that periodically connected the charger to one battery, then to the other.

12.5.1 Two means for recharging storage batteries shall be provided.

12.5.2 One method shall be the generator or alternator furnished with the engine.

The engine-driven generator/alternator is completely independent of other energy sources. This equipment recharges the batteries, after cranking, while the engine is running much faster than the ac-powered, automatically controlled charger.

12.5.3 The other method shall be an automatically controlled charger taking power from an ac power source.

To recharge storage batteries, one means is the static (ac powered) battery charger, and the other is the belt-driven battery charging alternator (or generator on older engines). Engines have used alternators instead of generators since the 1970s, although some generators are still in service. Controllers were equipped with optional battery chargers starting in the late 1960s or 1970s. It was a significant cost item that slowly became popular. Now the charger is pretty much standard equipment in modern controllers.

12.5.4 If an ac power source is not available or is not reliable, another charging method, in addition to the generator or alternator furnished with the engine, shall be provided.

Although uncommon, an inverter powered by 48 Vdc telephone batteries or 120 Vdc or higher station batteries are examples of alternative charging methods.

12.6 Battery Chargers.

The requirements for battery chargers shall be as follows:

(1) Chargers shall be specifically listed for fire pump service and be part of the diesel fire pump controller.

Chargers were not always specifically listed for fire pump service and there used to be wide variations in charger quality, performance, and reliability.

Exhibit II.12.12 shows a modern dual fire pump battery charger, while Exhibit II.12.13 shows the schematic diagram for it.

EXHIBIT II.12.12 Dual 24 Vdc 10A dc Battery Charger. (Photograph courtesy of Master Control Systems, Inc.)

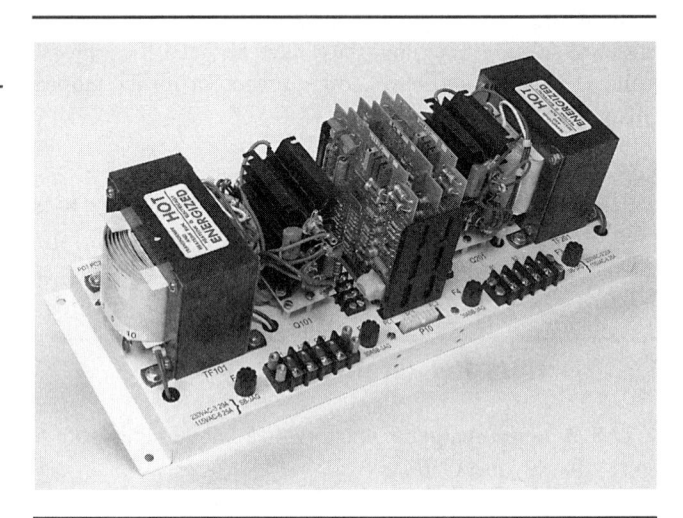

(2) The rectifier shall be a semiconductor type.

A semiconductor type rectifier is specified as opposed to a vacuum tube–type, which were still in use when this requirement was added.

(3) The charger for a lead-acid battery shall be a type that automatically reduces the charging rate to less than 500 mA when the battery reaches a full charge condition.

A common problem with historic chargers was "battery boiling." This phenomenon was dissolution of water in the electrolyte well past the time when the battery was fully charged. This was due to the poor regulation of these chargers, or the complete lack of any voltage regulation, which was the norm in that era. Modern chargers are regulated well enough that this requirement is easily met if the charger is properly adjusted for the battery type involved. For this reason, field adjustment of battery charger settings is not recommended without the specific approval and guidance of the controller manufacturer. The amount of battery water consumption is directly proportional to this "final" current flow.

(4) The battery charger at its rated voltage shall be capable of delivering energy into a fully discharged battery in such a manner that it will not damage the battery.

Modern battery chargers are of the "current limiting" type to avoid excessive initial charging currents. This was not the case with historic unregulated chargers.

(5) The battery charger shall restore to the battery 100 percent of the battery's reserve capacity or ampere-hour rating within 24 hours.

Part II • NFPA 20   291

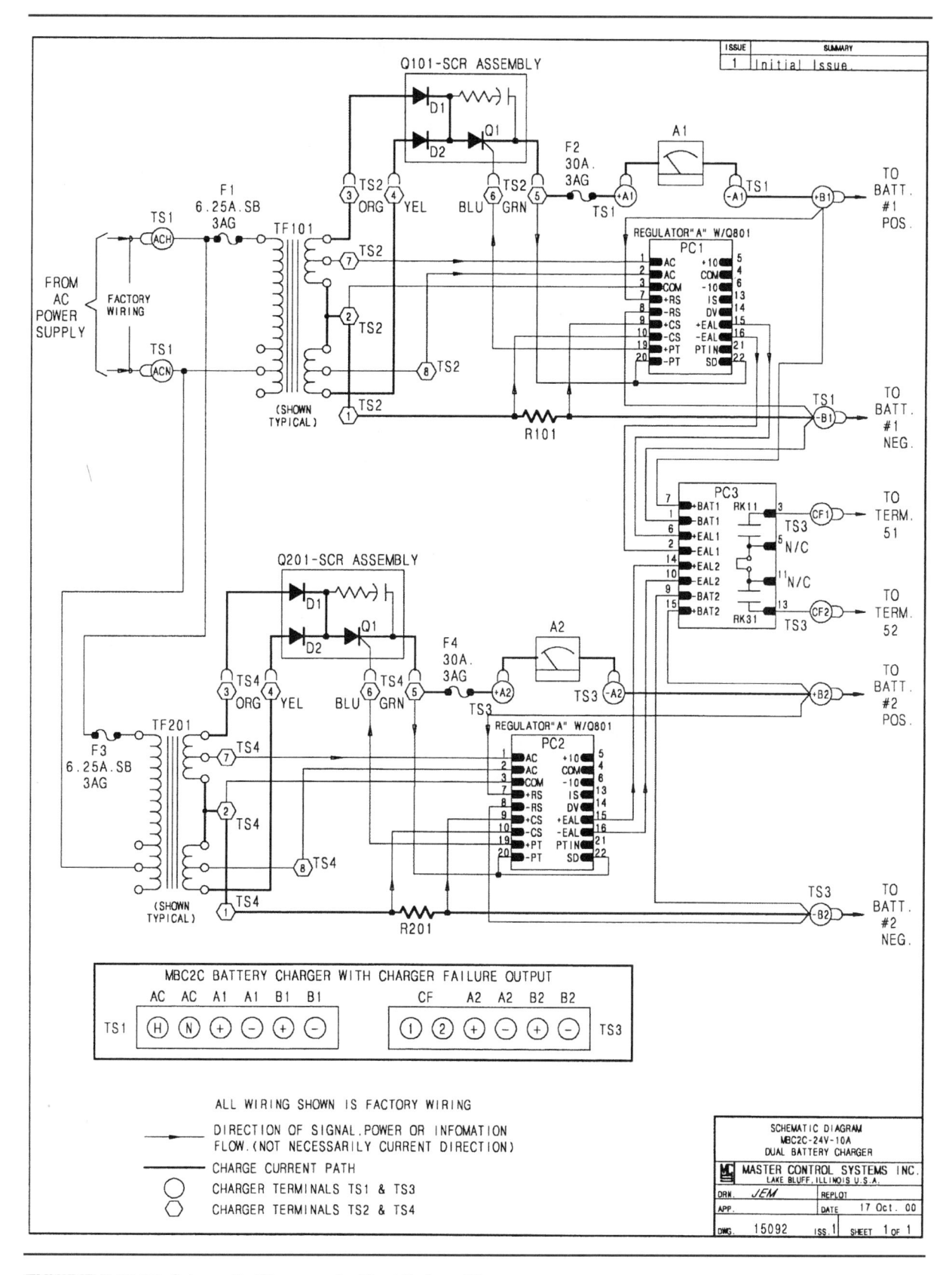

EXHIBIT II.12.13 Schematic Diagram of a Dual Battery Charger.

Stationary Fire Pumps Handbook 2010

In order to fulfill this requirement, all modern fire pump battery chargers are of the mode switching type—that is, they operate in two different modes, namely float mode (low rate), and equalize mode (high rate). The float mode is a lower voltage maintaining mode that allows the charger to meet the maximum 500 mA (½ amp) maintaining current. Better chargers reduce this maintaining current to 100 mA or less on batteries in good condition.

It is vitally important that the battery size and type match the battery charger. If the battery type does not match the type for which the charger is built and adjusted, one of two failure conditions will very likely occur: (1) the charger will not be able to meet the 24-hour recharge time requirement, or (2) the charger will wind up in the mode lock-up condition sooner or later.

This requirement was added in the 1972 edition to avoid the case in which the batteries were depleted after acceptance testing, leaving the premises with questionable, unreliable, or inoperative fire protection.

(6) The charger shall be marked with the reserve capacity or ampere-hour rating of the largest capacity battery that it can recharge in compliance with 12.6(4).

A 10 amp charger can supply no more than approximately 240 amp-hours in a 24-hour period. This is sufficient for most SAE Group 8D or 4D sized batteries but not for parallel battery banks. When batteries are connected in parallel, either 4D or 8D, the battery charger has to be a 20 amp rated unit. Marking the charger with the maximum battery size allows a field inspector to determine immediately if the charger is not the correct size for the installation involved. Undersized chargers are highly vulnerable to the mode lock-up phenomenon described in the commentary for 12.6(5).

(7) An ammeter with an accuracy of ±5 percent of the normal charging rate shall be furnished to indicate the operation of the charger.

While most modern controllers make use of digital ammeters, there is no assurance that the readings will be within ½ amp correct for a 10 amp charger unless this is checked as part of the listing (conformity assessment) process.

(8) The charger shall be designed such that it will not be damaged or blow fuses during the cranking cycle of the engine when operated by an automatic or manual controller.
(9) The charger shall automatically charge at the maximum rate whenever required by the state of charge of the battery.
(10) The battery charger shall be arranged to indicate loss of current output on the load side of the direct current (dc) overcurrent protective device where not connected through a control panel. *[See 12.4.1.4(2).]*

During starter motor breakaway, the battery voltage can drop to one-half of nominal voltage. This puts an immediate large demand on the battery charger. If the current limiting circuitry is too slow or imprecise, of if the overcurrent protection (fuses or circuit breakers) is too small, blown fuses or tripped circuit breakers will result.

The battery charger is required to be of the well-regulated type that will sense small changes in the battery from the desired voltage set point. As such, the charger will go to full output when it detects insufficient battery voltage.

It is an important requirement that the battery charger detect loss of ac (mains) power, malfunction of the charger, and blown fuses or tripped circuit breakers, or any loss of connection between the battery and the charger.

(11) The charger(s) shall remain in float mode or switch from equalize to float mode while the batteries are under the loads in 12.5.2.

This requirement is to help ensure that the charger does not become stuck in the equalize (high rate) mode.

For a more detailed discussion of battery charger theory and operation, refer to *Pumps for Fire Protection Systems*.

12.7* Starting and Control.

A.12.7 The following definitions are derived from *NFPA 70, National Electrical Code:*

(1) *Automatic.* Self-acting, operating by its own mechanism when actuated by some impersonal influence (e.g., a change in current strength, pressure, temperature, or mechanical configuration).
(2) *Nonautomatic.* The implied action requires personal intervention for its control. As applied to an electric controller, nonautomatic control does not necessarily imply a manual controller, but only that personal intervention is necessary.

12.7.1 Automatic and Nonautomatic.

12.7.1.1 An automatic controller shall be operable also as a nonautomatic controller.

There are four main modes of operation of these controllers:

1. *Emergency control:* Strictly manual operation (see 12.7.6).
2. *Test mode:* Starting is initiated manually by draining water through a test solenoid valve, but the controller initiates starting and running automatically.
3. *Automatic operation:* The normal standby fire protection mode. It starts on pressure loss (drop) or other signal.
4. *Manual starting:* A manual local or remote switch initiates starting similar to the test mode, but the signal is electrical rather than hydraulic.

12.7.1.2 The controller's primary source of power shall not be ac electric power.

All power needed for controller operation, as well as engine operation, is derived from the engine batteries. The ac (mains) power is used for ancillary functions, such as the battery charger, cabinet heater, and engine heater; pump room loads (heat, light, sump pumps); and so forth. Older version controllers also used the ac power for timers, such as the weekly test timer and the minimum running (automatic stop) timer.

When both the controller dc and the engine dc standby power use (battery drain) are low enough, the weekly 30-minute run can be enough to replenish the charge in the battery. In this case, the fire pump can remain in service despite extended periods of ac power (mains) loss.

Diesel drive fire pumps are often specified due to the lack of reliable ac power needed for an electric motor–driven pump. Also, some diesel drive fire pump controllers are equipped with an ac "power fail start" option.

◀ **FAQ**
Why does the controller need to be independent of ac (mains) power?

12.7.2 Automatic Operation of Controller.

12.7.2.1 Water Pressure Control.

Most controllers monitor the water pressure in a pressurized sprinkler system and start the pump (engine) when the pressure drops below a specific set point. Controllers make use of a pressure switch or pressure transducer with suitable circuitry to take these measurements and determine when to start the pump.

12.7.2.1.1 Pressure-Actuated Switch.

12.7.2.1.1.1 Unless the requirements of 12.7.2.1.1.2 are met, there shall be provided a pressure-actuated switch having adjustable high- and low-calibrated set-points as part of the controller.

Exhibits II.12.14 and II.12.15 show pressure switches mounted inside of the controller.

At one time, all controllers used similar bourdon type pressure switches. These had independent high and low set points and had a means for locking the settings. Exhibit II.12.15

EXHIBIT II.12.14 Pressure Switch Mounted Inside of Controller.

EXHIBIT II.12.15 Bourdon Tube–Type Pressure Switch. (Photograph courtesy of Dwyer Instruments, Inc.)

shows one example. Although traditionally these switches used a mercury "tilt bottle" switch element for sensitivity, the one shown uses a magnetic reed type switch element instead.

12.7.2.1.1.2 The requirements of 12.7.2.1.1.1 shall not apply to a non-pressure-actuated controller, where the pressure-actuated switch shall not be required.

12.7.2.1.2 There shall be no pressure snubber or restrictive orifice employed within the pressure switch.

A pressure snubber was once an option with bourdon type pressure switches, but was a problem for fire water installations since the switch could, under certain circumstances, trap pressure and thus prevent the pressure switch from responding to a drop in pressure. This would then prevent the pump from starting when it should.

◀ **FAQ**
Why is a pressure snubber prohibited?

12.7.2.1.3 This switch shall be responsive to water pressure in the fire protection system.

As opposed to other types of sensing, such as flow switches, the switch is responsive only to water pressure, not water flow.

12.7.2.1.4 The pressure-sensing element of the switch shall be capable of a momentary surge pressure of 400 psi (27.6 bar) or 133 percent of fire pump controller rated operating pressure, whichever is higher, without losing its accuracy.

12.7.2.1.5 Suitable provision shall be made for relieving pressure to the pressure-actuated switch to allow testing of the operation of the controller and the pumping unit. *[See Figure A.4.30(a) and Figure A.4.30(b).]*

The provisions consist of a solenoid valve that is actuated in the "test" mode of the controller, and also by the weekly test timer. Note that modern controllers re-close (de-energize) the valve once the controller receives the signal that the engine is running. This avoids constant draining of the water. Better controllers have no valves or solenoids between the pressure-sensing line connection and the pressure switch or transducer.

◀ **FAQ**
What provisions should be made to allow testing of the controller and pumping unit?

12.7.2.1.6 Water pressure control shall be as follows:

(1) There shall be no shutoff valve in the pressure-sensing line.

This requirement is to avoid valves of any type, including solenoid valves, in the pressure-sensing line or in the path of the pressure switch or transducer, either outside of or within the controller.

(2) Pressure switch actuation at the low adjustment setting shall initiate the pump starting sequence if the pump is not already in operation.

When the system fire water pressure drops to the start pressure set point of the pressure-sensing means (either pressure switch or pressure transducer circuitry), the controller needs to start the engine, either immediately or in accordance with allowed or mandated delays—that is, sequence start delay or high zone delay.

12.7.2.2 Fire Protection Equipment Control.

12.7.2.2.1 Where the pump supplies special water control equipment (e.g., deluge valves, dry-pipe valves), the engine shall be permitted to start before the pressure-actuated switch(es) would do so.

12.7.2.2.2 Under such conditions, the controller shall be equipped to start the engine upon operation of the fire protection equipment.

This equipment includes remote manual start (push button) stations, fire detection equipment as in aircraft hangers, or deluge valve systems. Most, but not all, of these systems will be pressurized systems. Any such circuits should be of the normally open ("drop-out") type so

◀ **FAQ**
What are examples of fire protection control equipment?

that opening the circuit results in starting the engine. See 10.5.2.3.3 for a more information on open circuits.

Since the inputs for external demand (start) signals have been, and still often are, optional, the controller has to be ordered with the proper options specified, where and when required.

12.7.2.3 Manual Electric Control at Remote Station. Where additional control stations for causing nonautomatic continuous operation of the pumping unit, independent of the pressure-actuated switch or control valve, are provided at locations remote from the controller, such stations shall not be operable to stop the engine.

Once a pump is started from one of these remote start (push button) stations, the controller is required to keep the pump running regardless of external circuitry or signals. This is because shutdown of the pump is to occur only by operating the switch on the controller within the pump room or pump house.

Some controllers provide a separate terminal for this input; some actuate on closing this remote start circuit, others actuate on opening it. Better controllers latch this signal since it is often from a momentary pushbutton station. If the sequence start delay is used, better controllers route this signal through the high zone delay.

12.7.2.4 Sequence Starting of Pumps.

This paragraph applies to pumps operating in parallel or in series. Pumps operating in parallel are those that take suction directly from the suction supply and discharge into the same system (same header or manifold). Parallel pumps are used either to provide more capacity than a single pump can deliver or to increase the reliability of the system.

The sequence-starting requirement prevents excessive hydraulic stress to piping, control valves, and other system components during pump start-up. The time delay in each unit allows independent controller operation without the need for electrical interconnections. The lead (first to start) controller may not be equipped with sequential starting, especially if there are no electric motor drive fire pumps in the system. When the lead controller is equipped with sequence starting, the time delay setting of the first pump is usually set to zero, unless there are electric motor–driven pumps required to start before the diesel driven pumps. Better installations equip all controllers with sequential starting, since this allows adjustment of which are the lead and lagging pumps in the system. Many modern controllers come equipped with this optional mode of operation as standard equipment.

The pressure switch (transducer) settings of all the pumps in parallel may be set to the same set point, or to different set points. However, upon a sudden demand for fire water, all pressure switches will trip in unison. This is why sequential starting is absolutely necessary for fire protection systems with multiple pumps. Where this rule is not adhered to, pump houses have been destroyed when the surge of tons of water breaks the headers or elbows in the underground piping. See Figure A.4.20.1.2(a) for a parallel pump arrangement.

FAQ ▶
Why not stagger the pressure switch start settings?

Staggering of the pressure switches will result in sequential starting of the pumps *only* when a very low and slowly rising demand for fire water exists. With a sudden sharp drop in pressure, all the switches will trip at the same time.

Pumps operating in series have at least one pump that takes suction from the discharge of another pump. This arrangement is often used in high-rise installations to achieve the pressure required for the upper floors or zone(s). The starting of the higher zone pump is delayed to allow the lower zone pump to start first. The delay prevents cavitation and/or dry running of the higher zone pump or pumps. Exhibit II.12.16 shows a diagram of two pumps in series.

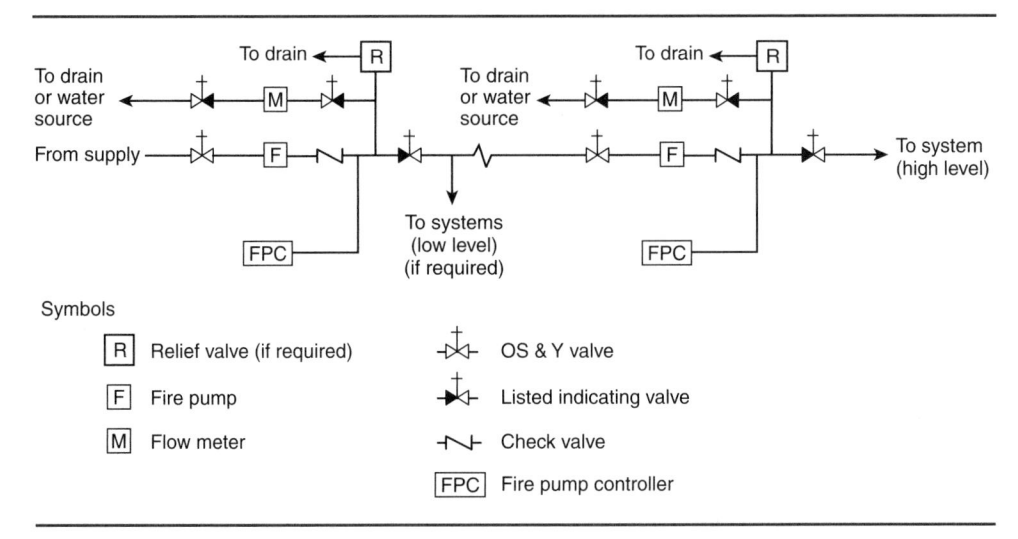

EXHIBIT II.12.16 *Two Fire Pumps in Series.*

In addition to provisions for delaying the starting of higher zone pumps, the lower zone controller(s) must be equipped with remote starting capabilities to receive the start signal sent by higher zone controllers. In three-zone systems, the mid zone controller must be equipped both ways—that is, with both high zone delay and with remote starting capabilities.

Also note that sometimes high-rise buildings make use of multiple pumps in each zone for reliability, with one pump acting as a back-up for the other. These cases require an additional special interlocking scheme to avoid false cycling or shutdown and restarting of pumps. While it is common for all pumps of a series pumping arrangement to be in the same room (basement), this is not always the case. Also, the highest zone in a high-rise building may be an independent pump with its own suction supply, such as a swimming pool.

12.7.2.4.1 The controller for each unit of multiple pump units shall incorporate a sequential timing device to prevent any one driver from starting simultaneously with any other driver.

The controller timing feature for pumps in parallel is usually called "sequence start" or "sequential starting." The controller timing for pumps in series is usually called "high zone delay." Better controllers apply these time delay functions differently with different demand signals.

12.7.2.4.2 Each pump supplying suction pressure to another pump shall be arranged to start within 10 seconds before the pump it supplies.

Better controllers for high zone operation delay all starting demands including local test start. This provides the signal to start lower zone pump(s) and delay long enough for suction water to arrive. The only exception will be manual (emergency) starting, since this should bypass all automatic circuitry. However, better controllers will still send the starting signal to the lower zone controller(s) to start the pump(s). Better controllers also latch in the starting demand signal and start the pump even if the demand clears during the delay interval.

EXAMPLE

With the high zone pump delay set to 15 seconds and the mid zone delay set to 7.5 seconds, and with a pump demand in the high zone, the sequence would be:

Low zone start: no delay (immediate starting)
Mid zone start: 7.5 second total delay
High zone start: 15 second total delay

12.7.2.4.2.1 The controllers for pumps arranged in series shall be interlocked to ensure the correct pump starting sequence.

12.7.2.4.3 If water requirements call for more than one pumping unit to operate, the units shall start at intervals of 5 to 10 seconds.

Better controllers do not latch in a pressure demand until after the sequence time interval has elapsed. This way, only the number of pumps actually needed start and run, while the remaining pumps remain in standby. Better controllers apply this sequential starting delay only to so-called "static demands" such as pressure, deluge, or ac power loss start, among others. Remote manual start (push-button) signals start the pump immediately to avoid an unexpected delay for fire water.

EXAMPLE

For four pumps in parallel, set at 0, 10, 20 and 30 seconds with a sudden large pressure drop:

Lead pump: No delay (immediate starting); if the pressure remains low, then
Second pump: 10 seconds; if the pressure still remains low, then
Third pump: 20 seconds; if the pressure still remains low, then
Fourth pump: 30 seconds.

FAQ ▶
Why wait out the sequence start delay instead of latching the demand signal right away?

By waiting to latch the demand, these controllers will start only as many pumps as needed to supply the required pressure and flow. This is because if the pressure rises enough to reset the pressure switch before the sequence start delay times out, the controller will let the demand expire at the end of the delay period if it no longer exists, and therefore it will not start a pump that is not needed.

12.7.2.4.4 Failure of a leading driver to start shall not prevent subsequent drivers from starting.

Properly designed sequential start controllers have no interwiring and thus operate independently of one another. Therefore, failure of any other pumps (engines or controller) does not affect starting and running of the other units.

For pumps in series, high zone units will start the pump regardless of the state of lower zone units—that is, whether the lower zone units work or not.

12.7.2.5 External Circuits Connected to Controllers.

12.7.2.5.1 With pumping units operating singly or in parallel, the control conductors entering or leaving the fire pump controller and extending outside the fire pump room shall be so arranged as to prevent failure to start due to fault.

Listed controllers are available with a "lockout" ("interlock") option, when and where an authority having jurisdiction requires this. This requirement is especially important to prevent defective external wiring from inhibiting starting of the pump when it is needed.

12.7.2.5.2 Breakage, disconnecting, shorting of the wires, or loss of power to these circuits shall be permitted to cause continuous running of the fire pump but shall not prevent the controller(s) from starting the fire pump(s) due to causes other than these external circuits.

This subparagraph allows the controller to employ "fail safe" type design in order to comply with 12.7.2.5.1. Better controllers include circuitry that detects a failure in the controlling transducer. These failures include a short circuit, an open circuit, or a disconnection. Upon detecting any such fault, a fail safe type controller will start the pump. The only viable alternative is a signal and alarm to monitor system pressure and report if the pump fails to respond. If the controller is the heart of a sprinkler system, the transducer is the heart of the controller.

12.7.2.5.3 All control conductors within the fire pump room that are not fault tolerant shall be protected against mechanical injury.

Any such wiring will be best protected by running the wiring in RMC rigid metal (threaded) conduit or raceway.

12.7.2.5.4 When a diesel driver is used in conjunction with a positive displacement pump, the diesel controller shall provide a circuit and timer to actuate and then close the dump valve after engine start is finished.

This requirement is for the controller to employ circuitry to actuate an "unloader" (dump) valve to allow the starter motor to crank the engine. A timer de-energizes the valve after a few seconds after receipt of the crank terminate (engine running) signal from the engine. Without this provision, the starter motor would attempt to crank against water-filled cylinders.

12.7.2.6 Sole Supply Pumps.

12.7.2.6.1 Shutdown shall be accomplished by manual or automatic means.

12.7.2.6.2 Automatic shutdown shall not be permitted where the pump constitutes the sole source of supply of a fire sprinkler or standpipe system or where the authority having jurisdiction has required manual shutdown.

Note that at the present time, Factory Mutual does not allow automatic shutdown on their installations without special exemption.

12.7.2.7 Weekly Program Timer.

12.7.2.7.1 To ensure dependable operation of the engine and its controller, the controller equipment shall be arranged to automatically start and run the engine for at least 30 minutes once a week.

The 30-minute running time is normally sufficient to heat up the lube oil enough to drive out any accumulated moisture and to prevent "wet stacking" of the exhaust system. Wet stacking occurs when the engine does not run long enough for the entire exhaust piping arrangement to be heated above the boiling point. In this case, water will gradually accumulate in low points. This weekly run also consumes some of the stored fuel, which helps avoid excessive aging of the fuel and also works to establish routine fuel replacement as part of the maintenance program.

It is most important that the weekly test be attended by appropriate personnel to observe cranking times and voltages, water temperature, oil pressure, running speed, and all of the other required inspections and observations delineated in NFPA 25, *Standard for the Inspection, Testing, and Maintenance of Water-Based Fire Protection Systems*.

12.7.2.7.2 The controller shall use the opposite battery bank (every other bank) for cranking on subsequent weeks.

This requirement ensures that each battery unit (battery bank) is exercised every other week, assuming that the engine starts during the first cranking cycle. If it does not, corrective action should be taken.

12.7.2.7.3 Means shall be permitted within the controller to manually terminate the weekly test provided a minimum of 30 minutes has expired.

12.7.2.7.4 A solenoid valve drain on the pressure control line shall be the initiating means.

This requirement ensures that the pump is started by actuation of the pressure sensing means—hydraulically, for example.

12.7.2.7.5 Performance of this weekly program timer shall be recorded as a pressure drop indication on the pressure recorder. *(See 12.4.4.)*

12.7.2.7.6 In a non-pressure-actuated controller, the weekly test shall be permitted to be initiated by means other than a solenoid valve.

This provision permits electric-only initiation of the cranking and starting in the controller when there are no hydraulic (pressure) components. For example, for non-pressurized applications, neither the pressure switch (transducer), the pressure recorder, nor the weekly test drain solenoid valve is needed.

12.7.3 Nonautomatic Operation of Controller.

12.7.3.1 Manual Control at Controller.

12.7.3.1.1 There shall be a manually operated switch or valve on the controller panel.

This switch is usually the mode or control selector switch, which is behind a break glass window or other means if it is part of the "automatic" control switch. See 12.3.4.

12.7.3.1.2 This switch or valve shall be so arranged that operation of the engine, when manually started, cannot be affected by the pressure-actuated switch.

12.7.3.1.3 The arrangement shall also provide that the unit will remain in operation until manually shut down.

All automatic circuitry should be completely bypassed when the controller is in the manual position(s) (MAN 1 or MAN 2 on some controllers). See 12.7.3.1.4.

12.7.3.1.4 Failure of any of the automatic circuits shall not affect the manual operation.

12.7.3.2 Manual Testing of Automatic Operation.

12.7.3.2.1 The controller shall be arranged to manually start the engine by opening the solenoid valve drain when so initiated by the operator.

12.7.3.2.2 In a non-pressure-actuated controller, the manual test shall be permitted to be initiated by means other than a solenoid valve.

FAQ ▶
How are the manual test valves used?

One use of manual test valves is to determine and/or verify the start and reset pressures of the pressure switch or pressure transducer circuitry. This is done by slowly opening one or the other of the test valves so as to slowly bleed down the pressure read by the controller, until the start set point pressure is reached as read on a calibrated precision pressure gauge attached to the controller pressure sensing line. The valve is slowly closed to cause (allow) the system pressure to slowly rise until the stop (reset) pressure set point is reached.

12.7.4 Starting Equipment Arrangement.
The requirements for starting equipment arrangement shall be as follows:

(1) Two storage battery units, each complying with the requirements of 11.2.7.2, shall be provided and so arranged that manual and automatic starting of the engine can be accomplished with either battery unit.
(2) The starting current shall be furnished by first one battery and then the other on successive operations of the starter.
(3) The battery changeover shall be made automatically, except for manual start.
(4) In the event that the engine does not start after completion of its attempt-to-start cycle, the controller shall stop all further cranking and operate a visible indicator and audible fire pump alarm on the controller.
(5) The attempt-to-start cycle shall be fixed and shall consist of six crank periods of approximately 15-second duration separated by five rest periods of approximately 15-second duration.

NFPA 20 requires fire pump engines to be "dual battery" types, with the assumption that one of the two batteries, or its contactors, wiring, or other components, may be defective. As explained in Chapter 11, each battery unit must have "twice the capacity" for the full "3-minute attempt-to-start cycle, which is six consecutive cycles of 15 seconds of cranking and 15 seconds of rest."

Follow the manufacturer's instruction manual to perform the crank cycle test and check the temperature of the starter motor(s) after each crank cycle to avoid overheating. The cranking is to be on alternate batteries, for a maximum of three cycles on each battery. The controller will perform the alternating of the batteries.

At the end of the attempt-to-start cycle, the controller de-energizes terminal #1 to de-energize the water and/or fuel solenoids. Besides the local visible and audible alarm, the controller will transfer the engine failure alarm contacts to notify the remote monitoring alarm set or alarm service.

The six cranks and five rest periods add up to a total of 90 seconds of cranking and 75 seconds of resting or 2 ¾ minutes total.

(6) In the event that one battery is inoperative or missing, the control shall lock in on the remaining battery unit during the cranking sequence.

Since each battery unit must be sized (have capacity) for twelve 15 second cranks, one complete automatic crank cycle will consume around half of the cranking capacity of the battery. This leaves six more cranking attempts of 15 seconds left and available for manual intervention and starting attempts when one battery unit is inoperative, missing, or defective.

12.7.5 Methods of Stopping.

12.7.5.1 Manual Electric Shutdown. Manual shutdown shall be accomplished by either of the following:

(1) Operation of the main switch or stop valve inside the controller
(2) Operation of a stop button or stop valve on the outside of the controller enclosure as follows:
    (a) The stop button or stop valve shall cause engine shutdown through the automatic circuits only if all starting causes have been returned to normal.
    (b) The controller shall then return to the full automatic position.

There are two methods for shutting down the controller. One is an emergency stopping function, where the main selector control switch, located behind the break glass window or equivalent, is switched to the "off" position. A second method is to utilize the external stop pushbutton on the outside of the controller. It stops the engine by releasing the running latched circuitry if, and only if, all start demands (call to start) have cleared to the reset (normal) condition.

◀ FAQ
How is the controller shut down?

If any demands remain active, this external stop pushbutton is inoperative and the engine remains running without interruption. (Note that this operation is different from the stop pushbutton on electric motor–driven fire pump controllers.)

It is not permitted to press the stop pushbutton to take the controller out of service whether it is effective in shutting down the engine or not.

Exhibit II.12.17 shows an interior view of a modern diesel drive controller. Various plug-in modules in the right-hand chassis determine which options and type(s) of remote operation are installed and operative.

12.7.5.2* Automatic Shutdown After Automatic Start. The requirements for automatic shutdown after automatic start shall be as follows:

(1) If the controller is set up for automatic engine shutdown, the controller shall shut down the engine only after all starting causes have returned to normal and a 30-minute minimum run time has elapsed.

FAQ ▶
Why do some authorities having jurisdiction prohibit automatic stopping of the fire pump?

Note that automatic stop (shutdown) is not permitted for "sole supply pumps," and by some authorities having jurisdiction. See 12.7.2.6 for more information on sole supply pumps. Stopping a running engine presents a possibility, regardless of how remote, that the engine will not restart if it is needed to fight a fire. If the fire pump is fighting a fire and the water supply to the sprinkler system is interrupted for a few seconds only, the system will most likely continue to contain the fire after the water supply is restored. If the water supply is interrupted for more than a few seconds, it is likely that a fire plume or heat plume will fuse (open) more sprinklers. Also, the fire will resume growing during the interruption. A fire plume or heat plume can overwhelm the water supply or render the sprinkler system incapable of containing the fire, even if the fire was being contained. One should never purposely shut down a running fire pump without being absolutely certain that it is safe to do so.

(2) When the engine overspeed shutdown device operates, the controller shall remove power from the engine running devices, prevent further cranking, energize the overspeed fire pump alarm, and lock out until manually reset.

(3) Resetting of the overspeed circuit shall be required at the engine and by resetting the controller main switch to the off position.

EXHIBIT II.12.17 Components and Modules of a Diesel Drive Controller. (Photograph courtesy of Master Control Systems, Inc.)

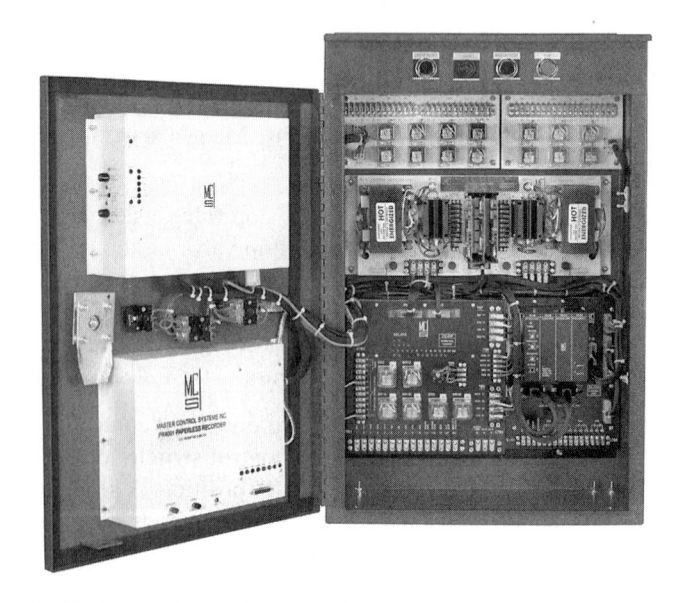

This requirement is both for personnel safety and to preserve the start of the local and remote alarms to ensure that manual intervention or repair occurs, since the engine will be off-line (out of service) until both the engine device and the controller are reset.

(4) The engine shall not shut down automatically on high water temperature or low oil pressure when any automatic starting or running cause exists, and the following also shall apply:
  (a) If no other starting or running cause exists during engine test, the engine shall shut down automatically on high water temperature or low oil pressure.
  (b) If after shutdown a starting cause occurs, the controller shall restart the engine and override the high water temperature and low oil shutdowns for the remainder of the test period.

◀ **FAQ**
Why must the resetting of the overspeed circuit be done at the main controller switch?

If the engine is running under a true "demand" condition, such as a pressure loss start, the controller will not shut down the engine. The engine will be run to destruction if it is fighting a fire. The fire pump and driver are considered sacrificial if running under a true demand.

The controller will shut down the engine during a test run only. This preserves the engine rather than allowing it to destroy itself during a test run. This also leaves the pump in service, even if the engine is in need of attention. If there is a small demand for water, such as a few sprinkler heads controlling a fire, the pump discharge pressure will normally reset the pressure switch (or transducer circuit), which will clear the demand. If the controller is set for automatic stop (shutdown) during the test, the controller will shut down the engine on low oil pressure or high water temperature conditions. If a demand occurs (reoccurs) after shutdown on low oil pressure or high water temperature conditions, it is likely that the pump is fighting a fire. Thus, the controller is required to change to the "run it to destruction" mode to keep fire water flowing as long as possible.

◀ **FAQ**
Why will the engine not shut down due to a low oil pressure or a high water temperature condition?

(5) The controller shall not be capable of being reset until the engine overspeed shutdown device is manually reset.

Modern engines supply the speed switch and its contacts directly from the batteries. Hence, if the controller selector switch is cycled to the "off" position (to reset the visible, audible, and/or remote alarms), the alarm will reactivate as soon as the controller's selector switch is set back to the auto position.

A.12.7.5.2 Manual shutdown of fire pumps is preferred. Automatic fire pump shutdown can occur during an actual fire condition when relatively low-flow conditions signal the controller that pressure requirements have been satisfied.

Shutting down a running engine is not advisable due to the possibility of having problems restarting the engine, especially while hot. If the engine is "distressed" in one way or another, it may not restart.

12.7.6 Emergency Control. Automatic control circuits, the failure of which could prevent engine starting and running, shall be completely bypassed during manual start and run.

This requirement is to ensure that any failed, removed, or unplugged component, including computer board assemblies or modules, does not affect the ability to start the engine using the manual control switch position(s). See also 12.7.3.1.4.

◀ **FAQ**
Why are automatic controls bypassed during a manual start?

12.8 Air-Starting Engine Controllers.

12.8.1 Existing Requirements. In addition to the requirements in Sections 12.1 through 12.7, the requirements in Section 12.8 shall apply.

12.8.2 Starting Equipment Arrangement. The requirements for starting equipment arrangement shall be as follows:

(1) The air supply container, complying with the requirements of 11.2.7.4.4, shall be provided and so arranged that manual and automatic starting of the engine can be accomplished.

This provision is to match the requirements and ability of electrically (battery) cranked engines.

(2) In the event that the engine does not start after completion of its attempt-to-start cycle, the controller shall stop all further cranking and operate the audible and visible fire pump alarms.
(3) The attempt-to-start cycle shall be fixed and shall consist of one crank period of an approximately 90-second duration.

This requirement replaces the six-time 15 second electrical starting scheme. One reason for intermittent cranking of electric-cranked engines is to prevent overheating of the starter motor. This is not applicable to air (or hydraulic) started (cranked) engines. The air or hydraulic motor does not generate significant heat while cranking.

12.8.3 Manual Shutdown. Manual shutdown shall be accomplished by either of the following:

(1) Operation of a stop valve or switch on the controller panel
(2) Operation of a stop valve or switch on the outside of the controller enclosure

12.8.3.1 The stop valve shall cause engine shutdown through the automatic circuits only after starting causes have been returned to normal.

12.8.3.2 This action shall return the controller to full automatic position.

REFERENCES CITED IN COMMENTARY

National Fire Protection Association, 1 Batterymarch Park, Quincy, MA 02169-7471.

Isman, K. E., and Puchovsky, M. T, *Pumps for Fire Protection Systems,* 2002.
NFPA 25, *Standard for the Inspection, Testing, and Maintenance of Water-Based Fire Protection Systems,* 2008 edition.
NFPA 70®, National Electrical Code®, 2008 edition.
NFPA 72®, National Fire Alarm and Signaling Code, 2010 edition.

Underwriters Laboratories Inc., 333 Pfingsten Road, Northbrook, IL 60062-2096.

ANSI/UL 50, *Enclosures for Electrical Equipment, Non-Environmental Considerations,* 2003.

Steam Turbine Drive

CHAPTER 13

Steam-driven fire pump technology has evolved to its current state over many years. The application of steam as a driver for fire pumps is not as common as that of electric or diesel engine drive because steam is not as readily available as it once was. Further, steam generation is not very energy efficient, so other forms of heating have been developed for most modern buildings. In fact, the number of new installations of steam-driven fire pumps is decreasing steadily in the world each year. Chapter 13 covers the requirements for the use of steam-driven fire pumps, and the code and commentary presented in this chapter reflect the current technology used for this type of pump.

13.1 General.

Although the use of steam turbines to drive fire pumps is rare, they are permitted by NFPA 20 and are likely to be used where steam is produced for manufacturing or industrial uses.

When selecting a turbine, it is important to analyze the total steam consumption and the demand of the driver and compare these factors to the steam supply that is available. The supply must be verified to be greater than the demand. See Exhibits II.13.1 through II.13.3 for illustrations of steam-driven fire pumps.

◀ **FAQ**
Does NFPA 20 permit the use of steam turbines to drive fire pumps?

EXHIBIT II.13.1 Steam Turbine–Driven Fire Pump. (Photograph courtesy of Larry Wenzel)

EXHIBIT II.13.2 Fire Pump Information for Steam Turbine Pump. (Photograph courtesy of Larry Wenzel)

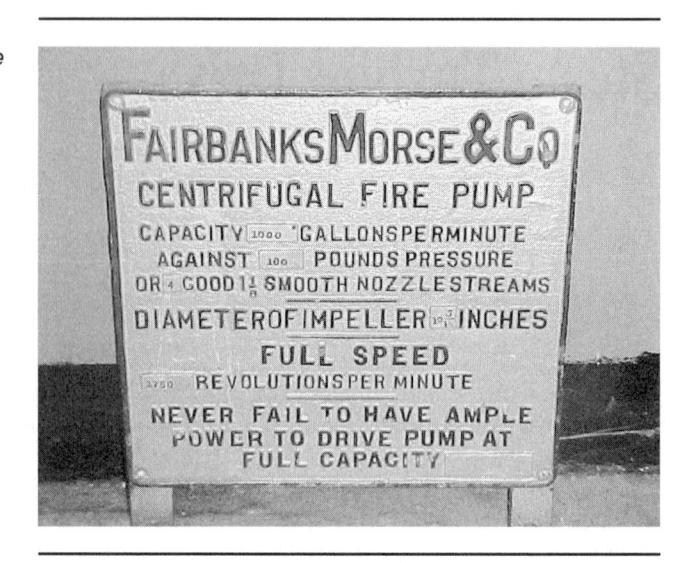

EXHIBIT II.13.3 Reciprocating Steam Engine–Driven Fire Pump. (Photograph courtesy of Larry Wenzel)

Installation of a steam turbine driver for a fire pump should be performed by a person knowledgeable in boilers and turbines as well as fire pumps.

13.1.1 Acceptability.

13.1.1.1 Steam turbines of adequate power are acceptable prime movers for driving fire pumps. Reliability of the turbines shall have been proved in commercial work.

FAQ ▶
Are any steam turbine drivers listed?

At the present time, no steam turbine drivers are listed. The approval of the individual driver must be made using sound engineering principles by the authority having jurisdiction.

As with other drivers, the reliability of the power source is one of the main considerations when choosing a means of supplying power to the fire pump. Very often, use of a steam turbine–driven fire pump is in a remote location where steam is used to generate the electrical power that operates the entire facility.

In most cases the steam turbine–driven fire pump will not be the primary automatic fire pump. The steam turbine–driven units normally are used as a secondary supply. The evolution of today's steam turbine drivers versus the old reciprocating steam pumps is like comparing night and day. The newer equipment has automatic pressure-regulating control valves that improved the safety features of the turbine drive operations.

However, if for some reason the steam supply is shut down, the power to the steam turbine is also shut down. In such situations an alternate source of steam or a pump with a different source of power (diesel) should be installed or provided as a backup.

13.1.1.2 The steam turbine shall be directly connected to the fire pump.

Direct connection to a fire pump includes a flexible coupling that is used for alignment. The intent of NFPA 20 is not to require the installation of a gear to reduce or increase transmission. The pump and the driver should run at the same rpm.

13.1.2 Turbine Capacity.

13.1.2.1 For steam boiler pressures not exceeding 120 psi (8.3 bar), the turbine shall be capable of driving the pump at its rated speed and maximum pump load with a pressure as low as 80 psi (5.5 bar) at the turbine throttle when exhausting against atmospheric back pressure with the hand valve open.

The pressure difference identified in 13.1.2.1 represents the loss of steam due to friction in the steam piping system while the steam turbine is running.

13.1.2.2 For steam boiler pressures exceeding 120 psi (8.3 bar), where steam is continuously maintained, a pressure 70 percent of the usual boiler pressure shall take the place of the 80 psi (5.5 bar) pressure required in 13.1.2.1.

13.1.2.3 In ordering turbines for stationary fire pumps, the purchaser shall specify the rated and maximum pump loads at rated speed, the rated speed, the boiler pressure, the steam pressure at the turbine throttle (if possible), and the steam superheat.

13.1.3* Steam Consumption.

A.13.1.3 Single-stage turbines of maximum reliability and simplicity are recommended where the available steam supply will permit.

13.1.3.1 Prime consideration shall be given to the selection of a turbine having a total steam consumption commensurate with the steam supply available.

As with other power supplies, the normal steam supply must be maintained for other on-line equipment while maintaining the quantity of steam to operate the steam turbine drive. The turbine must be able to come up to speed instantly when required by the fire water demand. If load shedding of other steam-driven equipment is required, the action must be completed automatically.

The steam turbine should always be in a ready state. The turbine drive unit should always be kept warm to minimize the buildup of steam condensate when steam is injected into the unit. Manufacturers' requirements for static mode and operating mode functions should be followed. Full open steam pressure should not be injected to cold equipment.

The available steam supply should always be capable of supplying the steam at the minimum pressure required to operate the steam turbine for the minimum period of time required by the design of the fire protection system. A reserve supply of steam should be anticipated in the design to handle operation of the fire pump at 150 percent of its capacity.

13.1.3.2 Where multistage turbines are used, they shall be so designed that the pump can be brought up to speed without a warmup time requirement.

13.2 Turbine.

FAQ ▶
Where can more information regarding steam turbines be found?

A recognized standard for the manufacture of steam turbines is API Standard 611, *General-Purpose Steam Turbines for Petroleum, Chemical and Gas Industry Services,* which outlines the minimum manufacturing tolerances that should be considered in approval of a turbine.

13.2.1 Casing and Other Parts.

13.2.1.1* The casing shall be designed to permit access with the least possible removal of parts or piping.

A.13.2.1.1 The casing can be of cast iron.

Some applications can require a turbine-driven fire pump to start automatically but not require the turbine to be on pressure control after starting. In such cases, a satisfactory quick-opening manual reset valve installed in a bypass of the steam feeder line around a manual control valve can be used.

Where the application requires the pump unit to start automatically and after starting continue to operate by means of a pressure signal, the use of a satisfactory pilot-type pressure control valve is recommended. This valve should be located in the bypass around the manual control valve in the steam feeder line. The turbine governor control valve, when set at approximately 5 percent above the normal full-load speed of the pump under automatic control, would act as a pre-emergency control.

In the arrangements set forth in the two preceding paragraphs, the automatic valve should be located in the bypass around the manual control valve, which would normally be kept in the closed position. In the event of failure of the automatic valve, this manual valve could be opened, allowing the turbine to come to speed and be controlled by the turbine governor control valve(s).

The use of a direct acting pressure regulator operating on the control valve(s) of a steam turbine is not recommended.

13.2.1.2 A safety valve shall be connected directly to the turbine casing to relieve high steam pressure in the casing.

13.2.1.3 Main Throttle Valve.

13.2.1.3.1 The main throttle valve shall be located in a horizontal run of pipe connected directly to the turbine.

13.2.1.3.2 There shall be a water leg on the supply side of the throttle valve.

13.2.1.3.3 This leg shall be connected to a suitable steam trap to automatically drain all condensate from the line supplying steam to the turbine.

13.2.1.3.4 Steam and exhaust chambers shall be equipped with suitable condensate drains.

13.2.1.3.5 Where the turbine is automatically controlled, these drains shall discharge through adequate traps.

13.2.1.3.6 In addition, if the exhaust pipe discharges vertically, there shall be an open drain at the bottom elbow.

13.2.1.3.7 This drain shall not be valved but shall discharge to a safe location.

13.2.1.4 The nozzle chamber, governor-valve body, pressure regulator, and other parts through which steam passes shall be made of a metal able to withstand the maximum temperatures involved.

13.2.2 Speed Governor.

13.2.2.1 The steam turbine shall be equipped with a speed governor set to maintain rated speed at maximum pump load.

13.2.2.2 The governor shall be capable of maintaining, at all loads, the rated speed within a total range of approximately 8 percent from no turbine load to full-rated turbine load by either of the following methods:

(1) With normal steam pressure and with hand valve closed
(2) With steam pressures down to 80 psi (5.5 bar) [or down to 70 percent of full pressure where this is in excess of 120 psi (8.3 bar)] and with hand valve open

13.2.2.3 While the turbine is running at rated pump load, the speed governor shall be capable of adjustment to secure speeds of approximately 5 percent above and 5 percent below the rated speed of the pump.

13.2.2.4 There shall also be provided an independent emergency governing device.

13.2.2.5 The independent emergency governing device shall be arranged to shut off the steam supply at a turbine speed approximately 20 percent higher than the rated pump speed.

13.2.3 Gauge and Gauge Connections.

13.2.3.1 A listed steam pressure gauge shall be provided on the entrance side of the speed governor.

13.2.3.2 A 0.25 in. (6 mm) pipe tap for a gauge connection shall be provided on the nozzle chamber of the turbine.

13.2.3.3 The gauge shall indicate pressures not less than one and one-half times the boiler pressure and in no case less than 240 psi (16.5 bar).

13.2.3.4 The gauge shall be marked "Steam."

13.2.4 Rotor.

13.2.4.1 The rotor of the turbine shall be of suitable material.

13.2.4.2 The first unit of a rotor design shall be type tested in the manufacturer's shop at 40 percent above rated speed.

13.2.4.3 All subsequent units of the same design shall be tested at 25 percent above rated speed.

13.2.5 Shaft.

13.2.5.1 The shaft of the turbine shall be of high-grade steel, such as open-hearth carbon steel or nickel steel.

13.2.5.2 Where the pump and turbine are assembled as independent units, a flexible coupling shall be provided between the two units.

13.2.5.3 Where an overhung rotor is used, the shaft for the combined unit shall be in one piece with only two bearings.

13.2.5.4 The critical speed of the shaft shall be well above the highest speed of the turbine so that the turbine will operate at all speeds up to 120 percent of rated speed without objectionable vibration.

13.2.6 Bearings.

13.2.6.1 Sleeve Bearings. Turbines having sleeve bearings shall have split-type bearing shells and caps.

13.2.6.2 Ball Bearings.

13.2.6.2.1 Turbines having ball bearings shall be acceptable after they have established a satisfactory record in the commercial field.

13.2.6.2.2 Means shall be provided to give visible indication of the oil level.

13.3* Installation.

Details of steam supply, exhaust, and boiler feed shall be carefully planned to provide reliability and effective operation of a steam turbine–driven fire pump.

Steam can be provided in a loop to the machine requiring the power. This piping should be designed with the same care that is required for fire sprinkler systems. The piping in earthquake-prone areas should be properly supported and braced. The boiler in earthquake-prone areas must be able to operate and resist the minimum horizontal and vertical force conditions as required by the authority having jurisdiction.

In addition to the requirements set forth in Chapter 13, regulatory requirements for the installation of steam piping, control valves, and special fabrication methods fall under the purview of the ASME codes. Refer to ASME B31, *Standards of Pressure Piping Installation,* ASME B31.1–2001, *Power Piping,* and ASME B31.9–1998, *Building Services Piping.* All steam piping work is required to be performed by individuals or companies certified for ASME work.

A.13.3 The following information should be taken into consideration when planning a steam supply, exhaust, and boiler feed for a steam turbine–driven fire pump.

The steam supply for the fire pump should preferably be an independent line from the boilers. It should be run so as not to be liable to damage in case of fire in any part of the property. The other steam lines from the boilers should be controlled by valves located in the boiler room. In an emergency, steam can be promptly shut off from these lines, leaving the steam supply entirely available for the fire pump. Strainers in steam lines to turbines are recommended.

The steam throttle at the pump should close against the steam pressure. It should preferably be of the globe pattern with a solid disc. If, however, the valve used has a removable composition ring, the disc should be of bronze and the ring made of sufficiently hard and durable material, and so held in place in the disc as to satisfactorily meet severe service conditions. Gate valves are undesirable for this service because they cannot readily be made leaktight, as is possible with the globe type of valve. The steam piping should be so arranged and trapped that the pipes can be kept free of condensed steam.

In general, a pressure-reducing valve should not be placed in the steam pipe supplying the fire pump. There is no difficulty in designing turbines for modern high-pressure steam, and this gives the simplest and most dependable unit. A pressure-reducing valve introduces a possible obstruction in the steam line in case it becomes deranged. In most cases, the turbines can be protected by making the safety valve required by 13.2.1.2 of such size that the pressure in the casing will not exceed 25 psi (1.7 bar). This valve should be piped outside of the pump room and, if possible, to some point where the discharge could be seen by the pump attendant. Where a pressure-reducing valve is used, the following points should be carefully considered:

(1) *Pressure-Reducing Valve.*
   (a) The pressure-reducing valve should not contain a stuffing box or a piston working in a cylinder.
   (b) The pressure-reducing valve should be provided with a bypass containing a globe valve to be opened in case of an emergency. The bypass and stop valves should be one pipe size smaller than the reducing valve, and they should be located so as to be readily accessible. This bypass should be arranged to prevent the accumulation of condensate above the reducing valve.
   (c) The pressure-reducing valve should be smaller than the steam pipe required by the specifications for the turbine.
(2) *Exhaust Pipe.* The exhaust pipe should run directly to the atmosphere and should not contain valves of any type. It should not be connected with any condenser, heater, or other system of exhaust piping.
(3) *Emergency Boiler Feed.* A convenient method of ensuring a supply of steam for the fire pump unit, in case the usual boiler feed fails, is to provide an emergency connection from the discharge of the fire pump. This connection should have a controlling valve at the fire pump and also, if desired, an additional valve located in the boiler room. A check valve also should be located in this connection, preferably in the boiler room. This emergency connection should be about 2 in. (50 mm) in diameter.

This method should not be used when there is any danger of contaminating a potable water supply. In situations where the fire pump is handling salt or brackish water, it might also be undesirable to make this emergency boiler feed connection. In such situations, an effort should be made to secure some other secondary boiler feed supply that will always be available.

REFERENCES CITED IN COMMENTARY

American Petroleum Institute, 1220 L Street, N.W., Washington, DC 20005-4070.

API Standard 611, *General-Purpose Steam Turbines for Petroleum, Chemical and Gas Industry Services,* 4th edition, 1997.

The American Society of Mechanical Engineers, Three Park Avenue, New York, NY 10016-5990

ASME B31, *Standards of Pressure Piping Installation:* ASME B31.1-2001, *Power Piping,* and ASME B31.9-1998, *Building Services Piping.*

Acceptance Testing, Performance, and Maintenance

CHAPTER 14

The final step in the design and installation of a fire pump system is to demonstrate that the fire pump functions as intended, the installation meets the requirements of NFPA 20, and the work has been completed in a professional manner. The various tests and system checks required in Chapter 14 are intended to show that all of the critical components of the system are in place and the fire pump performs in accordance with design requirements. The fire pump should perform in accordance with the shop test curve as described in Chapter 6. However, if the pump's performance does not match the shop test curve, as a minimum the pump is expected to meet the fire protection system demand. In a case where the pump's performance does not match the shop test curve, the pump can still be considered to be meeting the intent of the standard and should be accepted.

This chapter also deals with maintenance of the pump and related equipment. For more detailed information regarding fire pump maintenance, see NFPA 25, *Standard for the Inspection, Testing, and Maintenance of Water-Based Fire Protection Systems,* and *Water-Based Fire Protection Systems Handbook.*

The acceptance test evaluates the pump's performance over a range of conditions in order to evaluate the performance of the installation so that the pump will perform as needed during an actual fire event.

The acceptance test includes the following steps:

◀ **FAQ**
What are the steps of the acceptance test?

1. Demonstration of the adequacy of the pump and its ability to deliver water in accordance with the manufacturer's certified shop test characteristic curve, as described in 14.2.4
2. Operation of the pump driver under various conditions to confirm satisfactory performance under all expected conditions
3. Repeated operations of the primary and alternative power supply equipment (if supplied) to demonstrate satisfactory operation under all expected conditions from either automatic or manual operation of the controller (see 14.2.7)
4. Verification of the operation and pressure settings of the variable speed pressure limiting control system (see 14.2.5.3)

Any defects, faults, or other performance problems discovered during the acceptance test must be corrected before final acceptance is granted.

Note that Annex B, Possible Causes of Pump Troubles, of NFPA 20 contains a partial list of pump problems, their possible causes, and some suggested remedies. Annex B may be useful to help identify and correct problems discovered during fire pump acceptance tests.

When defective components are replaced, NFPA 20 requires that Table 14.5.2.4, Summary of Component Replacement Testing Requirements, be used to recommission the fire pump. An identical table is included in NFPA 25.

14.1 Hydrostatic Tests and Flushing.

14.1.1 Flushing.

Stones, gravel, blocks of wood, bottles, work gloves, and other foreign objects have been found in underground piping and piping taking suction from municipal or private water mains or water storage tanks when flushing and/or flow testing procedures are performed. Also, objects in underground piping that are quite remote from the fire pump installation and that would otherwise remain stationary during low-flow conditions may sometimes be carried into the pump during the high-volume flows required during the fire pump acceptance test. Because many of these objects can cause damage to the fire pump, flushing of suction piping prior to connection to the fire pump minimizes potential damage from foreign objects during the fire pump acceptance flow tests. See Exhibit II.14.1 for sample objects removed from underground pipe.

For booster pumps taking suction from underground mains, the flushing flow rate should be at least equal to that in Table 14.1.1.1(a) or (b), or the maximum water demand, whichever is greater. The flushing flow rates given in Table 14.1.1.1(a) or (b) are based on 15 ft/sec (4.6 m/sec) water flow velocity, which is the basis for the minimum suction inlet pipe sizes in Tables 4.25(a) and (b). The minimum suction pipe inlet sizes in Tables 4.25(a) and (b) are the same for pumps of several different capacities.

Experience has shown that the flow rates established in Tables 14.1.1.1(a) and (b) are of sufficient velocity to clear the suction piping of debris. Additionally, most of these flow rates exceed the 150 percent point of the pump ratings where these pipe sizes are the minimum mandated in Tables 4.25(a) and (b). For instance, the largest pump allowed to use a minimum 5 in. (125 mm) suction pipe is one rated at 500 gpm (1893 L/min). The peak (150 percent) point for a 500 gpm (1893 L/min) pump is 750 gpm (2839 L/min). For 5 in. (125 mm) suction pipes, the flush rate is 920 gpm (3482 L/min). In another instance, the largest pump allowed to use an 8 in. (200 mm) suction pipe is one rated at 1500 gpm (5678 L/min). The peak (150 percent) point for a 1500 gpm (5678 L/min) pump is 2250 gpm (8516 L/min). For 8 in. (200 mm) suction pipes, the flush rate is 2350 gpm (8895 L/min).

For fire pumps taking suction from a storage tank, reservoir, or wet pit, the following should be determined prior to the pump acceptance flow test:

1. No obstructing materials are in the water main, tank, reservoir, or pit.
2. Any piping to the pump suction inlet is free from obstructing material.

EXHIBIT II.14.1 Objects Found in a Fire Pump Impeller During Acceptance Testing.

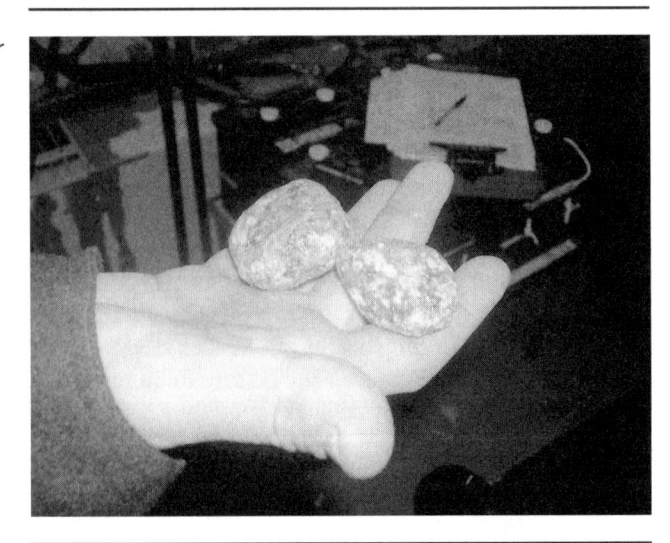

Although field experience is such that oftentimes actual flushing is done without a measurement of the flow rate, verification of the flow rate should be determined. The actual flow rate can be determined using methods such as pitot readings or flowmeters described in NFPA 291, *Recommended Practice for Fire Flow Testing and Marking of Hydrants* (see Part III of this handbook). Flushing is normally accomplished at the maximum flow rate available from the water supply. In some jurisdictions, restrictions may be on the minimum suction pressures allowed. In most cases, the flow at 150 percent of the fire pump rating should not result in unacceptable minimum residual suction pressures and should serve as an appropriate flushing flow rate. Where this scenario is not the case, actual flushing flow rates should at least be equal to the maximum water demand for the fire protection system. Also important to know is that the flushing rates required by NFPA 24, *Standard for the Installation of Private Fire Service Mains and Their Appurtenances,* are based on a velocity of 10 ft/sec (3.1 m/sec) and not the rates required by NFPA 20.

Regardless of the flushing methods used, the installing contractor should never connect piping to the underground stub-up without obtaining a copy of the Contractor's Material and Test Certificate for Underground Piping from the site utilities contractor or the underground piping installing contractor. This certificate provides verification that the underground piping has been flushed in accordance with NFPA 24 and that no debris was present in the piping system at the time of the test.

14.1.1.1 Suction piping shall be flushed at a flow rate not less than indicated in Table 14.1.1.1(a) and Table 14.1.1.1(b) or at the hydraulically calculated water demand rate of the system, whichever is greater.

TABLE 14.1.1.1(a) Flow Rates for Stationary Pumps

U.S. Customary Units		Metric Units	
Pipe Size (in.)	Flow Rate (gpm)	Pipe Size (mm)	Flow Rate (L/min)
4	590	100	2,233
5	920	125	3,482
6	1,360	150	5,148
8	2,350	200	8,895
10	3,670	250	13,891
12	5,290	300	20,023

TABLE 14.1.1.1(b) Flush Rates for Positive Displacement Pumps

U.S. Customary Units		Metric Units	
Pipe Size (in.)	Flow (gpm)	Pipe Size (mm)	Flow (L/min)
1½	100	40	378.50
2	250	50	945.25
3	400	80	1514.00
4	450	100	1703.25
6	500	150	1892.50

14.1.1.2 Flushing shall occur prior to hydrostatic test.

14.1.2 Hydrostatic Test.

14.1.2.1 Suction and discharge piping shall be hydrostatically tested at not less than 200 psi (13.8 bar) pressure or at 50 psi (3.4 bar) in excess of the maximum pressure to be maintained in the system, whichever is greater.

All new systems are required to be tested hydrostatically at a minimum pressure of either 200 psi (13.8 bar), or 50 psi (3.4 bar) above the maximum discharge pressure, whichever is greater, at the fire pump discharge flange. Verification that the proper pressure rating for pipe and fittings is used in the installation is important. This value is set to ensure that all pipe joints and other equipment are installed properly to withstand that pressure without coming apart or leaking. Although the test is primarily a quality control test on the installation and not a materials performance test, damaged materials (e.g., cracked fittings, leaky valves, bad joints, etc.) are routinely discovered during the hydrostatic test.

The measurement of the hydrostatic test pressure is taken at the lowest elevation within the system or portion of the system being tested. Testing at the high point of the system, which, due to static head, could increase the test pressure at lower points in the system to exceptionally high pressures, is not considered necessary.

The hydrostatic test should include all suction and discharge piping between the suction pipe control valve and the discharge pipe control valve. Also, any bypass, meter, jockey pump, cooling water piping, and hose header piping should be included in the test. The normally closed control valve for the hose header should be open for this test. Suction piping on the upstream side of the suction piping control valve on booster pumps, and discharge piping on the system side of the discharge piping control valve, which feeds into underground mains, is required to be hydrostatically tested in accordance with NFPA 24. For discharge piping on the system side of the discharge piping control valves, which feeds directly into the building's fire protection systems, hydrostatic testing is required in accordance with other standards such as NFPA 13, *Standard for Installation of Sprinkler Systems,* and NFPA 14, *Standard for the Installation of Standpipe and Hose Systems.* The coordination of hydrostatic test procedures to facilitate testing of piping for various portions of the entire fire protection system simultaneously may be practical. However, when testing underground piping, NFPA 24 permits a minimal amount of leakage. As such, underground and aboveground piping should be isolated for the hydrostatic test.

Where the fire pump takes suction from a storage tank or reservoir, suction piping upstream of the suction piping control valve must be in accordance with the requirements of NFPA 22, *Standard for Water Tanks for Private Fire Protection.*

Where a main pressure relief valve is provided on the discharge side of the fire pump, and the relief valve setting is less than the required hydrostatic test pressure, suitable means should be employed to allow full hydrostatic testing without discharging water through the relief valve. One method is to temporarily set the relief valve at or above the hydrostatic test pressure, then reset to the planned set pressure and test its performance at that pressure during the fire pump acceptance flow test.

Another method is to isolate the relief valve connection prior to conducting the hydrostatic test. Once the hydrostatic test is completed, the relief valve can be reconnected and its performance verified during the fire pump flow test.

14.1.2.2 The pressure required in 14.1.2.1 shall be maintained for 2 hours.

FAQ ▶
What are the minimum acceptance criteria for the hydrostatic test?

The minimum acceptance criteria for the hydrostatic test are no visible leaks and no drop in the test pressure for aboveground piping systems. The intention of this standard is to require the same procedure for hydrostatic testing as that of Chapter 24 of NFPA 13, 2010 edition.

14.1.3* The installing contractor shall furnish a certificate for flushing and hydrostatic test prior to the start of the fire pump field acceptance test.

A.14.1.3 See Figure A.14.1.3(a) for a sample of a contractor's material and test certificate for fire pumps and Figure A.14.1.3(b) for a sample certificate for private fire service mains.

14.2 Field Acceptance Tests.

The acceptance test evaluates the pump performance over a range of conditions in order to evaluate the performance of the installation so that the pump will perform as needed during an actual fire event.

14.2.1 The pump manufacturer, the engine manufacturer (when supplied), the controller manufacturer, and the transfer switch manufacturer (when supplied) or their factory-authorized representatives shall be present for the field acceptance test. *(See Section 4.4.)*

Contractor's Material and Test Certificate for Fire Pump Systems

PROCEDURE Upon completion of work, inspection and tests shall be made by the contractor's representative and witnessed by an owner's representative. All defects shall be corrected and system left in service before contractor's personnel finally leave the job.

A certificate shall be filled out and signed by both representatives. Copies shall be prepared for approving authorities, owners, and contractor. It is understood the owner's representative's signature in no way prejudices any claim against contractor for faulty material, poor workmanship, or failure to comply with approving authority's requirements or local ordinances.

PROPERTY NAME	DATE
PROPERTY ADDRESS	

PLANS	ACCEPTED BY APPROVING AUTHORITIES (NAMES)		
	ADDRESS		
	INSTALLATION CONFORMS TO ACCEPTED PLANS	☐ YES	☐ NO
	ALL EQUIPMENT USED IS APPROVED FOR FIRE SYSTEM SERVICE IF NO, STATE DEVIATIONS	☐ YES	☐ NO
INSTRUCTIONS	HAS PERSON IN CHARGE OF FIRE PUMP EQUIPMENT BEEN INSTRUCTED AS TO LOCATION OF SYSTEM CONTROL VALVES AND CARE AND MAINTENANCE OF THIS NEW EQUIPMENT? IF NO, EXPLAIN	☐ YES	☐ NO
	HAVE COPIES OF APPROPRIATE INSTRUCTIONS AND CARE AND MAINTENANCE CHARTS BEEN LEFT ON PREMISES? IF NO, EXPLAIN	☐ YES	☐ NO
LOCATION	SUPPLIES BUILDING(S) (CAMPUS, WAREHOUSE, HIGH RISE) EXPLAIN		
PUMP ROOM EQUIPMENT	IS THE PUMP ROOM EQUIPMENT PER THE PLANS AND SPECS?	☐ YES	☐ NO
	IS THE FIRE PUMP PROPERLY MOUNTED AND ANCHORED TO THE FOUNDATION? IF NO, EXPLAIN	☐ YES	☐ NO
	IS THE FIRE PUMP BASE PROPERLY GROUTED? IF NO, EXPLAIN	☐ YES	☐ NO
	DOES THE PUMP ROOM HAVE THE PROPER FLOOR DRAINS? IF NO, EXPLAIN	☐ YES	☐ NO
	IS THE SUCTION AND DISCHARGE PIPING PROPERLY SUPPORTED?	☐ YES	☐ NO
	IS THE PUMP ROOM HEATED AND VENTILATED PER NFPA 20?	☐ YES	☐ NO
PIPES AND FITTINGS	PIPE TYPES AND CLASS		
	PIPE CONFORMS TO _____ STANDARD FITTINGS CONFORM TO _____ STANDARD IF NO, EXPLAIN	☐ YES ☐ YES	☐ NO ☐ NO
	SUCTION AND DISCHARGE PIPING ANCHORED OR RESTRAINED?:	☐ YES	☐ NO
PRE-PACKAGED PUMP HOUSE	IS THIS A PACKAGE OR SKID MOUNTED PUMP?	☐ YES	☐ NO
	IS THE PACKAGE/SKID PROPERLY ANCHORED TO A CONCRETE FOUNDATION? IF NO, EXPLAIN	☐ YES	☐ NO
	IS THE STRUCTURAL FOUNDATION FRAME FILLED WITH CONCRETE TO FORM A FINISHED FLOOR?	☐ YES	☐ NO
	IS THERE A FLOOR DRAIN INSTALLED?	☐ YES	☐ NO
TEST DESCRIPTION	HYDROSTATIC: Hydrostatic tests shall be made at not less than 200 psi (13.8 bar) for 2 hours or 50 psi (3.4 bar) above static pressure in excess of 200 psi (13.8 bar) for 2 hours.		
	HYDROSTATIC TEST: ALL NEW PIPING HYDROSTATICALLY TESTED AT: _____ PSI/BAR FOR _____ HOURS	NO LEAKAGE ALLOWED	
FLUSHING TESTS	FLUSHING: Flow the required rate until water is clear as indicated by no collection of foreign material in burlap bags at outlets such as hydrants and blowoffs. Flush at flows not less than 390 gpm (1476 L/min) for 4 in. pipe, 610 gpm (2309 L/min) for 5 in. pipe, 880 gpm (3331 L/min) for 6 in. pipe, 1560 gpm (5905 L/min) for 8 in. pipe, 2440 gpm (9235 L/min) for 10 in. pipe, and 3520 gpm (13,323 L/min) for 12 in. pipe. When supply cannot produce stipulated flow rates, obtain maximum available.		

© 2009 National Fire Protection Association

FIGURE A.14.1.3(a) *Sample of Contractor's Material Test Certificate for Fire Pump Systems.*

FLUSHING TESTS (continued)	NEW PIPING FLUSHED ACCORDING TO _____ STANDARD BY (COMPANY) _____ IF NO, EXPLAIN	☐ YES	☐ NO
	HOW FLUSHING FLOW WAS OBTAINED ☐ PUBLIC WATER ☐ TANK OR RESERVOIR ☐ OTHER (EXPLAIN)	THROUGH WHAT TYPE OPENING ☐ TEST HEADER ☐ OPEN PIPE	
	LEAD-INS FLUSHED ACCORDING TO _____ STANDARD BY (COMPANY) _____ IF NO, EXPLAIN	☐ YES	☐ NO
	HOW FLUSHING FLOW WAS OBTAINED ☐ PUBLIC WATER ☐ TANK OR RESERVOIR ☐ OTHER (EXPLAIN)	THROUGH WHAT TYPE OPENING ☐ Y CONNECTION TO FLANGE & SPIGOT ☐ OPEN PIPE	
FIELD ACCEPTANCE TEST	ALL EQUIPMENT APPROVED?	☐ YES	☐ NO
	ALL REQUIRED REPRESENTATIVES PRESENT FOR TEST	☐ YES	☐ NO
	AHJ AND OWNER'S REPRESENTATIVE PRESENT FOR TEST IF NO, EXPLAIN	☐ YES	☐ NO
	ALL ELECTRICAL WIRING COMPLETE AND PER *NFPA 70* AND NFPA 20 IF NO, EXPLAIN	☐ YES	☐ NO
	CALIBRATE TEST EQUIPMENT USED CALIBRATION DATE _____	☐ YES	☐ NO
	FLOW TESTS PUMP DESIGN _____ GPM _____ PSI DOES THE PUMP MEET OR EXCEED THE CERTIFIED CURVE? PUMP TYPE ☐ HORIZONTAL ☐ VERTICAL TURBINE ☐ OTHER PUMP MAKE _____ MODEL # _____ SERIAL # _____ COMMENTS _____	☐ YES	☐ NO
	ELECTRIC DRIVER OPERATIONAL TEST SATISFACTORY ELEC. DRIVER _____ MODEL # _____ SERIAL # _____ VOLTAGE _____ VAC @ _____ HP _____ RPM _____ FLA	☐ YES	☐ NO
	ENGINE DRIVEN ENGINE MAKE _____ MODEL # _____ SERIAL # _____ _____ HP _____ RPM SPEED	☐ YES	☐ NO
	DIESEL DRIVER OPERATIONAL TEST SATISFACTORY? OTHER EXPLAIN	☐ YES	☐ NO
	CONTROLLER MAKE _____ MODEL # _____ SERIAL # _____		
	VARIABLE SPEED PRESSURE LIMITING CONTROL	☐ YES	☐ NO
	TESTED AT MINIMUM, RATED, AND PEAK FLOW	☐ YES	☐ NO

	CONTROLLER TEST:
	   SIX AUTO STARTS
	   SIX MANUAL STARTS
		☐ YES
		☐ YES
			☐ NO
			☐ NO

	PHASE REVERSAL TEST PERFORMED (ELECTRIC ONLY)	☐ YES	☐ NO
	ALTERNATE POWER SOURCE TESTED (ELECTRIC ONLY)	☐ YES	☐ NO
	ELECTRONIC FUEL MANAGEMENT (ECM) FUNCTION TEST PERFORMED (DIESEL ONLY)	☐ YES	☐ NO
CONTROL VALVES	SYSTEM CONTROL VALVES LEFT WIDE OPEN IF NO, STATE REASON	☐ YES	☐ NO
	HOSE THREADS OF FIRE DEPARTMENT CONNECTIONS AND HYDRANTS INTERCHANGEABLE WITH THOSE OF FIRE DEPARTMENT ANSWERING ALARM	☐ YES	☐ NO
REMARKS	DATE LEFT IN SERVICE _____ ADDITIONAL COMMENTS:		
SIGNATURES	NAME OF INSTALLING CONTRACTOR		
	TESTS WITNESSED BY FOR PROPERTY OWNER (SIGNED) _____ TITLE _____ DATE _____		
	FOR INSTALLING CONTRACTOR (SIGNED) _____ TITLE _____ DATE _____		
ADDITIONAL COMMENTS AND NOTES:			

© 2009 National Fire Protection Association (NFPA 20, 2 of 2)

FIGURE A.14.1.3(a) *Continued*

Contractor's Material and Test Certificate for Private Fire Service Mains

PROCEDURE Upon completion of work, inspection and tests shall be made by the contractor's representative and witnessed by an owner's representative. All defects shall be corrected and system left in service before contractor's personnel finally leave the job.

A certificate shall be filled out and signed by both representatives. Copies shall be prepared for approving authorities, owners, and contractor. It is understood the owner's representative's signature in no way prejudices any claim against contractor for faulty material, poor workmanship, or failure to comply with approving authority's requirements or local ordinances.

PROPERTY NAME		DATE	
PROPERTY ADDRESS			

PLANS	ACCEPTED BY APPROVING AUTHORITIES (NAMES)			
	ADDRESS			
	INSTALLATION CONFORMS TO ACCEPTED PLANS		☐ YES	☐ NO
	EQUIPMENT USED IS APPROVED IF NO, STATE DEVIATIONS		☐ YES	☐ NO
INSTRUCTIONS	HAS PERSON IN CHARGE OF FIRE EQUIPMENT BEEN INSTRUCTED AS TO LOCATION OF CONTROL VALVES AND CARE AND MAINTENANCE OF THIS NEW EQUIPMENT? IF NO, EXPLAIN		☐ YES	☐ NO
	HAVE COPIES OF APPROPRIATE INSTRUCTIONS AND CARE AND MAINTENANCE CHARTS BEEN LEFT ON PREMISES? IF NO, EXPLAIN		☐ YES	☐ NO
LOCATION	SUPPLIES BUILDINGS			
PIPES AND JOINTS	PIPE TYPES AND CLASS	TYPE JOINT		
	PIPE CONFORMS TO ———— STANDARD		☐ YES	☐ NO
	FITTINGS CONFORM TO ———— STANDARD IF NO, EXPLAIN		☐ YES	☐ NO
	BURIED JOINTS NEEDING ANCHORAGE CLAMPED, STRAPPED, OR BLOCKED IN ACCORDANCE WITH ———— STANDARD IF NO, EXPLAIN		☐ YES	☐ NO
TEST DESCRIPTION	FLUSHING: Flow the required rate until water is clear as indicated by no collection of foreign material in burlap bags at outlets such as hydrants and blowoffs. Flush at flows not less than 390 gpm (1476 L/min) for 4 in. pipe, 610 gpm (2309 L/min) for 5 in. pipe, 880 gpm (3331 L/min) for 6 in. pipe, 1560 gpm (5905 L/min) for 8 in. pipe, 2440 gpm (9235 L/min) for 10 in. pipe, and 3520 gpm (13323 L/min) for 12 in. pipe. When supply cannot produce stipulated flow rates, obtain maximum available. HYDROSTATIC: Hydrostatic tests shall be made at not less than 200 psi (13.8 bar) for 2 hours or 50 psi (3.4 bar) above static pressure in excess of 150 psi (10.3 bar) for 2 hours. LEAKAGE: New pipe laid with rubber gasketed joints shall, if the workmanship is satisfactory, have little or no leakage at the joints. The amount of leakage at the joints shall not exceed 2 qt/hr (1.89 L/hr) per 100 joints irrespective of pipe diameter. The amount of allowable leakage specified above can be increased by 1 fl oz per inch valve diameter per hour (30 mL/25 mm/hr) for each metal seated valve isolating the test section. If dry barrel hydrants are tested with the main valve open, so the hydrants are under pressure, an additional 5 oz per minute (150 mL/min) leakage is permitted for each hydrant.			
FLUSHING TESTS	NEW PIPING FLUSHED ACCORDING TO ———— STANDARD BY (COMPANY) IF NO, EXPLAIN		☐ YES	☐ NO
	HOW FLUSHING FLOW WAS OBTAINED ☐ PUBLIC WATER ☐ TANK OR RESERVOIR ☐ FIRE PUMP	THROUGH WHAT TYPE OPENING ☐ HYDRANT BUTT ☐ OPEN PIPE		
	LEAD-INS FLUSHED ACCORDING TO ———— STANDARD BY (COMPANY) IF NO, EXPLAIN		☐ YES	☐ NO
	HOW FLUSHING FLOW WAS OBTAINED ☐ PUBLIC WATER ☐ TANK OR RESERVOIR ☐ FIRE PUMP	THROUGH WHAT TYPE OPENING ☐ Y CONNECTION TO FLANGE & SPIGOT ☐ OPEN PIPE		

© 2009 National Fire Protection Association   (NFPA 20, 1 of 2)

FIGURE A.14.1.3(b) *Sample of Contractor's Material and Test Certificate for Private Fire Service Mains.*

HYDROSTATIC TEST	ALL NEW PIPING HYDROSTATICALLY TESTED AT _____ PSI FOR _____ HOURS		BURIED JOINTS COVERED ☐ YES ☐ NO
LEAKAGE TEST	TOTAL AMOUNT OF LEAKAGE MEASURED _____ GALLONS	NO LEAKAGE ALLOWED FOR VISIBLE JOINTS _____ HOURS	
	ALLOWABLE LEAKAGE (BURIED) _____ GALLONS	NO LEAKAGE ALLOWED FOR VISIBLE JOINTS _____ HOURS	
HYDRANTS	NUMBER INSTALLED	TYPE AND MAKE	ALL OPERATE SATISFACTORILY ☐ YES ☐ NO
CONTROL VALVES	WATER CONTROL VALVES LEFT WIDE OPEN IF NO, STATE REASON		☐ YES ☐ NO
	HOSE THREADS OF FIRE DEPARTMENT CONNECTIONS AND HYDRANTS INTERCHANGEABLE WITH THOSE OF FIRE DEPARTMENT ANSWERING ALARM		☐ YES ☐ NO
REMARKS	DATE LEFT IN SERVICE _____ ADDITIONAL COMMENTS: _____		
SIGNATURES	NAME OF INSTALLING CONTRACTOR		
	TESTS WITNESSED BY		
	FOR PROPERTY OWNER (SIGNED)	TITLE	DATE
	FOR INSTALLING CONTRACTOR (SIGNED)	TITLE	DATE
	ADDITIONAL EXPLANATION AND NOTES		

© 2009 National Fire Protection Association (NFPA 20, 2 of 2)

FIGURE A.14.1.3(b) Continued

As noted in 4.4.1 and A.4.4.1, NFPA 20 requires one single entity to have the responsibility to ensure a properly completed, tested, and accepted fire pump installation. To this end, 14.2.1 requires that all key component manufacturers (or their representatives) be present at the field acceptance test. The presence of the manufacturer's representatives allows easy resolution of any problems with the quality of the installation, the equipment, and the performance of the completed fire pump installation to be effectively identified and corrected to the satisfaction of the owner, design entity, authority having jurisdiction, the installing contractor, and any other involved parties.

14.2.2* All the authorities having jurisdiction shall be notified as to the time and place of the field acceptance test.

FAQ ▶ Who should be notified of the field acceptance test?

In many cases, the local fire/building official ultimately has legal jurisdictional responsibility and may be needed to approve an installation or even issue a valid Certificate of Occupancy. As such, adequate notice needs to be given to arrange the testing. See Exhibit II.14.2 for a sample Certificate of Occupancy.

While the authority having jurisdiction can represent a number of different individuals, such as various code enforcement officers, it is best to have all involved parties, such as the owner, tenant, or insurance companies, notified and present rather than have to repeat the acceptance test multiple times.

CERTIFICATE OF OCCUPANCY

Building Permit Number _____ **Date** _____

The undersigned hereby applies for a permit of occupancy in accordance with 780 CMR 120, sixth edition:

1. Location of building _____
              Street Address   *Unit Number*
2. Applicant _____
3. Owner _____

   Address _____
4. Occupant _____
5. Use group _____ Occupancy _____
6. Construction type _____ Occupant load _____
7. Special stipulations or conditions _____

   _____
   _____
   _____
   _____
   _____
   _____
   _____

Plumbing/gas _____ Fire _____

Electrical _____ Water and sewer _____

Health _____ Public works _____

I hereby certify that the work specified by the above named building permit has been completed and is ready for occupancy.

Building Inspector _____

Inspection Director _____

Date _____

© 2005 National Fire Protection Association

EXHIBIT II.14.2 *Sample Certificate of Occupancy.*

A.14.2.2 In addition, representatives of the installing contractor and owner should be present.

Sometimes, the owner can assist in notifying other interested parties. The owner may also be needed to provide access to other important areas, such as electrical rooms.

14.2.3 Pump Room Electrical Wiring. All electric wiring to the fire pump motor(s), including control (multiple pumps) interwiring, normal power supply, alternate power supply where provided, and jockey pump, shall be completed and checked by the electrical contractor prior to the initial startup and acceptance test.

The acceptance test is intended to evaluate the performance of a fully completed installation over a range of conditions, including a minimum flow rate that equals or exceeds the maximum fire protection demand, so that the pump will perform as needed during an actual fire event. Therefore, it is important to verify that all wiring and connections are completed and tested prior to acceptance testing and the pump rotation direction verified so that any problems associated with incorrect or incomplete connections, unacceptable quality, incompatibilities due to mismatched components, or any other performance problems can be readily identified and corrected. Acceptance should not be granted on the assumption that an incomplete installation will be correctly completed at some future time.

Most instances of driver failures during the acceptance testing of electric motor–driven fire pumps are usually traced back to noncompliant wiring methods, materials, or practices. By checking the entire installation prior to the acceptance test, unnecessary and time-consuming troubleshooting efforts can be avoided.

14.2.4* Certified Pump Curve.

The manufacturer's certified pump test characteristic curve provides a graphical representation of the pump's performance under controlled conditions, prior to any damage that occurs in transit and prior to the completion of the installation. The curve, therefore, provides a benchmark to which installed pumps can be compared (see Section 4.5).

A.14.2.4 If a complete fire pump submittal package is available, it should provide for comparison of the equipment specified. Such a package should include an approved copy of the fire pump room general arrangement drawings, including the electrical layout, the layout of the pump and water source, the layout of the pump room drainage details, the pump foundation layout, and the mechanical layout for heat and ventilation.

14.2.4.1 A copy of the manufacturer's certified pump test characteristic curve shall be available for comparison of the results of the field acceptance test.

14.2.4.2 The fire pump as installed shall equal the performance as indicated on the manufacturer's certified shop test characteristic curve within the accuracy limits of the test equipment.

Exhibit II.14.3 shows fire pump test data obtained from a fire pump acceptance test, and Exhibit II.14.4 shows the plots of the test data with respect to the manufacturer's certified test curve.

14.2.5* Field Acceptance Test Procedures.

A.14.2.5 The fire pump operation is as follows:

(1) *Motor-Driven Pump.* To start a motor-driven pump, the following steps should be taken in the following order:
   (a) See that pump is completely primed.
   (b) Close isolating switch and then close circuit breaker.

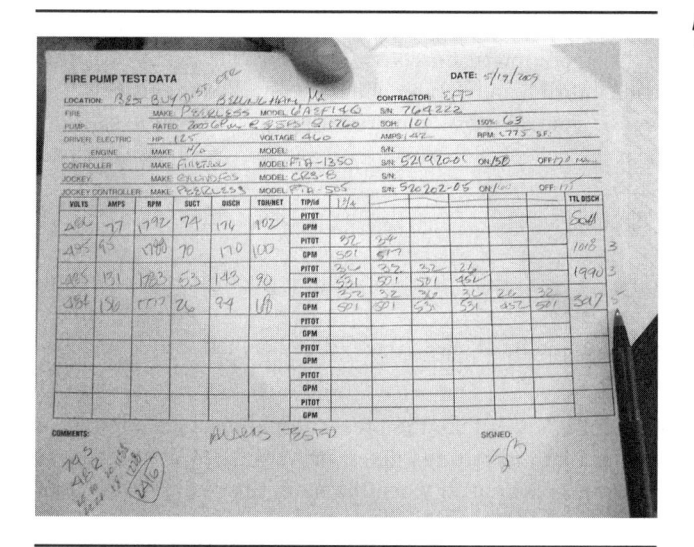

EXHIBIT II.14.3 Fire Pump Test Data.

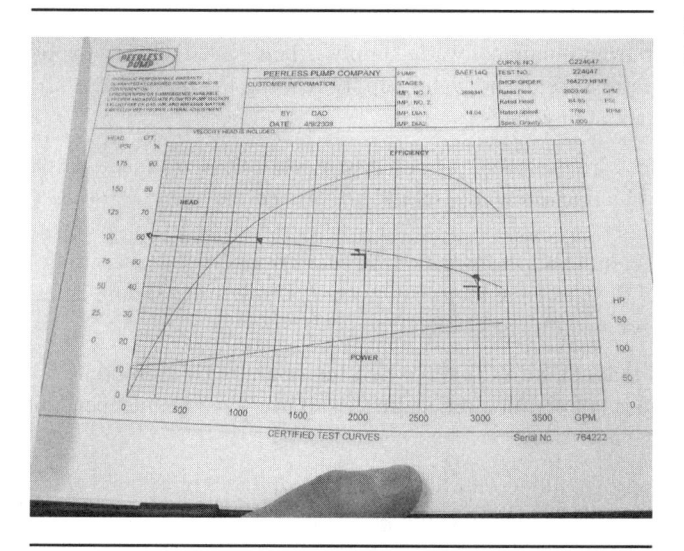

EXHIBIT II.14.4 Manufacturer Certified Test Curve.

(c) Automatic controller will start pump if system demand is not satisfied (e.g., pressure low, deluge tripped).

(d) For manual operation, activate switch, pushbutton, or manual start handle. Circuit breaker tripping mechanism should be set so that it will not operate when current in circuit is excessively large.

(2) *Steam-Driven Pump.* A steam turbine driving a fire pump should always be kept warmed up to permit instant operation at full-rated speed. The automatic starting of the turbine should not be dependent on any manual valve operation or period of low-speed operation. If the pop safety valve on the casing blows, steam should be shut off and the exhaust piping examined for a possible closed valve or an obstructed portion of piping. Steam turbines are provided with governors to maintain speed at a predetermined point, with some adjustment for higher or lower speeds. Desired speeds below this range can be obtained by throttling the main throttle valve.

(3) *Diesel Engine–Driven Pump.* To start a diesel engine–driven pump, the operator should be familiar beforehand with the operation of this type of equipment. The instruction books issued by the engine and control manufacturer should be studied to this end. The storage batteries should always be maintained in good order to ensure prompt, satisfactory operation of this equipment (i.e., check electrolyte level and specific gravity, inspect cable conditions, corrosion, etc.).

(4) *Fire Pump Settings.* The fire pump system, when started by pressure drop, should be arranged as follows:
   (a) The jockey pump stop point should equal the pump churn pressure plus the minimum static supply pressure.
   (b) The jockey pump start point should be at least 10 psi (0.68 bar) less than the jockey pump stop point.

Because the main fire pump start point is lower than the pressure maintenance pump start point, care should be taken to ensure that the settings given in A.14.2.5(4)(b) are not so low that a pressure shock (water hammer) is created when the main fire pump finally does come on.

   (c) The fire pump start point should be 5 psi (0.34 bar) less than the jockey pump start point. Use 10 psi (0.68 bar) increments for each additional pump.

In the case where a booster pump from municipal water supply is being used, the start point should not be set so low that the pressure of the water supply feeding the pump keeps the pump from starting, even at substantial flow rates.

   (d) Where minimum run times are provided, the pump will continue to operate after attaining these pressures. The final pressures should not exceed the pressure rating of the system.
   (e) Where the operating differential of pressure switches does not permit these settings, the settings should be as close as equipment will permit. The settings should be established by pressures observed on test gauges.

An alternative to A.14.2.5(4)(e) that can be explored when the operating differential of a pressure switch (or transducer) does not permit close settings is to use separate, independent switches for start and stop settings.

   (f) Examples of fire pump settings follow (for SI units, 1 psi = 0.0689 bar):
      i. Pump: 1000 gpm, 100 psi pump with churn pressure of 115 psi
      ii. Suction supply: 50 psi from city—minimum static; 60 psi from city—maximum static
      iii. Jockey pump stop = 115 psi + 50 psi = 165 psi
      iv. Jockey pump start = 165 psi - 10 psi = 155 psi
      v. Fire pump stop = 115 psi + 50 psi = 165 psi
      vi. Fire pump start = 155 psi - 5 psi = 150 psi
      vii. Fire pump maximum churn = 115 psi + 60 psi = 175 psi

Fire pumps should not be stopped automatically at the shutoff (or churn) pressure. In many cases, the authority having jurisdiction requires that the pump be manually stopped (see 10.5.4.1 and 12.7.2.6.2).

   (g) Where minimum-run timers are provided, the pumps will continue to operate at churn pressure beyond the stop setting. The final pressures should not exceed the pressure rating of the system components.

(5) *Automatic Recorder.* The performance of all fire pumps should be automatically indicated on a pressure recorder to provide a record of pump operation and assistance in fire loss investigation.

14.2.5.1* Test Equipment.

A.14.2.5.1 The test equipment should be furnished by either the authority having jurisdiction or the installing contractor or the pump manufacturer, depending upon the prevailing arrangements made between the aforementioned parties. The equipment should include, but not necessarily be limited to, the following:

(1) *Equipment for Use with Test Valve Header.* 50 ft (15 m) lengths of 2½ in. (65 mm) lined hose should be provided including Underwriters Laboratories' play pipe nozzles as needed to flow required volume of water. Where test meter is provided, however, these might not be needed.
(2) *Instrumentation.* The following test instruments should be of high quality, accurate, and in good repair:
 (a) Clamp-on volt/ammeter
 (b) Test gauges
 (c) Tachometer

Exhibit II.14.5 shows application of a typical tachometer for measuring pump speed.

EXHIBIT II.14.5 *Tachometer Used to Measure Engine Speed.*

(d) Pitot tube with gauge (for use with hose and nozzle)

Exhibit II.14.6 (top) illustrates a Pitot tube, which is used to measure the amount of water flowing out of an open nozzle. Exhibit II.14.6 (bottom) illustrates another type of flow measuring device that is commonly used for the purpose of measuring flow during hydrant flow tests of fire pump acceptance tests. The end with the orifice is inserted into the center of the stream of flowing water (orifice facing the nozzle opening) approximately one-half the diameter of the nozzle away from the opening of the nozzle. The pressure registered on the gauge corresponds to the amount of water flowing. The following equation is used to determine the water flow in gpm:

$$Q = 29.83 c d^2 \sqrt{P}$$

where:

Q = flow (gpm)

c = coefficient of the nozzle opening (provided by the manufacturer)

EXHIBIT II.14.6 Pitot Tube (top) and Flow Measuring Device (bottom).

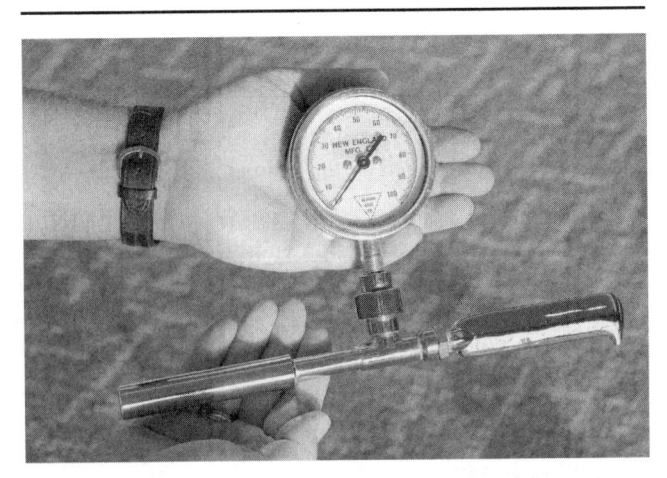

d = actual diameter of the nozzle opening (in.)

P = pressure registered on the Pitot tube gauge (psi)

(3) *Instrumentation Calibration.* All test instrumentation should be calibrated by an approved testing and calibration facility within the 12 months prior to the test. Calibration documentation should be available for review by the authority having jurisdiction.

FAQ ▶
Why does NFPA 20 recommend that all test instrumentation be calibrated?

It is a waste of time and effort to go through the motions of an acceptance test, only to have someone question the accuracy of the results. The use of calibrated equipment can reduce the possibility of measurement errors and having to conduct the test again at a later date.

A majority of the test equipment used for acceptance and annual testing has never been calibrated. This equipment can have errors of 15 to 30 percent in readings. The use of uncalibrated test equipment can lead to inaccurately reported test results.

While it is desirable to achieve a true churn condition (no flow) during the test for comparison to the manufacturer's certified pump test characteristic curve, it might not be possible in all circumstances. Pumps with circulation relief valves will discharge a small amount of water, even when no water is flowing into the fire protection system. The small discharge through the circulation relief valve should not be shut off during the test since it is necessary to keep the pump from overheating. For pumps with circulation relief valves, the minimum flow condition in the test is expected to be the situation where no water is flowing to the fire protection system but a small flow is present through the circulation relief valve. During a test on a pump with a pressure relief valve, the pressure relief valve should not open because these valves are installed purely as a safety precaution to prevent overpressurization during overspeed conditions.

Overspeed conditions should not be present during the test, so the pressure relief valve should not open. When pressure relief valves are installed on systems to relieve pressure under normal operating conditions, and if a true churn condition is desired during the acceptance test, the system discharge valve can be closed and the pressure relief valve can be adjusted to eliminate the flow. The pressure readings can be quickly noted and the pressure relief valve adjusted again to allow flow and relief of pressure. After this one-time test, a reference net pressure can be noted with the relief valve open so that the relief valve can remain open during subsequent annual tests with the comparison back to the reference residual net pressure rather than the manufacturer's curve.

14.2.5.1.1 Calibrated test equipment shall be provided to determine net pump pressures, rate of flow through the pump, volts and amperes for electric motor–driven pumps, and speed.

14.2.5.1.2 Calibrated test gauges shall be used and bear a label with the latest date of calibration. Gauges shall be calibrated a minimum of annually. Calibration of test gauges shall be maintained at an accuracy level of ±1 percent.

The intent of NFPA 20 is to require calibrated, listed, or approved test equipment to determine the performance of the fire pump system. Shop fabricated or untested equipment is not permitted.

In many cases, the installing contractor is held responsible for an acceptable installation, including a satisfactory acceptance test. Therefore, the contractor should have the test equipment available, even when the authority having jurisdiction has made arrangements to furnish the equipment. A good practice is to have a backup set of equipment available in the event of test equipment breakdown. This spare set can reduce the possibility of having to make arrangements for a later, second test.

14.2.5.2 Fire Pump Flow Testing(s).

See Exhibits II.14.7 and II.14.8 for examples of flow testing.

14.2.5.2.1 The fire pump shall perform at minimum, rated, and peak loads without objectionable overheating of any component.

Temperatures should be monitored at regular intervals during the entire test to ensure that any overheating can be detected and corrected before damage occurs to any component. If overheating does occur during the acceptance test, the test should be stopped until the

◀ **FAQ**
What happens if overheating occurs during the acceptance test?

EXHIBIT II.14.7 Determining Flow Using a Flowmeter.

EXHIBIT II.14.8 Determining Flow Using a Pitot Gauge.

problem is remedied. Once the problem is remedied, a retest should include a full 1-hour duration without overheating. Other successfully completed tests on the controller or other components do not need to be repeated.

14.2.5.2.2 Vibrations of the fire pump assembly shall not be of a magnitude to pose potential damage to any fire pump component.

14.2.5.2.3* The minimum, rated, and peak loads of the fire pump shall be determined by controlling the quantity of water discharged through approved test devices.

A.14.2.5.2.3 Where a hose valve header is used, it should be located where a limited [approximately 100 ft (30 m)] amount of hose is used to discharge water safely.

Where a flow test meter is used in a closed loop according to manufacturer's instructions, additional outlets such as hydrants, hose valves, and so forth should be available to determine the accuracy of the metering device.

14.2.5.2.3.1 Where simultaneous operation of multiple pumps is possible or required as part of a system design, the acceptance test shall include a flow test of all pumps operating simultaneously.

14.2.5.2.4 Where the maximum flow available from the water supply cannot provide a flow of 150 percent of the rated flow of the pump, the fire pump shall be operated at the greater of 100 percent of rated flow or the maximum flow demand of the fire protection system(s) maximum allowable discharge to determine its acceptance. This reduced capacity shall constitute an acceptable test, provided that the pump discharge exceeds the fire protection system design and flow rate.

FAQ ▶
Does NFPA 20 require pumps to flow 150 percent of the rated capacity during tests?

A test including a minimum flow rate of of 150 percent of the rated pump capacity is required if it can be completed without creating a dangerously low suction pressure. If the 150 percent of rated pump capacity cannot be achieved at the initial acceptance test of a new installation, the pump may be oversized relative to the available water supply.

Nevertheless, when pump flow tests are conducted at maximum rates less than 150 percent of the pump's rated capacity, the maximum flow should at least exceed the greatest demand of all fire protection systems supplied by the respective pump. Based on 14.2.5.2.4, the inability to operate the fire pump at 150 percent of the pump's rated capacity should not result in an unacceptable installation. However, the pump will not be properly tested for pump performance.

14.2.5.2.5 Where the suction to the fire pump is from a break tank, the tank refill rate shall be tested and recorded. The refill device shall be operated a minimum of five times.

14.2.5.2.6 Water Level Detection. Water level detection is required for all vertical turbine pumps installed in wells to determine the suction pressure available at the shutoff, 100 percent flow, and 150 percent flow points to determine if the pump is operating within its design conditions.

14.2.5.3 Variable Speed Pressure Limiting Control.

14.2.5.3.1 Pumps with variable speed pressure limiting control shall be tested at minimum, rated, and peak loads, with both the variable speed pressure limiting control operational and the fire pump operating at rated speed.

Paragraph 14.2.5.3.1 mandates a fire pump test with the simultaneous operation of the variable speed pressure limiting control and the fire pump at rated speed. This requirement is intended to ensure that both the pump and the variable speed pressure limiting control are operating correctly. When testing the performance of the pump without the variable speed pressure limiting control and pressure relief valve in operation, the fire protection system components could be overpressurized, so the discharge isolation valve should be closed for these tests.

14.2.5.3.2 The fire protection system shall be isolated and the pressure relief valve closed for the tests required in 14.2.5.3.1.

14.2.5.4* Measurement Procedure.

In order to produce the most accurate curve and to discover an occasional invalid reading, gathering readings at numerous flow rates including at least 0 percent, 100 percent, and 150 percent of the rated capacity is prudent. Flow testing at low-flow conditions for the pump may lead to false readings; therefore, flow testing should not be attempted below 50 percent of the rated flow. However, testing at 0 percent or no flow, normally referred to as churn or shutoff, is necessary to verify the correct operating pressure of the fire pump.

◀ FAQ
At what flow rates should readings be recorded?

A.14.2.5.4 The test procedure is as follows:

(1) Make a visual check of the unit. If hose and nozzles are used, see that they are securely tied down. See that the hose valves are closed. If a test meter is used, the valve on the discharge side of the meter should be closed.
(2) Start the pump.
(3) Partially open one or two hose valves, or slightly open the meter discharge valve.
(4) Check the general operation of the unit. Watch for vibration, leaks (oil or water), unusual noises, and general operation. Adjust packing glands.
(5) Measure water discharge. The steps to do so are as follows:
   (a) Where a test valve header is used, regulate the discharge by means of the hose valves and a selection of the nozzle tips. It will be noticed that the play pipe has a removable tip. This tip has a 1⅛ in. (28.6 mm) nozzle, and when the tip is removed, the play pipe has a 1¾ in. (44.4 mm) nozzle. Hose valves should be shut off before removing or putting on the 1⅛ in. (28.6 mm) tip.
   (b) Where a test meter is used, regulate the discharge valve to achieve various flow readings.
   (c) Important test points are at 150 percent rated capacity, rated capacity, and shutoff. Intermediate points can be taken if desired to help develop the performance curve.
(6) Record the following data at each test point (see Figure A.14.2.5.4):
   (a) Pump rpm
   (b) Suction pressure
   (c) Discharge pressure
   (d) Number and size of hose nozzles, pitot pressure for each nozzle, and total gpm (L/min); for test meter, simply a record of gpm (L/min)

Centrifugal Fire Pump Acceptance Test Form

Information on this form covers the minimum requirements of NFPA 20-2007 for performing acceptance tests on pumps with electric motors or diesel engine drivers. Other forms are available for periodic inspection, testing and maintenance.

Owner: _____
Owner's Address: _____

Property on which pump is installed: _____

Property Address: _____

Date of Test: _____
Demand(s) of Fire Protection Systems Supplied By Pump: _____

Pump: ❑ Horizontal ❑ Vertical
Manufacturer: _____ Shop/Serial Number: _____
Model or Type: _____
Rated GPM _____ Rated Pressure _____ Rated RPM _____
Suction From _____ If Tank, Size and Height _____
Driver: ❑ Electric Motor ❑ Diesel Engine ❑ Steam Turbine
Manufacturer: _____ Shop/Serial Number: _____
Model or Type: _____
Rated Horsepower: _____ Rated Speed: _____
If Electric Motor, Rated Voltage _____ Operating Voltage _____
Rated Amps _____ Phase Cycles _____ Service Factor _____
Controller Manufacturer: _____
Shop/Serial Number: _____ Model or Type: _____
Jockey Pump on System? ❑ Yes ❑ No   Settings: On ___ Off ___

Note: All questions are to be answered Yes, No or Not Applicable. All "No" answers are to be explained in the comments portion of this form.

I. Flush Test *(Conduct before Hydrostatic Test)*
Suction piping was flushed at _____ gpm? ❑ Yes ❑ No ❑ N/A
(See Table 14.1.1.1(a) of NFPA 20.)
Certificate presented showing flush test? ❑ Yes ❑ No ❑ N/A

II. Hydrostatic Test
Piping tested at _____ psi for 2 hours? ❑ Yes ❑ No ❑ N/A
(Note: NFPA 20 requires 200 psi or 50 psi above maximum system pressure whichever is greater.)
Piping passed test? ❑ Yes ❑ No ❑ N/A
Certificate presented showing test? ❑ Yes ❑ No ❑ N/A

III. People Present
Were the following present to witness the test:
A. Pump manufacturer/representative ❑ Yes ❑ No ❑ N/A
B. Engine manufacturer/representative ❑ Yes ❑ No ❑ N/A
C. Controller manufacturer/representative ❑ Yes ❑ No ❑ N/A
D. Transfer switch manufacturer/rep. ❑ Yes ❑ No ❑ N/A
E. Authority having jurisdiction/rep. ❑ Yes ❑ No ❑ N/A

IV. Electric Wiring
Was all electric wiring including control interwiring for multiple pumps, emergency power supply, and the jockey pump completed and checked by the electrical contractor prior to the initial start-up and acceptance test? ❑ Yes ❑ No ❑ N/A

V. Flow Test
Run the pump at no-load, rated load and peak load (usually 150% of rated load) conditions. For variable speed drivers, run the test with the pressure limiting control "on" and then again at rated speed with the pump isolated from the fire protection system and the relief valve closed.

A. Was a copy of the manufacturers' certified pump test characteristic curve available for comparison to the results of the acceptance test? ❑ Yes ❑ No ❑ N/A
B. Equipment and gages calibrated? ❑ Yes ❑ No ❑ N/A
C. No vibrations that could potentially damage any fire pump component? ❑ Yes ❑ No ❑ N/A
D. The fire pump performed at all conditions without objectionable overheating of any component? ❑ Yes ❑ No ❑ N/A

E. For each test, record the following for each load condition:

Test	Driver Speed	Suction Pressure	Discharge Pressure	Nozzle Size	Pitot Readings or Flow					
	rpm	psi	psi	inch	1	2	3	4	5	6
0										
100%										
150%										

F. For electric motor driven pumps also record:

Test	Voltage	Amperes
0		
100%		
150%		

E. For each test, record the following for each load condition:
$P_{Net} = P_{Discharge} - P_{Suction}$ $Q = 29.83 \, cd^2$

Test	Net Pressure	Flow						Total Flow
		1	2	3	4	5	6	
0		0	0	0	0	0	0	0
100%								
150%								

H. For electric motors operating at rated voltage and frequency, is the ampere demand less than or equal to the product of the full load ampere rating times the allowable service factor as stamped on the motor nameplate? ❑ Yes ❑ No ❑ N/A

I. For electric motors operating under varying voltage:
1. Was the product of the actual voltage and current demand less than or equal to the product of the rated full load current times the rated voltage times the allowable service factor? ❑ Yes ❑ No ❑ N/A
2. Was the voltage always less than 5% below the rated voltage during the test? ❑ Yes ❑ No ❑ N/A
3. Was the voltage always less than 10% above the rated voltage during the test? ❑ Yes ❑ No ❑ N/A

J. Did engine-driven units show no signs of overload or stress? ❑ Yes ❑ No ❑ N/A
K. Was the governor set to properly regulate the engine speed at rated pump speed? ❑ Yes ❑ No ❑ N/A
L. Did the gear drive assembly operate without excessive objectionable noise, vibration or heating? ❑ Yes ❑ No ❑ N/A
M. Was the fire pump unit started and brought up to rated speed without interruption under the conditions of a discharge equal to peak load? ❑ Yes ❑ No ❑ N/A
N. Did the fire pump perform equal to the manufacturer's characteristic curve within the accuracy limits of the test equipment? ❑ Yes ❑ No ❑ N/A
O. Electric motor pumps passed phase reversal test on normal and alternate (if provided) power? ❑ Yes ❑ No ❑ N/A

© 2005 National Fire Sprinkler Association, P.O. Box 1000, Patterson, NY 12563 (845) 878-4200 (used with permission)   Form 20-A Sheet 1 of 2

FIGURE A.14.2.5.4 Centifugal Fire Pump Acceptance Test Form. (Source: National Fire Sprinkler Association, Inc.)

VI. Controller Test

A. Did the pump start at least 6 times from automatic sources? ❏ Yes ❏ No ❏ N/A

B. Was each automatic starting feature tested at least once? ❏ Yes ❏ No ❏ N/A

C. Did the pump start at least 6 times manually? ❏ Yes ❏ No ❏ N/A

D. Was the pump run for at least 5 minutes during each of the operations in Parts A, B and C above? ❏ Yes ❏ No ❏ N/A

(Note: An engine driver is not required to run for 5 minutes at full speed between successive starts until the cumulative cranking time of successive starts reaches 45 seconds.)

E. Were the starting operations divided between both sets of batteries for engine-driven controllers? ❏ Yes ❏ No ❏ N/A

F. Electric Driven Pump Controllers

1. Were all overcurrent protective devices (including the controller circuit-breaker) selected, sized and set in accordance with NFPA 20? ❏ Yes ❏ No ❏ N/A

2. Was the fire pump started at least once from each power service and run for at least 5 minutes? ❏ Yes ❏ No ❏ N/A

3. Upon simulation of a power failure, while the pump is operating at peak load, did the transfer switch transfer from the normal to the emergency source without opening overcurrent protection devices on either line? ❏ Yes ❏ No ❏ N/A

4. When normal power was restored, did retransfer from emergency to normal power occur without overcurrent protection devices opening on either line? ❏ Yes ❏ No ❏ N/A

5. Were at least half of the automatic and manual starts required by Parts A and C performed with the pump connected to the alternate source? ❏ Yes ❏ No ❏ N/A

G. Were all signal conditions simulated demonstrating satisfactory operation? ❏ Yes ❏ No ❏ N/A

H. Was the pump run for at least 1 hour total during all of the above tests? ❏ Yes ❏ No ❏ N/A

I. For engines with ECM fuel management systems, primary and alternate ECM passed function test? ❏ Yes ❏ No ❏ N/A

VII. Information For Owner

Was the owner given all of the following? ❏ Yes ❏ No ❏ N/A

A. A manual explaining the operation of all components.
B. Instructions for routine maintenance and repairs.
C. Parts list and parts identification.
D. Schematic electrical drawings of controller, transfer switch and alarm panels.
E. Manufacturer's Certified Shop Test curve or Acceptance Test Curve.

VIII. Tester Information

Tester: _____

Company: _____

Company Address: _____

I state that the information on this form is correct at the time and place of my test, and that all equipment tested was left in operational condition upon completion of this test except as noted in the comments section below.

Signature of Tester: _____ Date: _____

License or Certification Number if Applicable: _____

IX. Comments

(Any "No" answers, test failures, or other problems must be explained here.)

Pump Test Results

Pressure (psi, fill in scale) — Flow (gpm, fill in scale) — Amperes (for electric driven pumps) (fill in scale)

© 2005 National Fire Sprinkler Association, P.O. Box 1000, Patterson, NY 12563 (845) 878-4200 (used with permission)   Form 20-A Sheet 2 of 2

FIGURE A.14.2.5.4 Continued

(e) Amperes (each phase)
(f) Volts (phase to phase)
(7) Calculation of test results is as follows:
  (a) *Rated Speed.* Determine that pump is operating at rated rpm.
  (b) *Capacity.* For hose valve header, using a fire stream table, determine the gpm (L/min) for each nozzle at each pitot reading. For example, 16 psi (1.1 bar) pitot pressure with 1¾ in. (44.4 mm) nozzle indicates 364 gpm (1378 L/min). Add the gpm for each hose line to determine total volume. For test meter, the total gpm (L/min) is read directly.
  (c) *Total Head for Horizontal Pump.* Total head is the sum of the following:
    i. Pressure measured by the discharge gauge at pump discharge flange
    ii. Velocity head difference, pump discharge, and pump suction
    iii. Gauge elevation corrections to pump centerline (plus or minus)
    iv. Pressure measured by suction gauge at pump suction flange—negative value when pressure is above zero
  (d) *Total Head for Vertical Pump.* Total head is the sum of the following:
    i. Pressure measured by the discharge gauge at pump discharge flange
    ii. Velocity head at the discharge
    iii. Distance to the supply water level
    iv. Discharge gauge elevation correction to centerline of discharge
  (e) *Electrical Input.* Voltage and amperes are read directly from the volt/ammeter. This reading is compared to the motor nameplate full-load amperes. The only general calculation is to determine the maximum amperes allowed due to the motor service factor. In the case of 1.15 service factor, the maximum amperes are approximately 1.15 times motor amperes, because changes in power factor and efficiency are not considered. If the maximum amperes recorded on the test do not exceed this figure, the motor and pump will be judged satisfactory. It is most important to measure voltage and amperes accurately on each phase should the maximum amperes logged on the test exceed the calculated maximum amperes. This measurement is important because a poor power supply with low voltage will cause a high ampere reading. This condition can be corrected only by improvement in the power supply. There is nothing that can be done to the motor or the pump.
  (f) *Correction to Rated Speed.* For purposes of plotting, the capacity, head, and power should be corrected from the test values at test speed to the rated speed of the pump. The corrections are made as follows.

Capacity:

$$Q_2 = \left(\frac{N_2}{N_1}\right) Q_1$$

where:
Q_1 = capacity at test speed in gpm (L/min)
Q_2 = capacity at rated speed in gpm (L/min)
N_1 = test speed in rpm
N_2 = rated speed in rpm

Head:

$$H_2 = \left(\frac{N_2}{N_1}\right)^2 H_1$$

where:

H_1 = head at test speed in ft (m)
H_2 = head at rated speed in ft (m)

Horsepower:

$$hp_2 = \left(\frac{N_2}{N_1}\right)^3 hp_1$$

where:

hp_1 = kW (horsepower) at test speed
hp_2 = kW (horsepower) at rated speed

The water supply to the fire pump should be tested before the fire pump test. Significant deviations between flow test results and the design analysis should also be investigated and resolved before the pump test. Partially closed valves in the underground piping can be identified prior to the pump test. In some jurisdictions, where a pump is fed from the public mains, the local code/fire official may want to see a hydrant test at a rate in excess of 150 percent of the pump's rated capacity. This test is conducted in order to eliminate any questions regarding the capability of the water supply prior to setting up the meters, gauges, and hoses. When the pump test begins, the capability of the water supply should already be known. In this instance, any deficiency will most likely be related to the fire pump, not the attached water supply. This testing should be limited to 165 percent of the rated flow because of the discharge curve of the pump.

An example of the information obtained during a pump acceptance test is shown in Exhibit II.14.9.

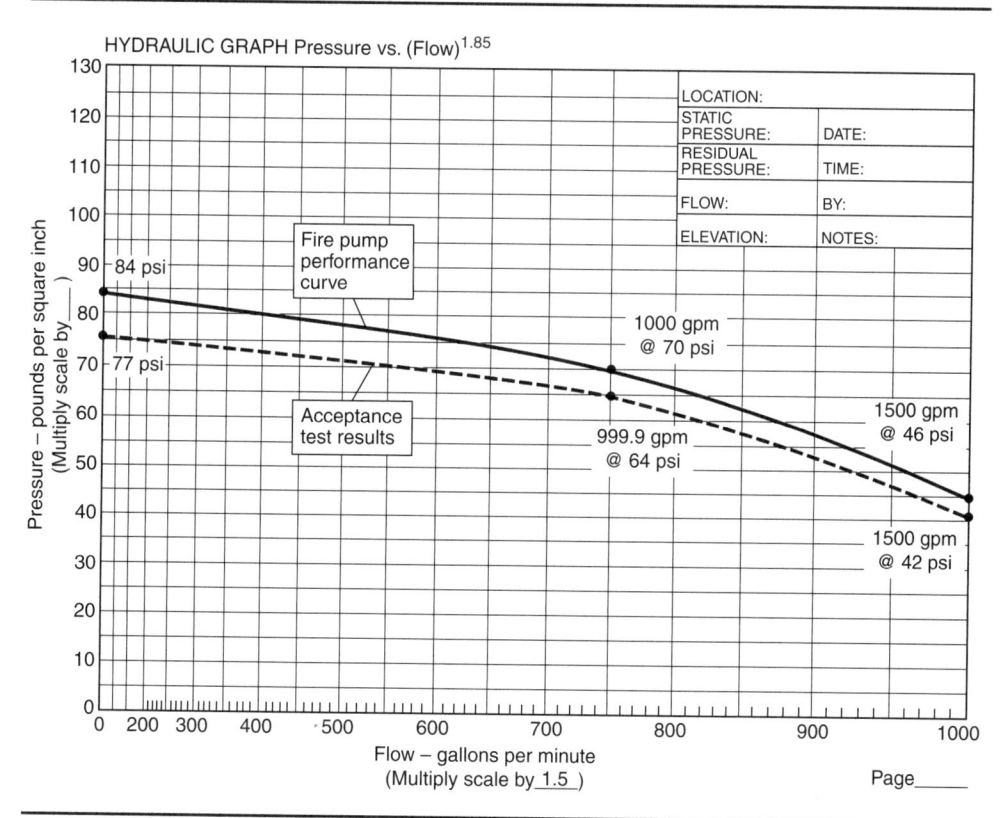

EXHIBIT II.14.9 Example of Pump Acceptance Test Data.

Usually, test results with pump speed variations of 1 or 2 percent from rated speed are disregarded because the capacity, head, and horsepower differences are insignificant. The following example illustrates how to correct pump performance for its rated speed.

EXAMPLE

A pump rated for 80 psi at 750 gpm when turning at 1780 rpm is operating at 1700 rpm during the acceptance test. The manufacturer's certified shop test indicates the following operational characteristics when the pump is operating at 1780 rpm:

	Churn	Rated	Overload
Flow	0 gpm (0 L/min)	750 gpm (2839 L/min)	1125 gpm (4258 L/min)
Net pressure	96 psi (6.7 bar)	80 psi (5.6 bar)	52 psi (3.6 bar)

The acceptance test shows the following results when the pump is operating at 1700 rpm:

	Churn	Rated	Overload
Flow	0 gpm (0 L/min)	716 gpm (2710 L/min)	1075 gpm (4069 L/min)
Net pressure	88 psi (6.1 bar)	73 psi (5.0 bar)	47 psi (3.2 bar)

Solution: At first, the results when the pump is operating at 1700 rpm may seem to reflect a deficiency in the pump performance when compared to the shop curve. However, by adjusting for the proper speed, the corrected flow (Q_c) and pressure (H_c) results are obtained. Using the equations presented in A.14.2.5.4(f), the corrected flow and pressure can be obtained.

$$Q_c = \left(\frac{N_2}{N_1}\right) \times Q_1$$

$$H_c = \left(\frac{N_2}{n_1}\right)^2 \times H_1$$

For churn,

$$Q_c = (1780/1700) \times 0 = 1.047 \times 0 = 0 \text{ gpm (0 L/min)}$$
$$H_c = (1780/1700)^2 \times 88 = (1.047)^2 \times 88 = 1.096 \times 88 = 96 \text{ psi (6.6 bar)}$$

For rated,

$$Q_c = (1780/1700) \times 716 = 1.047 \times 716 = 750 \text{ gpm (2839 L/min)}$$
$$H_c = (1780/1700)^2 \times 73 = (1.047)^2 \times 73 = 1.096 \times 73 = 80 \text{ psi (5.5 bar)}$$

For overload,

$$Q_c = (1780/1700) \times 1075 = 1.047 \times 1075 = 1125 \text{ gpm (4258 L/min)}$$
$$H_c = (1780/1700)^2 \times 47 = (1.047)^2 \times 47 = 1.096 \times 47 = 52 \text{ psi (3.6 bar)}$$

Exhibit II.14.10 shows the plots of these results. By plotting these results on hydraulic graph paper, overall pump performance can be better evaluated.

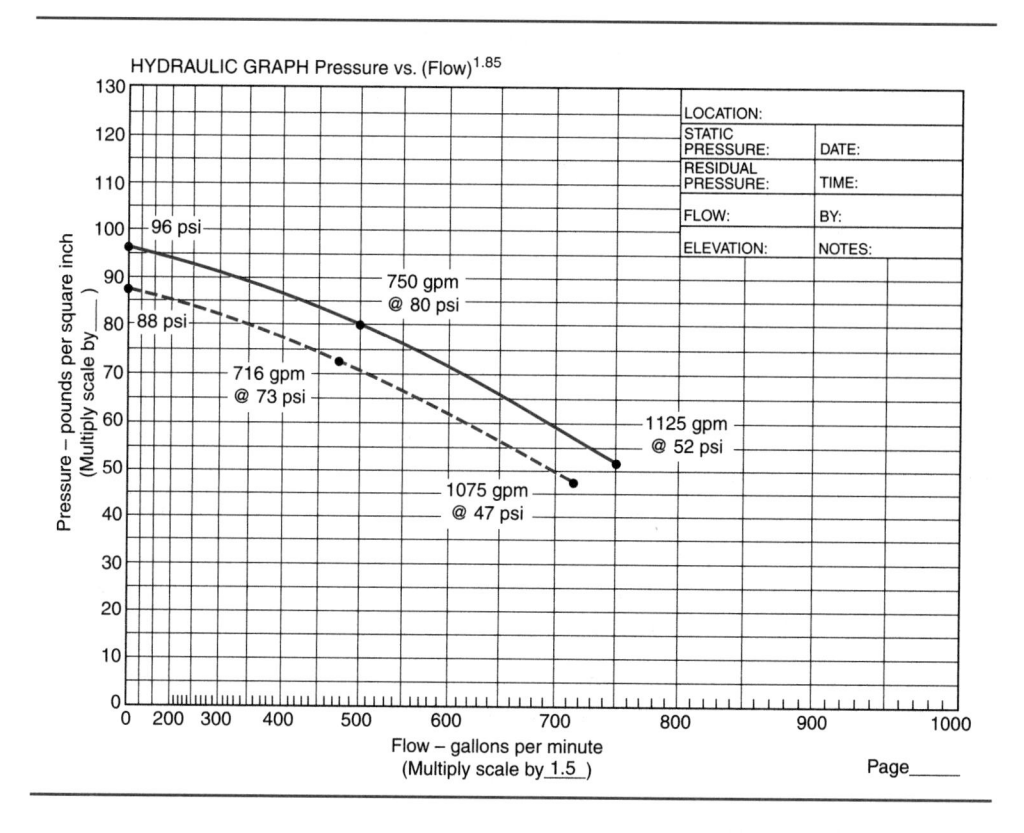

EXHIBIT II.14.10 Plot of Pump Performance Corrected for Its Rated Speed.

As a minimum, the pump as tested (without an rpm adjustment) must deliver the maximum flow and pressure required by any fire protection system it supplies. If the rpm adjusted curve indicates that the fire pump is performing satisfactorily, but the actual performance is below the fire protection requirement, the driver must investigate to determine why it is operating below rated speed.

(g) *Conclusion.* The final step in the test calculation is generally a plot of test points. A head-capacity curve is plotted, and an ampere-capacity curve is plotted. A study of these curves will show the performance picture of the pump as it was tested.

14.2.5.4.1 The quantity of water discharging from the fire pump assembly shall be determined and stabilized.

14.2.5.4.2 Immediately thereafter, the operating conditions of the fire pump and driver shall be measured.

14.2.5.4.3 Positive Displacement Pumps.

14.2.5.4.3.1 The pump flow for positive displacement pumps shall be tested and determined to meet the specified rated performance criteria where only one performance point is required to establish positive displacement pump acceptability.

14.2.5.4.3.2 The pump flow test for positive displacement pumps shall be accomplished using a flowmeter or orifice plate installed in a test loop back to the supply tank, inlet side of a positive displacement water pump, or to drain.

14.2.5.4.3.3 The flowmeter reading or discharge pressure shall be recorded and shall be in accordance with the pump manufacturer's flow performance data.

14.2.5.4.3.4 If orifice plates are used, the orifice size and corresponding discharge pressure to be maintained on the upstream side of the orifice plate shall be made available to the authority having jurisdiction.

14.2.5.4.3.5 Flow rates shall be as specified while operating at the system design pressure. Tests shall be performed in accordance with HI 3.6, *Rotary Pump Tests*.

14.2.5.4.3.6 Positive displacement pumps intended to pump liquids other than water shall be permitted to be tested with water; however, the pump performance will be affected, and manufacturer's calculations shall be provided showing the difference in viscosity between water and the system liquid.

14.2.5.4.4 Electric Motor–Driven Units. For electric motors operating at rated voltage and frequency, the ampere demand on each phase shall not exceed the product of the full-load ampere rating times the allowable service factor as stamped on the motor nameplate.

14.2.5.4.5 For electric motors operating under varying voltage, the product of the actual voltage and current demand on each phase shall not exceed the product of the rated voltage and rated full-load current times the allowable service factor.

14.2.5.4.6 The voltage at the motor shall not vary more than 5 percent below or 10 percent above rated (nameplate) voltage during the test. *(See Section 9.4.)*

When recording the ampere reading, the difference in current between the phases should normally not be more than 10 percent. Regardless of the flow rate during the tests, the average current readings in the three phases should not exceed the nameplate current by more than 10 percent.

14.2.5.4.7 Engine-Driven Units.

14.2.5.4.7.1 When dry charge batteries have been supplied, electrolyte shall be added to the batteries a minimum of 24 hours prior to the time the engine is to be started and the batteries given a conditioning charge.

14.2.5.4.7.2 Engine-driven units shall not show signs of overload or stress.

14.2.5.4.7.3 The governor of such units shall be set at the time of the test to properly regulate the engine speed at rated pump speed. *(See 11.2.4.1.)*

Many engines require special tools to reset speed in the field. While in some cases the fire pump installer may be able to accomplish this action, in other cases the engine manufacturer may need to be present. For this reason, the governor must be set at the time of the test when all appropriate parties are present per the requirements of 14.2.5.

14.2.5.4.7.4 Engines equipped with a variable speed control shall have the variable speed control device nonfunctioning when the governor field adjustment in 11.2.4.1 is set and secured.

14.2.5.4.8 Steam Turbine–Driven Units. The steam turbine shall maintain its speed within the limits specified in 13.2.2.

14.2.5.4.9 Right Angle Gear Drive Units. The gear drive assembly shall operate without excessive objectionable noise, vibration, or heating.

14.2.5.5 Loads Start Test. The fire pump unit shall be started and brought up to rated speed without interruption under the conditions of a discharge equal to peak load.

For centrifugal pumps, peak load (or maximum brake horsepower) usually occurs at approximately 145 percent of rated pump capacity.

14.2.5.6* Phase Reversal Test. For electric motors, a test shall be performed to ensure that there is not a phase reversal condition in either the normal power supply configuration or from the alternate power supply (where provided).

Phase reversal is a continuing problem that has not always been verified during pump acceptance testing when the alternate power supply may not yet be in service. Paragraph 14.2.5.6 requires a suitable test to be conducted for both supplies (when appropriate) even if performed at two separate times.

A.14.2.5.6 A simulated test of the phase reversal device is an acceptable test method.

14.2.6 Controller Acceptance Test for Electric and Diesel Driven Units.

14.2.6.1* Fire pump controllers shall be tested in accordance with the manufacturer's recommended test procedure.

Some electric motor manufacturers suggest that starting and stopping large (over 250 hp) motors in rapid sequence could adversely affect motor life, especially motors installed prior to 1986. For proper cooling, the motor should be run for intervals of at least 10 minutes.

A.14.2.6.1 All controller starts required for tests described in 14.2.5 through 14.2.8 should accrue respectively to this number of tests.

The minimum required number of starts is six. If an alternate power source is provided, then three of these six starts should be from the primary power supply and three from the alternate power source (see 14.2.7). Tests required to satisfy 14.2.5.5, 14.2.5.6, 14.2.6.1, and 14.2.6.9 should be added to the six minimum required starts. Note that satisfying several of these additional tests simultaneously may be possible. For example, once the loads start test required by 14.2.5.5 is satisfied, the pump and motor may be left running in order to test the power transfer and retransfer between the primary and alternate power supplies as required by 14.2.7.

14.2.6.2 As a minimum, no fewer than six automatic and six manual operations shall be performed during the acceptance test.

14.2.6.3 An electric driven fire pump shall be operated for a period of at least 5 minutes at full speed during each of the operations required in 14.2.5.

14.2.6.4 An engine driver shall not be required to run for 5 minutes at full speed between successive starts until the cumulative cranking time of successive starts reaches 45 seconds.

14.2.6.5 The automatic operation sequence of the controller shall start the pump from all provided starting features.

14.2.6.6 This sequence shall include pressure switches or remote starting signals.

14.2.6.7 Tests of engine-driven controllers shall be divided between both sets of batteries.

14.2.6.8 The selection, size, and setting of all overcurrent protective devices, including fire pump controller circuit breaker, shall be confirmed to be in accordance with this standard.

14.2.6.9 The fire pump shall be started once from each power service and run for a minimum of 5 minutes.

> **CAUTION:** Manual emergency operation shall be accomplished by manual actuation of the emergency handle to the fully latched position in a continuous motion. The handle shall be latched for the duration of this test run.

If the emergency handle cannot be placed in the fully latched position in one continuous motion, single phasing and severe arcing can occur.

14.2.7 Alternate Power Supply.

14.2.7.1 On installations with an alternate source of power and an automatic transfer switch, loss of primary source shall be simulated and transfer shall occur while the pump is operating at peak load.

When the primary power source is interrupted while flowing 150 percent of rated pump capacity, the transfer switch and the alternate power source should be achieved within 10 seconds and the peak flow should be successfully redelivered within 20 to 30 seconds.

14.2.7.2 Transfer from normal to alternate source and retransfer from alternate to normal source shall not cause opening of overcurrent protection devices in either line.

14.2.7.3 At least half of the manual and automatic operations of 14.2.6.2 shall be performed with the fire pump connected to the alternate source.

14.2.7.4 If the alternate power source is a generator set required by 9.3.2, installation acceptance shall be in accordance with NFPA 110, *Standard for Emergency and Standby Power Systems.*

14.2.8 Emergency Governor for Steam Driven Units.

14.2.8.1 Emergency governor valve for steam shall be operated to demonstrate satisfactory performance of the assembly.

14.2.8.2 Hand tripping shall be acceptable.

14.2.9 Simulated Conditions. Both local and remote signals and fire pump alarm conditions shall be simulated to demonstrate satisfactory operation.

14.2.10 Test Duration. The fire pump or foam concentrate pump shall be in operation for not less than 1 hour total time during all of the foregoing tests.

14.2.11* Electronic Fuel Management (ECM). For engines with electronic fuel management (ECM) control systems, a function test of both the primary and the alternate ECM shall be conducted.

A.14.2.11 To verify the operation of the alternate ECM, with the motor stopped, move the ECM selector switch to the alternate ECM position. Repositioning of the selector switch should cause a signal on the fire pump controller. Start the engine; it should operate normally with all functions. Shut engine down, switch back to the primary ECM, and restart the engine briefly to verify that correct switchback has been accomplished.

To verify the operation of the redundant sensor, with the engine running, disconnect the wires from the primary sensor. There should be no change in the engine operation. Reconnect the wires to the sensor. Next, disconnect the wires from the redundant sensor. There should be no change in the engine operation. Reconnect the wires to the sensor. Repeat this process for all primary and redundant sensors on the engines. *Note:* If desired, the disconnecting and reconnecting of wires to the sensors can be done while the engine is not running, then starting the engine after each disconnection and reconnection of the wires to verify engine operation.

14.3 Manuals, Special Tools, and Spare Parts.

14.3.1 A minimum of one set of instruction manuals for all major components of the fire pump system shall be supplied by the manufacturer of each major component.

14.3.2 The manual shall contain the following:

(1) A detailed explanation of the operation of the component
(2) Instructions for routine maintenance
(3) Detailed instructions concerning repairs
(4) Parts list and parts identification
(5) Schematic electrical drawings of controller, transfer switch, and fire pump control panels

Ordinarily, the operation and maintenance manual is submitted by the installing contractor for review and approval by the building owner or his or her representative, which is usually the project engineer. Multiple copies need to be maintained on file for future reference. This manual should be made available at the acceptance test for review of systems and components. Prior to final acceptance, the manual should be used for training of operating personnel.

In addition to the required information listed in 14.3.2, the manual should contain a complete list of parts and suppliers including contact information. The manual should also list any recommended spare parts that should be kept on site for immediate use if the need arises.

◀ **FAQ**
What information should the manual contain?

14.3.3 Any special tools and testing devices required for routine maintenance shall be available for inspection by the authority having jurisdiction at the time of the field acceptance test.

14.3.4 Consideration shall be given to stocking spare parts for critical items not readily available.

14.4 Periodic Inspection, Testing, and Maintenance.

Fire pumps shall be inspected, tested, and maintained in accordance with NFPA 25, *Standard for the Inspection, Testing, and Maintenance of Water-Based Fire Protection Systems.*

Starting the pump weekly and running it at full speed as required by NFPA 25 helps to ensure that the pump installation is always ready for service.

When conducting annual flow tests, a reduction in capacity of up to 10 percent over the life of the pump due to normal wear can be expected. More pronounced reduction in capacity may be due to one or more causes as shown in Annex B. A plugged impeller, worn wear rings, and obstructed suction are among the most common causes. In some cases, the reading of the suction gauge indicates whether the trouble is due to an obstruction in the water supply or suction pipe. In other cases, the difference between the suction gauge and the discharge gauge is the indicator of a plugged impeller or overly worn wear rings.

When testing a pump by discharging through a listed meter back to the pump suction, a hydraulic imbalance within the pump is possible. Additionally, this test method is not able to check the suction supply and piping upstream of the connection to the pump suction. If possible, the flowmeter should not be piped to the pump suction. Piping the flow back to the water storage tank (with an air gap) or to the pump house exterior results in accurate measurements.

At least once a year, diesel engine–driven pumps should be checked for overheating by running them with the pump discharging 150 percent or more of its rated capacity. Once the engine temperature has stabilized, the engine should be run at least an additional 15 minutes. If the engine overheats, a blockage of the cooling system is probably the cause, such as inadequate cooling water flow caused by an obstruction in the cooling water system, plugged strainer, or a partly closed valve.

In addition to the requirements of NFPA 25 on engine-driven pumps, the engine should be kept clean and dry. The fuel tank should be kept at least at a level capable of running the engine for 8 hours at peak load. For instance, if tanks are not refilled and they drop to the two-thirds level, the tank size should be for a 12-hour supply when full. The crankcase oil should be checked to see that it is at the proper level, that it has not become fouled, and that it has not lost its viscosity. Additionally, the strainers in the cooling water system should be cleaned, and the specific gravity of the battery electrolyte should be checked monthly.

Even though NFPA 25 specifies the records that should be kept, it is also beneficial to maintain a specific record of the temperature and tightness of the glands, the readings of the suction and discharge gauges, and the condition of the suction supply.

14.5 Component Replacement.

14.5.1 Positive Displacement Pumps.

14.5.1.1 Whenever a critical path component in a positive displacement fire pump is replaced, as defined in 14.5.2.5, a field test of the pump shall be performed.

14.5.1.2 If components that do not affect performance are replaced, such as shafts, then only a functional test shall be required to ensure proper installation and reassembly.

14.5.1.3 If components that affect performance are replaced, such as rotors, plungers, and so forth, then a retest shall be conducted by the pump manufacturer or designated representative or qualified persons acceptable to the authority having jurisdiction.

14.5.1.4 Field Retest Results.

14.5.1.4.1 The field retest results shall be compared to the original pump performance as indicated by the original factory-certified test curve, whenever it is available.

14.5.1.4.2 The field retest results shall meet or exceed the performance characteristics as indicated on the pump nameplate, and the results shall be within the accuracy limits of field testing as stated elsewhere in this standard.

14.5.2 Centrifugal Pumps.

The replacement of components in a listed fire pump, fire pump controller, or driver should be performed by the manufacturer's factory-trained representative. All replacement parts should be supplied by the original equipment manufacturer and should be listed for fire pump service. When the replacement parts are no longer available, the equipment should be replaced with new listed and approved components.

The replacement of any of the components listed in 14.5.2.4 requires a complete performance retest, as defined in Table 14.5.2.4, by the manufacturer's representative or other qualified person. The field retest should equal or exceed the original factory-certified test curve, controller starting and trip curves, and material specifications. The tests should be within the accuracy limits established in 14.5.2.7.

The major components of the fire pump system and fire pump room should be returned to service as quickly as possible when system impairment occurs. This work should be performed by a factory-trained and qualified technician who has established competency through training.

14.5.2.1 Whenever a critical path component in a piece of centrifugal pump equipment is replaced, changed, or modified, a field/on-site retest shall be performed.

14.5.2.2 The replacement of components in fire pumps, fire pump controllers, and drivers shall be performed by factory-authorized representatives or qualified persons acceptable to the authority having jurisdiction.

14.5.2.3* When an ECM on an electronic fuel management–controlled engine is replaced, the replacement ECM shall include the same software programming that was in the original ECM.

TABLE 14.5.2.4 Summary of Component Replacement Testing Requirements

	Component	Adjust	Repair	Rebuild	Replace	Test Criteria
A. Fire Pump Systems						
1	Entire pump assembly				X	Perform acceptance test in accordance with NFPA 20, 14.5.2.7.2, 14.5.1.4
2	Impeller/rotating assembly		X		X	Perform acceptance test in accordance with NFPA 20, 14.5.2.7.2, 14.5.1.4
3	Casing		X		X	Perform acceptance test in accordance with NFPA 20, 14.5.2.7.2, 14.5.1.4
4	Bearings				X	Perform annual test in accordance with NFPA 25, 8.3.3 with alignment
5	Sleeves				X	Perform annual test in accordance with NFPA 25, 8.3.3 with alignment
6	Wear rings				X	Perform annual test in accordance with NFPA 25, 8.3.3 with alignment
7	Main shaft		X		X	Perform annual test in accordance with NFPA 25, 8.3.3 with alignment
8	Packing	X			X	Perform weekly test in accordance with NFPA 25, 8.3.2
B. Mechanical Transmission						
1	Gear right angle drives		X	X	X	Perform acceptance test in accordance with NFPA 20, 14.5.2.7.2
2	Drive coupling	X	X	X	X	Perform annual test in accordance with NFPA 20, 14.2.5
C. Electrical System/Controller						
1	Entire controller		X	X	X	Perform acceptance test with NFPA 20, 14.2.6
2	Isolating switch				X	Perform acceptance test with NFPA 25, 8.3.2 and exercise six times
3	Circuit breaker	X				Perform six momentary starts in accordance with NFPA 20, 14.2.6.9
4	Circuit breaker				X	Perform a 1 hour full load current test
5	Electrical connections	X				Perform weekly test in accordance with NFPA 25, 8.3.2
6	Main contactor			X		Perform weekly test in accordance with NFPA 25, 8.3.2
7	Main contactor				X	Perform acceptance test in accordance with NFPA 20, 14.2.6
8	Power monitor				X	Perform weekly test in accordance with NFPA 25, 8.3.2
9	Start relay				X	Perform weekly test in accordance with NFPA 25, 8.3.2
10	Pressure switch	X			X	Perform acceptance test in accordance with NFPA 20, 14.2.6.9
11	Pressure transducer	X			X	Perform acceptance test in accordance with NFPA 20, 14.2.6.9
12	Manual start or stop switch				X	Perform six operations under load with NFPA 25, 8.3.2
13	Transfer switch—load carrying parts		X	X	X	Perform acceptance test and transfer from normal power to emergency power and back one time with NFPA 20, 14.5.2
14	Transfer switch—no load parts		X	X	X	Perform six no load operations of transfer of power with NFPA 20, 14.5.2.5

(continued)

TABLE 14.5.2.4 (continued)

	Component	Adjust	Repair	Rebuild	Replace	Test Criteria
D. Electric Motor Driver						
1	Electric motor		X	X	X	Perform acceptance test in accordance with NFPA 20, 14.5.2.7.2, 14.5.2.6
2	Motor bearings				X	Perform annual test in accordance with NFPA 25, 8.3.3 with alignment
3	Incoming power conductors/disconnects				X	Perform acceptance test NFPA 20, 14.5.2
E. Diesel Engine Driver						
1	Entire engine			X	X	Perform annual test in accordance with NFPA 25, 8.3.3, NFPA 20, 14.5.2.4.1
2	Fuel transfer pump	X		X	X	Perform weekly test in accordance with NFPA 25, 8.3.2
3	Fuel injector pump or ECM	X			X	Perform annual test in accordance with NFPA 25, 8.3.3
4	Fuel system filter		X		X	Perform weekly test in accordance with NFPA 25, 8.3.2
5	Combustion air intake system		X		X	Perform weekly test in accordance with NFPA 25, 8.3.2
6	Fuel tank		X		X	Perform weekly test in accordance with NFPA 25, 8.3.2
7	Cooling system		X	X	X	Perform weekly test in accordance with NFPA 25, 8.3.2
8	Batteries				X	Perform weekly test in accordance with NFPA 25, 8.3.2
9	Battery charger		X		X	Perform a start/stop sequence with NFPA 25, 8.3.2
10	Electrical system		X		X	Perform weekly test in accordance with NFPA 25, 8.3.2
11	Lubrication filter/oil service		X		X	Perform weekly test in accordance with NFPA 25, 8.3.2
F. Steam Turbines						
1	Steam turbine		X		X	Perform annual test in accordance with NFPA 25, 8.3.3
2	Steam regulator or source upgrade		X		X	Perform annual test in accordance with NFPA 20, 14.5.2.7.2
G. Positive Displacement Pumps						
1	Entire pump				X	Perform annual test in accordance with NFPA 20, 14.5.2.7.2
2	Rotors				X	Perform annual test in accordance with NFPA 25, 8.3.3
3	Plungers				X	Perform annual test in accordance with NFPA 25, 8.3.3
4	Shaft				X	Perform annual test in accordance with NFPA 25, 8.3.3
5	Driver		X	X	X	Perform annual test in accordance with NFPA 25, 8.3.3
6	Bearings				X	Perform weekly test in accordance with NFPA 25, 8.3.2
7	Seals				X	Perform weekly test in accordance with NFPA 25, 8.3.2
H. Pump House/Room and Misc Components						
1	Base plate		X		X	Perform weekly test in accordance with NFPA 25, 8.3.2 with alignment check
2	Foundation		X	X	X	Perform weekly test in accordance with NFPA 25, 8.3.2 with alignment check
3	Suction/discharge pipe		X		X	Hydrostatic test in accordance with NFPA 13, 24.2.1
4	Suction/discharge fittings		X		X	Hydrostatic test in accordance with NFPA 13, 24.2.1
5	Suction/discharge valves		X	X	X	Hydrostatic test in accordance with NFPA 13, 24.2.1

A.14.5.2.3 Fire pump engines have unique features compared to standard industrial engines. The standard industrial ECM programming can result in the reduction of power to self-protect the engine during a fire or the inability to accelerate the pump to rated speed in rated flow condition.

14.5.2.4 Replacement Parts. Table 14.5.2.4 shall be used for component replacement testing requirements.

14.5.2.4.1 Replacement parts shall be provided that will maintain the listing for the fire pump component whenever possible.

14.5.2.4.2 If it is not possible to maintain the listing of a component or if the component was not originally listed for fire protection use, the replacement parts shall meet or exceed the quality of the parts being replaced.

14.5.2.5 Critical path components include the following features of the pump equipment:

(1) Fire pumps
    (a) Impeller, casing
    (b) Gear drives
(2) Fire pump controllers (electric or diesel): total replacement
(3) Electric motor, steam turbines, or diesel engine drivers
    (a) Electric motor replacement
    (b) Steam turbine replacement or rebuild
    (c) Steam regulator or source upgrade
    (d) Engine replacement or engine rebuild

14.5.2.6 Whenever replacement, change, or modification to a critical path component is performed on a fire pump, driver, or controller, as described in 14.5.2.5, a retest shall be conducted as indicated in Table 14.5.2.4 by the pump manufacturer, factory-authorized representative, or qualified persons acceptable to the authority having jurisdiction.

14.5.2.7 Field Retests.

14.5.2.7.1 The field retest results shall be compared to the original pump performance as indicated by the original factory-certified test curve, whenever it is available.

14.5.2.7.2 The field retest results shall meet or exceed the performance characteristics as indicated on the pump nameplate, and they shall be within the accuracy limits of field testing as stated elsewhere in this standard.

REFERENCES CITED IN COMMENTARY

National Fire Protection Association, 1 Batterymarch Park, Quincy, MA 02169-7471.

Hague, D. R., ed., *Water-Based Fire Protection Systems Handbook,* 2008 edition.
NFPA 13, *Standard for the Installation of Sprinkler Systems,* 2009 edition.
NFPA 14, *Standard for the Installation of Standpipe and Hose Systems,* 2007 edition.
NFPA 22, *Standard for Water Tanks for Private Fire Protection,* 2008 edition.
NFPA 24, *Standard for Installation of Private Fire Service Mains and Their Appurtenances,* 2010 edition.
NFPA 25, *Standard for the Inspection, Testing, and Maintenance of Water-Based Fire Protection Systems,* 2008 edition.
NFPA 291, *Recommended Practice for Fire Flow Testing and Marking of Hydrants,* 2010 edition.

Explanatory Material

ANNEX A

The material contained in Annex A of NFPA 20 is included within the text of this handbook and, therefore, is not repeated here.

Possible Causes of Pump Troubles

ANNEX B

This annex is not a part of the requirements of this NFPA document but is included for informational purposes only.

B.1 Causes of Pump Troubles.

This annex contains a partial guide for locating pump troubles and their possible causes *(see Figure B.1)*. It also contains a partial list of suggested remedies. *(For other information on this subject, see Hydraulic Institute Standards for Centrifugal, Rotary and Reciprocating Pumps.)*

The causes listed here are in addition to possible mechanical breakage that would be obvious on visual inspection. In case of trouble, it is suggested that those troubles that can be checked easily should be corrected first or eliminated as possibilities.

B.1.1 Air Drawn into Suction Connection Through Leak(s). Air drawn into suction line through leaks causes a pump to lose suction or fail to maintain its discharge pressure. Uncover suction pipe and locate and repair leak(s).

B.1.2 Suction Connection Obstructed. Examine suction intake, screen, and suction pipe and remove obstruction. Repair or provide screens to prevent recurrence. *(See 4.14.8.)*

B.1.3 Air Pocket in Suction Pipe. Air pockets cause a reduction in delivery and pressure similar to an obstructed pipe. Uncover suction pipe and rearrange to eliminate pocket. *(See 4.14.6.)*

B.1.4 Well Collapsed or Serious Misalignment. Consult a reliable well drilling company and the pump manufacturer regarding recommended repairs.

B.1.5 Stuffing Box Too Tight or Packing Improperly Installed, Worn, Defective, Too Tight, or of Incorrect Type. Loosen gland swing bolts and remove stuffing box gland halves. Replace packing.

B.1.6 Water Seal or Pipe to Seal Obstructed. Loosen gland swing bolt and remove stuffing box gland halves along with the water seal ring and packing. Clean the water passage to and in the water seal ring. Replace water seal ring, packing gland, and packing in accordance with manufacturer's instructions.

B.1.7 Air Leak into Pump Through Stuffing Boxes. Same as the possible cause in B.1.6.

B.1.8 Impeller Obstructed. Does not show on any one instrument, but pressures fall off rapidly when an attempt is made to draw a large amount of water.

For horizontal split-case pumps, remove upper case of pump and remove obstruction from impeller. Repair or provide screens on suction intake to prevent recurrence.

For vertical shaft turbine–type pumps, lift out column pipe *(see Figure A.7.2.2.1 and Figure A.7.2.2.2)* and pump bowls from wet pit or well and disassemble pump bowl to remove obstruction from impeller.

Fire pump Troubles	Suction				Pump																Driver and/or Pump					Driver						
	Air drawn into suction connection through leak(s)	Suction connection obstructed	Air pocket in suction pipe	Well collapsed or serious misalignment	Stuffing box too tight or packing improperly installed, worn, defective, too tight, or incorrect type	Water seal or pipe to seal obstructed	Air leak into pump through stuffing boxes	Impeller obstructed	Wearing rings worn	Impeller damaged	Wrong diameter impeller	Actual net head lower than rated	Casing gasket defective, permitting internal leakage (single-stage and multistage pumps)	Pressure gauge is on top of pump casing	Incorrect impeller adjustment (vertical shaft turbine-type pump only)	Impellers locked	Pump is frozen	Pump shaft or shaft sleeve scored, bent, or worn	Pump not primed	Seal ring improperly located in stuffing box, preventing water from entering space to form seal	Excess bearing friction due to lack of lubrication, wear, dirt, rusting, failure, or improper installation	Rotating element binds against stationary element	Pump and driver misaligned	Foundation not rigid	Engine-cooling system obstructed	Faulty driver	Lack of lubrication	Speed too low	Wrong direction of rotation	Speed too high	Rated motor voltage different from line voltage	Faulty electric circuit, obstructed fuel system, obstructed steam pipe, or dead battery
	1	2	3	4	5	6	7	8	9	10	11	12	13	14	15	16	17	18	19	20	21	22	23	24	25	26	27	28	29	30	31	32
Excessive leakage at stuffing box					X													X					X									
Pump or driver overheats				X	X	X		X			X							X	X	X	X	X	X	X	X		X		X	X	X	
Pump unit will not start				X	X										X	X	X				X					X	X					X
No water discharge	X	X	X					X									X															
Pump is noisy or vibrates				X	X			X		X								X			X	X	X	X		X						
Too much power required				X	X			X	X		X		X		X			X			X	X	X	X		X			X	X	X	
Discharge pressure not constant for same gpm	X				X	X	X																									
Pump loses suction after starting	X	X	X		X	X														X												
Insufficient water discharge	X	X	X		X	X	X	X	X	X	X	X			X													X	X		X	
Discharge pressure too low for gpm discharge	X	X	X		X	X	X	X	X	X	X	X																X	X		X	

FIGURE B.1 *Possible Causes of Fire Pump Troubles.*

For close-coupled, vertical in-line pumps, lift motor on top pull-out design and remove obstruction from impeller.

B.1.9 Wearing Rings Worn. Remove upper case and insert feeler gauge between case wearing ring and impeller wearing ring. Clearance when new is 0.0075 in. (0.19 mm). Clearances of more than 0.015 in. (0.38 mm) are excessive.

B.1.10 Impeller Damaged. Make minor repairs or return to manufacturer for replacement. If defect is not too serious, order new impeller and use damaged one until replacement arrives.

B.1.11 Wrong Diameter Impeller. Replace with impeller of proper diameter.

B.1.12 Actual Net Head Lower than Rated. Check impeller diameter and number and pump model number to make sure correct head curve is being used.

B.1.13 Casing Gasket Defective, Permitting Internal Leakage (Single-Stage and Multistage Pumps). Replace defective gasket. Check manufacturer's drawing to see whether gasket is required.

B.1.14 Pressure Gauge Is on Top of Pump Casing. Place gauges in correct location. *[See Figure A.6.3.1(a).]*

B.1.15 Incorrect Impeller Adjustment (Vertical Shaft Turbine–Type Pump Only). Adjust impellers according to manufacturer's instructions.

B.1.16 Impellers Locked. For vertical shaft turbine–type pumps, raise and lower impellers by the top shaft adjusting nut. If this adjustment is not successful, follow the manufacturer's instructions.

For horizontal split-case pumps, remove upper case and locate and eliminate obstruction.

B.1.17 Pump Is Frozen. Provide heat in the pump room. Disassemble pump and remove ice as necessary. Examine parts carefully for damage.

B.1.18 Pump Shaft or Shaft Sleeve Scored, Bent, or Worn. Replace shaft or shaft sleeve.

B.1.19 Pump Not Primed. If a pump is operated without water in its casing, the wearing rings are likely to seize. The first warning is a change in pitch of the sound of the driver. Shut down the pump.

For vertical shaft turbine–type pumps, check water level to determine whether pump bowls have proper submergence.

B.1.20 Seal Ring Improperly Located in Stuffing Box, Preventing Water from Entering Space to Form Seal. Loosen gland swing bolt and remove stuffing box gland halves along with the water-seal ring and packing. Replace, putting seal ring in proper location.

B.1.21 Excess Bearing Friction Due to Lack of Lubrication, Wear, Dirt, Rusting, Failure, or Improper Installation. Remove bearings and clean, lubricate, or replace as necessary.

B.1.22 Rotating Element Binding Against Stationary Element. Check clearances and lubrication and replace or repair the defective part.

B.1.23 Pump and Driver Misaligned. Shaft running off center because of worn bearings or misalignment. Align pump and driver according to manufacturer's instructions. Replace bearings according to manufacturer's instructions.*(See Section 6.5.)*

B.1.24 Foundation Not Rigid. Tighten foundation bolts or replace foundation if necessary. *(See Section 6.4.)*

B.1.25 Engine Cooling System Obstructed. Heat exchanger or cooling water systems too small or cooling pump faulty. Remove thermostats. Open bypass around regulator valve and strainer. Check regulator valve operation. Check strainer. Clean and repair if necessary. Disconnect sections of cooling system to locate and remove possible obstruction. Adjust engine cooling water circulating pump belt to obtain proper speed without binding. Lubricate bearings of this pump.

If overheating still occurs at loads up to 150 percent of rated capacity, contact pump or engine manufacturer so that necessary steps can be taken to eliminate overheating.

B.1.26 Faulty Driver. Check electric motor, internal combustion engine, or steam turbine, in accordance with manufacturer's instructions, to locate reason for failure to start.

B.1.27 Lack of Lubrication. If parts have seized, replace damaged parts and provide proper lubrication. If not, stop pump and provide proper lubrication.

B.1.28 Speed Too Low. For electric motor drive, check that rated motor speed corresponds to rated speed of pump, voltage is correct, and starting equipment is operating properly.

Low frequency and low voltage in the electric power supply prevent a motor from running at rated speed. Low voltage can be due to excessive loads and inadequate feeder capacity or (with private generating plants) low generator voltage. The generator voltage of private generating plants can be corrected by changing the field excitation. When low voltage is from the other causes mentioned, it might be necessary to change transformer taps or increase feeder capacity.

Low frequency usually occurs with a private generating plant and should be corrected at the source. Low speed can result in older type squirrel-cage-type motors if fastenings of copper bars to end rings become loose. The remedy is to weld or braze these joints.

For steam turbine drive, check that valves in steam supply pipe are wide open; boiler steam pressure is adequate; steam pressure is adequate at the turbine; strainer in the steam supply pipe is not plugged; steam supply pipe is of adequate size; condensate is removed from steam supply pipe, trap, and turbine; turbine nozzles are not plugged; and setting of speed and emergency governor is correct.

For internal combustion engine drive, check that setting of speed governor is correct; hand throttle is opened wide; and there are no mechanical defects such as sticking valves, timing off, or spark plugs fouled, and so forth. These problems might require the services of a trained mechanic.

B.1.29 Wrong Direction of Rotation. Instances of an impeller turning backward are rare but are clearly recognizable by the extreme deficiency of pump delivery. Wrong direction of rotation can be determined by comparing the direction in which the flexible coupling is turning with the directional arrow on the pump casing.

With a polyphase electric motor drive, two wires must be reversed; with a dc driver, the armature connections must be reversed with respect to the field connections. Where two sources of electrical current are available, the direction of rotation produced by each should be checked.

B.1.30 Speed Too High. See that pump- and driver-rated speed correspond. Replace electric motor with one of correct rated speed. Set governors of drivers for correct speed. Frequency at private generating stations might be too high.

B.1.31 Rated Motor Voltage Different from Line Voltage. For example, a 220 V or 440 V motor on 208 V or 416 V line. Obtain motor of correct rated voltage or larger size motor. *(See Section 9.4.)*

B.1.32 Faulty Electric Circuit, Obstructed Fuel System, Obstructed Steam Pipe, or Dead Battery. Check for break in wiring open switch, open circuit breaker, or dead battery. If circuit breaker in controller trips for no apparent reason, make sure oil is in dash pots in accordance with manufacturer's specifications. Make sure fuel pipe is clear, strainers are clean, and control valves are open in fuel system to internal combustion engine. Make sure all valves are open and strainer is clean in steam line to turbine.

B.2 Warning.

Chapters 9 and 10 include electrical requirements that discourage the installation of disconnect means in the power supply to electric motor–driven fire pumps. This requirement is intended to ensure the availability of power to the fire pumps. When equipment connected to those circuits is serviced or maintained, the employee can have unusual exposure to electrical and other hazards. It can be necessary to require special safe work practices and special safeguards, personal protective clothing, or both.

B.3 Maintenance of Fire Pump Controllers After a Fault Condition.

B.3.1 Introduction. In a fire pump motor circuit that has been properly installed, coordinated, and in service prior to the fault, tripping of the circuit breaker or the isolating switch indicates a fault condition in excess of operating overload.

It is recommended that the following general procedures be observed by qualified personnel in the inspection and repair of the controller involved in the fault. These procedures are not intended to cover other elements of the circuit, such as wiring and motor, which can also require attention.

B.3.2 Caution. All inspections and tests are to be made on controllers that are de-energized at the line terminal, disconnected, locked out, and tagged so that accidental contact cannot be made with live parts and so that all plant safety procedures will be observed.

B.3.2.1 Enclosure. Where substantial damage to the enclosure, such as deformation, displacement of parts, or burning has occurred, replace the entire controller.

B.3.2.2 Circuit Breaker and Isolating Switch. Examine the enclosure interior, circuit breaker, and isolating switch for evidence of possible damage. If evidence of damage is not apparent, the circuit breaker and isolating switch can continue to be used after closing the door.

If there is any indication that the circuit breaker has opened several short-circuit faults, or if signs of possible deterioration appear within either the enclosure, circuit breaker, or isolating switch (e.g., deposits on surface, surface discoloration, insulation cracking, or unusual toggle operation), replace the components. Verify that the external operating handle is capable of opening and closing the circuit breaker and isolating switch. If the handle fails to operate the device, this would also indicate the need for adjustment or replacement.

B.3.2.3 Terminals and Internal Conductors. Where there are indications of arcing damage, overheating, or both, such as discoloration and melting of insulation, replace the damaged parts.

B.3.2.4 Contactor. Replace contacts showing heat damage, displacement of metal, or loss of adequate wear allowance of the contacts. Replace the contact springs where applicable. If deterioration extends beyond the contacts, such as binding in the guides or evidence of insulation damage, replace the damaged parts or the entire contactor.

B.3.2.5 Return to Service. Before returning the controller to service, check for the tightness of electrical connections and for the absence of short circuits, ground faults, and leakage current.

Close and secure the enclosure before the controller circuit breaker and isolating switch are energized. Follow operating procedures on the controller to bring it into standby condition.

Informational References

ANNEX C

C.1 Referenced Publications.

The documents or portions thereof listed in this annex are referenced within the informational sections of this standard and are not part of the requirements of this document unless also listed in Chapter 2 for other reasons.

C.1.1 NFPA Publications. National Fire Protection Association, 1 Batterymarch Park, Quincy, MA 02169-7471.

NFPA 13, *Standard for the Installation of Sprinkler Systems,* 2010 edition.
NFPA 14, *Standard for the Installation of Standpipe and Hose Systems,* 2007 edition.
NFPA 15, *Standard for Water Spray Fixed Systems for Fire Protection,* 2007 edition.
NFPA 16, *Standard for the Installation of Foam-Water Sprinkler and Foam-Water Spray Systems,* 2007 edition.
NFPA 24, *Standard for the Installation of Private Fire Service Mains and Their Appurtenances,* 2010 edition.
NFPA 25, *Standard for the Inspection, Testing, and Maintenance of Water-Based Fire Protection Systems,* 2008 edition.
NFPA 30, *Flammable and Combustible Liquids Code,* 2008 edition.
NFPA 31, *Standard for the Installation of Oil-Burning Equipment,* 2006 edition.
NFPA 37, *Standard for the Installation and Use of Stationary Combustion Engines and Gas Turbines,* 2006 edition.
NFPA 70®, National Electrical Code®, 2008 edition.

C.1.2 Other Publications.

C.1.2.1 ANSI Publications. American National Standards Institute, Inc., 25 West 43rd Street, 4th Floor, New York, NY 10036.

ANSI/IEEE C62.11, *IEEE Standard for Metal-Oxide Surge Arresters for AC Power Circuits,* 1987.

C.1.2.2 AWWA Publications. American Water Works Association, 6666 West Quincy Avenue, Denver, CO 80235.

AWWA C104, *Cement-Mortar Lining for Cast-Iron and Ductile-Iron Pipe and Fittings for Water,* 1990.

C.1.2.3 HI Publications. Hydraulic Institute, 6 Campus Drive, First Floor North, Parsippany, NJ 07054-4406.

Hydraulic Institute Standards for Centrifugal, Rotary and Reciprocating Pumps, 14th ed., 1983.
HI 3.5, *Standard for Rotary Pumps for Nomenclature, Design, Application and Operation,* 1994.
HI 3.6, *Rotary Pump Tests,* 1994.

C.1.2.4 IEEE Publications. Institute of Electrical and Electronics Engineers, Three Park Avenue, 17th Floor, New York, NY 10016-5997.

IEEE 141, *Electric Power Distribution for Industrial Plants,* 1986.
IEEE 241, *Electric Systems for Commercial Buildings,* 1990.

C.1.2.5 NEMA Publications. National Electrical Manufacturers Association, 1300 North 17th Street, Suite 1847, Rosslyn, VA 22209.

NEMA ICS 2.2, *Maintenance of Motor Controllers After a Fault Condition,* 1983.
NEMA ICS 14, *Application Guide for Electric Pump Controllers,* 2001.
NEMA 250, *Enclosures for Electrical Equipment,* 1991.

C.1.2.6 SAE Publications. Society of Automotive Engineers, 400 Commonwealth Drive, Warrendale, PA 15096.

SAE J-1349, *Engine Power Test Code—Spark Ignition and Compression Engine,* 1990.

C.1.2.7 UL Publications. Underwriters Laboratories Inc., 333 Pfingsten Road, Northbrook, IL 60062-2096.

ANSI/UL 508, *Standard for Safety Industrial Control Equipment,* 2005.
ANSI/UL 1008, *Standard for Safety Automatic Transfer Switches,* 2007.

C.2 Informational References.

The following documents or portions thereof are listed here as informational resources only. They are not a part of the requirements of this document.

C.2.1 UL Publications. Underwriters Laboratories Inc., 333 Pfingsten Road, Northbrook, IL 60062-2096.

ANSI/UL 80, *Standard for Steel Tanks for Oil Burner Fuels and Other Combustible Liquids,* 2007.
UL 2080, *Standard for Fire Resistant Tanks for Flammable and Combustible Liquids,* 2000.
ANSI/UL 2085, *Standard for Protected Aboveground Tanks for Flammable and Combustible Liquids,* 1997.

C.3 References for Extracts in Informational Sections. (Reserved)

Material Extracted by NFPA 70, Article 695

ANNEX D

D.1 General.

Table D.1 indicates corresponding sections of *NFPA 70*, Article 695.

TABLE D.1 NFPA 70, National Electrical Code, Extracted Material

NFPA 20 2007 Edition	NFPA 20 2010 Edition	NFPA 70 Section 695 2005 Edition	NFPA 70 Section 695 2008 Edition	Section 695 Titles or Text
(Reference only)		695.2	Same	Definitions
3.3.7.2	Same	695.2	Same	Fault Tolerant External Control Circuits
3.3.34	Same	695.2	Same	On-Site Power Production Facility
3.3.35	Same	695.2	Same	On-Site Standby Generator
9.2.1	Same	695.3	Same	Power Source(s) for Electric Motor-Driven Fire Pumps
9.2	Same	695.3(A)	Same	Individual Sources
9.2.2	Same	695.3(A)(1)	Same	Electric Utility Service Connection (2nd sentence)
9.2.2	Same	695.3(A)(2)	Same	On-Site Power Production Facility
9.2.2	Same	695.3(B)	Same	Multiple Sources
9.6.1	Same	695.3(B)(1)	Same	Generator Capacity
9.3.4(1)	Same	695.3(B)(2)	Same	Feeder Sources
9.3.2	Same	695.3(B)(3)	Same	Arrangement
9.2.1	Same	695.4	Same	Continuity of Power
9.2.2	Same	695.4(A)	Same	Direct Connection
9.2	Same	695.4(B)	Same	Supervised Connection
9.2	Same	695.4(B)-(1)	Same	(listed fire pump controller)
9.2	Same	695.4(B)-(2)	Same	(listed fire pump power transfer switch)
9.2	Same	695.4(B)-(3)	Same	(listed combo. fire pump controller / transfer switch)
9.2(5)	9.2.2(5)	695.4(B)— continued	Same	For systems installed under the provisions of 695.3(B)(2) only
9.2.3.1	Same	695.4(B)— continued	Same	All disconnecting devices and overcurrent protective devices
9.2.3.4	Same	695.4(B)(1)	Same	Overcurrent Device Selection
9.2.3.2	Same	695.4(B)(2)	Same	Disconnecting Means
9.2.3.1	Same	695.4(B)(2)-(1)	Same	Be identified as suitable for use as service equipment
9.2.3.1(2)	Same	695.4(B)(2)-(2)	Same	Be lockable in the closed position
(Reference only)	9.2.3.1(3)	695.4(B)(2)-(3)	Same	Not be located within equipment that feeds loads other than
9.2.3.1(4)	Same	695.4(B)(2)-(4)	Same	Be located sufficiently remote from
9.2.3.1(5)	Same	695.4(B)(3)	Same	Disconnect Marking
10.1.2.2	Same	695.4(B)(4)	Same	Controller Marking
9.2.3.3	Same	695.4(B)(5)	Same	Supervision

(continued)

355

TABLE D.1 *Continued*

NFPA 20		NFPA 70 Section 695		
2007 Edition	2010 Edition	2005 Edition	2008 Edition	Section 695 Titles or Text
9.2.3.3(1)	Same	695.4(B)(5)-(1)	Same	Central station, proprietary, or remote station
9.2.3.3(2)	Same	695.4(B)(5)-(2)	Same	Local signaling service
9.2.3.3(3)	Same	695.4(B)(5)-(3)	Same	Locking the disconnecting means
9.2.3.3(4)	Same	695.4(B)(5)-(4)	Same	Sealing of disconnecting means
9.2.2(5)	Same	695.5	Same	Transformers
9.2.3.4	Same	695.5(B)	Same	Overcurrent Protection
(Reference only)		695.5(C)	Same	Feeder Source
(Reference only)		695.5(C)(1)	Same	Size
9.2.3	Same	695.5(C)(2)	Same	Overcurrent Protection
(Reference only)		695.6	Same	Power Wiring
10.3.4.7	10.3.4.5.1	695.6(F)	Same	Junction Points
10.3.4.8	10.3.4.6	695.6(F)	Same	Junction Points
9.4	Same	695.7	Same	Voltage Drop
9.5.1.1	Same	695.10	Same	Listed Equipment
10.1.2.1	Same	695.10	Same	Listed Equipment
10.8.3.1	10.7.3.1	695.10	Same	Listed Equipment
12.1.3.1	Same	695.10	Same	Listed Equipment
(Reference only)		695.12	Same	Equipment Location
10.2.1	Same	695.12(A)	Same	Controllers and Transfer Switches
12.2.1	Same	695.12(B)	Same	Engine-Drive Controllers
11.2.5.2.6	11.2.7.2.4	695.12(C)	Same	Storage Batteries
11.2.5.2.7	11.2.7.2.4.2	695.12(D)	Same	Energized Equipment
10.2.2	Same	695.12(E)	Same	Protection Against Pump Water
12.2.2	Same	695.12(E)	Same	Protection Against Pump Water
10.3.2	Same	695.12(F)	Same	Mounting
12.3.2	Same	695.12(F)	Same	Mounting
10.5.2.6	Same	695.14(A)	Same	Control Circuit Failures
12.5.2.5	12.7.2.5	695.14(A)	Same	Control Circuit Failures
10.4.5.7	Same	695.14(B)	Same	Sensor Functioning
10.8.1.3	10.7.1.3	695.14(C)	Same	Remote Device(s)
12.3.5.1	Same	695.14(D)	Same	Engine-Drive Control Wiring
(Reference only)		695.14(E)	Same	Electric Fire Pump Control Wiring Methods

Additional References—Informational Only

A.9.3.2(3)	Same	695.6(A)	Same	Service Conductors
9.6.4	Same	695.12(A)	Same	Controllers and Transfer Switches
A.9.3.2(3)	Same	695.14(F)	Same	Generator Control Wiring Methods

PART III

Private Water Supplies, Hydrants, Tanks, and Piping

Part III of this handbook contains a collection of code requirements, annexes, and associated commentary from several water supply documents, organized into four sections. Section 1 consists of selected extracts from NFPA 22, *Standard for Water Tanks for Private Fire Protection,* 2008 edition. Section 2 contains NFPA 24, *Standard for the Installation of Private Fire Service Mains and Their Appurtenances,* 2010 edition, in its entirety. Section 3 consists of Section 18.4 and Annex I from NFPA 1, *Fire Code,* 2009 edition. Section 4 contains NFPA 291, *Recommended Practice for Fire Flow Testing and Marking of Hydrants,* 2010 edition, in its entirety.

All of these documents contain requirements and recommendations on the various aspects of fire pumps. All four documents and the associated commentary are invaluable to the system designer or registered design professional in designing an adequate water supply system to meet the flow and pressure demands of the fire protection system installed in a building. For the reader's convenience, marginal icons indicate material of special interest with respect to design and calculations.

Extracts from NFPA® 22,

Standard for Water Tanks for Private Fire Protection, 2008 Edition

SECTION 1

Requirements from NFPA 22, *Standard for Water Tanks for Private Fire Protection,* are extracted in this section since water tanks are frequently used to supply water to fire pumps. Only requirements that directly relate to fire pumps are included. Omitted code paragraphs are indicated by ellipses. The user should note that this section of the handbook is not intended to be a design guide for water tanks.

CHAPTER 1 INTRODUCTION

1.1 Scope.

This standard provides the minimum requirements for the design, construction, installation, and maintenance of tanks and accessory equipment that supply water for private fire protection, including the following:

(1) Gravity tanks, suction tanks, pressure tanks, and embankment-supported coated fabric suction tanks
(2) Towers
(3) Foundations
(4) Pipe connections and fittings
(5) Valve enclosures
(6) Tank filling
(7) Protection against freezing

By specifying water tanks for private fire protection in its scope, NFPA 22 is stating its intended use as a standard for water tanks that serve fire protection systems only. The standard is not intended to provide water for domestic consumption and, therefore, does not include chlorination or disinfection requirements to achieve drinking water quality. Use of water tanks for other than fire protection is recognized by the standard but is limited to cases that are unavoidable and situations that do not affect the quantity of fire protection water.

. . .

1.5 Types of Tanks.

This standard addresses elevated tanks on towers or building structures, water storage tanks that are at grade or below grade, and pressure tanks.

Section 1.5 describes in general terms the types of tanks normally used for fire protection. Of these types of tanks, those that are installed at or below grade are normally used in series with fire pumps.

. . .

CHAPTER 4 GENERAL INFORMATION

4.1 Capacity and Elevation.

4.1.1* The size and elevation of the tank shall be determined by conditions at each individual property after due consideration of all factors involved.

A.4.1.1 Where tanks are to supply sprinklers, see separately published NFPA standards; also see NFPA 13.

As discussed in Part II of this handbook, other systems such as standpipe systems or private water supplies (including master streams) may require higher flow rates for longer durations than most sprinkler systems. When serving more than one system, the size and capacity of the water tank should be determined based on the highest system demand for the longest required duration of flow. The installation standard for each type of system (i.e., sprinkler, standpipe, or master stream) specifies the water supply duration. The water supply duration multiplied by the flow in gallons per minute (liters per minute) determines the water tank capacity.

4.1.2 Wherever possible, standard sizes of tanks and heights of towers shall be as specified in 5.1.3, 6.1.2, 8.1.3, and Section 9.2.

4.1.3 For suction tanks, the net capacity shall be the number of U.S. gallons (cubic meters) between the inlet of the overflow and the level of the vortex plate.

The amount of water between the overflow and the anti-vortex plate constitutes the amount of usable water in the storage tank. The tank size as determined in 4.1.2 may be slightly larger than the quantity needed by the fire protection system. For example, a standpipe system that requires 1000 gpm (3785 L/min) for the required 30-minute duration needs a total quantity of 30,000 gal (113.55 m^3) of fire protection water. In this case, the next size tank may be needed [i.e., 40,000 gal (151.40 m^3)] in order to provide the required 30,000 gal (113.55 m^3) of fire protection water. This requirement applies to all types of tanks except break tanks. For the required size of break tanks, see NFPA 20, *Standard for the Installation of Stationary Pumps for Fire Protection,* Section 4.31.2.

4.2 Location of Tanks.

4.2.1 The location of tanks shall be such that the tank and structure are not subject to fire exposure.

Like fire pumps, water tanks should be protected from exposure to fire. Protection of water tanks usually involves physical separation from buildings or other exposures for ground-mounted suction tanks or elevated gravity tanks. Tanks that are installed inside the protected building should be located in the fire pump room or enclosed by sufficient fire-rated construction to prevent exposure. Buried tanks are considered to be protected.

. . .

4.3 Tank Materials.

4.3.1 Materials shall be limited to steel, wood, concrete, coated fabrics, and fiberglass-reinforced plastic tanks.

The 2008 edition of NFPA 22 recognizes that fiberglass-reinforced tanks are allowed by the standard, so 4.3.1 has been revised to include fiberglass-reinforced plastic tanks in the list of permitted materials. A new chapter has also been developed on fiberglass-reinforced plastic; see the extracts from Chapter 11.

4.3.2 The elevated wood and steel tanks shall be supported on steel towers or reinforced concrete towers.

...

4.5 Plans.

4.5.1 The contractor shall furnish stress sheets and plans required by the purchaser and the authority having jurisdiction for approval or for obtaining building permits and licenses for the erection of the structure.

4.5.2 Approval of Layouts.

Construction should not proceed until plans and calculations have been reviewed and approved and all permits and fees have been applied for and paid. This documentation should be modified by the installing contractor to reflect field conditions and submitted to the authority having jurisdiction and building owner for final review upon the completion of work. In addition, the manufacturer's inspection, testing, and maintenance instructions and a copy of NFPA 25, *Standard for the Inspection, Testing, and Maintenance of Water-Based Fire Protection Systems,* should be provided in the final submission package.

Complete information regarding the tank piping on the tank side of the connection to the yard or sprinkler system is required to be submitted by the installing contractor to the authority having juridiction for approval. This submittal should be made before work begins and should accompany the submittal requirements referenced in NFPA 22. No work should begin until these plans have been accepted by all authorities having jurisdiction involved in the project. In addition to these items, complete details of the heating system (where required) should be provided and approved prior to construction.

4.5.2.1 Complete information regarding the tank piping on the tank side of the connection to the yard or sprinkler system shall be submitted to the authority having jurisdiction for approval.

4.5.2.2 The information submitted shall include the following:

(1) Size and arrangement of all pipes
(2) Size, location, and type of all valves, tank heater, and other accessories
(3) Steam pressures available at the heater
(4) Arrangement of, and full information regarding, the steam supply and return system together with pipe sizes
(5) Details of construction of the frostproof casing
(6) Where heating is required, heat loss calculations
(7) Structural drawings and calculations
(8) Seismic bracing details and calculations
(9) Operational settings and sequence of operation
(10) Monitoring equipment and connections
(11) Underground details including foundations, compaction, and backfill details and calculations
(12) Buoyancy calculations for buried tanks

4.6 Tank Contractor Responsibility.

4.6.1 Any necessary work shall be handled by experienced contractors.

4.6.1.1 Careful workmanship and expert supervision shall be employed.

4.6.1.2 The manufacturer shall warranty the tank for at least 1 year from the date of completion and final customer acceptance.

4.6.2 Upon completion of the tank construction contract, and after the contractor has tested the tank and made it watertight, the tank contractor shall notify the authority having jurisdiction so that the tank can be inspected and approved.

A test report should be completed and signed by the installing contractor, the registered design professional (RDP), and the authority having jurisdiction. This test report should include a description of the installation and tests conducted and should be included in the final submission package.

4.6.3 Cleaning Up.

4.6.3.1 During and upon completion of the work, the contractor shall remove or dispose of all rubbish and other unsightly material in accordance with NFPA 241.

4.6.3.2 The condition of the premises shall be as it was before tank construction.

In addition to construction debris, NFPA 241, *Standard for Safeguarding Construction, Alteration, and Demolition Operations,* addresses issues such as temporary construction; equipment and storage; process and hazards; utilities; fire protection; safeguarding construction and alteration operations; and safeguarding roofing, demolition, and underground operations. Any one of these items can cause an exposure to the tank installation and should be remedied before demobilization.

• • •

CHAPTER 11 FIBERGLASS-REINFORCED PLASTIC TANKS

11.1 General.

Fiberglass-reinforced plastic tanks shall be permitted to be used for fire protection systems when installed in accordance with this standard.

11.2* Application.

Fiberglass-reinforced plastic tanks shall be permitted only for storage of water at atmospheric pressure.

Fiberglass-reinforced plastic (FRP) tanks are only permitted for storage of water at atmospheric pressure and should not be used as pressure tanks.

A.11.2 See Figure A.11.2 for an example of a fiberglass tank being used under ground as a cistern to supply fire flow for fire department apparatus in a rural area.

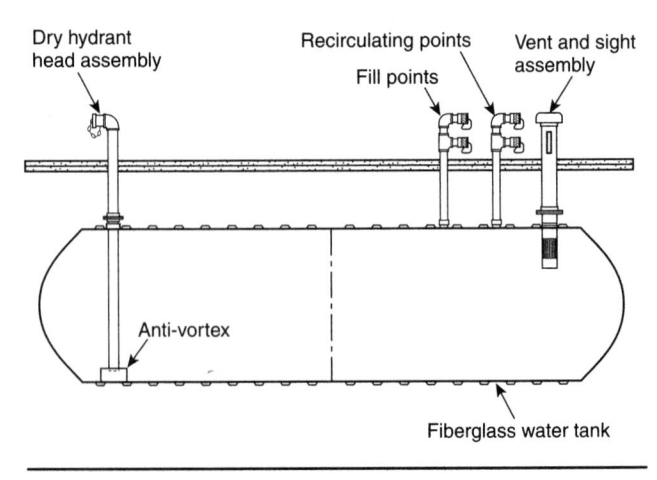

FIGURE A.11.2 Fiberglass Tank as an Underground Cistern.

FRP tanks are permitted for use only when buried for protection of the tank against mechanical and fire damage. NFPA 22 does not address the use of tanks inside buildings. However, when proposed for use inside buildings, if the same precautions are used for protection from mechanical and fire damage, and the proposed installation is approved by the authority having jurisdiction, NFPA 22 does permit this type of tank to be used for this application. The same protection methods used for the fire pumps that are listed in Section 4.12 in NFPA 20 should be used for the protection of FRP tanks that are installed indoors.

11.3* Tank Specification.

Fiberglass-reinforced plastic tanks shall meet the requirements of AWWA D120.

A.11.3 The standard capacities shall be from 2,000 gal to 50,000 gal (7.6 m^3 to 190 m^3). Tanks of other capacities are permitted.

11.4 Monolithic Tanks.

Monolithic tanks shall be tested for leakage by the manufacturer prior to shipment.

11.4.1 Tanks that are assembled on site shall be tested for leakage by the manufacturer.

11.5 Protection of Buried Tanks.

11.5.1 Tanks shall be designed to resist the pressure of earth against them.

11.5.2 Tanks shall meet local building code requirements for resisting earthquake damage.

11.5.3 Tanks shall be installed in accordance with the manufacturer's instructions and 11.5.4 through 11.5.13.

11.5.4 Bedding and backfill shall be noncorrosive inert material, of a type recommended by the tank manufacturer, such as crushed stone or pea gravel that is properly compacted.

11.5.5 Tanks shall be set on the minimum depth of bedding, as recommended by the tank manufacturer, that extends 1 ft (0.3 m) beyond the end and sides of the tank.

11.5.6 Tanks shall be located completely below the frost line to protect against freezing.

Tank installation below the frost line prevents freezing and negates the need to install a heating system and temperature alarm.

11.5.7 Where tanks are buried below railroad tracks, the minimum depth of cover shall be 4 ft (1.2 m).

11.5.8 Where the tanks are not subjected to traffic, tanks shall be covered with not less than 12 in. (305 mm) of compacted backfill and topped with up to 18 in. (457 mm) of compacted backfill or with not less than 12 in. (305 mm) of compacted backfill, on top of which shall be placed a slab of reinforced concrete not less than 4 in. (100 mm) thick.

11.5.9 Where tanks are, or are likely to be, subjected to traffic, they shall be protected from vehicles passing over them by at least 36 in. (914 mm) of backfill, or 18 in. (457 mm) of compacted backfill, of a type recommended by the tank manufacturer, plus either 6 in. (152 mm) of reinforced concrete or 9 in. (229 mm) of asphaltic concrete or greater where specified by the tank manufacturer.

11.5.10 Where asphaltic or reinforced concrete paving is used as part of the protection, it shall extend at least 12 in. (305 mm) horizontally beyond the outline of the tank in all directions.

11.5.11 Tanks shall be safeguarded against movement when exposed to high groundwater or floodwater by anchoring with non-metallic straps to a bottom hold-down pad or deadman anchors with fittings built up or protected to prevent corrosion failure over the life of the tank or by securing by other equivalent means using recognized engineering standards.

11.5.12 The depth of cover shall be measured from the top of the tank to the finished grade, and due consideration shall be given to future or final grade and the nature of the soil.

11.5.13 Maximum burial depths, measured from the top of the tank, are established by underground tank manufacturers and independent testing laboratories. Maximum burial depth shall be specified by the tank manufacturer and shall be marked on the tank.

The requirements in 11.5.7 through 11.5.10 and 11.5.12 are based on NFPA 24, *Standard for the Installation of Private Fire Service Mains and Their Appurtenances,* Section 10.4, and are consistent with the burial depths for underground piping systems. These measures are meant to protect the tank from damage when heavy objects roll over the surface above the tank location. It is preferable that tanks be located where railroad tracks and roadways do not pass over them.

This information is provided to standardize the burial depth for the tank, considering finished grade and the potential for soil settlement.

11.6 Protection of Aboveground Tanks

11.6.1 Tanks shall meet local building code requirements for resisting earthquake damage.

11.6.2 Tanks shall be installed in accordance with the manufacturer's instructions and 11.6.3 through 11.5.13.

11.6.3 Fiberglass-reinforced plastic (FRP) tanks located inside a building shall be protected by automatic sprinklers in accordance with ordinary hazard Group 2 occupancies.

Where FRP tanks are used to supply fire pumps and/or used as break tanks and are installed in the immediate vicinity of fire pumps, the tank should be protected in the same fashion as the fire pump. NFPA 20, Section 4.12, provides guidance on the type and extent of protection needed (see Part II of this handbook). This protection should include separation from the hazard by means of at least 1-hour fire resistance rating for buildings that are completely sprinklered and 2-hour fire resistance rating for buildings that are only partially sprinklered. Where the hazard is greater than ordinary hazard Group II, tank protection should be in accordance with NFPA 13, *Standard for the Installation of Sprinkler Systems.*

11.6.3.1 Where the hazard is greater than OH2, protection shall be in accordance with NFPA 13.

11.6.4 Horizontal fiberglass-reinforced plastic tanks that are greater than 4 ft (1.2 m) in diameter and are positioned 18 in. (457 mm) or greater above finished floor shall be protected in accordance with the obstruction rules of NFPA 13.

When the size of the tank exceeds 4 ft (1.2 m) and the tank is installed more than 18 in. (457 mm) above the finished floor, the tank becomes an obstruction to the sprinkler spray pattern. To solve this problem, a line of sprinklers must be installed below the tank in order to comply with NFPA 13.

11.6.5 Fiberglass tanks installed outdoors shall be protected from freezing and mechanical and UV damage.

Outdoor use of FRP tanks is permitted provided the tanks are protected from freezing, mechanical, and ultraviolet (UV) damage. For outdoor installation, the tank should be placed in a protective enclosure to prevent UV exposure and access should be limited to prevent mechanical damage.

Electronic supervision of the tank water temperature should be provided and can be accomplished by the installation of a temperature supervisory switch. This switch must be installed in accordance with *NFPA 72®, National Fire Alarm and Signaling Code,* and should

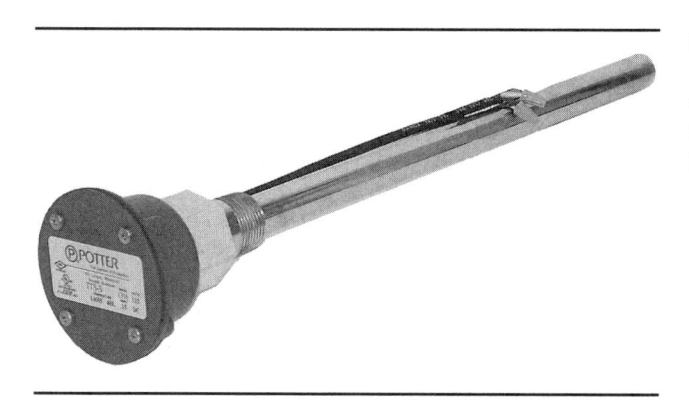

EXHIBIT III.1 *Tank Water Temperature Supervisory Switch. (Photograph courtesy of Potter Electric Signal Company, LLC)*

produce two separate and distinct signals. One signal should indicate a decrease in water temperature to 40°F (4.4°C), and the other signal should indicate a return to water temperature above 40°F (4.4°C). A single switch can be used to accomplish this task. See Exhibit III.1 for a typical tank water temperature supervisory switch.

11.7 Tank Connections.

11.7.1 Tanks shall have a vent that extends above the ground to prevent against pressurization of the tank during filling and creation of a vacuum during use. Tank venting systems shall be provided with a minimum 2.0 in. (50 mm) nominal inside diameter.

11.7.2* For underground tanks, water level monitoring required by 14.1.8 shall be capable of being read above ground.

A.11.7.2 See Figure A.11.7.2 for an example of a combination vent and sight assembly, which allows the tank to stay at atmospheric pressure while allowing the user to know the water

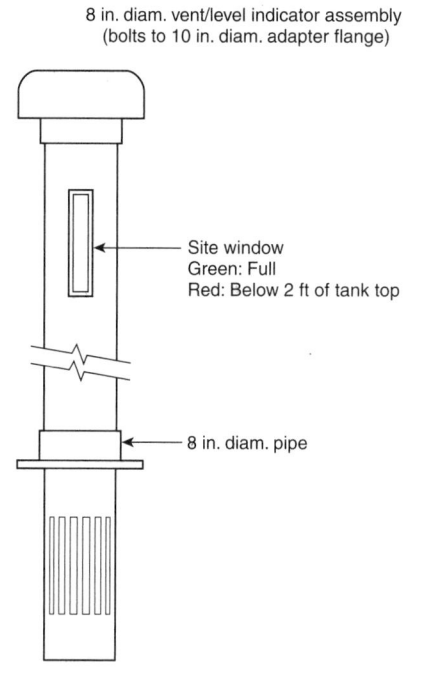

FIGURE A.11.7.2 Typical Combination Vent and Sight Assembly.

Stationary Fire Pumps Handbook 2010

level in the tank. While these two devices are not required to be combined, it is convenient since they are both required to be above ground.

Water level monitoring should be capable of being read above ground. The water level in a tank must be monitored for obvious reasons. According to NFPA 25, the water level should be inspected monthly when not supervised by an alarm connected to a constantly attended location, and it should be inspected quarterly when supervised by an alarm that is connected to a constantly attended location. An important factor to emphasize is that the quantity of water needed is determined by the required flow rate times a specified duration; anything less than this quantity in the tank constitutes an insufficient amount of water for fire fighting and a code deficiency. For tanks with an automatic fill connection, the flow rate of the fill connection is frequently included in the tank capacity calculation, thus reducing the overall size of the tank.

. . .

CHAPTER 14 PIPE CONNECTIONS AND FITTINGS

14.1* General Information.

A.14.1 For embankment-supported coated fabric suction tanks, see Section 9.6.

14.1.1 Watertight Intersections at Roofs and Floors.

14.1.1.1 The intersections of all tank pipes with roofs and concrete or waterproof floors of buildings shall be watertight.

14.1.1.2 Where tank pipes pass through concrete roofs, a watertight intersection shall be obtained by using fittings that are caulked with oakum or by pouring the concrete solidly around the pipes, which first shall be wrapped with two or three thicknesses of building paper.

14.1.1.3 Where concrete is used, the upper side of the intersection shall be well flashed with a suitable, firm, waterproof material that is noncracking and that retains its adhesion and flexibility.

14.1.1.4 Wood roofs also shall be built tightly around the pipes and shall be made watertight by means of fittings that are caulked with oakum or by using flashing.

14.1.1.5 Where tank pipes pass through a concrete or waterproof floor, a watertight intersection, as described in 14.1.1.1, shall be obtained so that water from above cannot follow down the pipe to the lower floors or to the basement.

14.1.2 Rigid connections to steel tanks shall be made by means of a welded joint with approval by the authority having jurisdiction.

14.1.2.1 A rigid connection to a wood tank shall be made by means of a running nipple or by means of threaded flanges, one inside the tank and one outside the tank, bolted together through the wood with movable nuts outside.

14.1.3* Placing Tank in Service. All tank piping shall be installed immediately after completion of the tank and tower construction so that the tank can be filled and placed in service promptly.

A.14.1.3 Wood tanks can be extensively damaged by shrinkage if left empty after they are erected.

A water storage tank is a critical component of a fire protection system. Should fire protection be needed for the construction site, the tank should be scheduled for early completion so it can be placed in service early on during construction.

14.1.4 The Contract. To ensure the installation of equipment, the contract shall specify that the finished work shall conform with this standard in all respects.

14.1.5 Precautions During Repairs. The authority having jurisdiction shall be notified well in advance when the tank is to be drained. The precautions required by 14.1.5.1 through 14.1.5.5 shall be observed.

Rarely is a facility more exposed to a potentially catastrophic fire loss than when a fire protection system is impaired or out of service, particularly when it is the water supply, as in the case of water tanks. Experience has shown time and again that if a proper impairment program had been in place and adhered to, many devastating fires could have been mitigated.

Whether an impairment is preplanned or occurs in an emergency situation seems to have little bearing on the propensity for catastrophe. Failure to take the following measures has been identified as a major factor leading to large fire losses:

1. Recognize the exposures created by the impairment.
2. Establish a fire watch and provide backup fire protection.
3. Control ignition sources and/or discontinue hazardous processes.
4. Expedite repairs.
5. Verify that fire protection systems are properly placed back into service.

Chapter 15 of NFPA 25, 2008 edition, should be followed any time a water storage tank is taken out of service for maintenance or repair. Note that 15.5.2(3) of NFPA 25 requires a comprehensive impairment procedure any time a fire protection system is removed from service for more than 10 hours in a 24-hour period. Subsection 15.5.2 is extracted as follows:

> Before authorization is given, the impairment coordinator shall be responsible for verifying that the following procedures have been implemented:
>
> (1) The extent and expected duration of the impairment have been determined.
> (2) The areas or buildings involved have been inspected and the increased risks determined.
> (3) Recommendations have been submitted to management or building owner/manager. Where a required fire protection system is out of service for more than 10 hours in a 24-hour period, the impairment coordinator shall arrange for one of the following:
>   (a)* Evacuation of the building or portion of the building affected by the system out of service
>   (b)* An approved fire watch
>   (c)* Establishment of a temporary water supply
>   (d)* Establishment and implementation of an approved program to eliminate potential ignition sources and limit the amount of fuel available to the fire
> (4) The fire department has been notified.
> (5) The insurance carrier, the alarm company, building owner/manager, and other authorities having jurisdiction have been notified.
> (6) The supervisors in the areas to be affected have been notified.
> (7) A tag impairment system has been implemented. (See Section 15.3.)
> (8) All necessary tools and materials have been assembled on the impairment site.
>
> [**NFPA 25,** 15.5.2]

14.1.5.1 Work shall be planned carefully to enable its completion in the shortest possible time.

14.1.5.2 Where available, a second, reasonably reliable water supply with constant suitable pressure and volume, usually public water, shall be connected to the system.

14.1.5.3 Where such a supply is not available, the fire pump shall be started and kept running to maintain suitable pressure in the system.

14.1.5.4 Additional portable fire extinguishers shall be placed in buildings where protection is impaired, and extra, well-instructed watch personnel shall be continuously on duty.

14.1.5.5 The members of the private fire brigade, as well as the public fire department, shall be familiar with conditions that affect repairs.

14.1.6* Heater Thermometer.

A.14.1.6 One of the chief advantages of the gravity circulation system of heating tanks is that it enables convenient observation of the temperature of the coldest water at a thermometer located in the cold-water return pipe near the heater. Failure to provide an accurate thermometer at this point or failure to observe it daily and ensure that it registers the proper temperature forfeits this advantage and can result in the freezing of the equipment. *[See Figure 16.1.7.5.5(a) and Figure 16.1.7.5.5(b), Figure B.1(i) through Figure B.1(k), and Figure B.1(s), Figure B.1(t), and Figure B.1(v).]*

Note that, for the purposes of this handbook, only Figures 16.1.7.5.5(a) and 16.1.7.5.5(b) are reprinted here.

For SI units, 1 in. = 25.4 mm; 1 ft = 0.3048 m; 1 psi = 0.0689 bar; °C = 5/9 (°F−32).

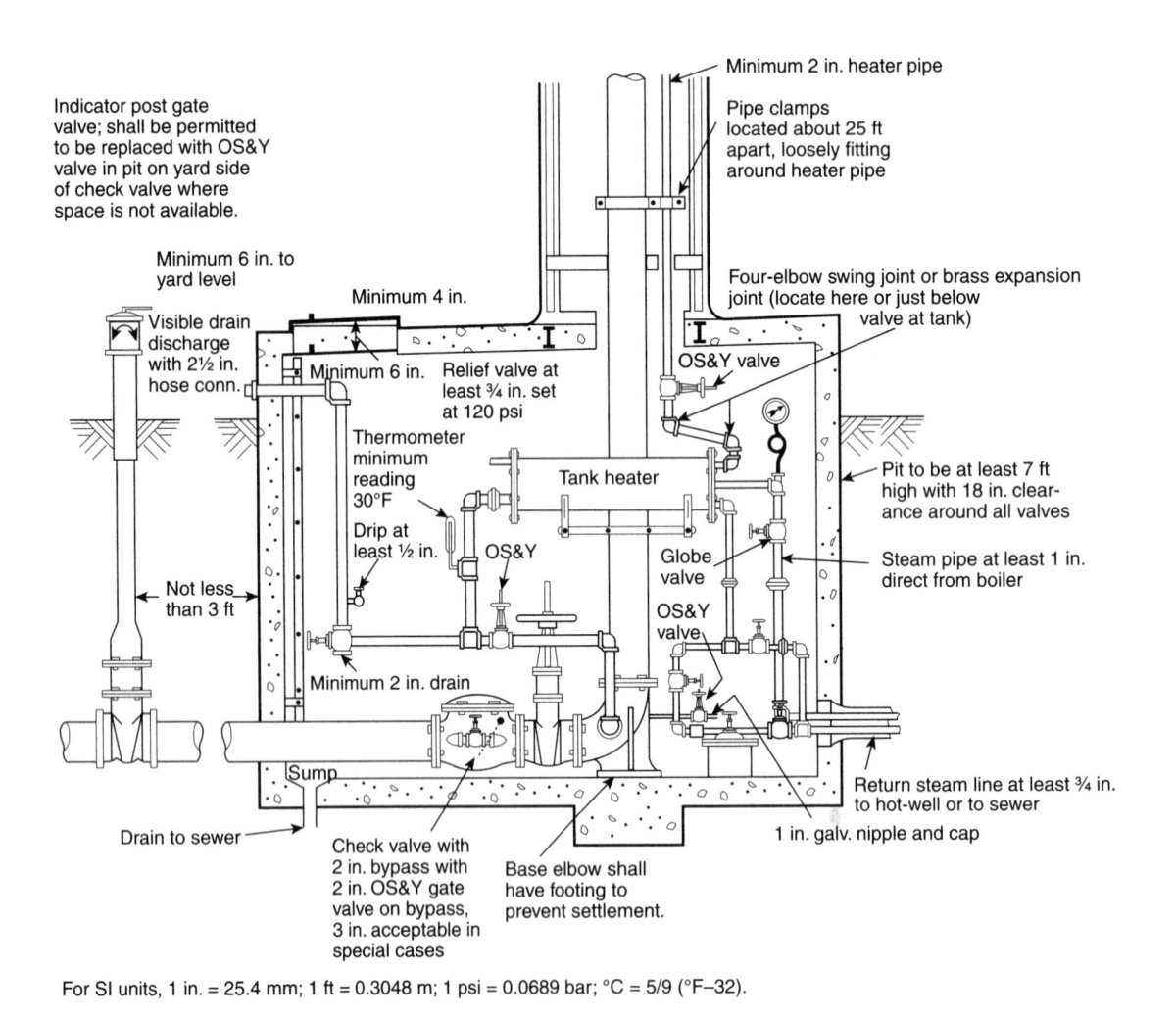

FIGURE 16.1.7.5.5(a) Tank Heater Drain Arrangement at Base of Riser.

FIGURE 16.1.7.5.5(b) *Tank Heater Drain Arrangement.*

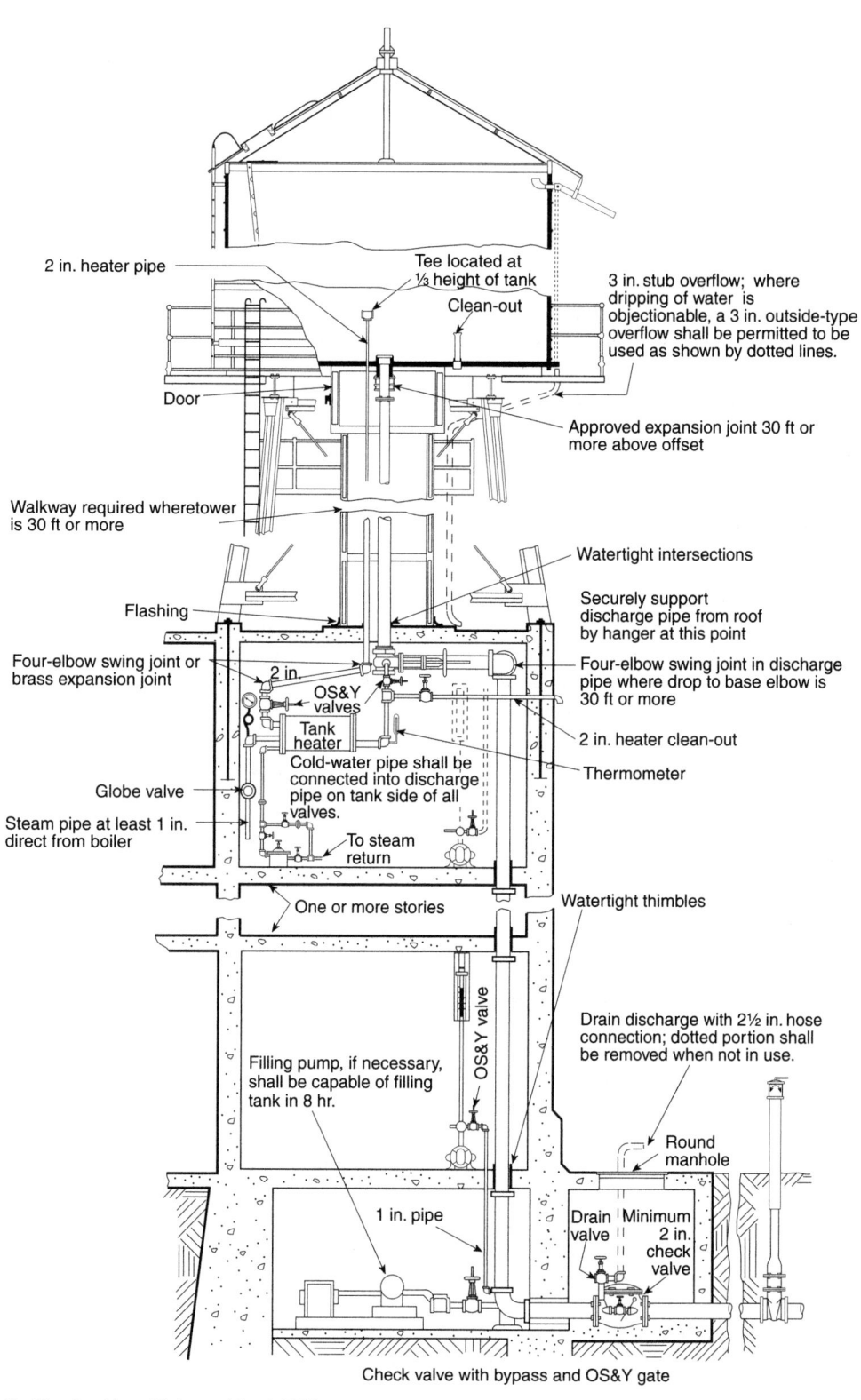

14.1.6.1 In the case of a gravity circulating heating system, an accurate thermometer shall be located as specified in 16.1.7.5.

Subparagraph 16.1.7.5.2 requires an accurate thermometer that is graduated at least as low as 30°F (−1.1°C) to be placed in the cold-water pipe at a point where it registers the temperature of the coldest water in the system. This requirement is clearly illustrated in Figures 16.1.7.5.5(a) and (b).

14.1.6.2 Where a tank contains a radiator steam heater, an accurate socket thermometer shall be located as specified in 16.3.6.

Subsection 16.3.6 requires an accurate angle socket thermometer that has at least a 6 in. (152 mm) stem and that is calibrated as low as 30°F (−1.1) to be permanently inserted through the plate or standpipe and as far from the heating unit as possible. An angle socket thermometer is not required for suction tanks with a maximum height of 25 ft (7.6 m).

14.1.7* Connections for Use Other Than for Fire Protection. The authority having jurisdiction shall be consulted before the tank is designed where water for other than fire protection purposes is to be drawn from the tank.

A.14.1.7 The circulation of water through the tank causes an accumulation of sediment that can obstruct the piping or sprinklers. A leak or break in a pipe for use other than fire protection may seriously impair the fire protection by partly or completely draining the elevated tank.

When a buildup of sediment occurs, the tank must be drained and cleaned, creating an impaired system and potentially expensive cleaning and refilling process. A buildup of sediment can be avoided by dedicating the water tank to fire protection use only. Any multiple-use water tank should be closely monitored for increases in the domestic or industrial consumption to prevent drawing too much water from the tank. The connections for use other than for fire protection should be made above the full level for the fire protection system to prevent depleting the stored fire protection supply.

14.1.8* Water-Level Gauge. A water-level gauge of suitable design shall be provided. It shall be carefully installed, adjusted, and properly maintained.

A.14.1.8 Water-Level Gauges. The following information is provided for existing installations where mercury gauges are in use. Mercury gauges are no longer permitted for new installations.

(1) *Mercury Gauge Materials.* Pipe and fittings that contain mercury should be iron or steel. Brass, copper, or galvanized parts, if in contact with mercury, are amalgamated, and leaks result.
(2) *Water Pipe.* The water pipe to the mercury gauge should be 1 in. (25 mm) galvanized throughout and connected into the discharge pipe on the tank side of the check valve. Where possible, the pipe should be short, should be run with a continual upward pitch toward the tank piping, and should be without air pockets to avoid false readings. The pipe should be buried well below the frost line or located in a heated conduit.
(3) *Valves.* The valve at the mercury gauge should be a listed OS&Y gate valve. An additional listed OS&Y gate valve should be installed close to the discharge pipe where the distance to the mercury gauge exceeds 50 ft (15.2 m).
(4) *Mercury Catcher.* Occasionally, fluctuating water pressures require a mercury catcher at the top of the gauge glass to prevent loss of mercury. The catcher is not a standard part of the equipment and is not furnished by the gauge manufacturer unless specially ordered.
(5) *Extension Piece.* Where the mercury catcher is not needed, it can be replaced by approximately a 3 ft (0.91 m) extension of $1/8$ in. (3 mm) pipe, vented at the top.

(6) *Water-Drain Plug.* A plugged tee should be provided in the mercury pipe between the mercury pot and the gauge glass to allow water that sometimes accumulates on top of the mercury column to drain off.

(7) *Location.* The gauge should be installed in a heated room such as a boiler room, engine room, or office, where it is readily accessible for reading, testing, and maintenance. It should be so located that it is not liable to break or to be damaged. The column of mercury, extending from the mercury pot to the top, is roughly $1/13$ the height from the mercury pot to the top of the tank. This fact should be considered when planning a location for the instrument.

(8) *Cleaning.* Before installing the gauge, all grease, dirt, and moisture should be removed from the pot and piping that are to contain mercury, and it should be ensured that the mercury itself is clean. Warm water that contains a small amount of washing soda is a good cleaning agent.

(9) *Installing.* The gauge should be accurately installed so that when the tank is filled to the level of the overflow, the mercury level is opposite the FULL mark on the gauge board.

(10) *Testing.* To determine that it is accurate, the instrument should be tested occasionally as follows:

  (a) Overflow the tank.
  (b) Close the OS&Y valve. Open the test cock. The mercury should quickly drop into the mercury pot. If it does not, there is an obstruction that must be removed from the pipe or pot between the test cock and the gauge glass.
  (c) If the mercury lowers at once, as expected, close the test cock and open the OS&Y valve. If the mercury responds immediately and comes to rest promptly opposite the FULL mark on the gauge board, the instrument is operating properly.
  (d) If the mercury column does not respond promptly and read correctly during the test specified in A.14.1.8(10)(c), there are probably air pockets or possibly obstructions in the water-connecting pipe. Open the test cock. Water should flow out forcefully. Allow water to flow through the test cock until all air is expelled and rusty water from the tank riser appears. Then close the test cock. The gauge should now read correctly. If air separates from the water in the 1 in. (25 mm) pipe due to being enclosed in a buried tile conduit with steam pipes, the air can be automatically removed by installing a ¾ in. (19 mm) air trap at the high point of the piping. The air trap can usually be best installed in a tee connected by a short piece of pipe at a location between the OS&Y valve and the test cock, using a plug in the top of the tee, so that mercury can be added in the future, if necessary, without removing the trap. If there are inaccessible pockets in the piping, such as locations below grade or under concrete floors, the air can be removed only through the test cock.
  (e) If, in the test specified in A.14.1.8(10)(d), the water does not flow forcefully through the test cock, there is an obstruction that must be removed from the outlet of the test cock or from the waterpipe between the test cock and the tank riser.
  (f) If there is water on top of the mercury column in the gauge glass, it will cause inaccurate readings and must be removed. First lower the mercury into the pot as in the test specified in A.14.1.8(10)(b). Close the test cock and remove the plug at the base of the mercury gauge. Open the OS&Y valve very slowly, causing mercury to rise slowly and the water above it to drain through the plug at the base of the mercury gauge. Close the OS&Y valve quickly when mercury appears at the outlet at the base of the mercury gauge, but have a receptacle ready to catch any mercury that drains out. Replace the plug. Replace any escaped mercury in the pot by removing the plug between the OS&Y valve and the test cock, and with the OS&Y valve closed, fill the pot with mercury to the mark on the cover corresponding to the height above the pot that indicates the full water level in the tank. Replace the plug.
  (g) After testing leave the OS&Y valve open, except as noted in A.14.1.8(11).

(11) *Excessive Water Pressures.* If found necessary, to prevent forcing mercury and water into the mercury catcher, the controlling OS&Y valve may be closed when filling the tank but should be left open after the tank is filled, except when the gauge is subjected to continual fluctuation of pressure, when it may be necessary to keep the gauge shut off, except when it is being read. Otherwise it may be necessary to frequently remove water from the top of the mercury column as in A.14.18(10).

Since mercury is a hazardous substance, its use as a liquid level gauge for new installations has come to an end. The material in A.14.1.8 remains in the standard for those installations that still use this technology. New installations rely on other means to indicate water level, such as an altitude gauge or float mechanism in the tank. Exhibit III.2 illustrates a liquid level gauge on a wood tank.

EXHIBIT III.2 Liquid Level Gauge on a Wood Tank.

14.1.8.1 Where an altitude gauge is used, it shall be at least 6 in. (152 mm) in diameter and shall be of noncorrodable construction.

14.1.8.2 The gauge shall be located to prevent it from freezing.

14.1.8.2.1 If necessary, it shall be located in a heated building or enclosure.

14.1.8.2.2 A blow-off cock shall be located between the gauge and the connection to the tank.

14.1.8.3 A listed, closed-circuit, high-water and low-water level electric alarm shall be permitted to be used in place of the gauge where acceptable to the authority having jurisdiction.

When a water level switch is used, the requirements of *NFPA 72* state that the switch must provide two separate and distinct signals—one indicating that the required water level has been raised or lowered (off-normal), and the other indicating restoration of the normal water level. See Exhibit III.3 for an example of a water level supervisory switch.

14.1.8.3.1 Provisions shall be made for the attachment of a calibrated test gauge.

14.1.8.4 For underground tanks, water-level monitoring shall be capable of being read and/or supervised above ground.

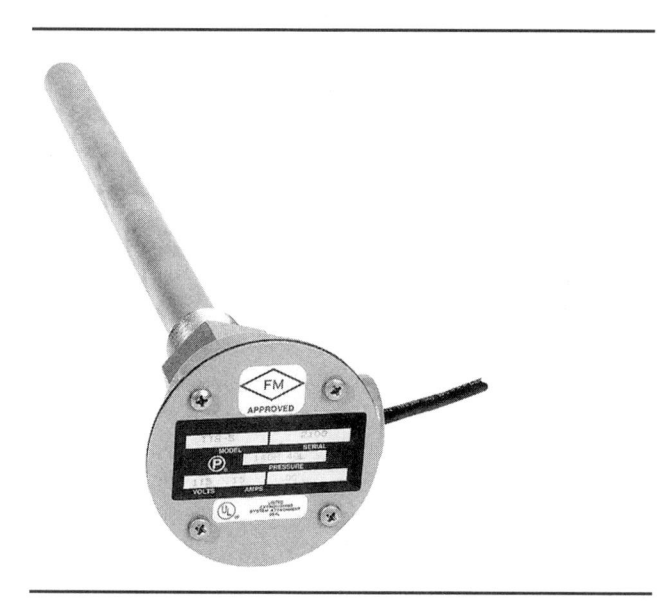

EXHIBIT III.3 Tank Water Level Supervisory Switch. (Photograph courtesy of Potter Electric Signal Company, LLC)

14.1.9* Frostproof Casing. The frostproof casing shall be maintained in good repair and shall be weathertight throughout.

A.14.1.9 The insulating qualities of frostproof casing are seriously impaired if joints spring open, if the casing settles away from the tank, or if rotting occurs around the base.

14.1.10 Tanks with Large Risers.

14.1.10.1* Large steel-plate riser pipes of 3 ft (0.91 m) or more in diameter and without frostproof casing shall be acceptable where properly heated.

A.14.1.10.1 By heating the large steel-plate riser pipes, the fire hazard and upkeep of the frostproof casing and the provision of an expansion joint or walkway are avoided. However, painting and heating the larger riser and building the stronger and larger valve pit cost more than the equipment for smaller risers.

A blow-off valve is sometimes furnished near the base of the larger riser.

A check valve and gates in the discharge pipe, filling arrangement, overflow, and drain are generally provided.

14.1.10.2 A manhole at least 12 in. × 16 in. (305 mm × 406 mm) shall be provided, and its lower edge shall be level with the discharge piping protection specified.

14.1.11 Discharge Piping Protection.

14.1.11.1 In the case of tanks with a large steel-plate riser [3 ft (0.91 m) diameter or larger], the inlet to the vertical discharge pipe that is located within the large riser shall be protected against the entry of foreign material.

14.1.11.2 The inlet can be done with an American National Standards Institute 125 lb/in. (8.6 bar) flanged tee, with the "run" of the reducing tee placed horizontally and with horizontal outlets one pipe size smaller than the discharge pipe, or with a fabricated plate extending at least 4 in. (102 mm) beyond the outside diameter of the pipe.

14.1.11.3 The plate shall be supported by at least three supporting bars 1½ in. × ¼ in. (38.1 mm × 6.4 mm), by ⅝ in. (15.9 mm) round rods, or by the equivalent, that elevate all portions of the plate at a height at least equal to the pipe diameter located above the discharge pipe inlet.

14.1.11.4 The attachment of the supports to the discharge pipe shall be made directly by welding or bolting or by means of a ¼ in. (6.4 mm) thick tightly fitting sectional clamp or collar that has ⅝ in. (15.9 mm) bolts in the outstanding legs of the clamps or collar.

14.1.11.5 A clearance of at least 6 in. (152 mm) shall be provided between all portions of the flanges of a tee or fabricated plate and the large riser plate.

14.1.12 Steel Pipe.

14.1.12.1 Steel pipe shall conform to ASTM A 53, Type E, Type F, Type S, Grade A, or Grade B, manufactured by the open-hearth, electric furnace, or basic oxygen process, or it shall conform to ASTM A 106, Grade A or Grade B.

14.1.12.2 Paragraphs 14.1.12.2.1 through 14.1.12.2.3 shall apply to steel pipe that is in contact with storage water.

14.1.12.2.1 Steel pipe smaller than 2 in. (50 mm) shall not be used.

14.1.12.2.2 Steel pipe of 2 in. to 5 in. (50 mm to 125 mm) to 5 in. (125 mm) shall be extra-strong weight.

Piping in fire protection systems is ordinarily Schedule 10 (light wall) or Schedule 40 (standard weight). Occasionally, extra strong (Schedule 80) pipe is used. In this application due to continual contact with storage water, pipe with a thicker wall is required. A comparison of wall thickness for Schedules 10, 40, and 80 pipe is shown in Commentary Table III.1.

COMMENTARY TABLE III.1 Comparison of Wall Thickness for Schedules 10, 40, and 80 Pipe

Pipe Size		Schedule Number	Outside Diameter		Inside Diameter		Wall Thickness	
in.	mm		in.	mm	in.	mm	in.	mm
2	50	10	2.375	60.3	2.157	54.8	0.109	2.8
2	50	40	2.375	60.3	2.067	52.5	0.154	3.9
2	50	80	2.375	60.3	1.939	49.3	0.218	5.5

14.1.12.2.3 All steel pipe 6 in. (150 mm) and larger shall be standard weight.

14.2 Discharge Pipe.

14.2.1 At Roofs and Floors. The intersection of discharge pipes, as well as the intersection of all other tank pipes, with roofs or with waterproof or concrete floors shall be watertight.

14.2.2 Size.

14.2.2.1 The conditions at each plant shall determine the size of the discharge pipe that is needed.

The factor that determines the size of the discharge pipe is the fire protection system demand. In the case of sprinkler systems, the discharge density, system size, and distance from the tank have a definite impact on the diameter of the discharge pipe. Where elevated tanks provide the sole supply of water, the hydraulic calculations should terminate at the base of the tank. The calculations must include flow through the discharge pipe and the elevation difference. The elevation to the top of the full water level is not considered because this level decreases during discharge. Sizes other than those specified can be used provided they are approved by the authority having jurisdiction. However, selection of the discharge pipe diameter should also take into consideration future demand due to the possibility of expansion.

14.2.2.2 The size shall not be less than 6 in. (150 mm) for tanks up to and including a 25,000 gal (94.63 m^3) capacity and shall not be less than 8 in. (200 mm) for capacities of 30,000 gal to 100,000 gal (113.55 m^3 to 378.50 m^3) inclusive, or 10 in. (250 mm) for greater capacities.

Where tanks supply fire pumps directly, the size of the discharge pipe when located within 10 pipe diameters of the pump suction flange should not be smaller than that specified in Tables 4.26(a) and (b) of NFPA 20 (see Part II of this handbook).

14.2.2.3 Pipe that is smaller than specified in 14.2.2.2 [not less than 6 in. (150 mm)] shall be permitted in some cases where conditions are favorable and large flows of water are not needed.

14.2.2.3.1 Larger pipe shall be required where deemed necessary because of the location and arrangement of piping, height of buildings, or other conditions.

14.2.2.3.2 In all cases, approval of the pipe sizes shall be obtained from the authority having jurisdiction.

14.2.3 Pipe Material.

14.2.3.1 Underground Pipe Material. Piping shall be in accordance with NFPA 24.

14.2.3.2 Aboveground Pipe Material. Aboveground pipe material shall be in accordance with NFPA 13 and NFPA 20.

14.2.4 Braces.

14.2.4.1 Either the pipe or the large steel-plate riser pipe, or both, shall be braced laterally by rods of not less than ⅝ in. (15.9 mm) in diameter and shall be connected to the tower columns near each panel point.

14.2.4.2 The end connection of braces shall be by means of eyes or shackles; open hooks shall not be permitted.

14.2.5 Support.

14.2.5.1 The discharge pipe shall be supported at its base by a double-flanged base elbow that rests on a concrete or masonry foundation.

14.2.5.1.1 The base elbow of tanks with steel-plate risers, of suction tanks, or of standpipes shall have bell ends.

14.2.5.2 The joint at the connection of yard piping to the base elbow shall be strapped, or the base elbow shall be backed up by concrete.

14.2.5.2.1 If the discharge pipe is offset inside a building, it shall be supported at the offset by suitable hangers that extend from the roof or floors, in which case the base elbow might not be required.

14.2.5.2.2 Large steel riser pipes shall be supported on a reinforced concrete pier that is designed to support the load specified in Section 13.3.

14.2.5.2.3 Concrete grout shall be provided beneath the large riser to furnish uniform bearing when the tank is empty.

14.2.6 Offsets.

14.2.6.1 The discharge pipe outside of buildings shall extend vertically to the base elbow or building roof without offsets where possible.

Offsets in the discharge pipe increase the need for additional bracing. They also add to the friction loss in the hydraulic calculations and should be avoided.

14.2.6.2 If an offset is unavoidable, it shall be supported at the offsetting elbows and at intermediate points not over 12 ft (3.7 m) apart, and it also shall be rigidly braced laterally.

14.2.6.3 The supports shall consist of steel beams that run across the tower struts or of steel rods from the tower columns arranged so that there is no slipping or loosening.

14.2.7 Expansion Joint.

14.2.7.1 Tanks with flanged or welded pipe risers [12 in. (250 mm) and under] shall have a listed expansion joint on the fire-service discharge pipe where the tank is on a tower that elevates the bottom 30 ft (9.1 m) or more above the base elbow or any offset in the discharge pipe.

Due to the rigid connections of the discharge pipe at the top and bottom of the pipe, an expansion joint is needed. Any fluctuations in the length of the piping material due to temperature variations could cause stress at either connection that could result in connection failure.

14.2.7.2 Expansion joints shall be built to conform to Section 14.3.

14.2.8 Rigid Connection.

14.2.8.1 When the distance between the tank bottom and the base elbow or supporting hanger is less than 30 ft (9.1 m), the discharge pipe shall be connected by an expansion joint that is built to conform to Section 14.3 or shall be rigidly connected in accordance with 14.1.2.

14.2.8.2 The top of the pipe (or the fitting attached to the top) shall extend above the inside of the tank bottom or base of a steel-plate riser to form a settling basin.

14.2.8.2.1 The top of a steel-plate riser shall be connected rigidly to the suspended bottom of the tank.

14.2.8.2.2 The discharge pipe from a steel plate riser of a tank that is located over a building shall be connected rigidly to the base of the larger riser.

14.2.8.2.3 A rigid flanged connection or welded joint shall be permitted to be used between the discharge pipe and the bottom of a suction tank, a standpipe, or the base of a steel-plate riser of a tank that is located on an independent tower where special approval is obtained from the authority having jurisdiction.

14.2.8.2.4 When the base of a steel-plate riser is in its final position on a concrete support, it shall be grouted to obtain complete bearing.

14.2.9 Swing Joints.
Where the vertical length of a discharge pipe that is located below an offset, either inside or outside a building, is 30 ft (9.1 m) or more, a four-elbow swing joint that is formed, in part, by the offset shall be provided in the pipe.

Because of the four elbows in an offset, thereby forming a swing joint, the riser can tolerate movement in any direction without producing any stress on the end connections.

14.2.10 Settling Basin.

14.2.10.1 The depth of the settling basin in the tank bottom shall be 4 in. (102 mm) for a flat-bottom tank and 18 in. (457 mm) for a suspended-bottom tank.

14.2.10.2 The settling basin at the base of a large steel-plate riser shall be at least 3 ft (0.91 m) deep.

EXHIBIT III.4 Material Found During an Interior Tank Inspection.

The settling basin is intended to permit potentially obstructing material to settle on the tank bottom without entering the discharge pipe. Obstructing material can be in the form of paint chips or scale from steel tanks or steel piping, rocks and other debris from the fill connection, construction debris that has not been removed, or aquatic growth. See Exhibit III.4 for an example of potentially obstructing material.

14.2.11 Check Valve.

A control valve is needed to isolate the tank or check valve for inspection and maintenance purposes. When tanks are used to supply water to a fire pump, devices in the suction pipe in close proximity to the fire pump suction flange are of concern. NFPA 20 permits the suction pressure to a fire pump to be as low as 0 psi (0 bar). When the water supply is a suction tank and the base of the tank is even with or above the elevation of the fire pump, the suction pressure is permitted to be as low as –3 psi (–0.2 bar). The friction loss of the devices required in NFPA 22 and any offsets in the tank discharge pipe must be carefully calculated to verify that these pressure limitations are not exceeded. (See 4.14.3 of NFPA 20 in Part II of this handbook.)

14.2.11.1 A listed check valve shall be placed horizontally in the discharge pipe and shall be located in a pit under the tank where the tank is located on an independent tower.

The check valve prevents water from flowing back into the tank or discharge pipe when sprinklers are located above the elevation of the discharge pipe or tank bottom.

14.2.11.2 Where the tank is located over a building, the check valve shall ordinarily be placed in a pit, preferably outside the building.

14.2.11.3 Where yard room is not available, the check valve shall be located on the ground floor or in the basement of a building, provided it is protected against breakage.

14.2.12 Controlling Valves.

14.2.12.1 A listed gate valve shall be placed in the discharge pipe on the yard side of the check valve between the check valve and any connection of the tank discharge to other piping.

A gate valve enables the isolation of the tank for the purpose of filling where a separate fill pipe or fill pump is used. The valve should be equipped with a seal, lock, or tamper switch with the appropriate inspection frequency based upon the method of supervision practiced on a routine basis. The inspection procedure and frequency should be carried out in accordance with NFPA 25.

14.2.12.1.1 The listed gate valve shall be permitted to be equipped with an indicating post.

14.2.12.2 Where yard room for an indicator post is not available, a listed outside screw and yoke (OS&Y) gate valve that is of similar arrangement, but that is located inside the valve pit or room, shall be used.

14.2.12.3 A listed indicating control valve shall be placed in the discharge pipe on the tank side of the check valve.

An indicating control valve is needed to isolate the discharge check valve for maintenance or repair. Note that according to NFPA 25, each system check valve must be inspected internally every 5 years, and therefore isolation of the check valve is critical in order to comply with this requirement. Isolation of the check valve may also be needed at some point for replacement or repair.

14.2.12.3.1 Where the tank is on an independent tower, the valve shall be placed in the pit with the check valve, preferably on the yard side of the base elbow.

14.2.12.3.2 Where a tank is used as a suction source for a fire pump, the listed indicating control valve shall be of the OS&Y type.

NFPA 20 requires any valve within 50 ft (15.3 m) of the pump suction flange to be an OS&Y-type valve.

14.2.12.3.3 Where the tank is located over a building, the valve shall be placed under the roof near the point where the discharge pipe enters the building.

14.2.12.3.4 For suction tanks, the valve shall be as close to the tank as possible.

14.2.13 Anti-Vortex Plate Assembly.

14.2.13.1 Where a tank is used as the suction source for a fire pump, the discharge outlet shall be equipped with an assembly that controls vortex flow.

14.2.13.2* The assembly shall consist of a horizontal steel plate that is at least twice the diameter of the outlet on an elbow fitting, where required, mounted at the outlet a distance above the bottom of the tank equal to one-half the diameter of the discharge pipe.

As its name suggests, the anti-vortex plate assembly is intended to prevent the development of a vortex when the fire pump is in operation. A vortex, if present, entrains air and could cause the fire pump to cavitate. The inspection required by 17.1.1 should include the anti-vortex plate to verify that it is in place and has been installed correctly. The inspection report should indicate that the anti-vortex plate was installed and found to be acceptable.

A.14.2.13.2 Large, standard size anti-vortex plates [48 in. × 48 in. (1219 mm × 1219 mm)] are desirable, as they are adequate for all sizes of pump suction pipes normally used. Smaller plates may be used; however, they should comply with 14.2.13.

14.2.13.3* The minimum distance above the bottom of the tank shall be 6 in. (152 mm).

A.14.2.13.3 See Figure B.1(o).

14.4 Filling.

14.4.1 A permanent pipe connected to a water supply shall be provided to fill the tank.

14.4.2 The means to fill the tank shall be sized to fill the tank in a maximum time of 8 hours.

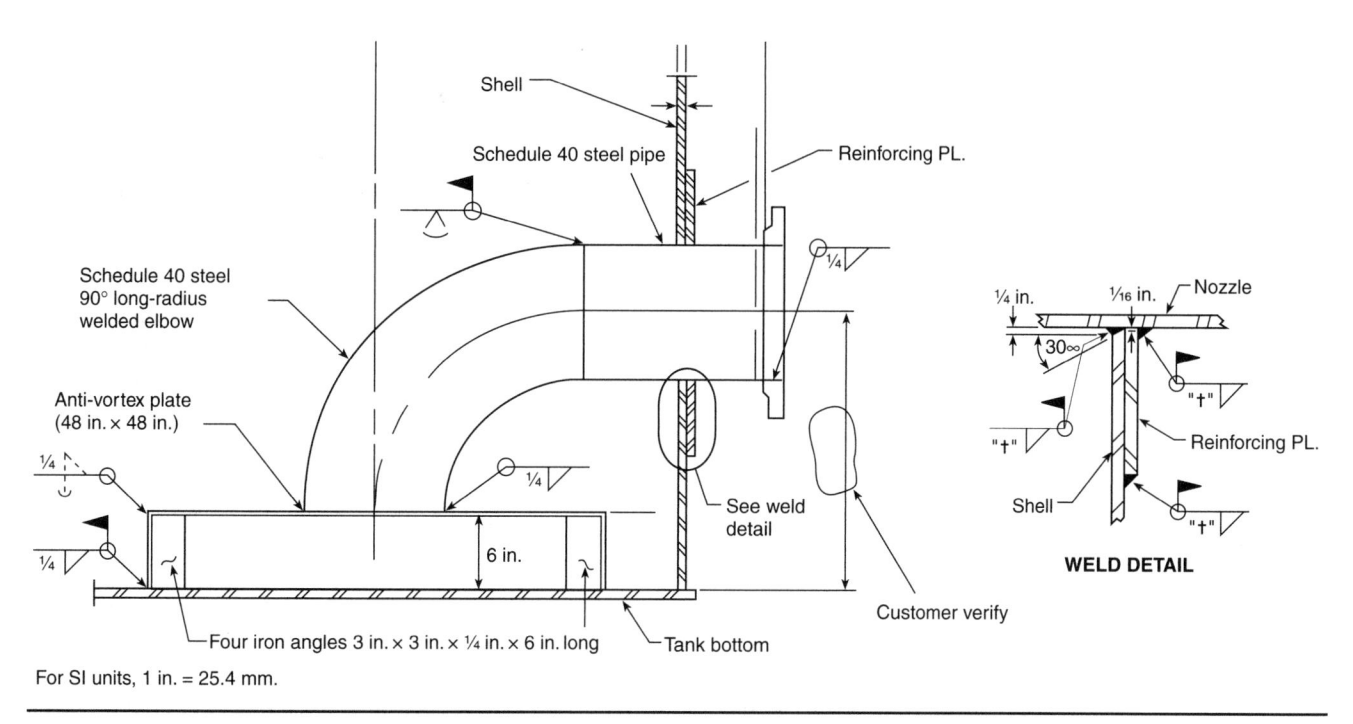

FIGURE B.1(o) *Suction Nozzle with Anti-Vortex Plate for Welded Suction Tanks.*

14.4.3 The tank shall be kept filled, and the water level shall never be more than 4 in. (102 mm) below the designated fire service level.

14.4.4 The filling bypass shall be kept closed when not in use.

Section 14.4 establishes the requirement for a permanently installed filling connection. The intent of the standard is to prohibit the filling of a tank from either a hydrant and hose arrangement or a filling truck.

This section has been revised to clarify the requirement to design and install a fill connection that is capable of filling the tank in not more than 8 hours.

A tolerance of 4 in. (102 mm) is permitted by the standard. Failure to keep this bypass closed results in wasting water by causing a constant discharge into the overflow pipe.

14.4.5 Bypass Around Check Valve.

14.4.5.1 Where the tank is to be filled from the fire protection system under city or fire-pump pressure, the filling pipe shall be a bypass around the check.

The filling arrangement specified in 14.4.5.1 is allowed only when the tank is used to supplement another water supply such as a city supply or when the tank connects into the fire protection main on the discharge side of the fire pump. With this arrangement, the control valve in the bypass is opened only for filling the tank and relies on city or fire pump pressure to fill the tank.

14.4.5.2 The bypass shall be connected into tapped bosses on the check valve or into the discharge pipe between the check valve and all other valves.

Most large bore [6 in. (150 mm) and larger] check valves can be tapped for up to 2 in. (50 mm) connections on each side of the clapper. The bypass is required to be connected by means of this tap in the check valve body. This method provides a convenient connection for the bypass. The bypass consists of a few short lengths of pipe, two 90 degree elbows, and a threaded

EXHIBIT III.5 Check Valve with Bypass.

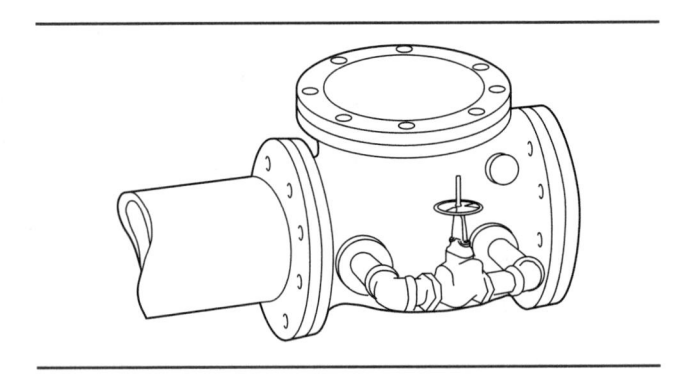

gate valve. The gate valve is normally closed and is only opened to fill the tank. Exhibit III.5 illustrates a typical fill connection.

14.4.5.3 The bypass shall be sized to fill the tank in accordance with 13.4.2 but shall not be smaller than 2 in. (50 mm).

14.4.5.4 A listed indicating control valve shall be placed in the bypass and shall be kept closed except when the tank is being filled.

14.4.6 Filling Pumps.

14.4.6.1 When the tank is to be filled by a filling pump, the pump and connections shall be of such size that the tank can be filled in accordance with 13.4.2.

The 8-hour requirement is a maximum fill time. Any arrangement that fills the tank sooner should be considered to be acceptable.

14.4.6.2 The filling pipe shall be of at least 2 in. (50 mm) and, except as noted in 14.4.7, shall be connected directly into the tank discharge pipe, in which case a listed indicating control valve and a check valve shall be placed in the filling pipe near the tank discharge pipe, with the check valve located on the pump side of the listed indicating valve.

14.4.6.3 The filling pump suction pipe shall not be connected to a fire service main that is supplied from the tank. The filling valve shall be open only when the tank is being filled.

14.4.7 Where a separate fill pipe is used, automatic filling shall be permitted.

For automatic filling, the tank water level is monitored and the fill pump is turned on when the water level drops. With this arrangement, the fill valve is normally open. If the exact fill rate is known and the reliability of the fill pump actuation mechanism is verified, the fill rate can be considered when the tank capacity is calculated. This approach to filling must be presented to the authority having jurisdiction for approval prior to the purchase and installation of the tank.

14.4.8 Filling from Drinking Water Supply. Where the water in the fire protection system is not suitable for drinking purposes and the tank is filled from a potable water supply, the filling pipe shall be installed in accordance with the regulations of the local health authority.

When filling from a drinking water supply, the fill connection may create a cross-connection that could ultimately contaminate the potable water supply. When this arrangement is proposed, a backflow preventer should be installed to prevent contamination of the potable water supply. The friction loss through the backflow preventer must be considered when calculating the flow capacity of the fill connection in order to meet the 8-hour fill requirement.

14.4.9 Filling Pipe at Roofs and Floors. The intersection of a separate filling pipe with a roof or a waterproof or concrete floor shall be watertight.

14.4.10 Suction Tanks.

14.4.10.1 Pipes for the automatic filling of suction tanks shall discharge into the opposite half of the tanks from the pump suction pipe.

14.4.10.2 Where an over-the-top fill line is used, the outlet shall be directed downward.

14.5 Overflow.

14.5.1 Size. The overflow pipe shall be of adequate capacity for the operating conditions and shall be of not less than 3 in. (75 mm) throughout.

In cases where the fill valve is left open or the automatic fill connection fails to open, an overflow connection is required to prevent overflowing of the tank and potential damage to the tank roof. The following subsection 14.5.2 is intended to ensure that the overflow pipe is properly pitched to automatically drain and that the overflow piping capacity exceeds that of the fill connection.

14.5.2 Inlet.

14.5.2.1 The inlet of the overflow pipe shall be located at the top capacity line or high waterline.

14.5.2.2 The inlet also shall be located at least 1 in. (25 mm) below the bottom of the flat cover joists in a wood tank, but shall never be closer than 2 in. (50 mm) to the top of the tank.

14.5.2.3 Unless the maximum fill capacity is known and the overflow capacity is calculated to be at least equal to the fill capacity, the overflow pipe shall be at least one pipe size larger than the fill line and shall be equipped with an inlet such as a concentric reducer, or equivalent, that is at least 2 in. (50 mm) larger in diameter.

14.5.2.4 The inlet shall be arranged so that the flow of water is not retarded by any obstruction.

14.5.2.5 An overflow pipe that is cut with the opening to fit the roof shall be used on a steel tank, provided a suitable horizontal suction plate and vortex breaker are used to ensure full capacity flow for the overflow.

14.5.3* Stub Pipe.

A.14.5.3 On column-supported tanks with outside overflow, vertical extensions of the pipe that is located below the balcony are not recommended, as they can become plugged with ice.

14.5.3.1 Where dripping water or a small accumulation of ice is not objectionable, the overflow shall be permitted, at the discretion of the owner, to pass through the side of the tank near the top.

14.5.3.2 The pipe shall be extended with a slight downward pitch to discharge beyond the tank or balcony and away from the ladders and shall be adequately supported.

14.5.3.3 Overflows for pedestal tanks shall be extended to ground level within the access tube and pedestal.

14.5.4 Inside Pipe.

14.5.4.1 Where a stub pipe is undesirable, the overflow pipe shall extend down through the tank bottom and inside the frostproof casing or steel-plate riser and shall discharge through the casing near the ground or roof level.

14.5.4.2* The section of the pipe inside the tank shall be of brass, flanged cast-iron, or steel.

A.14.5.4.2 See 14.1.12.

14.5.4.2.1 Inside overflow pipes shall be braced by substantial clamps to tank and riser plates at points not over 25 ft (7.6 m) apart.

14.5.4.2.2 The discharge shall be visible, and the pipe shall be pitched to drain.

14.5.4.2.3 Where the discharge is exposed, the exposed length shall not exceed 4 ft (1.2 m) and shall avoid the entrance to the valve pit or house.

14.6 Clean-Out and Drain.

14.6.1 Handhole. A standard handhole, with a minimum dimension of 3 in. (76 mm), or a manhole shall be provided in the saucer plate outside of the frostproof casing and at the bottom of an elevated steel tank with a suspended bottom unless the tank has a large riser pipe 3 ft (0.91 m) or more in diameter.

14.6.2 Manholes.

14.6.2.1 Two manholes shall be provided in the first ring of the steel suction tank shell at locations to be designated by the purchaser.

The manholes can serve the following two purposes:

1. Accommodate confined space entry requirements, thus providing a separate exit route
2. Permit the temporary installation of an exhaust fan to introduce fresh air when the tank is drained and opened for maintenance

14.6.2.1.1 The design of the manholes for steel tanks shall be in accordance with AWWA D100 for welded steel tanks, and AWWA D103 for bolted steel tanks.

14.6.3 For Elevated Flat-Bottom Tanks.

14.6.3.1 Where elevated, at least a 2 in. (50 mm) pipe clean-out also shall be provided outside of the frostproof casing in the bottom of a wood tank or a flat-bottom steel tank.

14.6.3.2 The clean-out connection for wood tanks shall consist of a special screw fitting with a gasket or a pair of 2 in. (50 mm) pipe flanges.

14.6.3.3 The connection for steel tanks shall consist of an extra-heavy coupling welded to the bottom plate.

14.6.3.4 The coupling shall be welded to both sides of the tank plates.

14.6.3.5 A piece of 2 in. (50 mm) brass pipe about 5 in. (127 mm) long that is capped at the top with a brass cap shall be screwed into the inner fitting or flange.

14.6.3.6* The clean-out shall be watertight.

A.14.6.3.6 See Figure B.1(k).

14.6.4 Riser Drain.

14.6.4.1 A drain pipe of at least 2 in. (50 mm) that is fitted with a controlling valve and a ½ in. (12 mm) drip valve shall be connected into the tank discharge pipe near its base and, where possible, on the tank side of all valves.

14.6.4.2 Where the outlet is an open end outlet, it shall be fitted with a 2½ in. (65 mm) hose connection unless it discharges into a funnel or cistern piped to a sewer.

14.6.4.3 Where the drain is piped directly to a sewer, a sight glass or a ¾ in. (19.1 mm) test valve on the underside of the pipe shall be provided.

14.6.4.4 Where the drain pipe is to be used for a hose stream, the controlling valve shall be a listed gate valve or angle valve.

14.6.4.5* Where a circulation-tank heater is located near the base of the tank riser, the drain pipe shall, if possible, be connected from the cold-water return pipe between the cold-water valve and the heater in order to permit flushing water from the tank through the hot-water pipe heater and drain for clean-out purposes.

A.14.6.4.5 See Figure B.1(i).

14.7 Connections for Other Than Fire Protection.

14.7.1* Dual-Service Tanks. Where dual service is necessary, an adequate supply of water shall be constantly and automatically reserved in the tank for fire protection purposes.

A.14.7.1 The use of an elevated tank, in part, for purposes other than fire protection, is not advised. Frequent circulation of the water results in an accumulation of sediment that can obstruct the piping of sprinklers, and a fluctuating water level hastens decaying of wood and corrosion of steel.

14.7.2 Pipe for Other Than Fire Protection Purposes.

14.7.2.1 Pipe used for other than fire protection purposes shall be entirely separate from fire-service pipes and shall extend to an elevation inside the tank below which an adequate quantity of water is constantly available for fire protection.

14.7.2.2 Pipe inside the tank that is used for other than fire protection purposes shall be brass.

14.7.2.2.1 Steel pipe shall be permitted to be used where the pipe is larger than 3 in. (75 mm), or cast-iron shall be permitted where the pipe is 6 in. (150 mm) or larger.

14.7.2.3 Pipe inside the tank shall be braced near the top and at points not over 25 ft (7.6 m) apart.

14.7.2.4* Where an expansion joint exists, it shall be of the standard type, shall be located below the tank, and shall be without connection to the tank plates.

A.14.7.2.4 See 14.3.8.

14.7.3* At Roofs and Floors. Where a pipe used for other than fire protection purposes intersects with a building roof or a waterproof or concrete floor, the intersection shall be watertight.

A.14.7.3 See 14.1.1.

14.8* Sensors.

A.14.8 It is not the intent of this standard to require the electronic supervision of tanks; however, where such supervision is required in accordance with *NFPA 72*, the following alarms should be required:

(1) Water temperature below 40°F (4.4°C)
(2) Return of water temperature to 40°F (4.4°C)
(3) Water level 3 in. (76.2 mm) (pressure tanks) or 5 in. (127 mm) (all other tanks) below normal

(4) Return of water level to normal
(5) Pressure in pressure tank 10 psi (0.48 kPa) below normal
(6) Pressure in pressure tank 10 psi (0.48 kPa) above normal

14.8.1 Provisions shall be made for the installation of sensors in accordance with *NFPA 72* for two critical water temperatures and two critical water levels.

14.8.2 Pressure Tanks. In addition to the requirements of 14.8.1, pressure tanks shall be provided with connections for the installation of high- and low-water pressure supervisory signals in accordance with *NFPA 72*.

Wood tanks should be filled and tested for liquid tightness for 48 hours. The leakage test for wood tanks must last for 48 hours in order to allow the wood to become completely wet and swell. Dry wood, when first installed, will most likely leak. This test should be done under the supervision of a qualified wood tank specialist. The test should be in accordance with the National Wood Tank Institute Bulletin S82, "Specification for Tanks and Pipes."

For fiberglass-reinforced plastic (FRP) tanks, a 4 in. (100 mm) standpipe should be attached to the tank that extends 4 ft (1.2 m) above the top of the tank. The tank and standpipe should be filled with water and allowed to stand for 24 hours. The tank should be examined for leakage or drop in water elevation in the standpipe. The tank should show no visible signs of leakage, and the water level should not fall more than ½ in. (13 mm) within the 24-hour test period.

The hydrostatic test for an FRP tank should be completed before backfill in order to inspect the tank for leaks. The tank should be in place and properly secured before filling with water.

• • •

CHAPTER 17 ACCEPTANCE TEST REQUIREMENTS

17.1 Inspection of Completed Equipment.

The installing contractor should notify the local authority having jurisdiction and encourage their involvement in the inspection of the completed tank and accessories. In some jurisdictions, work or commissioning cannot proceed without a written authorization from the authority having jurisdiction. On large projects, this inspection should also involve the general contractor and insurance company representative for their approval. Verification in writing that the inspection has been completed and the work has been accepted is important before concealment of any work or portion of the system and equipment, since once concealed, inspection is difficult.

17.1.1 Prior to placing the tank in service, a representative of the tank contractor and a representative of the owner shall conduct a joint inspection of the completed equipment.

Immediately following construction of the tank, an inspection of the joints provides a training opportunity for the operating personnel to become familiar with the operation and maintenance of the tank and related accessories. Operation and maintenance personnel should become familiar with NFPA 25 at this time, and the critical inspection, test, and maintenance requirements for tanks and valves should be highlighted during this training period.

17.1.2* Written reports of completed equipment inspections shall be made in triplicate, and a copy that has been signed by the contractors and the owners shall be sent to the authority having jurisdiction.

These final written inspection reports can be part of the test report previously mentioned and should be included in the final submission as an "as-built" record of completion of work. The tank installation specification and contract should be written in such a fashion so as to withhold final payment until all of the aforementioned documentation has been submitted and approved by the authority having jurisdiction and RDP.

A.17.1.2 This joint inspection reasonably ensures that there are no defects in the work of sufficient importance to prevent the system from being put into service immediately. The inspection also permits the owner's representatives to become more familiar with the system and equipment.

17.2 Testing.

17.2.1 All coated steel tanks shall be tested for holidays and coating thickness.

17.2.2 Corrective action shall be completed prior to acceptance.

The paint system on steel tanks must be inspected to verify that it has been applied correctly. Failure of the paint system results in excessive corrosion and potential failure of the tank.

17.3 Welded Steel Tanks.

17.3.1 Flat Bottoms. Upon completion of the welding of the tank bottom, it shall be tested by one of the following methods and shall be made entirely tight:

(1) Air pressure or vacuum applied to the joints, using soap suds, linseed oil, or other suitable material for the detection of leaks
(2) Joints tested by the magnetic particle method

17.3.2 General. Upon completion of the tank construction, it shall be filled with water furnished at the tank site by the owner's representative using the pressure necessary to fill the tank to the maximum working water level.

The fill test described in 17.3.2, although directed primarily towards welded-steel tanks, should be conducted on all tanks prior to placing them in service.

17.3.3 Any leaks in the shell, bottom, or roof (if the roof contains water) that are disclosed by the test shall be repaired by chipping or melting out any defective welds and then rewelding.

17.3.4 Repair work shall be done on joints only when the water in the tank is a minimum of 2 ft (0.6 m) below the point under repair.

17.3.5 The tank shall be tested as watertight to the satisfaction of the authority having jurisdiction and/or the owner's representative.

Although documentation of the fill test is not required by NFPA 22, the fill test should be documented and signed by the owner's representative, RDP, and authority having jurisdiction. A copy of this test report should be included in the final submission package.

17.4* Bolted Steel Tanks.

The completed tank shall be tested by filling it with water, and any detected leaks shall be repaired in accordance with AWWA D103.

A.17.4 Care should be taken when retorquing bolts in leaking areas. Overtorqued bolts can cause linings to crack, to splinter, or to be otherwise damaged. Manufacturers' recommendations for the repair or replacement of panels should be followed.

17.5 Pressure Tanks.

Tests shall be performed according to 17.5.1 through 17.5.4.

17.5.1 Each pressure tank shall be tested in accordance with the ASME *Boiler and Pressure Vessel Code,* "Rules for the Construction of Unfired Pressure Vessels," before painting.

17.5.1.1 The hydrostatic test pressure shall be a minimum of 150 lb/in.^{2} (10.3 bar).

17.5.2 In addition to the ASME tests, each pressure tank shall be filled to two-thirds of its capacity and tested at the normal working pressure with all valves closed and shall not lose more than ½ psi (0.03 bar) pressure in 24 hours.

17.5.3 A certificate signed by the manufacturer that certifies that the foregoing tests have been made shall be filed with the authority having jurisdiction.

The certificate should also be signed by the owner's representative, RDP, and authority having jurisdiction. The completed certificate or test report should be included in the final submission package.

17.5.4 A repetition of the tests specified in 17.5.1 through 17.5.3 shall be required after the tank has been set in place and connected. Where conditions do not allow shipping the tank after it is assembled, these tests shall be conducted following its assembly in the presence of a representative of the authority having jurisdiction.

17.6 Embankment-Supported Coated Fabric Tanks.

17.6.1 The tank shall be tested for leakage prior to shipment.

The "shop test" report should accompany shipment of the tank and should be included in the final submission package. This test is conducted in the shop prior to shipment.

17.6.2 The tank also shall be tested for leakage after installation.

The field test report should be signed by the owner's representative, RDP, and authority having jurisdiction and should be included in the final submission package.

17.7 Concrete Tanks.

17.7.1 Leakage Testing. On completion of the tank and prior to any specified backfill placement at the footing or wall, the test specified in 17.7.2 through 17.7.4 shall be applied to ensure watertightness.

17.7.2 Preparation. The tank shall be filled with water to the maximum level and left to stand for at least 24 hours.

17.7.3 Measurement. The drop in liquid level shall be measured over the next 72-hour period to determine the liquid volume loss. Evaporative losses shall be measured or calculated and shall be deducted from the measured loss to determine whether there is net leakage.

17.7.4 There shall be no measurable leakage after the tank is placed in service.

A test report should be completed and signed by the installing contractor, owner's representative, RDP, and authority having jurisdiction and should be included in the final submission package.

17.8 Wood Tanks.

17.8.1 Wood tanks shall be filled and tested for liquid tightness for 48 hours.

17.8.2 Testing shall be done under the supervision of a qualified wood tank specialist.

17.8.3 Tests shall be in accordance with the National Wood Tank Institute Bulletin S82.

17.9 Fiberglass-Reinforced Plastic Tanks—Hydrostatic Test.

17.9.1 After the excavation hole is backfilled to the bottom of the influent and effluent piping, influent and effluent piping shall be sealed off with watertight caps or plugs.

17.9.2 The tank shall be filled with water up to 3 in. (76 mm) into the access openings.

17.9.3 The water shall be allowed to stand in the tank for a minimum of 2 hours.

17.9.4 The tank shall be examined for leakage or drop in water elevation.

17.9.5 If the water level drops, plugs or caps sealing off piping shall be checked to see that they are tight.

17.9.6 If tightening is required, more water shall be added to fill air voids back to the standard testing level.

17.9.7 The tank shall show no visible signs of leakage, and the water level shall stabilize within a 2-hour test period.

17.10 Disposal of Test Water.

The owner's representative shall provide a means for disposing of test water up to the tank inlet or drain pipe.

CHAPTER 18 INSPECTION, TESTING, AND MAINTENANCE OF WATER TANKS

18.1 General.

Tanks shall be periodically inspected, tested, and maintained in accordance with NFPA 25.

. . .

SUMMARY

Portions of NFPA 22 have been included in this handbook because water storage tanks are frequently used to supply water to fire pumps. Interfacing the installation of water storage tanks and fire pumps can be difficult if the installation requirements of each system are not fully understood. Water storage tanks are frequently used as "break tanks" to provide a separation between the attached water supply and the fire pump. This arrangement stabilizes the supply pressure to fire pumps, thus preventing the fire pump from overpressurizing the fire protection system due to varying pressures in the attached water supply. The intention of this handbook is to provide the proper guidance for the design, installation, and acceptance testing for both tanks and pumps in order to improve their overall system performance.

REFERENCES CITED IN COMMENTARY

National Fire Protection Association, 1 Batterymarch Park, Quincy, MA 02169-7471.

NFPA 13, *Standard for the Installation of Sprinkler Systems,* 2010 edition.
NFPA 20, *Standard for the Installation of Stationary Pumps for Fire Protection,* 2010 edition.
NFPA 24, *Standard for the Installation of Private Fire Service Mains and Their Appurtenances,* 2010 edition.
NFPA 25, *Standard for the Inspection, Testing, and Maintenance of Water-Based Fire Protection Systems,* 2008 edition.
NFPA 72®, National Fire Alarm and Signaling Code, 2010 edition.
NFPA 241, *Standard for Safeguarding Construction, Alteration, and Demolition Operations,* 2009 edition.

National Wood Tank Institute, 5500 N. Water Street, Philadelphia, PA 19120.

NWTI Bulletin S82, "Specification for Tanks and Pipes."

Complete Text of NFPA® 24,

SECTION 2

Standard for the Installation of Private Fire Service Mains and Their Appurtenances, 2010 Edition

Because most fire pumps are served by a private fire service main, the fire pump designer and installer need to recognize the requirements for the installation of a private fire service main in order to properly interface with the water supply piping. In many cases, the installation of a new private water supply includes the design and installation of a private fire service main and fire pump. NFPA 24 is included in its entirety in this handbook to assist in the design, installation, and acceptance testing of a complete fire protection water supply.

CHAPTER 1 ADMINISTRATION

1.1 Scope.

1.1.1 This standard shall cover the minimum requirements for the installation of private fire service mains and their appurtenances supplying the following:

(1) Automatic sprinkler systems
(2) Open sprinkler systems
(3) Water spray fixed systems
(4) Foam systems
(5) Private hydrants
(6) Monitor nozzles or standpipe systems with reference to water supplies
(7) Hose houses

The scope of this standard establishes the requirements for the installation of private fire service mains, which distinguishes NFPA 24 from other standards that address public service mains, such as those published by the American Water Works Association (AWWA). An important distinction to remember is that NFPA 24 deals with the underground main once that pipe enters private property and becomes the responsibility of the building owner. See the definition *private fire service main* in 3.3.11 and accompanying annex text and commentary.

1.1.2 This standard shall apply to combined service mains used to carry water for fire service and other uses.

1.1.3 This standard shall not apply to the following situations:

(1) Mains under the control of a water utility
(2) Mains providing fire protection and/or domestic water that are privately owned but are operated as a water utility

1.1.4 This standard shall not apply to underground mains serving sprinkler systems designed and installed in accordance with NFPA 13R that are under 4 in. (102 mm) in size.

1.1.5 This standard shall not apply to underground mains serving sprinkler systems designed and installed in accordance with NFPA 13D.

The 2010 edition of NFPA 24 has been revised to clarify that the requirements of NFPA 24 do not apply to public mains or to mains owned by private companies that operate as water utilities. The technical committee believes that requirements for listed pipe and 150 psi (10 bar) rated piping are unnecessary and unrealistic for some utilities to meet. NFPA 13D, *Standard for the Installation of Sprinkler Systems in One- and Two-Family Dwellings and Manufactured Homes,* and NFPA 13R, *Standard for the Installation of Sprinkler Systems in Residential Occupancies up to and Including Four Stories in Height,* adequately address underground piping for systems.

1.2 Purpose.

The purpose of this standard shall be to provide a reasonable degree of protection for life and property from fire through installation requirements for private fire service main systems based on sound engineering principles, test data, and field experience.

1.3 Retroactivity.

The provisions of this standard reflect a consensus of what is necessary to provide an acceptable degree of protection from the hazards addressed in this standard at the time the standard was issued.

1.3.1 Unless otherwise specified, the provisions of this standard shall not apply to facilities, equipment, structures, or installations that existed or were approved for construction or installation prior to the effective date of the standard. Where specified, the provisions of this standard shall be retroactive.

1.3.2 In those cases where the authority having jurisdiction determines that the existing situation presents an unacceptable degree of risk, the authority having jurisdiction shall be permitted to apply retroactively any portions of this standard deemed appropriate.

1.3.3 The retroactive requirements of this standard shall be permitted to be modified if their application clearly would be impractical in the judgment of the authority having jurisdiction and only where it is clearly evident that a reasonable degree of safety is provided.

1.4 Equivalency.

Nothing in this standard is intended to prevent the use of systems, methods, or devices of equivalent or superior quality, strength, fire resistance, effectiveness, durability, and safety over those prescribed by this standard. Technical documentation shall be submitted to the authority having jurisdiction to demonstrate equivalency. The system, method, or device shall be approved for the intended purpose by the authority having jurisdiction.

1.5 Units.

1.5.1 Metric units of measurement in this standard shall be in accordance with the modernized metric system known as the International System of Units (SI). Liter and bar units are not part of, but are recognized by, SI and are commonly used in international fire protection. These units are shown in Table 1.5.1 with conversion factors.

1.5.2 If a value for measurement as given in this standard is followed by an equivalent value in other units, the first stated is to be regarded as the requirement. A given equivalent value might be approximate.

1.5.3 SI units have been converted by multiplying the quantity by the conversion factor and then rounding the result to the appropriate number of significant digits.

TABLE 1.5.1 Conversion Table for SI Units

Name of Unit	Unit Symbol	Conversion Factor
Liter	L	1 gal = 3.785 L
Liter per minute per square meter	(L/min)/m^2	1 gpm/ft^2 = (40.746 L/min)/m^2
Cubic decimeter	dm^3	1 gal = 3.785 dm^3
Pascal	Pa	1 psi = 6894.757 Pa
Bar	bar	1 psi = 0.0689 bar
Bar	bar	1 bar = 10^5 Pa

Note: For additional conversions and information, see IEEE/ASTM-SI-10.

CHAPTER 2 REFERENCED PUBLICATIONS

2.1 General.

The documents or portions thereof listed in this chapter are referenced within this standard and shall be considered part of the requirements of this document.

2.2 NFPA Publications.

National Fire Protection Association, 1 Batterymarch Park, Quincy, MA 02169-7471.

NFPA 13, *Standard for the Installation of Sprinkler Systems,* 2010 edition.
NFPA 13D, *Standard for the Installation of Sprinkler Systems in One- and Two-Family Dwellings and Manufactured Homes,* 2010 edition.
NFPA 13R, *Standard for the Installation of Sprinkler Systems in Residential Occupancies up to and Including Four Stories in Height,* 2010 edition.
NFPA 20, *Standard for the Installation of Stationary Pumps for Fire Protection,* 2010 edition.
NFPA 22, *Standard for Water Tanks for Private Fire Protection,* 2008 edition.
NFPA 25, *Standard for the Inspection, Testing, and Maintenance of Water-Based Fire Protection Systems,* 2008 edition.
NFPA 780, *Standard for the Installation of Lightning Protection Systems,* 2008 edition.
NFPA 1961, *Standard on Fire Hose,* 2007 edition.
NFPA 1963, *Standard for Fire Hose Connections,* 2009 edition.

2.3 Other Publications.

2.3.1 ASME Publications.

American Society of Mechanical Engineers, Three Park Avenue, New York, NY 10016-5990.

ASME B1.20.1, *Pipe Threads, General Purpose (Inch),* 2001.
ASME B16.1, *Gray Iron Pipe Flanges and Flanged Fittings, Classes 25, 125, and 250,* 2005.
ASME B16.3, *Malleable Iron Threaded Fittings, Classes 150 and 300,* 2006.
ASME B16.4, *Gray Iron Threaded Fittings, Classes 125 and 250,* 2006.
ASME B16.5, *Pipe Flanges and Flanged Fittings NPS ½ through 24,* 2003.
ASME B16.9, *Factory-Made Wrought Steel Buttweld Fittings,* 2007.

ASME B16.11, *Forged Steel Fittings, Socket Welded and Threaded*, 2005.
ASME B16.18, *Cast Bronze Solder Joint Pressure Fittings*, 2001.
ASME B16.22, *Wrought Copper and Bronze Solder Joint Pressure Fittings*, 2001.
ASME B16.25, *Buttwelding Ends*, 2007.

2.3.2 ASTM Publications.
ASTM International, 100 Barr Harbor Drive, P.O. Box C700, West Conshohocken, PA 19428-2959.

ASTM A 234, *Specification for Piping Fittings of Wrought Carbon Steel and Alloy Steel for Moderate and Elevated Temperatures*, 2007.
ASTM B 75, *Specification for Seamless Copper Tube*, 2002.
ASTM B 88, *Specification for Seamless Copper Water Tube*, 2003.
ASTM B 251, *Requirements for Wrought Seamless Copper and Copper-Alloy Tube*, 2002.
ASTM F 437, *Chlorinated Polyvinyl Chloride (CPVC) Specification for Schedule 80 CPVC Threaded Fittings*, 2006.
ASTM F 438, *Specification for Schedule 40 CPVC Socket-Type Fittings*, 2004.
ASTM F 439, *Specification for Schedule 80 CPVC Socket-Type Fittings*, 2006.
IEEE/ASTM-SI-10, *Standard for Use of the International System of Units (SI): The Modern Metric System*, 2002.

2.3.3 AWWA Publications.
American Water Works Association, 6666 West Quincy Avenue, Denver, CO 80235.

AWWA C104, *Cement Mortar Lining for Ductile Iron Pipe and Fittings for Water*, 2008.
AWWA C105, *Polyethylene Encasement for Ductile Iron Pipe Systems*, 2005.
AWWA C110, *Ductile Iron and Gray Iron Fittings*, 2008.
AWWA C111, *Rubber-Gasket Joints for Ductile Iron Pressure Pipe and Fittings*, 2000.
AWWA C115, *Flanged Ductile Iron Pipe with Ductile Iron or Gray Iron Threaded Flanges*, 2005.
AWWA C116, *Protective Fusion-Bonded Epoxy Coatings for the Interior and Exterior Surfaces of Ductile-Iron and Gray-Iron Fittings for Water Supply Service*, 2003.
AWWA C150, *Thickness Design of Ductile Iron Pipe*, 2008.
AWWA C151, *Ductile Iron Pipe, Centrifugally Cast for Water*, 2002.
AWWA C153, *Ductile-Iron Compact Fittings for Water Service*, 2006.
AWWA C200, *Steel Water Pipe 6 in. and Larger*, 2005.
AWWA C203, *Coal-Tar Protective Coatings and Linings for Steel Water Pipelines Enamel and Tape—Hot Applied*, 2002.
AWWA C205, *Cement-Mortar Protective Lining and Coating for Steel Water Pipe 4 in. and Larger—Shop Applied*, 2007.
AWWA C206, *Field Welding of Steel Water Pipe*, 2003.
AWWA C207, *Steel Pipe Flanges for Waterworks Service—Sizes 4 in. Through 144 in.*, 2007.
AWWA C208, *Dimensions for Fabricated Steel Water Pipe Fittings*, 2007.
AWWA C300, *Reinforced Concrete Pressure Pipe, Steel-Cylinder Type*, 2004.
AWWA C301, *Prestressed Concrete Pressure Pipe, Steel-Cylinder Type*, 2007.
AWWA C302, *Reinforced Concrete Pressure Pipe, Non-Cylinder Type*, 2004.
AWWA C303, *Reinforced Concrete Pressure Pipe, Steel-Cylinder Type, Pretensioned*, 2002.
AWWA C400, *Standard for Asbestos-Cement Distribution Pipe, 4 in. Through 16 in. (100 mm through 400 mm), for Water Distribution Systems*, 2003.
AWWA C401, *Standard for the Selection of Asbestos-Cement Pressure Pipe 4 in. through 16 in. (100 mm through 400 mm)*, 2003.
AWWA C600, *Standard for the Installation of Ductile Iron Water Mains and Their Appurtenances*, 2005.
AWWA C602, *Cement-Mortar Lining of Water Pipe Lines 4 in. and Larger—in Place*, 2006.
AWWA C603, *Standard for the Installation of Asbestos-Cement Pressure Pipe*, 2005.

AWWA C900, *Polyvinyl Chloride (PVC) Pressure Pipe, 4 in. Through 12 in., for Water Distribution,* 2007.

AWWA C906, *Polyethylene (PE) Pressure Pipe and Fittings, 4 in. (100 mm) Through 63 in. (1575 mm) for Water Distribution,* 2007.

AWWA M11, *A Guide for Steel Pipe Design and Installation,* 4th edition, 2004.

2.3.4 Other Publications.

Merriam-Webster's Collegiate Dictionary, 11th edition, Merriam-Webster, Inc., Springfield, MA, 2003.

2.4 References for Extracts in Mandatory Sections.

NFPA 20, *Standard for the Installation of Stationary Pumps for Fire Protection,* 2010 edition.

CHAPTER 3 DEFINITIONS

3.1 General.

The definitions contained in this chapter shall apply to the terms used in this standard. Where terms are not defined in this chapter or within another chapter, they shall be defined using their ordinarily accepted meanings within the context in which they are used. *Merriam-Webster's Collegiate Dictionary,* 11th edition, shall be the source for the ordinarily accepted meaning.

3.2 NFPA Official Definitions.

3.2.1* Approved. Acceptable to the authority having jurisdiction.

A.3.2.1 Approved. The National Fire Protection Association does not approve, inspect, or certify any installations, procedures, equipment, or materials; nor does it approve or evaluate testing laboratories. In determining the acceptability of installations, procedures, equipment, or materials, the authority having jurisdiction may base acceptance on compliance with NFPA or other appropriate standards. In the absence of such standards, said authority may require evidence of proper installation, procedure, or use. The authority having jurisdiction may also refer to the listings or labeling practices of an organization that is concerned with product evaluations and is thus in a position to determine compliance with appropriate standards for the current production of listed items.

3.2.2* Authority Having Jurisdiction (AHJ). An organization, office, or individual responsible for enforcing the requirements of a code or standard, or for approving equipment, materials, an installation, or a procedure.

A.3.2.2 Authority Having Jurisdiction (AHJ). The phrase "authority having jurisdiction," or its acronym AHJ, is used in NFPA documents in a broad manner, since jurisdictions and approval agencies vary, as do their responsibilities. Where public safety is primary, the authority having jurisdiction may be a federal, state, local, or other regional department or individual such as a fire chief; fire marshal; chief of a fire prevention bureau, labor department, or health department; building official; electrical inspector; or others having statutory authority. For insurance purposes, an insurance inspection department, rating bureau, or other insurance company representative may be the authority having jurisdiction. In many circumstances, the property owner or his or her designated agent assumes the role of the authority having jurisdiction; at government installations, the commanding officer or departmental official may be the authority having jurisdiction.

3.2.3 Labeled. Equipment or materials to which has been attached a label, symbol, or other identifying mark of an organization that is acceptable to the authority having jurisdiction and concerned with product evaluation, that maintains periodic inspection of production of labeled equipment or materials, and by whose labeling the manufacturer indicates compliance with appropriate standards or performance in a specified manner.

3.2.4* Listed. Equipment, materials, or services included in a list published by an organization that is acceptable to the authority having jurisdiction and concerned with evaluation of products or services, that maintains periodic inspection of production of listed equipment or materials or periodic evaluation of services, and whose listing states that either the equipment, material, or service meets appropriate designated standards or has been tested and found suitable for a specified purpose.

A.3.2.4 Listed. The means for identifying listed equipment may vary for each organization concerned with product evaluation; some organizations do not recognize equipment as listed unless it is also labeled. The authority having jurisdiction should utilize the system employed by the listing organization to identify a listed product.

3.2.5 Shall. Indicates a mandatory requirement.

3.2.6 Should. Indicates a recommendation or that which is advised but not required.

3.2.7 Standard. A document, the main text of which contains only mandatory provisions using the word "shall" to indicate requirements and which is in a form generally suitable for mandatory reference by another standard or code or for adoption into law. Nonmandatory provisions shall be located in an appendix or annex, footnote, or fine-print note and are not to be considered a part of the requirements of a standard.

3.3 General Definitions.

3.3.1 Appurtenance. An accessory or attachment that enables the private fire service main to perform its intended function.

As used in NFPA 24, an appurtenance can be a hydrant, valve, monitor nozzle (for master streams), or a fire department connection (FDC).

3.3.2 Corrosion Resistant Piping. Piping that has the property of being able to withstand deterioration of its surface or its properties when exposed to its environment.

Corrosion resistant piping is often equated with galvanized steel piping. However, this standard restricts the use of galvanized steel piping to fire department connections. Piping such as polyvinyl chloride (PVC) can be considered corrosion resistant.

3.3.3 Corrosion Retardant Material. A lining or coating material that when applied to piping or appurtenances has the property of reducing or slowing the deterioration of the object's surface or properties when exposed to its environment.

Piping used for underground applications is frequently coated with corrosion retardant materials such as cement mortar lining, epoxy, petroleum asphalt, or bituminous coatings. The coatings increase the survivability of the piping when buried. Some soil conditions can create a corrosive environment, and in some cases, stray electric currents can accelerate corrosion of pipe and fittings.

3.3.4 Fire Department Connection. A connection through which the fire department can pump supplemental water into the sprinkler system, standpipe, or other system, furnishing water for fire extinguishment to supplement existing water supplies.

3.3.5 Fire Pump. A pump that is a provider of liquid flow and pressure dedicated to fire protection. [20, 2010]

3.3.6 Hose House. An enclosure located over or adjacent to a hydrant or other water supply designed to contain the necessary hose nozzles, hose wrenches, gaskets, and spanners to be used in fire fighting in conjunction with and to provide aid to the local fire department.

3.3.7 Hydrant Butt. The hose connection outlet of a hydrant.

3.3.8 Hydraulically Calculated Water Demand Flow Rate. The water flow rate for a system or hose stream that has been calculated using accepted engineering practices.

3.3.9 Pressure.

3.3.9.1 Residual Pressure. The pressure that exists in the distribution system, measured at the residual hydrant at the time the flow readings are taken at the flow hydrants.

The residual pressure is measured when the flow hydrant is open and discharging through at least one outlet. This pressure is measured at the pressure or test hydrant when fire flow is occurring. See Exhibit III.6.

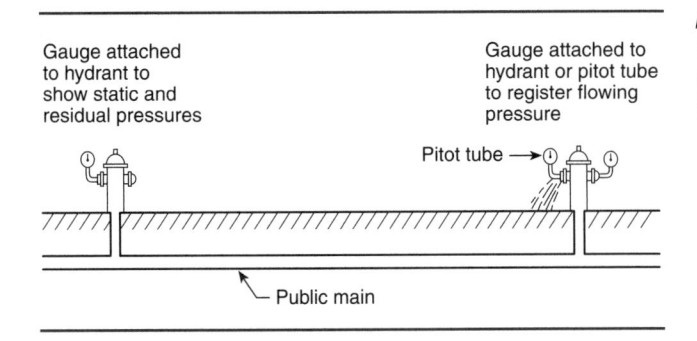

EXHIBIT III.6 Method of Conducting Flow Tests. (Source: NFPA 13, 2010, Figure A.23.2.1)

3.3.9.2 Static Pressure. The pressure that exists at a given point under normal distribution system conditions measured at the residual hydrant with no hydrants flowing.

Static pressure is the pressure measured at the pressure or test hydrant when no fire flow is occurring.

3.3.10* Pressure Regulating Device. A device designed for the purpose of reducing, regulating, controlling, or restricting water pressure.

Pressure regulating devices are valves normally used only when pressures in the system exceed that for which the system components are rated [usually 175 psi (12.1 bar)]. The exposure of a private fire service main to such high pressures is unusual unless the private fire service main is located on the discharge side of a fire pump.

A.3.3.10 Pressure Regulating Device. Examples include pressure-reducing valves, pressure-control valves, and pressure-restricting devices.

3.3.11* Private Fire Service Main. Private fire service main, as used in this standard, is that pipe and its appurtenances on private property (1) between a source of water and the base of the system riser for water-based fire protection systems, (2) between a source of water and inlets to foam-making systems, (3) between a source of water and the base elbow of private hydrants or monitor nozzles, and (4) used as fire pump suction and discharge piping, (5) beginning at the inlet side of the check valve on a gravity or pressure tank. [**13,** 2010]

A.3.3.11 Private Fire Service Main. See Figure A.3.3.11.

Figure A.3.3.11 shows a private fire service main that is fed from a public water supply. As shown in the figure, the piping becomes the responsibility of the building owner when the

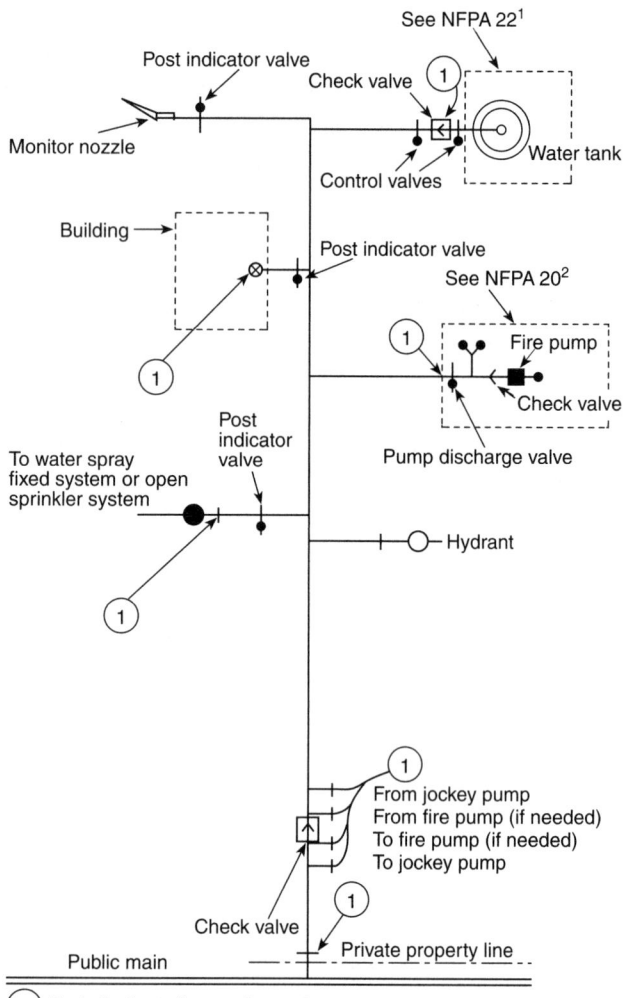

FIGURE A.3.3.11 Typical Private Fire Service Main.

pipe crosses the property line from public to private land. The transition from public to private has implications for acceptance testing, because public mains and accessories are tested to a different standard and a different hydrostatic pressure. This difference in standards can create problems in the field when the private main needs to be tested and isolated from the public main. The two must be separated by a valve that holds the higher pressure, which is sometimes difficult because the valve may not hold the higher test pressure.

Also, note in Figure A.3.3.11 that a fire pump is supplied by a private fire service main. The requirements in NFPA 24 for pipe, fittings, and valves are frequently used to design and install a water supply that serves a fire pump. The piping arrangement can take many forms, such as a direct connection from a public supply to a fire pump, as illustrated in Figure

A.3.3.11. NFPA 24 requirements can also be used for piping between a private water tank and a fire pump and for piping from a fire pump to one or more buildings in a multi-building complex.

3.3.12 Pumper Outlet. The hydrant outlet intended for use by fire departments for taking supply from the hydrant for pumpers.

3.3.13 Rated Capacity. The flow available from a hydrant at the designated residual pressure (rated pressure) either measured or calculated.

The rated capacity and color code of a hydrant as stated in Section D.4 are based on the flow available at a residual pressure of 20 psi (1.4 bar). While not restricted in NFPA 24, some jurisdictions restrict the use of pressures below 20 psi (1.4 bar), thus rating the hydrant at that pressure. By restricting the residual pressure to 20 psi (1.4 bar), problems such as drawing a vacuum on the main and inducing a backflow can be avoided. In some cases, a vacuum can cause the main to collapse, thus damaging the pipe.

3.3.14 Test.

3.3.14.1 Flow Test. A test performed by the flow and measurement of water from one hydrant and the static and residual pressures from an adjacent hydrant for the purpose of determining the available water supply at that location.

NFPA 14, *Standard for the Installation of Standpipe and Hose Systems,* requires that a flow test be conducted not more than 1 year prior to the commencement of the installation of the standpipe system. In most circumstances, if the water supply data are known at a given location in the system, water supply data obtained within the last 5 years can be used. When water supply data older than 1 year are used, consideration should be given to conditions that could affect the existing test data. Examples of such conditions are changes or modifications to the distribution system or increased development in the general area, which would place a higher demand on the system. Evidence of rapid deterioration of the system due to tuberculation, microbiologically influenced corrosion (MIC), or other potential obstruction sources is another condition that could impact the test data.

3.3.14.2 Flushing Test. A test of a piping system using high velocity flows to remove debris from the piping system prior to it being placed in service.

Dirt, rocks, and other obstructing material can enter piping during installation, creating a potential obstruction for sprinklers and valves. For this reason, the flushing test is intended to be conducted prior to connecting the underground system to any aboveground system.

3.3.14.3 Hydrostatic Test. A test of a closed piping system and its attached appurtenances consisting of subjecting the piping to an increased internal pressure for a specified period of duration to verify system integrity and leak rates.

The hydrostatic test is a "stress" test for the system to reveal leaks prior to placing the system in service.

3.3.15 Valve.

3.3.15.1 Check Valve. A valve that allows flow in one direction only.

3.3.15.2 Indicating Valve. A valve that has components that show if the valve is open or closed. Examples are outside screw and yoke (OS&Y) gate valves and underground gate valves with indicator posts.

All NFPA installation standards for fire protection systems require that all valves on fire protection systems be indicating valves. Because the leading cause of sprinkler system failure is a shut valve (over 35 percent according to the NFPA report, "U.S. Experience with Sprinklers and Other Fire Extinguishing Equipment"), indicating valves are required for all systems so that an inspector can easily determine whether the valve is open or closed.

3.4 Hydrant Definitions.

3.4.1 Hydrant. An exterior valved connection to a water supply system that provides hose connections.

3.4.1.1* Dry Barrel Hydrant. This is the most common type of hydrant; it has a control valve below the frost line between the footpiece and the barrel.

A dry barrel hydrant is used in climates where freezing weather occurs. As the name suggests, the barrel or body of the hydrant does not contain water. When the operating nut on top of the hydrant is used to open the hydrant, the barrel is flooded with water. The drain at the bottom of the hydrant should cause the hydrant to drain completely of water within 60 minutes. Hydrants that do not drain within 60 minutes should be marked as requiring pumping to drain.

A.3.4.1.1 Dry Barrel Hydrant. A drain is located at the bottom of the barrel above the control valve seat for proper drainage after operation.

3.4.1.2 Flow Hydrant. The hydrant that is used for the flow and flow measurement of water during a flow test.

3.4.1.3* Private Fire Hydrant. A valved connection on a water supply system having one or more outlets and that is used to supply hose and fire department pumpers with water on private property.

A private hydrant may or may not have a 4 in. (100 mm) connection and is not normally color coded as are public hydrants. See NFPA 291, *Recommended Practice for Fire Flow Testing and Marking of Hydrants,* for hydrant color codes.

A.3.4.1.3 Private Fire Hydrant. Where connected to a public water system, the private hydrants are supplied by a private service main that begins at the point of service designated by the AHJ, usually at a manually operated valve near the property line.

3.4.1.4 Public Hydrant. A valved connection on a water supply system having one or more outlets and that is used to supply hose and fire department pumpers with water.

Public hydrants are normally found on sidewalks or on the side of public streets or roads. These hydrants are usually color coded to indicate their flow capacity at 20 psi (1.4 bar) residual pressure. See Exhibit III.7 for a typical dry barrel hydrant.

3.4.1.5 Residual Hydrant. The hydrant that is used for measuring static and residual pressures during a flow test.

3.4.1.6 Wet Barrel Hydrant. A type of hydrant that sometimes is used where there is no danger of freezing weather. Each outlet on a wet barrel hydrant is provided with a valved outlet threaded for fire hose.

Sometimes referred to as the "California hydrant," the wet barrel hydrant is used in climates that do not experience freezing temperatures. As the name suggests, the barrel is wet or full of water to the outlet caps. Each outlet has its own valve mechanism and can be independently controlled. See Exhibit III.8 for a typical wet barrel hydrant.

CHAPTER 4 GENERAL REQUIREMENTS

4.1* Plans.

A.4.1 Underground mains should be designed so that the system can be extended with a minimum of expense. Possible future plant expansion should also be considered and the piping designed so that it is not covered by buildings.

EXHIBIT III.7 Dry Barrel Hydrant.

EXHIBIT III.8 Wet Barrel Hydrant.

When designing the piping layout, tees with plugs, and perhaps isolation valves, should be installed to accommodate future expansion. The pipe diameter should be increased in order to permit additional waterflow from future expansions.

4.1.1 Working plans shall be submitted for approval to the authority having jurisdiction before any equipment is installed or remodeled.

By submitting plans before installation, the designer and installing contractor will avoid making modifications to installed equipment based on the authority having jurisdiction's plan review. A permit to install such equipment may also be required in some jurisdictions.

4.1.2 Deviation from approved plans shall require permission of the authority having jurisdiction.

4.1.3 Working plans shall be drawn to an indicated scale on sheets of uniform size, with a plan of each floor as applicable, and shall include the following items that pertain to the design of the system:

(1) Name of owner
(2) Location, including street address
(3) Point of compass
(4) A graphic representation of the scale used on all plans
(5) Name and address of contractor
(6) Size and location of all water supplies
(7) Size and location of standpipe risers, hose outlets, hand hose, monitor nozzles, and related equipment
(8) The following items that pertain to private fire service mains:
   (a) Size
   (b) Length
   (c) Location
   (d) Weight
   (e) Material
   (f) Point of connection to city main
   (g) Sizes, types, and locations of valves, valve indicators, regulators, meters, and valve pits
   (h) Depth at which the top of the pipe is laid below grade
   (i) Method of restraint
(9) The following items that pertain to hydrants:
   (a) Size and location, including size and number of outlets and whether outlets are to be equipped with independent gate valves
   (b) Whether hose houses and equipment are to be provided, and by whom
   (c) Static and residual hydrants used in flow
   (d) Method of restraint
(10) Size, location, and piping arrangement of fire department connections

4.1.4 The working plan submittal shall include the manufacturer's installation instructions for any specially listed equipment, including descriptions, applications, and limitations for any devices, piping, or fittings.

4.2 Installation Work.

4.2.1 Installation work shall be performed by fully experienced and responsible persons.

4.2.2 The authority having jurisdiction shall always be consulted before the installation or remodeling of private fire service mains.

CHAPTER 5 WATER SUPPLIES

5.1* Connection to Waterworks Systems.

A.5.1 If possible, dead-end mains should be avoided by arranging for mains to be supplied from both directions. Where private fire service mains are connected to dead-end public mains, each situation should be examined to determine if it is practical to request the water utility to loop the mains to obtain a more reliable supply.

5.1.1 A connection to a reliable waterworks system shall be an acceptable water supply source.

5.1.2 The volume and pressure of a public water supply shall be determined from waterflow test data.

5.1.3 An adjustment to the waterflow test data to account for the following shall be made, as appropriate:

(1) Daily and seasonal fluctuations
(2) Possible interruption by flood or ice conditions
(3) Large simultaneous industrial use
(4) Future demand on the water supply system
(5) Other conditions that could affect the water supply

Water supplies must be evaluated for all of the conditions listed in 5.1.3. The water supply test must be conducted during normal business hours to include the effects of simultaneous business and industrial demand. Investigation into the possibility of future expansion should be discussed with the authority having jurisdiction and accounted for in the system demand, since expansion can be expected to create additional demand on the water supply.

5.2 Size of Fire Mains.

5.2.1 Private Fire Service Mains. Pipe smaller than 6 in. (152.4 mm) in diameter shall not be installed as a private service main supplying hydrants.

Pipes smaller than 6 in. (150 mm) do not permit the full flow capabilities for the hydrant.

5.2.2 Mains Not Supplying Hydrants. For mains that do not supply hydrants, sizes smaller than 6 in. (152.4 mm) shall be permitted to be used subject to the following restrictions:

(1) The main shall supply only the following types of systems:
   (a) Automatic sprinkler systems
   (b) Open sprinkler systems
   (c) Water spray fixed systems
   (d) Foam systems
   (e) Class II standpipe systems
(2) Hydraulic calculations shall show that the main is able to supply the total demand at the appropriate pressure.
(3) Systems that are not hydraulically calculated shall have a main at least as large as the riser.

When private fire service mains supply water to a fire pump, the pipe should be hydraulically calculated to provide not less than 0 psi (0 bar) at the pump suction flange. Regardless of the resulting calculations, the last few feet of piping at the pump suction flange (exactly 10 pipe diameters) must be sized in accordance with Tables 4.26(a) and 4.26(b) in NFPA 20, *Standard for the Installation of Stationary Pumps for Fire Protection* (see Part II of this handbook). The intent of pipe sizing in the aforementioned tables is to limit water velocity to not more than 15 ft/sec (4.57 m/sec).

Private fire service main piping supplying Class I and Class III standpipe systems (not referenced in 5.2.2) must be sized in accordance with NFPA 14 due to the high flows and pressure normally experienced in these systems.

5.3 Pressure-Regulating Devices and Meters.

5.3.1 No pressure-regulating valve shall be used in the water supply, except by special permission of the authority having jurisdiction.

The piping and fittings for private fire service mains are required in 10.1.5 to be rated for a pressure of at least 150 psi (10 bar). Pressures in excess of this value in a private fire service main are unusual and should be avoided by the system designer. The 150 psi (10 bar) pressure limitation may be exceeded when a fire pump is supplying pressure to a private fire service main, although even this situation should be avoided if at all possible. When pressures must exceed this limitation, a pressure-regulating device should be used. This situation is the only time a pressure-regulating device is permitted to be used.

Subsection 5.3.1 is intended to discourage the use of high pressures in this type of piping system and only permits the use of a pressure-regulating device with the approval of the authority having jurisdiction. See Exhibit III.9 for an example of a pressure-reducing valve.

5.3.2 Where meters are required by other authorities, they shall be listed.

5.4* Connection from Waterworks Systems.

A.5.4 Where connections are made from public waterworks systems, such systems should be guarded against possible contamination as follows *(see AWWA M14)*:

(1) For private fire service mains with direct connections from public waterworks mains only or with booster pumps installed in the connections from the street mains, no tanks or reservoirs, no physical connection from other water supplies, no antifreeze or other additives of any kind, and with all drains discharging to atmosphere, dry well, or other safe outlets, no backflow protection is recommended at the service connection.

(2) For private fire service mains with direct connection from the public water supply main plus one or more elevated storage tanks or fire pumps taking suction from aboveground covered reservoirs or tanks (all storage facilities are filled or connected to public water only, and the water in the tanks is to be maintained in a potable condition), an approved double check valve assembly is recommended.

EXHIBIT III.9 Pressure-Reducing Valve.

(3) For private fire service mains directly supplied from public mains with an auxiliary water supply, such as a pond or river on or available to the premises and dedicated to fire department use; or for systems supplied from public mains and interconnected with auxiliary supplies, such as pumps taking suction from reservoirs exposed to contamination or rivers and ponds; driven wells, mills, or other industrial water systems; or for systems or portions of systems where antifreeze or other solutions are used, an approved reduced pressure zone-type backflow preventer is recommended.

5.4.1 The requirements of the public health authority having jurisdiction shall be determined and followed.

5.4.2 Where equipment is installed to guard against possible contamination of the public water system, such equipment and devices shall be listed for fire protection service.

The following extract is a sample requirement for the installation of a double check valve assembly (DCVA):

> A double check is the minimum type of backflow protection required to protect the public water supply at the water user connection for the following situations:
>
> 1. The public water system serves water user premises where the fire protection system is directly supplied from the public water system and an unapproved auxiliary water supply is on the premises or accessible to the premises that is not connected to the public water system.
> 2. The public water system serves water user premises where the fire protection system is supplied from the public water system and where either elevated storage tanks or fire pumps that take suction from private reservoirs or tanks are on the user's premises.

In situation number one, the fire department connection (FDC) is the "unapproved auxiliary water supply." When an FDC is used, the recommended procedure for supplementing the sprinkler system is to connect a hose line to the FDC and pressurize the system to 150 psi (10 bar), unless the FDC is posted for a different pressure. When the fire department pumps into an FDC, a higher pressure exists in the system than in the supply and a backflow situation can occur that can contaminate the potable water supply. Since fire protection systems are not sanitized for human consumption, protection against backflow conditions by means of a backflow preventer must be installed. Exhibit III.10 illustrates the correct location for the installation of a backflow preventer.

In situation number two, either an elevated tank or a fire pump can pressurize a fire protection system so that a higher pressure exists in the system than in the water supply. Again,

EXHIBIT III.10 A Typical Layout of a Water Supply System for Automatic Sprinkler Systems. (Source: NFPA 13E, 2005, Figure A.4.2)

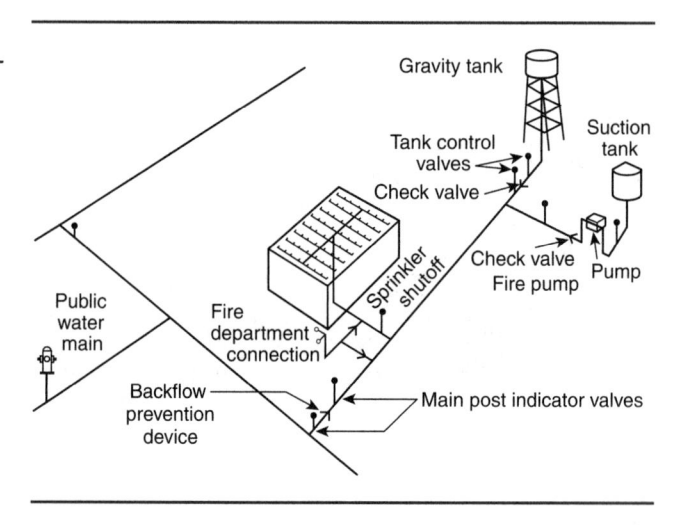

this case also creates the potential for a backflow and subsequent contamination of the potable water.

5.5 Connections to Public Water Systems.

Connections to public water systems shall be arranged to be isolated by one of the methods permitted in 6.2.11.

The 2010 edition of NFPA 24 has been revised to recognize that there are many arrangements that will satisfy the performance requirement of providing a safe way to turn off water to public water systems. See 6.2.11 for these methods.

5.6* Pumps.

A single, automatically controlled fire pump installed in accordance with NFPA 20 shall be an acceptable water supply source.

A.5.6 A fire pump installation consisting of pump, driver, and suction supply, when of adequate capacity and reliability and properly located, makes a good supply. An automatically controlled fire pump taking water from a water main of adequate capacity, or taking draft under a head from a reliable storage of adequate capacity, are permitted to be, under certain conditions, accepted by the authority having jurisdiction as a single supply.

5.7 Tanks.

Tanks shall be installed in accordance with NFPA 22.

5.8 Penstocks, Flumes, Rivers, Lakes, or Reservoirs.

Water supply connections from penstocks, flumes, rivers, lakes, or reservoirs shall be arranged to avoid mud and sediment and shall be provided with approved, double, removable screens or approved strainers installed in an approved manner.

Extreme caution should be exercised before using a natural body of water to supply a fire protection system. Screens and strainers may remove larger obstructing material but do not remove mud or sediment. Fire protection strainers typically have $\frac{1}{8}$ in. (3 mm) perforations in the strainer basket, which are too large to remove fine particulate matter. See Exhibit III.11 for an illustration of a typical fire protection strainer. Sprinkler systems that are supplied from natural sources of water must be installed with return bends or goosenecks to prevent the settlement of mud or other sediment in the bottom of the sprinkler drop, thus obstructing the sprinkler orifice. See Exhibit III.12 for an illustration of a return bend.

EXHIBIT III.11 Fire Protection Strainer.

EXHIBIT III.12 *Return Bend Arrangement. (Source: NFPA 13, 2010, Figure 8.15.18.2)*

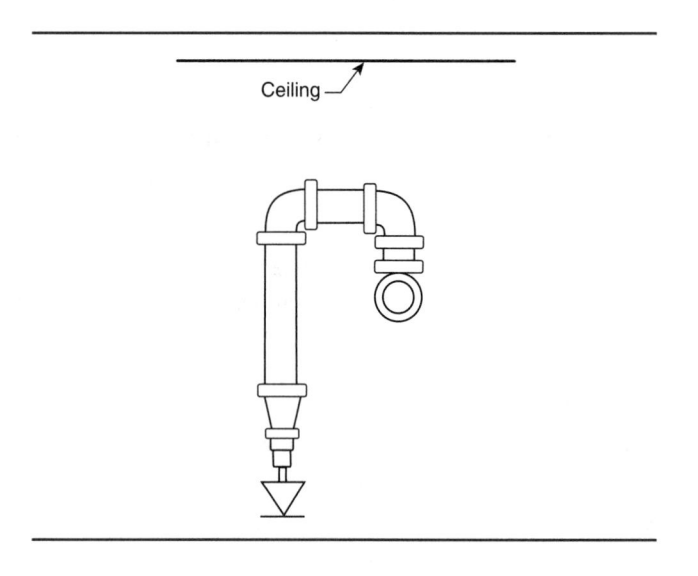

5.9* Fire Department Connections.

A.5.9 The fire department connection should be located not less than 18 in. (457 mm) and not more than 4 ft (1.2 m) above the level of the adjacent grade or access level. Typical fire department connections are shown in Figure A.5.9(a) and Figure A.5.9(b). Where a hydrant is not available, other water supply sources such as a natural body of water, a tank, or a reservoir should be utilized. The water authority should be consulted when a nonpotable water supply is proposed as a suction source for the fire department.

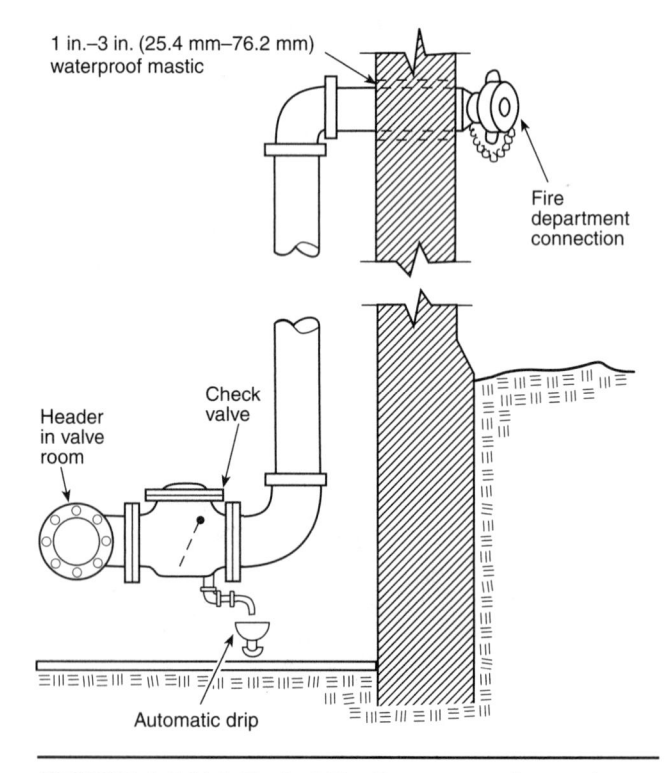

FIGURE A.5.9(a) *Typical Fire Department Connection.*

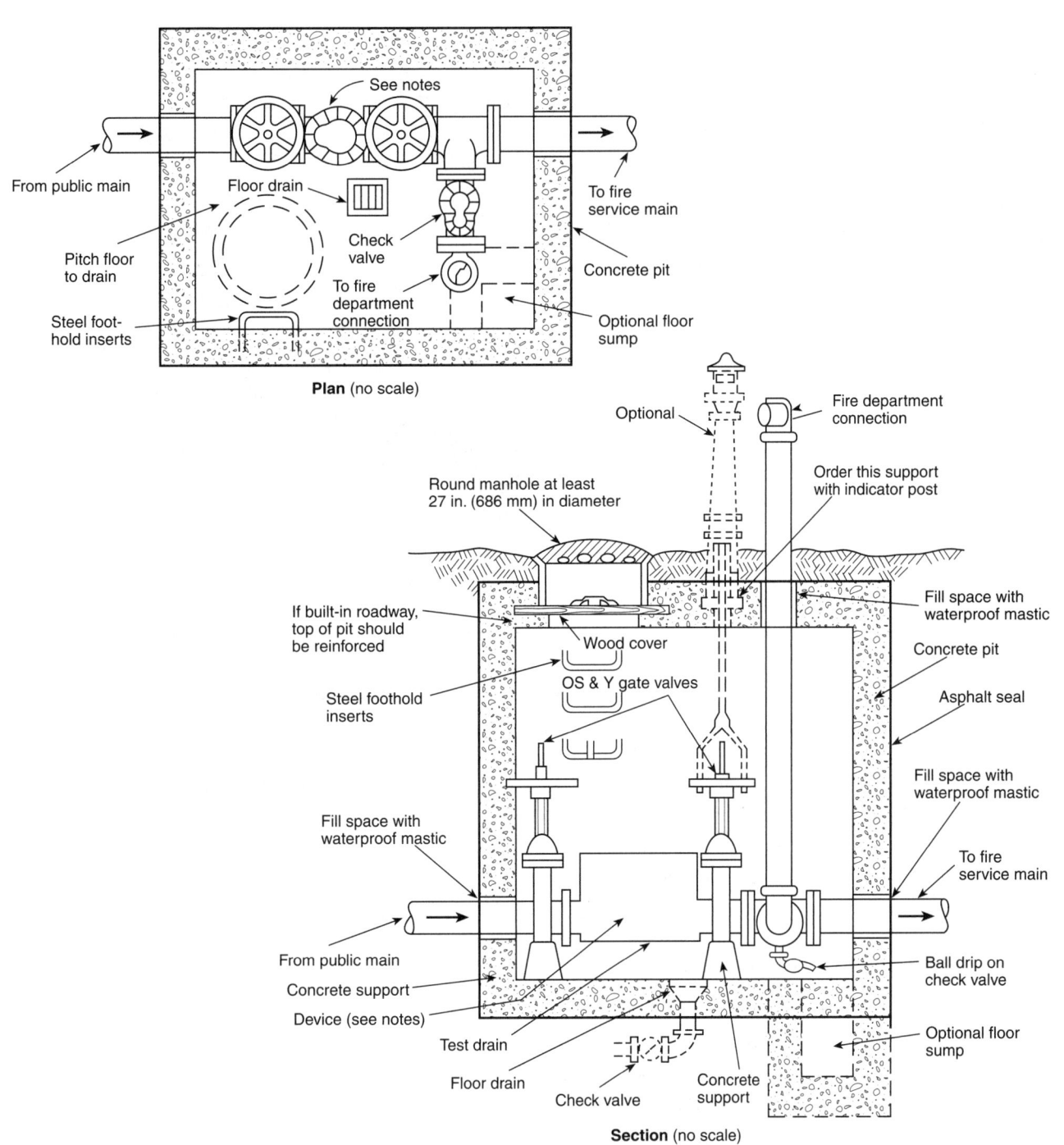

Notes:
1. Various backflow prevention regulations accept different devices at the connection between public water mains and private fire service mains.
2. The device shown in the pit could be any or a combination of the following:
   (a) Gravity check valve
   (b) Detector check valve
   (c) Double check valve assembly
   (d) Reduced pressure zone (RPZ) device
   (e) Vacuum breaker
3. Some backflow prevention regulations prohibit these devices from being installed in a pit.
4. In all cases, the device(s) in the pit should be approved or listed as necessary. The requirements of the local or municipal water department should be reviewed prior to design or installation of the connection.
5. Pressure drop should be considered prior to the installation of any backflow prevention device.

FIGURE A.5.9(b) *Typical City Water Pit—Valve Arrangement.*

EXHIBIT III.13 Freestanding FDC.

5.9.1 General. A fire department connection shall be provided as described in Section 5.9.

The FDC may serve more than one building. The underground piping, as illustrated in Figure A.3.3.11, can be pressurized through the FDC, and therefore some guidelines on the design and installation of the FDC are needed in Section 5.9. See Exhibit III.13 for an illustration of a freestanding FDC supplying an underground private fire service main.

5.9.1.1 Fire department connections shall not be required where approved by the authority having jurisdiction.

The authority having jurisdiction in this case is the local fire department. Since the local fire department is the end user, the fire marshal should be consulted for the location and number of FDCs needed.

5.9.1.2 Fire department connections shall be properly supported.

5.9.1.3 Fire department connections shall be of an approved type.

5.9.1.4 Fire department connections shall be equipped with listed plugs or caps that are secured and arranged for easy removal by fire departments.

5.9.2 Couplings.

5.9.2.1 The fire department connection(s) shall use an NH internal threaded swivel fitting(s) with an NH standard thread(s).

5.9.2.2 At least one of the connections shall be the 2.5–7.5 NH standard thread specified in NFPA 1963.

5.9.2.3 Where local fire department connections do not conform to NFPA 1963, the authority having jurisdiction shall designate the connection to be used.

5.9.2.4 The use of threadless couplings shall be permitted where required by the authority having jurisdiction and where listed for such use.

An important requirement to point out is that the type of threads must match that of the local fire department. Ideally, all fire departments should be using national standard hose threads, but this standardization is not the case in all areas of the United States and other countries. Paragraph 5.9.2.4 provides some flexibility for jurisdictions that use threads or connections other than the national standard. See Exhibit III.14 for an illustration of a nonthreaded hose connection.

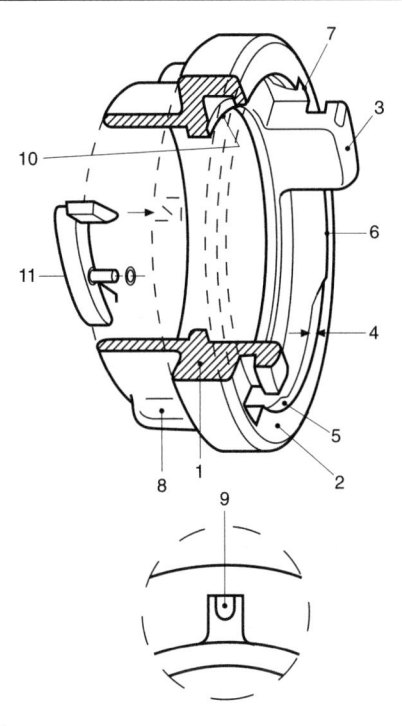

EXHIBIT III.14 Diagram of a Nonthreaded Connection. (Source: NFPA 1963, 2009, Figure A.6.1)

Legend:
1. Circumferential O.D.: The largest outer diameter of connection that protects the connection from damage.
2. Coupling face: The front part of the connection from which dimensions are developed.
3. Internal lug: The two internal lugs with recesses that fit on the ramp under the face of the cam head.
4. Ramp: The inclined plane under the face of the cam head that, when turned clockwise, increases pressure on lip seals.
5. Lug stop: The stop at the end of the ramp that the internal lug comes against.
6. Lug recess: Recessed area where opposite internal lugs enter the ramp.
7. Cleaning port: Area on end of connection face where dirt is pushed in by mating lug.
8. External wrenching lug: The external ribs or lugs on back diameter of connection head.
9. External wrenching lug indicator: The identification on rib or lug that, when lined up together, indicates the connection is fully engaged.
10. Tail place recess: The recess counterbore on the interface of the cam head that the tail piece rides in.
11. Lock: To keep the connection from becoming unintentionally disengaged.

5.9.3 Valves.

5.9.3.1 A listed check valve shall be installed in each fire department connection.

5.9.3.2 No shutoff valve shall be permitted in the fire department connection piping.

Due to the presence of the check valve in the FDC, a shutoff valve is not needed. Installing a shutoff valve where it is not needed could increase the likelihood that the valve will be closed (rendering the FDC unusable) when needed.

5.9.4 Drainage.

5.9.4.1 The pipe between the check valve and the outside hose coupling shall be equipped with an approved automatic drip.

For freestanding FDCs the requirement in 5.9.4.1 presents a problem. Usually the check valve in a freestanding FDC is buried or may be installed in a valve pit. If installed in a valve pit, the usual method of draining, which is a ball-drip valve, can be used. See Exhibit III.15 for a typical freestanding FDC pit arrangement.

EXHIBIT III.15 Pit for Gate Valve, Check Valve, and Fire Department Connection. (Source: NFPA 13, 2010, Figure A.8.16.1.1.4)

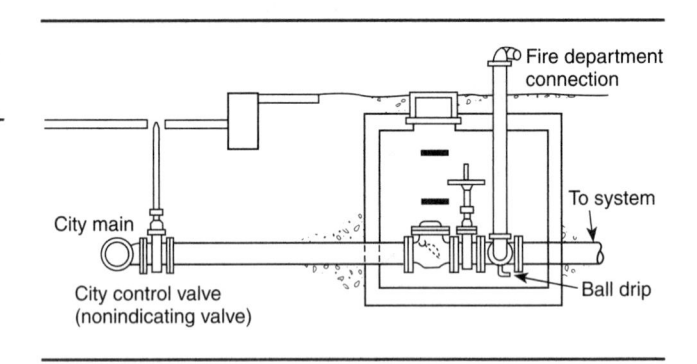

5.9.4.2 An automatic drip shall not be required in areas not subject to freezing.

5.9.5 Location and Signage.

5.9.5.1* Fire department connections shall be located at the nearest point of fire department apparatus accessibility or at a location approved by the authority having jurisdiction.

Careful consideration should be given to the location of FDCs that supply a standpipe system or combined sprinkler/standpipe system. NFPA 14 requires that an FDC be located within 100 ft (30.5 m) of a hydrant. When designing a private fire service main, the designer should consider this requirement when determining the location of all hydrants and FDCs.

A.5.9.5.1 The requirement in 5.9.5.1 applies to fire department connections attached to underground piping. If the fire department connection is attached directly to a system riser, the requirements of the appropriate installation standard apply.

5.9.5.2* Fire department connections shall be located and arranged so that hose lines can be attached to the inlets without interference.

A clearance of 4 ft (1.2 m) is recommended to be maintained around all FDCs in order to permit fire fighters to connect and use FDCs. Protective bollards should also not be located within the recommended 4 ft (1.2 m) radius. See Exhibit III.16 for a recommended FDC arrangement.

A.5.9.5.2 Obstructions to fire department connections include, but are not limited to, buildings, fences, posts, landscaping, other fire department connections, gas meters, and electrical equipment.

5.9.5.3* Each fire department connection shall be designated by a sign as follows:

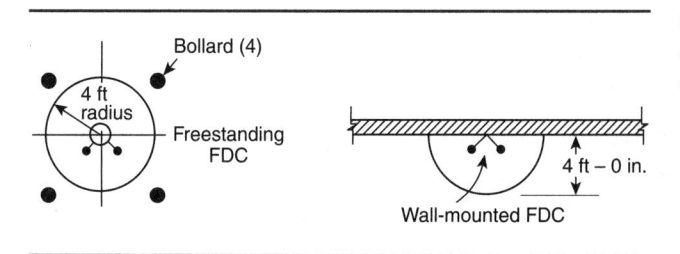

EXHIBIT III.16 Recommended Arrangement for an FDC.

(1) The sign shall have raised or engraved letters at least 1 in. (25.4 mm) in height on a plate or fitting.

(2)* The sign shall indicate the type of system for which the connection is intended.

A.5.9.5.3 Where a fire department connection services multiple buildings, structures, or locations, a sign should be provided indicating the buildings, structures, or locations served.

A.5.9.5.3(2) Examples for wording of signs are:

AUTOSPKR
OPEN SPKR
AND STANDPIPE

The sign required in 5.9.5.3 should also include a graphic representation of the type of FDC as provided in NFPA 170, *Standard for Fire Safety and Emergency Symbols:*

FDC—Automatic Sprinkler System

FDC—Standpipe System Fire Department

FDC—Combined Automatic Sprinkler/Standpipe System

5.9.5.4 Where the system demand pressure exceeds 150 psi (10.3 bar), the sign required by 5.9.5.3 shall indicate the required design pressure.

5.9.5.5 Where a fire department connection only supplies a portion(s) of the building, a sign shall be attached to indicate the portion(s) of the building supplied.

CHAPTER 6 VALVES

6.1 Types of Valves.

6.1.1 All valves controlling connections to water supplies and to supply pipes to sprinklers shall be listed indicating valves.

All valves controlling fire protection systems must be listed indicating valves. As mentioned in Chapter 3, a shut valve is the leading cause of system failure. An indicating valve, as its name suggests, indicates its status (open or shut) by quick visual observation.

6.1.2 Indicating valves shall not close in less than 5 seconds when operated at maximum possible speed from the fully open position.

The requirement in 6.1.2 is intended to reduce the possibility of water hammer in the system when water is flowing. Water hammer, which sounds like knocking in the pipes, is caused by

EXHIBIT III.17 Post Indicator Valve (PIV).

quickly stopping the flow of water, resulting in a pressure surge. Water hammer may instantaneously raise the pressure in the system beyond the pressure rating of the system components, which may result in damage or leaks.

6.1.3 A listed underground gate valve equipped with a listed indicator post shall be permitted.

6.1.4 A listed water control valve assembly with a reliable position indication connected to a remote supervisory station shall be permitted.

Although in 6.1.3 and 6.1.4 valves are permitted to be nonindicating, attachments are needed to provide an indication of their status. A post indicator valve (PIV) offers its status through a window in the body of the post and indicates "open" or "shut." A position indicator electronically signals a fire alarm panel displaying the status of the valve. See Exhibit III.17 for an illustration of a PIV.

6.1.5* A nonindicating valve, such as an underground gate valve with approved roadway box, complete with T-wrench, and accepted by the authority having jurisdiction, shall be permitted.

Although a nonindicating valve with a roadway box is permitted by 6.1.5, this design should be avoided since it offers no easy means of inspection or indication of the valve status.

A.6.1.5 A valve wrench with a long handle should be provided at a convenient location on the premises.

6.2 Valves Controlling Water Supplies.

6.2.1 At least one listed indicating valve shall be installed in each source of water supply.

6.2.2 No shutoff valve shall be permitted in the fire department connection.

6.2.3 Where more than one source of water supply exists, a check valve shall be installed in each connection.

6.2.4 Where break tanks are used with automatic fire pumps, a check valve shall not be required in the break tank connection.

6.2.5* In a connection serving as one source of supply, listed indicating valves or post indicator valves shall be installed on both sides of all check valves required in 6.2.3.

The installation of two valves on each side of a check valve allows the check valve to be isolated for internal inspection, maintenance, or repair.

A.6.2.5 See Figure A.6.2.5. For additional information on controlling valves, see NFPA 22.

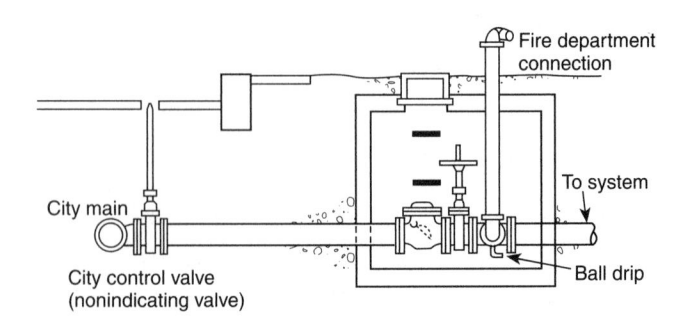

FIGURE A.6.2.5 *Pit for Gate Valve, Check Valve, and Fire Department Connection.*

6.2.6 In the discharge pipe from a pressure tank or a gravity tank of less than 15,000 gal (56.78 m3) capacity, a control valve shall not be required to be installed on the tank side of the check valve.

6.2.7* The following requirements shall apply where a gravity tank is located on a tower in the yard:

(1) The control valve on the tank side of the check valve shall be an outside screw and yoke or a listed indicating valve.
(2) The other control valve shall be either an outside screw and yoke, a listed indicating valve, or a listed valve having a post-type indicator.

A.6.2.7 For additional information on controlling valves, see NFPA 22.

6.2.8* The following requirements shall apply where a gravity tank is located on a building:

(1) Both control valves shall be outside screw and yoke or listed indicating valves.
(2) All fittings inside the building, except the drain tee and heater connections, shall be under the control of a listed valve.

A.6.2.8 For additional information on controlling valves, see NFPA 22.

6.2.9 One of the following requirements shall be met where a pump is located in a combustible pump house or exposed to danger from fire or falling walls, or where a tank discharges into a private fire service main fed by another supply:

(1)* The check valve in the connection shall be located in a pit.
(2) The control valve shall be of the post indicator type and located a safe distance outside buildings.

A.6.2.9(1) Where located underground, check valves on tank or pump connections can be placed inside of buildings and at a safe distance from the tank riser or pump, except in cases where the building is entirely of one fire area. Where the building is one fire area, it is ordinarily considered satisfactory to locate the check valve overhead in the lowest level.

6.2.10* All control valves shall be located where readily accessible and free of obstructions.

A.6.2.10 It might be necessary to provide valves located in pits with an indicator post extending above grade or other means so that the valve can be operated without entering the pit.

6.2.11 All connections to private fire service mains for fire protection systems shall be arranged in accordance with one of the following so that they can be isolated:

(1)* A post indicator valve installed not less than 40 ft (12 m) from the building
    (a) For buildings less than 40 ft (12 m) in height, a post indicator valve shall be permitted to be installed closer than 40 ft (12 m) but at least as far from the building as the height of the wall facing the post indicator valve.
(2) A wall post indicator valve
(3) An indicating valve in a pit, installed in accordance with Section 6.4
(4)* A backflow preventer with at least one indicating valve not less than 40 ft (12 m) from the building
    (a) For buildings less than 40 ft (12 m) in height, a backflow preventer with at least one indicating valve shall be permitted to be installed closer than 40 ft (12 m) but at least as far from the building as the height of the wall facing the backflow preventer.
(5)* A nonindicating valve, such as an underground gate valve with an approved roadway box, complete with T-wrench, located not less than 40 ft (12 m) from the building

(a) For buildings less than 40 ft (12 m) in height, a nonindicating valve, such as an underground gate valve with an approved roadway box, complete with T-wrench, shall be permitted to be installed closer than 40 ft (12 m) but at least as far from the building as the height of the wall facing the backflow preventer.
(6) Control valves installed in a fire-rated room accessible from the exterior
(7) Control valves in a fire-rated stair enclosure accessible from the exterior as permitted by the authority having jurisdiction

The 2010 edition now contains requirements offering many arrangements that satisfy the performance requirement of providing a safe way to turn off water. The standard now permits a variety of reasonable alternatives with less liability for the authority having jurisdiction.

A.6.2.11(1) Distances greater than 40 ft (12 m) are not required but can be permitted regardless of the building height.

A.6.2.11(4) Distances greater than 40 ft (12 m) are not required but can be permitted regardless of the building height.

A.6.2.11(5) Distances greater than 40 ft (12 m) are not required but can be permitted regardless of the building height.

6.3 Post Indicator Valves.

6.3.1 Where post indicator valves are used, they shall be set so that the top of the post is 32 in. to 40 in. (0.8 m to 1.0 m) above the final grade.

6.3.2 Where post indicator valves are used, they shall be protected against mechanical damage where needed.

The PIV should be installed in such a location that a 4 ft (1.2 m) radius around the valve is provided for proper access. Bollards for mechanical protection should also fall outside of this 4 ft (1.2 m) protective space.

6.3.3 Arrangement.

6.3.3.1 Post indicator valves shall be set so that the top of the post is 36 in. (0.9 m) above the final grade.

6.3.3.2 Post indicator valves shall be protected against mechanical damage where needed.

6.4 Valves in Pits.

6.4.1 Valve pits located at or near the base of the riser of an elevated tank shall be designed in accordance with Chapter 14 of NFPA 22.

6.4.2 Where used, valve pits shall be of adequate size and readily accessible for inspection, operation, testing, maintenance, and removal of equipment contained therein.

6.4.3 Valve pits shall be constructed and arranged to properly protect the installed equipment from movement of earth, freezing, and accumulation of water.

6.4.3.1 Depending on soil conditions and the size of the pit, valve pits shall be permitted to be constructed of any of the following materials:

(1) Poured-in-place or precast concrete, with or without reinforcement
(2) Brick
(3) Other approved materials

6.4.3.2 Where the water table is low and the soil is porous, crushed stone or gravel shall be permitted to be used for the floor of the pit.

6.4.4 The location of the valve shall be marked, and the cover of the pit shall be kept free of obstructions.

6.5 Backflow Prevention Assemblies

6.5.1 Where used in accordance with 6.2.11(4), backflow prevention assemblies shall be installed in accordance with their installation instructions.

6.5.2 Where backflow prevention assemblies are used, they shall be protected against mechanical damage where needed.

6.6 Sectional Valves.

6.6.1 Large, private, fire service main systems shall have sectional controlling valves at appropriate points to permit sectionalizing the system in the event of a break or to make repairs or extensions.

The purpose of a sectional control valve is to have the ability to isolate a segment of the underground main for maintenance or repair while impacting as few systems as possible. A common practice is to provide a sectional valve for every four to six pieces of equipment, such as sprinkler systems or hydrants. While calculating flow to an underground loop that is provided with sectional valves, a common practice is to assume the shorter leg of the loop is out of service. This plan ensures a design that is capable of supplying the system demand even when a portion of the underground loop is out of service. Exhibit III.18 is an example of a private fire service main that has been divided into segments by valves A, B, C, E, and F.

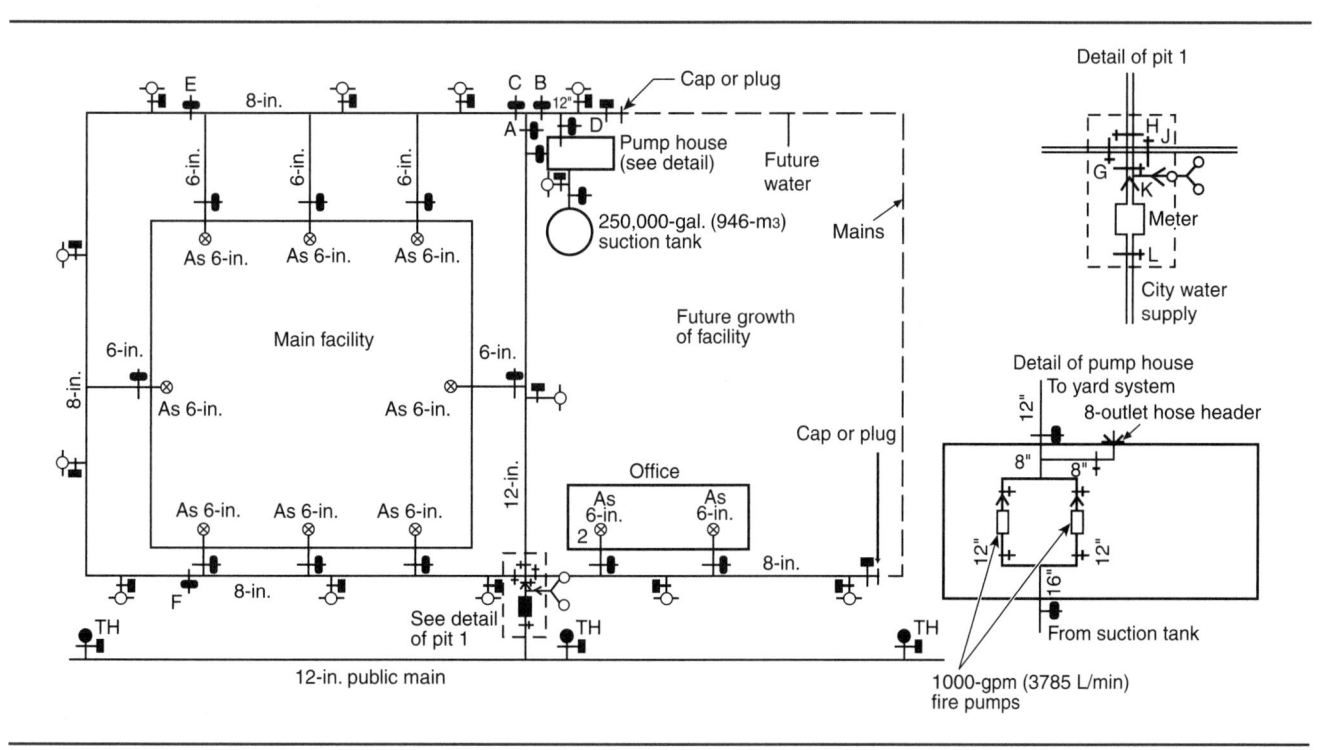

EXHIBIT III.18 *Water Piping for Fire Protection of an Industrial Site. (Source: Fire Protection Handbook®, NFPA, 2003, Figure 10.3.30)*

6.6.2 A sectional valve shall be provided at the following locations:

(1) On each bank where a main crosses water
(2) Outside the building foundation(s) where a main or a section of a main runs under a building

6.7 Identifying and Securing Valves.

6.7.1 Identification signs shall be provided at each valve to indicate its function and what it controls.

The valve sign should include a graphical symbol for the valve function, such as the following provided in NFPA 170:

Automatic Sprinkler Control Valve

6.7.1.1 Identification signs in 6.6.1 shall not be required for underground gate valves with roadway boxes.

6.7.2* Control valves shall be supervised by one of the following methods:

(1) Central station, proprietary, or remote station signaling service
(2) Local signaling service that causes the sounding of an audible signal at a constantly attended location
(3) An approved procedure to ensure that valves are locked in the correct position
(4) An approved procedure to ensure that valves are located within fenced enclosures under the control of the owner, sealed in the open position, and inspected weekly

A.6.7.2 See Annex B.

6.7.3 Supervision of underground gate valves with roadway boxes shall not be required.

6.8 Check Valves.

Check valves shall be installed in a vertical or horizontal position in accordance with their listing.

CHAPTER 7 HYDRANTS

7.1* General.

A.7.1 For information regarding identification and marking of hydrants, see Annex D.

7.1.1 Hydrants shall be of an approved type and have not less than a 6 in. (152 mm) diameter connection with the mains.

7.1.1.1 A valve shall be installed in the hydrant connection.

7.1.1.2* The number, size, and arrangement of outlets; the size of the main valve opening; and the size of the barrel shall be suitable for the protection to be provided and shall be approved by the authority having jurisdiction.

A.7.1.1.2 The flows required for private fire protection service mains are determined by system installation standards or fire codes. The impact of the number and size of hydrant outlets on the fire protection system demand is not addressed in this standard. The appropriate code or standard should be consulted for the requirements for calculating system demand.

7.1.1.3 Independent gate valves on 2½ in. (64 mm) outlets shall be permitted.

7.1.2 Hydrant outlet threads shall have NHS external threads for the size outlet(s) supplied as specified in NFPA 1963.

7.1.3 Where local fire department connections do not conform to NFPA 1963, the authority having jurisdiction shall designate the connection to be used.

7.2 Number and Location.

7.2.1* Hydrants shall be provided and spaced in accordance with the requirements of the authority having jurisdiction.

Section 18.3 and Annex I from NFPA 1, *Fire Code,* provide guidance on the number and location of hydrants (see Part III, Section 3, of this handbook). The authority having jurisdiction may have requirements to meet specific local issues regarding the location of hydrants.

A.7.2.1 Fire department pumpers will normally be required to augment the pressure available from public hydrants.

7.2.2 Public hydrants shall be permitted to be recognized as meeting all or part of the requirements of Section 7.2.

7.2.3* Hydrants shall be located not less than 40 ft (12.2 m) from the buildings to be protected.

Hydrants in parking areas or in any area in which they are exposed to potential damage should be protected by bollards on all sides. The bollards should not be closer than 4 ft (1.2 m) to the protected hydrant. When located near a roadway, the hydrant should be placed within 6 ft (1.8 m) of the pavement unless the authority having jurisdiction determines another location is more acceptable.

A.7.2.3 Where wall hydrants are used, the authority having jurisdiction should be consulted regarding the necessary water supply and arrangement of control valves at the point of supply in each individual case. *(See Figure A.7.2.3.)*

7.2.4 Where hydrants cannot be located in accordance with 7.2.3, locations closer than 40 ft (12.2 m) from the building or wall hydrants shall be permitted to be used where approved by the authority having jurisdiction.

7.2.5 Hydrants shall not be installed at less than the equivalent depth of burial from retaining walls where there is danger of frost through the walls.

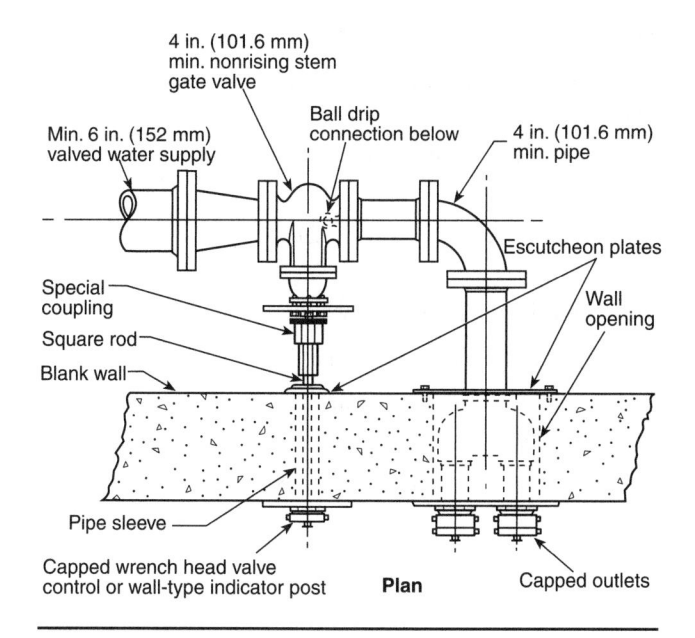

FIGURE A.7.2.3 Typical Wall Fire Hydrant Installation.

7.3 Installation.

7.3.1* Hydrants shall be set on flat stones or concrete slabs and shall be provided with small stones (or the equivalent) placed about the drain to ensure drainage.

In cold climates, the hydrant barrel is expected to drain within 60 minutes. Where soil conditions do not permit this drainage or where groundwater is above the hydrant drain, the hydrant drain should be plugged and provisions made for pumping out the barrel.

A.7.3.1 See Figure A.7.3.1(a) and Figure A.7.3.1(b).

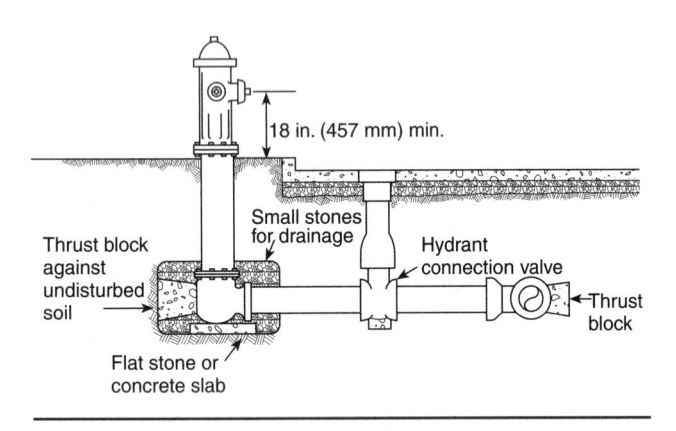

FIGURE A.7.3.1(a) Typical Hydrant Connection with a Minimum Height Requirement.

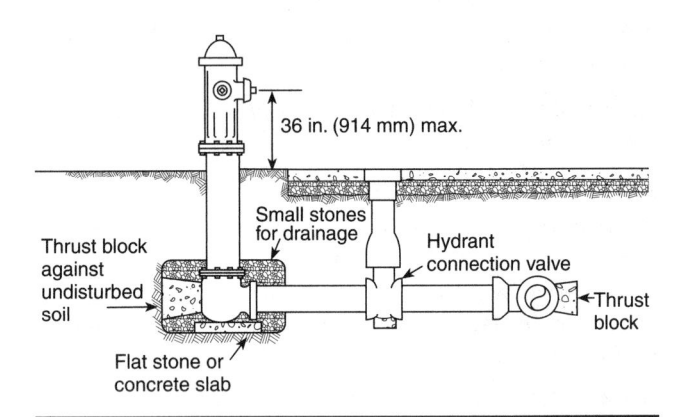

FIGURE A.7.3.1(b) Typical Hydrant Connection with Maximum Height Requirement.

7.3.2 Where soil is of such a nature that the hydrants will not drain properly with the arrangement specified in 7.3.1, or where groundwater stands at levels above that of the drain, the hydrant drain shall be plugged at the time of installation.

7.3.2.1 If the drain is plugged, hydrants in service in cold climates shall be pumped out after usage.

7.3.2.2 Such hydrants shall be marked to indicate the need for pumping out after usage.

7.3.3* The center of a hose outlet shall be not less than 18 in. (457 mm) above final grade or, where located in a hose house, 12 in. (305 mm) above the floor.

The center of the hose outlet should not be more than 36 in. (914.4 mm) above final grade [see Figure A.7.3.1(b)].

7.3.4 Hydrants shall be fastened to piping and anchored in accordance with the requirements of Chapter 10.

A.7.3.3 When setting hydrants, due regard should be given to the final grade line.

7.3.5 Hydrants shall be protected if subject to mechanical damage.

7.3.6 The means of hydrant protection shall be arranged in a manner that does not interfere with the connection to, or operation of, hydrants.

7.3.7 The following shall not be installed in the service stub between a fire hydrant and private water supply piping:

(1) Check valves
(2) Detector check valves
(3) Backflow prevention valves
(4) Other similar appurtenances

CHAPTER 8 HOSE HOUSES AND EQUIPMENT

8.1 General.

8.1.1* A supply of hose and equipment shall be provided where hydrants are intended for use by plant personnel or a fire brigade.

Only where a trained industrial fire brigade is present should this equipment be installed. Industrial fire brigades should be trained in accordance with NFPA 600, *Standard on Industrial Fire Brigades.*

A.8.1.1 All hose should not be removed from a hose house for testing at the same time, since the time taken to return the hose in case of fire could allow a fire to spread beyond control. *(See NFPA 1962.)*

8.1.1.1 The quantity and type of hose and equipment shall depend on the following:

(1) Number and location of hydrants relative to the protected property
(2) Extent of the hazard
(3) Fire-fighting capabilities of potential users

8.1.1.2 The authority having jurisdiction shall be consulted regarding quantity and type of hose.

8.1.2 Hose shall be stored so it is readily accessible and is protected from the weather by storing in hose houses or by placing hose reels or hose carriers in weatherproof enclosures.

8.1.3* Hose shall conform to NFPA 1961.

A.8.1.3 Where hose will be subjected to acids, acid fumes, or other corrosive materials, as in chemical plants, the purchase of approved rubber-covered, rubber-lined hose is advised. For hose used in plant yards containing rough surfaces that cause heavy wear or used where working pressures are above 150 psi (10.3 bar), double-jacketed hose should be considered.

8.1.4 Hose Connections.

8.1.4.1 Hose connections shall have external national hose standard (NHS) threads, for the valve size specified, in accordance with NFPA 1963.

8.1.4.2 Hose connections shall be equipped with caps to protect the hose threads.

8.1.4.3 Where local fire department hose threads do not conform to NFPA 1963, the authority having jurisdiction shall designate the hose threads to be used.

8.2 Location.

8.2.1 Where hose houses are utilized, they shall be located over, or immediately adjacent to, the hydrant.

8.2.2 Hydrants within hose houses shall be as close to the front of the house as possible and still allow sufficient room in back of the doors for the hose gates and the attached hose.

8.2.3 Where hose reels or hose carriers are utilized, they shall be located so that the hose can be brought into use at a hydrant.

8.3 Construction.

8.3.1 Hose houses shall be of substantial construction on foundations.

8.3.2 The construction shall protect the hose from weather and vermin and shall be designed so that hose lines can be brought into use.

8.3.3 Clearance shall be provided for operation of the hydrant wrench.

8.3.4 Ventilation shall be provided.

8.3.5 The exterior shall be painted or otherwise protected against deterioration.

8.4* Size and Arrangement.

Hose houses shall be of a size and arrangement that provide shelves or racks for the hose and equipment.

A.8.4 Typical hose houses are shown in Figure A.8.4(a) through Figure A.8.4(c).

8.5 Marking.

Hose houses shall be plainly identified.

FIGURE A.8.4(a) Hose House of Five-Sided Design for Installation over a Private Hydrant.

FIGURE A.8.4(b) Closed Steel Hose House of Compact Dimensions for Installation over a Private Hydrant, in Which Top Lifts Up and Doors on Front Open for Complete Accessibility.

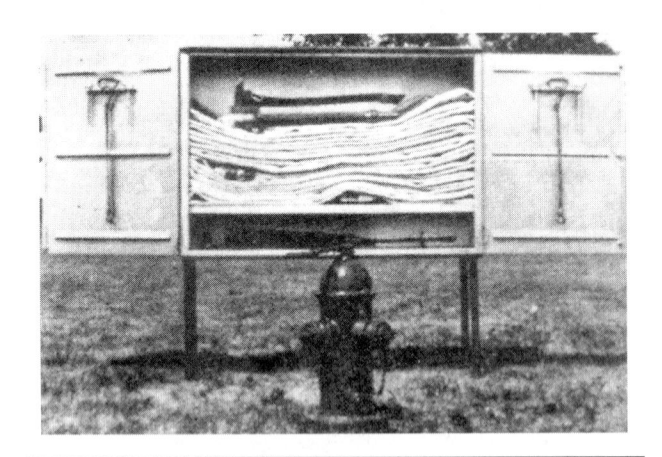

FIGURE A.8.4(c) Hose House That Can Be Installed on Legs, as Shown, or Installed on a Wall near, but Not Directly over, a Private Hydrant.

8.6 General Equipment.

8.6.1* Where hose houses are used in addition to the hose, each shall be equipped with the following:

(1) Two approved adjustable spray–solid stream nozzles equipped with shutoffs for each size of hose provided
(2) One hydrant wrench (in addition to wrench on hydrant)
(3) Four coupling spanners for each size hose provided
(4) Two hose coupling gaskets for each size hose

A.8.6.1 All hose should not be removed from a hose house for testing at the same time, since the time taken to return the hose in case of fire could allow a fire to spread beyond control. *(See NFPA 1962.)*

8.6.2 Where two sizes of hose and nozzles are provided, reducers or gated wyes shall be included in the hose house equipment.

8.7 Domestic Service Use Prohibited.

The use of hydrants and hose for purposes other than fire-related services shall be prohibited.

CHAPTER 9 MASTER STREAMS

9.1* Master Streams.

Master streams shall be delivered by monitor nozzles, hydrant-mounted monitor nozzles, and similar master stream equipment capable of delivering more than 250 gpm (946 L/min).

Some standards may specify higher flows for master streams. In any case, the equipment used to provide a master stream can be arranged to flow from 250 gpm (946 L) to 2000 gpm (7570 L) or more. See Exhibit III.19 for typical master stream equipment.

A.9.1 For typical master stream devices, see Figure A.9.1(a) and Figure A.9.1(b). Gear control nozzles are acceptable for use as monitor nozzles.

9.2 Application and Special Considerations.

Master streams shall be provided as protection for the following:

(1) Large amounts of combustible materials located in yards
(2) Average amounts of combustible materials in inaccessible locations
(3) Occupancies presenting special hazards, as required by the authority having jurisdiction

CHAPTER 10 UNDERGROUND PIPING

10.1* Piping Materials.

A.10.1 The term *underground* is intended to mean direct buried piping. For example, piping installed in trenches and tunnels but exposed should be treated as aboveground piping. Loop systems for yard piping are recommended for increased reliability and improved hydraulics. Loop systems should be sectionalized by placing valves at branches and at strategic locations to minimize the extent of impairments.

10.1.1* Listing. Piping shall be listed for fire protection service or shall comply with the standards in Table 10.1.1.

A.10.1.1 Copper tubing (Type K) with brazed joints conforming to Table 10.1.1 and Table 10.2.1(a) is acceptable for underground service. Listing and labeling information, along with applicable publications for reference, is as follows:

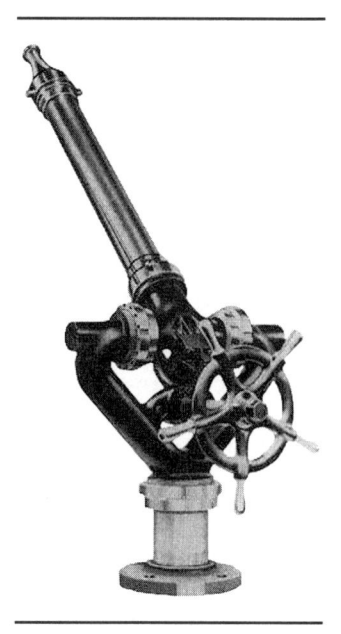

EXHIBIT III.19 *Master Stream Equipment. (Courtesy of Potter Electric Signal Company, LLC)*

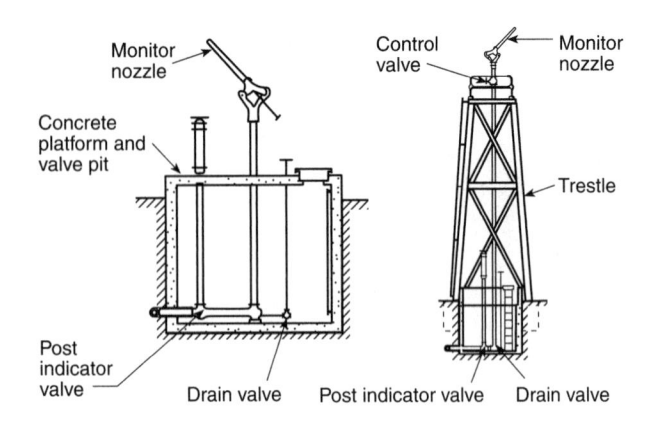

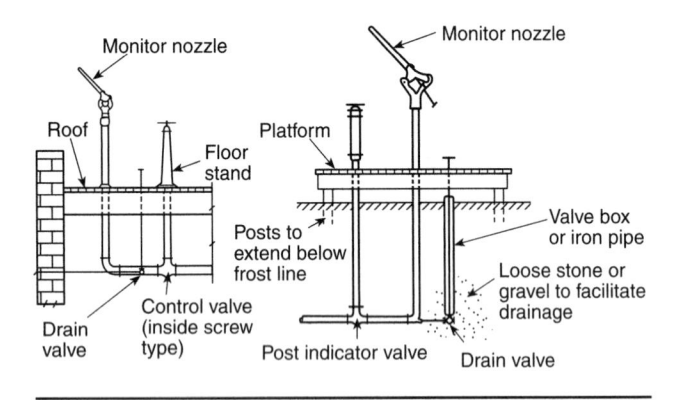

FIGURE A.9.1(a) Standard Monitor Nozzles.

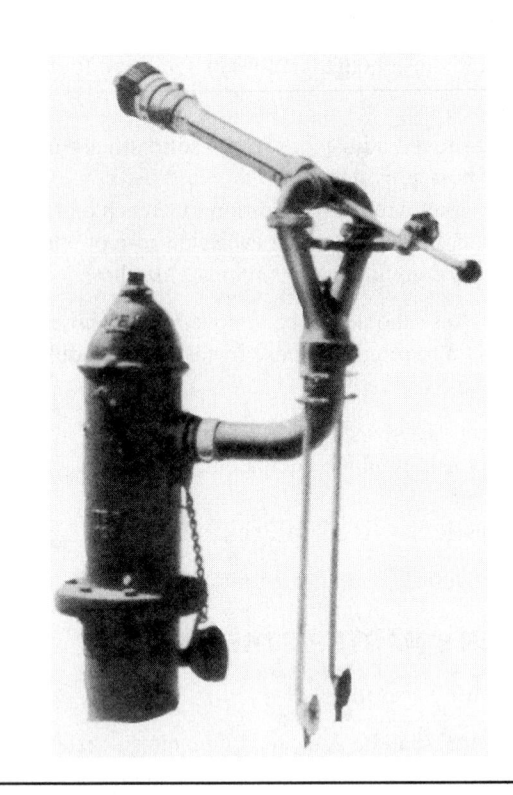

FIGURE A.9.1(b) Typical Hydrant-Mounted Monitor Nozzle.

(1) *Listing and Labeling.* Testing laboratories list or label the following:
   (a) Cast iron and ductile iron pipe (cement-lined and unlined, coated and uncoated)
   (b) Asbestos-cement pipe and couplings
   (c) Steel pipe
   (d) Copper pipe
   (e) Fiberglass filament-wound epoxy pipe and couplings
   (f) Polyethylene pipe
   (g) Polyvinyl chloride (PVC) pipe and couplings
   (h) Underwriters Laboratories Inc. lists, under re-examination service, reinforced concrete pipe (cylinder pipe, nonprestressed and prestressed)

(2) *Pipe Standards.* The various types of pipe are usually manufactured to one of the following standards:
   (a) ASTM C 296, *Standard Specification for Asbestos-Cement Pressure Pipe*
   (b) AWWA C151, *Ductile Iron Pipe, Centrifugally Cast for Water*
   (c) AWWA C300, *Reinforced Concrete Pressure Pipe, Steel-Cylinder Type*
   (d) AWWA C301, *Prestressed Concrete Pressure Pipe, Steel-Cylinder Type*
   (e) AWWA C302, *Reinforced Concrete Pressure Pipe, Non-Cylinder Type*
   (f) AWWA C303, *Reinforced Concrete Pressure Pipe, Steel-Cylinder Type, Pretensioned*
   (g) AWWA C400, *Standard for Asbestos-Cement Distribution Pipe, 4 in. Through 16 in. (100 mm through 400 mm), for Water Distribution Systems*
   (h) AWWA C900, *Polyvinyl Chloride (PVC) Pressure Pipe, 4 in. Through 12 in., for Water Distribution*

TABLE 10.1.1 *Manufacturing Standards for Underground Pipe*

Materials and Dimensions	Standard	Materials and Dimensions	Standard
Ductile Iron		**Concrete**	
Cement Mortar Lining for Ductile Iron Pipe and Fittings for Water	AWWA C104	Reinforced Concrete Pressure Pipe, Steel-Cylinder Type	AWWA C300
Polyethylene Encasement for Ductile Iron Pipe Systems	AWWA C105	Prestressed Concrete Pressure Pipe, Steel-Cylinder Type	AWWA C301
Ductile Iron and Gray Iron Fittings, 3 in. Through 48 in., for Water and Other Liquids	AWWA C110	Reinforced Concrete Pressure Pipe, Non-Cylinder Type	AWWA C302
Rubber-Gasket Joints for Ductile Iron Pressure Pipe and Fittings	AWWA C111	Reinforced Concrete Pressure Pipe, Steel-Cylinder Type, Pretensioned	AWWA C303
Flanged Ductile Iron Pipe with Ductile Iron or Gray Iron Threaded Flanges	AWWA C115	Standard for Asbestos-Cement Distribution Pipe, 4 in. Through 16 in., for Water Distribution Systems	AWWA C400
Protective Fusion-Bonded Epoxy Coatings for the Interior and Exterior Surfaces of Ductile-Iron and Gray-Iron Fittings for Water Supply Service	AWWA C116	Standard for the Selection of Asbestos-Cement Pressure Pipe	AWWA C401
		Cement-Mortar Lining of Water Pipe Lines 4 in. and Larger—in Place	AWWA C602
Thickness Design of Ductile Iron Pipe	AWWA C150	Standard for the Installation of Asbestos-Cement Water Pipe	AWWA C603
Ductile Iron Pipe, Centrifugally Cast for Water	AWWA C151		
Ductile-Iron Compact Fittings for Water Service	AWWA C153	**Plastic**	
Standard for the Installation of Ductile Iron Water Mains and Their Appurtenances	AWWA C600	Polyvinyl Chloride (PVC) Pressure Pipe, 4 in. Through 12 in., for Water Distribution	AWWA C900
		Polyethylene (PE) Pressure Pipe and Fittings, 4 in. (100 mm) Through 63 in. (1575 mm) for Water Distribution	AWWA C906
Steel			
Steel Water Pipe 6 in. and Larger	AWWA C200	**Copper**	
Coal-Tar Protective Coatings and Linings for Steel Water Pipelines Enamel and Tape—Hot Applied	AWWA C203	Specification for Seamless Copper Tube	ASTM B 75
		Specification for Seamless Copper Water Tube	ASTM B 88
Cement-Mortar Protective Lining and Coating for Steel Water Pipe 4 in. and Larger—Shop Applied	AWWA C205	Requirements for Wrought Seamless Copper and Copper-Alloy Tube	ASTM B 251
Field Welding of Steel Water Pipe	AWWA C206		
Steel Pipe Flanges for Waterworks Service—Sizes 4 in. Through 144 in.	AWWA C207		
Dimensions for Fabricated Steel Water Pipe Fittings	AWWA C208		
A Guide for Steel Pipe Design and Installation	AWWA M11		

Underground pipe must be listed for fire protection service or must comply with the American Water Works Association (AWWA) standards specified in Table 10.1.1. Many piping products that are manufactured in accordance with the AWWA standards are also listed for fire protection service.

Although several of these standards deal with steel pipe, flanges, welding, and installation, steel pipe is currently prohibited from use in underground fire protection service except for fire department connections or unless specially listed for this service. (See 10.1.2 and 10.1.3.)

Although the only plastic pipe included in Table 10.1.1 is polyvinyl chloride (PVC) pipe manufactured in accordance with AWWA C900, *Polyvinyl Chloride (PVC) Pressure Pipe, 4* in. Through 12 in. (100 mm Through 300 mm), for Water and Other Liquids, with diameters up to 12 in. (300 mm), listings are available for PVC with nominal diameters up to 30 in. (750 mm). Listings are also available for chlorinated polyvinyl chloride (CPVC) in the ¾ in. (20 mm) to 3 in. (75 mm) size, and for fiberglass-wound epoxy pipe in sizes 2 in. (50 mm) through 16 in. (400 mm) nominal diameter.

An often overlooked materials issue is that most nonmetallic underground piping brought up through the floor of a building is vulnerable to fire exposure and spills of corrosive liquids. A transition should be made to metallic pipe to avoid problems, and this transition is essentially required by the listing of the nonmetallic pipe, since such use is no longer underground.

10.1.2 Steel Pipe. Steel piping shall not be used for general underground service unless specifically listed for such service.

Past experience and loss data indicate that a high potential for failure of buried steel pipe exists even where the pipe is externally coated and wrapped and internally galvanized. The integrity of the coating, wrapping, and galvanizing can be easily compromised in many underground applications. Accordingly, steel pipe can no longer be installed for general underground service unless it is specially listed for this purpose. At this time, no such listings are available.

10.1.3 Steel Pipe Used with Fire Department Connections. Where externally coated and wrapped and internally galvanized, steel pipe shall be permitted to be used between the check valve and the outside hose coupling for the fire department connection.

Steel pipe can be used for piping to a fire department connection, because the fire department connection is an auxiliary rather than a primary water supply for the sprinkler system. Both coating and wrapping are required when internally galvanized steel pipe is used between an exterior fire department connection and the check valve where the pipe connects to the system. Coating is also required for clamps, tie rods, and all bolted joint accessories. (See 10.3.5.2 and 10.8.3.5.)

AWWA C203, *Coal-Tar Protective Coatings and Linings for Steel Water Pipelines Enamel and Tape—Hot Applied,* is a referenced standard for coating materials in Table 10.1.1, but other bituminous (coal-tar) coatings are available in the marketplace.

AWWA C105, *Polyethylene Encasement for Ductile Iron Pipe Systems,* another of the referenced standards in Table 10.1.1, provides useful installation guidance where wrapping is required. The standard basically requires that high-density polyethylene tubes or sheets with a minimum nominal thickness of 0.004 in. (0.1 mm) or low-density polyethylene tubes or sheets with a minimum nominal thickness of 0.008 in. (0.2 mm) be used for the wrapping. The minimum width of high-density polyethylene tubes or sheets to be used is based on the nominal pipe diameter as shown in Commentary Table III.2.

The polyethylene is not intended to create an airtight or watertight enclosure but to prevent contact between the pipe and surrounding backfill/bedding material. Slack must be provided to prevent stretching the polyethylene where it bridges irregular surfaces such as bolted joints, anticipating backfilling operations. Cuts, tears, and punctures must be repaired. Overlaps must be a minimum of 1 ft (0.3 m) and must be secured using tape, string, or tie straps. Overlaps to pipe areas not wrapped must extend a minimum of 3 ft (1.0 m) and be circumferentially taped. Where piping is installed below the water table or in areas affected by tidal flooding, the wrapping must be circumferentially taped every 2 ft (0.6 m).

10.1.4* Pipe Type and Class. The type and class of pipe for a particular underground installation shall be determined through consideration of the following factors:

(1) Fire resistance of the pipe

COMMENTARY TABLE III.2 Minimum Width of High-Density Polyethylene Flat Tubes and Sheets

Nominal Pipe Diameter		Minimum Width of Flat Tube		Minimum Width of Sheet	
in.	mm	in.	mm	in.	mm
3	75	14	350	28	700
4	100	14	350	28	700
6	150	16	400	32	800
8	200	20	500	40	1000
10	250	24	600	48	1200
12	300	27	670	54	1350
14	350	30	750	60	1500

(2) Maximum system working pressure
(3) Depth at which the pipe is to be installed
(4) Soil conditions
(5) Corrosion
(6) Susceptibility of pipe to other external loads, including earth loads, installation beneath buildings, and traffic or vehicle loads

A.10.1.4 The following pipe design manuals can be used as guides:

(1) AWWA C150, *Thickness Design of Ductile Iron Pipe*
(2) AWWA C401, *Standard Practice for the Selection of Asbestos-Cement Water Pipe*
(3) AWWA M41, *Ductile Iron Pipe and Fittings*
(4) *Concrete Pipe Handbook,* American Concrete Pipe Association

10.1.5* Working Pressure. Piping, fittings, and other system components shall be rated for the maximum system working pressure to which they are exposed but shall not be rated at less than 150 psi (10.3 bar).

The 2010 edition has been revised to recognize that an underground main may have working pressures in excess of 150 psi (10 bar), such as an early suppression fast-response (ESFR) application. Higher-rated piping may be needed on the discharge side of the pump for the private fire service mains.

A.10.1.5 For underground system components, a minimum system pressure rating of 150 psi (10.3 bar) is specified in 10.1.5, based on satisfactory historical performance. Also, this pressure rating reflects that of the components typically used underground, such as piping, valves, and fittings. Where system pressures are expected to exceed pressures of 150 psi (10.3 bar), system components and materials manufactured and listed for higher pressures should be used. Systems that do not incorporate a fire pump or are not part of a combined standpipe system do not typically experience pressures exceeding 150 psi (10.3 bar) in underground piping. However, each system should be evaluated on an individual basis, because the presence of a fire department connection introduces the possibility of high pressures being applied by fire department apparatus. It is not the intent of this section to include the pressures generated through fire department connections as part of the maximum working pressure.

10.1.6* Lining of Buried Pipe.

Nominal diameters of available listed, lined ferrous metal pipe range from 3 in. (75 mm) to 24 in. (600 mm). Ductile iron pipe is generally cement mortar lined in accordance with AWWA C104, *Cement Mortar Lining for Ductile Iron Pipe and Fittings for Water.* Cement mortar linings were first used with cast iron pipe in 1922, and help prevent tuberculation of ferrous pipe through the creation of a high pH at the pipe wall as well as by providing a physical barrier to the water. The minimum thickness of the lining is 1/16 in. (1.6 mm) for pipe sizes 3 in. (75 mm) to 12 in. (300 mm) nominal diameter and greater for larger diameter pipe.

Care should be taken in hydraulic calculations of underground piping since manufacturers generally do not include the thickness of linings when providing information on internal diameters of their piping products. Therefore, utilizing the actual internal diameter throughout the calculations, accounting for any lining material, is imperative when calculating underground lined piping.

A.10.1.6 The following standards apply to the application of coating and linings:

(1) AWWA C104, *Cement Mortar Lining for Ductile Iron Pipe and Fittings for Water*
(2) AWWA C105, *Polyethylene Encasement for Ductile Iron Pipe Systems*
(3) AWWA C203, *Coal-Tar Protective Coatings and Linings for Steel Water Pipelines Enamel and Tape—Hot Applied*
(4) AWWA C205, *Cement-Mortar Protective Lining and Coating for Steel Water Pipe 4 in. and Larger—Shop Applied*
(5) AWWA C602, *Cement-Mortar Lining of Water Pipe Lines 4 in. and Larger—in Place*
(6) AWWA C116, *Protective Fusion-Bonded Epoxy Coatings for the Interior and Exterior Surfaces of Ductile-Iron and Gray-Iron Fittings for Water Supply Service*

For internal diameters of cement-lined ductile iron pipe, see Table A.10.1.6.

10.1.6.1 Unless the requirements of 10.1.6.2 are met, all ferrous metal pipe shall be lined in accordance with the applicable standards in Table 10.1.1.

10.1.6.2 Steel pipe utilized in fire department connections and protected in accordance with the requirements of 10.1.3 shall not be additionally required to be lined.

10.2 Fittings.

10.2.1 Standard Fittings. Fittings shall meet the standards in Table 10.2.1(a) or shall be in accordance with 10.2.2. In addition to the standards in Table 10.2.1(b), CPVC fittings shall also be in accordance with 10.2.2 and with the portions of the ASTM standards specified in Table 10.2.1(b) that apply to fire protection service.

10.2.2 Special Listed Fittings. Other types of fittings investigated for suitability in automatic sprinkler installations and listed for this service, including, but not limited to, polybutylene, CPVC, and steel differing from that provided in Table 10.2.1(a), shall be permitted when installed in accordance with their listing limitations, including installation instructions.

10.2.3 Pressure Limits. Listed fittings shall be permitted for the system pressures as specified in their listings, but not less than 150 psi (10 bar).

10.2.4* Buried Joints. Joints shall be approved.

A.10.2.4 The following standards apply to joints used with the various types of pipe:

(1) ASME B16.1, *Cast Iron Pipe Flanges and Flanged Fittings*
(2) AWWA C111, *Rubber-Gasket Joints for Ductile Iron Pressure Pipe and Fittings*
(3) AWWA C115, *Flanged Ductile Iron Pipe with Ductile Iron or Gray Iron Threaded Flanges*
(4) AWWA C206, *Field Welding of Steel Water Pipe*
(5) AWWA C606, *Grooved and Shouldered Joints*

10.2.5* Buried Fittings. Fittings shall be of an approved type with joints and pressure class ratings compatible with the pipe used.

TABLE A.10.1.6 *IDs for Cement-Lined Ductile Iron Pipe*

Pipe Size (in.)	OD (in.)	Pressure Class	Thickness Class	Wall Thickness	Minimum Lining Thickness*	ID (in.) with Lining
3	3.96	350		0.25	1/16	3.34
3	3.96		51	0.25	1/16	3.34
3	3.96		52	0.28	1/16	3.28
3	3.96		53	0.31	1/16	3.22
3	3.96		54	0.34	1/16	3.16
3	3.96		55	0.37	1/16	3.10
3	3.96		56	0.40	1/16	3.04
4	4.80	350		0.25	1/16	4.18
4	4.80		51	0.26	1/16	4.16
4	4.80		52	0.29	1/16	4.10
4	4.80		53	0.32	1/16	4.04
4	4.80		54	0.35	1/16	3.98
4	4.80		55	0.38	1/16	3.92
4	4.80		56	0.41	1/16	3.86
6	6.90	350		0.25	1/16	6.28
6	6.90		50	0.25	1/16	6.28
6	6.90		51	0.28	1/16	6.22
6	6.90		52	0.31	1/16	6.16
6	6.90		53	0.34	1/16	6.10
6	6.90		54	0.37	1/16	6.04
6	6.90		55	0.40	1/16	5.98
6	6.90		56	0.43	1/16	5.92
8	9.05	350		0.25	1/16	8.43
8	9.05		50	0.27	1/16	8.39
8	9.05		51	0.30	1/16	8.33
8	9.05		52	0.33	1/16	8.27
8	9.05		53	0.36	1/16	8.21
8	9.05		54	0.39	1/16	8.15
8	9.05		55	0.42	1/16	8.09
8	9.05		56	0.45	1/16	8.03
10	11.10	350		0.26	1/16	10.46
10	11.10		50	0.29	1/16	10.40
10	11.10		51	0.32	1/16	10.34
10	11.10		52	0.35	1/16	10.28
10	11.10		53	0.38	1/16	10.22
10	11.10		54	0.41	1/16	10.16
10	11.10		55	0.44	1/16	10.10
10	11.10		56	0.47	1/16	10.04
12	13.20	350		0.28	1/16	12.52
12	13.20		50	0.31	1/16	12.46
12	13.20		51	0.34	1/16	12.40
12	13.20		52	0.37	1/16	12.34
12	13.20		53	0.40	1/16	12.28
12	13.20		54	0.43	1/16	12.22
12	13.20		55	0.46	1/16	12.16
12	13.20		56	0.49	1/16	12.10

(continued)

TABLE A.10.1.6 Continued

Pipe Size (in.)	OD (in.)	Pressure Class	Thickness Class	Wall Thickness	Minimum Lining Thickness*	ID (in.) with Lining
14	15.30	250		0.28	3/32	14.55
14	15.30	300		0.30	3/32	14.51
14	15.30	350		0.31	3/32	14.49
14	15.30		50	0.33	3/32	14.45
14	15.30		51	0.36	3/32	14.39
14	15.30		52	0.39	3/32	14.33
14	15.30		53	0.42	3/32	14.27
14	15.30		54	0.45	3/32	14.21
14	15.30		55	0.48	3/32	14.15
14	15.30		56	0.51	3/32	14.09
16	17.40	250		0.30	3/32	16.61
16	17.40	300		0.32	3/32	16.57
16	17.40	350		0.34	3/32	16.53
16	17.40		50	0.34	3/32	16.53
16	17.40		51	0.37	3/32	16.47
16	17.40		52	0.40	3/32	16.41
16	17.40		53	0.43	3/32	16.35
16	17.40		54	0.46	3/32	16.29
16	17.40		55	0.49	3/32	16.23
16	17.40		56	0.52	3/32	16.17
18	19.50	250		0.31	3/32	18.69
18	19.50	300		0.34	3/32	18.63
18	19.50	350		0.36	3/32	18.59
18	19.50		50	0.35	3/32	18.61
18	19.50		51	0.35	3/32	18.61
18	19.50		52	0.41	3/32	18.49
18	19.50		53	0.44	3/32	18.43
18	19.50		54	0.47	3/32	18.37
18	19.50		55	0.50	3/32	18.31
18	19.50		56	0.53	3/32	18.25
20	21.60	250		0.33	3/32	20.75
20	21.60	300		0.36	3/32	20.69
20	21.60	350		0.38	3/32	20.65
20	21.60		50	0.36	3/32	20.69
20	21.60		51	0.39	3/32	20.63
20	21.60		52	0.42	3/32	20.57
20	21.60		53	0.45	3/32	20.51
20	21.60		54	0.48	3/32	20.45
20	21.60		55	0.51	3/32	20.39
20	21.60		56	0.54	3/32	20.33
24	25.80	200		0.33	3/32	24.95
24	25.80	250		0.37	3/32	24.87
24	25.80	300		0.40	3/32	24.81
24	25.80	350		0.43	3/32	24.75
24	25.80		50	0.38	3/32	24.85
24	25.80		51	0.41	3/32	24.79

TABLE A.10.1.6 Continued

Pipe Size (in.)	OD (in.)	Pressure Class	Thickness Class	Wall Thickness	Minimum Lining Thickness*	ID (in.) with Lining
24	25.80		52	0.44	$3/32$	24.73
24	25.80		53	0.47	$3/32$	24.67
24	25.80		54	0.50	$3/32$	24.61
24	25.80		55	0.53	$3/32$	24.55
24	25.80		56	0.56	$3/32$	24.49

TABLE 10.2.1(a) Fittings Materials and Dimensions

Materials and Dimensions	Standard
Cast Iron	
Gray Iron Threaded Fittings, Classes 125 and 250	ASME B16.4
Gray Iron Pipe Flanges and Flanged Fittings, Classes 12, 125, and 250	ASME B16.1
Malleable Iron	
Malleable Iron Threaded Fittings, Class 150 and 300	ASME B16.3
Steel	
Factory-Made Wrought Steel Buttweld Fittings	ASME B16.9
Buttwelding Ends	ASME B16.25
Specification for Piping Fittings of Wrought Carbon Steel and Alloy Steel for Moderate and Elevated Temperatures	ASTM A 234
Pipe Flanges and Flanged Fittings, NPS ½ Through 24	ASME B16.5
Forged Steel Fittings, Socket Welded and Threaded	ASME B16.11
Copper	
Wrought Copper and Bronze Solder Joint Pressure Fittings	ASME B16.22
Cast Bronze Solder Joint Pressure Fittings	ASME B16.18

TABLE 10.2.1(b) Specially Listed Fittings Materials and Dimensions

Materials and Dimensions	Standard
Chlorinated Polyvinyl Chloride (CPVC) Specification for Schedule 80 CPVC Threaded Fittings	ASTM F 437
Specification for Schedule 40 CPVC Socket-Type Fittings	ASTM F 438
Specification for Schedule 80 CPVC Socket-Type Fittings	ASTM F 439

A.10.2.5 Fittings generally used are cast iron with joints made to the specifications of the manufacturer of the particular type of pipe *(see the standards listed in A.10.2.4)*. Steel fittings also have some applications. The following standards apply to fittings:

(1) ASME B16.1, *Cast Iron Pipe Flanges and Flanged Fittings*
(2) AWWA C110, *Ductile Iron and Gray Iron Fittings, 3-in. Through 48-in., for Water and Other Liquids*
(3) AWWA C153, *Ductile Iron Compact Fittings, 3 in. through 24 in. and 54 in. through 64 in. for Water Service*

(4) AWWA C208, *Dimensions for Fabricated Steel Water Pipe Fittings*

10.3 Joining of Pipe and Fittings.

Although 10.2.4 and 10.2.5 simply require that buried joints be approved and that buried fittings be of an approved type, acceptable joining methods for listed piping are generally controlled as part of the pipe listing. Both ductile iron and PVC pipe are generally joined by bell and spigot ends in conjunction with rubber gaskets or by listed mechanical joints. A simple rubber gasket joint is termed a *push-on joint,* in which a single rubber ring gasket is fitted into the recess of a bell end of a length of pipe and is compressed by the plain end of the entering pipe, forming a seal. The gasket and the annular recess are specially shaped to lock the gasket in place against displacement. AWWA C111, *Rubber-Gasket Joints for Ductile Iron Pressure Pipe and Fittings,* one of the standards referenced in Table 10.1.1, requires that lubricants used in conjunction with push-on joints be labeled with the trade name or trademark and the pipe manufacturer's name to ensure compatibility with gasket material.

10.3.1 Threaded Pipe and Fittings. All threaded steel pipe and fittings shall have threads cut in accordance with ASME B1.20.1.

10.3.2 Groove Joining Methods. Pipes joined with grooved fittings shall be joined by a listed combination of fittings, gaskets, and grooves.

10.3.3 Brazed and Pressure Fitting Methods. Joints for the connection of copper tube shall be brazed or joined using pressure fittings as specified in Table 10.2.1(a).

10.3.4 Other Joining Methods. Other joining methods listed for this service shall be permitted where installed in accordance with their listing limitations.

10.3.5 Pipe Joint Assembly.

10.3.5.1 Joints shall be assembled by persons familiar with the particular materials being used and in accordance with the manufacturer's instructions and specifications.

10.3.5.2 All bolted joint accessories shall be cleaned and thoroughly coated with asphalt or other corrosion-retarding material after installation.

10.4 Depth of Cover.

Buried piping must be located below the frost line to prevent freezing during the winter months. The piping must also be located deep enough to be protected from other surface loads that could cause mechanical damage. Piping located under driveways, roads, and railroad tracks must be buried deeper to prevent undue loads on the sprinkler piping.

10.4.1* The depth of cover over water pipes shall be determined by the maximum depth of frost penetration in the locality where the pipe is laid.

A.10.4.1 The following documents apply to the installation of pipe and fittings:

(1) AWWA C603, *Standard for the Installation of Asbestos-Cement Water Pipe*
(2) AWWA C600, *Standard for the Installation of Ductile-Iron Water Mains and Their Appurtenances*
(3) AWWA M11, *A Guide for Steel Pipe Design and Installation*
(4) AWWA M41, *Ductile Iron Pipe and Fittings*
(5) *Concrete Pipe Handbook,* American Concrete Pipe Association
(6) *Handbook of PVC Pipe,* Uni-Bell PVC Pipe Association
(7) *Installation Guide for Ductile Iron Pipe,* Ductile Iron Pipe Research Association
(8) *Thrust Restraint Design for Ductile Iron Pipe,* Ductile Iron Pipe Research Association

As there is normally no circulation of water in private fire mains, they require greater depth of covering than do public mains. Greater depth is required in a loose gravelly soil (or in rock) than in compact soil containing large quantities of clay. The recommended depth of cover above the top of underground yard mains is shown in Figure A.10.4.1.

10.4.2 The top of the pipe shall be buried not less than 1 ft (0.3 m) below the frost line for the locality.

10.4.3 In those locations where frost is not a factor, the depth of cover shall be not less than 2½ ft (0.8 m) to prevent mechanical damage.

10.4.4 Pipe under driveways shall be buried at a minimum depth of 3 ft (0.9 m).

10.4.5 Pipe under railroad tracks shall be buried at a minimum depth of 4 ft (1.2 m).

10.4.6 The depth of cover shall be measured from the top of the pipe to finished grade, and due consideration shall always be given to future or final grade and nature of soil.

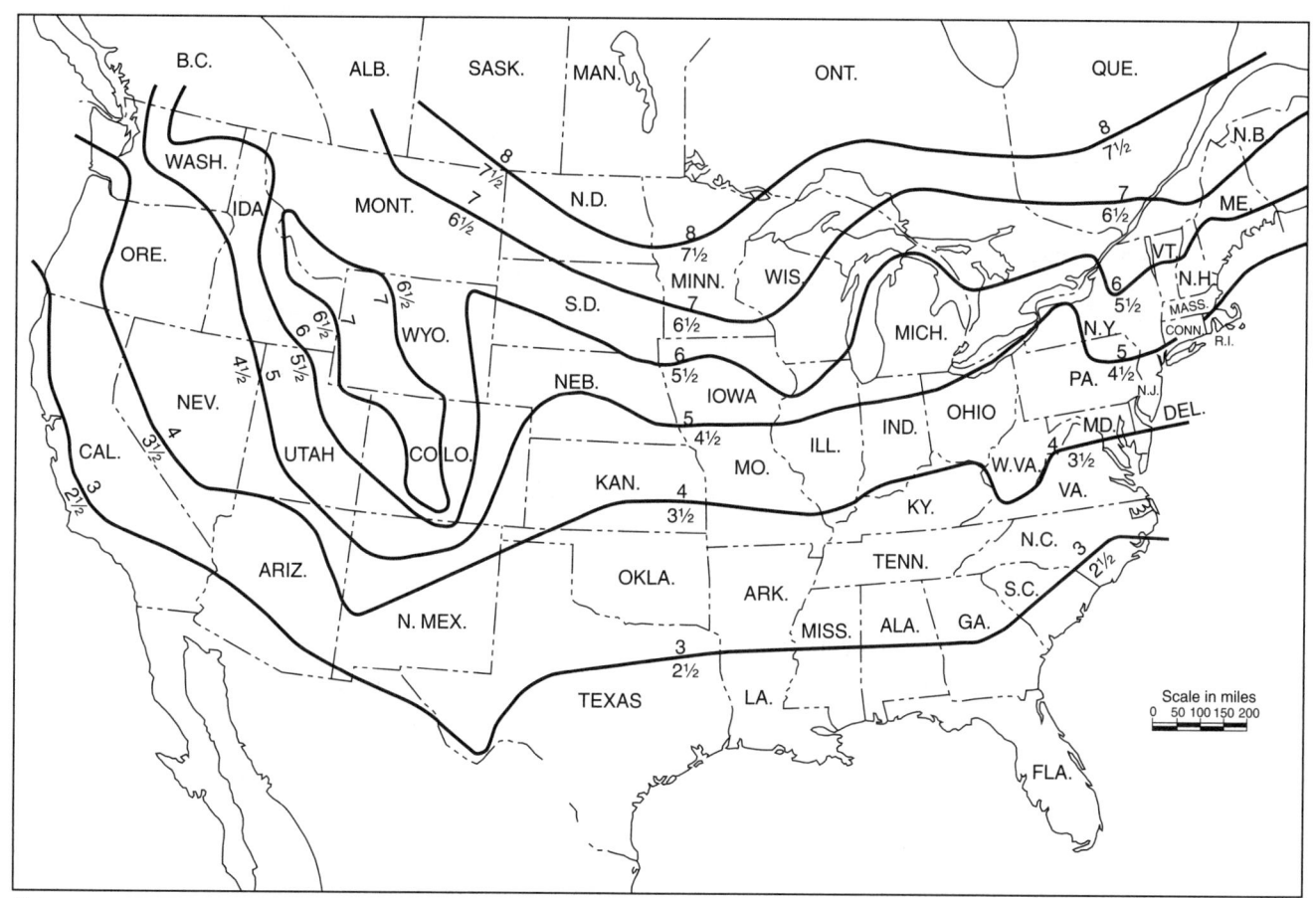

Notes:
1. For SI Units, 1 in. = 25.4 mm; 1 ft = 0.304 m.
2. Where frost penetration is a factor, the depth of cover shown averages 6 in. greater than that usually provided by the municipal waterworks. Greater depth is needed because of the absence of flow in yard mains.

FIGURE A.10.4.1 Recommended Depth of Cover (in feet) Above Top of Underground Yard Mains.

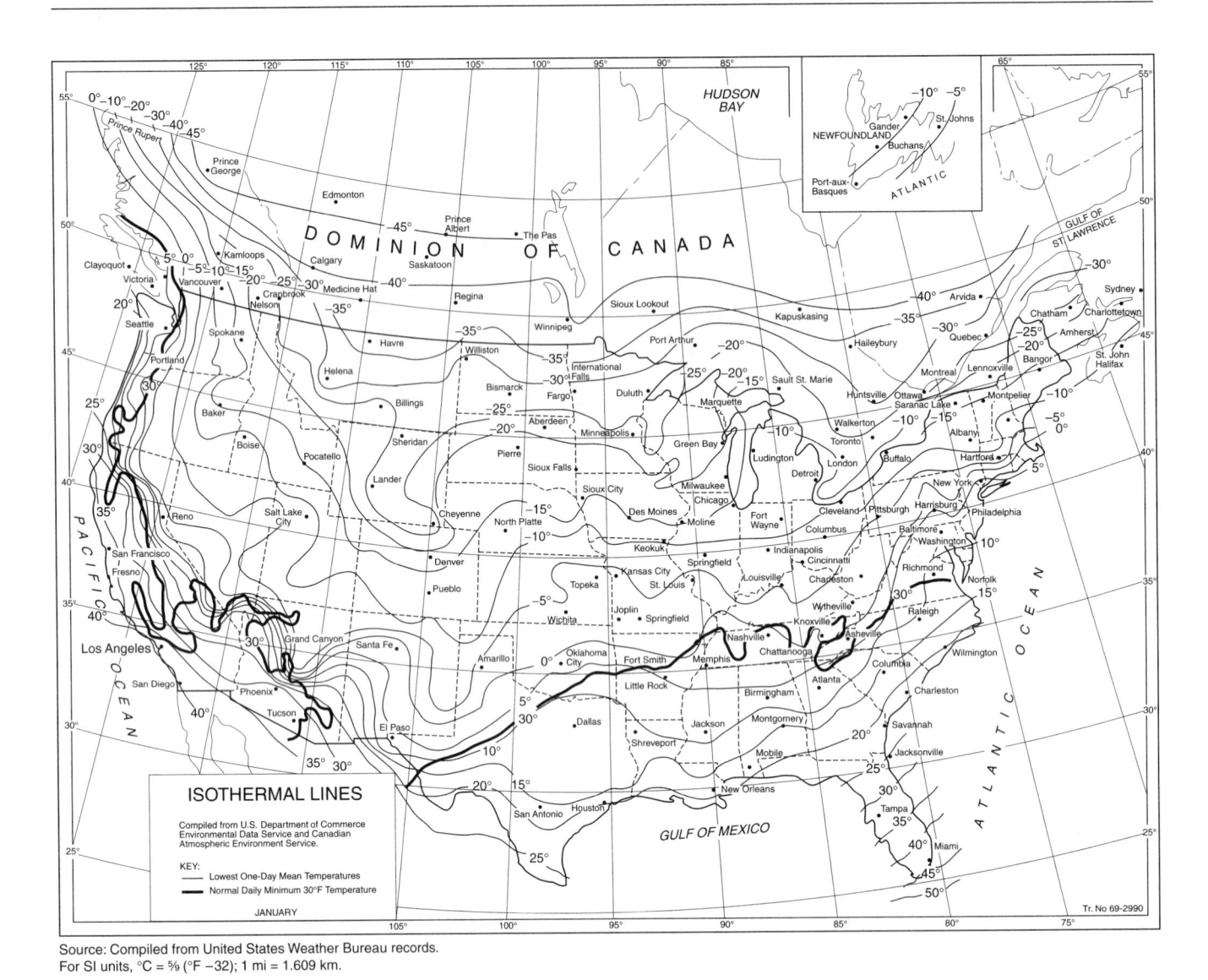

FIGURE A.10.5.1 *Isothermal Lines—Lowest One-Day Mean Temperature (°F).*

10.5 Protection against Freezing.

10.5.1* Where it is impracticable to bury pipe, pipe shall be permitted to be laid aboveground, provided that the pipe is protected against freezing and mechanical damage.

A.10.5.1 In determining the need to protect aboveground piping from freezing, the lowest mean temperature should be considered, as shown in Figure A.10.5.1.

10.5.2 Pipe shall be buried below the frost line where entering streams and other bodies of water.

10.5.3 Where pipe is laid in water raceways or shallow streams, care shall be taken that there will be sufficient depth of running water between the pipe and the frost line during all seasons of frost; a safer method is to bury the pipe 1 ft (0.3048 m) or more under the bed of the waterway.

10.5.4 Pipe shall be located at a distance from stream banks and embankment walls that prevents danger of freezing through the side of the bank.

10.6 Protection against Damage.

10.6.1 Pipe shall not be run under buildings.

Piping located under buildings is extremely difficult to repair, which is one of the reasons that 10.6.1 prohibits the installation of pipe under buildings. When buildings are built over existing underground piping, the piping should be rerouted around the new building. When piping under buildings requires repair, operations in the building must be curtailed, equipment may need to be moved, and the floor must be excavated. Leaks in the buried piping underneath the building can go undetected for long periods, and they can also undermine the building support. The location of system valves in the center of a building also places the system controls in the center of the building, which may not be desirable during a fire event.

10.6.2 Where pipe must be run under buildings, special precautions shall be taken, including the following:

(1) Arching the foundation walls over the pipe
(2) Running pipe in covered trenches
(3) Providing valves to isolate sections of pipe under buildings

Subsection 10.6.2 outlines some of the measures that can be taken to minimize the exposure to damage but not completely eliminate it. Where the main penetrates the foundation wall, arching of the foundation wall over the main or other means of ensuring system and structural integrity needs to be provided.

10.6.3 Fire service mains shall be permitted to enter the building adjacent to the foundation.

Subsection 10.6.3 permits the fire main to run underneath the building's footing or to penetrate the foundation wall if the main rises immediately inside the building adjacent to the exterior wall.

10.6.3.1 Where fire service mains enter the building adjacent to the foundation, the requirements of 10.6.2(2) and 10.6.2(3) shall not apply.

10.6.4* Where adjacent structures or physical conditions make it impractical to locate risers immediately inside an exterior wall, such risers shall be permitted to be located to minimize underground piping under the building.

A.10.6.4 It is not the intent of this provision to prohibit piping from passing under or through the foundation. Best practices are to locate risers to minimize the run of pipe below the building.

10.6.4.1 Where risers are located according to 10.6.4, the requirements of 10.6.2(2) and 10.6.2(3) shall not apply.

Where locating risers adjacent to exterior walls is not practical, 10.6.4 nevertheless requires pipe runs to be minimized.

10.6.5* Pipe joints shall not be located under foundation footings.

A.10.6.5 The individual piping standards should be followed for load and bury depth, accounting for the load and stresses imposed by the building foundation.

Figure A.10.6.5 shows location where pipe joints would be prohibited.

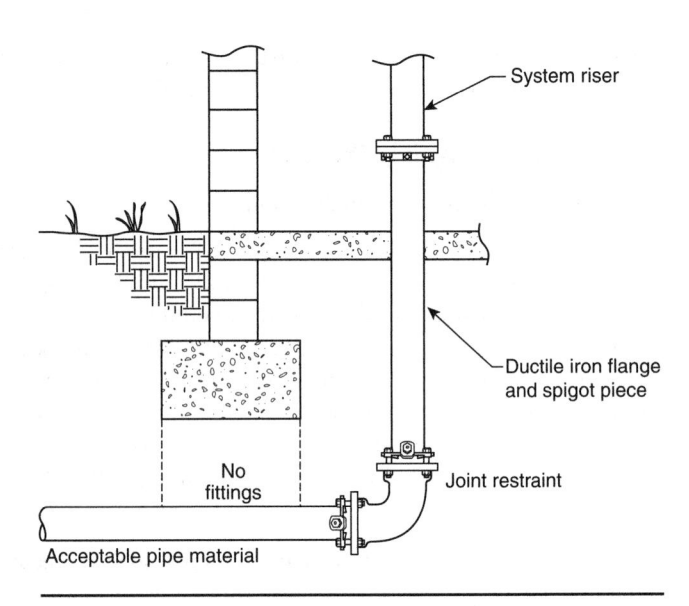

FIGURE A.10.6.5 *Pipe Joint Location in Relation to Foundation Footings.*

10.6.6 Mains shall be subjected to an evaluation of the following specific loading conditions and protected, if necessary:

(1) Mains running under railroads carrying heavy cargo
(2) Mains running under large piles of heavy commodities
(3) Mains located in areas that subject the main to heavy shock and vibrations

10.6.7* Where it is necessary to join metal pipe with pipe of dissimilar metal, the joint shall be insulated against the passage of an electric current using an approved method.

A.10.6.7 Gray cast iron is not considered galvanically dissimilar to ductile iron. Rubber gasket joints (unrestrained push-on or mechanical joints) are not considered connected electrically. Metal thickness should not be considered a protection against corrosive environments. In the case of cast iron or ductile iron pipe for soil evaluation and external protection systems, see Appendix A of AWWA C105.

10.6.8* In no case shall the underground piping be used as a grounding electrode for electrical systems.

The use of underground fire protection piping for electrical grounding increases the potential for stray currents and increased galvanic corrosion, which is why this use is prohibited by 10.6.8. Grounding to piping containing nonconductive joints is especially dangerous since it may not provide the expected ground. In no case should the underground piping be used as a grounding electrode for electrical systems. Electrical equipment should be grounded in accordance with *NFPA 70®*, *National Electrical Code®*.

A.10.6.8 Where lightning protection is provided for a structure, NFPA 780, Section 4.14, requires that all grounding media, including underground metallic piping systems, be interconnected to provide a common ground potential. These underground piping systems are not permitted to be substituted for grounding electrodes but must be bonded to the lightning protection grounding system. Where galvanic corrosion is of concern, this bond can be made via a spark gap or gas discharge tube.

10.6.8.1* The requirement of 10.6.8 shall not preclude the bonding of the underground piping to the lightning protection grounding system as required by NFPA 780 in those cases where lightning protection is provided for the structure.

A.10.6.8.1 While the use of the underground fire protection piping as the grounding electrode for the building is prohibited, *NFPA 70* requires that all metallic piping systems be bonded and grounded to disperse stray electrical currents. Therefore, the fire protection piping will be bonded to other metallic systems and grounded, but the electrical system will need an additional ground for its operation.

10.7 Requirement for Laying Pipe.

The precautions that must be taken to minimize damage to underground piping, eliminate stresses, and ensure a long service life are outlined in Section 10.7.

In the United States, the federal Occupational Safety and Health Administration (OSHA) has a number of requirements relating to safety issues involved in underground piping installation. These OSHA regulations can be found in standards 29 Code of Federal Regulations (CFR) 1926, Subpart P, Appendix F, on excavations, and 29 CFR 1910.146, Appendix A, on confined spaces.

For trenching operations involving excavations exceeding 5 ft (1.5 m) in depth, sloping, shoring, or shielding is required unless the excavation is entirely in stable rock. Additional soil classification is needed to select options, or, in the absence of soil classification, sloping of the excavation wall must be performed with the slope limited to 1½ horizontal to 1 vertical, which is a maximum 34 degrees from horizontal.

Confined spaces regulations can be invoked when valves on underground piping are located in pits. For this reason, valves are usually brought above ground in areas where freezing is not a concern, and special prefabricated heated enclosures (see Exhibit III.20) are now marketed to allow aboveground valve installations in other areas as well.

EXHIBIT III.20 *Heated Enclosure. (Photograph courtesy of AquaSHIELD)*

10.7.1 Pipes, valves, hydrants, gaskets, and fittings shall be inspected for damage when received and shall be inspected prior to installation. *(See Figure 10.10.1.)*

10.7.2 The torquing of bolted joints shall be checked.

10.7.3 Pipe, valves, hydrants, and fittings shall be clean inside.

10.7.4 When work is stopped, the open ends of pipe, valves, hydrants, and fittings shall be plugged to prevent stones and foreign materials from entering.

Precautions must be taken to prevent rocks and other foreign materials from entering piping during installation, as stated in 10.7.4. These materials can be carried into fire protection system piping and cause blockages and system failure during an emergency. Care in installation is necessary but does not eliminate the need to flush the piping as part of the acceptance testing.

10.7.5 All pipe, fittings, valves, and hydrants shall be carefully lowered into the trench using appropriate equipment and carefully examined for cracks or other defects while suspended above the trench.

10.7.6 Plain ends shall be inspected for signs of damage prior to installation.

10.7.7 Under no circumstances shall water main materials be dropped or dumped.

10.7.8 Pipe shall not be rolled or skidded against other pipe materials.

10.7.9 Pipes shall bear throughout their full length and shall not be supported by the bell ends only or by blocks.

10.7.10 If the ground is soft or of a quicksand nature, special provisions shall be made for supporting pipe.

10.7.11 Valves and fittings used with nonmetallic pipe shall be supported and restrained in accordance with the manufacturer's specifications.

10.8 Joint Restraint.

The requirements of Section 10.8 address the need to provide thrust blocks only at turns in piping, because the soil surrounding underground pipe is generally satisfactory to hold the piping in place along straight runs.

10.8.1 General.

10.8.1.1* All tees, plugs, caps, bends, reducers, valves, and hydrant branches shall be restrained against movement by using thrust blocks in accordance with 10.8.2 or restrained joint systems in accordance with 10.8.3.

A.10.8.1.1 It is a fundamental design principle of fluid mechanics that dynamic and static pressures, acting at changes in size or direction of a pipe, produce unbalanced thrust forces at locations such as bends, tees, wyes, dead ends, and reducer offsets. This design principle includes consideration of lateral soil pressure and pipe/soil friction, variables that can be reliably determined using current soil engineering knowledge. Refer to A.10.8.3 for a list of references for use in calculating and determining joint restraint systems.

Except for the case of welded joints and approved special restrained joints, such as is provided by approved mechanical joint retainer glands or locked mechanical and push-on joints, the usual joints for underground pipe are expected to be held in place by the soil in which the pipe is buried. Gasketed push-on and mechanical joints without special locking devices have limited ability to resist separation due to movement of the pipe.

10.8.1.2 Piping with fused, threaded, grooved, or welded joints shall not require additional restraining, provided that such joints can pass the hydrostatic test of 10.10.2.2 without shifting of piping or leakage in excess of permitted amounts.

10.8.1.3 Steep Grades. On steep grades, mains shall be additionally restrained to prevent slipping.

10.8.1.3.1 Pipe shall be restrained at the bottom of a hill and at any turns (lateral or vertical).

10.8.1.3.2 The restraint specified in 10.8.1.3.1 shall be to natural rock or to suitable piers built on the downhill side of the bell.

10.8.1.3.3 Bell ends shall be installed facing uphill.

10.8.1.3.4 Straight runs on hills shall be restrained as determined by the design engineer.

10.8.2* Thrust Blocks.

Thrust blocks are covered in 10.8.2. Exhibit III.21 illustrates a typical thrust block arrangement.

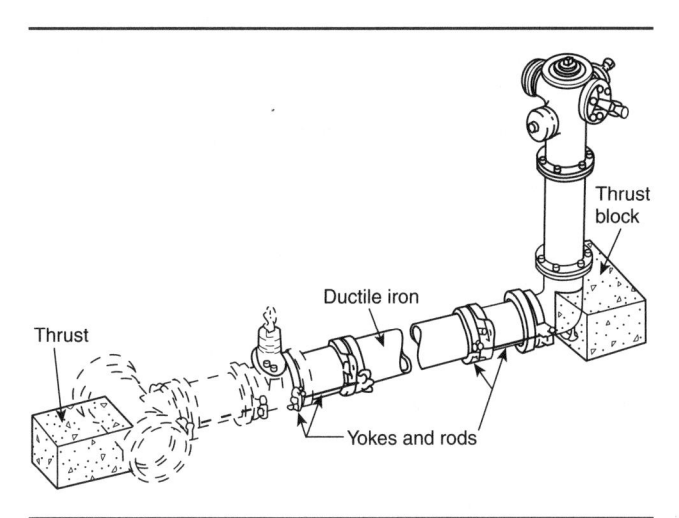

EXHIBIT III.21 Thrust Block Arrangement. (Courtesy of Los Angeles Department of Water and Power)

A.10.8.2 Thrust Blocks. Concrete thrust blocks are one of the methods of restraint now in use, provided that stable soil conditions prevail and space requirements permit placement. Successful blocking is dependent upon factors such as location, availability and placement of concrete, and possibility of disturbance by future excavations.

Resistance is provided by transferring the thrust force to the soil through the larger bearing area of the block such that the resultant pressure against the soil does not exceed the horizontal bearing strength of the soil. The design of thrust blocks consists of determining the appropriate bearing area of the block for a particular set of conditions. The parameters involved in the design include pipe size, design pressure, angle of the bend (or configuration of the fitting involved), and the horizontal bearing strength of the soil.

Table A.10.8.2(a) gives the nominal thrust at fittings for various sizes of ductile iron and PVC piping. Figure A.10.8.2(a) shows an example of how thrust forces act on a piping bend.

Thrust blocks are generally categorized into two groups—bearing and gravity blocks. Figure A.10.8.2(b) depicts a typical bearing thrust block on a horizontal bend.

The following are general criteria for bearing block design:

(1) The bearing surface should, where possible, be placed against undisturbed soil.
(2) Where it is not possible to place the bearing surface against undisturbed soil, the fill between the bearing surface and undisturbed soil must be compacted to at least 90 percent Standard Proctor density.

TABLE A.10.8.2(a) Thrust at Fittings at 100 psi (6.9 bar) Water Pressure for Ductile Iron and PVC Pipe

	Total Pounds					
Nominal Pipe Diameter (in.)	**Dead End**	**90 Degree Bend**	**45 Degree Bend**	**22½ Degree Bend**	**11¼ Degree Bend**	**5⅛ Degree Bend**
4	1,810	2,559	1,385	706	355	162
6	3,739	5,288	2,862	1,459	733	334
8	6,433	9,097	4,923	2,510	1,261	575
10	9,677	13,685	7,406	3,776	1,897	865
12	13,685	19,353	10,474	5,340	2,683	1,224
14	18,385	26,001	14,072	7,174	3,604	1,644
16	23,779	33,628	18,199	9,278	4,661	2,126
18	29,865	42,235	22,858	11,653	5,855	2,670
20	36,644	51,822	28,046	14,298	7,183	3,277
24	52,279	73,934	40,013	20,398	10,249	4,675
30	80,425	113,738	61,554	31,380	15,766	7,191
36	115,209	162,931	88,177	44,952	22,585	10,302
42	155,528	219,950	119,036	60,684	30,489	13,907
48	202,683	286,637	155,127	79,083	39,733	18,124

Notes:
(1) For SI units, 1 lb = 0.454 kg; 1 in. = 25.4 mm.
(2) To determine thrust at pressure other than 100 psi (6.9 bar), multiply the thrust obtained in the table by the ratio of the pressure to 100 psi (6.9 bar). For example, the thrust on a 12 in. (305 mm), 90 degree bend at 125 psi (8.6 bar) is 19,353 × 125/100 = 24,191 lb (10,973 kg).

(3) Block height (h) should be equal to or less than one-half the total depth to the bottom of the block (H_t) but not less than the pipe diameter (D).
(4) Block height (h) should be chosen such that the calculated block width (b) varies between one and two times the height.
(5) Gravity thrust blocks can be used to resist thrust at vertical down bends. In a gravity thrust block, the weight of the block is the force providing equilibrium with the thrust force. The design problem is then to calculate the required volume of the thrust block of a known density. The vertical component of the thrust force in Figure A.10.8.2(c) is balanced by the weight of the block. For required horizontal bearing block areas, see Table A.10.8.2(b).

The required block area (A_b) is as follows:

$$A_b = (h)(b) = \frac{T(S_f)}{S_b}$$

where:
A_b = required block area (ft²)
h = block height (ft)
b = calculated block width (ft)
T = thrust force (lbf)
S_f = safety factor (usually 1.5)
S_b = bearing strength (lb/ft²)

TABLE A.10.8.2(b) Required Horizontal Bearing Block Area

Nominal Pipe Diameter (in.)	Bearing Block Area (ft²)	Nominal Pipe Diameter (in.)	Bearing Block Area (ft²)	Nominal Pipe Diameter (in.)	Bearing Block Area (ft²)
3	2.6	12	29.0	24	110.9
4	3.8	14	39.0	30	170.6
6	7.9	16	50.4	36	244.4
8	13.6	18	63.3	42	329.9
10	20.5	20	77.7	48	430.0

Notes:
(1) Although the bearing strength values in this table have been used successfully in the design of thrust blocks and are considered to be conservative, their accuracy is totally dependent on accurate soil identification and evaluation. The ultimate responsibility for selecting the proper bearing strength of a particular soil type must rest with the design engineer.
(2) Values listed are based on a 90 degree horizontal bend, an internal pressure of 100 psi, a soil horizontal bearing strength of 1000 lb/ft², a safety factor of 1.5, and ductile iron pipe outside diameters.
(a) For other horizontal bends, multiply by the following coefficients: for 45 degrees, 0.414; for 22½ degrees, 0.199; for 11¼ degrees, 0.098.
(b) For other internal pressures, multiply by ratio to 100 psi.
(c) For other soil horizontal bearing strengths, divide by ratio to 1000 lb/ft2.
(d) For other safety factors, multiply by ratio to 1.5.

Example: Using Table A.10.8.2(b), find the horizontal bearing block area for a 6 in. diameter, 45 degree bend with an internal pressure of 150 psi. The soil bearing strength is 3000 lb/ft², and the safety factor is 1.5.

From Table A.10.8.2(b), the required bearing block area for a 6 in. diameter, 90 degree bend with an internal pressure of 100 psi and a soil horizontal bearing strength of 1000 psi is 7.9 ft².

For example:

$$\text{Area} = \frac{7.9 \text{ ft}^2 (0.414)\left(\frac{150}{100}\right)}{\left(\frac{3000}{1000}\right)} = 1.64 \text{ ft}^2$$

TABLE A.10.8.2(c) Horizontal Bearing Strengths

Soil	Bearing Strength (Sb) lb/ft²	kN/m²
Muck	0	0
Soft clay	1000	47.9
Silt	1500	71.8
Sandy silt	3000	143.6
Sand	4000	191.5
Sand clay	6000	287.3
Hard clay	9000	430.9

Note: Although the bearing strength values in this table have been used successfully in the design of thrust blocks and are considered to be conservative, their accuracy is totally dependent on accurate soil identification and evaluation. The ultimate responsibility for selecting the proper bearing strength of a particular soil type must rest with the design engineer.

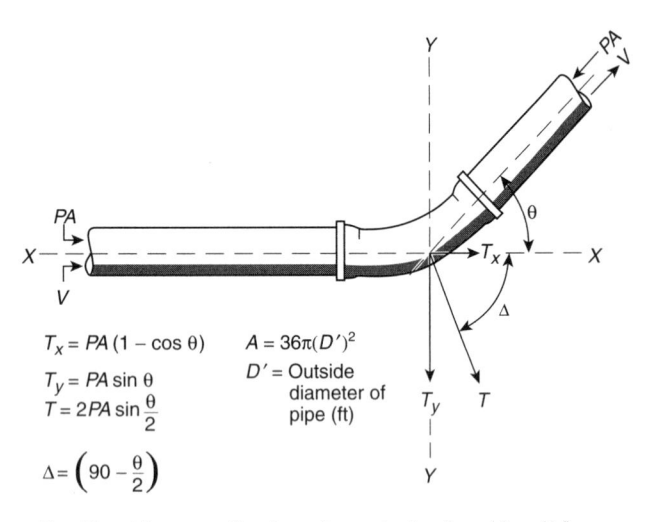

$T_x = PA(1 - \cos \theta)$ $A = 36\pi(D')^2$
$T_y = PA \sin \theta$ D' = Outside diameter of pipe (ft)
$T = 2PA \sin \dfrac{\theta}{2}$

$\Delta = \left(90 - \dfrac{\theta}{2}\right)$

T = Thrust force resulting from change in direction of flow (lbf)
T_x = Component of the thrust force acting parallel to the original direction of flow (lbf)
T_y = Component of the thrust force acting perpendicular to the original direction of flow (lbf)
P = Water pressure (psi^2)
A = Cross-sectional area of the pipe based on outside diameter (in.^{2})
V = Velocity in direction of flow

FIGURE A.10.8.2(a) *Thrust Forces Acting on a Bend.*

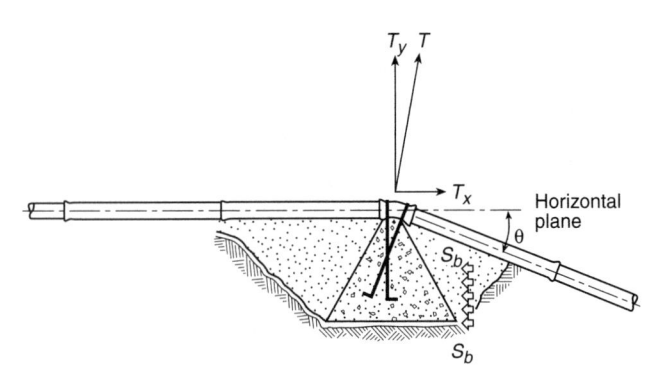

T = thrust force resulting from the change of direction of flow
T_x = horizontal component of the thrust force
T_y = vertical component of the thrust force
S_b = horizontal bearing strength of the soil

FIGURE A.10.8.2(c) *Gravity Thrust Block.*

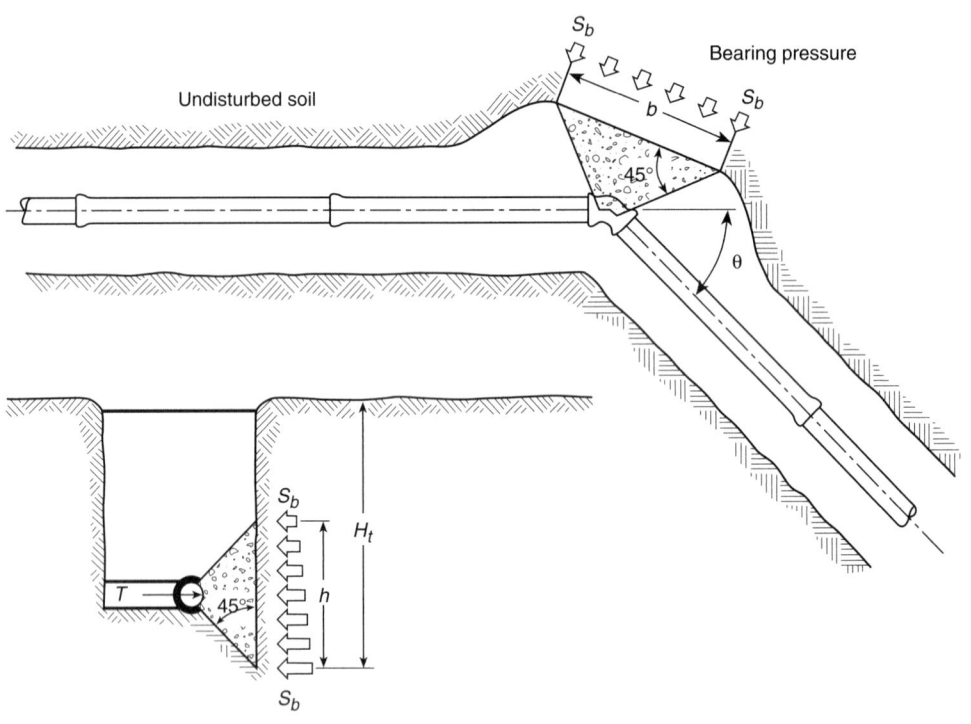

T = thrust force resulting from the change in direction of flow
S_b = horizontal bearing strength of the soil
h = block height
H_t = total depth to bottom of block

FIGURE A.10.8.2(b) *Bearing Thrust Block.*

Then, for a horizontal bend, the following formula is used:

$$b = \frac{2(S_f)(P)(A) \sin\left(\frac{\theta}{2}\right)}{(h)(S_b)}$$

where:

b = calculated block width (ft)

S_f = safety factor (usually 1.5 for thrust block design)

P = water pressure (lb/in.^{2})

A = cross-sectional area of the pipe based on outside diameter

h = block height (ft)

S_b = horizontal bearing strength of the soil (lb/ft^2)(in.^{2})

A similar approach can be used to design bearing blocks to resist the thrust forces at locations such as tees and dead ends. Typical values for conservative horizontal bearing strengths of various soil types are listed in Table A.10.8.2(c).

In lieu of the values for soil bearing strength shown in Table A.10.8.2(c), a designer might choose to use calculated Rankine passive pressure (Pp) or other determination of soil bearing strength based on actual soil properties.

It can be easily shown that $T_y = PA \sin \theta$. The required volume of the block is as follows:

$$V_g = \frac{S_f PA \sin \theta}{W_m}$$

where:

V_g = block volume (ft^3)

S_f = safety factor

P = water pressure (psi)

A = cross-sectional area of the pipe interior

W_m = density of the block material (lb/ft^3)

In a case such as the one shown, the horizontal component of thrust force is calculated as follows:

$$T_x = PA (1 - \cos \theta)$$

where:

T_x = horizontal component of the thrust force

P = water pressure (psi)

A = cross-sectional area of the pipe interior

The horizontal component of thrust force must be resisted by the bearing of the right side of the block against the soil. Analysis of this aspect follows the same principles as the previous section on bearing blocks.

The following is a sample calculation to determine the size of a thrust block using the formula in A.10.8.2(5), Figure A.10.8.2(b), and Table A.10.8.2(b):

Design parameters: 12 in. ductile iron pipe $(A) = 136.8$ in.^{2}
45° bend (θ)
85 psi water pressure at bend (P)
Soil type = sandy silt
Safety factor $(S_f) = 1.5$
Block height $(h) = 1.5$ ft

The cross-sectional area of the pipe is determined by the following formula:

$$A = 36\pi(D')^2$$

where:

D' = outside diameter of the pipe (ft)

Therefore:

$$A = 36\pi (1.1)^2$$
$$A = 136.8 \text{ in.}^2$$

Using the formula in A.10.8.2(5):

$$b = \frac{2(1.5)(85)(136.8) \sin (45/2)}{de(1.5)(3000)}$$

$$b = \frac{13325.688}{4500}$$

$$b = 2.96 \text{ ft}$$

Note that the result of 2.96 ft is consistent with A.10.8.2(4) in that the block base ($b = 2.96$ ft) is between one to two times the height ($h = 1.5$ ft). This method is the correct method for selecting the size and shape of a thrust block.

10.8.2.1 Thrust blocks shall be considered satisfactory where soil is suitable for their use.

10.8.2.2 Thrust blocks shall be of a concrete mix not leaner than one part cement, two and one-half parts sand, and five parts stone.

10.8.2.3 Thrust blocks shall be placed between undisturbed earth and the fitting to be restrained and shall be capable of resisting the calculated thrust forces.

10.8.2.4 Wherever possible, thrust blocks shall be placed so that the joints are accessible for repair.

10.8.3* Restrained Joint Systems. Fire mains utilizing restrained joint systems shall include one or more of the following:

(1) Locking mechanical or push-on joints
(2) Mechanical joints utilizing setscrew retainer glands
(3) Bolted flange joints
(4) Heat-fused or welded joints
(5) Pipe clamps and tie rods
(6) Other approved methods or devices

Any one of the restraining systems covered in 10.8.3 is considered acceptable. Flange adapter fittings and other mechanical joints do not necessarily provide joint restraint. For example, flange adapter fittings are listed by Underwriters Laboratories Inc. as "fittings, retainer type" in UL 194, *Gasketed Joints for Ductile-Iron Pipe and Fittings for Fire Protection Service.* UL 194 further categorizes such fittings as either "gasketed joints with self-restraining feature" or "gasketed joints without self-restraining feature." The latter includes gasketed joints consisting of merely a pipe or fitting bell and a spigot end without a feature for joint restraint. Under the provisions of UL 194, all gasketed joints are provided with a pressure test at twice the rated working pressure with samples deflected to the maximum angle specified by the manufacturer. Joints for which the manufacturer claims a self-restraining feature are not provided with external restraint during the test. The manufacturer's literature should specify whether external restraint is required. An example of a restrained joint system is shown in Exhibit III.22.

Rods and clamps can be used as an alternative to the restrained joint systems shown in Figure A.10.8.3.

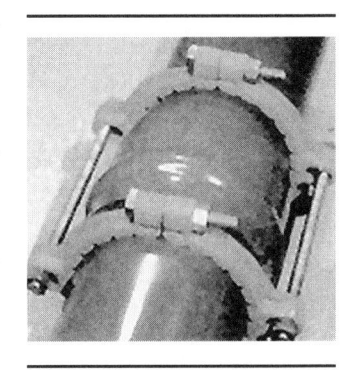

EXHIBIT III.22 Restrained Joint System.

A.10.8.3 Restrained Joint Systems. A method for providing thrust restraint is the use of restrained joints. A restrained joint is a special type of joint that is designed to provide longitudinal restraint. Restrained joint systems function in a manner similar to that of thrust blocks, insofar as the reaction of the entire restrained unit of piping with the soil balances the thrust forces.

The objective in designing a restrained joint thrust restraint system is to determine the length of pipe that must be restrained on each side of the focus of the thrust force. This will be a function of the pipe size, the internal pressure, the depth of cover, and the characteristics of the solid surrounding the pipe.

The following documents apply to the design, calculation, and determination of restrained joint systems:

(1) *Thrust Restraint Design for Ductile Iron Pipe,* Ductile Iron Pipe Research Association
(2) AWWA M41, *Ductile Iron Pipe and Fittings*
(3) AWWA M9, *Concrete Pressure Pipe*
(4) AWWA M11, *A Guide for Steel Pipe Design and Installation*
(5) *Thrust Restraint Design Equations and Tables for Ductile Iron and PVC Pipe,* EBAA Iron, Inc.

Figure A.10.8.3 shows an example of a typical connection to a fire protection system riser utilizing restrained joint pipe.

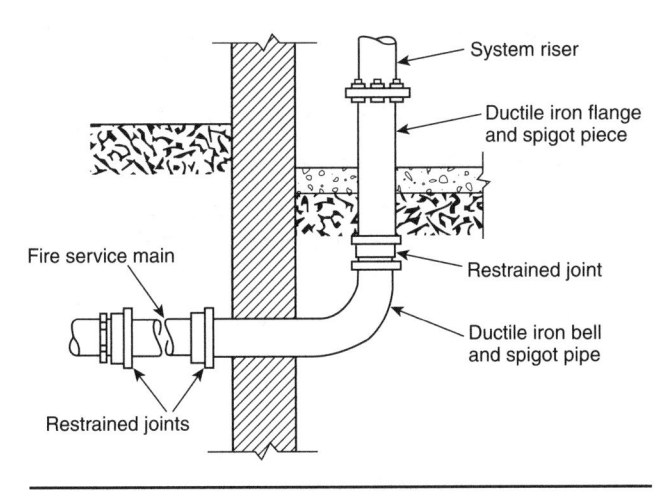

FIGURE A.10.8.3 *Typical Connection to a Fire Protection System Riser Illustrating Restrained Joints.*

10.8.3.1 Sizing Clamps, Rods, Bolts, and Washers.

10.8.3.1.1 Clamps.

10.8.3.1.1.1 Clamps shall have the following dimensions:

(1) ½ in. × 2 in. (12.7 mm × 50.8 mm) for pipe 4 in. to 6 in.
(2) ⅝ in. × 2½ in. (15.9 mm × 63.5 mm) for pipe 8 in. to 10 in.
(3) ⅝ in. × 3 in. (15.9 mm × 76.2 mm) for 12 in. pipe

10.8.3.1.1.2 The diameter of a bolt hole shall be ¹⁄₁₆ in. (1.6 mm) larger than that of the corresponding bolt.

10.8.3.1.2 Rods.

10.8.3.1.2.1 Rods shall be not less than ⅝ in. (15.9 mm) in diameter.

10.8.3.1.2.2 Table 10.8.3.1.2.2 provides the numbers of various diameter rods that shall be used for a given pipe size.

TABLE 10.8.3.1.2.2 Rod Number—Diameter Combinations

Nominal Pipe Size (in.)	⅝ in. (15.9 mm)	¾ in. (19.1 mm)	⅞ in. (22.2 mm)	1 in. (25.4 mm)
4	2	—	—	—
6	2	—	—	—
8	3	2	—	—
10	4	3	2	—
12	6	4	3	2
14	8	5	4	3
16	10	7	5	4

Note: This table has been derived using pressure of 225 psi (15.5 bar) and design stress of 25,000 psi (172.4 MPa).

10.8.3.1.2.3 Where using bolting rods, the diameter of mechanical joint bolts shall limit the diameter of rods to ¾ in. (19.1 mm).

10.8.3.1.2.4 Threaded sections of rods shall not be formed or bent.

10.8.3.1.2.5 Where using clamps, rods shall be used in pairs for each clamp.

10.8.3.1.2.6 Assemblies in which a restraint is made by means of two clamps canted on the barrel of the pipe shall be permitted to use one rod per clamp if approved for the specific installation by the authority having jurisdiction.

10.8.3.1.2.7 Where using combinations of rods, the rods shall be symmetrically spaced.

10.8.3.1.3 Clamp Bolts. Clamp bolts shall have the following diameters:

(1) ⅝ in. (15.9 mm) for pipe 4 in., 6 in., and 8 in.
(2) ¾ in. (19.1 mm) for pipe 10 in.
(3) ⅞ in. (22.2 mm) for 12 in. pipe

10.8.3.1.4 Washers.

10.8.3.1.4.1 Washers shall be permitted to be cast iron or steel and round or square.

10.8.3.1.4.2 Cast iron washers shall have the following dimensions:

(1) ⅝ in. × 3 in. (15.9 mm × 76.2 mm) for 4 in., 6 in., 8 in., and 10 in. pipe
(2) ¾ in. × 3½ in. (19.1 mm × 88.9 mm) for 12 in. pipe

10.8.3.1.4.3 Steel washers shall have the following dimensions:

(1) ½ in. × 3 in. (12.7 mm × 76.2 mm) for 4 in., 6 in., 8 in., and 10 in. pipe
(2) ½ in. × 3½ in. (12.7 mm × 88.9 mm) for 12 in. pipe

10.8.3.1.4.4 The diameter of holes shall be ⅛ in. (3.2 mm) larger than that of rods.

10.8.3.2 Sizes of Restraint Straps for Tees.

10.8.3.2.1 Restraint straps for tees shall have the following dimensions:

(1) ⅝ in. (15.9 mm) thick and 2½ in. (63.5 mm) wide for 4 in. (102 mm), 6 in. (152 mm), 8 in. (204 mm), and 10 in. (254 mm) pipe
(2) ⅝ in. (15.9 mm) thick and 3 in. (76.2 mm) wide for 12 in. (305 mm) pipe

10.8.3.2.2 The diameter of rod holes shall be 1/16 in. (1.6 mm) larger than that of rods.

10.8.3.2.3 Figure 10.8.3.2.3 and Table 10.8.3.2.3 shall be used in sizing the restraint straps for both mechanical and push-on joint tee fittings.

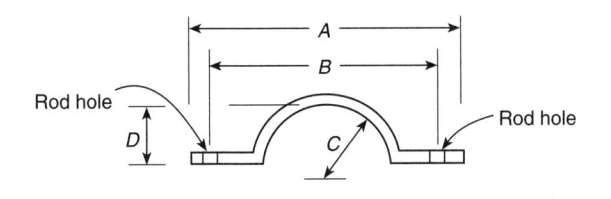

FIGURE 10.8.3.2.3 Restraint Straps for Tees.

TABLE 10.8.3.2.3 Restraint Straps for Tees

Nominal Pipe Size (in.)	A in.	A mm	B in.	B mm	C in.	C mm	D in.	D mm
4	12½	318	10⅛	257	2½	64	1¾	44
6	14½	368	12⅛	308	3 9/16	90	2 13/16	71
8	16¾	425	14⅜	365	4 21/32	118	3 29/32	99
10	19 1/16	484	16 11/16	424	5¾	146	5	127
12	22 5/16	567	19 3/16	487	6¾	171	5⅞	149

10.8.3.3 Sizes of Plug Strap for Bell End of Pipe.

10.8.3.3.1 The strap shall be ¾ in. (19.1 mm) thick and 2½ in. (63.5 mm) wide.

10.8.3.3.2 The strap length shall be the same as dimension *A* for tee straps as shown in Figure 10.8.3.2.3.

10.8.3.3.3 The distance between the centers of rod holes shall be the same as dimension *B* for tee straps as shown in Figure 10.8.3.2.3.

10.8.3.4 Material. Clamps, rods, rod couplings or turnbuckles, bolts, washers, restraint straps, and plug straps shall be of a material that has physical and chemical characteristics that indicate its deterioration under stress can be predicted with reliability.

10.8.3.5* Corrosion Resistance. After installation, rods, nuts, bolts, washers, clamps, and other restraining devices shall be cleaned and thoroughly coated with a bituminous or other acceptable corrosion-retarding material.

A.10.8.3.5 Examples of materials and the standards covering these materials are as follows:

(1) Clamps, steel *(see discussion on steel in the following paragraph)*
(2) Rods, steel *(see discussion on steel in the following paragraph)*
(3) Bolts, steel (ASTM A 307)
(4) Washers, steel *(see discussion on steel in the following paragraph)*; cast iron (Class A cast iron as defined by ASTM A 126)
(5) Anchor straps and plug straps, steel *(see discussion on steel in the following paragraph)*
(6) Rod couplings or turnbuckles, malleable iron (ASTM A 197)

Steel of modified range merchant quality as defined in U.S. Federal Standard No. 66C, April 18, 1967, change notice No. 2, April 16, 1970, as promulgated by the U.S. Federal Government General Services Administration.

The materials specified in A.10.8.3.5(1) through (6) do not preclude the use of other materials that also satisfy the requirements of this section.

10.9 Backfilling.

10.9.1 Backfill shall be tamped in layers or puddled under and around pipes to prevent settlement or lateral movement and shall contain no ashes, cinders, refuse, organic matter, or other corrosive materials.

Improper backfill is a major cause of underground piping failures. Proper consolidation of backfill can prevent voids that eventually place stress on the piping and joints.

Underground piping should be laid on a firm bed of earth for its entire length with the earth scooped out at the bells. Clean earth or screened gravel should be tamped under, around, and above the pipe to a level of 1 ft (304 mm) above. The excavation should then be backfilled to grade, compacting the fill in layers of 2 ft (608 mm) or less, unless otherwise directed by the project soil engineer.

Puddling, which involves the use of water to help consolidate soil around and below the piping, is occasionally used to assist the backfill compaction effort.

10.9.2 Rocks shall not be placed in trenches.

Subsection 10.9.2 prohibits the placement of rocks in trenches. Clean fill cushions the pipe and distributes the load to the surrounding earth. The presence of cinders, refuse, and other organic matter can create points of accelerated corrosion that can reduce the life of underground piping.

10.9.3 Frozen earth shall not be used for backfilling.

10.9.4 In trenches cut through rock, tamped backfill shall be used for at least 6 in. (150 mm) under and around the pipe and for at least 2 ft (0.6 m) above the pipe.

10.10 Testing and Acceptance.

10.10.1 Approval of Underground Piping. The installing contractor shall be responsible for the following:

(1) Notifying the authority having jurisdiction and the owner's representative of the time and date testing is to be performed
(2) Performing all required acceptance tests
(3) Completing and signing the contractor's material and test certificate(s) shown in Figure 10.10.1.

Contractor's Material and Test Certificate for Underground Piping

PROCEDURE
Upon completion of work, inspection and tests shall be made by the contractor's representative and witnessed by an owner's representative. All defects shall be corrected and system left in service before contractor's personnel finally leave the job.

A certificate shall be filled out and signed by both representatives. Copies shall be prepared for approving authorities, owners, and contractor. It is understood the owner's representative's signature in no way prejudices any claim against contractor for faulty material, poor workmanship, or failure to comply with approving authority's requirements or local ordinances.

Property name	Date
Property address	

Plans	Accepted by approving authorities (names)		
	Address		
	Installation conforms to accepted plans	☐ Yes	☐ No
	Equipment used is approved	☐ Yes	☐ No
	If no, state deviations		
Instructions	Has person in charge of fire equipment been instructed as to location of control valves and care and maintenance of this new equipment? If no, explain	☐ Yes	☐ No
	Have copies of appropriate instructions and care and maintenance charts been left on premises? If no, explain	☐ Yes	☐ No
Location	Supplies buildings		

Underground pipes and joints	Pipe types and class		Type joint	
	Pipe conforms to _____ standard		☐ Yes	☐ No
	Fittings conform to _____ standard		☐ Yes	☐ No
	If no, explain			
	Joints needing anchorage clamped, strapped, or blocked in accordance with _____ standard		☐ Yes	☐ No
	If no, explain			

Test description

<u>Flushing:</u> Flow the required rate until water is clear as indicated by no collection of foreign material in burlap bags at outlets such as hydrants and blow-offs. Flush at flows not less than 390 gpm (1476 L/min) for 4 in. pipe, 880 gpm (3331 L/min) for 6 in. pipe, 1560 gpm (5905 L/min) for 8 in. pipe, 2440 gpm (9235 L/min) for 10 in. pipe, and 3520 gpm (13,323 L/min) for 12 in. pipe. When supply cannot produce stipulated flow rates, obtain maximum available.

<u>Hydrostatic:</u> All piping and attached appurtenances subjected to system working pressure shall be hydrostatically tested at 200 psi (13.8 bar) or 50 psi (3.5 bar) in excess of the system working pressure, whichever is greater, and shall maintain that pressure ± 5 psi (0.35 bar) for 2 hours.

<u>Hydrostatic Testing Allowance:</u> Where additional water is added to the system to maintain the test pressures required by 10.10.2.2.1, the amount of water shall be measured and shall not exceed the limits of the following equation (For metric equation, see 10.10.2.2.6):

$$L = \frac{SD\sqrt{P}}{148,000}$$

L = testing allowance (makeup water), in gallons per hour
S = length of pipe tested, in feet
D = nominal diameter of the pipe, in inches
P = average test pressure during the hydrostatic test, in pounds per square inch (gauge)

Flushing tests

New underground piping flushed according to _____ standard by (company) ☐ Yes ☐ No
If no, explain

How flushing flow was obtained: ☐ Public water ☐ Tank or reservoir ☐ Fire pump
Through what type opening: ☐ Hydrant butt ☐ Open pipe

Lead-ins flushed according to _____ standard by (company) ☐ Yes ☐ No
If no, explain

How flushing flow was obtained: ☐ Public water ☐ Tank or reservoir ☐ Fire pump
Through what type opening: ☐ Y connection to flange and spigot ☐ Open pipe

© 2009 National Fire Protection Association NFPA 24 (p. 1 of 2)

FIGURE 10.10.1 *Sample of Contractor's Material and Test Certificate for Underground Piping.*

Hydrostatic test	All new underground piping hydrostatically tested at _____ psi for _____ hours		Joints covered ☐ Yes ☐ No
Leakage test	Total amount of leakage measured _____ gallons _____ hours		
	Allowable leakage _____ gallons _____ hours		
Hydrants	Number installed	Type and make	All operate satisfactorily ☐ Yes ☐ No
Control valves	Water control valves left wide open If no, state reason		☐ Yes ☐ No
	Hose threads of fire department connections and hydrants interchangeable with those of fire department answering alarm		☐ Yes ☐ No
Remarks	Date left in service		
Signatures	Name of installing contractor		
	Tests witnessed by		
	For property owner (signed)	Title	Date
	For installing contractor (signed)	Title	Date

Additional explanation and notes

© 2009 National Fire Protection Association NFPA 24 (p. 2 of 2)

FIGURE 10.10.1 Continued

10.10.2 Acceptance Requirements.

10.10.2.1* Flushing of Piping.

A.10.10.2.1 Underground mains and lead-in connections to system risers should be flushed through hydrants at dead ends of the system or through accessible aboveground flushing outlets allowing the water to run until clear. Figure A.10.10.2.1 shows acceptable examples of flushing the system. If water is supplied from more than one source or from a looped system, divisional valves should be closed to produce a high-velocity flow through each single line. The flows specified in Table 10.10.2.1.3 will produce a velocity of at least 10 ft/sec (3 m/sec), which is necessary for cleaning the pipe and for lifting foreign material to an aboveground flushing outlet.

10.10.2.1.1 Underground piping, from the water supply to the system riser, and lead-in connections to the system riser shall be completely flushed before the connection is made to downstream fire protection system piping.

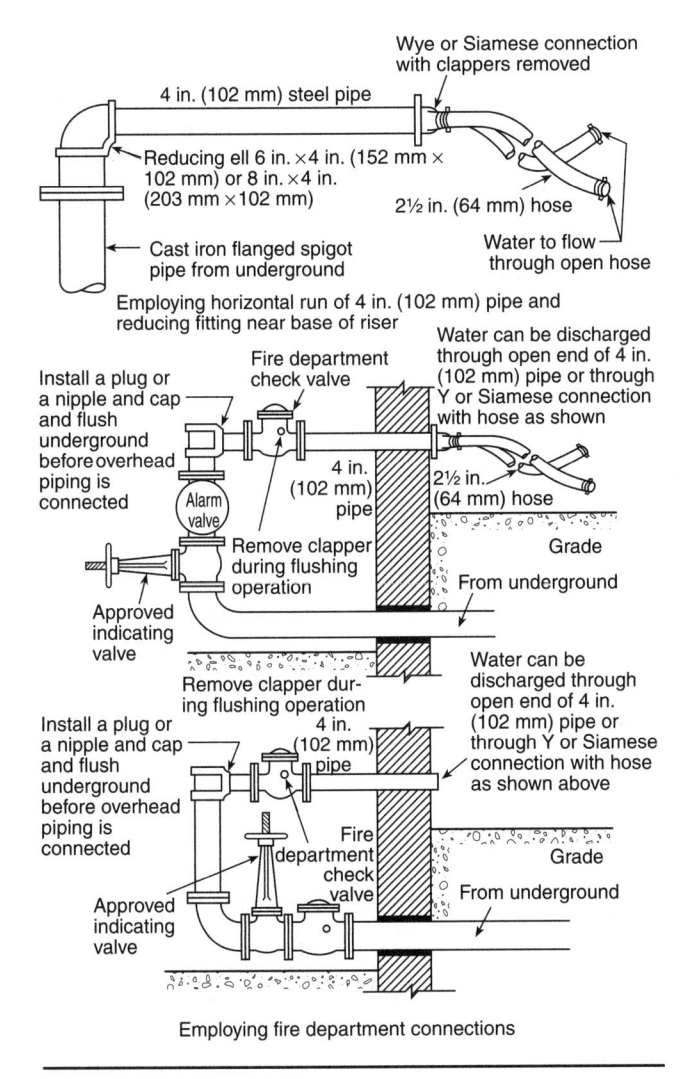

FIGURE A.10.10.2.1 *Methods of Flushing Water Supply Connections.*

TABLE 10.10.2.1.3 *Flow Required to Produce a Velocity of 10 ft/sec (3 m/sec) in Pipes*

Pipe Size		Flow Rate	
in.	mm	gpm	L/min
4	102	390	1,476
6	152	880	3,331
8	203	1,560	5,905
10	254	2,440	9,235
12	305	3,520	13,323

10.10.2.1.2 The flushing operation shall be continued for a sufficient time to ensure thorough cleaning.

10.10.2.1.3 The minimum rate of flow shall be not less than one of the following:

(1) Hydraulically calculated water demand flow rate of the system, including any hose requirements
(2) Flow necessary to provide a velocity of 10 ft/sec (3.1 m/sec) in accordance with Table 10.10.2.1.3
(3) Maximum flow rate available to the system under fire conditions

Flushing the underground piping before connection to the fire protection system piping is critical for removing potentially obstructing material. Dirt, rocks, and other debris can easily enter piping as it is laid in trenches during installation. Frequently, a burlap bag is placed over the flushing connection to capture any obstructing material and to verify, after flushing for a period of time, that all obstructing material has been removed from the system. Exhibit III.23 illustrates material captured while flushing an underground main.

10.10.2.1.4 Provision shall be made for the proper disposal of water used for flushing or testing.

10.10.2.2 Hydrostatic Test.

The trench must be backfilled prior to hydrostatic testing to prevent movement of underground piping. The backfilling can take place between joints if observing the joints for leakage is desired and if the backfill depth is sufficient to prevent movement. As an alternative, the joints can also be covered, but the contractor remains responsible for locating and correcting excessive leakage. The latter is usually preferred where a delay is expected in scheduling the hydrostatic test, since leaving trenches open for any length of time is not desirable from a safety standpoint.

A 2-hour hydrostatic test is required at not less than 200 psi (13.8 bar) but at least 50 psi (3.5 bar) above the maximum expected static pressure. The piping between an exterior fire department connection and the check valve in the connection's inlet pipe must also be hydrostatically tested. All thrust blocks should be hardened before testing.

Allowable leakage rates during hydrostatic testing were revised in the 2007 edition and are now based on the total length and diameter of the underground piping. Previously, the allowable leakage was based on the number of gaskets or joints, regardless of pipe diameter.

When underground piping is repaired, the piping should be tested as a new installation. If the repaired piping can be isolated, all of the underground piping is not required to be subjected to the hydrostatic test. Blind flanges or skillets can be used for isolating the re-

EXHIBIT III.23 Pipe Scale (Rust) and Work Gloves, Found When Flushing an Underground Main. (Photograph courtesy of John Jensen)

paired piping. Procedures must be implemented to remove these devices following hydrostatic testing.

10.10.2.2.1* All piping and attached appurtenances subjected to system working pressure shall be hydrostatically tested at 200 psi (13.8 bar) or 50 psi (3.5 bar) in excess of the system working pressure, whichever is greater, and shall maintain that pressure at ±5 psi (0.35 bar) for 2 hours.

The removal of all air from the system piping when conducting the hydrostatic test is very important. Air in the piping causes a fluctuation in the test pressure. Any time the test pressure drops below the minimum specified [200 psi (13.8 bar) or 50 psi (3.5 bar) in excess of the system working pressure], the test must be repeated and the correct pressure held for the 2-hour duration.

A.10.10.2.2.1 A sprinkler system has for its water supply a connection to a public water service main. A 100 psi (6.9 bar) rated pump is installed in the connection. With a maximum normal public water supply of 70 psi (4.8 bar), at the low elevation point of the individual system or portion of the system being tested and a 120 psi (8.3 bar) pump (churn) pressure, the hydrostatic test pressure is 70 psi (4.8 bar) + 120 psi (8.3 bar) + 50 psi (3.5 bar), or 240 psi (16.5 bar).

To reduce the possibility of serious water damage in case of a break, pressure can be maintained by a small pump, the main controlling gate meanwhile being kept shut during the test.

Polybutylene pipe will undergo expansion during initial pressurization. In this case, a reduction in gauge pressure might not necessarily indicate a leak. The pressure reduction should not exceed the manufacturer's specifications and listing criteria.

When systems having rigid thermoplastic piping such as CPVC are pressure tested, the sprinkler system should be filled with water. The air should be bled from the highest and farthest sprinklers. Compressed air or compressed gas should never be used to test systems with rigid thermoplastic pipe.

A recommended test procedure is as follows: The water pressure is to be increased in 50 psi (3.5 bar) increments until the test pressure described in 10.10.2.2.1 is attained. After each increase in pressure, observations are to be made of the stability of the joints. These observations are to include such items as protrusion or extrusion of the gasket, leakage, or other factors likely to affect the continued use of a pipe in service. During the test, the pressure is not to be increased by the next increment until the joint has become stable. This applies particularly to movement of the gasket. After the pressure has been increased to the required maximum value and held for 1 hour, the pressure is to be decreased to 0 psi while observations are made for leakage. The pressure is again to be slowly increased to the value specified in 10.10.2.2.1 and held for 1 more hour while observations are made for leakage and the leakage measurement is made.

10.10.2.2.2 Pressure loss shall be determined by a drop in gauge pressure or visual leakage.

10.10.2.2.3 The test pressure shall be read from one of the following, located at the lowest elevation of the system or the portion of the system being tested:

(1) A gauge located at one of the hydrant outlets
(2) A gauge located at the lowest point where no hydrants are provided

Note that the gauge is required to be located at the lowest portion of the system. If located at the highest portion of the system, in some cases, an exceptionally high test pressure would result in the lower portions of the system. Excessively high pressures could cause failure of some system components.

10.10.2.2.4* The trench shall be backfilled between joints before testing to prevent movement of pipe.

Backfilling the trench between the pipe joints helps keep the pipe in place while the trench (or the piping) is being filled and pressurized.

A.10.10.2.2.4 Hydrostatic tests should be made before the joints are covered, so that any leaks can be readily detected. Thrust blocks should be sufficiently hardened before hydrostatic testing is begun. If the joints are covered with backfill prior to testing, the contractor remains responsible for locating and correcting any leakage in excess of that permitted.

10.10.2.2.5 Where required for safety measures presented by the hazards of open trenches, the pipe and joints shall be permitted to be backfilled, provided the installing contractor takes the responsibility for locating and correcting leakage.

10.10.2.2.6* Hydrostatic Testing Allowance. Where additional water is added to the system to maintain the test pressures required by 10.10.2.2.1, the amount of water shall be measured and shall not exceed the limits of Table 10.10.2.2.6, which are based upon the following equations:

U.S. Customary Units:

$$L = \frac{SD\sqrt{P}}{148{,}000} \qquad [10.10.2.2.6(a)]$$

where:
L = testing allowance (makeup water) [gph (gal/hr)]
S = length of pipe tested (ft)
D = nominal diameter of the pipe (in.)
P = average test pressure during the hydrostatic test (gauge psi) Metric Units:

TABLE 10.10.2.2.6 Hydrostatic Testing Allowance at 200 psi per 100 ft of Pipe

Nominal Pipe Diameter (in.)	Testing Allowance in Gallons per Hour (gph) per 100 Feet of Pipe
2	0.019
4	0.038
6	0.057
8	0.076
10	0.096
12	0.115
14	0.134
16	0.153
18	0.172
20	0.191
24	0.229

Notes:
(1) For other length, diameters, and pressures, utilize Equation 10.10.2.2.6(a) or 10.10.2.2.6(b) to determine the appropriate testing allowance.
(2) For test sections that contain various sizes and sections of pipe, the testing allowance is the sum of the testing allowances for each size and section.

$$L = \frac{SD\sqrt{P}}{794{,}797} \quad\quad [10.10.2.2.6(b)]$$

where:
L = testing allowance (makeup water) (L/hr)
S = length of pipe tested (m)
D = nominal diameter of the pipe (mm)
P = average test pressure during the hydrostatic test (kPa)

Using equation 10.10.2.2.6(a), if testing 350 ft of a 10 in. main at an average of 203 psi during the duration of the test,

$$L = \frac{(350)(10)\sqrt{203}}{148{,}000}$$

L = 0.3369 gph (from Table 10.10.2.2.6: 0.096 gph loss is permitted per 100 ft of pipe.)

$$0.096 \times 3.5 = 0.336 \text{ gph}$$

When testing 350 ft of 10 in. pipe at an average test pressure of 203 psi, a total of 0.67 gal of water is permitted to be lost.

A.10.10.2.2.6 One acceptable means of completing this test is to utilize a pressure pump that draws its water supply from a full container. At the completion of the 2-hour test, the amount of water to refill the container can be measured to determine the amount of makeup water. In order to minimize pressure loss, the piping should be flushed to remove any trapped air. Additionally, the piping should be pressurized for 1 day prior to the hydrostatic test to account for expansion, absorption, entrapped air, and so on.

The use of a blind flange or skillet is preferred for hydrostatically testing segments of new work. Metal-seated valves are susceptible to developing slight imperfections during transport, installation, and operation and thus can be likely to leak more than 1 fl oz/in. (1.2 mL/mm) of valve diameter per hour. For this reason, the blind flange should be used when hydrostatically testing.

10.10.2.3 Other Means of Hydrostatic Tests. Where required by the authority having jurisdiction, hydrostatic tests shall be permitted to be completed in accordance with the requirements of AWWA C600, AWWA C602, AWWA C603, and AWWA C900.

10.10.2.4 Operating Test.

10.10.2.4.1 Each hydrant shall be fully opened and closed under system water pressure.

Exercising the hydrant in this fashion should be done annually after the hydrant is placed into service. Threaded outlets should be lubricated and checked for burrs or potential cross-threading at this time.

10.10.2.4.2 Dry barrel hydrants shall be checked for proper drainage.

The hydrant should completely drain within 60 minutes.

10.10.2.4.3 All control valves shall be fully closed and opened under system water pressure to ensure proper operation.

Post indicator valves should be opened completely until pressure is felt in the connecting rod. This pressure is an indication that the rod is still connected to the valve. Additional pressure should be exerted on the handle at this time and then suddenly released to determine if the handle "springs" back. The spring indicates that the valve is fully open and verifies that the gate is still attached to the handle. If the gate were jammed, the handle would not be likely to spring back. If the gate were loose or detached from the handle, the handle would continue to turn with little or no resistance.

Post indicator and gate valves should be turned back one-quarter turn from fully open to prevent jamming.

10.10.2.4.4 Where fire pumps are available, the operating tests required by 10.10.2.4 shall be completed with the pumps running.

10.10.2.5 Backflow Prevention Assemblies.

10.10.2.5.1 The backflow prevention assembly shall be forward flow tested to ensure proper operation.

10.10.2.5.2 The minimum flow rate required by 10.10.2.5.1 shall be the system demand, including hose stream demand where applicable.

In addition to the forward flow test, local regulations may require a backflow test at this time. The backflow test should also be documented and submitted to the building owner along with the contractor's material and test certificate.

CHAPTER 11 HYDRAULIC CALCULATIONS

When calculating friction loss through underground piping, the Hazen–Williams formula must be used. This formula is easily converted to table format to simplify the calculation process. The C value used must correspond to those listed in Commentary Table III.3.

COMMENTARY TABLE III.3 Hazen–Williams C Values*

Pipe or Tube	C Value*
Unlined cast or ductile iron	100
Black steel (dry systems including preaction)	100
Black steel (wet systems including deluge)	120
Galvanized (all)	120
Plastic (listed) all	150
Cement-lined cast or ductile iron	140
Copper tube or stainless steel	150
Asbestos cement	140
Concrete	140

*The authority having jurisdiction is permitted to consider other C values.
Source: NFPA 13, 2010, Table 22.4.4.7.

11.1* Calculations in English Units.

Pipe friction losses shall be determined based on the Hazen–Williams formula, as follows:

$$p = \frac{4.52 Q^{1.85}}{C^{1.85} d^{4.87}}$$

where:

p = frictional resistance (psi/ft of pipe)

Q = flow (gpm)

C = friction loss coefficient

d = actual internal diameter of pipe (in.)

A.11.1 When calculating the actual inside diameter of cement-mortar lined pipe, twice the thickness of the pipe wall and twice the thickness of the lining need to be subtracted from the outside diameter of the pipe. The actual lining thickness should be obtained from the manufacturer.

Table A.11.1(a) and Table A.11.1(b) indicate the minimum lining thickness.

TABLE A.11.1(a) Table for Minimum Thickness of Lining for Ductile Iron Pipe and Fittings

Pipe and Fitting Size		Thickness of Lining	
in.	mm	in.	mm
3–12	76–305	1/16	1.6
14–24	356–610	3/32	2.4
30–64	762–1600	1/8	3.2

Source: AWWA C104.

TABLE A.11.1(b) Table for Minimum Thickness of Lining for Steel Pipe

Nominal Pipe Size		Thickness of Lining		Tolerance	
in.	mm	in.	mm	in.	mm
4–10	100–250	1/4	6	–1/16, +1/8	–1.6, +3.2
11–23	280–580	5/16	8	–1/16, +1/8	–1.6, +3.2
24–36	600–900	3/8	10	–1/16, +1/8	–1.6, +3.2
>36	>900	1/2	13	–1/16, +3/16	–1.6, +4.8

Source: AWWA C205.

11.2 Calculations in SI Units.

Pipe friction losses shall be determined based on the Hazen–Williams formula in SI units, as follows:

$$p_m = 6.05 \left(\frac{Q_m^{1.85}}{C^{1.85} d_m^{4.87}} \right) 10^5$$

where:

p_m = frictional resistance (bar/m of pipe)
Q_m = flow (L/min)
C = friction loss coefficient
d_m = actual internal diameter of pipe (mm)

CHAPTER 12 ABOVEGROUND PIPE AND FITTINGS

12.1 General.

Aboveground pipe and fittings shall comply with the applicable sections of Chapters 6 and 8 of NFPA 13 that address pipe, fittings, joining methods, hangers, and installation.

12.2 Protection of Piping.

12.2.1 Aboveground piping for private fire service mains shall not pass through hazardous areas and shall be located so that it is protected from mechanical and fire damage.

12.2.2 Aboveground piping shall be permitted to be located in hazardous areas protected by an automatic sprinkler system.

12.2.3 Where aboveground water-filled supply pipes, risers, system risers, or feed mains pass through open areas, cold rooms, passageways, or other areas exposed to freezing temperatures, the pipe shall be protected against freezing by the following:

(1) Insulating coverings
(2) Frostproof casings
(3) Other reliable means capable of maintaining a minimum temperature between 40°F and 120°F (4°C and 48.9°C)

12.2.4 Where corrosive conditions exist or piping is exposed to the weather, corrosion-resistant types of pipe, fittings, and hangers or protective corrosion-resistant coatings shall be used.

12.2.5 To minimize or prevent pipe breakage where subject to earthquakes, aboveground pipe shall be protected in accordance with the seismic requirements of NFPA 13.

12.2.6 Mains that pass through walls, floors, and ceilings shall be provided with clearances in accordance with NFPA 13.

CHAPTER 13 SIZES OF ABOVEGROUND AND BURIED PIPE

13.1 Private Service Mains.

Pipe smaller than 6 in. (152 mm) in diameter shall not be installed as a private service main supplying hydrants.

13.2 Mains Not Supplying Hydrants.

For mains that do not supply hydrants, sizes smaller than 6 in. (152 mm) shall be permitted to be used, subject to the following restrictions:

(1) The main shall supply only the following types of systems:
    (a) Automatic sprinkler systems
    (b) Open sprinkler systems
    (c) Water spray fixed systems
    (d) Foam systems
    (e) Class II standpipe systems
(2) Hydraulic calculations shall show that the main is able to supply the total demand at the appropriate pressure.
(3) Systems that are not hydraulically calculated shall have a main at least as large as the riser.

13.3 Mains Supplying Fire Protection Systems.

The size of private fire service mains supplying fire protection systems shall be approved by the authority having jurisdiction, and the following factors shall be considered:

(1) Construction and occupancy of the plant
(2) Fire flow and pressure of the water required
(3) Adequacy of the water supply

CHAPTER 14 SYSTEM INSPECTION, TESTING, AND MAINTENANCE

14.1 General.

A private fire service main and its appurtenances installed in accordance with this standard shall be properly inspected, tested, and maintained in accordance with NFPA 25 to provide at least the same level of performance and protection as designed.

ANNEX B VALVE SUPERVISION ISSUES

B.1 Responsibility.

The management is responsible for the supervision of valves controlling the water supply for fire protection and should exert every effort to see that the valves are maintained in the normally open position. This effort includes special precautions to ensure that protection is promptly restored by completely opening valves that are necessarily closed during repairs or alterations. The precautions apply equally to the following:

(1) Valves controlling sprinklers and other fixed water-based fire suppression systems
(2) Hydrants
(3) Tanks
(4) Standpipes
(5) Pumps
(6) Street connections
(7) Sectional valves

Central station supervisory service systems or proprietary supervisory service systems, or a combination of these methods of valve supervision, as described in the following paragraphs, are considered essential to ensure that the valves controlling fire protection systems are in the normally open position. The methods described are intended as an aid to the person responsible for developing a systematic method of determining that the valves controlling sprinkler systems and other fire protection devices are open.

Continual vigilance is necessary if valves are to be kept in the open position. Responsible day and night employees should be familiar with the location of all valves and their proper use.

The authority having jurisdiction should be consulted as to the type of valve supervision required. Contracts for equipment should specify that all details are to be subject to the approval of the authority having jurisdiction.

B.2 Central Station Supervisory Service Systems.

Central station supervisory service systems involve complete, constant, and automatic supervision of valves by electrically operated devices and circuits. The devices and circuits are continually under test and operate through an approved outside central station in compliance with *NFPA 72*. It is understood that only the portions of *NFPA 72* that relate to valve supervision should apply.

B.3 Proprietary Supervisory Service Systems.

Proprietary supervisory service systems include systems in which the operation of a valve produces some form of signal and record at a common point by electrically operated devices and circuits. The device and circuits are continually under test and operate through a central supervising station at the protected property in compliance with the standards for the installation, maintenance, and use of local protective, auxiliary protective, remote-station protective, and proprietary signaling systems. It is understood that only the portions of the standards that relate to valve supervision should apply.

B.4 Locking and Sealing.

The standard method of locking, sealing, and tagging valves to prevent, as far as possible, their unnecessary closing, to obtain notification of such closing, and to aid in restoring the valve to normal condition is a satisfactory alternative to valve supervision. The authority having jurisdiction should be consulted for details for specific cases.

Where electrical supervision is not provided, locks or seals should be provided on all valves and should be of a type acceptable to the authority having jurisdiction.

Seals can be marked to indicate the organization under whose jurisdiction the sealing is conducted. All seals should be attached to the valve in such a manner that the valves cannot be operated without breaking the seals. Seals should be of a character that prevents injury in handling and that prevents reassembly when broken. Where seals are used, valves should be inspected weekly. The authority having jurisdiction can require a valve tag to be used in conjunction with the sealing.

A padlock, with a chain where necessary, is especially desirable to prevent unauthorized closing of valves in areas where valves are subject to tampering. Where such locks are employed, valves should be inspected monthly.

If valves are locked, any distribution of keys should be restricted to only those directly responsible for the fire protection system. Multiple valves should not be locked together; they should be individually locked.

The individual performing inspections should determine that each valve is in the normal position and properly locked or sealed, and so noted on an appropriate record form while still at the valve. The authority having jurisdiction should be consulted for assistance in preparing a suitable report form for this activity.

Identification signs should be provided at each valve to indicate its function and what it controls.

The position of the spindle of OS&Y valves or the target on the indicator valves cannot be accepted as conclusive proof that the valve is fully open. The opening of the valve should be followed by a test to determine that the operating parts have functioned properly.

The test consists of opening the main drain valve and allowing a free flow of water until the gauge reading becomes stationary. If the pressure drop is excessive for the water supply involved, the cause should be determined immediately and the proper remedies taken. Where sectional valves or other special conditions are encountered, other methods of testing should be used.

If it becomes necessary to break a seal for emergency reasons, the valve, following the emergency, should be opened by the individual responsible for the fire protection of the plant or his or her designated representative. The responsible individual should apply a seal at the time of the valve opening. The seal should be maintained in place until such time as the authority having jurisdiction can replace it with a seal of its own.

Seals or locks should not be applied to valves that have been reopened after closure until such time as the inspection procedure is carried out.

Where water is shut off to the sprinkler or other fixed water-based fire suppression systems, a guard or other qualified person should be placed on duty and required to continuously patrol the affected sections of the premises until such time as protection is restored.

During specific critical situations, a responsible individual should be stationed at the valve so that the valve can be reopened promptly if necessary. It is the intent of this recommendation that the individual remain within sight of the valve and have no additional duties. This recommendation is considered imperative when fire protection is shut off immediately following a fire.

An inspection of all other fire protection equipment should be made prior to shutting off water in order to ensure that it is in operative condition.

Where changes to fire protection equipment are to be made, as much work as possible should be done in advance of shutting off the water, so that final connections can be made quickly and protection restored promptly. With careful planning, open outlets often can be plugged and protection can be restored on a portion of the equipment while the alterations are being made.

Where changes are to be made in underground piping, as much piping as possible should be laid before shutting off the water for final connections. Where possible, temporary feed lines, such as temporary piping for reconnection of risers by hose lines, should be used to afford maximum protection. The plant, public fire department, and other authorities having jurisdiction should be notified of all impairments to fire protection equipment.

ANNEX C RECOMMENDED PRACTICE FOR FIRE FLOW TESTING

Annex C of NFPA 24 is based on Chapter 4 of NFPA 291, *Recommended Practice for Fire Flow Testing and Marking of Hydrants,* 2010 edition, and is therefore not reprinted here. See Part III, Section 4, of this handbook for NFPA 291, Chapter 4, and associated commentary.

ANNEX D RECOMMENDED PRACTICE FOR MARKING OF HYDRANTS

Annex D of NFPA 24 is based on Chapter 5 of NFPA 291 and is therefore not reprinted here. See Part III, Section 4, of this handbook for NFPA 291, Chapter 5, and associated commentary.

Stationary Fire Pumps Handbook 2010

ANNEX E INFORMATIONAL REFERENCES

E.1 Referenced Publications.

The documents or portions thereof listed in this annex are referenced within the informational sections of this standard and are not part of the requirements of this document unless also listed in Chapter 2 for other reasons.

E.1.1 NFPA Publications. National Fire Protection Association, 1 Batterymarch Park, Quincy, MA 02169-7471.

NFPA 20, *Standard for the Installation of Stationary Pumps for Fire Protection,* 2010 edition.
NFPA 22, *Standard for Water Tanks for Private Fire Protection,* 2008 edition.
NFPA 70®, *National Electrical Code®,* 2008 edition.
NFPA 72®, *National Fire Alarm and Signaling Code,* 2010 edition.
NFPA 291, *Recommended Practice for Fire Flow Testing and Marking of Hydrants,* 2010 edition.
NFPA 780, *Standard for the Installation of Lightning Protection Systems,* 2008 edition.
NFPA 1962, *Standard for the Inspection, Care, and Use of Fire Hose, Couplings, and Nozzles and the Service Testing of Fire Hose,* 2008 edition.

E.1.2 Other Publications.

E.1.2.1 ACPA Publications. American Concrete Pipe Association, 1303 West Walnut Hill Lane, Suite 305, Irving, TX 75038-3008.

Concrete Pipe Handbook.

E.1.2.2 ASME Publications. American Society of Mechanical Engineers, Three Park Avenue, New York, NY 10016-5990.

ASME B16.1, *Cast Iron Pipe Flanges and Flanged Fittings,* 1989.

E.1.2.3 ASTM Publications.

ASTM International, 100 Barr Harbor Drive, P.O. Box C700, West Conshohocken, PA 19428-2959.
ASTM A 126, *Standard Specification for Gray Iron Castings for Valves, Flanges and Pipe Fittings,* 1993.
ASTM A 197, *Standard Specification for Cupola Malleable Iron,* 1987.
ASTM A 307, *Standard Specification for Carbon Steel Bolts and Studs,* 1994.
ASTM C 296, *Standard Specification for Asbestos-Cement Pressure Pipe,* 1988.
IEEE/ASTM-SI-10, *Standard for Use of the International System of Units (SI): The Modern Metric System,* 1997.

E.1.2.4 AWWA Publications.

American Water Works Association, 6666 West Quincy Avenue, Denver, CO 80235.

AWWA C104, *Cement Mortar Lining for Ductile Iron Pipe and Fittings for Water,* 2008.
AWWA C105, *Polyethylene Encasement for Ductile Iron Pipe Systems,* 2005.
AWWA C110, *Ductile Iron and Gray Iron Fittings,* 2008.
AWWA C111, *Rubber-Gasket Joints for Ductile Iron Pressure Pipe and Fittings,* 2000.
AWWA C115, *Flanged Ductile Iron Pipe with Ductile Iron or Gray Iron Threaded Flanges,* 2005.

AWWA C116, *Protective Fusion-Bonded Epoxy Coatings for the Interior and Exterior Surfaces of Ductile-Iron and Gray-Iron Fittings for Water Supply Service,* 2003.
AWWA C150, *Thickness Design of Ductile Iron Pipe,* 2008.
AWWA C151, *Ductile Iron Pipe, Centrifugally Cast for Water,* 2002.
AWWA C153, *Ductile Iron Compact Fittings, 3 in. through 24 in. and 54 in. through 64 in. for Water Service,* 2006.
AWWA C203, *Coal-Tar Protective Coatings and Linings for Steel Water Pipelines Enamel and Tape—Hot Applied,* 2002.
AWWA C205, *Cement-Mortar Protective Lining and Coating for Steel Water Pipe 4 in. and Larger—Shop Applied,* 2007.
AWWA C206, *Field Welding of Steel Water Pipe,* 2003.
AWWA C208, *Dimensions for Fabricated Steel Water Pipe Fittings,* 2007.
AWWA C300, *Reinforced Concrete Pressure Pipe, Steel-Cylinder Type,* 2004.
AWWA C301, *Prestressed Concrete Pressure Pipe, Steel-Cylinder Type,* 2007.
AWWA C302, *Reinforced Concrete Pressure Pipe, Non-Cylinder Type,* 2004.
AWWA C303, *Reinforced Concrete Pressure Pipe, Steel-Cylinder Type, Pretensioned,* 2002.
AWWA C400, *Standard for Asbestos-Cement Distribution Pipe, 4 in. Through 16 in. (100 mm through 400 mm) for Water Distribution Systems,* 2003.
AWWA C401, *Standard for the Selection of Asbestos-Cement Pressure Pipe 4 in. through 16 in. (100 mm through 400 mm),* 2003.
AWWA C600, *Standard for the Installation of Ductile-Iron Water Mains and Their Appurtenances,* 2005.
AWWA C602, *Cement-Mortar Lining of Water Pipe Lines 4 in. and Larger—in Place,* 2006.
AWWA C603, *Standard for the Installation of Asbestos-Cement Pressure Pipe,* 2005.
AWWA C606, *Grooved and Shouldered Joints,* 1997.
AWWA C900, *Polyvinyl Chloride (PVC) Pressure Pipe, 4 in. Through 12 in., for Water Distribution,* 2007.
AWWA M9, *Concrete Pressure Pipe,* 2008.
AWWA M11, *A Guide for Steel Pipe Design and Installation,* 4th edition, 2004.
AWWA M14, *Recommended Practice for Backflow Prevention and Cross Connection Control,* 2004.
AWWA M41, *Ductile Iron Pipe and Fittings,* 2003.

E.1.2.5 DIPRA Publications. Ductile Iron Pipe Research Association, 245 Riverchase Parkway East, Suite O, Birmingham, AL 35244.

Installation Guide for Ductile Iron Pipe.
Thrust Restraint Design for Ductile Iron Pipe.

E.1.2.6 EBAA Iron Publications. EBAA Iron, Inc., P.O. Box 857, Eastland, TX 76448.

Thrust Restraint Design Equations and Tables for Ductile Iron and PVC Pipe.

E.1.2.7 UBPPA Publications. Uni-Bell PVC Pipe Association, 2655 Villa Creek Drive, Dallas, TX 75234.

Handbook of PVC Pipe.

E.1.2.8 U.S. Government Publication. U.S. Government Printing Office, Washington, DC 20402.

U.S. Federal Standard No. 66C, *Standard for Steel Chemical Composition and Harden Ability,* April 18, 1967 change notice No. 2, April 16, 1970, as promulgated by the U.S. Federal Government General Services Administration.

E.2 Informational References.

The following documents or portions thereof are listed here as informational resources only. They are not a part of the requirements of this document.

AWWA M17, *Installation, Field Testing and Maintenance of Fire Hydrants,* 1989.

E.3 References for Extracts in Informational Sections. (Reserved)

SUMMARY

NFPA 24 provides the complete information needed to design and install a private fire service water supply. In many cases, a private fire service main is used to supply water to a fire pump and must be properly interfaced with other systems designed and installed by other standards.

REFERENCES CITED IN COMMENTARY

National Fire Protection Association, 1 Batterymarch Park, Quincy, MA 02169-7471.

NFPA 1, *Fire Code,* 2009 edition.
NFPA 13, *Standard for the Installation of Sprinkler Systems,* 2010 edition.
NFPA 13D, *Standard for the Installation of Sprinkler Systems in One- and Two-Family Dwellings and Manufactured Homes,* 2010 edition.
NFPA 13E, *Recommended Practice for Fire Department Operations in Properties Protected by Sprinkler and Standpipe Systems,* 2005 edition.
NFPA 13R, *Standard for the Installation of Sprinkler Systems in Residential Occupancies up to and Including Four Stories in Height,* 2010 edition.
NFPA 14, *Standard for the Installation of Standpipe and Hose Systems,* 2007 edition.
NFPA 20, *Standard for the Installation of Stationary Pumps for Fire Protection,* 2010 edition.
NFPA 70®, National Electrical Code®, 2008 edition.
NFPA 170, *Standard for Fire Safety and Emergency Symbols,* 2009 edition.
NFPA 291, *Recommended Practice for Fire Flow Testing and Marking of Hydrants,* 2007 edition.
NFPA 600, *Standard on Industrial Fire Brigades,* 2005 edition.
NFPA 1963, *Standard for Fire Hose Connections,* 2009 edition.
Cote, A. E., ed., *Fire Protection Handbook®,* 19th edition, 2003.
Hall, John R., Jr., "U.S. Experience with Sprinklers and Other Fire Extinguishing Equipment," Fire Analysis and Research Division, 2005.

American Water Works Association, 6666 West Quincy Avenue, Denver, CO 80235.

AWWA C104, *Cement Mortar Lining for Ductile Iron Pipe and Fittings for Water,* 2003 edition.
AWWA C105, *Polyethylene Encasement for Ductile Iron Pipe Systems,* 2005 edition.
AWWA C111, *Rubber-Gasket Joints for Ductile Iron Pressure Pipe and Fittings,* 2000 edition.
AWWA C203, *Coal-Tar Protective Coatings and Linings for Steel Water Pipelines Enamel and Tape—Hot Applied,* 2002 edition.
AWWA C900, *Polyvinyl Chloride (PVC) Pressure Pipe, 4 in. Through 12 in. (100 mm Through 300 mm), for Water and Other Liquids,* 1997 edition.

Underwriters Laboratories Inc., 333 Pfingsten Road, Northbrook, IL 60062-2096.

UL 194, *Gasketed Joints for Ductile-Iron Pipe and Fittings for Fire Protection Service,* 1996 edition.

U.S. Government Printing Office, Washington, DC 20402.

Title 29, Code of Federal Regulations, Part 1926, Subpart P, Appendix F.
Title 29, Code of Federal Regulations, Part 1910.146, Appendix A.

Section 18.4 and Annex I

NFPA® 1, *Fire Code,* 2009 Edition

SECTION 3

In some cases, fire pumps may be needed to provide flow and pressure to an entire industrial complex, including private fire hydrants, as well as to fire protection systems. Section 18.4 from NFPA 1, *Fire Code,* provides the authority having jurisdiction and the design engineer with one method for determining the required fire flow for manual fire-fighting purposes. Annex I of NFPA 1 provides useful information in determining the location and distribution of hydrants. The method referenced in Section 18.4 does not take into consideration any fire flow from the fire suppression system or hose stream requirements from system installation standards such as NFPA 13, *Standard for the Installation of Sprinkler Systems,* and NFPA 14, *Standard for the Installation of Standpipe and Hose Systems.* However, the method does take into consideration the installation of an automatic sprinkler system. This method is used solely for fire department fire flow requirements. When fire pumps are used to supply flow and pressure to a system of fire hydrants, this method can be used to determine the correct size and capacity of the fire pump. Annex I provides valuable information regarding the number of hydrants and their distribution. Once the total flow requirements are determined from Section 18.4, Annex I can then be used to determine how many hydrants are needed to achieve that flow. For installations involving multiple hydrants, Annex I provides the average spacing between hydrants for uniform distribution.

SECTION 18.4 FIRE FLOW REQUIREMENTS FOR BUILDINGS

18.4.1* Scope.

A.18.4.1 Section 18.4 and the associated tables are only applicable for determining minimum water supplies for manual fire suppression efforts. Water supplies for fire protection systems are not addressed by this section. It is not the intent to add the minimum fire protection water supplies, such as for a fire sprinkler system, to the minimum fire flow for manual fire suppression purposes required by this section.

18.4.1.1* The procedure determining fire flow requirements for buildings hereafter constructed shall be in accordance with Section 18.4.

A.18.4.1.1 For the purpose of this section, a building subdivided by fire walls constructed in accordance with the building code is considered to be a separate building.

18.4.1.2 Section 18.4 does not apply to structures other than buildings.

18.4.2 Definitions. See definitions 3.3.13.6 (Fire Flow Area) and 3.3.108 (Fire Flow).

The definition of *fire flow area* is as follows:

> **3.3.13.6 Fire Flow Area.** The floor area, in square feet, used to determine the required fire flow.

The fire area should be determined based on the area within the surrounding exterior walls and 4-hour rated fire walls, exclusive of courts. Areas of the building without surrounding exterior walls should be included in the fire area, if such areas are within the horizontal projection of the roof or floor above.

The definition of *fire flow* is as follows:

3.3.108 Fire Flow. The flow rate of a water supply, measured at 20 psi (137.9 kPa) residual pressure, that is available for fire fighting.

According to NFPA 291, *Recommended Practice for Fire Flow Testing and Marking of Hydrants* (see Part III, Section 4, of this handbook), hydrant output should be identified by means of a color code that is based on the available flow at a residual pressure of 20 psi (1.4 bar). Hydrants on particularly weak systems—that is, those producing less than 40 psi (2.8 bar) static pressure—should be identified based on the available flow at one-half of the static pressure.

Generally, a minimum residual pressure of 20 psi (1.4 bar) should be maintained within the water supply system at all times. Fire department pumpers can operate at pressures of less than 20 psi (1.4 bar) but only do so with difficulty. Operating at a minimum pressure of 20 psi (1.4 bar) also prevents the development of a negative pressure at any point in the system. Negative pressures in an underground piping system can cause the collapse of the water main and can create a backflow of contaminated water from sources that may be cross-connected with the potable water supply system. Paragraph 4.14.3.1 of NFPA 20, *Standard for the Installation of Stationary Pumps for Fire Protection* (see Part II of this handbook), permits the suction pressure in the fire pump suction pipe to drop as low as 0 psi (0 bar); however, many water purveyors will not permit pressures below 20 psi (1.4 bar) in their piping system.

18.4.3 Modifications.

18.4.3.1 Decreases. Fire flow requirements shall be permitted to be modified downward by the AHJ for isolated buildings or a group of buildings in rural areas or small communities where the development of full fire flow requirements is impractical.

18.4.3.2 Increases. Fire flow shall be permitted to be modified upward by the AHJ where conditions indicate an unusual susceptibility to group fires or conflagrations. An upward modification shall not be more than twice that required for the building under consideration.

An evaluation of issues such as separation distances and other means of protection should be completed before making a determination of required fire flow. NFPA 80A, *Recommended Practice for Protection of Buildings from Exterior Fire Exposures,* should be used to determine the appropriate separation distances and other means of protection, such as clear space between buildings, blank walls, or walls with protected openings and/or water curtain protection.

18.4.4 Fire Flow Area.

18.4.4.1 General. The fire flow area shall be the total floor area of all floor levels of a building except as modified in 18.4.4.1.1.

18.4.4.1.1 Type I (443), Type I (332), and Type II (222) Construction. The fire flow area of a building constructed of Type I (443), Type I (332), and Type II (222) construction shall be the area of the three largest successive floors.

18.4.5 Fire Flow Requirements for Buildings.

18.4.5.1 One- and Two-Family Dwellings.

18.4.5.1.1 The minimum fire flow and flow duration requirements for one- and two-family dwellings having a fire flow area that does not exceed 5000 ft^2 (334.5 m^2) shall be 1000 gpm (3785 L/min) for 1 hour.

18.4.5.1.1.1 A reduction in required fire flow of 50 percent shall be permitted when the building is provided with an approved automatic sprinkler system.

18.4.5.1.1.2 A reduction in the required fire flow of 25 percent shall be permitted when the building is separated from other buildings by a minimum of 30 ft (9.1 m).

18.4.5.1.1.3 The reduction in 18.4.5.1.1.1 and 18.4.5.1.1.2 shall not reduce the required fire flow to less than 500 gpm (1900 L/min).

18.4.5.1.2 Fire flow and flow duration for dwellings having a fire flow area in excess of 5000 ft2 (334.5 m2) shall not be less than that specified in Table 18.4.5.1.2.

18.4.5.1.2.1 A reduction in required fire flow of 50 percent shall be permitted when the building is provided with an approved automatic sprinkler system.

18.4.5.2 Buildings Other Than One- and Two-Family Dwellings. The minimum fire flow and flow duration for buildings other than one- and two-family dwellings shall be as specified in Table 18.4.5.1.2.

18.4.5.2.1 A reduction in required fire flow of 75 percent shall be permitted when the building is protected throughout by an approved automatic sprinkler system. The resulting fire flow shall not be less than 1000 gpm (3785 L/min).

18.4.5.2.2 A reduction in required fire flow of 75 percent shall be permitted when the building is protected throughout by an approved automatic sprinkler system, which utilizes quick response sprinklers throughout. The resulting fire flow shall not be less than 600 gpm (2270 L/min).

The following example illustrates the method used to calculate the required fire flow for a building other than a one- or two-family dwelling, including the allowed reduction for the installation of an automatic sprinkler system.

EXAMPLE

Referring to Exhibit III.24, the building is a six-story mill of construction Type V. The main building measures 75 ft by 370 ft, for a total floor area per floor of 27,750 ft^2.

$$27,750 \text{ ft}^2 \times 6 \text{ floors} = 166,500 \text{ ft}^2 \text{ (per 18.4.4.1)}$$
$$8000 \text{ gpm (converted from Table 18.4.5.1.2)}$$
$$8000 \text{ gpm} \times 0.25 = 2000 \text{ gpm (per 18.4.5.2.1)}$$

For metric equivalents, the main building measures 22.9 m by 112.8 m, for a total floor area per floor of 2583.12 m^2.

$$2583.12 \text{ m}^2 \times 6 \text{ floors} = 15,498.72 \text{ m}^2 \text{ (per 18.4.4.1)}$$
$$30,280 \text{ L/min (converted from Table 18.4.5.1.2)}$$
$$30,280 \text{ L/min} \times 0.25 = 7570 \text{ L/min (per 18.4.5.2.1)}$$

Based on this approach, the minimum number of fire hydrants would be two and the average spacing between them would be 450 ft (137 m).

This example is taking the maximum credit for the installation of automatic sprinklers. In this particular example, since 18.4.5.2.1 permits a reduction of up to 75 percent, a reduction of something less can be expected due to a number of factors, including the judgment of the authority having jurisdiction. The building in Exhibit III.24 is an existing mill building that has been converted to office space. The required fire flow is considerably higher

TABLE 18.4.5.1.2 Minimum Required Fire Flow and Flow Duration for Buildings

\multicolumn{5}{c	}{Fire Flow Area ft² (× 0.0929 for m²)}	Fire Flow gpm† (× 3.785 for L/min)	Flow Duration (hours)
I(443), I(332), II(222)*	II(111), III(211)*	IV(2HH), V(111)*	II(000), III(200)*	V(000)*		
0–22,700	0–12,700	0–8200	0–5900	0–3600	1500	2
22,701–30,200	12,701–17,000	8201–10,900	5901–7900	3601–4800	1750	
30,201–38,700	17,001–21,800	10,901–12,900	7901–9800	4801–6200	2000	
38,701–48,300	21,801–24,200	12,901–17,400	9801–12,600	6201–7700	2250	
48,301–59,000	24,201–33,200	17,401–21,300	12,601–15,400	7701–9400	2500	
59,001–70,900	33,201–39,700	21,301–25,500	15,401–18,400	9401–11,300	2750	
70,901–83,700	39,701–47,100	25,501–30,100	18,401–21,800	11,301–13,400	3000	3
83,701–97,700	47,101–54,900	30,101–35,200	21,801–25,900	13,401–15,600	3250	
97,701–112,700	54,901–63,400	35,201–40,600	25,901–29,300	15,601–18,000	3500	
112,701–128,700	63,401–72,400	40,601–46,400	29,301–33,500	18,001–20,600	3750	
128,701–145,900	72,401–82,100	46,401–52,500	33,501–37,900	20,601–23,300	4000	4
145,901–164,200	82,101–92,400	52,501–59,100	37,901–42,700	23,301–26,300	4250	
164,201–183,400	92,401–103,100	59,101–66,000	42,701–47,700	26,301–29,300	4500	
183,401–203,700	103,101–114,600	66,001–73,300	47,701–53,000	29,301–32,600	4750	
203,701–225,200	114,601–126,700	73,301–81,100	53,001–58,600	32,601–36,000	5000	
225,201–247,700	126,701–139,400	81,101–89,200	58,601–65,400	36,001–39,600	5250	
247,701–271,200	139,401–152,600	89,201–97,700	65,401–70,600	39,601–43,400	5500	
271,201–295,900	152,601–166,500	97,701–106,500	70,601–77,000	43,401–47,400	5750	
Greater than 295,900	Greater than 166,500	106,501–115,800	77,001–83,700	47,401–51,500	6000	
		115,801–125,500	83,701–90,600	51,501–55,700	6250	
		125,501–135,500	90,601–97,900	55,701–60,200	6500	
		135,501–145,800	97,901–106,800	60,201–64,800	6750	
		145,801–156,700	106,801–113,200	64,801–69,600	7000	
		156,701–167,900	113,201–121,300	69,601–74,600	7250	
		167,901–179,400	121,301–129,600	74,601–79,800	7500	
		179,401–191,400	129,601–138,300	79,801–85,100	7750	
		Greater than 191,400	Greater than 138,300	Greater than 85,100	8000	

*Types of construction are based on NFPA 220.
†Measured at 20 psi (139.9 kPa).

than that for the typical modern office building, because the mill is still of Type V construction. However, taking into consideration the previous use of the building, which was textile manufacturing, the credit for sprinklers should be considerably less, perhaps on the order of 25 percent to 50 percent. Applying a reduction of only 25 percent would result as follows:

$$8000 \text{ gpm} \times 0.75 = 6000 \text{ gpm (per 18.4.5.2.1)}$$
$$30{,}278 \text{ L/min} \times 0.75 = 22{,}709 \text{ L/min}$$

Based on these results, the minimum number of hydrants per Table I.3 would be six and the average spacing between each would be 250 ft (76.2 m). Given the nature of the hazard, even considering the building's present use, the higher number of hydrants may be more appropriate for the hazard.

In cases where a fire pump is used to supply the hydrants for a facility, the pump capacity would be selected based on the flow of 6000 gpm (22,712 L/min) and sized sufficiently to produce at least 20 psi (1.4 bar) at the most hydraulically demanding hydrant.

Determination of the available water supply using this method should only be conducted by those having extensive experience in fire fighting and/or fire protection system design and knowledge of the construction of the building and its intended use.

ANNEX I FIRE HYDRANT LOCATIONS AND DISTRIBUTION

This annex is not a part of the requirements of this NFPA document unless specifically adopted by the jurisdiction.

I.1 Scope.

Fire hydrants shall be provided in accordance with Annex I for the protection of buildings, or portions of buildings, hereafter constructed.

I.2 Location.

Fire hydrants shall be provided along required fire apparatus access roads and adjacent public streets.

To be usable, fire hydrants must be reasonably accessible to the fire department, in accordance with Section I.2. Hydrants should be located a minimum of 40 ft (12.1 m) away from the building to provide some protection for the end user. Hydrants should also be sufficient in number and location such that they are within 100 ft (30.5 m) of the fire department connection serving the building's sprinkler system. Where hydrants cannot be located at the minimum distance from protected buildings, the application of wall hydrants should be considered. A wall hydrant should only be used where approved by the authority having jurisdiction.

I.3 Number of Fire Hydrants.

The minimum number of fire hydrants available to a building shall not be less than that listed in Table I.3. The number of fire hydrants available to a complex or subdivision shall not be less than that determined by spacing requirements listed in Table I.3 when applied to fire apparatus access roads and perimeter public streets from which fire operations could be conducted.

Section I.3 and Table I.3 do not require hydrants to be within a minimum distance of the building. The minimum distance to a building is determined by other issues such as fire department access road design and the proximity to a fire department connection. When the layout of the fire department access road is determined and separation between protected buildings is determined, the number of hydrants can be selected from Table I.3 based on the average spacing requirement between hydrants. In addition to this quantity, each building should be provided with the minimum number of hydrants as stated in the table.

I.4 Consideration of Existing Fire Hydrants.

Existing fire hydrants on public streets shall be permitted to be considered as available. Existing fire hydrants on adjacent properties shall not be considered available unless fire apparatus

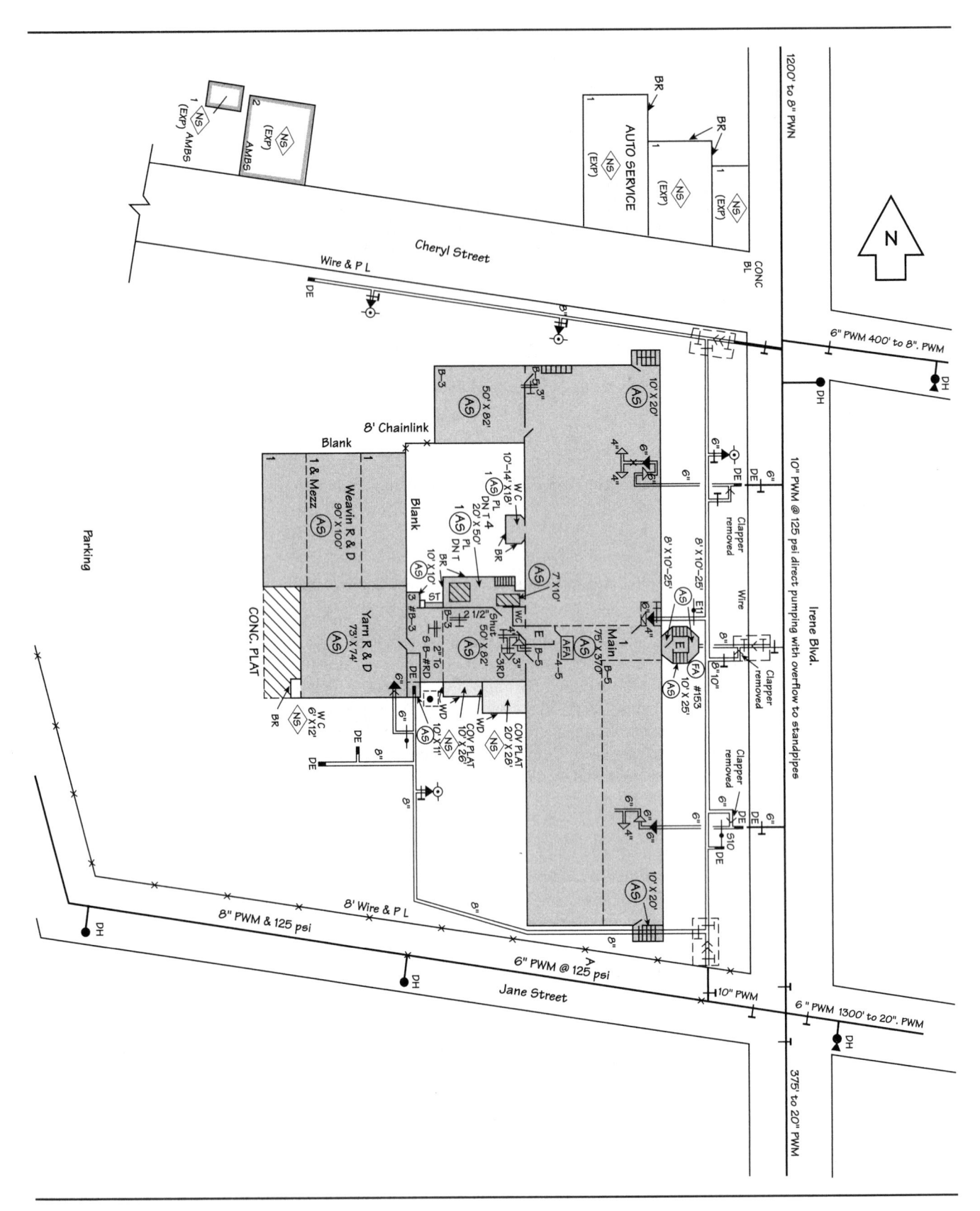

EXHIBIT III.24 Six-Story Mill Building—Construction Type V.

TABLE I.3 Number and Distribution of Fire Hydrants

Fire Flow Requirements		Minimum Number of Hydrants	Average Spacing Between Hydrants[1,2,3]		Maximum Distance from Any Point on Street or Road Frontage to a Hydrant[4]	
gpm	L/min		ft	m	ft	m
1750 or less	6650 or less	1	500	152	250	76
2000–2250	7600–8550	2	450	137	225	69
2500	9500	3	450	137	225	69
3000	11,400	3	400	122	225	69
3500–4000	13,300–15,200	4	350	107	210	64
4500–5000	17,100–19,000	5	300	91	180	55
5500	20,900	6	300	91	180	55
6000	22,800	6	250	76	150	46
6500–7000	24,700–26,500	7	250	76	150	46
7500 or more	28,500 or more	8 or more[5]	200	61	120	37

Note: 1 gpm = 3.8 L/min; 1 ft = 0.3 m.

[1] Reduce by 100 ft (30.5 m) for dead-end streets or roads.

[2] Where streets are provided with median dividers that can be crossed by fire fighters pulling hose lines, or arterial streets are provided with four or more traffic lanes and have a traffic count of more than 30,000 vehicles per day, hydrant spacing shall average 500 ft (152.4 m) on each side of the street and be arranged on an alternating basis up to a fire flow requirement of 7000 gpm (26,500 L/min) and 400 ft (122 m) or higher fire flow requirements.

[3] Where new water mains are extended along streets where hydrants are not needed for protection of structures or similar fire problems, fire hydrants shall be provided at spacing not to exceed 1000 ft (305 m) to provide for transportation hazards.

[4] Reduce by 50 ft (15.2 m) for dead-end streets or roads.

[5] One hydrant for each 1000 gpm (3785 L/min) or fraction thereof.

access roads extend between properties and easements are established to prevent obstruction of such roads.

As illustrated in Exhibit III.24, only four private hydrants are provided. However, four public hydrants are located in the immediate vicinity and would be available for fire fighting if needed. Section I.4 permits the use of the public hydrants in Exhibit III.24.

I.5 Distribution of Fire Hydrants.

The average spacing between fire hydrants shall not exceed that listed in Table I.3.

Exception: The AHJ shall be permitted to accept a deficiency of up to 10 percent where existing fire hydrants provide all or a portion of the required fire hydrant service. Regardless of the average spacing, fire hydrants shall be located such that all points on streets and access roads adjacent to a building are within the distances listed in Table I.3.

SUMMARY

Section 18.4 and Annex I from NFPA 1 are included to assist the designer in determining the correct number of hydrants needed to properly design a private fire service system. Hydrants are a critical component of any water supply, and their proper number and distribution are necessary to provide adequate fire protection. The information provided by Section 18.4 and Annex I is needed to coordinate the design and installation of underground piping and other aspects of a complete water supply system.

REFERENCES CITED IN COMMENTARY

National Fire Protection Association, 1 Batterymarch Park, Quincy, MA 02169-7471.

NFPA 13, *Standard for the Installation of Sprinkler Systems,* 2010 edition.

NFPA 14, *Standard for the Installation of Standpipe and Hose Systems,* 2007 edition.

NFPA 20, *Standard for the Installation of Stationary Pumps for Fire Protection,* 2010 edition.

NFPA 80A, *Recommended Practice for Protection of Buildings from Exterior Fire Exposures,* 2007 edition.

NFPA 291, *Recommended Practice for Fire Flow Testing and Marking of Hydrants,* 2010 edition.

Complete Text of NFPA® 291,

Recommended Practice for Fire Flow Testing and Marking of Hydrants, **2010 Edition**

SECTION 4

NFPA 291, *Recommended Practice for Fire Flow Testing and Marking of Hydrants,* is included in this handbook because water supply information is critical to the design of all types of water-based systems except water mist. A waterflow test should be conducted prior to the commencement of system design for every type of water-based system covered in this handbook. NFPA 291 outlines the correct procedure for obtaining water supply information and also provides the necessary information in the form of flow tables for calculating the flow test results.

NFPA 291 is a recommended practice and as such contains no mandatory language that is normally found in an NFPA code or standard. Instead, NFPA 291 uses the word *should* rather than *shall*. This difference is important because a variety of methods and equipment can be employed to correctly perform a hydrant flow test. NFPA 291 also recommends a color coding system for hydrants that provides a quick visual indication of the flow capacity of each hydrant.

CHAPTER 1 ADMINISTRATION

1.1 Scope.

The scope of this document is fire flow testing and marking of hydrants.

NFPA 291 outlines the procedures necessary for conducting fire flow testing and color coding of hydrants.

1.2 Purpose.

Fire flow tests are conducted on water distribution systems to determine the rate of flow available at various locations for fire-fighting purposes.

In addition to fire-fighting purposes, fire flow testing is also necessary for determining the available flow and pressure for fixed fire protection systems including sprinklers, standpipes, and fire pumps. In all cases, the available flow and pressure from a public or private water supply must be determined prior to the commencement of any system design. NFPA 13, *Standard for the Installation of Sprinkler Systems,* does not address the issue of when flow testing is needed and the issue of how old water supply information can be before retesting is needed. However, NFPA 14, *Standard for the Installation of Standpipe and Hose Systems,* requires a fire flow test not more than 1 year prior to the commencement of system design. This requirement from NFPA 14 is intended to make certain that the fire flow and pressures available are as accurate and as up-to-date as possible. A general recommendation is that water supply data should be not more than 5 years old before retesting is conducted.

Whenever modifications to a water supply system are made or when substantial development is experienced in an area, retesting the available water supply is recommended.

1.3 Application.

A certain residual pressure in the mains is specified at which the rate of flow should be available. Additional benefit is derived from fire flow tests by the indication of possible deficiencies, such as tuberculation of piping or closed valves or both, which could be corrected to ensure adequate fire flows as needed.

1.4 Units.

Metric units of measurement in this recommended practice are in accordance with the modernized metric system known as the International System of Units (SI). Two units (liter and bar), outside of but recognized by SI, are commonly used in international fire protection. These units are listed in Table 1.4 with conversion factors.

TABLE 1.4 SI Units and Conversion Factors

Unit Name	Unit Symbol	Conversion Factor
liter	L	1 gal = 3.785 L
liter per minute per square meter	(L/min)/m^2	1 gpm ft^2 = (40.746 L/min)/m^2
cubic decimeter	dm^3	1 gal = 3.785 dm^3
pascal	Pa	1 psi = 6894.757 Pa
bar	bar	1 psi = 0.0689 bar
bar	bar	1 bar = 10^5 Pa

1.4.1 If a value for measurement as given in this recommended practice is followed by an equivalent value in other units, the first value stated is to be regarded as the recommendation. A given equivalent value might be approximate.

CHAPTER 2 REFERENCED PUBLICATIONS

2.1 General.

The documents or portions thereof listed in this chapter are referenced within this recommended practice and shall be considered part of the recommendations of this document.

2.2 NFPA Publications. (Reserved)

2.3 Other Publications.

2.3.1 IEEE Publications.

Institute of Electrical and Electronics Engineers, Three Park Avenue, 17th Floor, New York, NY 10016-5997.

IEEE/ASTM-SI-10, *Standard for Use of the International System of Units (SI): The Modern Metric System,* 1992.

2.3.2 Other Publications.

Merriam-Webster's Collegiate Dictionary, 11th edition, Merriam-Webster, Inc., Springfield, MA, 2003.

2.4 References for Extracts in Recommendations Sections. (Reserved)

CHAPTER 3 DEFINITIONS

3.1 General.

The definitions contained in this chapter shall apply to the terms used in this recommended practice. Where terms are not defined in this chapter or within another chapter, they shall be defined using their ordinarily accepted meanings within the context in which they are used. *Merriam-Webster's Collegiate Dictionary,* 11th edition, shall be the source for the ordinarily accepted meaning.

3.2 NFPA Official Definitions.

3.2.1* Authority Having Jurisdiction (AHJ). An organization, office, or individual responsible for enforcing the requirements of a code or standard, or for approving equipment, materials, an installation, or a procedure.

A.3.2.1 Authority Having Jurisdiction (AHJ). The phrase "authority having jurisdiction," or its acronym AHJ, is used in NFPA documents in a broad manner, since jurisdictions and approval agencies vary, as do their responsibilities. Where public safety is primary, the authority having jurisdiction may be a federal, state, local, or other regional department or individual such as a fire chief; fire marshal; chief of a fire prevention bureau, labor department, or health department; building official; electrical inspector; or others having statutory authority. For insurance purposes, an insurance inspection department, rating bureau, or other insurance company representative may be the authority having jurisdiction. In many circumstances, the property owner or his or her designated agent assumes the role of the authority having jurisdiction; at government installations, the commanding officer or departmental official may be the authority having jurisdiction.

3.2.2* Listed. Equipment, materials, or services included in a list published by an organization that is acceptable to the authority having jurisdiction and concerned with evaluation of products or services, that maintains periodic inspection of production of listed equipment or materials or periodic evaluation of services, and whose listing states that either the equipment, material, or service meets appropriate designated standards or has been tested and found suitable for a specified purpose.

A.3.2.2 Listed. The means for identifying listed equipment may vary for each organization concerned with product evaluation; some organizations do not recognize equipment as listed unless it is also labeled. The authority having jurisdiction should utilize the system employed by the listing organization to identify a listed product.

3.2.3 Should. Indicates a recommendation or that which is advised but not required.

3.3 General Definitions.

3.3.1 Rated Capacity. The flow available from a hydrant at the designated residual pressure (rated pressure), either measured or calculated.

The rated capacity of a hydrant also indicates the flow capabilities of the underground main. The flow capacity of the underground main can be compared to the flow and pressure demand of an attached fire protection system. The flow and pressure demand of the attached fire protection system must be less than that available from the underground main.

3.3.2 Residual Pressure. The pressure that exists in the distribution system, measured at the residual hydrant at the time the flow readings are taken at the flow hydrants.

3.3.3 Static Pressure. The pressure that exists at a given point under normal distribution system conditions measured at the residual hydrant with no hydrants flowing.

CHAPTER 4 FLOW TESTING

4.1 Rating Pressure.

4.1.1 For the purpose of uniform marking of fire hydrants, the ratings should be based on a residual pressure of 20 psi (1.4 bar) for all hydrants having a static pressure in excess of 40 psi (2.8 bar).

The minimum pressure required by many water purveyors is 20 psi (1.4 bar). Pressures less than 20 psi (1.4 bar) may create a vacuum in the piping system, causing collapse of the piping. Low pressures such as 20 psi (1.4 bar) can also cause a backflow of nonpotable water contaminating a potable source. While fire apparatus can operate at such low pressures, they do so with difficulty.

4.1.2 Hydrants having a static pressure of less than 40 psi (2.8 bar) should be rated at one-half of the static pressure.

4.1.3 It is generally recommended that a minimum residual pressure of 20 psi (1.4 bar) should be maintained at hydrants when delivering the fire flow. Fire department pumpers can be operated where hydrant pressures are less, but with difficulty.

4.1.4 Where hydrants are well distributed and of the proper size and type (so that friction losses in the hydrant and suction line are not excessive), it might be possible to set a lesser pressure as the minimum pressure.

4.1.5 A primary concern should be the ability to maintain sufficient residual pressure to prevent developing a negative pressure at any point in the street mains, which could result in the collapse of the mains or other water system components or back-siphonage of polluted water from some other interconnected source.

4.1.6 It should be noted that the use of residual pressures of less than 20 psi (1.4 bar) is not permitted by many state health departments.

4.2 Procedure.

4.2.1 Tests should be made during a period of ordinary demand.

"Periods of ordinary demand" should include testing during normal business hours and during seasons of high usage, such as summertime. Testing at times of high use, however, can cause problems during periods of low demand when a water main supplies a fire pump. During periods of low demand, the available pressure can be considerably higher than at periods of high demand. If pressures fluctuate such that excessively high pressures risk causing system damage, components to control such high pressure must be installed. These controls can be either pressure-reducing valves or break tanks. Chapter 4 of NFPA 20, *Standard for the Installation of Stationary Pumps for Fire Protection,* addresses the proper applications of these devices (see Part II of this handbook).

4.2.2 The procedure consists of discharging water at a measured rate of flow from the system at a given location and observing the corresponding pressure drop in the mains.

4.3 Layout of Test.

4.3.1 After the location where the test is to be run has been determined, a group of test hydrants in the vicinity is selected.

4.3.2 Once selected, due consideration should be given to potential interference with traffic flow patterns, damage to surroundings (e.g., roadways, sidewalks, landscapes, vehicles, and pedestrians), and potential flooding problems both local and remote from the test site.

In addition to the considerations listed in 4.3.2, the local health authority should be consulted regarding the proper disposal of test water.

4.3.3 One hydrant, designated the residual hydrant, is chosen to be the hydrant where the normal static pressure will be observed with the other hydrants in the group closed, and where the residual pressure will be observed with the other hydrants flowing.

The point at which the residual hydrant connects to the underground main is where the flow and pressure (determined during the test) is considered to be available. For example, the static and residual pressures and flow are assumed to be available at the point of connection to the underground main from the residual hydrant.

4.3.4 This hydrant is chosen so it will be located between the hydrant to be flowed and the large mains that constitute the immediate sources of water supply in the area. In Figure 4.3.4, test layouts are indicated showing the residual hydrant designated with the letter R and hydrants to be flowed with the letter F.

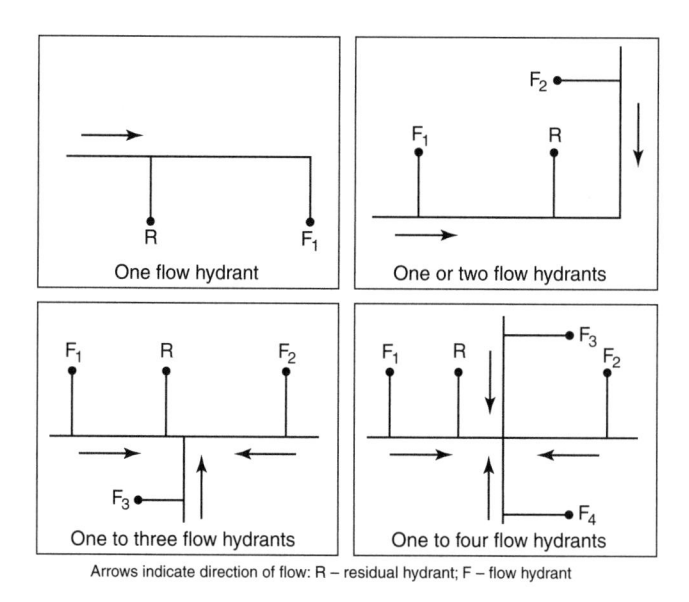

FIGURE 4.3.4 Suggested Test Layout for Hydrants.

The residual hydrant should also be the hydrant closest to the property to be protected. This designation simplifies hydraulic calculation of the sprinkler or standpipe system in the protected property.

4.3.5 The number of hydrants to be used in any test depends upon the strength of the distribution system in the vicinity of the test location.

4.3.6 To obtain satisfactory test results of theoretical calculation of expected flows or rated capacities, sufficient discharge should be achieved to cause a drop in pressure at the residual hydrant of at least 25 percent, or to flow the total demand necessary for fire-fighting purposes.

In order to produce a pressure drop of 25 percent, multiple outlets on each flow hydrant should be opened and flowed during the test. In some cases, several hydrants may need to be flowing to produce a sufficient pressure drop.

4.3.7 If the mains are small and the system weak, only one or two hydrants need to be flowed.

4.3.8 If, on the other hand, the mains are large and the system strong, it may be necessary to flow as many as seven or eight hydrants.

4.4 Equipment.

4.4.1 The equipment necessary for field work consists of the following:

(1) A single 200 psi (14 bar) bourdon pressure gauge with 1 psi (0.0689 bar) graduations.
(2) A number of pitot tubes.

(3) Hydrant wrenches.
(4) 50 or 60 psi (3.5 or 4.0 bar) bourdon pressure gauges with 1 psi (0.0689 bar) graduations, and scales with 1/16 in. (1.6 mm) graduations [One pitot tube, a 50 or 60 psi (3.5 or 4.0 bar) gauge, a hydrant wrench, a scale for each hydrant to be flowed].
(5) A special hydrant cap tapped with a hole into which a short length of ¼ in. (6.35 mm) brass pipe is fitted; this pipe is provided with a T connection for the 200 psi (14 bar) gauge and a cock at the end for relieving air pressure.

4.4.2 All pressure gauges should be calibrated at least every 12 months, or more frequently depending on use.

Each calibrated gauge should be identified by a sticker from the company or person performing the calibration. Gauges might lose their calibration after heavy use; therefore, recalibration every 12 months is recommended.

4.4.3 When more than one hydrant is flowed, it is desirable and could be necessary to use portable radios to facilitate communication between team members.

Some hydrants may be located several hundred feet apart, and therefore portable radios are needed for proper communication. Testing personnel need to communicate so that measurements can be taken at the appropriate time. For example, when hydrants are fully opened residual pressures should be taken. Exhibit III.25 illustrates a portable radio in use during a flow test.

4.4.4 It is preferred to use stream straightener with a known coefficient of discharge when testing hydrants due to a more streamlined flow and more accurate pitot reading.

In addition to improving the accuracy of test results, a diffuser also diffuses the energy of the stream, improves the safety of the test personnel, and helps to avoid property damage. Exhibit

EXHIBIT III.25 Portable Radio in Use During a Hydrant Flow Test.

EXHIBIT III.26 Diffuser in Use During a Hydrant Flow Test.

III.26 illustrates the use of a diffuser. A stream straightener helps to improve accuracy of the pressure reading but does not diffuse the energy of the water stream.

4.5 Test Procedure.

4.5.1 In a typical test, the 200 psi (14 bar) gauge is attached to one of the 2½ in. (6.4 cm) outlets of the residual hydrant using the special cap.

4.5.2 The cock on the gauge piping is opened, and the hydrant valve is opened full.

4.5.3 As soon as the air is exhausted from the barrel, the cock is closed.

4.5.4 A reading (static pressure) is taken when the needle comes to rest.

Before attaching test equipment to any hydrant, the hydrant should be flushed to remove any debris within the hydrant barrel. Debris from the hydrant barrel can damage test gauges or Pitot tubes. After flushing, the test equipment can be attached to the hydrant and the air bled off. At this point in the test, data such as date, time of day, weather conditions, and static pressure should be recorded. Exhibit III.27 illustrates flushing of the residual hydrant and attachment of test equipment.

EXHIBIT III.27 Flushing of Residual Hydrant (left) and Attachment of Test Equipment (right).

4.5.5 At a given signal, each of the other hydrants is opened in succession, with discharge taking place directly from the open hydrant butts.

4.5.6 Hydrants should be opened one at a time.

4.5.7 With all hydrants flowing, water should be allowed to flow for a sufficient time to clear all debris and foreign substances from the stream(s).

4.5.8 At that time, a signal is given to the people at the hydrants to read the pitot pressure of the streams simultaneously while the residual pressure is being read.

4.5.9 The final magnitude of the pressure drop can be controlled by the number of hydrants used and the number of outlets opened on each.

4.5.10 After the readings have been taken, hydrants should be shut down slowly, one at a time, to prevent undue surges in the system.

4.6 Pitot Readings.

4.6.1 When measuring discharge from open hydrant butts, it is always preferable from the standpoint of accuracy to use 2½ in. (6.4 cm) outlets rather than pumper outlets.

4.6.2 In practically all cases, the 2½ in. (6.4 cm) outlets are filled across the entire cross-section during flow, while in the case of the larger outlets there is very frequently a void near the bottom.

4.6.3 When measuring the pitot pressure of a stream of practically uniform velocity, the orifice in the pitot tube is held downstream approximately one-half the diameter of the hydrant outlet or nozzle opening, and in the center of the stream.

4.6.4 The center line of the orifice should be at right angles to the plane of the face of the hydrant outlet.

The use of a diffuser removes the difficulty in holding the Pitot tube at the one-half diameter position referenced in 4.6.3 and maintaining the centerline of the Pitot tube orifice at a right angle to the plane of the hydrant outlet. Exhibit III.28 illustrates the Pitot tube opening or slot for correct positioning of the Pitot tube in the water stream.

4.6.5 The air chamber on the pitot tube should be kept elevated.

4.6.6 Pitot readings of less than 10 psi (0.7 bar) and more than 30 psi (2.0 bar) should be avoided, if possible.

Pitot readings of less than 10 psi (0.7 bar) and more than 30 psi (2.0 bar) can be inaccurate and should be avoided.

4.6.7 Opening additional hydrant outlets will aid in controlling the pitot reading.

4.6.8 With dry barrel hydrants, the hydrant valve should be wide open to minimize problems with underground drain valves.

4.6.9 With wet barrel hydrants, the valve for the flowing outlet should be wide open to give a more streamlined flow and a more accurate pitot reading. *(See Figure 4.6.9.)*

4.7 Determination of Discharge.

4.7.1 At the hydrants used for flow during the test, the discharges from the open butts are determined from measurements of the diameter of the outlets flowed, the pitot pressure (velocity head) of the streams as indicated by the pitot gauge readings, and the coefficient of the outlet being flowed as determined from Figure 4.7.1.

4.7.2 If flow tubes (stream straighteners) are being utilized, a coefficient of 0.95 is suggested unless the coefficient of the tube is known.

EXHIBIT III.28 *Diffuser—Slot for Pitot Tube Insertion.*

4.7.3 The formula used to compute the discharge, Q, in gpm from these measurements is as follows:

$$Q = 29.84cd^2 \times \sqrt{p} \qquad (4.7.3)$$

where:

c = coefficient of discharge (*see Figure 4.7.1*)
d = diameter of the outlet in inches
p = pitot pressure (velocity head) in psi

4.8 Use of Pumper Outlets.

4.8.1 If it is necessary to use a pumper outlet, and flow tubes (stream straighteners) are not available, the best results are obtained with the pitot pressure (velocity head) maintained between 5 psi and 10 psi (0.3 bar and 0.7 bar).

4.8.2 For pumper outlets, the approximate discharge can be computed from Equation 4.7.3 using the pitot pressure (velocity head) at the center of the stream and multiplying the result by one of the coefficients in Table 4.8.2, depending upon the pitot pressure (velocity head).

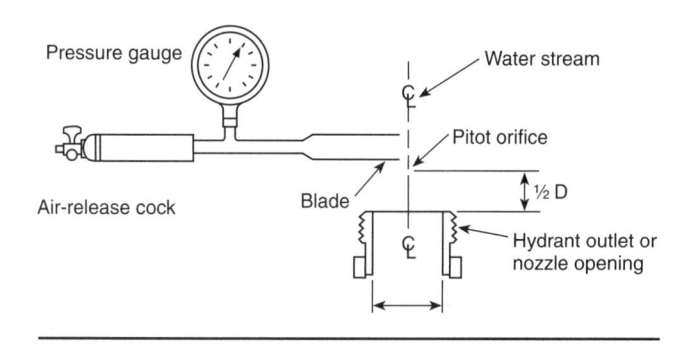

FIGURE 4.6.9 Pitot Tube Position.

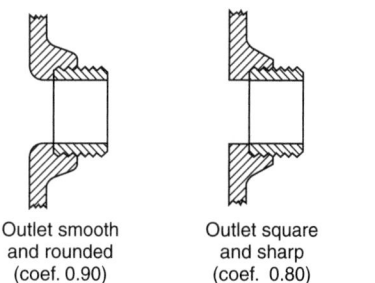

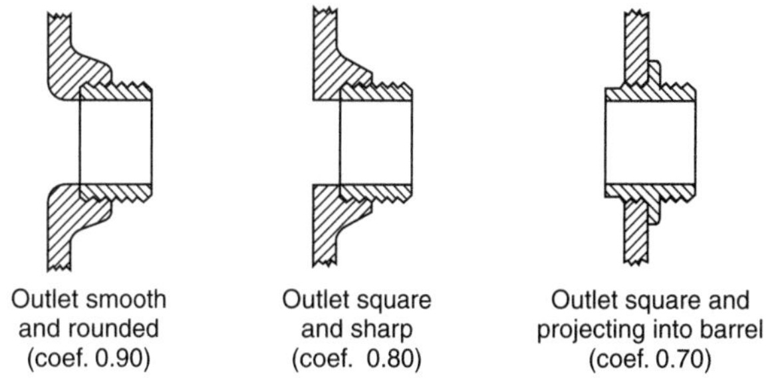

FIGURE 4.7.1 Three General Types of Hydrant Outlets and Their Coefficients of Discharge.

Outlet smooth and rounded (coef. 0.90)

Outlet square and sharp (coef. 0.80)

Outlet square and projecting into barrel (coef. 0.70)

4.8.3 These coefficients are applied in addition to the coefficient in Equation 4.7.3 and are for average-type hydrants.

4.9 Determination of Discharge Without a Pitot.

4.9.1 If a pitot tube is not available for use to measure the hydrant discharge, a 50 or 60 psi (3.5 or 4.0 bar) gauge tapped into a hydrant cap can be used.

4.9.2 The hydrant cap with gauge attached is placed on one outlet, and the flow is allowed to take place through the other outlet at the same elevation.

4.9.3 The readings obtained from a gauge so located, and the readings obtained from a gauge on a pitot tube held in the stream, are approximately the same.

4.10 Calculation Results.

4.10.1 The discharge in gpm (L/min) for each outlet flowed is obtained from Table 4.10.1(a) and Table 4.10.1(b) or by the use of Equation 4.7.3.

TABLE 4.8.2 Pumper Outlet Coefficients

Pitot Pressure (Velocity Head)		
psi	bar	Coefficient
2	0.14	0.97
3	0.21	0.92
4	0.28	0.89
5	0.35	0.86
6	0.41	0.84
7 and over	0.48 and over	0.83

To determine the discharge, the Pitot pressure and orifice diameter must be known. The Pitot pressure measured during the flow test in Exhibit III.29 is 50 psi (3.5 bar). From Table 4.10.1(a) and Table 4.10.1(b), a Pitot pressure of 50 psi (3.5 bar) is 1319 gpm (4992 L/min) theoretical flow. After applying the outlet coefficient of 0.9, the actual flow measured during this flow test is 1187 gpm (4493 L/min).

4.10.1.1 If more than one outlet is used, the discharges from all are added to obtain the total discharge.

4.10.1.2 The formula that is generally used to compute the discharge at the specified residual pressure or for any desired pressure drop is Equation 4.10.1.2:

$$Q_R = Q_F \times \frac{h_r^{0.54}}{h_f^{0.54}} \qquad (4.10.1.2)$$

where:

Q_R = flow predicted at desired residual pressure

Q_F = total flow measured during test

h_r = pressure drop to desired residual pressure

h_f = pressure drop measured during test

Applying the formula in 4.10.1.2, the total flow measured during the test is as previously calculated: 1187 gpm (4493 L/min). The pressure drop to the desired pressure in this case

TABLE 4.10.1(a) *Theoretical Discharge Through Circular Orifices (U.S. Gallons of Water per Minute)*

Pitot Pressure* (psi)	Feet†	Velocity Discharge (ft/sec)	2	2.25	2.375	2.5	2.625	2.75	3	3.25	3.5	3.75	4	4.5
1	2.31	12.20	119	151	168	187	206	226	269	315	366	420	477	604
2	4.61	17.25	169	214	238	264	291	319	380	446	517	593	675	855
3	6.92	21.13	207	262	292	323	356	391	465	546	633	727	827	1047
4	9.23	24.39	239	302	337	373	411	451	537	630	731	839	955	1209
5	11.54	27.26	267	338	376	417	460	505	601	705	817	938	1068	1351
6	13.84	29.87	292	370	412	457	504	553	658	772	895	1028	1169	1480
7	16.15	32.26	316	400	445	493	544	597	711	834	967	1110	1263	1599
8	18.46	34.49	338	427	476	528	582	638	760	891	1034	1187	1350	1709
9	20.76	36.58	358	453	505	560	617	677	806	946	1097	1259	1432	1813
10	23.07	38.56	377	478	532	590	650	714	849	997	1156	1327	1510	1911
11	25.38	40.45	396	501	558	619	682	748	891	1045	1212	1392	1583	2004
12	27.68	42.24	413	523	583	646	712	782	930	1092	1266	1454	1654	2093
13	29.99	43.97	430	545	607	672	741	814	968	1136	1318	1513	1721	2179
14	32.30	45.63	447	565	630	698	769	844	1005	1179	1368	1570	1786	2261
15	34.61	47.22	462	585	652	722	796	874	1040	1221	1416	1625	1849	2340
16	36.91	48.78	477	604	673	746	822	903	1074	1261	1462	1679	1910	2417
17	39.22	50.28	492	623	694	769	848	930	1107	1300	1507	1730	1969	2491
18	41.53	51.73	506	641	714	791	872	957	1139	1337	1551	1780	2026	2564
19	43.83	53.15	520	658	734	813	896	984	1171	1374	1593	1829	2081	2634
20	46.14	54.54	534	676	753	834	920	1009	1201	1410	1635	1877	2135	2702
22	50.75	57.19	560	709	789	875	964	1058	1260	1478	1715	1968	2239	2834
24	55.37	59.74	585	740	825	914	1007	1106	1316	1544	1791	2056	2339	2960
26	59.98	62.18	609	770	858	951	1048	1151	1369	1607	1864	2140	2434	3081
28	64.60	64.52	632	799	891	987	1088	1194	1421	1668	1934	2220	2526	3197
30	69.21	66.79	654	827	922	1022	1126	1236	1471	1726	2002	2298	2615	3310
32	73.82	68.98	675	855	952	1055	1163	1277	1519	1783	2068	2374	2701	3418
34	78.44	71.10	696	881	981	1087	1199	1316	1566	1838	2131	2447	2784	3523
36	83.05	73.16	716	906	1010	1119	1234	1354	1611	1891	2193	2518	2865	3626
38	87.67	75.17	736	931	1038	1150	1268	1391	1656	1943	2253	2587	2943	3725
40	92.28	77.11	755	955	1065	1180	1300	1427	1699	1993	2312	2654	3020	3822
42	96.89	79.03	774	979	1091	1209	1333	1462	1740	2043	2369	2719	3094	3916
44	101.51	80.88	792	1002	1116	1237	1364	1497	1781	2091	2425	2783	3167	4008
46	106.12	82.70	810	1025	1142	1265	1395	1531	1821	2138	2479	2846	3238	4098
48	110.74	84.48	827	1047	1166	1292	1425	1563	1861	2184	2533	2907	3308	4186
50	115.35	86.22	844	1068	1190	1319	1454	1596	1899	2229	2585	2967	3376	4273
52	119.96	87.93	861	1089	1214	1345	1483	1627	1937	2273	2636	3026	3443	4357
54	124.58	89.61	877	1110	1237	1370	1511	1658	1974	2316	2686	3084	3508	4440
56	129.19	91.20	893	1130	1260	1396	1539	1689	2010	2359	2735	3140	3573	4522
58	133.81	92.87	909	1150	1282	1420	1566	1719	2045	2400	2784	3196	3636	4602
60	138.42	94.45	925	1170	1304	1445	1593	1748	2080	2441	2831	3250	3698	4681
62	143.03	96.01	940	1189	1325	1469	1619	1777	2115	2482	2878	3304	3759	4758
64	147.65	97.55	955	1209	1347	1492	1645	1805	2148	2521	2924	3357	3820	4834
66	152.26	99.07	970	1227	1367	1515	1670	1833	2182	2561	2970	3409	3879	4909
68	156.88	100.55	984	1246	1388	1538	1696	1861	2215	2599	3014	3460	3937	4983
70	161.49	102.03	999	1264	1408	1560	1720	1888	2247	2637	3058	3511	3995	5056

(continued)

TABLE 4.10.1(a) *Continued*

Pitot Pressure* (psi)	Feet†	Velocity Discharge (ft/sec)	\multicolumn{12}{c}{Orifice Size (in.)}
			2	2.25	2.375	2.5	2.625	2.75	3	3.25	3.5	3.75	4	4.5
72	166.10	103.47	1013	1282	1428	1583	1745	1915	2279	2674	3102	3561	4051	5127
74	170.72	104.90	1027	1300	1448	1604	1769	1941	2310	2711	3144	3610	4107	5198
76	175.33	106.30	1041	1317	1467	1626	1793	1967	2341	2748	3187	3658	4162	5268
78	179.95	107.69	1054	1334	1487	1647	1816	1993	2372	2784	3228	3706	4217	5337
80	184.56	109.08	1068	1351	1505	1668	1839	2018	2402	2819	3269	3753	4270	5405
82	189.17	110.42	1081	1368	1524	1689	1862	2043	2432	2854	3310	3800	4323	5472
84	193.79	111.76	1094	1385	1543	1709	1885	2068	2461	2889	3350	3846	4376	5538
86	198.40	113.08	1107	1401	1561	1730	1907	2093	2491	2923	3390	3891	4428	5604
88	203.02	114.39	1120	1417	1579	1750	1929	2117	2519	2957	3429	3936	4479	5668
90	207.63	115.68	1132	1433	1597	1769	1951	2141	2548	2990	3468	3981	4529	5733
92	212.24	116.96	1145	1449	1614	1789	1972	2165	2576	3023	3506	4025	4579	5796
94	216.86	118.23	1157	1465	1632	1808	1994	2188	2604	3056	3544	4068	4629	5859
96	221.47	119.48	1169	1480	1649	1827	2015	2211	2631	3088	3582	4111	4678	5921
98	226.09	120.71	1182	1495	1666	1846	2035	2234	2659	3120	3619	4154	4726	5982
100	230.70	121.94	1194	1511	1683	1865	2056	2257	2686	3152	3655	4196	4774	6043
102	235.31	123.15	1205	1526	1700	1884	2077	2279	2712	3183	3692	4238	4822	6103
104	239.93	124.35	1217	1541	1716	1902	2097	2301	2739	3214	3728	4279	4869	6162
106	244.54	125.55	1229	1555	1733	1920	2117	2323	2765	3245	3763	4320	4916	6221
108	249.16	126.73	1240	1570	1749	1938	2137	2345	2791	3275	3799	4361	4962	6280
110	253.77	127.89	1252	1584	1765	1956	2157	2367	2817	3306	3834	4401	5007	6338
112	258.38	129.05	1263	1599	1781	1974	2176	2388	2842	3336	3869	4441	5053	6395
114	263.00	130.20	1274	1613	1797	1991	2195	2409	2867	3365	3903	4480	5098	6452
116	267.61	131.33	1286	1627	1813	2009	2215	2430	2892	3395	3937	4519	5142	6508
118	272.23	132.46	1297	1641	1828	2026	2234	2451	2917	3424	3971	4558	5186	6564
120	276.84	133.57	1308	1655	1844	2043	2252	2472	2942	3453	4004	4597	5230	6619
122	281.45	134.69	1318	1669	1859	2060	2271	2493	2966	3481	4038	4635	5273	6674
124	286.07	135.79	1329	1682	1874	2077	2290	2513	2991	3510	4070	4673	5317	6729
126	290.68	136.88	1340	1696	1889	2093	2308	2533	3015	3538	4103	4710	5359	6783
128	295.30	137.96	1350	1709	1904	2110	2326	2553	3038	3566	4136	4748	5402	6836
130	299.91	139.03	1361	1722	1919	2126	2344	2573	3062	3594	4168	4784	5444	6890
132	304.52	140.10	1371	1736	1934	2143	2362	2593	3086	3621	4200	4821	5485	6942
134	309.14	141.16	1382	1749	1948	2159	2380	2612	3109	3649	4231	4858	5527	6995
136	313.75	142.21	1392	1762	1963	2175	2398	2632	3132	3676	4263	4894	5568	7047

Notes:

(1) This table is computed from the formula $Q = 29.84cd^2 \times \sqrt{p}$, with $c = 1.00$. The theoretical discharge of seawater, as from fireboat nozzles, can be found by subtracting 1 percent from the figures in Table 4.10.2.1, or from the formula $Q = 29.84cd^2 \times \sqrt{p}$.

(2) Appropriate coefficient should be applied where it is read from hydrant outlet. Where more accurate results are required, a coefficient appropriate on the particular nozzle must be selected and applied to the figures of the table. The discharge from circular openings of sizes other than those in the table can readily be computed by applying the principle that quantity discharged under a given head varies as the square of the diameter of the opening.

*This pressure corresponds to velocity head.

†1 psi = 2.307 ft of water. For pressure in bars, multiply by 0.01.

TABLE 4.10.1(b) *Theoretical Discharge Through Circular Orifices (Liters of Water per Minute)*

Pitot Pressure* (psi)	Feet†	Velocity Discharge (ft/sec)	Orifice Size (in.)										
			51	57	60	64	67	70	76	83	89	95	101	114
6.89	0.70	3.72	455	568	629	716	785	857	1010	1204	1385	1578	1783	2272
13.8	1.41	5.26	644	804	891	1013	1111	1212	1429	1704	1960	2233	2524	3215
20.7	2.11	6.44	788	984	1091	1241	1360	1485	1750	2087	2400	2735	3091	3938
27.6	2.81	7.43	910	1137	1260	1433	1571	1714	2021	2410	2771	3158	3569	4547
34.5	3.52	8.31	1017	1271	1408	1602	1756	1917	2259	2695	3099	3530	3990	5084
41.4	4.22	9.10	1115	1392	1543	1755	1924	2100	2475	2952	3394	3867	4371	5569
48.3	4.92	9.83	1204	1504	1666	1896	2078	2268	2673	3189	3666	4177	4722	6015
55.2	5.63	10.51	1287	1608	1781	2027	2221	2425	2858	3409	3919	4466	5048	6431
62.0	6.33	11.15	1364	1704	1888	2148	2354	2570	3029	3613	4154	4733	5349	6815
68.9	7.03	11.75	1438	1796	1990	2264	2482	2709	3193	3808	4379	4989	5639	7184
75.8	7.73	12.33	1508	1884	2087	2375	2603	2841	3349	3995	4593	5233	5915	7536
82.7	8.44	12.87	1575	1968	2180	2481	2719	2968	3498	4172	4797	5466	6178	7871
89.6	9.14	13.40	1640	2048	2270	2582	2830	3089	3641	4343	4994	5690	6431	8193
96.5	9.84	13.91	1702	2126	2355	2680	2937	3206	3779	4507	5182	5905	6674	8503
103	10.55	14.39	1758	2196	2433	2769	3034	3312	3904	4656	5354	6100	6895	8784
110	11.25	14.87	1817	2269	2515	2861	3136	3423	4035	4812	5533	6304	7125	9078
117	11.95	15.33	1874	2341	2593	2951	3234	3530	4161	4963	5706	6502	7349	9362
124	12.66	15.77	1929	2410	2670	3038	3329	3634	4284	5109	5874	6693	7565	9638
131	13.36	16.20	1983	2477	2744	3122	3422	3735	4403	5251	6038	6880	7776	9906
138	14.06	16.62	2035	2542	2817	3205	3512	3834	4519	5390	6197	7061	7981	10168
152	15.47	17.43	2136	2668	2956	3363	3686	4023	4743	5657	6504	7410	8376	10671
165	16.88	18.21	2225	2779	3080	3504	3840	4192	4941	5893	6776	7721	8727	11118
179	18.28	18.95	2318	2895	3208	3650	4000	4366	5147	6138	7058	8042	9090	11580
193	19.69	19.67	2407	3006	3331	3790	4153	4534	5344	6374	7329	8350	9438	12024
207	21.10	20.36	2492	3113	3450	3925	4301	4695	5535	6601	7590	8648	9775	12453
221	22.50	21.03	2575	3217	3564	4055	4444	4851	5719	6821	7842	8935	10100	12867
234	23.91	21.67	2650	3310	3668	4173	4573	4992	5884	7018	8070	9195	10393	13240
248	25.31	22.30	2728	3408	3776	4296	4708	5139	6058	7225	8308	9466	10699	13630
262	26.72	22.91	2804	3502	3881	4416	4839	5282	6227	7426	8539	9729	10997	14010
276	28.13	23.50	2878	3595	3983	4532	4967	5422	6391	7622	8764	9986	11287	14379
290	29.53	24.09	2950	3685	4083	4646	5091	5557	6551	7813	8984	10236	11570	14740
303	30.94	24.65	3015	3767	4173	4748	5204	5681	6696	7986	9183	10463	11826	15066
317	32.35	25.21	3084	3853	4269	4857	5323	5810	6849	8169	9393	10702	12096	15410
331	33.75	25.75	3152	3937	4362	4963	5439	5937	6999	8347	9598	10935	12360	15747
345	35.16	26.28	3218	4019	4453	5067	5553	6061	7145	8522	9799	11164	12619	16077
358	36.57	26.80	3278	4094	4536	5161	5657	6175	7279	8681	9981	11373	12855	16377
372	37.97	27.31	3341	4173	4624	5261	5766	6294	7419	8849	10175	11593	13104	16694
386	39.38	27.80	3403	4251	4711	5360	5874	6412	7558	9014	10364	11809	13348	17005
400	40.78	28.31	3465	4328	4795	5456	5979	6527	7694	9176	10551	12021	13588	17311
414	42.19	28.79	3525	4403	4878	5551	6083	6640	7827	9335	10734	12230	13823	17611
427	43.60	29.26	3580	4471	4954	5637	6178	6743	7949	9481	10901	12420	14039	17885
441	45.00	29.73	3638	4544	5035	5729	6278	6853	8078	9635	11078	12622	14267	18176
455	46.41	30.20	3695	4616	5114	5819	6377	6961	8206	9787	11253	12821	14492	18462
469	47.82	30.65	3751	4686	5192	5908	6475	7067	8331	9936	11425	13017	14713	18744
483	49.22	31.10	3807	4756	5269	5995	6570	7172	8454	10083	11594	13210	14931	19022
496	50.63	31.54	3858	4819	5340	6075	6658	7268	8567	10218	11749	13386	15131	19276
510	52.03	31.97	3912	4887	5415	6161	6752	7370	8687	10361	11913	13574	15343	19547

(continued)

TABLE 4.10.1(b) Continued

| Pitot Pressure* (psi) | Feet† | Velocity Discharge (ft/sec) | \multicolumn{12}{c}{Orifice Size (in.)} |
			51	57	60	64	67	70	76	83	89	95	101	114
524	53.44	32.71	3965	4953	5488	6245	6844	7470	8806	10503	12076	13759	15552	19813
538	54.85	32.82	4018	5019	5561	6327	6934	7569	8923	10642	12236	13942	15758	20076
552	56.25	33.25	4070	5084	5633	6409	7024	7667	9038	10780	12394	14122	15962	20335
565	57.66	33.66	4118	5143	5699	6484	7106	7757	9144	10906	12539	14287	16149	20573
579	59.07	34.06	4168	5207	5769	6564	7194	7853	9256	11040	12694	14463	16348	20827
593	60.47	34.47	4218	5269	5839	6643	7280	7947	9368	11173	12846	14637	16544	21077
607	61.88	34.87	4268	5331	5907	6721	7366	8040	9478	11304	12997	14809	16738	21324
620	63.29	35.26	4313	5388	5970	6793	7444	8126	9578	11424	13136	14966	16917	21552
634	64.69	35.65	4362	5448	6037	6869	7528	8217	9686	11552	13283	15134	17107	21794
648	66.10	36.04	4410	5508	6103	6944	7610	8307	9792	11679	13429	15301	17294	22033
662	67.50	36.42	4457	5567	6169	7019	7692	8397	9898	11805	13573	15465	17480	22270
676	68.91	36.79	4504	5626	6234	7093	7773	8485	10002	11929	13716	15628	17664	22504
689	70.32	37.17	4547	5680	6293	7161	7848	8566	10097	12043	13847	15777	17833	22719
703	71.72	37.54	4593	5737	6357	7233	7927	8653	10200	12165	13987	15937	18013	22949
717	73.13	37.90	4638	5794	6420	7305	8005	8738	10301	12285	14126	16095	18192	23176
731	74.54	38.27	4684	5850	6482	7376	8083	8823	10401	12405	14263	16251	18369	23401
745	75.94	38.63	4728	5906	6544	7446	8160	8907	10500	12523	14399	16406	18544	23624
758	77.35	38.98	4769	5957	6601	7510	8231	8985	10591	12632	14524	16548	18705	23830
772	78.76	39.33	4813	6012	6662	7580	8307	9067	10688	12748	14658	16701	18877	24049
786	80.16	39.68	4857	6066	6722	7648	8382	9149	10785	12863	14790	16851	19047	24266
800	81.57	40.03	4900	6120	6781	7716	8456	9230	10880	12977	14921	17001	19216	24481
813	82.97	40.37	4939	6170	6836	7778	8525	9305	10968	13082	15042	17138	19371	24679
827	84.38	40.71	4982	6223	6895	7845	8598	9385	11063	13194	15171	17285	19538	24891
841	85.79	41.05	5024	6275	6953	7911	8670	9464	11156	13305	15299	17431	19702	25100
855	87.19	41.39	5065	6327	7011	7977	8742	9542	11248	13416	15425	17575	19866	25309
869	88.60	41.72	5107	6379	7068	8042	8813	9620	11340	13525	15551	17719	20028	25515
882	90.01	42.05	5145	6426	7121	8102	8879	9692	11424	13626	15667	17851	20177	25705
896	91.41	42.38	5185	6477	7177	8166	8949	9768	11515	13734	15791	17992	20336	25908
910	92.82	42.70	5226	6527	7233	8229	9019	9844	11604	13840	15914	18132	20495	26110
924	94.23	43.03	5266	6577	7288	8292	9088	9920	11693	13947	16036	18271	20652	26310
938	95.63	43.35	5305	6627	7343	8355	9156	9995	11782	14052	16157	18409	20807	26509

Notes:

(1) This table is computed from the formula $Q = 0.0666cd^2 \times \sqrt{p_m}$, with $c = 1.00$. The theoretical discharge of seawater, as from fireboat nozzles, can be found by subtracting 1 percent from the figures in Table 4.10.2.1, or from the formula $Q = 0.0666cd^2 \times \sqrt{p_m}$.

(2) Appropriate coefficient should be applied where it is read from the hydrant outlet. Where more accurate results are required, a coefficient appropriate on the particular nozzle must be selected and applied to the figures of the table. The discharge from circular openings of sizes other than those in the table can readily be computed by applying the principle that quantity discharged under a given head varies as the square of the diameter of the opening.

*This pressure corresponds to velocity head.

†1 kPa = 0.102 m of water. For pressure in bars, multiply by 0.01.

FLOW TEST REPORT

Date: __6/2/10__

Inspector: __John Doe__

Residual hydrant location: __Main Street (see map on p. 2)__

Flow hydrant location: __"   "__

Static pressure (residual hydrant): __75__ psi

Residual pressure (residual hydrant): __52__ psi

Pitot pressure (flow hydrant): __50__ psi

Nozzle size (flow hydrant): __2.5__ in.
From Table 4.10.1(a) of NFPA 291,
Recommended Practice for Fire Flow Testing and Marking of Hydrants, 2010 edition: __1319__ gpm

Nozzle coefficient (flow hydrant): __0.9__

$$\underline{1319} \times \underline{0.9} = \underline{1187}$$

Available water flow: __1187__ gpm at __52__ psi

Compute discharge:

$$Q_R = Q_F \times h_r^{0.54}/h_f^{0.54}$$

$$Q_R = \underline{1187} \times (\underline{55})^{0.54} / (\underline{23})^{0.54}$$

$$Q_R = \underline{1901} \text{ gpm at 20 psi}$$

(p. 1 of 3)

EXHIBIT III.29 Sample Flow Test Report.

Water Supply Data

Static:	75 psi
Residual:	52 psi
Flow:	1220 gpm

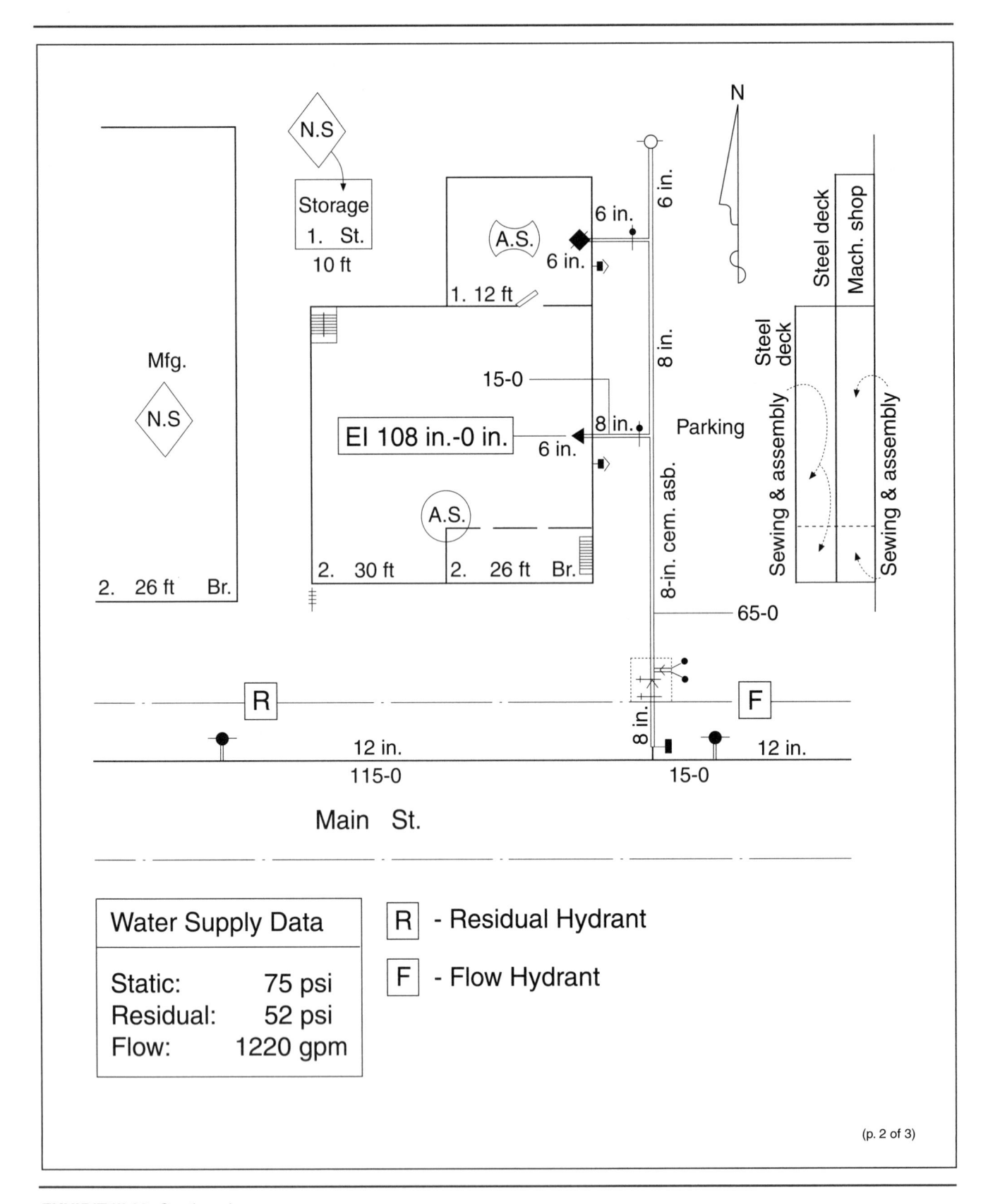

R - Residual Hydrant

F - Flow Hydrant

EXHIBIT III.29 Continued.

Part III • Complete Text of NFPA 291   487

HYDRAULIC GRAPH Pressure vs. (Flow)$^{1.85}$

LOCATION:
STATIC PRESSURE: 75
RESIDUAL PRESSURE: 52
FLOW: 1187
ELEVATION: 108'
DATE: 6-2-10
TIME: 9:40 AM
BY: J.D.
NOTES:

STATIC 75 PSI
RESIDUAL 52 PSI @ 1187 GPM

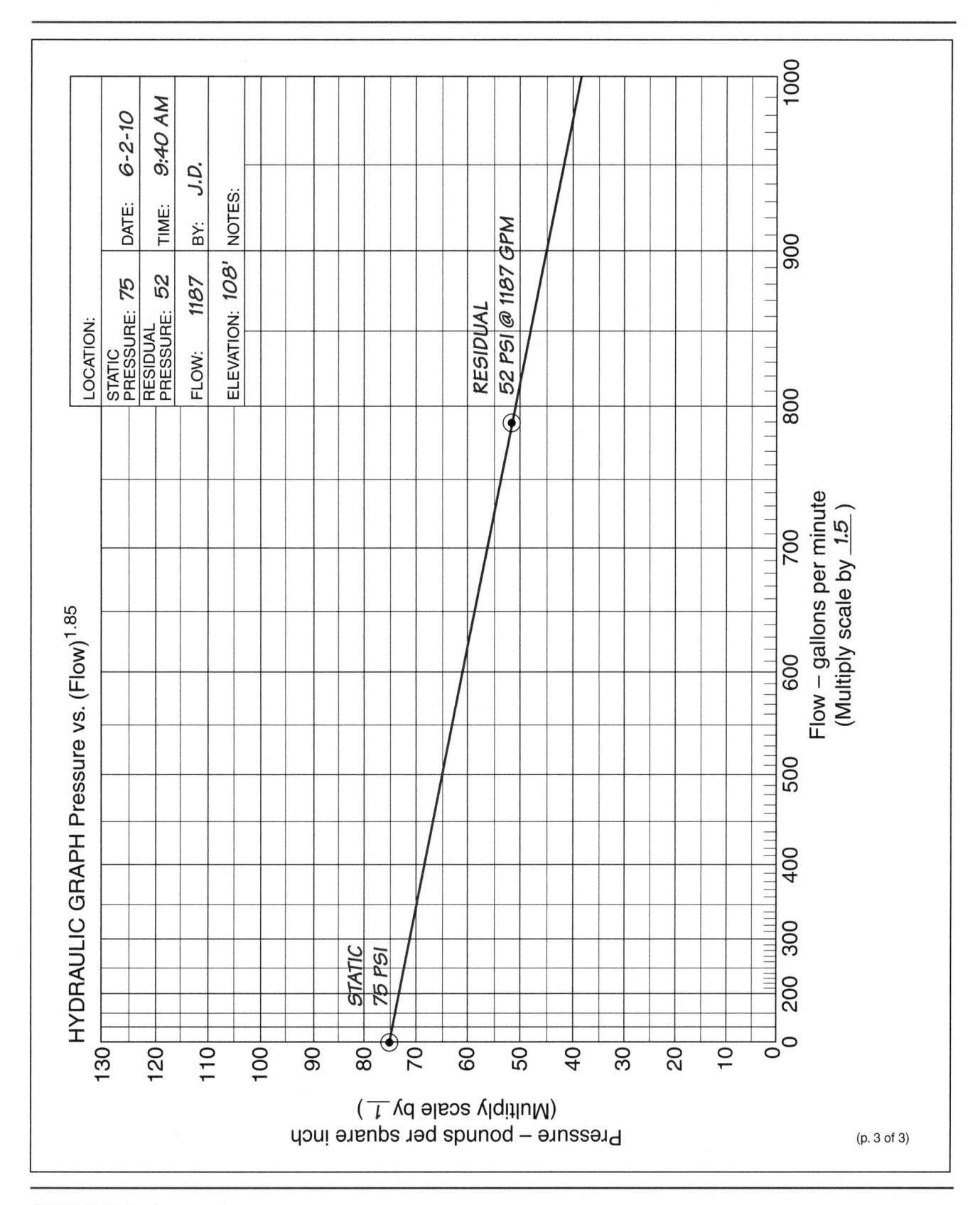

Pressure – pounds per square inch (Multiply scale by 1)
Flow – gallons per minute (Multiply scale by 1.5)

(p. 3 of 3)

EXHIBIT III.29 Continued.

Stationary Fire Pumps Handbook   2010

is 55 psi (3.8 bar) [75 psi − 20 psi = 55 psi (5.2 bar − 1.4 bar = 3.8 bar)]. The pressure drop measured during the test was 23 psi (1.6 bar) [75 psi − 52 psi = 23 psi (5.2 bar − 3.6 (bar) = 1.6 bar)]. The result of this calculation predicts a flow of 1901 gpm (7195.3 L/min) when the residual pressure is drawn down to the minimum recommended 20 psi (1.4 bar). As stated in 4.1.1, the ratings and uniform marking of hydrants should be based on a residual pressure of 20 psi (1.4 bar). Therefore, the rating for the hydrant in this example should be based on a flow of 1901 gpm (7195.3 L/min) not the 1187 gpm (4493 L/min) measured during the test.

4.10.1.3 In this equation, any units of discharge or pressure drop may be used as long as the same units are used for each value of the same variable.

4.10.1.4 In other words, if Q_R is expressed in gpm, Q_F must be in gpm, and if hr is expressed in psi, h_f must be expressed in psi.

4.10.1.5 These are the units that are normally used in applying Equation 4.10.1.2 to fire flow test computations.

4.10.2 Discharge Calculations from Table.

4.10.2.1 One means of solving this equation without the use of logarithms is by using Table 4.10.2.1, which gives the values of the 0.54 power of the numbers from 1 to 175.

4.10.2.2 Knowing the values of h_f, h_r, and QF, the values of $h_f^{0.54}$ and $h_r^{0.54}$ can be read from the table and Equation 4.10.1.2 solved for QR.

4.10.2.3 Results are usually carried to the nearest 100 gpm (380 L/min) for discharges of 1000 gpm (3800 L/min) or more, and to the nearest 50 gpm (190 L/min) for smaller discharges, which is as close as can be justified by the degree of accuracy of the field observations.

4.10.2.4 Insert in Equation 4.10.1.2 the values of $h_r^{0.54}$ and $h_f^{0.54}$ determined from the table and the value of Q_F, and solve the equation for QR.

4.11 Data Sheet.

4.11.1 The data secured during the testing of hydrants for uniform marking can be valuable for other purposes.

4.11.2 With this in mind, it is suggested that the form shown in Figure 4.11.2 be used to record information that is taken.

4.11.3 The back of the form should include a location sketch.

4.11.4 Results of the flow test should be indicated on a hydraulic graph, such as the one shown in Figure 4.11.4.

4.11.5 When the tests are complete, the forms should be filed for future reference by interested parties.

The data recorded in the forms may change over time, in which case the test should be repeated. Water flow test data should never be used if the data are more than 5 years old. Some installation standards may require more recent flow test data. For example, NFPA 14, Section 10.2 requires that a flow test be conducted not more than 1 year before the commencement of design of a standpipe system. Subsection 7.3.1 of NFPA 25, *Standard for the Inspection, Testing, and Maintenance of Water-Based Fire Protection Systems*, requires a flow test of underground piping every 5 years to evaluate the internal condition of the system piping. All subsequent testing should be documented and compared to previous results.

4.12 System Corrections.

4.12.1 It must be remembered that flow test results show the strength of the distribution system and do not necessarily indicate the degree of adequacy of the entire water works system.

TABLE 4.10.2.1 Values of h to the 0.54 Power

h	$h^{0.54}$	h	$h^{0.54}$	h	$h^{0.54}$	h	$h^{0.54}$	h	$h^{0.54}$
1	1.00	36	6.93	71	9.99	106	12.41	141	14.47
2	1.45	37	7.03	72	10.07	107	12.47	142	14.53
3	1.81	38	7.13	73	10.14	108	12.53	143	14.58
4	2.11	39	7.23	74	10.22	109	12.60	144	14.64
5	2.39	40	7.33	75	10.29	110	12.66	145	14.69
6	2.63	41	7.43	76	10.37	111	12.72	146	14.75
7	2.86	42	7.53	77	10.44	112	12.78	147	14.80
8	3.07	43	7.62	78	10.51	113	12.84	148	14.86
9	3.28	44	7.72	79	10.59	114	12.90	149	14.91
10	3.47	45	7.81	80	10.66	115	12.96	150	14.97
11	3.65	46	7.91	81	10.73	116	13.03	151	15.02
12	3.83	47	8.00	82	10.80	117	13.09	152	15.07
13	4.00	48	8.09	83	10.87	118	13.15	153	15.13
14	4.16	49	8.18	84	10.94	119	13.21	154	15.18
15	4.32	50	8.27	85	11.01	120	13.27	155	15.23
16	4.48	51	8.36	86	11.08	121	13.33	156	15.29
17	4.62	52	8.44	87	11.15	122	13.39	157	15.34
18	4.76	53	8.53	88	11.22	123	13.44	158	15.39
19	4.90	54	8.62	89	11.29	124	13.50	159	15.44
20	5.04	55	8.71	90	11.36	125	13.56	160	15.50
21	5.18	56	8.79	91	11.43	126	13.62	161	15.55
22	5.31	57	8.88	92	11.49	127	13.68	162	15.60
23	5.44	58	8.96	93	11.56	128	13.74	163	15.65
24	5.56	59	9.04	94	11.63	129	13.80	164	15.70
25	5.69	60	9.12	95	11.69	130	13.85	165	15.76
26	5.81	61	9.21	96	11.76	131	13.91	166	15.81
27	5.93	62	9.29	97	11.83	132	13.97	167	15.86
28	6.05	63	9.37	98	11.89	133	14.02	168	15.91
29	6.16	64	9.45	99	11.96	134	14.08	169	15.96
30	6.28	65	9.53	100	12.02	135	14.14	170	16.01
31	6.39	66	9.61	101	12.09	136	14.19	171	16.06
32	6.50	67	9.69	102	12.15	137	14.25	172	16.11
33	6.61	68	9.76	103	12.22	138	14.31	173	16.16
34	6.71	69	9.84	104	12.28	139	14.36	174	16.21
35	6.82	70	9.92	105	12.34	140	14.42	175	16.26

4.12.2 Consider a system supplied by pumps at one location and having no elevated storage.

4.12.3 If the pressure at the pump station drops during the test, it is an indication that the distribution system is capable of delivering more than the pumps can deliver at their normal operating pressure.

4.12.4 It is necessary to use a value for the drop in pressure for the test that is equal to the actual drop obtained in the field during the test, minus the drop in discharge pressure at the pumping station.

```
                    Hydrant Flow Test Report

    Location _____ Date _____
    Test made by _____ Time _____
    Representative of _____
    Witness _____
    State purpose of test _____
    _____
    Consumption rate during test _____
    If pumps affect test, indicate pumps operating _____
                          A₁      A₂      A₃      A₄
    Flow hydrants: _____
        Size nozzle       _____
        Pitot reading     _____
        Discharge coefficient _____ Total gpm
        gpm               _____

    Static B _____ psi    Residual B _____ psi
    Projected results  @20 psi Residual ___ gpm;  or @ ___ psi Residual ___ gpm
    Remarks: _____
    _____
    _____
    _____
    Location map: Show line sizes and distance to next cross-connected line. Show valves and
        hydrant branch size. Indicate north. Show flowing hydrants – Label A₁, A₂, A₃, A₄. Show
        location of static and residual – Label B.
    Indicate B  Hydrant _____ Sprinkler _____ Other (identify) _____
```

FIGURE 4.11.2 *Sample Report of a Hydrant Flow Test.*

4.12.5 If sufficient pumping capacity is available at the station and the discharge pressure could be maintained by operating additional pumps, the water system as a whole could deliver the computed quantity.

4.12.6 If, however, additional pumping units are not available, the distribution system would be capable of delivering the computed quantity, but the water system as a whole would be limited by the pumping capacity.

4.12.7 The portion of the pressure drop for which a correction can be made for tests on systems with storage is generally estimated upon the basis of a study of all the tests made and the pressure drops observed on the recording gauge at the station for each.

4.12.8 The corrections may vary from very substantial portions of the observed pressure drops for tests near the pumping station, to zero for tests remote from the station.

CHAPTER 5 MARKING OF HYDRANTS

5.1 Classification of Hydrants.

Hydrants should be classified in accordance with their rated capacities [at 20 psi (1.4 bar) residual pressure or other designated value] as follows:

(1) Class AA—Rated capacity of 1500 gpm (5680 L/min) or greater
(2) Class A—Rated capacity of 1000–1499 gpm (3785–5675 L/min)
(3) Class B—Rated capacity of 500–999 gpm (1900–3780 L/min)
(4) Class C—Rated capacity of less than 500 gpm (1900 L/min)

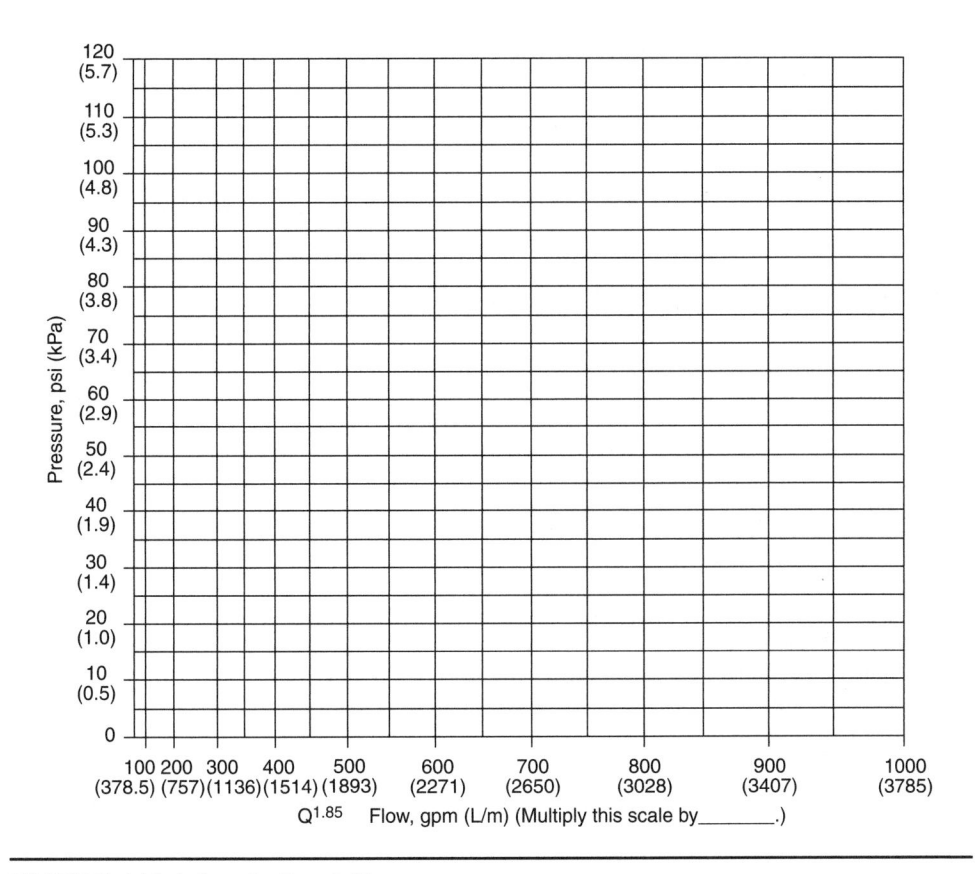

FIGURE 4.11.4 Sample Graph Sheet.

In accordance with Section 5.1, the hydrant in Exhibit III.29 should be rated as Class AA since its capacity is greater than 1500 gpm (5678 L/min).

5.2 Marking of Hydrants.

5.2.1 Public Hydrants.

5.2.1.1 All barrels are to be chrome yellow except in cases where another color has already been adopted.

5.2.1.2 The tops and nozzle caps should be painted with the following capacity-indicating color scheme to provide simplicity and consistency with colors used in signal work for safety, danger, and intermediate condition:

(1) Class AA—Light blue
(2) Class A—Green
(3) Class B—Orange
(4) Class C—Red

In accordance with 5.2.1.2, the bonnet and nozzle caps of the hydrant in Exhibit III.29 should be painted light blue to indicate that the hydrant is rated as a Class AA hydrant.

5.2.1.3 For rapid identification at night, it is recommended that the capacity colors be of a reflective-type paint.

5.2.1.4 Hydrants rated at less than 20 psi (1.4 bar) should have the rated pressure stenciled in black on the hydrant top.

5.2.1.5 In addition to the painted top and nozzle caps, it can be advantageous to stencil the rated capacity of high-volume hydrants on the top.

5.2.1.6 The classification and marking of hydrants provided for in this chapter anticipate determination based on individual flow test.

5.2.1.7 Where a group of hydrants can be used at the time of a fire, some special marking designating group-flow capacity may be desirable.

5.2.1.8 Marking on private hydrants within private enclosures is to be done at the owner's discretion.

5.2.1.9 When private hydrants are located on public streets, they should be painted red or another color to distinguish them from public hydrants.

5.2.2 Permanently Inoperative Hydrants. Fire hydrants that are permanently inoperative or unusable should be removed.

5.2.3 Temporarily Inoperative Hydrants. Fire hydrants that are temporarily inoperative or unusable should be wrapped or otherwise provided with temporary indication of their condition.

5.2.4 Flush Hydrants. Location markers for flush hydrants should carry the same background color as stated above for class indication, with such other data stenciled thereon as deemed necessary.

5.2.5 Private Hydrants.

5.2.5.1 Marking on private hydrants within private enclosures is to be at the owner's discretion.

5.2.5.2 When private hydrants are located on public streets, they should be painted red or some other color to distinguish them from public hydrants.

SUMMARY

Proper fire flow testing of hydrants provides the needed information for the design of any water-based fire protection system including fire pumps. By following the recommended practices of NFPA 291, the data obtained can be used for fire protection system design in addition to the proper color coding of hydrants.

REFERENCES CITED IN COMMENTARY

National Fire Protection Association, 1 Batterymarch Park, Quincy, MA 02169-7471.

NFPA 13, *Standard for the Installation of Sprinkler Systems,* 2010 edition.
NFPA 14, *Standard for the Installation of Standpipe and Hose Systems,* 2007 edition.
NFPA 20, *Standard for the Installation of Stationary Pumps for Fire Protection,* 2010 edition.
NFPA 25, *Standard for the Inspection, Testing, and Maintenance of Water-Based Fire Protection Systems,* 2008 edition.

PART IV

Supplements

In addition to the commentary presented in Parts I through III, the *Handbook for Stationary Fire Pumps* includes four supplements.

The supplements in Part IV explore the background of selected topics related to fire pumps in more detail than the commentary in Parts I through III. These supplements are not part of the standards; they are included as additional information for handbook users.

The four supplements in Part IV are as follows:

1. Fire Pump Installation from Design to Acceptance
2. Commissioning Forms for Fire Pumps
3. Article 695 from *NFPA 70*®, *National Electrical Code*®, 2008 Edition
4. Technical/Substantive Changes from the 2007 Edition to 2010 Edition of NFPA 20

SUPPLEMENT 1

Fire Pump Installation from Design to Acceptance

John Jensen and David R. Hague

Editor's Note: This supplement is a guide for the design of a fire pump installation. Material has been gathered from this handbook, the manufacturers' data sheets, and the author's experience to provide the design guide. The expectation of the author is that the supplement will be of value to those who are just getting started in the design of such systems and may be of assistance to those who are reviewing new and existing installations.

INTRODUCTION

In this supplement, the method used to explain the information is intended to briefly state what is being covered and then outline in color the issues that need to be considered. Most of the needed information is found in Parts I through III of this handbook. Items that are not covered in adequate detail elsewhere are presented in the outline segments or text of this supplement.

An important part of any fire protection installation is the follow-up inspection, testing, and maintenance of the system, which is not addressed in this supplement. Detailed information on that topic can be found in *The NFPA 25 Handbook*.

Fire Protection System Requirements

The fire protection system water demand requirements are most often obtained from NFPA standards or other standards listed in I(A) through I(C) of the outline. Part I of this handbook gives a detailed description of those requirements. The outline lists the most common fire protection systems in I(A) through I(C). Without exception, each fire protection system using water as the fire suppression agent requires that the water supply have adequate pressure and volume to control and/or extinguish the design fire. The need for a fire pump is only determined after an analysis of the existing water supply and a determination of whether it is adequate or if no water supply is available. The decision process can then be followed to design and install a fire pump that meets the water supply needs of the fire protection system.

If the building or structure is equipped with fire protection systems having different requirements, the water supply is designed based on the most demanding system. The most demanding system may be the fire sprinkler, standpipe, or outside fire hose system. The outline shows considerations for each type of system in I(A) through I(C). If multiple buildings with multiple hazards need to be protected, the water supply is designed for the most severe hazard in the building that presents the highest demand unless the authority having jurisdiction requires additional considerations. Therefore, when planning a water supply using a fire pump, the future addition of more hazardous operations must be considered in the sizing of the water supply and the fire pump.

I. Evaluation
   A. Sprinkler system (See NFPA 13, NFPA 15, NFPA 2001, insurance company interpretive guidelines, or NFPA's *Fire Protection Handbook*)
      1. Wet pipe
      2. Dry pipe
      3. Deluge or water spray
      4. Water mist

5. Preaction
   6. Authority having jurisdiction—see insurance company interpretive guidelines
   7. Select the most demanding waterflow requirement in a building as the minimum required quantity and estimate the required pressure by adding 15 psi to the pressure lost due to the elevation from the bottom to the highest point in the building. Then add 15 to 25 percent to the minimum flow required because of design inefficiencies in the actual piping configurations.
B. Inside fire hose demand
   1. Independent standpipe system (see NFPA 14, NFPA 1, *NFPA 5000®*, or the applicable building code) with separate piping
      a. Class I
      b. Class II
      c. Class III
   2. Supplied from sprinkler system (see NFPA 13)
   3. Authority having jurisdiction—see insurance company interpretive guidelines
C. Outside fire water requirements (see NFPA 24, NFPA's *Fire Protection Handbook, NFPA 5000,* or other building code)
   1. Private hydrant system
   2. Public fire hydrants
   3. Authority having jurisdiction

EXISTING AVAILABLE WATER SUPPLIES

One of the most common reasons a fire pump is used is to boost the pressure from an existing municipal water system. In this application, the municipal system must be reliable and must have an adequate volume of water available, but the pressure is inadequate to provide the pressure demand at the top or hydraulically remote part of the fire protection system. The municipal system is evaluated using the criteria listed in the outline in I(D)(1) through I(D)(5). If the system is not reliable or has insufficient volume, then an independent source of water storage must also be designed and installed with the fire pump.

   D. City water main and supply
      1. Reliability—Consider the following features:
         a. Seasonal and hourly differences in static pressure and volume of water available
         b. Gridding versus dead-end
         c. Water source
            i. Wells
            ii. Rivers
            iii. Lakes
            iv. Reserve capacity
         d. Storage capacity reserved for fire
         e. Maintenance
         f. Hydrant spacing
         g. Pipe sizing and condition
         h. Sectional control valves
         i. Insurance rating of the water supply
      2. Inspection, testing, and maintenance adequacy (see insurance or town grading report)
      3. Testing
         a. Conduct a water flow test in accordance with NFPA 291
         b. Use data collected within the past 12 months during the period of maximum usage

Note: Computer simulations should not be used because of the unknown condition of the piping, system control valves, water storage, and source of the water.

      4. Backflow prevention requirements
         a. Install for all systems connected to a potable water system
         b. Do not draw the system water pressure below 20 psi when the fire pump is operating at 150 percent of the rated waterflow
      5. Piping
         a. Limit the velocity of the water in the suction piping to the fire pump to not more than 15 fps (4.6 mps) while flowing 150 percent of the rated water demand of the fire pump
         b. Flush all suction piping prior to connection to the fire pump at the flow rates indicated in NFPA 20 for the size of pipe installed

NEW OR UPGRADED WATER SUPPLY

A new or upgraded water supply can be designed using many different systems and combinations of systems. The most common systems are listed in the outline in I(E) through I(G)(4), and the referenced NFPA standards contain the information on the design parameters and considerations.

   E. Gravity (see NFPA 22)
      1. Capacity
      2. Pressure
      3. Heating adequacy in cold climates
      4. Water fill line
   F. Grade level (see NFPA 22)
      1. Capacity

2. Pressure
3. Heating adequacy in cold climates
4. Water fill line
G. Subgrade level (see NFPA 20 and NFPA 22)
1. Capacity
2. Pressure
3. Depth of bury (freeze and load protection)
4. Water fill line

Gravity tanks on hills or towers are probably the oldest supplies that are still being used today. Old high-rise buildings, remote manufacturing plants, and industrial plants requiring a secondary supply to improve their water supply reliability use this type of system. A horizontal split-case, vertical in-line, or end suction pump is the most common type of pump used to boost the system pressure.

Grade level water storage tanks constructed of concrete or welded or bolted steel are still used extensively to provide a positive suction pressure to fire pumps. The tanks are also used to provide a constant pressure when municipal systems have a widely fluctuating pressure because of changing demand or limited sources of water. They may be used as a break tank to provide an even pressure at the pump discharge when the municipal water supply pressure is low or fluctuates widely. A horizontal split-case, vertical in-line, end suction, or positive displacement pump is the most common type of pump used to boost the pressure for these water tanks.

Subgrade level tanks are constructed of concrete or reinforced fiberglass. The tanks are used where space or heating of the water is a problem. They may be used as a break tank to provide an even pressure at the pump discharge when the municipal water supply pressure is low or fluctuates widely. Most of the installations are used when a full supply of water is needed to be stored and space and heating are the main issues. A vertical lineshaft turbine pump is normally used for this application.

WATER WELLS

Water wells are also used as a source of water for fire pumps. Chapter 7 of NFPA 20, *Standard for the Installation of Stationary Pumps for Fire Protection,* on vertical lineshaft turbine fire pumps, includes requirements pertaining to water wells (see Part II of this handbook).

The process of designing a fire well requires extensive preliminary work to be completed prior to the purchase and installation of the fire pump. The outline provides more detail in I(G)(5) through I(G)(6) on what items should be considered. Other good sources of information are the *Ground Water Handbook* and *Manual of Well Construction Practices,* published by the National Ground Water Association (NGWA), and individual state regulations.

5. Water source
   a. Depth to water (normally limit to less than 200 ft)
   b. History of water levels in area (10 years minimum)
   c. Other wells in vicinity
      i. Municipal
      ii. Industrial
      iii. Agricultural
      iv. Injection
   d. Seasonal effects
      i. Irrigation
      ii. Drought
      iii. Floods
6. Installation
   a. The proper submergence of the pump bowls must allow for operation of the fire pump under all circumstances. This requires that the second from the bottom pump bowl be installed so that it is 10 ft below the pumping water level when the pump is pumping at 150 percent of the rated waterflow.
   b. The well should be vertically straight for the installation of vertical lineshaft turbine pumps. The well straightness should be within 1 in. per 100 ft in depth and not bend back and forth, which causes jamming of the pump. The well should be gauged prior to installation of the pump by lowering a dummy assembly on a cable to check the potential binding.

II. Construction
   A. Unconsolidated formations (sand and gravel)
      1. Location of casing
      2. Minimum size of casing should be as follows:

Gallons per minute	Casing Wall Thickness	Outside: First 50 ft of Depth	50 ft to 200 ft
250	0.250	20	14
500	0.250	20	14
750	0.250	20	14
1000	0.312	22	16
1500	0.312	22	16
2000	0.312	24	18
2500	0.312	26	20
3000	0.312	30	24
3500	0.312	30	24
4000	0.375	32	26
5000	0.375	32	26

B. Material used for casing
 1. Well casings must be steel only
    a. ANSI/ASTM A 53, *Standard Specification for Pipe, Steel, Black and Hot-Dipped, Zinc-Coated, Welded and Seamless*
    b. Consolidated formations (rock)
       i. Location of casing
       ii. Minimum size of casing should be as follows:

Gallons per minute	Casing Wall Thickness	Outside: First 50 ft of Depth	50 ft to 200 ft
250	0.250	20	14
500	0.250	20	14
750	0.250	20	14
1000	0.312	22	16
1500	0.312	22	16
2000	0.312	24	18
2500	0.312	26	20
3000	0.312	30	24
3500	0.312	30	24
4000	0.375	32	26
5000	0.375	32	26

Currently, no nationally recognized consensus standard is written on the construction of wells. The specific requirements for water wells are regulated by individual state rules and laws and the U.S. Environmental Protection Agency. California and Utah, for example, have clear and easy-to-follow regulations. Most other states follow the national guidelines published by the NGWA.

The pump manufacturer's information provides design information, such as the size of the well and how far the pump must be submerged. The construction of a well needs to be completed by an experienced and trained contractor that specializes in the drilling of wells for water. Wells used for fire protection purposes should be fully cased when possible. The exception is when the well is drilled in solid rock. The well can be of any depth, but the preferable limit is 200 ft or less for pump lubrication purposes. Generally, the first 5 ft of depth are of a larger diameter because the well needs to be protected from surface pollution and to support the pumping equipment that will be on the surface. Currently, no listed submersible pumps for fire protection are available. If the water available from the well is inadequate, the well and submersible pump can be used to fill a storage tank above or below grade level.

WATER WELL CONSTRUCTION

Prior to testing, the quantity of water can be increased by breaking up the rock, drilling to multiple water strata, and other means. The outline identifies the major steps and considerations needed to develop a well in II(C).

C. Well development
 1. Reputation of well driller
    a. Certification (see National Ground Water Association)
       i. Certified well driller
       ii. Certified pump installer
       iii. Master ground water contractor
       iv. State licensed well contractor
    b. Experience
 2. Well designer
    a. Certification (see National Ground Water Association)
       i. Ground water professional
       ii. Professional engineer or geologist
 3. Well development
    a. The water capacity of the well may be increased by using one of the accepted methods, such as over pumping, surging with compressed air, jetting with water, hydraulic fracturing, explosives, or chemicals acceptable to the well designer.
 4. Waterflow test
    a. The waterflow test must be conducted at the time of the year when the water usage in the vicinity of the new well is at the maximum level to make sure that water levels are adequate when the worst conditions occur for the fire pump.
 5. Results of 24-hour pumping at 150 percent of water demand
    a. Actual drawdown of water level
    b. Recommended pump bowl location
    c. Sand pumped out during testing
    d. Capacity of pump used for development of well
    e. How straight is well (vertically) (1 in. deviation in 100 ft of depth is allowed)
    f. Availability of well geology log
 6. Well drillers log must be available prior to purchasing and setting of the pump

OTHER WATER SOURCES

Other sources of fire protection water are available. The information on how to utilize these sources is not located in a single reference or design standard. NFPA 1142, *Standard on Water Supplies for Suburban and Rural Fire Fighting,* contains useful information on the utilization of these water supplies as a suction source for fire department apparatus. The situation discussed in this section is similar, except it pertains to designing a stationary fire pump installation. These sources are normally nonpotable and require a means to remove trash and filter out sand and soil prior to the water being drawn into the fire pump.

The best type of pump for these purposes is a vertical lineshaft turbine pump installed in a suction crib. See II(D) through II(G) in the outline for design considerations for alternative sources of water. The design of the suction crib should be in accordance with the requirements of the Hydraulic Institute and NFPA 20.

D. River or creek
   1. Water level
      a. Seasonable fluctuations
      b. Freezing
      c. Measure flow per minute past pump suction pipe or crib
   2. Suction crib
      a. See Figure A.7.2.2.2 in NFPA 20 for a visual presentation of the terms associated with pumps in suction cribs
E. Lake or pond
   1. Water level
      a. Seasonal fluctuations
      b. Minimum quantity of water available
      c. Freezing
   2. Suction crib
      a. See Figure A.7.2.2.2 in NFPA 20 for a visual presentation of the terms associated with pumps in suction cribs
F. Embankment supported tank
G. Redundant reliable water supply and fire pump for fire protection in seismic areas or as required by the authority having jurisdiction

FIRE PUMP SELECTION

What kind of fire pump should be used? In order to make this determination, the designer must consider the design of the pump, foundation, capacity, pressure, and reliability issues. All fire pumps are required to provide performance that has the following characteristics: a maximum of 140 percent and a minimum of 100 percent of rated pressure when the flow is zero, 100 percent of the rated flow when the pressure is at 100 percent of the rated pressure, and 150 percent of the rated flow when the discharge pressure is at 65 percent of the rated pressure. For example, a pump having a rating of 1000 gpm at 150 psi (3785 L/min at 10.3 bar) would have a characteristic curve having a maximum shut off or churn pressure of 210 psi to 150 psi (14.5 bar to 10.3 bar) at zero flow, at 100 percent of the flow [1000 gpm at 150 psi (3785 L/min at 10.3 bar)], and at 150 percent of flow [1500 gpm (5678 L/min)] a pressure of at least 97.5 psi (6.7 bar), as indicated in Exhibit S1.1.

According to the Hydraulic Institute standards, the foundation of a pump should be designed for 5 times the weight of the pump package. For example, if a 1000 gpm at 150 psi (3785 L/min at 10.3 bar) pump was driven by a diesel engine having about 160 hp (119.36 kW) at 3000 rpm, the foundation would be required to support 5 times the weight of the pump, driver, and accessories [520 lb + 2290 lb = 2810 lb × 5 = 14,050 lb (235.8 kg + 1038.7 kg = 1274.5 kg × 5 = 6372.5 kg)] for the pump and driver or 94 ft^3 (2.7 m^3) of typical concrete at 150 lb/ft^3 (2403 kg/m^3). If the platform was 84 in. × 24 in. [7.17 ft × 2 ft (2137 mm × 610 mm)], the depth of the foundation would be 6.5 ft (1981 mm). Most pump manufacturers recommend 2.5 to 5 times the weight. NFPA 20 does not address this issue except to be in general compliance with the manufacturers and Hydraulic Institute standards. When the pump is installed inside of a building, the pump should also be isolated from the building so that it does not transmit the operational noise to the structure.

The horizontal split-case pump is the traditional favorite and has the highest reputation for reliability. Most pumps operated at 1750 rpm in the past, and many still operate at that speed. However, now diesel engines and electric motors are available at much higher speeds, extending the horsepower curve out and making a reliable system use a smaller engine, and thus reducing the cost of the installation. At 1750 rpm the pump is quieter, has less vibration, and will easily last for 20 to 40 years. See III(A) through III(E) in the outline for design attributes for horizontal shaft, vertical shaft, end suction, and vertical in-line fire pumps.

III. Pump Selection
   A. Vertical in-line centrifugal
      1. Design of pump and driver foundation 2.5 times weight of pumping unit
      2. Limited to electric motor drive. The electric motor is supported by the pump.
      3. Capacities of 35 gpm to 1500 gpm
      4. Pressures of 40 psi to 165 psi
   B. Positive displacement
      1. Design of pump and driver foundation 2.5 to 5 times weight of pumping unit
      2. Use the minimum piping requirements as described in Chapter 8 of NFPA 20 and as shown in Figure A.8.4.2.
   C. Horizontal split-case
      1. Design of pump foundation
         a. 2.5 to 5 times weight of pumping unit

EXHIBIT S1.1 Fire Pump Performance Curve for a Pump Rated for 1000 gpm at 150 psi (3785 L/min at 10.3 bar).

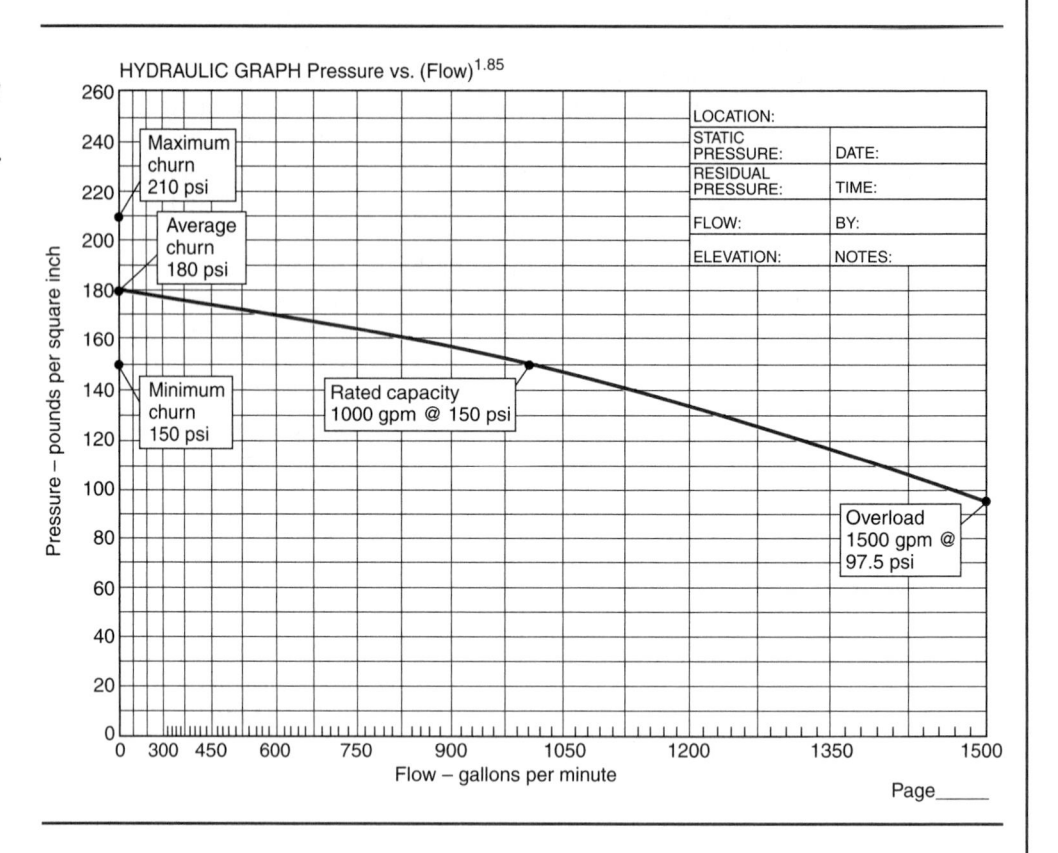

HYDRAULIC GRAPH Pressure vs. (Flow)$^{1.85}$

- Maximum churn 210 psi
- Average churn 180 psi
- Minimum churn 150 psi
- Rated capacity 1000 gpm @ 150 psi
- Overload 1500 gpm @ 97.5 psi

LOCATION:
STATIC PRESSURE:
DATE:
RESIDUAL PRESSURE:
TIME:
FLOW:
BY:
ELEVATION:
NOTES:

Page_____

      b. Isolate from the building, if installed inside a major building
2. Capacities of 25 gpm to 5000 gpm
3. Pressures of 40 psi to 640 psi
4. Single and multiple stage
D. Vertical split-case
  1. Design of pump and driver foundation
     a. 2.5 to 5 times weight of pumping unit
     b. Isolate from the building, if installed inside a major building
  2. Electric motor is supported separately from the pump
  3. Capacities of 25 gpm to 5000 gpm
  4. Pressures of 40 psi to 640 psi
  5. Single and multiple stage discharge
E. End suction
  1. Design of pump foundation 2.5 to 5 times weight of pumping unit
  2. Horizontal applications only
  3. Capacities of 50 gpm to 750 gpm
  4. Pressures of 40 psi to 160 psi

The vertical split-case has traditionally been used where space is a problem and the electrical supply is very reliable. Diesel engines and steam turbines are not used for this application.

The end suction pump has traditionally been used for commercial pump applications where limited quantities of water are needed. Recently, the size of these pumps has been increasing with new engineering. A typical application might be a small industrial plant or hospital where the quantity of water is limited.

The vertical in-line pump is relatively new in the fire protection market and specifically targets the smaller sprinkler-protected buildings such as light hazard, low- to mid-rise hotels, offices, and health care applications where there is a reliable municipal water supply and a need for limited standpipe applications. The pumps are usually used with electric motors.

Vertical lineshaft turbine pumps have been used for many fire protection applications. Where space is a problem the pumps can take suction from belowgrade water storage tanks and deep wells. Another frequent application is to take suction from a river, a pond, or a municipal water main. In cold climates they have an advantage because the water storage tank can be installed underground where the water doesn't need to be heated. In the case of a well, the earth serves as the water storage facility. They can be driven by diesel engines or electric motors and can be considered as reliable as any other kind of fire pump. The installations are usually more complicated because more equipment is involved and the coordination of that equipment requires an in-depth knowledge of system integration to maintain the reliability of the system. A wide variety of pumping capacities and pressures is available to the user.

Each vertical pump installation is required to include a method to measure the water level in the suction water source. The most common way for wells, underground tanks, and suction cribs is to use a ¼ in. (6.4 mm) OD copper air line extended into the water source to a point about 2 ft (610 mm) above the pump suction inlet and extended through the discharge head to a water level gauge. See Figure A.7.3.5.3 of NFPA 20 for a graphical presentation of this method and how to measure the water level (see Part II of this handbook). Other methods are available, but this method is the easiest to maintain and use during the required testing and inspection. See III(F)(1) through III(F)(6) in the outline for design considerations for vertical turbine fire pumps.

F. Vertical turbine
 1. Design of pump foundation
  a. 5 times weight of pumping unit
  b. Must include the weight of the lineshaft and pumping unit
  c. Consider downthrust and upthrust of pump during operation
 3. Capacities of 250 gpm to 5000 gpm
 4. Pressures of 50 psi to 350 psi
 5. Vertical electric motor drive
 6. Right angle drive for fire pumps with a diesel engine drive
  a. Listed
  b. Should be 1:1 ratio
  c. Provide adequate horsepower rating
  d. Listed drive lineshaft to diesel engine
  e. Listed couplings
 7. Suction water supply
  a. Underground tanks (break or full supply)
   i. See Figure A.7.2.2.1 in NFPA 20 for a visual presentation of the terms associated with pumps in underground tanks.
  b. As a booster pump installed in a "can": taking suction from a municipal water main. See Hydraulic Institute standards for "can" details.
  c. Wells
   i. See Figure A.7.2.2.1 in NFPA 20 for a visual presentation of the terms associated with pumps in wells.
   ii. What is the normal water level in the well?
   iii. What is the drawdown water level in the well?
   iv. If the difference is large (greater than 20 ft) the shutoff pressure may be in excess of the fitting rating planned for the fire protection systems.
   v. Is the well casing large enough for the vertical turbine pump?
   vi. What method will be used to lubricate the pump shaft (oil or water)? If oil, is the oil biodegradable?
   vii. Is the well casing vertically straight enough to allow installation of the lineshaft and pump?
   viii. Are provisions included in the fire pump enclosure to remove the lineshaft and pump for maintenance?
  d. Wet pits
   i. See Figure A.7.2.2.2 in NFPA 20 for a visual presentation of the terms associated with pumps in wet pits.
   ii. This method is normally used when taking suction from rivers, lakes, and ponds.
   iii. Construction of the wet pit should follow the design guidelines developed by the Hydraulic Institute.

Positive displacement pumps are used for foam-water sprinkler systems and for water mist water spray sprinkler systems. They should not be used for pressure maintenance pumps. See the applicable standards for the appropriate application such as NFPA 11, *Standard for Low-, Medium-, and High-Expansion Foam*; NFPA 16, *Standard for the Installation of Foam-Water Sprinkler and Foam-Water Spray Systems*; and NFPA 750, *Standard on Water Mist Fire Protection Systems*. The engineering associated with these pumps should only be undertaken by experienced and qualified engineers.

FIRE PUMP DRIVERS

The selection of the appropriate fire pump driver is one of the most important decisions that is made as a part of the design package. Electric motors have long been the pump driver of choice because of their reliability and long life. Each driver has a different set of considerations however.

Currently, the diesel engine is used more often, mainly because of the high cost of connecting to a reliable electrical power system and a decrease in reliability of electric motors. The diesel engine has been modernized by electronic controls and fuel management brought about by environmental considerations. If the diesel engine is properly maintained and tested, this power source is one of the most reliable sources of power to drive a fire pump. A diesel engine is available that automatically controls the pump discharge pressure for situations in which the suction pressure may vary a great deal or the discharge pressure needs to be controlled because of discharge pressures that may exceed the rating of the downstream equipment. For design considerations regarding diesel engine applications, see III(G)(1) through III(G)(9) in the outline.

G. Diesel engine (full speed or variable speed)–listed (compression ignition type only)
   1. The horsepower capability of the engine shall be not less than 10 percent greater than the listed horsepower on the engine nameplate.
   2. Design of foundations
      a. 2.5 to 5 times the weight of the pump/driver assembly
      b. Seismic
      c. Sound isolation
   3. Cooling requirements
      a. Water
         i. Pressure requirements of engine (maximum and minimum)
         ii. Radiator cooling
            a) Required to keep engine operating temperature at the combustion air inlet to 120°F.
         iii. Heat exchanger water supply
            a) Potable
               1. Backflow requirements [see Figure A.11.2.8.5.3.7(B) of NFPA 20]
            b) Raw water (nonpotable)
               1. Strainers (see Figure A.11.2.8.5 of NFPA 20)
            c) Water disposal
               1. Building drain
               2. French drain
               3. On to ground
               4. Return to tank
   4. Engine exhaust requirements
      a. Maximum back pressure
      b. Size and length
      c. Sound levels
   5. Derating parameters
      a. Altitude (above 300 ft). See Figure A.11.2.2.4 of NFPA 20 (3 percent per 1000 ft elevation above 300 ft elevation).
      b. Ambient temperature (77°F). See Figure A.11.2.2.5 of NFPA 20 (1 percent per 10°F above 77°F ambient temperature).
      c. Right angle drive (vertical turbine only). Increase horsepower for power loss in the gear drive.
   6. After derating, the engine shall have a 4-hour minimum horsepower rating required to drive the pump at its rated speed under any conditions of pump load.
   7. Engines may be electrical, hydraulic, or air started.
   8. Torsion analysis
      a. Horizontal pump
      b. Vertical turbine
   9. Emission standards
      a. EPA, 40 CFR 60, Part llll (NSPS Regulation)
      b. California Air Resources Board, Airborne Toxic Control Measures (CARB ATCM)

Steam turbine drives are not used often today unless a reliable supply of steam is available at all times. This supply may occur in private and public power plants or in manufacturing plants that use steam to power most of their manufacturing equipment. No listed steam turbine pumps are currently available in the marketplace, and their acceptance is a matter for the authority having jurisdiction to determine. Steam turbines of adequate power are an acceptable driver for a fire pump. For specific requirements, see Chapter 13 of NFPA 20 in Part II of this handbook.

At the present time gasoline, natural gas, or propane engines are not listed or approved for use as fire pump drivers. In many situations in which the reliability of an electric motor drive needs to be improved, an on-site backup electric generator is connected to the electric motor–driven fire pump by means of a two-source pump controller and transfer switch arrangement. Typical situations in which this arrangement is needed are high-rise buildings and health care facilities. In this case, caution should be taken to make sure the power supply is improved and does not decrease the overall system reliability.

Electric motors are listed in horsepower rating up to 500 hp (373 kW) in medium voltage. The listing organizations use NEMA MG-1, *Motors and Generators,* as their guideline. The circuit and protective device requirements are specified in NFPA 20 (sizing and what is allowed) and in Article 695 of *NFPA 70®, National Electrical Code®* (wiring, arrangements, sizing, short circuit ratings, etc.).

Both standards must be used when designing the power supply circuits (see Part II and Supplement 3 of this handbook).

For fire pump installations, all of the equipment and wiring is considered sacrificial when the fire pump is required to operate because of the importance of maintaining water for the fire protection system. Many engineers do not understand this principle and try to design the system to shut down and save the equipment. This design principle should not be followed because loss of life and property is at stake if the fire gets out of control and there is no water supply. As stated previously, the use of alternate power supplies to improve the reliability should be carefully considered during design to make sure that the reliability is improved, not decreased, in the installed system. For design considerations regarding electric motors for fire pump service, see III(H)(1) through III(H)(3) in the outline.

    H. Electric motors
        1. Power supply
           a. Normal
              i. Reliability
              ii. Voltage
           b. Alternate power
              i. Reliability
              ii. Voltage
           c. Generator power
              i. Capacity
              ii. Load shedding
              iii. Duration of fuel supply
              iv. Inspection, testing, and maintenance adequacy
        2. Voltage drop
           a. 15 percent maximum at controller line terminals
           b. 5 percent when motor operating at 115 percent full-load current rating
        3. Type
           a. General purpose, open or dripproof
           b. Totally enclosed, fan cooled (TEFC)
           c. Totally enclosed, nonventilated (TENV)
           Note: Derating is required for elevations above 3300 ft in accordance with NEMA MG-1.

FIRE PUMP CONTROLLERS

The controllers for fire pumps are designed with reliability and independence in mind. For instance, the diesel engine controller only takes battery charging current from the commercial power source. All operations are based on using redundant sets of batteries that are charged by the engine during operation. The controller is designed to shut down the engine only when overspeed conditions occur, otherwise the system will run to destruction to maintain the water supply for the fire protection system.

The electrical controllers are available in two types. The first is a limited-service controller designed for motors having less than 30 hp (22.38 kW) This type is a simplified controller originally designed for small booster pump applications where the municipal water supply was distributed at very low pressures in large water mains. Historically, this controller has been very reliable.

The other type of controller is a full-service controller available in all horsepower ratings. This controller is designed specifically for electric motor–driven fire pumps and is available for use with a single source or two power sources, with or without integral transfer switches. The full-service controller is available for all voltage applications. A variable speed controller is available for those situations requiring the speed to vary because of the need to control the discharge pressures produced by the pump. The outline lists the types of controllers in IV(A) through IV(B) that are available for electric and diesel fire pump service.

    IV. Fire Pump Controllers
        A. Diesel engine
           1. Electrical
           2. Pneumatic
           3. Hydraulic
        B. Electrical
           1. Full service
           2. Limited (30 hp or less)
           3. Variable speed
           4. Normal power
           5. Alternative power

PRESSURE MAINTENANCE (JOCKEY) PUMPS

Most fire pump installations include a pressure maintenance pump. This pump is installed to keep the fire protection system piping pressurized at a constant pressure just below the pump discharge pressure. The purpose of this pump is to prevent the fire pump from starting due to small leaks or air pockets in the system piping. If the fire pump installation is part of a potable water system, the pressure maintained on the system may be adequate so that a separate pressure maintenance pump is not required. When a pressure maintenance pump is installed, the pressure on the system should be kept high enough so that excessive water hammer is not applied to the system when the fire pumps starts. Preventing water hammer is particularly important when dry type sprinkler systems are connected to the water distribution system.

FIRE PUMP ENCLOSURES

The enclosure that contains the fire pump installation is an important part of the installation. The design must be in accordance with the local building code. In NFPA 20 the references to what is required to provide a satisfactory enclosure are very few. In areas where the weather is mild and the chance of freezing or adverse weather conditions is absent, the installation of the fire pump and its equipment outside is usually acceptable in some circumstances. Unfortunately, these conditions are very few and far between. The need to maintain the reliability of the installation and equipment usually requires that the fire pump be installed inside an enclosure. The following discussion looks at what is important for protection of the fire pump.

The enclosure can be part of a prepackaged system where the fire pump, driver accessories, enclosure, and building are assembled in a manufacturing facility, tested, and then shipped to the construction site for connection to the water supply, electrical power, fire protection system, and drainage system. This method is of particular value for fast-track projects and in areas where the construction season is limited. At this time this type of integration is not listed, and it is usually approved by the authority having jurisdiction for the particular project.

The outline in 4(a) through 4(p)(ii) can be used to make sure that the most important design items have been included in the installation.

V. Enclosure
  A. Is the enclosure constructed of noncombustible materials?
  B. Has all other equipment been excluded from the fire pump enclosure (e.g., mechanical equipment, chemical storage, maintenance supplies, electrical service equipment, and other potable water pumps)?
  C. Fire resistance requirements
    1. Is the enclosure "inside of" or "adjacent to" a high-rise building? If it is, the requirement is for a 2-hour rated enclosure.
    2. Other locations within a building require 1-hour fire rated construction.
    3. Enclosures adjacent to property lines usually require a 1-hour fire resistance rating on the sides exposed by the adjoining the property(ies).
    4. When the fire pump installation is exposed by a hazardous operation, the exposed sides require a fire resistance rating or protection by a fire protection system such as fire sprinklers.
  D. How are high noise levels mitigated?
    1. Sound deadening material on inside walls
    2. Foundation isolation
    3. Residential, critical, or hospital grade muffler for diesel engines
    4. Locate pump room away from protected building
  E. Has adequate space been provided for inspection, testing, and maintenance of the fire pump equipment?
  F. Heating and cooling system
    1. What is the minimum expected outside temperature? If less that 40°F, heat is required.
    2. What is the maximum outside and inside temperature expected? If above driver (diesel engine or electric motor) maximum allowable value (nominally 120°F), then supplemental cooling may be required.
  G. Combustion air
    1. Required for diesel engine pump drivers, see engine manufacturer's data for values. For example, a 210 hp diesel engine could require 416 [ft^3/min] at 1760 rpm and 580 [ft^3/min] at 2600 rpm. This requirement needs to be built into the enclosure.
  H. Electric power
    1. Separate service connection for the building from the power utility or campus distribution system if there is an electric motor driver.
    2. Subpanel from facility service connection is adequate for diesel engine–driven pumps.
  I. Exposure protection
    1. Location on property (a 50 ft separation is preferred)
    2. Location with respect to hazardous storage
  J. Waste water drainage. As a minimum, the waste water drain must accommodate the cooling water discharge from the diesel engine and any seal leakage from the pump. The anticipated amount is considered to be normally about or less than 50 gpm. In an electric pump installation, the water from the casing relief valve must be disposed of, which is usually less than 10 gpm.

  Another potential source of waste water is when a main relief valve is installed on the discharge of the fire pump to keep the discharge pressure to the fire protection system below the pressure ratings of the system. In this situation, the discharge from the relief valve needs to be piped back to the water storage tank or to a location that can store the full

flow of the pump that occurs during the 30-minute weekly operational testing of the fire pump. For example, consider a 1000 gpm at 150 psi pump with a shutoff pressure of 210 psi and a suction pressure of 10 psi. The relief valve would be set to open at 200 psi. The quantity of water discharged during the weekly operational test would be about 1100 gpm or 33,000 gal of water. In cold climate locations and where pumps are installed inside buildings or along streets, this quantity of water can be a major consideration. Some of the choices are as follows:
1. Municipal sewer system
2. Private sewer system
3. French drain
4. Septic tank
5. Surface drainage
6. Water storage tank

K. Access
1. Entry
2. Service and removal

L. Acceptance and annual testing. Any of the following methods can be installed for flow testing of the fire pump:
1. Fire hose header (1 in. to 2½ in. outlet for each 250 gpm of pump capacity)
2. Flowmeter
3. Calibrated nozzle

M. Fuel storage
1. Gravity feed to diesel engine injectors. The fuel outlet is located at the same elevation as the engine-mounted fuel pump.
2. An 8-hour supply is required. To determine this amount, the fuel consumption of the engine must be known. For example, a 210 hp diesel engine operating at 1760 rpm might consume 10.2 gal/hour (81.6 gal) and at 2600 rpm it might consume 11.6 gal/hour (92.8 gal). If the engine data are not available, the rule of thumb is 1 pint/hp/hour (210 hp × 8 hr/16 pints/gal = 105 gal).The tank is sized by adding a 5 percent allowance for sump and a 5 percent allowance for expansion. Therefore in this example, the minimum sized fuel tank (210 hp at 1760 rpm) is 81.6 + 8.1 = 89.7 gal or, to simplify, 100 gal.
3. A double contained tank is used for environmental containment of any fuel spill. (Locate inside enclosure.)
4. Listed
5. Vent line sizing and height above enclosure
6. Fill line location (outside of enclosure)
7. Fuel suction line (¼ in. to ½ in. minimum)
8. Flexible piping is required for the connection to diesel engine (vibration isolation).
9. Rotate or treat fuel to minimize biological growth. For example, using a 210 hp diesel engine at 1760 rpm and a fuel consumption of 10.2 gal/hour may result in a normal usage per year of 275 gal of fuel.
10. Design of tank support to consider earthquakes in seismic areas

N. If the pump to be used is a vertical turbine pump taking suction from a well, belowgrade tank, or a suction pit, is roof access available for installation of the pump and "lineshaft"? Is access available for installation of an "A" frame or truck-mounted crane to remove and replace the pump for maintenance purposes?

O. What fire protection is required for the fire pump room?
1. Fire sprinkler system
2. Portable fire extinguishers

P. How is the fire pump installation supervised?
1. Direct connection to constantly attended location on the premises
2. Fire alarm system with supervisory and fire alarm signals for monitoring protection and operations connected to facility central fire alarm system or to a monitoring system such as central station or remote station (see *NFPA 72®*)

Consider a standpipe system in a high-rise building consisting of 15 floors with four exit stairs. According to Section 7.4 of NFPA 14, *Standard for the Installation of Standpipe and Hose Systems* (see Part I of this handbook), each required exit stairway must be provided with a standpipe. In this case, the required flow would be 1250 gpm (4731 L/min)—500 gpm (1892 L/min) for the most demanding standpipe and 250 gpm (946 L/min) for each additional standpipe. Assuming a pressure demand of 210 psi (14.5 bar), what type of pump should be proposed and what should be its rated capacity?

Since the high-rise building is assumed to be located in a metropolitan area, the nearby electrical supply grid should be able to supply an electric fire pump with ease. Further, since the building is a high rise and the height of the structure is beyond the pumping capacity of the local fire department, at least one alternate source of power must be provided (see NFPA 20, 9.3.1). In addition, NFPA *101®, Life Safety Code®*, requires standby power in 11.8.4. The most reliable fire pump system in such a structure would be an electrically driven fire pump with a power transfer switch.

To determine the capacity of the fire pump, Section A.5.8 in NFPA 20 suggests that operation at capacities over 140 percent can be adversely affected by the available suction supply and operation at less than 90 percent is not recommended. In this case, a demand of 1250 gpm (4731 L/min) can be supplied by a fire pump rated for 1000 gpm (3785 L/min); 1250/1000 × 100 = 125 percent of rated capacity. A fire pump rated for 1250 gpm (4731 L/min) also fits this application; however, due to the added cost of a larger pump, the designer should consider using the smaller rated capacity. The pressure rating of this pump depends on basically two factors: total system demand pressure and available pressures from the attached water supply. Assuming a city water supply when adjusted for friction and head loss at the pump suction flange of 60 psi (4.1 bar) static pressure and 40 psi (2.8 bar) residual pressure when flowing 3000 gpm (11,355 L/min), a fire pump rated at 210 psi (14.5 bar) is needed. See Exhibit S1.2.

SUMMARY

This supplement is intended to provide the reader with useful information regarding the design and selection of a fire pump. Many factors must be considered when planning a fire pump system design: needed flow and pressure, availability of water, available power supply, and general site conditions. In addition to the requirements of NFPA 20 and related suppression systems design standards, other codes and standards may need to be reviewed before commencing design of a fire pump system such as water quality and air quality standards.

INTERNET LINKS TO MANUFACTURERS AND ORGANIZATIONS

Pump Manufacturers

Armstrong Pumps *www.armstrongpumps.com/fire_protection.asp*

Aurora *www.aurorapump.com/index.htm*

Fairbanks Morse *www.pentairpump.com/fr_fmtitle.htm*

Floway *www.weirclearliquid.com/weir/clearliquid/home.nsf/Page/Products_Vertical_Turbine_Fire_Pump*

ITT-AC *www.acfirepump.com*

Patterson *www.pattersonpumps.com/fire.html*

Peerless *www.peerlesspump.com/fire_pumps.aspx*

R-B Pumps *www.rbpump.com/*

SPP Pumps *www.spppumps.com/fire.htm*

EXHIBIT S1.2 *Fire Pump Performance Curve for a Pump Rated at 210 psi (14.5 bar).*

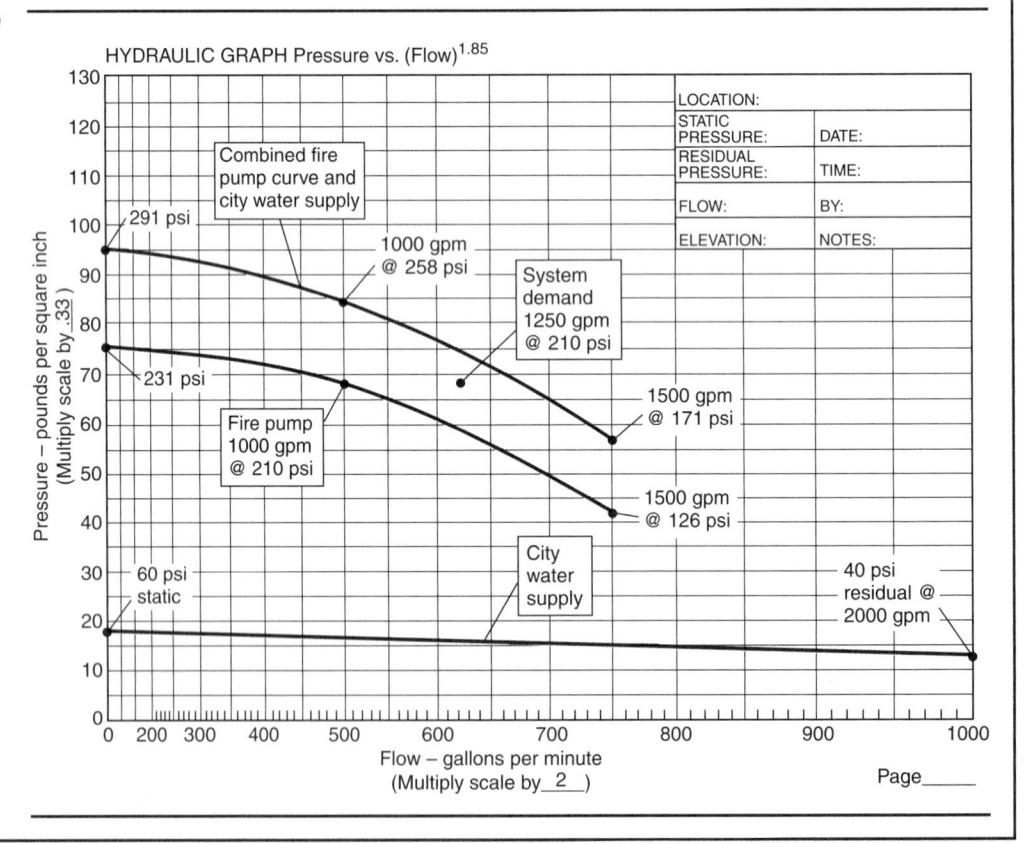

Controller Manufacturers

ASCO Power Technologies *www.firetrol.com*

Eaton Corporation *www.chfire.com*

Hubbell Industrial Controls *www.hubbell-icd.com/fpcIndex.asp*

Joslyn Clark Controls *www.joslynclark.com*

Master Control Systems *www.mastercontrols.com*

Metron *www.metroninc.com*

Tornatech *www.tornatech.com/contenu/index.cfm*

Diesel Engine Manufacturers

Caterpillar *www.cat.com/cda/layout?m=37557&x=7#firepump*

Clarke Fire Protection Products *www.clarkefire.com*

Cummins Fire Power *www.cumminsfirepower.com*

Gear Manufacturers

Randolph Gear Drive *www.randolphgear.com*

Electric Motor Manufacturers

Baldor *www.baldor.com*

Emerson Motor Technology *www.usmotors.com*

Marathon Electric *www.marathonelectric.com*

Valve and Accessory Manufacturers

CLA-VAL *www.cla-val.com*

Cummins Filtration (silencers and coolant) *www.fleetguard.com*

Gerand Engineering (flowmeters) *www.gerandengineeringco.com*

Goulds Pumps *www.goulds.com*

Kunkle Valves *www.kunklevalve.com*

NIBCO *www.nibco.com*

Preso Meters (flowmeters) *www.preso.com*

Watts *www.wattsreg.com*

Associated Organizations

American Fire Sprinkler Association *www.sprinklernet.org* or *www.firesprinkler.org*

American Water Works Association *www.awwa.org*

FM Global *www.fmglobal.com*

Hydraulic Institute *www.pumps.org*

National Electrical Manufacturers Association *www.nema.org*

National Fire Protection Association *www.nfpa.org*

National Fire Sprinkler Association *www.nfsa.org*

National Ground Water Association *www.ngwa.org*

Society of Fire Protection Engineers *www.sfpe.org*

Underwriters Laboratories Inc. *www.ul.com*

REFERENCES CITED

National Fire Protection Association, 1 Batterymarch Park, Quincy, MA 02169-7471.

Cote, A. E., ed., *Fire Protection Handbook®*, 20th edition, 2008.

Hague, D. R., ed., *Water-Based Fire Protection Systems Handbook*, 2008 edition.

NFPA 1, *Fire Code*, 2009 edition.

NFPA 11, *Standard for Low-, Medium-, and High-Expansion Foam*, 2005 edition.

NFPA 13, *Standard for the Installation of Sprinkler Systems*, 2010 edition.

NFPA 14, *Standard for the Installation of Standpipe and Hose Systems*, 2007 edition.

NFPA 15, *Standard for Water Spray Fixed Systems for Fire Protection*, 2007 edition.

NFPA 16, *Standard for the Installation of Foam-Water Sprinkler and Foam-Water Spray Systems*, 2007 edition.

NFPA 20, *Standard for the Installation of Stationary Pumps for Fire Protection*, 2010 edition.

NFPA 22, *Standard for Water Tanks for Private Fire Protection*, 2008 edition.

NFPA 24, *Standard for the Installation of Private Fire Service Mains and Their Appurtenances*, 2010 edition.

NFPA 70®, *National Electrical Code®*, 2008 edition.

NFPA 72®, *National Fire Alarm and Signaling Code*, 2010 edition.

NFPA 101®, *Life Safety Code®*, 2009 edition.

NFPA 291, *Recommended Practice for Fire Flow Testing and Marking of Hydrants*, 2010 edition.

NFPA 750, *Standard on Water Mist Fire Protection Systems*, 2006 edition.

NFPA 1142, *Standard on Water Supplies for Suburban and Rural Fire Fighting*, 2007 edition.

NFPA 2001, *Standard on Clean Agent Fire Extinguishing Systems*, 2008 edition.

NFPA 5000®, *Building Construction and Safety Code®*, 2009 edition.

ASTM International, 100 Barr Harbor Drive, P.O. Box C700, West Conshohocken, PA 19428-2959.

ANSI/ASTM A 53, *Standard Specification for Pipe, Steel, Black and Hot-Dipped, Zinc-Coated, Welded and Seamless,* 2004 edition.

California Air Resources Board, Sacramento, CA 95812.

Airborne Toxic Control Measures.

National Electrical Manufacturers Association, 1300 North 17th Street, Suite 1847, Rosslyn, VA 22209.

NEMA MG-1, *Motors and Generators,* 1998 edition.

National Ground Water Association, 601 Dempsey Road, Westerville, OH 43081-8978.
Ground Water Handbook, 1992.
Manual of Well Construction Practices, 1998.

U.S. Government Printing Office, Washington, DC 20402.

Title 40, Code of Federal Regulations, 60, Part 1111.

SUPPLEMENT 2

Commissioning Forms for Fire Pumps

Editor's Note: This supplement contains commissioning forms in a reproducible format for use by the fire pump designer and/or inspector. Duplication and use of these forms is encouraged. They can also be found on-line in an interactive format at www.nfpa.org/20handbook. They may also be used as a basis for developing special forms that are relevant to a particular jurisdiction.

This supplement consists of the following forms, which the reader may reproduce and use:

- Form S2.1 Sample Plan Review Checklist
- Form S2.2 Sample Acceptance Test Checklist
- Form S2.3 Sample Routine Inspection Checklist

PLAN REVIEW CHECKLIST

Date Documents Submitted: _____

Log No.: _____

File No.: _____

Property Information

Building Name: _____

Building Address: _____

Owner's Name: _____

Owner's Address: _____

Owner's Phone: _____ Fax: _____ E-mail: _____

System Designer/Contractor

Company Name: _____

Company Address: _____

Contact Person (Designer): _____

Designer Qualifications: _____

Phone: _____ Fax: _____ E-mail: _____

General

Building type:

❑ New   ❑ Existing   ❑ Renovation

Pump make: _____   Drive: ❑ Electric   ❑ Diesel

Model No.: _____   Pump rating: _____ gpm @ _____ psi

Rated speed: _____ rpm

❑ Yes   ❑ No   Area/building protected

What is fire pump feeding?

❑ Automatic sprinkler system   ❑ Standpipe system

❑ Fire hydrants   ❑ Other _____

Building Use and Occupancy Classification

Applicable building code: _____   Edition: _____

❑ NFPA 20, *Standard for the Illustration of Stationary Pumps for Fire Protection*   Edition: _____

❑ Other _____

Copyright © 2010 National Fire Protection Association.
This form may be copied for individual use other than for resale. It may not be copied for commercial sale or distribution.   (p. 1 of 4)

FORM S2.1 *Sample Plan Review Checklist.*

Building Use and Occupancy Classification *(continued)*

❏ Fire pump system required by building or fire code

❏ Fire pump system required by local ordinances

❏ Fire pump system required for equivalency, alternative level of protection, etc.

❏ Fire pump system not required, property owner voluntary safety improvements

❏ Other _____

Fire Pump Plan Review

Installation Features

❏ Yes ❏ No   Water supply to the fire pump adequate to meet fire pump requirements

❏ Yes ❏ No   Water supply to the fire pump added to the fire pump rating meets or exceeds the demands placed on it

❏ Yes ❏ No   Centrifugal fire pump listed for, and used exclusively for, fire protection service

❏ Yes ❏ No   Horizontal pump/driver on common base plate and connected by a listed flexible coupling

❏ Yes ❏ No   Indoor fire pump units separated from all other areas of the building by 2-hour fire-rated construction; 1-hour fire-rated construction if fire pump buildings is protected with an automatic sprinkler system

❏ Yes ❏ No   Fire pump units located outdoors and fire pump installations in buildings other than that building being protected by the fire pump located at least 50 feet away from the protected building

❏ Yes ❏ No   Pump room/house can be inundated by water

❏ Yes ❏ No   Suction piping the proper size: (5 inch for 500 gpm) (6 inch for 750 gpm) (8 inch for 1000 or 1500 gpm) (10 inch for 2000 or 2500 gpm)

❏ Yes ❏ No   Suction pipe galvanized or painted on the inside

❏ Yes ❏ No   An OS&Y valve provided in the suction piping *(Butterfly valves not permitted in suction piping)*

❏ Yes ❏ No   Backflow prevention or other device in the suction piping

❏ Yes ❏ No   Elbows perpendicular to impeller of horizontal pump within 10 pipe diameters of the intake flange

❏ Yes ❏ No   Reducer at pump intake, if provided, eccentric and installed with flat side up

❏ Yes ❏ No   If the suction supply is of sufficient pressure to be of material value without the pump, a bypass at least the required size of the discharge pipe provided

❏ Yes ❏ No   A listed indicating-type valve on each side of the check valve in the bypass and they are normally open

❏ Yes ❏ No   A 3½ inch compound gauge, having a rating of at least 100 psi and a range of at least twice the maximum suction pressure, provided on the suction piping

❏ Yes ❏ No   A 3½ inch pressure gauge, with a rating of at least 200 psi and a range of at least twice the working pressure of the pump, provided near the discharge casting

❏ Yes ❏ No   A ¾ inch circulating relief valve (1 inch if pump is rated over 2500 gpm) provided and piped to a drain *(Not needed for engine driven pumps cooled by water from pump discharge)*

Copyright © 2010 National Fire Protection Association.
This form may be copied for individual use other than for resale. It may not be copied for commercial sale or distribution.

FORM S2.1 *Continued.*

Installation Features (continued)

☐ Yes ☐ No   Listed, float-operated, automatic, air release valve (no less than ½ inch in size) provided

☐ Yes ☐ No   Discharge piping the proper size: (5 inch for 500 gpm) (6 inch for 750 or 1000 gpm) (8 inch for 1250 or 1500 gpm) (10 inch for 2000 or 2500 gpm)

☐ Yes ☐ No   A listed indicating valve installed on the fire protection system side of the pump

☐ Yes ☐ No   A check valve provided between discharge valve and the pump

☐ Yes ☐ No   Pump driver, regardless of diesel or electric, listed for fire pump service

☐ Yes ☐ No   Pump controller, regardless of diesel or electric, listed

☐ Yes ☐ No   If pump is electric motor driven, wiring, wiring elements, and components arranged in approved manner

☐ Yes ☐ No   If pump is diesel driven or if churn pressure can exceed rating of system components, a properly sized relief valve provided (5 inch for 500 gpm) (6 inch for 750 gpm) (8 inch for 1000 and 1500 gpm) (10 inch for 2000 gpm)

☐ Yes ☐ No   Relief valve piping installed with no valves

☐ Yes ☐ No   For a diesel engine driver, storage battery units provided with battery chargers specifically listed for fire pump service, arranged to automatically charge at the maximum rate whenever required by the state of charge of the battery, and arranged to indicate loss of current

☐ Yes ☐ No   For a diesel engine driver cooled by a heat exchanger, the cooling water supply from the discharge of the pump taken is prior to the discharge check valve

☐ Yes ☐ No   The heat exchanger piping for a diesel engine driver is equipped with an indicating manual shutoff valve, an approved flushing-type strainer, a pressure regulator, an automatic valve listed for fire protection service, and a second indicating manual shutoff valve

☐ Yes ☐ No   The heat exchanger piping for a diesel engine driver is equipped with a pressure gauge installed in the cooling water supply system on the engine side of the last manual valve

☐ Yes ☐ No   The heat exchanger piping for a diesel engine driver is provided with a bypass line

☐ Yes ☐ No   An outlet is provided for the wastewater line from the heat exchanger with a discharge line not less than one size larger than the inlet line, discharging into a visible open waste cone, and having no valves

☐ Yes ☐ No   A diesel fuel supply tank provided with a capacity of 1 gallon per engine horsepower plus 10%

☐ Yes ☐ No   A diesel fuel supply tank located aboveground

☐ Yes ☐ No   A test header or flowmeter tapped between the discharge check valve and the discharge valve is provided for annual fire pump flow testing *(If a flowmeter is used, it is to be arranged so as to test both pump performance and suction supply)*

☐ Yes ☐ No   Proper number of listed 2½ inch hose valves provided on test header (2 for 500 gpm) (3 for 750 gpm) (4 for 1000 gpm) (6 for up to 2500 gpm)

☐ Yes ☐ No   Test header piping is the proper size (4 inch for 500 gpm) (6 inch for 750 & 1000 gpm) (8 inch up to 2500 gpm) (10 inch for 2500 gpm)

☐ Yes ☐ No   For test header piping over 15 feet in length, next larger piping size is used

Copyright © 2010 National Fire Protection Association.
This form may be copied for individual use other than for resale. It may not be copied for commercial sale or distribution.

(p. 3 of 4)

FORM S2.1 Continued.

Installation Features (continued)

❏ Yes ❏ No		Drain valve located at a low point of the test header pipe between the normally closed test header valve and the test header
❏ Yes ❏ No		If a flowmeter provided, is the meter system piping is the proper size (5 inch for 500 and 750 GPM) (6 inch for 1000 and 1250 GPM) (8 inch for up to 2500 GPM) *(If the meter system piping exceeds 100 feet equivalent of pipe is next larger size pipe used)*
❏ Yes ❏ No		Jockey pump takes suction upstream of the main pump suction control valve and discharges downstream of the installation, typically the main pump discharge valve
❏ Yes ❏ No		Jockey pump provided with sensing line totally independent from that of main pump sensing line
❏ Yes ❏ No		Sensing lines both tap the discharge pipes between the check valve and the discharge control valve of the pumps they respectively serve
❏ Yes ❏ No		Both sensing lines ½ inch nominal and brass, copper, or series 300 stainless steel piping, tube, and fittings
❏ Yes ❏ No		Two check valves are installed in each pressure sensing line at least 5 feet apart with 3/32 inch holes drilled in the clappers
❏ Yes ❏ No		There are no shutoff valves in the sensing lines
❏ Yes ❏ No		Valves supervised open *(Test header and flow meter valves should be supervised shut)*

Location of test header: _____

Approval

Reviewer: _____ Date: _____

Approved ❏ Yes ❏ No

If no, reason(s): _____
_____
_____
_____

Notes:
_____
_____
_____
_____

Copyright © 2010 National Fire Protection Association.
This form may be copied for individual use other than for resale. It may not be copied for commercial sale or distribution.

FORM S2.1 *Continued.*

ACCEPTANCE TEST CHECKLIST

Date Documents Submitted: _____

Log No.: _____

File No.: _____

Plan Examiner: _____

Date of Approval: _____

Permit No.: _____

Property Information

Building Name: _____

Building Address: _____

Owner's Name: _____

Owner's Address: _____

Owner's Phone: _____ Fax: _____ E-mail: _____

System Designer/Contractor

Company Name: _____

Company Address: _____

Contact Person (Designer): _____

Designer Qualifications: _____

Phone: _____ Fax: _____ E-mail: _____

General

Type of building:

   ❏ New   ❏ Existing   ❏ Renovation

Pump make: _____ Drive: ❏ Electric   ❏ Diesel

Model No.: _____ Pump rating: _____ gpm @ _____ psi

                                           Rated speed: _____ rpm

What is fire pump feeding?

   ❏ Automatic sprinkler system   ❏ Standpipe system

   ❏ Fire hydrants   ❏ Other _____

Copyright © 2010 National Fire Protection Association.
This form may be copied for individual use other than for resale. It may not be copied for commercial sale or distribution.

FORM S2.2 *Sample Acceptance Test Checklist.*

General *(continued)*

Present at test:

	Authorized Representative	**Manufacturer**
Pump		
Engine (if diesel)		
Controller		
Transfer switch		

Date the suction piping was flushed prior to hydrostatic test: _____

Flow rate: _____ gpm

Pressure at which piping hydrostatic tested: _____ psi

Fire Pump Acceptance Test

Installation

❏ Yes ❏ No   Certificate for flushing and hydrostatic testing furnished

❏ Yes ❏ No   Centrifugal fire pump listed for fire protection service

❏ Yes ❏ No   Horizontal pump/driver on common base plate and connected by a listed flexible coupling

❏ Yes ❏ No   Guards provided for flexible couplings and flexible connecting shafts

❏ Yes ❏ No   Base plate securely attached to a solid reinforced concrete foundation

❏ Yes ❏ No   Indoor fire pump units separated from all other areas of the building by 2-hour fire-rated construction; 1-hour fire-rated construction in buildings protected with an automatic sprinkler system

❏ Yes ❏ No   If fire pump unit is located outdoors or if fire pump installation is in a building other than that building being protected by the fire pump, it is located at least 50 feet away from the protected building

❏ Yes ❏ No   A suitable means for maintaining 40°F provided; 70°F if driver is diesel engine. *(Portable units, plug-in units, and hardwired electric units without secured circuit breakers are not reliable)*

❏ Yes ❏ No   Pump room/house provided with normal lighting and emergency lighting

❏ Yes ❏ No   Pump room/house adequately ventilated and floor is pitched toward drain

❏ Yes ❏ No   Suction piping is the proper size. (5 inch for 500 gpm) (6 inch for 750 gpm) (8 inch for 1000 or 1500 gpm) (10 inch for 2000 or 2500 gpm)

❏ Yes ❏ No   OS&Y valve provided in the suction piping *(Butterfly valves not permitted in suction piping)*

❏ Yes ❏ No   No backflow prevention or other devices are in the suction piping

❏ Yes ❏ No   No elbows perpendicular to impeller of horizontal pump are within 10 pipe diameters of the intake flange

Copyright © 2010 National Fire Protection Association.
This form may be copied for individual use other than for resale. It may not be copied for commercial sale or distribution.

FORM S2.2 *Continued.*

Installation (continued)

☐ Yes ☐ No Reducer at pump intake is eccentric and installed with flat side up

☐ Yes ☐ No A bypass, at least the required size of the discharge pipe, is provided if the suction supply is of sufficient pressure to be of material value without the pump

☐ Yes ☐ No Listed indicating type valves are on each side of the check valve in the bypass and are normally open

☐ Yes ☐ No A 3½ inch compound gauge, having a rating of at least 100 psi and a range of at least twice the maximum suction pressure, is provided on the suction piping

☐ Yes ☐ No A 3½ inch pressure gauge, with a rating of at least 200 psi and a range of at least twice the working pressure of the pump, is provided near the discharge casting

☐ Yes ☐ No A ¾ inch circulating relief valve (1 inch if pump is rated over 2500 gpm) is provided and piped to a drain. *(Not needed for engine driven pumps cooled by water from pump discharge)*

☐ Yes ☐ No A listed, float-operated, automatic, air release valve (no less than ½ inch in size) is provided

☐ Yes ☐ No Discharge piping is of the proper size. (5 inch for 500 gpm) (6 inch for 750 or 1000 gpm) (8 inch for 1250 or 1500 gpm) (10 inch for 2000 or 2500 gpm)

☐ Yes ☐ No A listed indicating valve is installed on the fire protection system side of the pump

☐ Yes ☐ No A check valve is provided between the discharge valve and the pump

☐ Yes ☐ No The pump driver, regardless of diesel or electric, is listed for fire pump service

☐ Yes ☐ No A properly sized relief valve has been provided if pump is diesel driven or if churn pressure can exceed rating of system components. (5 inch for 500 gpm) (6 inch for 750 gpm) (8 inch for 1000 and 1500 gpm) (10 inch for 2000 gpm)

☐ Yes ☐ No No valves are installed in the relief valve piping

☐ Yes ☐ No For diesel engine drivers, there are two storage battery units provided and rack-supported above the floor, secured against displacement, and located where they are readily accessible for servicing and not subject to excessive temperature, vibration, mechanical injury, or flooding

☐ Yes ☐ No For diesel engine driver, storage battery units are provided with battery chargers specifically listed for fire pump service, arranged to automatically charge at the maximum rate whenever required by the state of charge of the battery, and arranged to indicate loss of current

☐ Yes ☐ No For diesel engine driver cooled by a heat exchanger, the cooling water supply is from the discharge of the pump and taken prior to the discharge check valve

☐ Yes ☐ No The heat exchanger piping for a diesel engine driver is equipped with an indicating manual shutoff valve, an approved flushing-type strainer, a pressure regulator, an automatic valve listed for fire protection service, and a second indicating manual shutoff valve

☐ Yes ☐ No Heat exchanger piping of a diesel engine driver is equipped with a pressure gauge installed in the cooling water supply system on the engine side of the last manual valve

☐ Yes ☐ No Heat exchanger piping of a diesel engine driver is provided with a bypass line

☐ Yes ☐ No The outlet provided for the wastewater line from the heat exchanger has a discharge line not less than one size larger than the inlet line, discharges into a visible open waste cone, and has no valves

Copyright © 2010 National Fire Protection Association.
This form may be copied for individual use other than for resale. It may not be copied for commercial sale or distribution.

FORM S2.2 Continued.

Installation (continued)

☐ Yes ☐ No Diesel fuel supply tank has a capacity of 1 gallon per engine horsepower plus 10%

☐ Yes ☐ No Diesel fuel supply tank is located aboveground

☐ Yes ☐ No Exposed fuel lines are provided with guard or protecting pipe

☐ Yes ☐ No The test header or flowmeter is tapped between the discharge check valve and the discharge valve provided for annual fire pump flow testing (If a flowmeter is used, verify that it is arranged so as to test both pump performance and suction supply)

☐ Yes ☐ No Proper number of listed 2½ inch hose valves is provided on test header (2 for 500 gpm) (3 for 750 gpm) (4 for 1000 gpm) (6 for up to 2500 gpm)

☐ Yes ☐ No Test header piping is of the proper size (4 inch for 500 gpm) (6 inch for 750 and 1000 gpm) (8 inch for up to 2500 gpm) (10 inch for 2500 gpm)

☐ Yes ☐ No If test header piping is over 15 feet in length, the next larger piping size is used

☐ Yes ☐ No A drain valve is located at a low point of the test header pipe between the normally closed test header valve and the test header

☐ Yes ☐ No If a flowmeter is provided, meter system piping is of the proper size. (5 inch for 500 and 750 gpm) (6 inch for 1000 and 1250 gpm) (8 inch for up to 2500 gpm)

☐ Yes ☐ No If the meter system piping exceeds 100 feet equivalent of pipe, the next larger size pipe is used

☐ Yes ☐ No Jockey pump is provided with sensing line totally independent from that of main pump sensing line

☐ Yes ☐ No The sensing lines both tap the discharge pipes between the check valve and the discharge control valve of the pumps they respectively serve

☐ Yes ☐ No Both sensing lines are ½ inch and brass, copper, or series 300 stainless steel piping, tube, and fittings

☐ Yes ☐ No Two check valves are installed in each pressure sensing line at least 5 feet apart

☐ Yes ☐ No No shut off valves in the sensing lines

For diesel driven pumps, verify that the following alarms are provided on the controller and operative:

☐ Low oil pressure   ☐ Battery failure/battery missing

☐ High engine temperature   ☐ Battery charger failure

☐ Failure to start   ☐ Low (less than ⅔) fuel level

☐ Shutdown on overspeed

For diesel driven pumps, verify that the following alarms are provided and transmit to a constantly attended location:

☐ Pump running

☐ Controller main switch in a position other than "AUTOMATIC"

☐ Trouble on controller or engine

FORM S2.2 Continued.

Installation (continued)

For electric driven pumps, verify that are the following alarms are operative:

- ❏ Loss of power
- ❏ Phase reversal
- ❏ Pump running
- ❏ Other _____

❏ Yes ❏ No Verify that the cut-in and cut-out of the jockey pump is properly set.

❏ Yes ❏ No Verify that the cut-in of the main pump is properly set.

❏ Yes ❏ No Verify that all valves are supervised open. *(Test header and flowmeter valves should be supervised shut)*

❏ Yes ❏ No Verify that the pump performance met or exceeds the demands of the systems supplied by pump. *(See results below.)*

Test Results

Test	Discharge Pressure	Intake pressure	Net pressure	Speed	Nozzle Size and Pitot Pressures	gpm
1						
2						
3						
4						
5						

Inspector: _____ Date: _____

Approval

Inspector: _____ Date: _____

Approved ❏ Yes ❏ No

If no, reason(s): _____
_____
_____

Notes:
_____
_____
_____
_____

Copyright © 2010 National Fire Protection Association.
This form may be copied for individual use other than for resale. It may not be copied for commercial sale or distribution.

FORM S2.2 Continued.

ROUTINE INSPECTION CHECKLIST

Date of Inspection: _____

Inspector: _____

Date of Last Inspection: _____

Property Information

Building Name: _____

Building Address: _____

Contact Person (Owner/Tenant): _____

Address: _____

Phone: _____ Fax: _____ E-mail: _____

General

Type of building:

❏ New   ❏ Existing   ❏ Renovation

Pump make: _____ Drive: ❏ Electric   ❏ Diesel

Model No.: _____ Pump rating: _____ gpm @ _____ psi

Rated speed: _____ rpm

What is fire pump feeding?

❏ Automatic sprinkler system   ❏ Standpipe system

❏ Fire hydrants   ❏ Other _____

❏ Yes ❏ No   Area protected

Fire Pump Inspection

Installation

❏ Yes ❏ No   Water supply to the fire pump adequate to meet fire pump requirements

❏ Yes ❏ No   Water supply to the fire pump added to the fire pump rating meets or exceeds the demands placed on it

❏ Yes ❏ No   Centrifugal fire pump listed for, and used exclusively for, fire protection service

❏ Yes ❏ No   Change in installation since last inspection

❏ Yes ❏ No   Guards provided for the flexible couplings and flexible connecting shafts in good order

❏ Yes ❏ No   Required rated building construction housing the fire pump intact

❏ Yes ❏ No   Suitable means for maintaining 40°F being provided; 70°F if driver is diesel engine *(Portable units, plug-in units, and hardwired electric units without secured circuit breakers are not reliable)*

Copyright © 2010 National Fire Protection Association.
This form may be copied for individual use other than for resale. It may not be copied for commercial sale or distribution.   (p. 1 of 2)

FORM S2.3 *Sample Routine Inspection Checklist.*

Installation (continued)

- ❏ Yes ❏ No Both normal lighting and emergency lighting maintained for pump room/house
- ❏ Yes ❏ No Pump room/house adequately ventilated
- ❏ Yes ❏ No All valves in the fire pump piping (except the test header valve) normally open
- ❏ Yes ❏ No Suction piping compound and the discharge pressure gauges appear operative
- ❏ Yes ❏ No Circulating relief valve functions properly
- ❏ Yes ❏ No For diesel engine driver, storage battery units maintained
- ❏ Yes ❏ No For diesel engine driver, battery charger units maintained
- ❏ Yes ❏ No For diesel driver cooled by heat exchanger, cooling water able to discharge through the waste cone, manual shutoff valves in the bypass line normally closed, and flushing-type strainer being maintained
- ❏ Yes ❏ No For diesel driven pumps, fuel level is appropriate
- ❏ Yes ❏ No All alarms functional
- ❏ Yes ❏ No Approved vendor serviced fire pump in the past 12 months
- ❏ Yes ❏ No Annual fire pump test conducted

    Date of last certification _____
- ❏ Yes ❏ No Copy of annual inspection by approved vendor provided
- ❏ Yes ❏ No Pump performance meets or exceeds the demands of the systems supplied by pump

Approval

Inspector: _____   Date: _____

System inspection considered satisfactory   ❏ Yes   ❏ No

If no, reason(s): _____
_____
_____
_____

Notes:
_____
_____
_____
_____

Copyright © 2010 National Fire Protection Association.
This form may be copied for individual use other than for resale. It may not be copied for commercial sale or distribution.

FORM S2.3 *Continued.*

SUPPLEMENT 3

Article 695 from *NFPA 70*, *National Electrical Code*, 2008 Edition

Editor's Note: Supplement 3 extracts Article 695 from the 2008 edition of NFPA 70®, National Electrical Code® (NEC®), which contains the installation requirements for electric power sources and interconnecting circuitry for electric motor–driven fire pumps and switching and control equipment for electric fire pump motors. Article 695 does not cover performance, maintenance, or acceptance testing of the fire pump system nor does it cover the internal wiring of the system components or wiring for diesel fire pumps or pressure maintenance pumps.

ARTICLE 695
Fire Pumps

695.1 Scope.

FPN: Rules that are followed by a reference in brackets contain text that has been extracted from NFPA 20-2007, *Standard for the Installation of Stationary Pumps for Fire Protection*. Only editorial changes were made to the extracted text to make it consistent with this *Code*.

(A) Covered. This article covers the installation of the following:

(1) Electric power sources and interconnecting circuits
(2) Switching and control equipment dedicated to fire pump drivers

(B) Not Covered. This article does not cover the following:

(1) The performance, maintenance, and acceptance testing of the fire pump system, and the internal wiring of the components of the system
(2) Pressure maintenance (jockey or makeup) pumps
   FPN: See NFPA 20-2007, *Standard for the Installation of Stationary Pumps for Fire Protection*, for further information.

695.2 Definitions.

Fault-Tolerant External Control Circuits. Those control circuits either entering or leaving the fire pump controller enclosure, which if broken, disconnected, or shorted will not prevent the controller from starting the fire pump from all other internal or external means and may cause the controller to start the pump under these conditions.

On-Site Power Production Facility. The normal supply of electric power for the site that is expected to be constantly producing power.

On-Site Standby Generator. A facility producing electric power on site as the alternate supply of electric power. It differs from an on-site power production facility, in that it is not constantly producing power.

695.3 Power Source(s) for Electric Motor-Driven Fire Pumps.
Electric motor-driven fire pumps shall have a reliable source of power.

(A) Individual Sources. Where reliable, and where capable of carrying indefinitely the sum of the locked-rotor current of the fire pump motor(s) and the pressure maintenance pump motor(s) and the full-load current of the associated fire pump accessory equipment when connected to this power supply, the power source for an electric motor driven fire pump shall be one or more of the following.

(1) Electric Utility Service Connection. A fire pump shall be permitted to be supplied by a separate service, or from a connection located ahead of and not within the same cabinet, enclosure, or vertical switchboard section as the service disconnecting means. The connection shall be located and arranged so as to minimize the possibility of damage by fire from within the premises and from exposing hazards. A tap ahead of the service disconnecting means shall comply with 230.82(5). The service equipment shall comply with the labeling requirements in 230.2 and the location requirements in 230.72(B). [**20**:9.2.2]

(2) On-Site Power Production Facility. A fire pump shall be permitted to be supplied by an on-site power production facility. The source facility shall be located and protected to minimize the possibility of damage by fire. [**20**:9.2.3]

(B) Multiple Sources. Where reliable power cannot be obtained from a source described in 695.3(A), power shall be supplied from an approved combination of two or more of either of such sources, or from an approved combination of feeders constituting two or more power sources as covered in 695.3(B)(2), or from an approved combination of one or more of such power sources in combination with an on-site standby generator complying with 695.3(B)(1) and (B)(3).

(1) Generator Capacity. An on-site generator(s) used to comply with this section shall be of sufficient capacity to allow normal starting and running of the motor(s) driving the fire pump(s) while supplying all other simultaneously operated load. Automatic shedding of one or more optional standby loads in order to comply with this capacity requirement shall be permitted. A tap ahead of the on-site generator disconnecting means shall not be required. The requirements of 430.113 shall not apply. [**20**:9.6.1]

(2) Feeder Sources. This section applies to multibuilding campus-style complexes with fire pumps at one or more buildings. Where sources in 695.3(A) are not practicable, and with the approval of the authority having jurisdiction, two or more feeder sources shall be permitted as one power source or as more than one power source where such feeders are connected to or derived from separate utility services. The connection(s), overcurrent protective device(s), and disconnecting means for such feeders shall meet the requirements of 695.4(B). [**20**:9.2.2(4)]

(3) Arrangement. The power sources shall be arranged so that a fire at one source will not cause an interruption at the other source.

695.4 Continuity of Power. Circuits that supply electric motor-driven fire pumps shall be supervised from inadvertent disconnection as covered in 695.4(A) or (B).

(A) Direct Connection. The supply conductors shall directly connect the power source to either a listed fire pump controller or listed combination fire pump controller and power transfer switch.

Where the power source is supplied by on-site generator(s), the supply conductors shall connect to a generator disconnecting means dedicated for the purposes of serving the fire pump. The disconnecting means shall be located in a separate enclosure from the other generator disconnecting means.

(B) Supervised Connection. A single disconnecting means and associated overcurrent protective device(s) shall be permitted to be installed between a remote power source and one of the following:

(1) A listed fire pump controller
(2) A listed fire pump power transfer switch
(3) A listed combination fire pump controller and power transfer switch

For systems installed under the provisions of 695.3(B)(2) only, such additional disconnecting means and associated overcurrent protective device(s) shall be permitted as required to comply with other provisions of this *Code*. Overcurrent protective devices between an on-site standby generator and a fire pump controller shall be selected and sized according to 430.62 to provide short-circuit protection only. All disconnecting devices and overcurrent protective devices that are unique to the fire pump loads shall comply with 695.4(B)(1) through (B)(5).

(1) Overcurrent Device Selection. The overcurrent protective device(s) shall be selected or set to carry indefinitely the sum of the locked-rotor current of the fire pump motor(s) and the pressure maintenance pump motor(s) and the full-load current of the associated fire pump accessory equipment when connected to this power supply. The next standard overcurrent device shall be used in accordance with 240.6. The requirement to carry the locked-rotor currents indefinitely shall not apply to conductors or devices other than overcurrent devices in the fire pump motor circuit(s).

(2) Disconnecting Means. The disconnecting means shall comply with all of the following:

(1) Be identified as suitable for use as service equipment

(2) Be lockable in the closed position
(3) Not be located within equipment that feeds loads other than the fire pump
(4) Be located sufficiently remote from other building or other fire pump source disconnecting means such that inadvertent operation at the same time would be unlikely

(3) Disconnect Marking. The disconnecting means shall be marked "Fire Pump Disconnecting Means." The letters shall be at least 25 mm (1 in.) in height, and they shall be visible without opening enclosure doors or covers.

(4) Controller Marking. A placard shall be placed adjacent to the fire pump controller, stating the location of this disconnecting means and the location of the key (if the disconnecting means is locked).

(5) Supervision. The disconnecting means shall be supervised in the closed position by one of the following methods:

(1) Central station, proprietary, or remote station signal device
(2) Local signaling service that causes the sounding of an audible signal at a constantly attended point
(3) Locking the disconnecting means in the closed position
(4) Sealing of disconnecting means and approved weekly recorded inspections when the disconnecting means are located within fenced enclosures or in buildings under the control of the owner [20:9.2.3.3]

695.5 Transformers. Where the service or system voltage is different from the utilization voltage of the fire pump motor, transformer(s) protected by disconnecting means and overcurrent protective devices shall be permitted to be installed between the system supply and the fire pump controller in accordance with 695.5(A) and (B), or with (C). Only transformers covered in 695.5(C) shall be permitted to supply loads not directly associated with the fire pump system.

(A) Size. Where a transformer supplies an electric motor driven fire pump, it shall be rated at a minimum of 125 percent of the sum of the fire pump motor(s) and pressure maintenance pump(s) motor loads, and 100 percent of the associated fire pump accessory equipment supplied by the transformer.

(B) Overcurrent Protection. The primary overcurrent protective device(s) shall be selected or set to carry indefinitely the sum of the locked-rotor current of the fire pump motor(s) and the pressure maintenance pump motor(s) and the full-load current of the associated fire pump accessory equipment when connected to this power supply. Secondary overcurrent protection shall not be permitted. The requirement to carry the locked-rotor currents indefinitely shall not apply to conductors or devices other than overcurrent devices in the fire pump motor circuit(s).

(C) Feeder Source. Where a feeder source is provided in accordance with 695.3(B)(2), transformers supplying the fire pump system shall be permitted to supply other loads. All other loads shall be calculated in accordance with Article 220, including demand factors as applicable.

(1) Size. Transformers shall be rated at a minimum of 125 percent of the sum of the fire pump motor(s) and pressure maintenance pump(s) motor loads, and 100 percent of the remaining load supplied by the transformer.

(2) Overcurrent Protection. The transformer size, the feeder size, and the overcurrent protective device(s) shall be coordinated such that overcurrent protection is provided for the transformer in accordance with 450.3 and for the feeder in accordance with 215.3, and such that the overcurrent protective device(s) is selected or set to carry indefinitely the sum of the locked-rotor current of the fire pump motor(s), the pressure maintenance pump motor(s), the full-load current of the associated fire pump accessory equipment, and 100 percent of the remaining loads supplied by the transformer. The requirement to carry the locked-rotor currents indefinitely shall not apply to conductors or devices other than overcurrent devices in the fire pump motor circuit(s).

695.6 Power Wiring. Power circuits and wiring methods shall comply with the requirements in 695.6(A) through (H), and as permitted in 230.90(A), Exception No. 4; 230.94, Exception No. 4; 240.13; 230.208; 240.4(A); and 430.31.

(A) Service Conductors. Supply conductors shall be physically routed outside a building(s) and shall be installed as service-entrance conductors in accordance with 230.6, 230.9, and Parts III and IV of Article 230. Where supply conductors cannot be physically routed outside buildings, they shall be permitted to be routed through buildings where installed in accordance with 230.6(1) or (2). Where a fire pump is wired under the provisions of 695.3(B)(2), this requirement shall apply to all supply conductors on the load side of the service disconnecting means that constitute the normal source of supply to that fire pump.

Exception: Where there are multiple sources of supply with means for automatic connection from one source to the other, the requirement shall apply only to those conductors on the load side of that point of automatic connection between sources.

(B) Circuit Conductors. Fire pump supply conductors on the load side of the final disconnecting means and overcurrent device(s) permitted by 695.4(B) shall be kept entirely independent of all other wiring. They shall supply only loads that are directly associated with the fire pump system, and they shall be protected to resist potential damage by fire, structural failure, or operational accident. They shall be permitted to be routed through a building(s) using one of the following methods:

(1) Be encased in a minimum 50 mm (2 in.) of concrete
(2) Be protected by a fire-rated assembly listed to achieve a minimum fire rating of 2 hours and dedicated to the fire pump circuit(s).
(3) Be a listed electrical circuit protective system with a minimum 2-hour fire rating

FPN: UL guide information for electrical circuit protective systems (FHIT) contains information on proper installation requirements to maintain the fire rating.

Exception: The supply conductors located in the electrical equipment room where they originate and in the fire pump room shall not be required to have the minimum 1-hour fire separation or fire resistance rating, unless otherwise required by 700.9(D) of this Code.

(C) Conductor Size.

(1) Fire Pump Motors and Other Equipment. Conductors supplying a fire pump motor(s), pressure maintenance pumps, and associated fire pump accessory equipment shall have a rating not less than 125 percent of the sum of the fire pump motor(s) and pressure maintenance motor(s) full-load current(s), and 100 percent of the associated fire pump accessory equipment.

(2) Fire Pump Motors Only. Conductors supplying only a fire pump motor shall have a minimum ampacity in accordance with 430.22 and shall comply with the voltage drop requirements in 695.7.

(D) Overload Protection. Power circuits shall not have automatic protection against overloads. Except for protection of transformer primaries provided in 695.5(C)(2), branch-circuit and feeder conductors shall be protected against short circuit only. Where a tap is made to supply a fire pump, the wiring shall be treated as service conductors in accordance with 230.6. The applicable distance and size restrictions in 240.21 shall not apply.

Exception No. 1: Conductors between storage batteries and the engine shall not require overcurrent protection or disconnecting means.

Exception No. 2: For on-site standby generator(s) rated to produce continuous current in excess of 225 percent of the full-load amperes of the fire pump motor, the conductors between the on-site generator(s) and the combination fire pump transfer switch controller or separately mounted transfer switch shall be installed in accordance with 695.6(B). The protection provided shall be in accordance with the short-circuit current rating of the combination fire pump transfer switch controller or separately mounted transfer switch.

(E) Pump Wiring. All wiring from the controllers to the pump motors shall be in rigid metal conduit, intermediate metal conduit, liquidtight flexible metal conduit, or liquidtight flexible nonmetallic conduit Type LFNC-B, listed Type MC cable with an impervious covering, or Type MI cable.

(F) Junction Points. Where wire connectors are used in the fire pump circuit, the connectors shall be listed. A fire pump controller or fire pump power transfer switch, where provided, shall not be used as a junction box to supply other equipment, including a pressure maintenance (jockey) pump(s). A fire pump controller and fire pump power transfer switch, where provided, shall not serve any load other than the fire pump for which it is intended.

(G) Mechanical Protection. All wiring from engine controllers and batteries shall be protected against physical damage and shall be installed in accordance with the controller and engine manufacturer's instructions.

(H) Ground-Fault Protection of Equipment. Ground-fault protection of equipment shall not be permitted for fire pumps.

695.7 Voltage Drop. The voltage at the controller line terminals shall not drop more than 15 percent below normal (controller-rated voltage) under motor starting conditions. The voltage at the motor terminals shall not drop more than 5 percent below the voltage rating of the motor when the motor is operating at 115 percent of the full-load current rating of the motor.

Exception: This limitation shall not apply for emergency run mechanical starting. [20:9.4.2]

695.10 Listed Equipment. Diesel engine fire pump controllers, electric fire pump controllers, electric motors, fire pump power transfer switches, foam pump controllers, and limited service controllers shall be listed for fire pump service. [20:9.5.1.1, 10.1.2.1, 12.1.3.1]

695.12 Equipment Location.

(A) Controllers and Transfer Switches. Electric motor-driven fire pump controllers and power transfer switches shall be located as close as practicable to, and within sight of, the motors that they control.

(B) Engine-Drive Controllers. Engine-drive fire pump controllers shall be located as close as is practical to, and within sight of, the engines that they control.

(C) Storage Batteries. Storage batteries for fire pump engine drives shall be supported above the floor, secured against displacement, and located where they are not subject to physical damage, flooding with water, excessive temperature, or excessive vibration.

(D) Energized Equipment. All energized equipment parts shall be located at least 300 mm (12 in.) above the floor level.

(E) Protection Against Pump Water. Fire pump controller and power transfer switches shall be located or protected so that they are not damaged by water escaping from pumps or pump connections.

(F) Mounting. All fire pump control equipment shall be mounted in a substantial manner on noncombustible supporting structures.

695.14 Control Wiring.

(A) Control Circuit Failures. External control circuits that extend outside the fire pump room shall be arranged so that failure of any external circuit (open or short circuit) shall not prevent the operation of a pump(s) from all other internal or external means. Breakage, disconnecting, shorting of the wires, or loss of power to these circuits could cause continuous running of the fire pump but shall not prevent the controller(s) from starting the fire pump(s) due to causes other than these external control circuits. All control conductors within the fire pump room that are not fault tolerant shall be protected against physical damage. [20:10.5.2.6, 12.5.2.5]

(B) Sensor Functioning. No undervoltage, phase-loss, frequency-sensitive, or other sensor(s) shall be installed that automatically or manually prohibits actuation of the motor contactor. [20:10.4.5.6]

Exception: A phase loss sensor(s) shall be permitted only as a part of a listed fire pump controller.

(C) Remote Device(s). No remote device(s) shall be installed that will prevent automatic operation of the transfer switch. [20:10.8.1.3]

(D) Engine-Drive Control Wiring. All wiring between the controller and the diesel engine shall be stranded and sized to continuously carry the charging or control currents as required by the controller manufacturer. Such wiring shall be protected against physical damage. Controller manufacturer's specifications for distance and wire size shall be followed. [20:12.3.5.1]

(E) Electric Fire Pump Control Wiring Methods. All electric motor–driven fire pump control wiring shall be in rigid metal conduit, intermediate metal conduit, liquidtight flexible metal conduit, liquidtight flexible nonmetallic conduit Type B (LFNC-B), listed Type MC cable with an impervious covering, or Type MI cable.

(F) Generator Control Wiring Methods. Control conductors installed between the fire pump power transfer switch and the standby generator supplying the fire pump during normal power loss shall be kept entirely independent of all other wiring. They shall be protected to resist potential damage by fire or structural failure. They shall be permitted to be routed through a building(s) encased in 50 mm (2 in.) of concrete or within enclosed construction dedicated to the fire pump circuits and having a minimum 1-hour fire resistance rating, or circuit protective systems with a minimum of 1-hour fire resistance. The installation shall comply with any restrictions provided in the listing of the electrical circuit protective system used.

SUPPLEMENT 4

Technical/Substantive Changes from the 2007 Edition to the 2010 Edition of NFPA 20

Editor's Note: Supplement 4 contains a useful table of major code changes from the 2007 to the 2010 edition of NFPA 20, Standard for the Installation of Stationary Pumps for Fire Protection.

Subject / 2010 Edition Text	Notes
Chapter 1 Administration	There were no revisions to Chapter 1 for the 2010 edition.
Chapter 2 Referenced Publications	Chapter was revised to reflect the documents referenced by other chapters of the standard; edition dates were updated.
Chapter 3 Definitions **3.3.37.11 Packaged Fire Pump Assembly.** Fire pump unit components assembled at a packaging facility and shipped as a unit to the installation site. The scope of listed components (where required to be listed by this standard) in a pre-assembled package includes the pump, driver, controller, and other accessories identified by the packager assembled onto a base with or without an enclosure.	This definition was added to provide needed clarification and clearly identifies the components included in a packaged fire pump assembly.
3.3.37.14 Pressure Maintenance (Jockey or Make-Up) Pump. A pump designed to maintain the pressure on the fire protection system(s) between preset limits when the system is not flowing water.	This definition was added to provide clarification.
3.3.40 Series Fire Pump Unit. All fire pump units that operate in a series arrangement where the first fire pump takes suction directly from a water supply and each sequential pump takes suction from the preceding pump; pumps taking suction from tanks or break tanks are not considered series fire pump units even if fire pumps at lower elevations are used to refill the tanks or break tanks.	This definition clarifies that the series fire pump unit takes suction from another pump, not a tank or break tank.
3.3.57 Variable Speed Suction Limiting Control. A speed control system used to maintain a minimum positive suction pressure at the pump inlet by reducing the pump driver speed while monitoring pressure in the suction piping through a sensing line.	Variable speed control has proven to be a reliable method for limiting excessive fire pump discharge pressures. The new definition is necessary since the term is used in the revised paragraphs on variable speed control.

Subject / 2010 Edition Text	Notes
Chapter 4 General Requirements	The 2007 edition of Chapter 5 (General Requirements) has become Chapter 4 of the 2010 edition. A new Chapter 5 on high-rise buildings has been added to NFPA 20 in 2010.
4.3.2 System Designer. The system designer shall be identified on system design documents. Acceptable minimum evidence of qualifications or certification shall be provided when requested by the authority having jurisdiction. Qualified personnel shall include, but not be limited to, one or more of the following: (1) Personnel who are factory trained and certified for fire pump system design of the specific type and brand of system being designed (2)* Personnel who are certified by a nationally recognized fire protection certification organization acceptable to the authority having jurisdiction (3) Personnel who are registered, licensed, or certified by a state or local authority **4.3.2.1** Additional evidence of qualification or certification shall be permitted to be required by the AHJ. **4.3.3 System Installer.** Installation personnel shall be qualified or shall be supervised by persons who are qualified in the installation, inspection, and testing of fire protection systems. Minimum evidence of qualifications or certification shall be provided when requested by the authority having jurisdiction. Qualified personnel shall include, but not be limited to, one or more of the following: (1) Personnel who are factory trained and certified for fire pump system design of the specific type and brand of system being designed (2)* Personnel who are certified by a nationally recognized fire protection certification organization acceptable to the authority having jurisdiction (3) Personnel who are registered, licensed, or certified by a state or local authority **4.3.3.1** Additional evidence of qualification or certification shall be permitted to be required by the AHJ. **4.3.4* Service Personnel Qualifications and Experience.** **4.3.4.1** Service personnel shall be qualified and experienced in the inspection, testing, and maintenance of fire protection systems. Qualified personnel shall include, but not be limited to, one or more of the following: (1) Personnel who are factory trained and certified for fire pump system design of the specific type and brand of system being designed (2)* Personnel who are certified by a nationally recognized fire protection certification organization acceptable to the authority having jurisdiction (3) Personnel who are registered, licensed, or certified by a state or local authority (4) Personnel who are employed and qualified by an organization listed by a nationally recognized testing laboratory for the servicing of fire protection systems **4.3.4.2** Additional evidence of qualification or certification shall be permitted to be required by the AHJ.	The technical committee found that many fire pumps have been installed by those who have little or no training or experience with fire protection electrical systems. As a result, the committee found that many fire pumps have been installed incorrectly. Subsections 4.3.2 through 4.3.4 now provide guidance for the qualifications of personnel to ensure that fire pumps are designed, installed, and serviced by qualified personnel.
4.6.2.3.1 Where the maximum flow available from the water supply cannot provide a flow of 150 percent of the rated flow of the pump, but the water supply can provide the greater of 100 percent of rated flow or the maximum flow demand of the fire protection system(s), the water supply shall be deemed to be adequate. In this case, the maximum flow shall be considered the highest flow that the water supply can achieve.	It is common for certain size fire pumps to be required due to the flow demand of the fire protection system and to have a water supply that is not capable of meeting the maximum flow of 150 percent of the rated flow of the pump. As long as the water supply can provide the flow demand of the fire protection system, and as long as the pump can be tested at some high flow so that the performance curve can be reproduced, there is no need to actually require the water supply to be capable of achieving the 150 percent of rated flow point.

Subject / 2010 Edition Text	Notes
4.6.2.3.2 Where the water supply cannot provide 150 percent of the rated flow of the pump, a placard shall be placed in the pump room indicating the minimum suction pressure that the fire pump is allowed to be tested at and also indicating the required flow rate.	This paragraph provides cautionary guidance for potential changes in the water supply.
4.6.5.1 Except as provided in 4.6.5.2, the head available from a water supply shall be figured on the basis of a flow of 150 percent of rated capacity of the fire pump.	It is common for certain size fire pumps to be required due to the flow demand of the fire protection system and to have a water supply that is not capable of meeting the maximum flow of 150 percent of the rated flow of the pump. As long as the water supply can provide the flow demand of the fire protection system, and as long as the pump can be tested at some high flow so that the performance curve can be reproduced, there is no need to actually require the water supply to be capable of achieving the 150 percent of rated flow point.
4.6.5.2 Where the water supply cannot provide a flow of 150 percent of the rated flow of the pump, but the water supply can provide the flow demand of the fire protection system, the head available from the water supply shall be permitted to be calculated on the basis of the maximum flow available as allowed by 4.6.2.3.1.	
4.7.6* The driver shall be selected in accordance with 9.5.2 (electric motors), 11.2.2 (diesel engines), or 13.1.2 (steam turbines) to provide the required power to operate the pump at rated speed and maximum pump load under any flow condition.	New text has been added to specify the selection of the driver for all types of fire pump installations.
4.8.1 A centrifugal fire pump for fire protection shall be selected so that the greatest single demand for any fire protection system connected to the pump is less than or equal to 150 percent of the rated capacity (flow) of the pump.	This subsection was revised in order to clarify the language in the requirement.
4.10.2.1 Unless the requirements of 4.10.2.4 are met, a gauge having a dial not less than 3.5 in. (89 mm) in diameter shall be connected to the suction pipe near the pump with a nominal 0.25 in. (6 mm) gauge valve.	Paragraphs 4.10.2.1 and 4.10.2.3 were revised and a new 4.10.2.1.1 was added in order to provide adequate pressure gauge range and more flexibility to match the gauge to the actual water supply.
4.10.2.1.1 Where the minimum pump suction pressure is below 20 psi (1.3 bar) under any flow condition, the suction gauge shall be a compound pressure and vacuum gauge.	
4.10.2.3 The gauge shall have a pressure range two times the rated maximum suction pressure of the pump.	
4.12.1.1.1 Fire pumps units serving high-rise buildings shall be protected from surrounding occupancies by a minimum of 2-hour fire-rated construction or physically separated from the protected building by a minimum of 50 ft (15.3 m).	Several of the requirements in Section 4.12 were revised for clarity. A new 4.12.1.1.5 was added to clarify that domestic water distribution is permitted in the fire pump room.
4.12.1.1.2 Indoor fire pump rooms in non-high-rise buildings or in separate fire pump buildings shall be physically separated or protected by fire-rated construction in accordance with Table 4.12.1.1.2.	
4.12.1.1.5 Equipment related to domestic water distribution shall be located within the same room as the fire pump equipment.	
4.12.1.2.1 Fire pumps units that are outdoors shall be located at least 50 ft (15.3 m) away from any buildings and other fire exposures exposing the building.	
4.12.2 Equipment Access.	These requirements were added to ensure that fire pump rooms are accessible under actual fire conditions.
4.12.2.1 Access to the fire pump room shall be pre-planned with the fire department.	
4.12.2.1.1 Fire pump rooms not directly accessible from the outside shall be accessible through an enclosed passageway from an enclosed stairway or exterior exit. The enclosed passageway shall have a minimum 2-hour fire-resistance rating.	
4.13.4 Drain Piping. Drain pipe and its fittings that discharge to atmosphere shall be permitted to be constructed of metallic or polymeric materials.	New text was added to clarify that drain piping can be constructed of plastic as well as metallic materials.
4.13.5 Piping, Hangers, and Seismic Bracing. Pipe, fittings, hangers, and seismic bracing for the fire pump unit, including the suction and discharge piping, shall comply with the applicable requirements of NFPA 13, *Standard for the Installation of Sprinkler Systems*.	New text was added to reference the requirements in NFPA 13 for pipe, hangers, and seismic bracing for all fire pump installations.

Subject / 2010 Edition Text	Notes
4.14.3.1* Unless the requirements of 4.14.3.2 are met, the size of the suction pipe for a single pump or of the suction header pipe for multiple pumps (operating together) shall be such that, with all pumps operating at maximum flow (150 percent of rated capacity or the maximum flow available from the water supply as discussed in 4.6.2.3.1), the gauge pressure at the pump suction flanges shall be 0 psi (0 bar) or higher.	This paragraph was revised to recognize that NFPA 20 does not require the water supply to go up to the flow of 150 percent of the water supply as long as it meets the conditions of 4.6.2.3.1.
4.14.9.2(2) Where the authority having jurisdiction requires positive pressure to be maintained on the suction piping, a pressure sensing line for a low suction pressure control, specifically listed for fire pump service, shall be permitted to be connected to the suction piping.	This paragraph was revised to agree with revised 4.15.9 on low suction pressure controls.
4.15.9 Low Suction Pressure Controls.	
4.15.9.1 Low suction throttling valves or variable speed suction limiting controls for pump drivers that are listed for fire pump service and that are suction pressure sensitive shall be permitted where the authority having jurisdiction requires positive pressure to be maintained on the suction piping.	Variable speed controlled fire pumps maintain the desired pressure limits without the use of in-flow devices, which introduce the risk of flow impediment and are therefore now permitted as an alternative to inline devices. This paragraph has been revised to allow for variable speed suction limiting controls.
4.15.9.2 When a low suction throttling valve is used, it shall be installed according to manufacturers' recommendations in the piping between the pump and the discharge check valve.	
4.15.9.3* The size of the low suction throttling valve shall not be less than that given for discharge piping in Section 4.26.	
4.18.6.2 The discharge pipe shall be permitted to be sized hydraulically to discharge sufficient water to prevent the pump discharge pressure, adjusted for elevation, from exceeding the pressure rating of the system components.	The intent of this new requirement is to allow a performance-based option.
4.19 Pumps Arranged in Series.	New requirements were added to address pumps arranged in series.
4.19.1 Series Fire Pump Unit Performance.	
4.19.1.1 A series fire pump unit (pumps, drivers, controllers, and accessories) shall perform in compliance with this standard as an entire unit.	
4.19.1.2 Within 20 seconds after a demand to start, pumps in series shall supply and maintain a stable discharge pressure (±10 percent) throughout the entire range of operation.	
4.19.1.2.1 The discharge pressure shall be permitted to be adjusted and restabilized whenever the flow condition changes.	
4.19.1.3 The complete series fire pump unit shall be field acceptance tested for proper performance in accordance with the provisions of this standard. (*See Section 14.2.*)	
4.19.2 Fire Pump Arrangement.	
4.19.2.1 No more than three pumps shall be allowed to operate in series.	
4.19.2.2 No pump in a series pump unit shall be shut down automatically for any condition of suction pressure.	
4.19.2.3 No pressure reducing or pressure regulating valves shall be installed between fire pumps arranged in series.	
4.19.2.4 The pressure at any point in any pump in a series fire pump unit, with all pumps running at shutoff and rated speed at the maximum static suction supply, shall not exceed any pump suction, discharge, or case working pressure rating.	
4.20.1.4 Where a test header is installed, it shall be installed on an exterior wall or in another location outside the pump room that allows for water discharge during testing in accordance with 14.2.7.2.	This is a new requirement that was added for fire pump testing.

Subject / 2010 Edition Text	Notes
4.20.2.9 When discharging back into a tank, the discharge nozzle(s) or pipe shall be located at a point as far from the pump suction as is necessary to prevent the pump from drafting air introduced by the discharge of test water into the tank.	This is a new requirement regarding the discharge of test water back to tanks.
4.20.3.1.3 Where outlets are being utilized as a means to test the fire pump in accordance with 4.20.1.1, one of the following methods shall be used:	New language provides further clarification.
(1)* Hose valves mounted on a hose valve header with supply pipe sized in accordance with 4.20.3.4 and Section 4.26	
(2) Wall hydrants, yard hydrants, or standpipe outlets of sufficient number and size to allow testing of the pump	
4.21 Steam Power Supply Dependability.	The section was revised to clarify the intention of the text.
4.21.1 Steam Supply.	
4.21.1.1 Careful consideration shall be given in each case to the dependability of the steam supply and the steam supply system.	
4.21.1.2 Consideration shall include the possible effect of interruption of transmission piping either on the property or in adjoining buildings that could threaten the property.	
4.25.1 Pressure maintenance pumps shall not be required to be listed. Pressure maintenance pumps shall be approved.	Clarifies that pressure maintenance pumps are not required to be listed.
4.25.1.1* The pressure maintenance pump shall be sized to replenish the fire protection system pressure due to allowable leakage and normal drops in pressure.	New language was added to include the use of pressure maintenance pumps for stabilizing the air pressure in fire protection systems for situations such as thermal changes and expansion of liquid to fill voids left by escaping pressurized air.
4.25.5 Piping and Components for Pressure Maintenance Pumps.	New requirements were added to address the piping and components for pressure maintenance pumps.
4.25.5.1 Steel pipe shall be used for suction and discharge piping on pressure maintenance pumps, which includes packaged prefabricated systems.	
4.25.5.2 Valves and components for the pressure maintenance pump shall not be required to be listed.	
4.25.5.3 An isolation valve shall be installed on the suction side of the pressure maintenance pump to isolate the pump for repair.	
4.25.5.4 A check valve and isolation valve shall be installed in the discharge pipe.	
4.25.5.5* Indicating valves shall be installed in such places as needed to make the pump, check valve, and miscellaneous fittings accessible for repair.	
4.25.5.6 The pressure sensing line for the pressure maintenance pump shall be in accordance with Section 4.30.	
4.25.5.7 The isolation valves serving the pressure maintenance pump shall not be required to be supervised.	
4.25.6 The primary or standby fire pump shall not be used as a pressure maintenance pump.	
4.25.7 The controller for a pressure maintenance pump shall be listed but shall not be required to be listed for fire pump service.	
4.25.8 The pressure maintenance pump is not required to have secondary or standby power.	
4.27.3.1 Where a backflow preventer with butterfly control valves is installed in the suction pipe, the backflow preventer is required to be at least 50 ft (15.2 m) from the pump suction flange (as measured along the route of pipe) in accordance with 4.14.5.2.	New requirement clarifies the distance required when a backflow preventer with butterfly control valves is installed.

Subject / 2010 Edition Text	Notes
4.29.9* The interior floor shall be solid with grading to provide for proper drainage for the fire pump components. The structural frame for a packaged pump shall be mounted to a properly engineered footing designed to withstand the live loads of the packaged unit and the applicable wind loading requirements. The foundation footings shall include the necessary anchor points required to secure the package to the foundation.	New requirements provide guidance in the handling, manufacture, and setting of prepackaged units.
4.29.10 A high skid-resistant, solid structural plate floor with grout holes shall be permitted to be used where protected from corrosion and drainage is provided for all incidental pump room spillage or leakage.	
4.30* Pressure Actuated Controller Pressure Sensing Lines.	Section title has been revised from Pressure-Sensing Lines.
4.30.3* The pressure sensing line shall be brass, rigid copper pipe Types K, L, or M, or Series 300 stainless steel pipe or tube, and the fittings shall be of ½ in. (15 mm) nominal size.	New requirement was added to prohibit the use of soft copper tubing in pressure sensing lines since that type of tubing can be easily damaged.
4.31.1 Application. Break tanks shall be used for one or more of the following reasons: (1) As a backflow prevention device between the water supply and the fire pump suction pipe (2) To eliminate fluctuations in the water supply pressure and provide a steady suction pressure to the fire pump (3) To provide a quantity of stored water on site where the normal water supply will not provide the required quantity of water required by the fire protection system	Revisions were made to clarify the existing language.
Chapter 5 Fire Pumps for High-Rise Buildings	The 2007 edition of Chapter 5 (General Requirements) has become Chapter 4 of the 2010 edition. A new Chapter 5 on high-rise buildings has been added to NFPA 20 in 2010.
5.1 General. **5.1.1 Application.** **5.1.1.1** This chapter applies to all fire pumps within a building wherever a building is defined as high-rise per 3.3.24. **5.1.1.2** The provisions of all other chapters of this standard shall apply unless specifically addressed by this chapter. **5.2 Types.** **5.2.1** Fire pumps used for high-rise application shall be of a type addressed by Chapter 6 or 7 (centrifugal pumps) of this standard. **5.2.2** Fire pumps complying with Chapter 8 (positive displacement pumps) shall be allowed for local applications. **5.3 Equipment Access.** Location and access to the fire pump room shall be pre-planned with the fire department. **5.4 Fire Pump Test Arrangement.** Where the water supply to a fire pump is a tank, a listed flowmeter or a test header discharging back into the tank with a calibrated nozzle(s) arranged for the attachment of a pressure gauge to determine pitot pressure shall be permitted. **5.5 Auxiliary Power.** Where electric motors are used and the height of the structure is beyond the pumping capability of the fire department apparatus, a reliable emergency source of power in accordance with Section 9.6 shall be provided for the fire pump installation.	Because some of the most complex fire pump installations exist in high-rise buildings, the technical committee has provided a new Chapter 5 containing special requirements for fire pumps in high-rise buildings. The requirements were developed after reviewing current code requirements, current design practices, fire-fighting operations, maintenance implications, and overall reliability and risk exposure.

Subject / 2010 Edition Text	Notes
5.6* Fire Pump Backup. Fire pumps serving zones that are partially or wholly beyond the pumping capability of the fire department apparatus shall be provided with an auxiliary means that is capable of providing the full fire protection demand. **5.7 Water Supply Tanks.** **5.7.1** Water tanks shall be installed in accordance with NFPA 22, *Standard for Water Tanks for Private Fire Protection*. **5.7.2** When a water tank serves domestic and fire protection systems, the domestic supply connection shall be connected above the level required for fire protection demand. **5.7.3** Water tanks supplying suction to fire pumps serving zones that are partially or wholly beyond the pumping capability of the fire department apparatus shall meet the requirements in 5.7.3.1 through 5.7.3.5 with a minimum of two automatic fill valves with separate piping connected to the zone below or primary water supply to the building. A manual fill valve shall also be provided. **5.7.3.1** Two or more water tanks shall be provided. Alternatively, a water tank shall be permitted to be divided into compartments such that the compartments function as individual tanks. **5.7.3.2** Water tank(s) shall be sized for the full fire protection demand and arranged so that at least 50 percent of the fire protection demand is stored with any one compartment or tank out of service. **5.7.3.3** An automatic refill valve shall be provided for each tank or tank compartment. **5.7.3.4** A manual refill valve shall be provided. **5.7.3.5** Each refill valve shall be sized and arranged to independently supply the system fire protection demand.	
Chapter 6 Centrifugal Pumps	No changes.
Chapter 7 Vertical Shaft Turbine–Type Pumps **7.3.5.3* Water Level Detection.** Water level detection shall be required for all vertical turbine pumps installed in wells to monitor the suction pressure available at the shutoff, 100 percent flow, and 150 percent flow points, to determine if the pump is operating within its design conditions. **7.5.1.3** Vertical shaft turbine pumps shall be driven by a vertical hollow shaft electric motor or vertical hollow shaft right-angle gear drive with diesel engine or steam turbine except as permitted in 7.5.1.4. **7.5.1.6.1** For drive systems that include a right angle gear drive, the pump manufacturer shall provide a complete mass elastic system torsional analysis to ensure there are no damaging stresses or critical speeds within 25 percent above and below the operating speed of the pump and drive. **7.5.1.6.3** For variable speed vertical hollow shaft electric motors, the pump manufacturer shall provide a complete mass elastic system torsional analysis to ensure there are no damaging stresses or critical speeds within 25 percent above and below the operating speed of the pump and drive. **7.5.1.8.2** The operating angle for the flexible connecting shaft shall not exceed the limits specified by the manufacturer for the speed and horsepower transmitted under any static or operating conditions. **7.5.3 Variable Speed Vertical Turbine Pumps.** **7.5.3.1** The pump supplier shall inform the controller manufacturer of any and all critical resonant speeds within the operating speed range of the pump, which is from zero up to full speed. **7.5.3.2** When water-lubricated pumps with line shaft bearings are installed, the pump manufacturer shall inform the controller manufacturer of the maximum allowed time for water to reach the top bearing under the condition of the lowest anticipated water level of the well or reservoir.	New text was added to 7.3.5.3 to mandate a method to determine the suction pressure of a vertical turbine pump installed in a well. The requirements in 7.5.1.3 and 7.5.1.6.1 were revised regarding the torsional analysis for right angle gear drives. A new 7.5.1.6.3 was added regarding the requirements for torsional analysis for variable speed drive pumps. This paragraph was revised to include requirements for rigidly mounted pumps and drivers. A new subsection on variable speed vertical turbine pumps was added. The two new requirements are essential for safe and reliable operation of vertical turbine pumps when operating at variable speeds.

Subject / 2010 Edition Text	Notes
Chapter 8 Positive Displacement Pumps **8.4.2.1** All pumps shall be equipped with a listed safety relief valve capable of relieving 100 percent of the rated pump capacity at a pressure not exceeding 125 percent of the relief valve set pressure. **8.4.2.2** The pressure relief valve shall be set such that the pressure required to discharge the rated pump capacity is at or below the lowest rated pressure of any component.	Paragraphs 8.4.2.1 and 8.4.2.2 were revised because relief valves used on positive displacement pumps can require a pressure up to 125 percent of valve set pressure to discharge the rated pump capacity.
8.5.2.4 For drive systems that include a gear case, the pump manufacturer shall provide a complete mass elastic system torsional analysis to ensure there are no damaging stresses or critical speeds within 25 percent above and below the operating speed of the pump(s) and driver. **8.5.2.4.1** For variable speed drives, the analysis of 8.5.2.4 shall include all speeds down to 25 percent below the lowest operating speed obtainable with the variable speed drive.	Paragraphs 8.5.2.4 and 8.5.2.4.1 are new requirements in NFPA 20. Positive displacement pumps require the ability to have gear case drives to provide drive for multiple pumps and/or speed correction. However, because of sensitive torsional issues to achieve a properly designed and reliable drive system utilizing a gear case, the system needs to have a full mass elastic torsional analysis considering all components in the system.
Chapter 9 Electric Drive for Pumps **9.1.7*** Phase converters shall not be used to supply power to a fire pump.	Phase converters that take single-phase power and convert it to three-phase power for the use of fire pump motors are not permitted because of the imbalance in the voltage between the phases when there is no load on the equipment.
9.2.3.4 Where the overcurrent protection permitted by 9.2.3 is installed, the overcurrent protection device shall be rated to carry indefinitely the sum of the locked rotor current of the fire pump motor(s) and the pressure maintenance pump motor(s) and the full-load current of the associated fire pump accessory equipment.	This revision clarifies the sizing of the overcurrent protection. Often circuit breakers are used for this purpose but are not sized thermally to carry the locked rotor current indefinitely.
9.3.6 Two or More Alternate Sources. Where the alternate source consists of two or more sources of power and one of the sources is a dedicated feeder derived from a utility service separate from that used by the normal source, the disconnecting means, overcurrent protective device, and conductors shall not be required to meet the requirements of Section 9.2 and shall be permitted to be installed in accordance with *NFPA 70, National Electrical Code*.	New requirement was added to clarify that the alternate source may consist of two or more sources of power. When one of those sources is a dedicated feeder derived from a utility service separate from that used by the normal source, the reliability of the power available to the fire pump controller is increased significantly and the special requirements of Section 9.2 are considered unnecessary.
9.4.3 The requirements of 9.4.1 shall not apply to the bypass mode of a variable speed pressure limiting control *(see 10.10.3)*, provided a successful start can be demonstrated on the standby gen-set.	New requirement was added to clarify that the voltage drop requirements of Section 9.4 do not apply to the bypass mode of a variable speed pressure limiting control.
9.5.1.3 Part-winding motors shall have a 50-50 winding ratio in order to have equal currents in both windings while running at nominal speed.	New requirement was added to specify the criteria needed for proper sizing of contactors in a part-winding fire pump controller.
9.6.2.2 The engine shall run and continue to produce rated nameplate power without shutdown or de-rate for alarms and warnings, or failed engine sensors, except for overspeed shutdown.	This requirement ensures that engines used as the electrical backup to electric drivers have a level of performance similar to what is required in Chapter 11, which includes a requirement that is more stringent than what is required for a Level 1 EPSS.

Subject / 2010 Edition Text	Notes
9.6.5* Protective Devices. Protective devices installed in the on-site power source circuits at the generator shall allow instantaneous pickup of the full pump room load and shall comply with *NFPA 70, National Electrical Code*, Section 700.27.	The wording of 9.6.5 was revised to clarify that these protective devices are not optional. Article 445 of the *NEC* requires that the load side of the generator be supplied with overcurrent protection. The reference to Section 700.27 of the *NEC* was added to ensure that these protective devices are selectively coordinated with all overcurrent devices in the alternate power circuit.
9.8.2* The raceway between a junction box and the fire pump controller shall be sealed at the junction box end with an identified compound and in accordance with the instructions of the electrical circuit protective systems if provided.	The provision in 9.8.2 was revised to require sealing of the raceway between a junction box and the fire pump controller in all installations where a junction box is provided. Previously, sealing was only necessary where required by the manufacturer of a listed electrical circuit protective system or by the *NEC*, or by the listing.
Chapter 10 Electric-Drive Controllers and Accessories	
10.1.2.3 Preshipment. All controllers shall be completely assembled, wired, and tested by the manufacturer before shipment from the factory. Controllers shipped in sections shall be completely assembled, wired, and tested by the manufacturer before shipment from the factory. Such controllers shall be reassembled in the field, and the proper assembly shall be verified by the manufacturer or designated representative.	This paragraph was revised to recognize that some fire pump controllers due to size and construction cannot be shipped as a complete assembly and need to be shipped in sections. Proper reassembly at the installation site is ensured by requiring the reassembly to be verified by the manufacturer or designated representative.
10.3.3.1* The structure or panel shall be securely mounted in, as a minimum, a National Electrical Manufacturers Association (NEMA) Type 2, dripproof enclosure(s) or an enclosure(s) with an ingress protection (IP) rating of IP31.	This paragraph was revised to recognize the ingress protection (IP) rating system for enclosures. An enclosure with an IP rating of IP31 is considered to provide protection equivalent to the protection provided by a Type 2 enclosure.
10.4.5.1.1 Running contactors shall be sized for both the locked rotor currents and the continuous running currents encountered. **10.4.5.1.2** Starting contactors shall be sized for both the locked rotor current and the acceleration (starting) encountered.	Paragraphs 10.4.5.1.1 and 10.4.5.1.2 are new requirements that address the sizing criteria for the selection of running and starting contactors in a fire pump controller. Across-the-line starting fire pump controllers generally employ one contactor that needs to meet the sizing criteria specified for a running contactor. Reduced-voltage starting fire pump controllers employ multiple contactors and the sizing criteria that an individual contactor needs to meet are based on its function in the controller.
10.5.2.1.1.1 There shall be provided a pressure-actuated switch or electronic pressure sensor having adjustable high- and low-calibrated set-points as part of the controller.	This requirement was revised to reflect current technology and practice in permitting the use of electronic pressure transducers.
10.5.2.1.2 There shall be no pressure snubber or restrictive orifice employed within the pressure switch or pressure responsive means.	Paragraph 10.5.2.1.2 was revised to recognize pressure responsive means.
10.5.2.1.3 There shall be no valve or other restrictions within the controller ahead of the pressure switch or pressure responsive means.	Paragraph 10.5.2.1.3 is a new requirement. Valves are an added and unnecessary item in the critical starting path of fire pump controllers.
10.5.2.1.7.5 For variable speed pressure limiting control, a ½ in. (15 mm) nominal size inside diameter pressure line shall be connected to the discharge piping at a point recommended by the variable speed control manufacturer. The connection shall be between the discharge check valve and the discharge control valve.	This paragraph was revised to clarify the location of the control line connection based on the type of equipment used.

Subject / 2010 Edition Text	Notes
10.5.2.5.2 Each pump supplying suction pressure to another pump shall be arranged to start within 10 seconds before the pump it supplies.	This paragraph was revised to clarify the time interval for starting pumps in series in order to ensure adequate water flow.
10.6.5.5 Current Transformers. Unless rated at the incoming line voltage, the secondaries of all current transformers used in the high voltage path shall be grounded.	This is a new requirement in NFPA 20. Unless the current transformer is rated for use at the incoming line (medium) voltage, the secondaries of all current transformers should be grounded to reduce the risk of electric shock to anyone who may come in contact with the fire pump controller.
10.8.2.1.3 Circuit Breaker. The transfer switch emergency side shall be provided with a circuit breaker complying with 10.4.3 and 10.4.4.	The requirements for equipment in the alternate source side of the transfer switch when supplied by a second utility were revised such that the requirements are now identical to those for equipment in the normal source side of the transfer switch. The essence of the change is that the transfer switch emergency side is now required to be provided with both an isolation switch and a circuit breaker.
10.10* Controllers with Variable Speed Pressure Limiting Control or Variable Speed Suction Limiting Control. **10.10.1.1** Controllers equipped with variable speed pressure limiting control or variable speed suction limiting control shall comply with the requirements of Chapter 10, except as provided in 10.10.1 through 10.10.11. **10.10.1.2** Controllers with variable speed pressure limiting control or variable speed suction limiting control shall be listed for fire service. **10.10.1.3** The variable speed pressure limiting control or variable speed suction limiting control shall have a horsepower rating at least equal to the motor horsepower or, where rated in amperes, shall have an ampere rating not less than the motor full-load current.	The title of Section 10.10 and the requirements of 10.10.1.1, 10.10.1.2, and 10.10.1.3 were revised to reflect current technology.
10.10.3.2 When the variable speed pressure limiting control is bypassed, the unit shall remain bypassed until manually restored.	This paragraph was revised to clarify that the minimum run timer is not to be used in the bypass mode.
10.10.12 Variable Speed Drives for Vertical Pumps. **10.10.12.1** The pump supplier shall inform the controller manufacturer of any and all critical resonant speeds within the operating speed range of the pump, which is from zero speed up to full speed. The controller shall avoid operating at or ramping through these speeds. The controller shall make use of skip frequencies with sufficient bandwidth to avoid exciting the pump into resonance. **10.10.12.2** When water-lubricated pumps with line shaft bearings are installed, the pump manufacturer shall inform the controller manufacturer of the maximum allowed time for water to reach the top bearing under the condition of the lowest anticipated water level of the well or reservoir. The controller shall provide a ramp up speed within this time period. **10.10.12.3** The ramp down time shall be approved or agreed to by the pump manufacturer. **10.10.12.4** Any skip frequencies employed and their bandwidth shall be included along with the information required in 10.10.11. **10.10.12.5** Ramp up and ramp down times for water-lubricated pumps shall be included along with the information required in 10.10.11.	New subsection was added to recognize the application of variable speed drives to vertical turbine pumps. This application is primarily for well pumps.

Subject / 2010 Edition Text	Notes
Chapter 11 Diesel Engine Drive	Many of the requirements in Chapter 11 were reorganized within the chapter for user friendliness.
11.2.2.2* The horsepower capability of the engine, when equipped for fire pump service, shall have a 4-hour minimum horsepower rating not less than 10 percent greater than the listed horsepower on the engine nameplate.	This revision clarifies that this is a fire pump driver certification test and not an individual installation test.
11.2.4.2.3.1 Operation. The transition from the primary ECM to the alternate ECM shall be accomplished automatically upon failure of the primary ECM, and a hand/automatic switch without an off position shall be provided.	This paragraph was revised to provide clarification.
11.2.4.2.3.3 Contacts. **(A)** The contacts for each circuit shall be rated for both the minimum and maximum current and voltage. **(B)** The total resistance of each ECM circuit through the selector switch shall be approved by the engine manufacturer.	The electrical characteristics of circuits between the engines and ECMs vary widely depending on function and manufacturer. The new requirements in 11.2.4.2.3.3 will be useful for the third party listing agency to ensure proper circuit design.
11.2.4.2.3.4 Enclosure. **(A)** The selector switch shall be enclosed in a NEMA Type 2 dripproof enclosure. **(B)** Where special environments exist, suitably rated enclosures shall be used.	These new requirements ensure proper protection.
11.2.4.2.3.5 Mounting. **(A)** The selector switch and enclosure shall be engine mounted. **(B)** The selector switch enclosure and/or the selector switch inside shall be isolated from engine vibration to prevent any deterioration of contact operation.	This paragraph provides new requirements for vibration isolation.
11.2.4.2.7 ECM and Engine Power Supply. **11.2.4.2.7.1** In the standby mode, the engine standby battery shall be used to power the ECM. **11.2.4.2.7.2** These engines shall not require more than 0.5 ampere from the battery or battery charger while the engine is not running.	The requirements in 11.2.4.2.7 are necessary to accommodate electronic controlled engines.
11.2.4.3 Variable Speed Pressure Limiting Control or Variable Speed Suction Limiting Control (Optional). **11.2.4.3.1** Variable speed pressure limiting control or variable speed suction limiting control systems used on diesel engines for fire pump drive shall be listed for fire pump service and be capable of limiting the pump output total rated head (pressure) or suction pressure by reducing pump speed. **11.2.4.3.2** Variable speed control systems shall not replace the engine governor as defined in 11.2.4.1. **11.2.4.3.3** In the event of a failure of the variable speeds control system, the engine shall be fully functional with the governor defined in 11.2.4.1.	The title of 11.2.4.3 and the requirements in 11.2.4.3.1, 11.2.4.3.2, and 11.2.4.3.3 were revised to reflect current technology.
11.2.4.4.2 The overspeed device shall be arranged to shut down the engine in a speed range of 10 to 20 percent above rated engine speed and to be manually reset.	This requirement was revised to permit an acceptable range. Set-point tolerances in the manufacturing setting along with the field verification lends itself better to an acceptable range than a single percentage.
11.2.4.4.4 Means shall be provided for verifying overspeed switch and circuitry shutdown function.	This is a new requirement that is needed to verify overspeed shutdown function.
11.2.7.2.1.5* Batteries shall be sized on a calculated capacity of 72 hours of standby power followed by three 15-second attempt-to-start cycles per battery unit, without ac power being available for battery charging.	New requirement provides performance criteria for batteries.

Subject / 2010 Edition Text	Notes
11.2.7.2.2* Battery Isolation. **11.2.7.2.2.1** Engines with only one cranking motor shall include a main battery contactor installed between each battery and the cranking motor for battery isolation. **(A)** Main battery contactors shall be listed for fire pump driver service. **(B)** Main battery contactors shall be rated for the cranking motor current. **(C)** Main battery contactors shall be capable of manual mechanical operation including positive methods such as spring-loaded, over-center operator to energize the starting motor in the event of controller circuit failure. **11.2.7.2.2.2** Engines with two cranking motors shall have one cranking motor dedicated to each battery. **(A)** Each cranking motor shall meet the cranking requirements of a single cranking motor system. **(B)** To activate cranking, each cranking motor shall have an integral solenoid relay to be operated by the pump set controller. **(C)** Each cranking motor integral solenoid relay shall be capable of being energized from a manual operator listed and rated for the cranking motor solenoid relay and include spring-loaded, over-center operation to energize the starting motor in the event of controller circuit failure.	Because of the unique role of main battery contactors, clarification has been provided to the standard regarding one and two starting motor systems.
11.2.7.2.3 Battery Loads. **11.2.7.2.3.1** Nonessential loads shall not be powered from the engine starting batteries. **11.2.7.2.3.2** Essential loads, including the engine, controller, and all pump equipment combined, shall not exceed 0.5 ampere each for a total of 1.5 amperes, on a continuous basis.	These requirements provide clarification of the allowed loads on the battery charger.
11.2.8.2 A means shall be provided to maintain 120°F (49°C) at the combustion chamber.	Revised text clarifies the intended performance of all engines, not just those provided with a jacket water heater.
11.3.3 The entire pump room shall be protected with fire sprinklers in accordance with NFPA 13, *Standard for the Installation of Sprinkler Systems,* as an Extra Hazard Group 2 space.	This is a new requirement for the pump room to be protected as an Extra Hazard Group 2 space. Since the fire pump is in the same room, the flow and pressure to protect this small room are most likely present, and this should help alleviate any of the fears regarding the installation of the fuel tank in the pump room.
11.4.1.2 Tank Construction. **11.4.1.2.1*** Tanks shall be designed and constructed in accordance with recognized engineering standards such as ANSI/UL 142, *Standard for Steel Aboveground Tanks for Flammable and Combustible Liquids.* **11.4.1.2.2** Tanks shall be securely mounted on noncombustible supports. **11.4.1.2.3** Tanks used in accordance with the rules of this standard shall be limited in size to 1320 gal (4996 L). For situations where fuel tanks in excess of 1320 gal (4996 L) are being used, the rules of NFPA 37, *Standard for the Installation and Use of Stationary Combustion Engines and Gas Turbines,* shall apply. **11.4.1.2.4** Fuel tanks shall be enclosed with a wall, curb, or dike sufficient to hold the entire capacity of the tank. **11.4.1.2.5** Each tank shall have suitable fill, drain, and vent connections. **11.4.1.2.6** Fill pipes that enter the top of the tank shall terminate within 6 in. (152 mm) of the bottom of the tank and shall be installed or arranged so that vibration is minimized.	New language was added by taking the applicable requirements of NFPA 30 and rewriting them so that they are clear, enforceable, and flow consistently with the already existing rules of NFPA 20.

Subject / 2010 Edition Text	Notes
11.4.1.2.7 The fuel tank shall have one 2 in. (50.8 mm) NPT threaded port in the top, near the center, of the tank to accommodate the low fuel level switch. **11.4.1.2.8 Tank Venting.** **11.4.1.2.8.1** Normal vents shall be sized in accordance with ANSI/UL 142, *Standard for Steel Aboveground Tanks for Flammable and Combustible Liquids,* or other approved standards. Alternatively, the normal vent shall be at least as large as the largest filling or withdrawal connection, but in no case shall it be less than 1¼ in. (32 mm) nominal inside diameter. **11.4.1.2.8.2** Vent piping shall be arranged so that the vapors are discharged upward or horizontally away from adjacent walls and so that vapors will not be trapped by eaves or other obstructions. Outlets shall terminate at least 5 ft (1.5 m) from building openings.	
11.4.2.4 A means shall be provided within the tank for low fuel level signal initiation.	New requirement was added for signaling to be provided upon low fuel supply level.
11.6.1.1 Engines shall be designed and installed so that they can be started no less than once a week and run for no less than 30 minutes to attain normal running temperature. **11.6.2* Engine Maintenance.** Engines shall be designed and installed so that they can be kept clean, dry, and well lubricated to ensure adequate performance. **11.6.3.1** Storage batteries shall be designed and installed so that they can be kept charged at all times. **11.6.3.2** Storage batteries shall be designed and installed so that they can be tested frequently to determine the condition of the battery cells and the amount of charge in the battery. **11.6.3.6** The battery and charger shall be designed and installed so that periodic inspection of both battery and charger is physically possible. **11.6.4.1** The fuel storage tanks shall be designed and installed so that they can be kept as full and maintained as practical at all times but never below 66 percent (two-thirds) of tank capacity. **11.6.4.2** The tanks shall be designed and installed so that they can always be filled by means that will ensure removal of all water and foreign material. **11.6.5.1** The temperature of the pump room, pump house, or area where engines are installed shall be designed so that the temperature is maintained at the minimum recommended by the engine manufacturer and is never less than the minimum recommended by the engine manufacturer.	The maintenance requirements were revised to take into account the initial design and installation of the fire pump and ancillary equipment. These provisions are found in NFPA 20 so that the installation can be made in such a manner that the system can be inspected, tested, and maintained in accordance with NFPA 25.
Chapter 12 Engine Drive Controllers	Chapter 12 was slightly reorganized to eliminate duplicate text within the chapter.
12.3.3.1.1 The structure or panel shall be securely mounted in, as a minimum, a NEMA Type 2 dripproof enclosure(s) or an enclosure(s) with an ingress protection (IP) rating of IP 31.	This paragraph was revised to recognize the ingress protection (IP) rating system for enclosures. An enclosure with an IP rating of IP31 is considered to provide protection equivalent to the protection provided by a Type 2 enclosure.
12.3.5.3.2 No undervoltage, phase loss, frequency sensitive, or other device(s) shall be field installed that automatically or manually prohibits electrical actuation of the motor contactor.	In the 2010 edition of NFPA 20, the wording change from *sensor(s)* to *device(s)* makes the requirement more general. Adding "field" installed clarifies the intent of no field modifications to the controller design or construction. Modifying the controller in any way could void the listing(s) and likely void the manufacturer's warranty.
12.4.1.4(9) Low engine temperature	Where engines rely on a jacket water heater to meet the requirements of 11.2.8.2, the fire pump starting is in jeopardy if the heater isn't working.

Subject / 2010 Edition Text	Notes
12.4.1.5 No audible signal silencing switch or valve, other than the controller main switch, shall be permitted for the conditions reflected in 12.4.1.3 and 12.4.1.4.	The words "or valve" were added in the 2010 edition to address controllers that operate pneumatically or use fluidic logic rather than traditional electrical circuitry.
12.4.1.5.1 A separate signal silencing switch shall be used for the conditions of 12.4.1.4(5), 12.4.1.4(7), and 12.4.1.4(8).	These three alarms — ECM selector switch in alternate ECM position, low fuel level, and low air pressure — may exist for extended periods of time. Making these alarms silenceable is for attended pump rooms and also to avoid unnecessary battery drain during power outage periods and/or to avoid someone taking the controller out of service to quiet the alarm.
12.4.1.5.2* The controller shall automatically return to the nonsilenced state when the alarm(s) have cleared (returned to normal). This switch shall be clearly marked as to its function.	This requirement was revised because it is important that the controller return to the reset state, ready to signal the next alarm, without intervention, when the initiating alarm condition has been cleared (reset to the normal condition).
12.4.6 Voltmeter. A voltmeter with an accuracy of ±5 percent shall be provided for each battery bank to indicate the voltage during cranking or to monitor the condition of batteries used with air-starting engine controllers.	The text was revised to address the condition of batteries used with air-starting engine controllers.
Section 12.5* Battery Recharging. **Section 12.6 Battery Chargers.**	The requirements for battery chargers are part of the diesel fire pump controller and as such the requirements for battery recharging and battery chargers have been deleted from Chapter 11 and are now in Chapter 12.
12.6(11) The charger(s) shall remain in float mode or switch from equalize to float mode while the batteries are under the loads in 12.5.2.	New requirement was added to ensure that the charger does not become stuck in the equalize (high rate) mode.
12.7.2.3 Manual Electric Control at Remote Station. Where additional control stations for causing nonautomatic continuous operation of the pumping unit, independent of the pressure-actuated switch or control valve, are provided at locations remote from the controller, such stations shall not be operable to stop the engine.	The text was revised to include control valve.
12.7.2.4.2 Each pump supplying suction pressure to another pump shall be arranged to start within 10 seconds before the pump it supplies.	This paragraph was revised to clarify the time interval for starting pumps in series in order to ensure adequate water flow.
12.7.2.4.2.1 The controllers for pumps arranged in series shall be interlocked to ensure the correct pump starting sequence.	New requirement was added addressing pumps in series.
12.7.2.7.2 The controller shall use the opposite battery bank (every other bank) for cranking on subsequent weeks.	This is a new requirement that is common practice with currently listed controllers.
Chapter 13 Steam Turbine Drive	No changes.
Chapter 14 Acceptance Testing, Performance, and Maintenance	Several formatting changes were made to Chapter 14 (such as adding additional section headings) to enhance user friendliness and provide clarification of the requirements.
14.2.5.2.3.1 Where simultaneous operation of multiple pumps is possible or required as part of a system design, the acceptance test shall include a flow test of all pumps operating simultaneously.	Chapter 4 recognizes the installation of multiple pumps operating together by requiring an appropriately sized suction header to limit velocity in the suction pipe to not more than 15 ft/sec with all pumps operating together. This new requirement in Chapter 14 verifies proper operation of such an arrangement.

Subject / 2010 Edition Text	Notes
14.2.5.2.4 Where the maximum flow available from the water supply cannot provide a flow of 150 percent of the rated flow of the pump, the fire pump shall be operated at the greater of 100 percent of rated flow or the maximum flow demand of the fire protection system(s) maximum allowable discharge to determine its acceptance. This reduced capacity shall constitute an acceptable test, provided that the pump discharge exceeds the fire protection system design and flow rate.	This requirement was revised to correlate with the language in 4.6.2.3.1.
14.2.5.2.6 Water Level Detection. Water level detection is required for all vertical turbine pumps installed in wells to determine the suction pressure available at the shutoff, 100 percent flow, and 150 percent flow points to determine if the pump is operating within its design conditions.	This new requirement originates from the new 7.3.5.3, which mandates a method to determine the suction pressure of a vertical turbine pump installed in a well.
14.2.5.4.3.1 The pump flow for positive displacement pumps shall be tested and determined to meet the specified rated performance criteria where only one performance point is required to establish positive displacement pump acceptability.	New requirement was added for positive displacement pumps where only one performance point is required to establish acceptability.
14.5.2.3* When an ECM on an electronic fuel management–controlled engine is replaced, the replacement ECM shall include the same software programming that was in the original ECM.	New requirement was added that addresses replacement of ECM. Fire pump engines have unique features when compared to standard industrial engines. The standard industrial ECM programming may result in the reduction of power to self protect the engine, the inability to accelerate the pump to rated speed in rated flow condition, or other unwanted operation during a fire.
Annex A Explanatory Material	
A.4.3.2(2) Nationally recognized fire protection certification programs include, but are not limited to, those programs offered by the International Municipal Signal Association (IMSA) and the National Institute for Certification in Engineering Technologies (NICET). Note: These organizations and the products or services offered by them have not been independently verified by the NFPA, nor have the products or services been endorsed or certified by the NFPA or any of its technical committees.	Annex text provides additional guidance for the qualifications of personnel.
A.4.3.4 Service personnel should be able to do the following: (1) Understand the requirements contained in this standard and in NFPA 25, *Standard for the Inspection, Testing, and Maintenance of Water-Based Fire Protection Systems*, and the fire pump requirements contained in *NFPA 70, National Electrical Code* (2) Understand basic job site safety laws and requirements (3) Apply troubleshooting techniques and determine the cause of fire protection system trouble conditions (4) Understand equipment-specific requirements, such as programming, application, and compatibility (5) Read and interpret fire protection system design documentation and manufacturers' inspection, testing, and maintenance guidelines (6) Properly use tools and equipment required for testing and maintenance of fire protection systems and their components (7) Properly apply the test methods required by this standard and NFPA 25, *Standard for the Inspection, Testing, and Maintenance of Water-Based Fire Protection Systems*	Annex text provides additional guidance for the qualifications of personnel.
A.4.14.3.1 It is permitted that the suction pressure drop to −3 psi for a horizontal pump that is taking suction from a grade level storage tank where the pump room elevation and bottom of the water storage tank are at the same elevation. This negative suction pressure is to allow for the friction loss in the suction piping when the pump is operating at 150 percent capacity.	This new annex material clarifies the intent of the suction sizing and tank elevation under maximum flow conditions.

Subject / 2010 Edition Text	Notes
A.4.15.9.3 The friction loss through a low suction throttling valve must be taken into account in the design of the fire protection system.	The proper sized valve is required to ensure that the system design is not compromised by the installation of an incorrect valve.
A.4.25.1.1 The sizing of the pressure maintenance pump requires a thorough analysis of the type and size of system the pressure maintenance pump will serve. Pressure maintenance pumps on fire protection systems that serve large underground mains need to be larger than pressure maintenance pumps that serve small aboveground fire protection systems. Underground mains are permitted by NFPA 24, *Standard for the Installation of Private Fire Service Mains and Their Appurtenances*, to have some leakage (*see 10.10.2.2.4 of NFPA 24 for allowable leakage rates*) while aboveground piping systems are required to be tight when new and should not have significant leakage. For situations where the pressure maintenance pump serves only aboveground piping for fire sprinkler and standpipe systems, the pressure maintenance pump should be sized to provide a flow less than a single fire sprinkler. The main fire pump should start and run (providing a pump running signal) for any waterflow situation where a sprinkler has opened, which will not happen if the pressure maintenance pump is too large. One guideline that has been successfully used to size pressure maintenance pumps is to select a pump that will make up the allowable leakage rate in 10 minutes or 1 gpm (3.8 L/min), whichever is larger.	New annex material addresses the concerns that need to be considered when sizing a pressure maintenance pump. The existing text from A.5.24 of the 2007 edition of NFPA 20 regarding the sizing of pressure maintenance pumps was incorporated into this annex note because this is a more specific and appropriate location.
A.4.29.9 Figure A.4.29.9 illustrates a typical foundation detail for a packaged fire pump assembly.	Annex figure provides an example of a typical foundation detail for a packaged fire pump assembly.
A.4.30.3 The use of soft copper tubing is not permitted for a pressure sensing line because it is easily damaged.	Annex text clarifies the intent of the requirement.
A.5.6 If backup fire pumps are required, they can be arranged to prevent both pumps from starting and running simultaneously and, if done, the following arrangement is suggested: (1) Turning off or disconnecting power to the primary fire pump controller should not prevent the redundant fire pump from starting or running. (2) Turning off or disconnecting power to the redundant fire pump controller should not prevent the primary fire pump from starting or running. (3) Once the primary fire pump is locked out, it should remain locked out until manual reset. (4) Once the redundant fire pump is running, it should remain running until manual reset. (5) Either controller should always be capable of being operated by local manual starting regardless of any lockout that has occurred. (6) A local visual alarm and remote contacts should be provided to indicate that the primary fire pump has been locked out. A fully independent and automatic backup fire pump unit(s) arranged so that all zones can be maintained in full service with any one pump out of service should be considered one such means.	Annex text provides guidance on backup fire pumps.
A.6.3.1 See Figure A.6.3.1(a) and Figure A.6.3.1(b).	New Figure A.6.3.1(b) was added to show a backflow preventer installation.
A.7.2.2.1.2 The acceptability of a well is determined by a 24-hour test that flows the well at 150 percent of the pump flow rating. This test should be reviewed by qualified personnel (usually a well drilling contractor or a person having experience in hydrology and geology). The adequacy and reliability of the water supply are critical to the successful operation of the fire pump and fire protection system.	New guidance has been provided for identifying the parameters desirable in the design of vertical turbine pumps in wells.

Subject / 2010 Edition Text	Notes
A 10 ft (3.05 m) submergence is considered the minimum acceptable level to provide proper pump operation in well applications. The increase of 1 ft (0.30 m) for each 1000 ft (305 m) increase in elevation is due to loss of atmospheric pressure that accompanies elevation. Therefore, the net positive suction head (NPSH) available must be considered in selection of the pump. For example, to obtain the equivalent of 10 ft (3.05 m) of NPSH available at an elevation of 1000 ft (305 m), approximately 11 ft (3.35 m) of water is required. Several other design parameters need to be considered in the selection of a vertical turbine pump, including the following: (1) *Lineshaft lubrication when the pump is installed in a well.* Bearings are required to have lubrication and are installed along the lineshaft to maintain alignment. Lubrication fluid is usually provided by a fluid reservoir located aboveground, and the fluid is supplied to each bearing by a copper tube or small pipe. This lubrication fluid should use a vegetable-based material that is approved by the federal Clean Water Act to minimize water contamination. (2) *Determination of the water level in the well.* When a vertical turbine pump is tested, the water level in the well needs to be known so that the suction pressure can be determined. Often the air line for determining the depth is omitted, so testing of the pump for performance is not possible. The arrangement of this device is shown in Figure A.7.3.5.3, and its installation should be included in the system design.	
A.11.2.2.2 For more information, see SAE J-1349, *Engine Power Test Code — Spark Ignition and Compression Engine.* The 4-hour minimum power requirement in NFPA 20 has been tested and witnessed during the engine listing process.	Annex text clarifies that this is a fire pump driver certification test and not an individual installation test.
A.11.2.7.2.1.5 The 72-hour requirement is intended to apply when batteries are new. Some degradation is expected as batteries age.	Annex text provides clarification of the 72-hour requirement.
A.11.3.2.3 When motor-operated dampers are used in the air supply path, they should be spring operated to the open position and motored closed. Motor-operated dampers should be signaled to open when or before the engine begins cranking to start. It is necessary that the maximum air flow restriction limit for the air supply ventilator be compatible with listed engines to ensure adequate air flow for cooling and combustion. This restriction typically includes louvers, bird screens, dampers, ducts, and anything else in the air supply path between the pump room and the outdoors. Motor-operated dampers are recommended for the heat exchanger–cooled engines to enhance convection circulation. Gravity-operated dampers are recommended for use with radiator-cooled engines to simplify their coordination with the air flow of the fan. Another method of designing the air supply ventilator in lieu of dampers is to use a vent duct (with rain cap), the top of which extends through the roof or outside wall of a pump house and the bottom of which is approximately 6 in. (152.4 mm) off the floor of the pump house. This passive method reduces heat loss in the winter. Sizing of this duct must meet the requirements of 11.3.2.1.	The last paragraph is new annex material and provides another method of designing the air supply ventilator.
A.11.4.5.1 Biodiesel and other alternative fuels are not recommended for diesel engines used for fire protection because of the unknown storage life issues. It is recommended that these engines use only petroleum fuels.	Because of the limited running time of fire protection engines and therefore the limited volume of fuel consumed by them, any environmental benefit from these fuels would also be limited. It is not in the best interest of quality fire protection to use fuels that would likely exhibit greater problems of degradation when storing the fuel for extended periods of time.
A.12.4.1.5.2 This automatic reset function can be accomplished by the use of a silence switch of the automatic reset type or of the self supervising type.	Annex text provides guidance on automatic reset function.

Subject / 2010 Edition Text	Notes
A.14.5.2.3 Fire pump engines have unique features compared to standard industrial engines. The standard industrial ECM programming can result in the reduction of power to self-protect the engine during a fire or the inability to accelerate the pump to rated speed in rated flow condition.	New requirement was added that addresses replacement of ECM. Fire pump engines have unique features when compared to standard industrial engines. The standard industrial ECM programming may result in the reduction of power to self protect the engine, the inability to accelerate the pump to rated speed in rated flow condition, or other unwanted operation during a fire.
Annex B Possible Causes of Pump Troubles	
Annex C Informational References	
Annex D Material Extracted by NFPA 70, Article 695 **D.1 General.** Table D.1 indicates corresponding sections of *NFPA 70*, Article 695.	This table was updated to the 2008 edition of *NFPA 70*.

NFPA 20 Index

-A-

Additive
  Definition, 3.3.1
  Piping, 4.13.3
Additive pumps, 8.2, 8.9.2, A.8.2.2 to A.8.2.6; *see also* Foam concentrate pumps
  Definition, 3.3.37.1
  Motors, controllers for, 10.9
Air leaks/pockets, B.1.1, B.1.3, B.1.7
Air release fittings, automatic, 6.3.1, 6.3.3, 7.3.5.1(1), 7.3.5.2, A.6.3.1
Air starting, 11.2.7.4, 12.4.1.4(3), 12.8, A.11.2.7.4.4
Alarms, fire pump, *see* Fire pump alarms
Application of standard, 1.3
Approved/approval
  Definition, 3.2.1, A.3.2.1
  Requirements, 4.2, A.4.2
Aquifer (definition), 3.3.2
Aquifer performance analysis, A.7.2.1.2
  Definition, 3.3.3
Authority having jurisdiction
  Definition, 3.2.2, A.3.2.2
  Fuel system plan review, 11.4.1.1
Automatic air release fittings, *see* Air release fittings, automatic
Automatic transfer switches, 10.8.1.3
  Definition, 3.3.48.2.1

-B-

Backflow preventers, 4.14.9.2(1), 4.15.6, 4.16.1, 4.27, 4.31.1, A.4.14.9, A.4.15.6, A.11.2.8.5.3.7(B)
Backup fire pump, high-rise buildings, 5.7, A.5.7
Batteries, storage, 11.2.7.2, 12.7.4, 14.2.5.4.7.1, 14.2.6.7, A.11.2.7.2.1.5 to A.11.2.7.2.4
  Failure indicators, 12.4.1.4(1), 12.4.1.4(2)
  Isolation, 11.2.7.2.2, A.11.2.7.2.2
  Load, 11.2.7.2.3
  Location, 11.2.7.2.4, A.11.2.7.2.4
  Maintenance, 11.6.3, B.1.32
  Recharging, 12.5, A.12.5
  Voltmeter, 12.4.5
Battery cables, 11.2.6.3
Battery chargers, 12.4.1.4(2), 12.5.3, 12.6
Bowl assembly, vertical shaft turbine pumps, 7.3.3
Branch circuit (definition), 3.3.7.1

Break tanks, 4.31
  Definition, 3.3.6
  Refill mechanism, 4.31.3
  Size, 4.31.2
Butterfly valves, 4.15.7, 4.15.8, 4.27.3.1, A.4.25.5.5
Bypass line, 4.14.4, 11.2.8.6, A.4.14.4, A.11.2.8.6, A.13.2.1.1, B.1.25
Bypass operation, variable speed pressure limiting control, 10.10.3, A.10.10.3
Bypass valves, 4.16.1

-C-

Can pump (definition), 3.3.37.2
Capacity, pump, 4.8, 6.2.1, A.4.8, A.14.2.5.4
Centrifugal pumps
  Capacity, 4.8, 6.2.1, A.4.8, A.14.2.5.4
  Component replacement, 14.5.2, A.14.5.2.3
  Connection to driver and alignment, 6.5, A.6.5, B.1.23
  Definition, 3.3.37.3
  End suction, 6.1.1.2
  Factory and field performance, 6.2, A.6.2
  High-rise buildings, 5.2.1
  In-line, 6.1.1.2
  Maximum pressure for, 4.7.7, A.4.7.7
  Pressure maintenance pumps, 4.25.4.1, A.25.4
  Relief valves, 4.18, A.4.18.1 to A.4.18.8
  Sole supply, 12.7.2.6
  Types, 6.1.1, A.6.1.1
Check valves, 4.14.9.2(1), 4.15.6, 4.25.5.4, 4.27, 4.30.2, 4.30.4.1, 11.2.8.5.3.2, A.4.15.6, A.4.25.5.5, A.11.2.8.5.3.7(A)
Circuit breakers, *see* Disconnecting means
Circuits, *see also* Fault tolerant external control circuits
  Branch circuit (definition), 3.3.7.1
  Internal conductors, B.3.2.3
  Protection of, *see* Protective devices
Circulation relief valves, *see* Relief valves
Columns, vertical shaft turbine pumps, 7.3.2, A.7.3.2.1
Contractor's material and test certificate, 14.1.3, A.14.1.3
Controllers, fire pump, *see* Fire pump controllers
Control valves, *see* Gate valves
Coolant, engine, 11.2.8.3, 11.2.8.4
Corrosion-resistant material (definition), 3.3.9
Couplings, flexible, *see* Flexible couplings
Cutting, torch, 4.13.6, A.4.13.6

-D-

Definitions, Chap. 3
Designer, system, 4.3.2, A.4.3.2(2)
Detectors, liquid level, *see* Liquid level
Diesel engines, 4.7.2, 4.7.6, Chap. 11, A.4.7.6; *see also* Engine drive controllers
   Connection to pump, 11.2.3
   Cooling, 11.2.8, A.11.2.8.5, A.11.2.8.6, B.1.25
   Definition, 3.3.13.1
   Emergency starting and stopping, 11.6.6
   Exhaust, 11.5, A.11.5.2
   Fire pump buildings or rooms with, 4.12.1.3
   Fuel supply and arrangement, 11.4, A.11.4.1.2.1 to A.11.4.5
   Instrumentation and control, 11.2.4, A.11.2.4.2
   Listing, 11.2.1
   Lubrication, 11.2.9
   Operation and maintenance, 11.6, A.11.6, A.14.2.5
   Pump room, 11.3, A.11.3
   Ratings, 11.2.2, A.11.2.2.2
   Redundant, 9.3.3
   Starting methods, 11.2.7, 11.6.6, A.11.2.7.2.1.5 to A.11.2.7.4.4, A.14.2.5
   Tests, 14.2.5 to 14.2.7, A.14.2.5
   Type, 11.1.3, A.11.1.3
   Vertical lineshaft turbine pumps, 7.5.1.3, 7.5.1.4, 7.5.2
Discharge, exhaust, 11.5.3
Discharge cones, 4.18.5.1, 4.18.5.4, 6.3.2(4), 7.3.5.1(4), A.4.18.5
Discharge pipe and fittings, 4.2.3, 4.15, A.4.15.3 to A.4.15.10; *see also* Drainage
   From dump valves, 8.1.6.5
   Pressure maintenance pumps, 4.25.5.1
   From relief valves, 4.18.5 to 4.18.9, A.4.18.5 to A.4.18.8
   Valves, 4.15.6 to 5.15.9, 4.16.1, 4.25.4.1, 4.30.2, A.4.15.6, A.4.15.9.3, A.4.25.4
Discharge pressure gauges, 4.10.1, 8.4.1
Disconnecting means
   Definition, 3.3.11
   Electric-drive controllers, 10.4.3, 10.6.3, 10.6.8, 10.8.2.1.3, A.10.4.3.1 to A.10.4.3.3.2, B.2, B.3.1, B.3.2.2
      Variable speed drives, 10.10.5, A.10.10.5
   Electric drivers, 9.2.3, 9.3.6, A.9.2.3.1(3), A.9.2.3.1(4), B.2
Domestic water supply, 4.12.1.1.5, 5.8.2
Drainage
   Pump room/house, 4.12.7, 11.3.1, A.4.12.7
   Relief valve, 4.27.2
   Water mist system pumps, 8.4.4.1
Drain piping, 4.13.4
Drawdown (definition), 3.3.12
Dripproof motors, 9.5.1.9.1, 9.5.2.4
   Definition, 3.3.32.2
   Guarded (definition), 3.3.32.1
Drivers, 4.2.3, 4.2.4, 4.4.1, 4.7, A.4.4.1, A.4.7.1 to A.4.7.7.3.2, B.1.23, B.1.26; *see also* Diesel engines; Electric motors (drivers); Steam turbines
   Earthquake protection, 4.28.1, A.4.28.1
   Positive displacement pumps, 8.5, 8.8, A.8.5.1
   Pump connection and alignment, 6.5, A.6.5, B.1.23

   Speed, B.1.28, B.1.30
   Vertical lineshaft turbine pumps, 7.5, 7.6.1.6
Dual-drive pump units, 4.7.3, A.4.7.3
Dump valves, 8.1.6
   Control, 8.1.6.3
   Definition, 3.3.55.1
Dust-ignition-proof motor (definition), 3.3.32.3

-E-

Earthquake protection, 4.12.1, 4.13.5, 4.28, A.4.12.1, A.4.28.1
Eccentric tapered reducer or increaser, 4.14.6.4, 6.3.2(1)
Electric-drive controllers, 9.1.2, Chap. 10; *see also* Electric motors (drivers)
   Additive pump motors, 10.9
   Automatic, 10.5.1, 10.5.2, 10.9.2, A.10.5.1, A.10.5.2.1, A.10.5.2.1.7.2
   Components, 10.4, A.10.4.1 to A.10.4.7
   Connections and wiring, 10.3.4
      Continuous-duty basis, 10.3.4.4
      Control circuits, protection of, 10.3.5
   Construction, 10.3, A.10.3.3.1 to A.10.3.7.3
   Design, 10.1.3, A.10.1.3
   Electrical diagrams and instructions, 10.3.7, A.10.3.7.3
   Electrical power supply connection, 9.6.4
   Emergency-run control, 10.5.3.2, 10.6.10, A.10.5.3.2
   External operations, 10.3.6, 10.5.2.6, A.10.3.6
   Indicating devices
      Available power, 10.6.6
      Power source, 10.8.3.8
      Remote indication, 10.4.8, 10.8.3.14, 10.10.9, A.10.4.5.7
      Variable speed drive operations, 10.10.8, 10.10.9
   Listing, 10.1.2.1, 10.1.2.4
   Location, 10.2, A.10.2.1
   Low-voltage control circuit, 10.6.5
   Nonautomatic, 10.5.1, 10.5.2.4, 10.5.3, A.10.5.1, A.10.5.3.2
   Power transfer for alternate power supply, 10.8, A.10.8
   Rated in excess of 600 Volts, 10.6
   Service arrangements, 10.1.2.6
   Starting and control, 10.5, A.10.5.1 to A.10.5.3.2
      Motor starting circuitry, 10.4.5, A.10.4.5.7
      Operating coils, 10.4.5.6
      Single-phase sensors in controller, 10.4.5.7, A.10.4.5.7
      Soft start units, 10.4.5.5
      Starting reactors and autotransformers, 10.4.5.4
      Starting resistors, 10.4.5.3
      Timed acceleration, 10.4.5.2
   State of readiness, 10.1.2.7
   Stopping methods, 10.5.4, 10.9.3
   Variable speed pressure limiting control, 10.5.2.1.7.5, 10.10, A.10.10
   Voltage drop, 9.1.6, 9.4, A.9.4
Electric motors (drivers), 4.7.2, 4.7.6, 6.5.1.1, Chap. 9, A.4.7.6; *see also* Electric-drive controllers
   Current limits, 9.5.2

Definition, 3.3.32.4
Operation, A.14.2.5
Phase reversal test, 14.2.5.6, A.14.2.5.6
Power sources and supply, *see* Power supply
Problems of, B.1.26, B.1.28 to B.1.30
Speed, B.1.28, B.1.30
Tests, 14.2.5 to 14.2.7, A.14.2.5
Vertical lineshaft turbine pumps, 7.5.1.3, 7.5.1.5, 7.5.2
Voltage drop, 9.1.6, 9.4, A.9.4

Electric starting, diesel engine, 11.2.7.2, A.11.2.7.2.1.5 to A.11.2.7.2.4

Electric supply, *see* Power supply

Electronic fuel management control, 11.2.4.2, 12.4.1.4(5), 12.4.1.4(6), 14.2.11, A.11.2.4.2, A.14.2.11

Emergency boiler feed, steam turbine, A.13.3

Emergency control for engine drive controllers, 12.7.6

Emergency governors, 13.2.2.4, 13.2.2.5, 14.2.8

Emergency lighting, 4.12.5

Emergency-run mechanical control, 10.5.3.2, 10.6.10, A.10.5.3.2

Enclosures, pump, *see* Pump rooms/houses

Enclosures for controllers, 9.7(2), 10.3.3, 12.3.3, A.9.7(2), A.10.3.3.1, A.12.3.3.1, B.3.2.1

End suction pumps, 6.1.1.2, Fig. A.6.1.1(a)
Definition, 3.3.37.4

Engine drive controllers, Chap. 12
Air starting, 11.2.7.4, 12.8, A.11.2.7.4.4
Automatic, 12.7.1, 12.7.2
  In-factory wiring, 11.2.6.1, 11.2.7.4.2
  In-field wiring, 11.2.6.2, A.11.2.6.2
Battery charging and recharging, 12.5, 12.6, A.12.5
Components, 12.4, A.12.4.1.2 to A.12.4.4
Connections and wiring, 11.2.6, 11.2.7.4.2, 12.3.5, A.11.2.6.2
Construction, 12.3, A.12.3.1.1 to A.12.3.8
Contacts, 12.4.3
Electrical diagrams and instructions, 12.3.6, 12.3.8, A.12.3.8
External operations, 12.3.6.3, 12.7.2.5
Indicators, 12.4.1, A.12.4.1.2, A.12.4.1.5.2
  Remote, 12.4.2, 12.4.3, 14.2.6.6, A.12.4.2.3(3)
Location, 12.2, A.12.2.1
Locked cabinet for switches, 12.3.4
Nonautomatic, 12.7.1, 12.7.2.3, 12.7.3
Starting and control, 12.7, 12.8.2, A.12.7
Stopping methods, 12.7.5, 12.8.3, A.12.7.5.2
  Automatic shutdown, 12.7.5.2, A.12.7.5.2
  Manual shutdown, 12.7.5.1, 12.8.3
Tests, 12.7.3.2, 14.2.6.7

Engines
Diesel, *see* Diesel engines
Internal combustion, *see* Internal combustion engines

Equivalency to standard, 1.5

Exhaust system
Diesel engine, 11.5, A.11.5.2
Steam turbines, A.13.3

Explosionproof motor (definition), 3.3.32.5

-F-

Fan-cooled motor, totally enclosed, 9.5.2.4
Definition, 3.3.32.8

Fault tolerant external control circuits, 10.5.2.6, 12.7.2.5
Definition, 3.3.7.2

Feeder, 9.2.2, 9.3.6
Definition, 3.3.15
Inadequate capacity, B.1.28

Field acceptance tests, 4.4.2, 4.32, 14.2, 14.3.3, 14.5, A.14.2.2 to A.14.2.11, A.14.5.2.3

Fill/refill valves, water tank, 5.8.3

Fire protection equipment control, 10.5.2.3, 12.7.2.2

Fire pump alarms
Definition, 3.3.16
Electric-drive controllers, 10.1.1.2, 10.4.7, A.10.4.7
Engine drive controllers, 12.1.1, 12.4.1.3 to 12.4.1.5, 12.4.2, 12.4.3, A.12.4.1.5.2, A.12.4.2.3(3)
Tests, 14.2.9

Fire pump controllers, 4.2.3, 4.2.4, 4.4.1, 8.6, A.4.4.1, A.8.6; *see also* Electric-drive controllers; Engine drive controllers
Acceptance test, 14.2.6, A.14.2.6.1
Additive pump motors, 10.9
Dedicated controller for each driver, 4.7.5
Definition, 3.3.17
Earthquake protection, 4.28.1, A.4.28.1
Maintenance, after fault condition, B.3
Positive displacement pump, 8.1.6.3
Pressure actuated controller pressure sensing lines, 4.30, A.4.30
Pressure maintenance pump, 4.25.7
Protection of, 4.12.1, A.4.12.1
Vertical lineshaft turbine pumps, 7.5.2, 10.10.12

Fire pumps, *see also* Pumps
Definition, 3.3.37.5
In high-rise buildings, *see* High-rise buildings
Operations, *see* Operations
Packaged fire pump assemblies, 4.29, A.4.29.9
Redundant, 4.25.6, 9.3.3
Summary of data, Table 4.26(a), Table 4.26(b)

Fire pump units
Definition, 3.3.18
Dual-drive, 4.7.3, A.4.7.3
Field acceptance tests, 4.32
Location and protection
  Indoor units, 4.12.1.1, A.4.12.1.1
  Outdoor units, 4.12.1.2
Performance, 4.4, 14.2.5.5, A.4.4.1
Series, 4.19, 5.3
  Definition, 3.3.40

Fittings, 4.2.3, 4.13, 6.3, A.4.13.1 to A.4.13.6, A.6.3.1
Discharge, 4.15, A.4.15.3 to A.4.15.10
Joining method, 4.13.2
Maintenance, B.1.13
Positive displacement pumps, 8.4, A.8.4.2 to A.8.4.5
Suction, 4.14, A.4.14.1 to A.4.14.10
Vertical lineshaft turbine pumps, 7.3.5, A.7.3.5.3

Flexible connecting shafts, 6.5.1.1, 11.2.3.1.1, 11.2.3.2.1, A.6.5
　Definition, 3.3.19
　Guards for, 4.12.8
　Vertical lineshaft turbine pumps, 7.5.1.7.1, 7.5.1.8
Flexible couplings, 6.5.1.1, Fig. A.6.1.1(e), A.6.4.1, A.6.5
　Definition, 3.3.20
　Diesel engine pump connection, 11.2.3.1
　Earthquake protection, 4.28.4
　Guards, 4.12.8
　Positive displacement pumps, 8.8.1, 8.8.2
Flooded suction (definition), 3.3.21
Flow-measuring device, 5.5, 6.3.2(3), 8.9.2, 8.9.3, 14.2.5.4.3.2, 14.2.7.4.3.3
Flow tests, *see* Tests
Foam concentrate piping, 4.13.3
Foam concentrate pumps, 8.2, 8.4.3, 14.2.10, A.8.2.2 to A.8.2.6, A.8.4.3; *see also* Additive pumps
Foundations, 6.4, A.6.4.1, A.6.4.4, B.1.24
　Positive displacement pumps, 8.7
　Vertical lineshaft turbine pumps, 7.4.3
Frequency-sensing devices, 10.8.3.7
Fuel supply, 11.4, A.11.4.1.2.1 to A.11.4.5; *see also* Electronic fuel management control
　Earthquake protection, 4.28.1, A.4.28.1
　Fuel types, 11.4.5, A.11.4.5
　Location, 11.4.3, A.11.4.3
　Maintenance, 11.6.4, A.11.6.4
　Obstructed system, B.1.32
　On-site standby generators, 9.6.2.3, A.9.6.2

-G-

Gate valves, 4.14.5, 4.15.7, A.4.14.5, A.4.25.5.5, A.13.3
Gear drives, 7.5.1.7, 11.2.3.2.1, 14.2.5.4.9
Gear pump (definition), 3.3.37.7
Generator, *see* On-site standby generator
Governors, B.1.30
　Diesel engine, 11.2.4.1, 14.2.5.4.7.3, 14.2.5.4.7.4
　Emergency, 13.2.2.4, 13.2.2.5, 14.2.8
　Speed, steam turbine, 13.2.2
Ground-face unions, 4.30.4.2
Grounding
　Electric drive controller enclosures, 10.3.3.3
　Engine drive controllers, 12.3.3.2
Groundwater (definition), 3.3.22
Guarded motors
　Definition, 3.3.32.6
　Dripproof (definition), 3.3.32.1
Guards for fuel lines, 11.4.4.5, A.11.4.4.5

-H-

Hangers, 4.13.5
Head, 4.5.1, 6.2.2; *see also* Net positive suction head (NPSH) (h_sv); Total head *(H)*
　Available from water supply, 4.6.5
　Definition, 3.3.23, A.3.3.23
　Net head lower than rated, B.1.12
　Static, A.6.1.2
　Total discharge head (h_d) (definition), 3.3.23.2
　Total rated head (definition), 3.3.23.4
　Total suction head (definition), 3.3.23.5
　Velocity head (h_v) (definition), 3.3.23.6
　Vertical turbine pump head component, 7.3.1, A.7.3.1
Heat exchangers, 11.2.8.1(1), 11.2.8.5, 11.2.8.6, A.11.2.8.5, A.11.2.8.6, A.11.3.2.3, A.11.3.2.4, B.1.25
Heat source, pump room/house, 4.12.3
High-rise buildings
　Definition, 3.3.24
　Fire pumps for, 4.12.1.1.1, Chap. 5, A.4.12.1.1.1
Horizontal pumps, 4.1.1
　Definition, 3.3.37.8
　Diesel engine drive connection, 11.2.3.1
　Installation, Fig. A.6.3.1
　Shaft rotation, A.4.23
　Split-case, 6.1.1.3, Fig. A.6.1.1(f), Fig. A.6.3.1
　　Definition, 3.3.37.9
　　Suction pipe and fittings, 4.14.6.3.1 to 4.14.6.3.3
　Total head *(H)*, A.14.2.5.4
　　Definition, 3.3.23.3.1, A.3.3.23.3.1
Hose valves, 4.20.3, 6.3.2(2), 7.3.5.1(5), A.4.20.3.1, A.14.2.5.2.3
Hydraulic starting, 11.2.7.3, 12.4.1.4(3)

-I-

Impellers
　Impeller between bearings design, 6.1.1.3, 6.4.1, A.6.1.1, A.6.4.1
　Overhung impeller design, 6.1.1.2, 6.4.1, 6.4.2, A.6.1.1, A.6.4.1
　Problems, B.1.8, B.1.10 to B.1.12, B.1.15, B.1.16, B.1.29
　Vertical lineshaft turbine pumps, 7.3.3.2, 7.5.1.1, 7.5.1.2, 7.5.1.6.2, 7.5.1.7.2, 7.6.1.3, A.7.6.1.1
Indicating valves, 4.15.7, 4.15.8, 4.25.5.5, A.4.25.5.5
In-line pumps, 6.1.1.2, Figs. A.6.1.1(c) to (e)
　Definition, 3.3.37.10
In-rush currents, 10.7.3.10
Installer, system, 4.3.3, A.4.3.3(2)
Instrument panel, 11.2.5.1
Internal combustion engines, B.1.26; *see also* Diesel engines
　Definition, 3.3.13.2
　Pump room or house, heat for, 4.12.3.2
　Spark-ignited, 11.1.3.2
　Speed, B.1.28
Isolating switches, 10.4.2, 10.8.2.1.2, 10.8.2.2, 10.8.3.12.1, A.10.4.2.1.2, A.10.4.2.3, B.3.1, B.3.2.2
　Definition, 3.3.48.1
Isolation valves, 4.16.1, 4.25.5.3, 4.25.5.4, 4.25.5.7

-J-

Jockey pumps, *see* Pressure maintenance (jockey or make-up) pumps
Junction boxes, 9.7, 10.3.4.5.1, A.9.7(2), A.9.7(3)

-L-

Lighting
   Emergency, 4.12.5
   Normal artificial, 4.12.4
Limited service controllers, 10.7, A.10.7
Liquid (definition), 3.3.27
Liquid level
   Detectors, 7.3.5.1(2), 7.3.5.3, A.7.3.5.3
   Pumping, 4.6.3, B.1.19
      Definition, 3.3.28.1
   Static, 6.1.2, A.6.1.2
      Definition, 3.3.28.2
   Well or wet pit, 4.6.3
Liquid supplies, 4.2.3, 4.2.4, 4.6, A.4.6.1 to A.4.6.4
   Domestic water supply, 4.12.1.1.5, 5.8.2
   Head, 4.6.5, A.6.1.2
   Heat exchanger, 11.2.8.5, 11.2.8.6, A.11.2.8.5, A.11.2.8.6
   Potable water, protection of, 8.4.6
   Pumps, priming of, B.1.19
   Reliability, 4.6.1, A.4.6.1
   Sources, 4.6.2, 7.2.1, A.4.6.2, A.7.1.1, A.7.2.1.1, A.7.2.1.2
      Discharge to, 4.18.7, A.4.18.7
   Stored supply, 4.6.4, 5.5, 5.8, A.4.6.4
   Vertical lineshaft turbine pumps, 7.1.1, 7.2, A.7.1.1, A.7.2.1.1 to A.7.2.7, B.1.19
Listed
   Break tank refill mechanism, 4.31.3
   Controllers and transfer switches, 4.25.7, 10.1.2.1, 10.1.2.4, 10.10.1.2, 12.1.3.1
   Definition, 3.2.3, A.3.2.3
   Dump valves, 8.1.6.4
   Electrical circuit protective system to controller wiring, 9.8, A.9.8.1, A.9.8.2
   Engines, 11.2.1
   Pumps, 4.1.2.1, 4.7.1, 6.5.1.2, A.4.7.1, A.6.2
Loads start test, 14.2.5.5
Locked rotor overcurrent protection, 10.4.4, 10.6.9, A.10.4.4(3)
Lockout, additive pump motors, 10.9.4
Loss of phase, 10.4.7.2.2
   Definition, 3.3.29
Low suction throttling valves, 4.14.9.2(2), 4.15.9
   Definition, 3.3.55.2
Lubrication, pump, B.1.21, B.1.27

-M-

Maintenance
   Batteries, storage, 11.6.3, B.1.32
   Controllers, B.3
   Diesel engines, 11.6, A.11.6
   Fittings, B.1.13
   Fuel supply, 11.6.4, A.11.6.4
   Pumps, 7.6.2, 14.4
   Water seals, B.1.6
Main throttle valve, steam turbine, 13.2.1.3

Make-up pumps, *see* Pressure maintenance (jockey or make-up) pumps
Manuals, instruction, 14.3.1, 14.3.2
Manual transfer switches, 10.8.1.2, 10.8.3.5
   Definition, 3.3.48.2.2
Marking
   Additive pump motor controllers, 10.9.5
   Disconnecting means, 9.2.3.1(5), 9.2.3.2
   Electric-drive controllers, 10.1.2.2, 10.1.2.5, 10.3.8, 10.9.5, 10.10.2, 10.10.11, A.10.1.2.2
   Electric drivers, 9.5.3
   Engine drive controllers, 12.1.3.3, 12.3.7
   Transfer switches, 10.1.2.2, A.10.1.2.2
Mass elastic system, 7.5.1.6
Maximum pump brake horsepower, 4.5.1
   Definition, 3.3.31
Measurement, units of, 1.6
Meters, 4.20.2, 8.9.2, 8.9.3, 14.2.7.4.3.2, 14.2.7.4.3.3, A.4.20.2.1.1
Motor contactors, 10.4.5.1, B.3.2.4
Motors
   Dripproof, *see* Dripproof motors
   Dust-ignition-proof (definition), 3.3.32.3
   Electric, *see* Electric motors (drivers)
   Explosionproof (definition), 3.3.32.5
   Guarded (definition), 3.3.32.6
   Open, 9.5.2.4
      Definition, 3.3.32.7
   Speed, *see* Speed
   Totally enclosed, *see* Totally enclosed motors
Multistage pumps, 4.1.1, 6.1.1.2, 6.1.1.3

-N-

Nameplates, on pumps, 4.9
Net positive suction head (NPSH) $(h_s v)$, 7.2.2.2.2, 8.2.2, 8.3.2, A.8.2.2
   Definition, 3.3.23.1
Nonreverse ratchets, 7.5.1.4, 7.5.1.7.3, 9.5.1.9.2, 11.2.3.2.2
Nonventilated motor, totally enclosed, 9.5.2.4
   Definition, 3.3.32.10

-O-

Oil pressure gauge, 11.2.5.3
On-site power production facility, 9.2.2(2), 9.2.3, A.9.2.3.1(3), A.9.2.3.1(4); *see also* Power supply
   Definition, 3.3.34
On-site standby generator, 9.3.4, 9.6, 10.8.3.6.2, 10.8.3.12, A.9.6.2, A.9.6.5; *see also* Power supply, Alternate power sources
   Definition, 3.3.35
Open motors, 9.5.2.4
   Definition, 3.3.32.7
Operations
   Controllers, external operations, 10.3.6, 10.5.2.6, 12.3.6.3, 12.7.2.5, A.10.3.6

Operations *(continued)*
    Diesel engines, 11.6, A.11.6
    Pumps, 4.3, 7.6.1, A.4.3.2(2) to A.4.3.4.1(2), A.7.6.1.1, A.7.6.1.4, A.14.2.5
Outdoor setting
    Fire pump units, 4.12.1.2, A.4.12.1
    Vertical lineshaft turbine pumps, 7.4.2
Outside screw and yoke gate valves, *see* Gate valves
Overcurrent protection, 9.2.2(4), 9.2.3, 9.3.6, 10.3.5.1, 10.8.2.2(3), 10.8.3.11, 14.2.6.8, A.9.2.3.1(3), A.9.2.3.1(4); *see also* Disconnecting means
    Isolating switch, 10.4.2.1.3
    Locked rotor, 10.4.4, 10.6.9, A.10.4.4(3)
Overspeed shutdown device, 11.2.4.4

-P-

Packaged fire pump assemblies, 4.29, A.4.29.9
    Definition, 3.3.37.11
Phase converters, 9.1.7, A.9.1.7
Phase reversal, 10.4.6.2, 10.4.7.2.3, 14.2.5.6, A.14.2.5.6
Pipe, 4.13, A.4.13.1 to A.4.13.6; *see also* Discharge pipe and fittings; Suction pipe and fittings
    Break tank refill, 4.31.3.2.3
    Exhaust, 11.5.2, A.11.5.2
    Flushing, 14.1, A.14.1.3 to A.14.5.2.3
    Fuel, 11.4.4, 11.4.4.5, A.11.4.4, A.11.4.4.5
    Joining method, 4.13.2
    Problems, causes of, B.1.1 to B.1.3, B.1.6
    Protection
        Against damage due to movement, 4.17, A.4.17
        For fuel line, 11.4.4.5, A.11.4.4.5
        Seismic bracing, 4.13.5
    Size
        Minimum pipe size data, Table 4.26(a), Table 4.26(b)
        Suction pipe, 4.14.3
    Steel, 4.13.1, 4.13.2.1, 4.25.5.1, 4.30.3, A.4.30.3
Pipeline strainers, 6.3.2(5)
Piston plunger pump (definition), 3.3.37.12
Positive displacement pumps, 4.1.1, Chap. 8
    Application, 8.1.3
    Component replacement, 14.5.1
    Connection to driver and alignment, 8.8
    Controllers, 8.6, A.8.6
    Definition, 3.3.37.13
    Drivers, 8.5, 8.8, A.8.5.1
    Fittings, 8.4, A.8.4.2 to A.8.4.5
    Flow tests, 8.9, 14.2.5.4.3
    Foam concentrate and additive pumps, 8.2, 8.4.3, 8.9.2, A.8.2.2 to A.8.2.6, A.8.4.3
    Foundation and setting, 8.7
    High-rise buildings, 5.2.2
    Materials, 8.1.5, A.8.1.5
    Seals, 8.1.4, 8.2.3
    Suitability, 8.1.2, A.8.1.2
    Types, 8.1.1
    Water mist system pumps, 8.3, A.8.3.1

Power supply, 4.2.3, 4.2.4
    Alternate power sources, 9.2.2(4), 9.3, 10.4.7.2.4, 10.8, 14.2.7, A.9.3.2, A.10.8; *see also* On-site standby generator
        Momentary test switch, 10.8.3.13
        Overcurrent protection, 10.8.3.11
        Transfer of power, 9.6.4, 10.8.3.9
    Electric drive for pumps, 9.1.4 to 9.1.6, 9.2, A.9.1.4, A.9.2.3.1(3), A.9.2.3.1(4), B.1.28, B.1.30
        Alternate power sources, *see subhead:* Alternate power sources
        Multiple power sources, 9.3.4
        Normal power, 9.2, A.9.2.3.1(3), A.9.2.3.1(4)
        On-site power production facility, A.9.2.3.1(3), A.9.2.3.1(4)
    Definition, 3.3.34
        Service-supplied, 9.2.2
        Variable speed drives, 10.10.6.1, A.10.10.6.3
    Steam supply, *see* Steam supply
    Tests, 10.8.3.13, 14.2.6.9, 14.2.7, A.14.2.5.4
Pressure control valves, A.13.2.1.1
    Definition, 3.3.55.3
Pressure gauges, 4.10, 5.4, 6.3.1, A.4.10.2, A.6.3.1, B.1.14
    Oil, 11.2.5.3
    Positive displacement pumps, 8.4.1
    Steam, 13.2.3
    Vertical lineshaft turbine pumps, 7.3.5.1(3)
Pressure maintenance (jockey or make-up) pumps, 4.25, 10.3.4.6, A.4.14.4, A.4.14.9, A.4.25
    Definition, 3.3.37.14
    Overcurrent protection, 9.2.3.4
    Pressure sensing lines, 4.30.1, 4.30.2
    Valves, 4.25.5.2 to 4.25.5.5, 4.25.5.7
Pressure recorders, 12.4.4, A.12.4.4
Pressure-reducing valves, A.13.3
    Definition, 3.3.55.4
Pressure regulating devices, 4.15.10, A.13.2.1.1
    Definition, 3.3.36
Pressure relief valves, *see* Relief valves
Pressure sensing lines, pressure actuated controllers, 4.30, A.4.30
Protection
    Equipment, 4.12, A.4.12
    Personnel, 10.6.7, B.2
    Piping
        Fuel line, 11.4.4.5, A.11.4.4.5
        Movement, damage due, 4.17, A.4.17
        Seismic bracing, 4.13.5
Protective devices, *see also* Overcurrent protection
    Control circuits, 10.3.5
    Controller, 10.4.2 to 10.4.4, 10.6.9, A.10.4.2.1.2 to A.10.4.4(3)
    Listed electrical circuit protective system to controller wiring, 9.8, A.9.8.1, A.9.8.2
    On-site power source circuits, 9.6.5, A.9.6.5
    Overspeed shutdown device, 11.2.4.4
    Variable speed drive circuit protection, 10.10.5, A.10.10.5
Pump brake horsepower, maximum (definition), 3.3.31
Pumping liquid level, 4.6.3, A.7.1.1
    Definition, 3.3.28.1
Pump manufacturers, 10.1.2.6, 12.1.4, 14.2.1

Index

Pump rooms/houses, 4.12, A.4.12
   Access to, 4.12.2, 5.3, A.4.12
   Controller, transfer of power to, 9.6.4
   Diesel engine drive, 11.3, A.11.3
   Drainage, 4.12.7, 11.3.1, A.4.12.7
   Lighting, 4.12.4, 4.12.5
   Temperature of, 4.12.3, 11.6.5, A.11.6.5, B.1.17
   Torch-cutting or welding in, 4.13.6, A.4.13.6
   Ventilation, 4.12.6, 11.3.2, A.11.3.2
   Vertical lineshaft turbine pumps, 7.4.1

Pumps
   Additive, *see* Additive pumps
   Bypass, with, 4.14.4, 11.2.8.6, A.4.14.4, A.11.2.8.6
   Can (definition), 3.3.37.2
   Centrifugal, *see* Centrifugal pumps
   End suction, *see* End suction pumps
   Fire, *see* Fire pumps
   Foam concentrate, *see* Foam concentrate pumps
   Gear (definition), 3.3.37.7
   Horizontal, *see* Horizontal pumps
   In-line, *see* In-line pumps
   Listed, 4.1.2.1, 4.7.1, A.4.7.1
   Lubrication, B.1.21, B.1.27
   Multiple, 4.14.3.1, 4.14.7, 12.1.3.3.2, A.4.14.3.1
      Sequence starting of, 9.6.3, 10.5.2.5, 12.7.2.4
   Multistage, 4.1.1, 6.1.1.2, 6.1.1.3
   Piston plunger (definition), 3.3.37.12
   Positive displacement, *see* Positive displacement pumps
   Pressure maintenance, *see* Pressure maintenance (jockey or make-up) pumps
   Pressure maintenance (jockey or make-up), *see* Pressure maintenance (jockey or make-up) pumps
   Priming, B.1.19
   Problems, causes of, Annex B
   Rotary lobe (definition), 3.3.37.15
   Rotary vane (definition), 3.3.37.16
   Single-stage, 4.1.1, 6.1.1.2, 6.1.1.3, A.6.1.1
   Speed, 4.7.6, A.4.7.6, B.1.30
   Vertical lineshaft turbine, *see* Vertical lineshaft turbine pumps

Pump shaft rotation, 4.23, A.4.23, B.1.18
Purpose of standard, 1.2

-Q-

Qualified person (definition), 3.3.39

-R-

Radiators, 11.2.8.8, 11.3.2.1(4), 11.3.2.4.3, A.11.3.2.3, A.11.3.2.4
References, Chap. 2, Annex C
Relief valves, 4.7.7.2, 4.18, 4.25.4.1, 6.3.1, 6.3.2(4), A.4.7.7, A.4.7.7.2, A.4.18.1 to A.4.18.8, A.6.3.1, A.14.2.5.1
   Circulation relief valves, 6.3.1, A.6.3.1, A.14.2.5.1
      Automatic, 4.11.1
      Definition, 3.3.55.5.1
   Definition, 3.3.55.5

   Drainage, backflow prevention device, 4.27.2
   Positive displacement pumps, 8.4.2 to 8.4.4, A.8.4.2 to A.8.4.4
   Vertical lineshaft turbine pumps, 7.3.5.1(4)

Remote station, manual control at, 10.5.2.4, 12.7.2.3
Retroactivity of standard, 1.4
Rotary lobe pump (definition), 3.3.37.15
Rotary vane pump (definition), 3.3.37.16

-S-

Scope of standard, 1.1, A.1.1
Screens
   Suction pipe, 4.14.8, 7.3.4.3, A.4.14.8
   Well, 7.2.4.5 to 7.2.4.11

Seals
   Positive displacement pumps, 8.1.4, 8.2.3
   Rings improperly located in stuffing box, B.1.20
   Water seals, maintenance of, B.1.6

Seismic bracing, 4.13.5
Sequence starting of pumps, 9.6.3, 10.5.2.5, 12.7.2.4
Series fire pump units, 4.19
   Definition, 3.3.40

Service (power source), 9.2.2
   Definition, 3.3.41, A.3.3.41

Service equipment, 9.2.3.1(1)
   Definition, 3.3.42, A.3.3.42

Service factor (definition), 3.3.43
Service personnel, 4.3.4, A.4.3.4
Set pressure, 4.7.7.3.2, A.4.7.7.3.2
   Definition, 3.3.44
   Variable speed pressure limiting control, 10.10.3.1, 10.10.8.3, A.10.10.3.1

Shaft rotation, *see* Pump shaft rotation
Shafts, flexible connecting, *see* Flexible connecting shafts
Shall (definition), 3.2.4
Shop tests, *see* Tests
Should (definition), 3.2.5
Shutoff valve, 4.18.9, 4.30.5, 11.2.8.5.3.2, 11.2.8.5.3.3, 11.2.8.6.2, 11.2.8.6.3

Signal devices
   Electric-drive controllers, 10.1.1.2, 10.4.6, 10.4.7, A.10.4.6, A.10.4.7
   Engine drive controllers, 12.1.2, 12.4.2, 12.4.3, A.12.4.2.3(3)
   Phase reversal indicator, 10.4.6.2, 10.4.7.2.3

Signals, 4.24, A.4.24
   Break tank low liquid level, 4.31.3.1.5
   Definition, 3.3.45, A.3.3.45
   Electric-drive controllers, 10.4.5.7.3, 10.4.7.1, 10.8.3.12.2
   Engine drive controllers, 12.4.1.3 to 12.4.1.5, 14.2.6.6, A.12.4.1.5.2
   Engine running and crank termination, 11.2.7.4.3
   Tests, 14.2.9
   Visible, 10.4.7.1; *see also* Visible indicators

Single-stage pumps, 4.1.1, 6.1.1.2, 6.1.1.3, A.6.1.1
SI units, 1.6
Spare parts, 14.3.4, 14.5.2, A.14.5.2.3

Stationary Fire Pumps Handbook 2010

Speed, *see also* Variable speed pressure limiting control
   Engine
      Definition, 3.3.46.1
      Internal combustion engine, B.1.28
      Overspeed shutdown device, 11.2.4.4
   Motor, B.1.28, B.1.30
      Definition, 3.3.46.2
   Rated, 4.7.6, A.4.7.6, A.14.2.5.4
      Definition, 3.3.46.3
   Steam turbine, B.1.28
Speed governor, steam turbine, 13.2.2
Split-case pumps, *see* Horizontal pumps
Sprinkler systems, 4.12.1.3
Standard (definition), 3.2.6
Static liquid level, 6.1.2, A.6.1.2
   Definition, 3.3.28.2
Steam supply, 4.21, 13.1.3.1, 13.3, A.13.3
Steam turbines, 4.7.2, 4.7.6, Chap. 13, A.4.7.6, B.1.26
   Acceptability, 13.1.1
   Bearings, 13.2.6
   Capacity, 13.1.2
   Casing and other parts, 13.2.1, A.13.2.1.1
   Gauge and gauge connections, 13.2.3
   Installation, 13.3, A.13.3
   Obstructed pipe, B.1.32
   Redundant, 9.3.3
   Rotor, 13.2.4
   Shaft, 13.2.5
   Speed, B.1.28
   Speed governor, 13.2.2
   Steam consumption, 13.1.3, A.13.1.3
   Tests, 14.2.5, 14.2.8.1, A.14.2.5
   Vertical lineshaft turbine pumps, 7.5.1.3, 7.5.1.4, 7.5.2
Steel pipe, *see* Pipe
Storage batteries, *see* Batteries, storage
Strainers, *see also* Suction strainers
   Engine cooling system, B.1.25
   Pipeline, 6.3.2(5)
   Suction, vertical shaft turbine pumps, 7.3.4
Stuffing boxes, B.1.5, B.1.7, B.1.20
Suction, *see also* Net positive suction head
   Backflow preventers, evaluation of, 4.27.4.1, 4.27.4.2
   Static suction lift, 6.1.2, A.6.1.2
   Total suction head (definition), 3.3.23.5
   Total suction lift (h_l) (definition), 3.3.54
Suction pipe and fittings, 4.2.3, 4.14, A.4.14.1 to A.4.14.10
   Devices in, 4.14.9, 4.27.3, 4.27.4.3, A.4.14.9
   Elbows and tees, 4.14.6.3
   Freeze protection, 4.14.6.2
   Pressure maintenance pumps, 4.25.5.1
   Problems, causes of, B.1.1 to B.1.3
   Strain relief, 4.14.6.5
   Valves, 4.16.1
Suction pressure gauges, 4.10.2, 4.27.4.2, 8.4.1, A.4.10.2
Suction reservoir, discharge to, 4.18.8, A.4.18.8
Suction screening, 4.14.8, A.4.14.8

Suction strainers
   Positive displacement pumps, 8.4.5, A.8.4.5
   Vertical shaft turbine pumps, 7.3.4
Sump, vertical shaft turbine-type pumps, 7.4.3.7
Switches, *see also* Isolating switches; Transfer switches
   Electronic control module (ECM) selector switch, 11.2.4.2.3, A.11.2.4.2.3.6
   Locked cabinets for, 12.3.4
   Pressure switch, pressure sensing line, 4.30.6
System designer, 4.3.2, A.4.3.2(2)
System installer, 4.3.3, A.4.3.3(2)

-T-

Tachometer, 11.2.5.2
Tanks
   Break, *see* Break tanks
   Fuel supply, 4.28.1, 11.4.1.2, 11.4.2, A.4.28.1, A.11.4.1.2.1, A.11.4.2
   Water supply, high-rise buildings, 5.4, 5.7
Temperature gauge, 11.2.5.4
Tests, *see also* Field acceptance tests; Water flow test devices
   Aquifer performance analysis, A.7.2.1.2
      Definition, 3.3.3
   Component replacement, 14.5, A.14.5.2.3
   Controllers, 10.1.2.3, 10.1.2.6
      Acceptance, 14.2.6, A.14.2.6.1
      Engine drive controllers, manual testing of, 12.7.3.2
      Pressure switch-actuated automatic controller, 10.5.2.1.6
      Rated in excess of 600 Volts, 10.6
   Duration, 14.2.10
   Flow, 5.4, 8.9, 14.2.5.2, 14.2.5.4.3, 14.2.7.3.1, A.14.2.5.2.3
   High-rise buildings, fire pumps for, 5.4
   Hydrostatic, 4.22.2, 14.1, A.14.1.3 to A.14.5.2.3
   Loads start test, 14.2.5.5
   Metering devices or fixed nozzles for, 4.20.2.1, 8.9.2, 8.9.3
   Momentary test switch, alternate power source, 10.8.3.13
   Periodic, 14.4
   Shop (preshipment), 4.5, 4.22, 10.1.2.3, 14.2.4.2
   Suction pipe, 4.14.2
   Vertical lineshaft turbine pumps, 7.6.1, A.7.6.1.1, A.7.6.1.4
   Vertical lineshaft turbine pump wells, 7.2.7, A.7.2.7
Timer, weekly program, 12.7.2.7
Tools, special, 14.3.3
Total discharge head (h_d) (definition), 3.3.23.2
Total head (H)
   Horizontal pumps, A.14.2.5.4
      Definition, 3.3.23.3.1, A.3.3.23.3.1
   Vertical turbine pumps, 7.1.2, A.14.2.5.4
      Definition, 3.3.23.3.2, A.3.3.23.3.2
Totally enclosed motors
   Definition, 3.3.32.9
   Fan-cooled, 9.5.2.4
      Definition, 3.3.32.8
   Nonventilated, 9.5.2.4
      Definition, 3.3.32.10

Index

Total rated head, 7.1.2
   Definition, 3.3.23.4
Total suction head, 4.22.2.2
   Definition, 3.3.23.5
Total suction lift (h_l) **(definition),** 3.3.54
Trade sizes, 1.6.5
Transfer switches, 9.1.2, 10.1.1.1, 10.4.8, 10.8, A.4.4.1, A.10.8
   Automatic, see Automatic transfer switches
   Listing, 10.1.2.1, 10.1.2.4
   Manual, see Manual transfer switches
   Marking, 10.1.2.2, A.10.1.2.2
   Nonpressure-actuated, 10.5.1.2, 10.5.2.2
   Pressure-actuated, 10.5.1.2, 10.5.2.1.1, 10.6.4, 12.7.2.1.1, 14.2.6.6
   Service arrangements, 10.1.2.6
Transformers, 9.2.2(5)
Troubleshooting, Annex B
Turbine pumps, vertical lineshaft, see Vertical lineshaft turbine pumps
Turbines, steam, see Steam turbines

-U-

Undervoltage-sensing devices, 10.8.3.6
Units, pump, see Fire pump units
Units of measurement, 1.6
Unloader valves, 8.3.4, 8.3.5
   Definition, 3.3.55.6

-V-

Valves, see also Butterfly valves; Check valves; Dump valves; Gate valves; Hose valves; Relief valves
   Bypass, 4.16.1
   Discharge pipe, 4.15.6 to 4.15.9, A.4.15.6, A.4.15.9.3
   Emergency governor, 14.2.8
   Fill/refill, water tank, 5.7.3
   Fuel solenoid, 11.4.4.6, 12.7.3.2
   Isolation, 4.16.1
   Low suction throttling, 4.14.9.2(2), 4.15.9
      Definition, 3.3.55.2
   Main throttle, steam turbine, 13.2.1.3
   Pressure control, A.13.2.1.1
      Definition, 3.3.55.3
   Pressure-reducing, A.13.3
      Definition, 3.3.55.4
   Shutoff, 4.18.9, 4.30.5, 11.2.8.5.3.2, 11.2.8.5.3.3, 11.2.8.6.2, 11.2.8.6.3
   Supervision of, 4.16, A.4.16
   Unloader, 8.3.4, 8.3.5
      Definition, 3.3.55.6
Variable speed pressure limiting control, 4.7.7.3, 4.18.1.3, 10.5.2.1.7.5, 10.10, 11.2.4.3, 14.2.5.3, 14.2.5.4.7.4, A.4.7.7.3.2, A.10.10, A.11.2.4.2.3.6
   Definition, 3.3.56
Variable speed suction limiting control, 10.10, 11.2.4.3, A.10.10, A.11.2.4.2.3.6

   Definition, 3.3.57
Variable speed vertical pumps, 7.5.3, 10.10.12
Velocity head (h_v) **(definition),** 3.3.23.6, A.3.3.23.6
Ventilation of pump room/house, 4.12.6, 11.3.2, A.11.3.2
Vertical hollow shaft motors, 7.5.1.5, 7.5.1.7.2
Vertical in-line pumps, 4.28.3
Vertical lineshaft turbine pumps, 4.1.1, Chap. 7; see also Wells, vertical shaft turbine pumps
   Bowl assembly, 7.3.3
   Characteristics, 7.1.2
   Column, 7.3.2, A.7.3.2.1
   Consolidated formations, 7.2.5, A.7.2.5
   Controllers, 7.5.2, 10.10.12
   Definition, 3.3.37.17
   Drivers, 7.5, 7.6.1.6, 9.5.1.9, 11.2.3.2
   Fittings, 7.3.5, A.7.3.5.3
   Head, 7.3.1, A.7.3.1
      Total head (H), 7.1.2, A.14.2.5.4
   Definition, 3.3.23.3.2, A.3.3.23.3.2
   Installation, 7.4, A.7.1, A.7.4
   Maintenance, 7.6.2
   Oil-lubricated type, 7.3.2.4 to 7.3.2.6, A.7.1.1
   Operation, 7.6.1, A.7.6.1.1, A.7.6.1.4
   Pump house, A.4.12
   Shaft rotation, A.4.23
   Shop test, 4.22.2.5
   Submergence, 7.2.2, A.7.2.2.1, A.7.2.2.2
   Suction strainer, 7.3.4
   Suitability, A.7.1.1
   Unconsolidated formations, 7.2.4
   Water supply, 7.1.1, 7.2, A.7.1.1, A.7.2.1.1 to A.7.2.7, B.1.19
Vibration, pump, 7.6.1.5
Visible indicators
   Electric-drive controllers, 10.4.5.7.3, 10.4.6.1, 10.6.6.2 to 10.6.6.4, 10.8.3.8, 10.8.3.12.2, A.10.4.5.7
   Engine drive controllers, 12.4.1.1 to 12.4.1.4, A.12.4.1.2
Voltage
   Low, B.1.28
   Rated motor voltage different from line voltage B.1.31
Voltage drop, 9.1.6, 9.4, A.9.4
Voltage-sensing devices, 10.8.3.7
Voltage surge arresters, 10.4.1, A.10.4.1
Voltmeter, 12.4.5, A.14.2.5.4
Vortex plate, 4.14.10, A.4.14.10

-W-

Waste outlet, heat exchanger, 11.2.8.7
Water flow test devices, 4.20, A.4.20.1.1 to A.4.20.3.4(2)
Water level, see Liquid level
Water mist system pumps, 8.3, 8.4.4, A.8.3.1, A.8.4.4
Water pressure control, 10.5.2.1, 10.6.4, 12.7.2.1, A.10.5.2.1
Water supplies, see Liquid supplies
Wearing rings, B.1.9, B.1.19
Weekly program timer, 12.7.2.7
Welding, 4.13.6, A.4.13.6

Wells, vertical shaft turbine pumps
  Construction, 7.2.3
  Developing, 7.2.6
  Installations, 7.2.2.1, Fig. A.7.1.1, A.7.2.2.1, Fig. A.7.2.2.1
  Problems, causes of, B.1.4
  Screens, 7.2.4.5 to 7.2.4.11
  Test and inspection, 7.2.7, A.7.2.7
  Tubular wells, 7.2.4.16
  In unconsolidated formations, 7.2.4
  Water level, 4.6.3

Wet pits
  Definition, 3.3.59
  Installation of vertical shaft turbine pumps, 7.2.2.2, 7.4.3.7, A.7.2.1.1, A.7.2.2.2, Fig. A.7.2.2.2
  Suction strainer requirement, 7.3.4.3
  Water level, 4.6.3

Wiring, *see also* Disconnecting means
  Electric-drive controllers, 10.3.4
  Engine drive controllers, 11.2.6, 12.3.5, A.11.2.6.2
  Field acceptance tests, 14.2.3
  Packaged fire pump assemblies, 4.29.2
  Problems of, B.1.31, B.1.32

IMPORTANT NOTICES AND DISCLAIMERS CONCERNING NFPA® DOCUMENTS

NOTICE AND DISCLAIMER OF LIABILITY CONCERNING THE USE OF NFPA DOCUMENTS

NFPA® codes, standards, recommended practices, and guides ("NFPA Documents"), including the NFPA Documents contained herein, are developed through a consensus standards development process approved by the American National Standards Institute. This process brings together volunteers representing varied viewpoints and interests to achieve consensus on fire and other safety issues. While the NFPA administers the process and establishes rules to promote fairness in the development of consensus, it does not independently test, evaluate, or verify the accuracy of any information or the soundness of any judgments contained in NFPA Documents.

The NFPA disclaims liability for any personal injury, property or other damages of any nature whatsoever, whether special, indirect, consequential or compensatory, directly or indirectly resulting from the publication, use of, or reliance on NFPA Documents. The NFPA also makes no guaranty or warranty as to the accuracy or completeness of any information published herein.

In issuing and making NFPA Documents available, the NFPA is not undertaking to render professional or other services for or on behalf of any person or entity. Nor is the NFPA undertaking to perform any duty owed by any person or entity to someone else. Anyone using this document should rely on his or her own independent judgment or, as appropriate, seek the advice of a competent professional in determining the exercise of reasonable care in any given circumstances.

The NFPA has no power, nor does it undertake, to police or enforce compliance with the contents of NFPA Documents. Nor does the NFPA list, certify, test, or inspect products, designs, or installations for compliance with this document. Any certification or other statement of compliance with the requirements of this document shall not be attributable to the NFPA and is solely the responsibility of the certifier or maker of the statement.

ADDITIONAL NOTICES AND DISCLAIMERS

Updating of NFPA Documents

Users of NFPA codes, standards, recommended practices, and guides ("NFPA Documents") should be aware that these documents may be superseded at any time by the issuance of new editions or may be amended from time to time through the issuance of Tentative Interim Amendments. An official NFPA Document at any point in time consists of the current edition of the document together with any Tentative Interim Amendments and any Errata then in effect. In order to determine whether a given document is the current edition and whether it has been amended through the issuance of Tentative Interim Amendments or corrected through the issuance of Errata, consult appropriate NFPA publications such as the National Fire Codes® Subscription Service, visit the NFPA website at www.nfpa.org, or contact the NFPA at the address listed below.

Interpretations of NFPA Documents

A statement, written or oral, that is not processed in accordance with Section 6 of the Regulations Governing Committee Projects shall not be considered the official position of NFPA or any of its Committees and shall not be considered to be, nor be relied upon as, a Formal Interpretation.

Patents

The NFPA does not take any position with respect to the validity of any patent rights referenced in, related to, or asserted in connection with an NFPA Document. The users of NFPA Documents bear the sole responsibility for determining the validity of any such patent rights, as well as the risk of infringement of such rights, and the NFPA disclaims liability for the infringement of any patent resulting from the use of or reliance on NFPA Documents.

NFPA adheres to the policy of the American National Standards Institute (ANSI) regarding the inclusion of patents in American National Standards ("the ANSI Patent Policy"), and hereby gives the following notice pursuant to that policy:

> **NOTICE:** The user's attention is called to the possibility that compliance with an NFPA Document may require use of an invention covered by patent rights. NFPA takes no position as to the validity of any such patent rights or as to whether such patent rights constitute or include essential patent claims under the ANSI Patent Policy. If, in connection with the ANSI Patent Policy, a patent holder has filed a statement of willingness to grant licenses under these rights on reasonable and nondiscriminatory terms and conditions to applicants desiring to obtain such a license, copies of such filed statements can be obtained, on request, from NFPA. For further information, contact the NFPA at the address listed below.

Law and Regulations

Users of NFPA Documents should consult applicable federal, state, and local laws and regulations. NFPA does not, by the publication of its codes, standards, recommended practices, and guides, intend to urge action that is not in compliance with applicable laws, and these documents may not be construed as doing so.

Copyrights

NFPA Documents are copyrighted by the NFPA. They are made available for a wide variety of both public and private uses. These include both use, by reference, in laws and regulations, and use in private self-regulation, standardization, and the promotion of safe practices and methods. By making these documents available for use and adoption by public authorities and private users, the NFPA does not waive any rights in copyright to these documents.

Use of NFPA Documents for regulatory purposes should be accomplished through adoption by reference. The term "adoption by reference" means the citing of title, edition, and publishing information only. Any deletions, additions, and changes desired by the adopting authority should be noted separately in the adopting instrument. In order to assist NFPA in following the uses made of its documents, adopting authorities are requested to notify the NFPA (Attention: Secretary, Standards Council) in writing of such use. For technical assistance and questions concerning adoption of NFPA Documents, contact NFPA at the address below.

For Further Information

All questions or other communications relating to NFPA Documents and all requests for information on NFPA procedures governing its codes and standards development process, including information on the procedures for requesting Formal Interpretations, for proposing Tentative Interim Amendments, and for proposing revisions to NFPA documents during regular revision cycles, should be sent to NFPA headquarters, addressed to the attention of the Secretary, Standards Council, NFPA, 1 Batterymarch Park, P.O. Box 9101, Quincy, MA 02269-9101; email: stds_admin@nfpa.org

For more information about NFPA, visit the NFPA website at www.nfpa.org.

A Guide to Using the Stationary Fire Pumps Handbook

This third edition of the *Stationary Fire Pumps Handbook* contains the complete text of the 2010 edition of NFPA 20, *Standard for the Installation of Stationary Pumps for Fire Protection.*

---

(1) Service connection dedicated to the fire pump installation

The requirement in 9.2.2(1) means that the service must supply no loads other than those associated with the fire pump. This requirement is normally met by providing a separate service or a tap ahead of the main service for the building in which the conductors supplying power to the fire pump enter. Frequently, the fire department shuts down power to the building during a fire event to provide for fire-fighter safety. In such situations, power is still needed for the fire pump.

(2) On-site power production facility connection dedicated to the fire pump installation

An on-site standby or emergency generator does not satisfy the requirement in 9.2.2(2). See 3.3.34 and 3.3.35 for the definitions of on-site power production facility and on-site standby generator. An on-site standby generator differs from an on-site power production facility in that it is not constantly producing power and is therefore not considered a private power production facility. In many cases, on-site power production sources are electric generating stations dedicated to a particular facility or to a particular campus-style distribution system. Information on fire protection systems for on-site generating stations can be found in NFPA 850, *Recommended Practice for Fire Protection for Electric Generating Plants and High Voltage Direct Current Converter Stations.*

(3) Dedicated feeder connection derived directly from the dedicated service to the fire pump installation

The dedicated service and feeder connection referred to in 9.2.2(3) may not supply loads other than those associated with the fire pump.

◀ FAQ
Can a fire pump feeder circuit supply loads other than those associated with the fire pump?

(4) As a feeder connection where all of the following conditions are met:

The feeder connection referred to in 9.2.2(4) may supply loads other than those associated with the fire pump. This practice is acceptable and not considered to reduce the reliability of the power supplied to the fire pump because a backup source of power is required in 9.2.2(4)(b) and all feeder circuits supplying fire pump(s) must be selectively coordinated.

◀ FAQ
Can a fire pump feeder circuit supply loads other than those associated with the fire pump if an alternate source of power is provided for the fire pump motor?

(a) The protected facility is part of a multibuilding campus-style arrangement.
(b) A backup source of power is provided from a source independent of the normal source of power.

In order to satisfy the requirement in 9.2.2(4)(b), the independent source must be a separate service, a separate feeder, or an on-site generator. If a separate feeder is provided, it cannot derive its power from the service providing the feeder connection for the normal supply.

◀ FAQ
What is meant by the term *independent*?

(c) It is impractical to supply the normal source of power through the arrangement in 9.2.2(1), 9.2.2(2), or 9.2.2(3).
(d) The arrangement is acceptable to the authority having jurisdiction.
(e) The overcurrent protection device(s) in each disconnecting means is selectively coordinated with any other supply side overcurrent protective device(s).

Selective coordination in the context of this chapter ensures that when an overcurrent condition occurs in a fire pump branch circuit, power is not interrupted to any loads served by upstream protective devices.

(5) Dedicated transformer connection directly from the service meeting the requirements of Article 695 of *NFPA 70, National Electrical Code*

A fire pump may be powered by a transformer provided that the transformer is dedicated to the fire pump and does not serve any other loads not associated with the fire pump. The exception to this is stated in *NEC* 695.5(C): "Where a feeder source is provided in accordance with 695.3(B)(2), transformers supplying the fire pump system shall be permitted to supply

◀ FAQ
What is n
transform

Stationary Fire Pumps Handbook 2010

The commentary text contains Frequently Asked Questions shown in the margin as FAQs. The FAQs answer the most commonly asked questions throughout the handbook.

SUPPLEMENT 1

Fire Pump Installation from Design to Acceptance

John Jensen and David R. Hague

Editor's Note: This supplement is a guide for the design of a fire pump installation. Material has been gathered from this handbook, the manufacturers' data sheets, and the author's experience to provide the design guide. The expectation of the author is that the supplement will be of value to those who are just getting started in the design of such systems and may be of assistance to those who are reviewing new and existing installations.

INTRODUCTION

In this supplement, the method used to explain the information is intended to briefly state what is being covered and then outline in color the issues that need to be considered. Most of the needed information is found in Parts I through III of this handbook. Items that are not covered in adequate detail elsewhere are presented in the outline segments or text of this supplement.

An important part of any fire protection installation is the follow-up inspection, testing, and maintenance of the system, which is not addressed in this supplement. Detailed information on that topic can be found in *The NFPA 25 Handbook.*

Fire Protection System Requirements

The fire protection system water demand requirements are most often obtained from NFPA standards or other standards listed in I(A) through I(C) of the outline. Part I of this handbook gives a detailed description of those requirements. The outline lists the most common fire protection systems in I(A) through I(C). Without exception, each fire protection system using water as the fire suppression agent requires that the water supply have adequate pressure and volume to control and/or extinguish the design fire. The need for a fire pump is only determined after an analysis of the existing water supply and a determination of whether it is adequate or if no water supply is available. The decision process can then be followed to design and install a fire pump that meets the water supply needs of the fire protection system.

If the building or structure is equipped with fire protection systems having different requirements, the water supply is designed based on the most demanding system. The most demanding system may be the fire sprinkler, standpipe, or outside fire hose system. The outline shows considerations for each type of system in I(A) through I(C). If multiple buildings with multiple hazards need to be protected, the water supply is designed for the most severe hazard in the building that presents the highest demand unless the authority having jurisdiction requires additional considerations. Therefore, when planning a water supply using a fire pump, the future addition of more hazardous operations must be considered in the sizing of the water supply and the fire pump.

I. Evaluation
  A. Sprinkler system (See NFPA 13, NFPA 15, NFPA 2001, insurance company interpretive guidelines, or NFPA's *Fire Protection Handbook*)
  1. Wet pipe
  2. Dry pipe
  3. Deluge or water spray
  4. Water mist

Four supplements appear at the end of the handbook and are not a part of the Standards or commentaries. They provide additional, useful information for NFPA 20 users and include drawings, photographs, tables, and extended text. The supplements appear in black type inside a red box.